Neurodegenerative Diseases

Molecular and Cellular Mechanisms and Therapeutic Advances

GWUMC Department of Biochemistry and Molecular Biology Annual Spring Symposia
Series Editors:
Allan L. Goldstein, Ajit Kumar, and J. Martyn Bailey
The George Washington University Medical Center

Recent volumes in this series:

ADVANCES IN MOLECULAR BIOLOGY AND TARGETED TREATMENT FOR AIDS
Edited by Ajit Kumar

BIOLOGY OF CELLULAR TRANSDUCING SIGNALS
Edited by Jack Y. Vanderhoek

BIOMEDICAL ADVANCES IN AGING
Edited by Allan L. Goldstein

CARDIOVASCULAR DISEASE
Molecular and Cellular Mechanisms, Prevention, and Treatment
Edited by Linda L. Gallo

CARDIOVASCULAR DISEASE 2
Cellular and Molecular Mechanisms, Prevention, and Treatment
Edited by Linda L. Gallo

CELL CALCIUM METABOLISM
Physiology, Biochemistry, Pharmacology, and Clinical Implications
Edited by Gary Fiskum

THE CELL CYCLE
Regulators, Targets, and Clinical Applications
Edited by Valerie W. Hu

GROWTH FACTORS, PEPTIDES, AND RECEPTORS
Edited by Terry W. Moody

NEURAL AND ENDOCRINE PEPTIDES AND RECEPTORS
Edited by Terry W. Moody

NEURODEGENERATIVE DISEASES
Molecular and Cellular Mechanisms and Therapeutic Advances
Edited by Gary Fiskum

PROSTAGLANDINS, LEUKOTRIENES, AND LIPOXINS
Biochemistry, Mechanism of Action, and Clinical Applications
Edited by J. Martyn Bailey

PROSTAGLANDINS, LEUKOTRIENES, LIPOXINS, AND PAF
Mechanism of Action, Molecular Biology, and Clinical Applications
Edited by J. Martyn Bailey

A Continuation Order Plan is available for this series. A continuation order will bring delivery of each new volume immediately upon publication. Volumes are billed only upon actual shipment. For further information please contact the publisher.

Neurodegenerative Diseases

Molecular and Cellular Mechanisms and Therapeutic Advances

Edited by

Gary Fiskum

The George Washington University Medical Center
Washington, D.C.

Plenum Press • New York and London

Library of Congress Cataloging in Publication Data

Neurodegenerative diseases: molecular and cellular mechanisms and therapeutic advances / edited by Gary Fiskum.

p. cm.—(GWUMC Department of Biochemistry and molecular biology annual spring symposia)

"Proceedings of the Fifteenth Washington International Spring Symposium at The George Washington University, held May 15–17, 1995, in Washington, D.C."—T.p. verso.

Includes bibliographical references and index.

ISBN 0-306-45298-7

1. Nervous system—Degeneration—Molecular aspects—Congresses. 2. Pathology, Cellular—Congresses. 3. Neurogenetics—Congresses. 4. Neurons—Congresses. 5. Nervous system—Metabolism—Congresses. 6. Amyloid—Congresses. I. Fiskum, Gary. II. International Spring Symposium on Health Sciences (15th: 1995: Washington, D.C.) III. Series.

RC365.N476 1996 96-5688

616.8'047—dc20 CIP

Proceedings of the Fifteenth Washington International Spring Symposium at The George Washington University, held May 15 – 17, 1995, in Washington, D.C.

ISBN 0-306-45298-7

A Division of Plenum Publishing Corporation
233 Spring Street, New York, N. Y. 10013

10 9 8 7 6 5 4 3 2 1

Printed in the United States of America

PREFACE

This book presents much of the exciting new information on mechanisms of neurodegenerative disorders that was presented at the XVth International Spring Symposium on Health Sciences at George Washington University in Washington, D.C. The organization of the symposium as well as the chapters within this book were based upon fundamental molecular and cellular mechanisms of neurodegeneration rather than upon different clinically defined disorders, in order to emphasize the commonality of cause rather than effect.

The first part of the book is devoted to the relationships between selective vulnerability of different neuronal cell types to injury and their functional characteristics related to the transport, binding, and responses for excitatory amino acids, e.g., glutamate, as well as other neurotransmitters involved in toxicity. These relationships are studied in the context of neurodegeneration associated with several disorders including Parkinson's disease, amyotrophic lateral sclerosis (ALS), cerebral ischemia, and AIDS dementia.

Part II emphasizes the role of amyloid proteins in neurodegeneration but also covers other molecular and genetic risk factors, e.g., expression of different apolipoprotein isoforms and the involvement of abnormal superoxide dismutase in neuropathology. These topics are applied primarily to the aging brain, Alzheimer's disease, and ALS.

The third part, Cellular Metabolism, provides an overview of how altered cerebral energy and amino acid metabolism and protein synthesis predispose the CNS to injury caused by other closely associated mechanisms, including oxidative stress and excitotoxicity. This concept is applied to a broad range of neurodegenerative disorders including Alzheimer's and Huntington's diseases as well as those caused by neurotoxins, poor nutrition, cerebral ischemia, and inborn errors of metabolism.

The fourth part of the book concentrates on the rapidly growing field of research addressing the roles that free radicals and oxidative stress play in cellular injury and death associated with neurological disorders, e.g., stroke, ALS, Alzheimer's and Parkinson's diseases, and meningitis. Important new information on the actions of nitric oxide in the nervous system is also covered in this section.

Part V elaborates on how calcium transport and calcium-dependent enzymes and regulatory proteins provide either protection against, or exacerbation of, the cascade of events leading to irreversible neuronal injury. In addition to the application of this information to mechanisms of neurodegeneration in Alzheimer's, ischemia, trauma, and AIDS dementia, this section of the book covers many of the pharmacological interventions being tested for therapeutic efficacy toward these disorders.

The final section of the book deals with the extremely important topics of neuronal growth regulation and apoptotic cell death. As these issues are considered to be of fundamental importance with regard to cellular mechanisms of neurodegeneration, it is not surprising that the chapters within this section touch upon many of the basic molecular mechanisms of neuronal injury covered in previous sections, including metabolic failure, excitotoxicity, and oxidative stress.

Gary Fiskum

CONTENTS

PART I - EXCITOTOXICITY AND SELECTIVE NEURONAL VULNERABILITY

PART II - MOLECULAR AND GENETIC RISK FACTORS

PART III - CELLULAR METABOLISM

PART IV - FREE RADICALS AND OXIDATIVE STRESS

PART V - CALCIUM TRANSPORT AND REGULATORY PROTEINS

PART VI - GROWTH REGULATION AND APOPTOSIS

INTRACELLULAR SIGNALLING IN GLUTAMATE EXCITOTOXICITY

Ian J. Reynolds, Kari R. Hoyt, R. James White and Amy K. Stout.

Department of Pharmacology
University of Pittsburgh
Pittsburgh, PA 15261

INTRODUCTION

The excitatory neurotransmitter glutamate is a potent and effective neurotoxin. When applied *in vitro*, a brief exposure to a moderate concentration of glutamate is sufficient to kill neurons.[1,2] *In vivo*, glutamate-induced neuronal injury probably contributes to damage that results from cerebrovascular accidents and trauma.[2-4] A number of important studies have characterized the temporal and pharmacological characteristics of glutamate excitotoxicity *in vitro*.[5,6] It is now clear that glutamate-induced activation of N-methyl-D-aspartate (NMDA) receptors for about 5 minutes is sufficient to kill neurons, and that death is expressed within 24 hours of glutamate application. Activation of non-NMDA receptors by, for example, kainate requires exposures of more than 30 minutes; death ensues over a similar time frame.

NMDA receptor-mediated injury is an interesting and convenient target for studies aimed at elucidating the critical events in glutamate excitotoxicity. In addition to the short exposure times necessary to lethally injure neurons, NMDA receptor mediated injury is substantially dependent on the presence of extracellular Ca^{2+} during the stimulus.[7] For this reason it seems likely that Ca^{2+}-triggered events are involved in the process of neuronal injury. Indeed, a critical role for a number of such events have been proposed, including the activation of nitric oxide synthase, proteases, phospholipases, protein kinases and phosphatases.[8-17] However, the sequence in which these processes are activated remains unclear, which is of concern if these mechanisms are to be exploited as therapeutic targets. It is also unclear which, if any, of these events are activated during the initial 5 minute period within which neurons are committed to die. Finally, the anomaly exists that a number of stimuli can increase intracellular free Ca^{2+} ($[Ca^{2+}]_i$) to similar concentrations as glutamate. These include depolarizing stimuli and non-NMDA receptor activation that activate voltage-sensitive Ca^{2+} channels. However, only $[Ca^{2+}]_i$ elevations resulting from NMDA receptor activation are capable of lethally injuring cells within the short exposure time.[18,19]

Neurodegenerative Diseases
Edited by Gary Fiskum, Plenum Press, New York, 1996

An important goal, then, in understanding glutamate excitotoxicity is the identification of cellular events that occur during and immediately after potentially lethal NMDA receptor activation *and* that are specific to injurious stimuli. In this chapter we will review studies that are in progress in this laboratory which have started to dissect out intracellular events specifically associated with neuronal injury. Ca^{2+} remains an important focus of these studies because of its role as the initial trigger for the subsequent pathology. However, there also appear to be a number of other signalling events that occur shortly after excitotoxic glutamate exposure, including changes in intracellular free Mg^{2+} ($[Mg^{2+}]_i$) and the production of reactive oxygen species that may also be of critical importance.

INTRACELLULAR Mg^{2+} AND EXCITOTOXICITY

Mg^{2+} is an abundant intracellular cation, and $[Mg^{2+}]_i$ is typically 0.5mM in most cells. However, total cellular Mg^{2+} is some 20-40 fold greater, suggesting that a substantial fraction is sequestered intracellularly via poorly understood mechanisms. It is generally believed that $[Mg^{2+}]_i$ is essentially constant. Our recent observation that glutamate could evoke rather large (>mM) changes in $[Mg^{2+}]_i$ was, therefore, a surprise.[20] Under normal buffer conditions most of the glutamate-induced change was mediated by NMDA receptors and required Ca^{2+} entry, thus suggesting that Ca^{2+} mobilized Mg^{2+} from an intracellular store, the location of which has yet to be determined. We believe that the $[Mg^{2+}]_i$ change stoichiometrically reflects the entry of Ca^{2+}, because the magnitude of the Mg^{2+} signal is likely to be similar to the size of the typical integrated NMDA receptor-mediated Ca^{2+} current in these neurons.[20] In this regard, Ca^{2+}-dependent increases in $[Mg^{2+}]_i$ appear to be an excitotoxicity-specific signal because other stimuli, such as kainate or veratridine that produce equivalent changes in $[Ca^{2+}]_i$ but are not acutely neurotoxic, are much less effective at increasing $[Mg^{2+}]_i$. The magnitude of the $[Mg^{2+}]_i$ change also corresponds with observations of the difference in the ability of agonists to induce $^{45}Ca^{2+}$ flux into neurons.[21,22] The similar pharmacology would support our contention that the $[Mg^{2+}]_i$ change is a marker

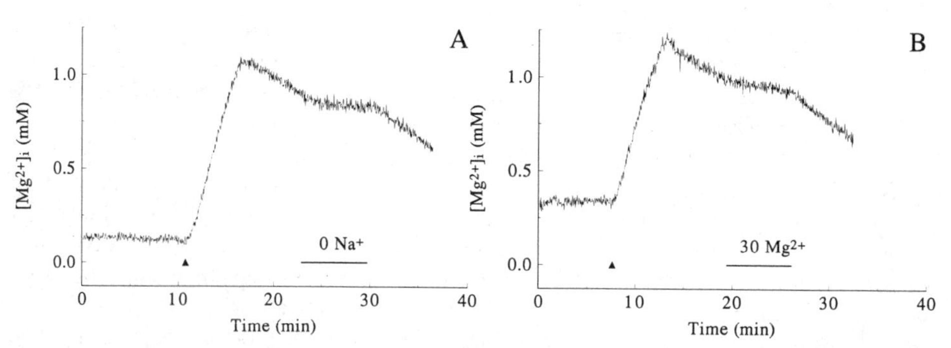

Figure 1. Effects of (A) decreased extracellular Na^+ and (B) increased extracellular Mg^{2+} on the recovery from a glutamate-induced elevation of $[Mg^{2+}]_i$. $[Mg^{2+}]_i$ was measured in single cultured forebrain neurons using magfura-2. Glutamate (100μM) and glycine (1μM) were added at the arrowhead in a Na^+- and Ca^{2+}-free buffer containing 134mM N-methyl-D-glucamine, 9mM Mg^{2+} and 20μM EGTA. At the end of the stimulus the buffer was replaced with Na^+-containing, Ca^{2+}-free buffer and the rate of reduction of $[Mg^{2+}]_i$ determined. The removal of Na^+ or the addition of 30mM Mg^{2+} during the recovery markedly decreased the rate of recovery, consistent with a role for Na^+/Mg^{2+} exchange in the recovery process.

for the total Ca^{2+} entry, a parameter distinct from the resulting $[Ca^{2+}]_i$ change reported by fura-2 and related agents.

Intracellular Mg^{2+} does not have a clear role in neuronal injury. We have demonstrated that elevating $[Mg^{2+}]_i$ independently of Ca^{2+} entry is sufficient to kill neurons in culture.[23] However, other studies in intact preparations have associated loss of tissue Mg^{2+} with injury.[24] Clearly, then, we must better define our understanding of Mg^{2+} homeostasis if we are to determine its role in neuronal injury. To approach this issue experimentally we have taken advantage of the observation that glutamate can evoke Mg^{2+} influx into neurons in the absence of extracellular Ca^{2+} and Na^+, apparently by NMDA receptor-mediated Mg^{2+} currents (A.K.S., Y. Li-Smerin, J.W. Johnson and I.J.R., manuscript submitted). This approach offers the advantage of raising $[Mg^{2+}]_i$ without altering $[Ca^{2+}]_i$, thus simplifying the interpretation of data. Using this approach we have established a role for a Na^+/Mg^{2+} exchange-like mechanism in the recovery from a glutamate-induced Mg^{2+} load (Figure 1). The rate of recovery of $[Mg^{2+}]_i$ to basal levels is significantly reduced by the removal of Na^+ during the recovery phase and is slowed to a similar extent by the addition of 30mM extracellular Mg^{2+}. The failure of these two manipulations to produce additive effects suggests that they are working on the same mechanism, and that the mechanism is consistent with Na^+-mediated Mg^{2+} extrusion. If loss of tissue Mg^{2+} is an important contributor to brain injury, the Na^+/Mg^{2+} exchange process in neurons would represent an important therapeutic target.

BUFFERING MECHANISMS FOR INTRACELLULAR Ca^{2+}

Notwithstanding the potentially injurious effects of altered $[Mg^{2+}]_i$ in neurons, Ca^{2+} still appears to be the critical trigger that initiates neuronal injury. For this reason it of interest to understand how neurons buffer glutamate-induced $[Ca^{2+}]_i$ changes and whether the function of those mechanisms is altered by the intensity of stimulus necessary to kill neurons. We recently demonstrated that both mitochondria and Na^+/Ca^{2+} exchange play an important role in buffering moderate $[Ca^{2+}]_i$ changes induced by brief exposures to glutamate concentrations that are ostensibly not toxic.[25] Progressively longer stimuli, as well as increasing glutamate concentrations, produce $[Ca^{2+}]_i$ transients that take much longer to return to baseline levels, a phenomenon that might broadly be interpreted as an increased Ca^{2+} load (Figure 2). Interestingly, mitochondria continue to play an important role during the recovery from a glutamate-induced load. The prolonged recovery associated with intense neuronal stimulation can be dramatically curtailed by selective inhibition of mitochondrial Na^+/Ca^{2+} exchange with CGP 37157 (Figure 2).[26,27] This finding demonstrates that the mitochondria contribute to the prolonged elevation of $[Ca^{2+}]_i$ by rapidly sequestering Ca^{2+} and then releasing it long after glutamate stimulation is terminated. A similar storage role for mitochondria has been described in dorsal root ganglion neurons.[28] The impact of mitochondrial Na^+/Ca^{2+} exchange inhibition on excitotoxicity has not yet been established, but this drug will allow us to distinguish between the possible toxic effects of mitochondrial Ca^{2+} retention compared to prolonged elevation of $[Ca^{2+}]_i$ by this putative cycling process.

REACTIVE OXYGEN SPECIES AND NEURONAL INJURY

Glutamate-induced changes in $[Ca^{2+}]_i$ and $[Mg^{2+}]_i$ are clearly and directly coupled to the activation of NMDA receptors and are therefore closely associated with the initial neurotoxic stimulus. The ever-expanding list of signal transduction events that putatively contribute to excitotoxic injury are less directly coupled to NMDA receptor activation, such

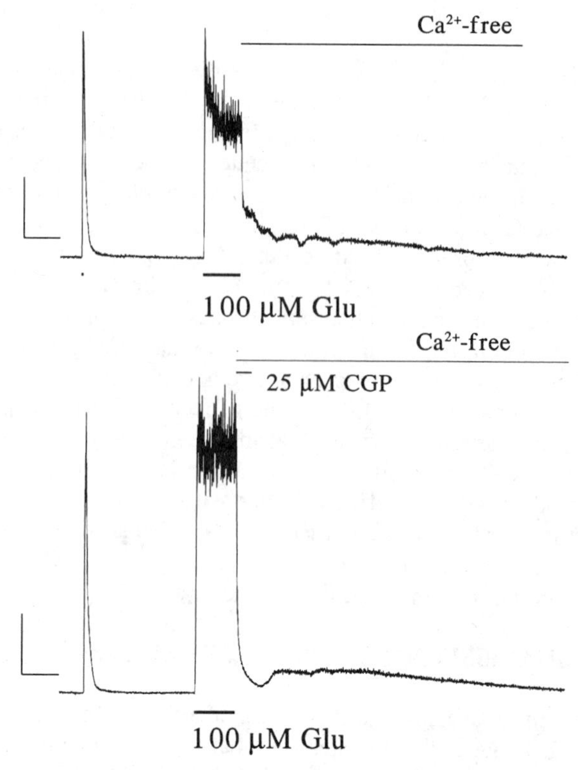

Figure 2. CGP 37157 reveals a large mitochondrial contribution to the prolonged elevation of $[Ca^{2+}]_i$ following an excitotoxic glutamate stimulus. (*Top*) After the cell had recovered from a reference glutamate stimulus (15 seconds, 3 μM; at the *thin box*), we applied 100 μM glutamate/1 μM glycine for 5 minutes. After return to normal buffer $[Ca^{2+}]_i$ begins to fall rapidly towards baseline. However $[Ca^{2+}]_i$ then remains elevated at several hundred nanomolar for some time. This plateau may reflect the mitochondrial set-point, a balance in which efflux (via mitochondrial Na^+/Ca^{2+}-exchange) and uptake (via the Ca^{2+} uniporter) are nearly equal. In the *lower* panel, the 5 minute glutamate stimulus was immediately followed by 25 μM CGP-37157 in Ca^{2+}-free buffer. In the presence of this inhibitor of mitochondrial Na^+/Ca^{2+}-exchange, the cell rapidly buffered $[Ca^{2+}]_i$ back to baseline; as the drug was washed out, $[Ca^{2+}]_i$ rose, presumably reflecting release of the mitochondrial matrix Ca^{2+} which was stored during and immediately after the glutamate stimulus. Scale bars in each figure represent 500 nM $[Ca^{2+}]_i$ and 5 minutes.

that it becomes necessary to very carefully establish their role in neuronal injury. If a given signal transduction pathway is causative in excitotoxic injury, rather than being a later consequence of deteriorating neurons, it should be possible to associate the activation of the signal transduction pathway specifically with the earliest phase of excitotoxic glutamate application. Recently it has become possible to demonstrate such an association betweeen ROS generation and NMDA receptor mediated excitotoxicity in cultured neurons.

Glutamate-Induced ROS Production.

There is ample evidence that ischemic injury *in vivo* is associated with the generation of ROS.[10,29] However, there is far less known about the mechanisms of ROS production in well-controlled systems, and evidence for a causative role in neuronal injury has not been presented. We monitored ROS production in neurons using the fluorescent probe 2',7'-dichlorodihydrofluorescein (DCF).[30] This probe is supplied in the reduced form and becomes fluorescent upon oxidation. By following the appearance of signal in neurons becomes fluorescent upon oxidation. By following the appearance of signal in neurons after treatment with glutamate we were able to observe ROS production within 2-4 minutes, thus associating ROS with the earliest events in excitotoxic injury. DCF is less than ideal as a marker for ROS as it leaks from neurons after oxidation and is also pH sensitive.[30]

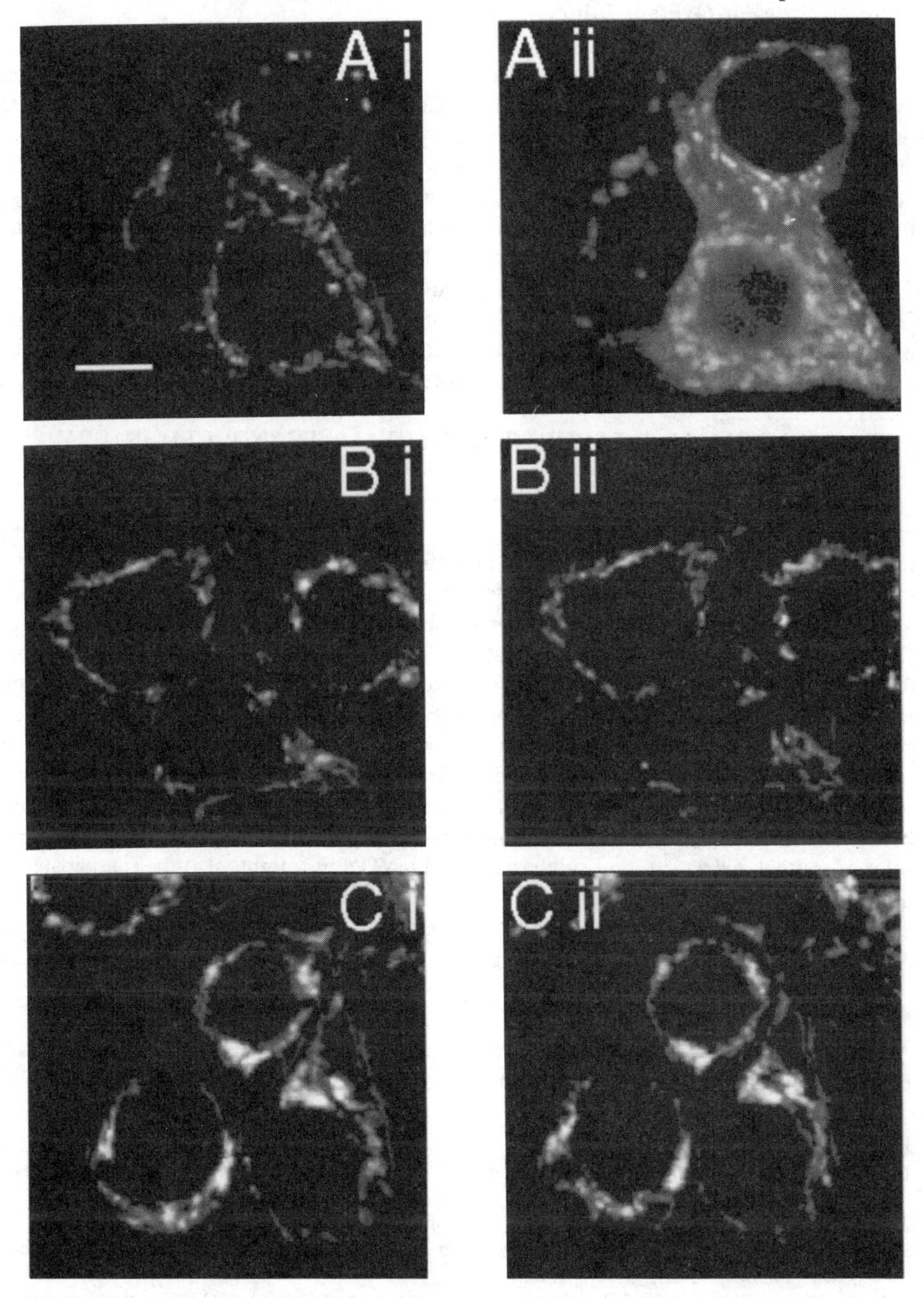

Figure 3. Excitotoxic concentrations of glutamate acutely depolarize mitochondria. Neurons were labelled with rhodamine 123 and then monitored using confocal microscopy. The localized areas of fluorescence reflect individual mitochondria. In each pair of images (i) represents an image taken before and (ii) an image taken 6 minutes after the application of (A) glutamate (100µM) and glycine (1µM), (B) glutamate (100µM) and glycine (1µM) in the absence of extracellular Ca^{2+}, and (C) kainate (100µM). Mitochondrial depolarization appears as a loss of punctate fluorescence (A ii) following dye leakage from the mitochondria. Scale bar represents 5µm.

Nevertheless it was possible to identify Ca^{2+} uptake by mitochondria as the likely source of ROS, while arachidonic acid metabolism inhibitors were much less effective in blocking the ROS signal.

Characteristics of Oxidant Injury.

Given that glutamate can induce the production of ROS during the initiation phase of excitotoxic injury, how much of the injurious effect of glutamate can be attributed to oxidant injury? This question could be addressed by simply adding an oxidant to neurons and observing the outcome. We used H_2O_2 and found that low concentrations ($\leq 25\mu M$) for <30 minutes were very effective at killing neurons within the following 24 hours. Interestingly, this form of neuronal injury exhibited many of the hallmarks of apoptosis, including shrunken cell bodies, condensed chromatin, and DNA damage as assayed by TUNEL labelling. Moreover, H_2O_2 injury did not produce a "glutamate-like" intracellular Ca^{2+} change. This pattern of injury was quite distinct from the necrotic appearance of glutamate-injured cells forcing the conclusion that oxidant injury may contribute to excitotoxicity but, under these conditions, does not dominate the excitotoxic process.

MITOCHONDRIA AS A TARGET FOR GLUTAMATE-INDUCED INJURY

Several independent experiments investigating excitotoxicity have indicated a role for mitochondria in glutamate pathophysiology. As described above, mitochondria represent an important intracellular Ca^{2+} buffering mechanism. Ca^{2+}-loaded mitochondria may also represent a source of ROS in glutamate-stimulated cells. Using rhodamine 123 we have also observed that mitochondria depolarize following NMDA receptor-induced Ca^{2+} entry (Figure 3) but not after stimulation by kainate. Although this dye clearly has limitations in terms of quantitation, the observation of mitochondrial depolarization is quite dramatic, rapid, and also explicitly limited to excitotoxic stimuli. Collectively, these data indicate that compromised ATP production, a decreased mitochondrial membrane potential, and increased mitochondrial ROS production may all be important hallmarks of excitotoxic neuronal injury. We lack sufficient data to determine whether mitochondria are actively involved in the injury process or are merely the targets of other very early pathologic signals. Given the critical role of mitochondria in the synthesis of Mg-ATP, it might also be anticipated that mitochondria have active mechanisms for Mg^{2+} transport. The possibility that mitochondria are, in fact, the source of the intracellular Mg^{2+} mobilized by Ca^{2+} influx is an attractive unifying hypothesis with respect to glutamate excitotoxicity that, at this time, lacks experimental evaluation.

Acknowledgements. These studies were supported by a Grant-in-Aid from the American Heart Association. I.J.R. is an established investigator of the American Heart Association.

REFERENCES

1. S.M. Rothman and J.W. Olney, Excitotoxicity and the NMDA receptor, *Trends Neurosci.*, 10:299 (1987).
2. D.W. Choi, Glutamate neurotoxicity and diseases of the nervous system, *Neuron*, 1:623 (1988).
3. B.S. Meldrum and J. Garthwaite, Excitatory amino acid neurotoxicity and neurodegenerative disease, *Trends Pharmacol. Sci.*, 11:379 (1990).
4. H. Benveniste, The excitotoxin hypothesis in relation to cerebral ischemia, *Cerebrovasc. Brain Metab. Rev.*, 3:213 (1991).

6. D.W. Choi, M. Maulucci-Gedde and A.R. Kriegstein, Glutamate neurotoxicity in cortical cell culture, *J. Neurosci.*, 7:357 (1987).
7. D.W. Choi, Ionic dependence of glutamate neurotoxicity, *J. Neurosci.*, 7:369 (1987).
8. M.L. Mayer and G.L. Westbrook, Cellular mechanisms underlying excitotoxicity, *Trends Neurosci.*, 10:59 (1987).
9. B.K. Siesjo and T. Wieloch, Cerebral metabolism in ischemia: neurochemical basis for therapy, *Br. J. Anaesth.*, 57:47 (1985).
10. E.D. Hall and J.M. Braughler, Central nervous system trauma and stroke: II. Physiological and pharmacological evidence for involvement of oxygen radicals and lipid peroxidation, *Free Radic. Biol. Med.*, 6:303 (1989).
11. M.E. Götz, G. Künig, P. Riederer and M.B.H. Youdim, Oxidative stress: Free radical production in neural degeneration, *Pharmacol. Ther.*, 63:37 (1994).
12. S.M. Rothman, K.A. Yamada and N. Lancaster, Nordihydroguaiaretic acid attenuates NMDA neurotoxicity--Action beyond the receptor, *Neuropharmacology*, 32:1279 (1993).
13. V.L. Dawson, T.M. Dawson, E.D. London, D.S. Bredt and S.H. Snyder, Nitric oxide mediates glutamate neurotoxicity in primary cortical cultures, *Proc. Natl. Acad. Sci. USA*, 88:6368 (1991).
14. H. Manev, M. Favaron, A. Guidotti and E. Costa, Delayed increase of Ca^{2+} influx elicited by glutamate: role in neuronal death, *Mol. Pharmacol.*, 36:106 (1989).
15. L.-Y. Wang, B.A. Orser, D.L. Brautigan and J.F. MacDonald, Regulation of NMDA receptors in cultured hippocampal neurons by protein phosphatases 1 and 2A, *Nature*, 369:230 (1994).
16. Y.T. Wang and M.W. Salter, Regulation of NMDA receptors by tyrosine kinases and phosphatases, *Nature*, 369:233 (1994).
17. D.N. Lieberman and I. Mody, Regulation of NMDA channel function by endogenous Ca^{2+}-dependent phosphatase, *Nature*, 369:235 (1994).
18. M. Tymianski, M.P. Charlton, P.L. Carlen and C.H. Tator, Source specificity of early calcium neurotoxicity in cultured embryonic spinal neurons, *J. Neurosci.*, 13:2085 (1993).
19. S. Rajdev and I.J. Reynolds, Glutamate-induced intracellular calcium changes and neurotoxicity *in vitro*: effects of chemical ischemia, *Neuroscience*, 62:667 (1994).
20. J.B. Brocard, S. Rajdev and I.J. Reynolds, Glutamate induced increases in intracellular free Mg^{2+} in cultured cortical neurons, *Neuron*, 11:751 (1993).
21. D.M. Hartley, M.C. Kurth, L. Bjerkness, J.H. Weiss and D.W. Choi, Glutamate receptor-induced $^{45}Ca^{2+}$ accumulation in cortical culture correlates with subsequent neuronal degeneration, *J. Neurosci.*, 13:1993 (1993).
22. S. Eimerl and M. Schramm, The quantity of calcium that appears to induce neuronal death, *J. Neurochem.*, 62:1223 (1994).
23. K.A. Hartnett, S. Rajdev, P.A. Rosenberg, E. Aizenman and I.J. Reynolds, A paradoxical requirement for extracellular Mg^{2+} in glutamate toxicity, *Soc. Neurosci.*, 19:1344 (1993).
24. R. Vink, T.K. McIntosh and A.I. Faden, Mg^{2+} in neurotrauma: its role and therapeutic implications, *in:* "Mg^{2+} and excitable membranes," P. Strata and E. Carbone (Eds.) Springer-Verlag, Berlin, (1991).
25. R.J. White and I.J. Reynolds, Mitochondria and Na^+/Ca^{2+} exchange buffer glutamate-induced calcium loads in cultured cortical neurons, *J. Neurosci.*, 15:1318 (1995).
26. D.A. Cox, L. Conforti, N. Sperelakis and M.A. Matlib, Selectivity of inhibition of Na^+-Ca^{2+} exchange of heart mitochondria by benzothiazepine CGP-37157, *J. Cardiovasc. Pharmacol.*, 21:595 (1993).
27. D.A. Cox and M.A. Matlib, Modulation of intramitochondrial free Ca^{2+} concentration by antagonists of Na^+-Ca^{2+} exchange, *Trends Pharmacol. Sci.*, 14:408 (1993).
28. S.A. Thayer and R.J. Miller, Regulation of the intracellular free calcium concentration in single rat dorsal root ganglion neurones in vitro, *J. Physiol.*, 425:85 (1990).
29. B. Halliwell, Reactive oxygen species in the central nervous system, *J. Neurochem.*, 59:1609 (1992).
30. I.J. Reynolds and T.G. Hastings, Glutamate induces the production of reactive oxygen species in cultured forebrain neurons following NMDA receptor activation, *J. Neurosci.*, 15:3318 (1995).

LONG-TERM EXPRESSION OF PROENKEPHALIN AND PRODYNORPHIN IN THE RAT BRAIN AFTER SYSTEMIC ADMINISTRATION OF KAINIC ACID ——AN *IN SITU* HYBRIDIZATION STUDY

Guoying Bing,[1] Belinda Wilson,[1] Michael McMillian,[1] Zhehui Feng,[1] Qiping Qi,[1] Hyoung-Chum Kim,[1] Wen Wang,[1] Karl Jensen,[2] and Jau-Shyong Hong[1]

[1]Laboratory of Environmental Neuroscience, NIEHS/NIH;
[2]Neurotoxicology Division Environmental Protection Agency,
Research Triangle Park, NC 27709

INTRODUCTION

Kainic acid (KA) is a glutamate analog that binds to and activates ionotropic glutamate receptors. Administration of KA causes robust and recurrent seizures in the rat, and produces permanent damage in the CNS, specifically in the limbic system. Systemic injection of kainate results in both short-term and long-term effects on the rat central nervous system (CNS). The excitatory neuronal stimulation by kainate increases the expression of a variety of genes including immediate-early genes, growth factors, and opioid peptides. The short-term effects of kainate in the rat brain have been well characterized[1-6]. However, the long-term effects of kainate, especially on the opioid peptides, have not been reported.

The opioid neuropeptides derived from the proenkephalin (PENK) and prodynorphin (PDYN) precursors are expressed at high levels in the CNS. These endogenous opioids can function as neurotransmitters, neurohormones, or neuromodulators (see Refs. 7 and 8 for reviews). Nevertheless, they may also serve as growth factors as well as exerting inhibitory influences on the developing nervous system[9, 10]. The present studies were designed to investigate the long-term effects of kainate on the expression of PENK and PDYN in regions of the rat brain known to be important in kindling and epilepsy, such as the dentate gyrus, and piriform and entorhinal cortices.

MATERIALS AND METHODS

Animals and Treatments

Adult male Fischer 344 rats (225-250 g body weight) were used throughout the study. Control animals (n = 8) were injected i.p. with physiological saline. Experimental animals (n = 36) were injected with KA (Sigma, 8 mg/kg, i.p.). The animals were rated according to the scale devised by Racine[11] for the initial 4 h following the KA injection. Only animals with full limbic seizures (forelimb clonus with rearing, stage 4) were chosen for further studies. Animals were perfused with 4% paraformaldehyde in PBS (pH 7.4) at different time points: 3 h, 6 h, 1 day, 3 days, 1 week, 2 weeks, 3 weeks, 1 month, 3 months, and 5 months.

Immunocytochemistry

The brains were sectioned at 35 μm on a horizontal sliding microtome. The sections were incubated in 4% normal goat serum in PBS for 30 min at room temperature to block non-specific immunostaining. After 3 washes with PBS (pH 7.4), the sections were incubated in PBS containing 0.025% Triton X-100, 1% normal goat serum, and the primary antiserum at 4°C overnight. The avidin-biotin immunoperoxidase method with 3,3'-diaminobenzidine tetrahydrochloride as the chromogen was used to visualize immunoreactive cells. The Fos-related antigen (FRA) antibody from Dr. M. Iadarola (NIDR/NIH) was used at a 1:2000 dilution. This antibody was raised against the M peptide region of c-Fos which is conserved among the FRA proteins[12].

In Situ Hybridization

The expression of PENK and PDYN mRNA was analyzed by *in situ* hybridization histochemistry. The hybridizations were performed using free-floating sections. After perfusion with 4% paraformaldehyde, the rat brains were frozen-sectioned on a sliding microtome. The 35 μm sections were collected in 24-well culture dishes containing a cryopreservative solution. Every 12th section was used for each specific probe. Both ^{35}S-labeled and non-radioactive digoxigenin-labeled cRNA probes were generated from PENK and PDYN cDNAs. The sections were hybridized with ^{35}S-labeled (2 x 10^6 cpm/well) probes at 50°C for 16 h. For post-hybridization washing, the sections were rinsed twice in

2 x SSC, 50% formamide, and 10 mM DTT at room temperature. After the incubation with RNase A (20 μg/ml) at 37^0C for 30 minutes, the sections were washed again with 2X SSC, 50% formamide and 10 mM DTT at room temperature followed by 2 washes with 0.25 x SSC with 10 mM DTT. Following post-hybridization, the sections were mounted on slides coated with poly-lysine and exposed to X-ray film for 3-4 days . The sections were then dipped in Kodak NTB-2 emulsion (2:1 with water) and exposed for 1 week before being developed in D-19 developer solution.

RESULTS

Behavioral and Morphological Effects of KA Treatment

All animals were closely observed and rated for seizure behavior according to the scale developed by Racine. Thirty min after treatment, the animals began to display wet dog shakes. Two h after injection, full limbic motor seizures were observed including rearing and loss of postural control. After 2.5-3 h the animals showed profuse salivation, circling and jumping, i.e., *status epilepticus*. Hyperexcitability was seen in most animals throughout the period of the experiment.

The sections from each time point were stained with cresyl violet to evaluate neuronal damage and reveal cellular architecture. Three h after KA injection, degeneration of pyramidal neurons in the hippocampal CA3 area became evident as shrunken cell bodies, condensed nuclei and more darkly stained cytoplasm. Cellular necrosis and infarction of the brain tissue were clearly observed 2 days after treatment. By 4 days, KA-induced damage in the brain became maximal. Most neuronal cells in the hippocampal CA1 and CA3 areas were no longer stained by cresyl violet and the pyramidal cell layers were not clearly visible. The loss of morphology and blood penetration into this area persisted for at least five months after KA treatment. The KA-treated rat brain showed permanent deformations, evident as enlarged ventricles, thinner cortical layers and smaller hippocampi 1 month after KA treatment. These deformations were more obvious five months after KA treatment.

Changes in the Expression of PENK and PDYN mRNA after KA Injection

The major findings of the present study are that KA produces biphasic increases in PENK and PDYN mRNAs in the rat hippocampus. The initial increase subsided for PENK mRNA around 4 days and for PDYN mRNA 7 days after KA injection. However, the mRNA levels for both PENK and PDYN rose again after 2-4 weeks and the elevated levels persisted for at least 5 months after the treatment.

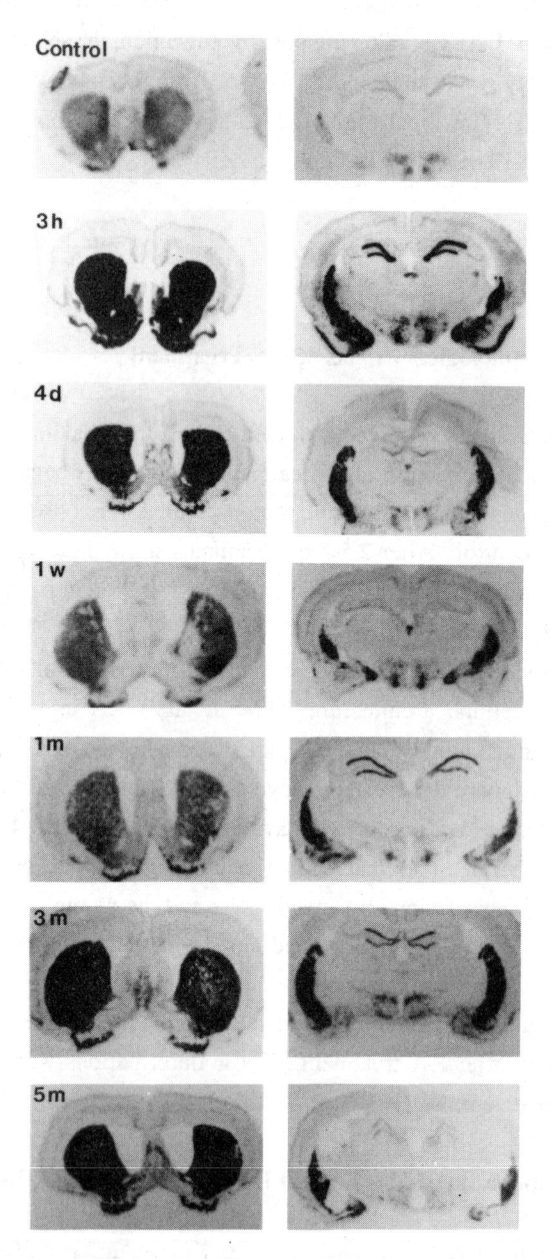

Figure 1. Autoradiographs of *in situ* hybridization of enkephalin (A) and dynorphin (B) on coronal sections of the rat brain. Two sections from +0.5 mm and -3.5 mm to bregma were taken to represent the brain structures of interest in the present study. The brains were taken from the control, 3 hours (h), 4 days (d), 1 week (w), 1 month (m), 3 months, and 5 months after KA (8 mg/kg, i.p.) treatment. The autoradiographs were taken from X-ray film 3 days after exposure to brain sections hybridized with ^{35}S-labeled probes.

B

DYN

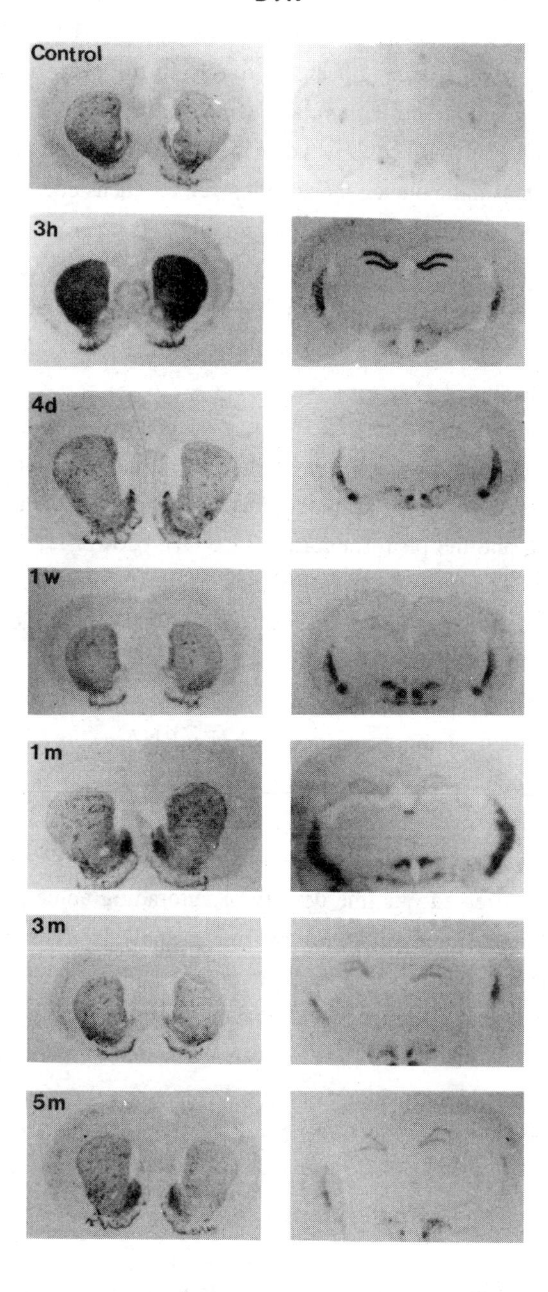

The initial increases in the levels of PENK and PDYN after KA are due to excitation of the granule cells resulted from the recurrent seizures[20]. However, the second phase increases of these two opioid peptides have not been reported.

The PENK mRNA expression, as revealed by *in situ* hybridization, is shown in Figure 1A. The dynamic alterations in PENK mRNA were primarily in the hippocampi and the striata. The peak increase in PENK mRNA levels was at 24 h. The mRNA hybridization signal as revealed by silver grains appeared strongly in the piriform cortex, entorhinal cortex, striatum, medial and lateral septum, and dentate granular layer of the hippocampus. The mRNAs were also visualized less heavily in the hypothalamic periventricular, supraoptic and suprachiasmatic nuclei, the zone inserta, the locus coeruleus, layers 4 and 5 of the frontal and parietal cortices, and in the granule cell layer of the cerebellum. All mRNA signals were decreased to control levels at 4 to 7 days after KA injection. However, the mRNA levels increased again 2 weeks after KA injection. This increase lasted for at least 5 months and was primarily localized in the dentate gyrus of the hippocampal formation (Fig. 1A).

The peak expression of PDYN mRNA was earlier than that of PENK mRNA, peaking at about 3-6 h after KA injection. The heavily labeled structures standing out at this time point were the granular layer of the dentate gyrus, striatum, hypothalamic periventricular and supraoptic nuclei, and the piriform cortex (Fig. 1B). Lower levels of PDYN mRNA were found in layers 2 and 4 of the frontal and parietal cortices and in some of the hippocampal CA1 and CA3 neurons. The expression of PDYN mRNA subsided 1 week after KA injection. However, the mRNA levels for PDYN rose again after 3 weeks and persisted for at least 5 months after treatment. Unlike the expression of PENK which was also persistently elevated in the striatum, PDYN mRNA expression only rose in the granule cell layer of the dentate gyrus. The expression of PDYN in the rest of the brain remained at control levels.

Since the major changes in the expression of PENK and PDYN mRNAs were in the hippocampi of the KA-treated rats, the density of autoradiographic grains overlying the hippocampi was estimated by computer-aided image analysis. The results showed that PENK mRNA levels were increased about 3 to 4-fold at 6 h after KA injection. These increases subsided by 4 days to control levels and they rose again at 3- to 4 weeks after KA injections (Fig. 2). The levels of PDYN mRNA in the dentate granule cells of the hippocampus also showed similar changes after KA injections (Fig. 2).

The specificity of the riboprobes for PENK were studied with a sense probe from the same region as the antisense probes. The sense probe revealed virtually no hybridization signals in the hippocampus 4 h after KA injections.

Expression of FRA Immunoreactivity in the Rat Hippocampus after KA Treatment

KA-induced epileptic seizures increase the expression of immediate-early genes, such as c-fos and FRAs, in the central nervous system with a specific pattern[13]. Systemic administration of KA elevated the 35-kDa FRA for up to two weeks in the rat hippocampus[14]. This long-term expression of FRA can also be induced by chronic electroconvulsive seizure[15]. Thus, the long-term consequences of KA treatment on the expression of immediate-early genes and damage in the limbic system can provide an opportunity to explore the role of FRAs in the long-term changes and in the expressions of PENK and PDYN mRNAs.

In order to examine the possibility that FRAs are involved in the regulation of the long-term changes in opioid peptide levels, we have processed alternate sections for immunocytochemistry with FRA antibodies and compared these change to those of PENK and PDYN expressions. Increased FRA immunoreactivity also persisted for at least 5 months and the spatiotemporal patterns of the FRA expression were closely correlated with PENK and PDYN mRNA expression (data not shown). This finding suggests that FRA is involved in the regulation of PENK and PDYN gene expression.

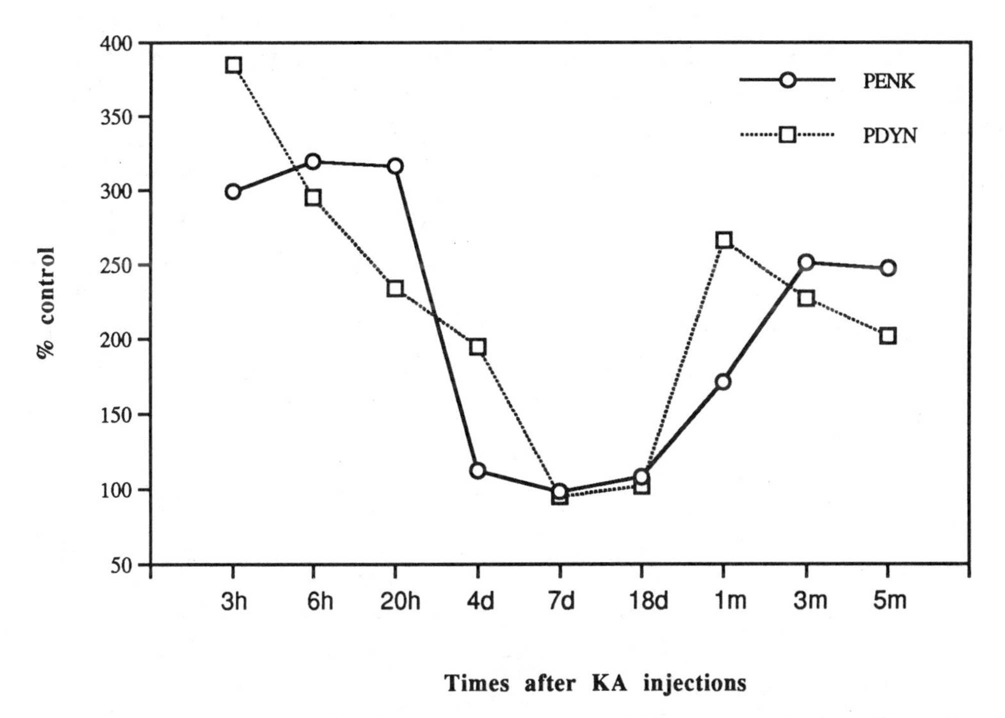

Figure 2. The changes of PENK and PDYN mRNA levels after KA injections. The density of silver grains overlying the dentate gyrus was estimated by digitizing the field of a 40X objective with the Presage Imaging System (ALC) and calculating the area occupied by silver grains. The results are expressed as percent of control density.

DISCUSSION

The present study demonstrates that the expression of the PENK and PDYN genes can be regulated by KA and that KA-induced increases in the expressions of these two genes persist for at least 5 months. The time courses of ENK and DYN expressions in the hippocampus are closely correlated with the expression of FRA immunoreactivity after KA injections. Since both PENK and PDYN have AP-1 recognition sites in the promoter region of their genes, FRA, an AP-1 component, may play important roles in regulating the expressions of PENK and PDYN in the rat brain.

The long-lasting increases in FRA immunoreactivity observed in this study and the AP-1 DNA binding activity reported by Pennypacker et al. [14] is surprising since most of the earlier reports indicated the transient nature of the charges of AP-1 DNA binding activity and the levels of the FRAs[15, 16]. The mechanism(s) underlying these long-term increases in the expression of FRAs is not presently known. We speculate that the degeneration of the hilar inhibitory interneurons in the hippocampus, such as the GABAergic neurons, after KA treatment may be closely associated with this phenomenon. Loss of these inhibitory neurons causes a disinhibition of the granule cells in the hippocampus which may induce the expression of FRAs and the AP-1 DNA binding activity. The increased AP-1 activity may in turn trigger the expression of some target genes which are involved in the sprouting of the mossy fibers to innervate the granule cells. This positive feedback loop may further excite the granule cells and exacerbate the sprouting phenomenon[16,17,18]. Since both PENK and PDYN genes are expressed in granule cells, it is likely that the long-lasting increase in the expression of both of these opioid genes after KA treatment is triggered by the increase in AP-1 activity. The significance of the persistent increase in the levels of these two opioid peptides after KA treatment is not clear. However, it has been documented that enkephalins exert proconvulsive effects, whereas dynorphins show anticonvulsive effects in the hippocampus[2, 7, 20]. The long-lasting increases in both PENK and PDYN may be associated with the permanent changes in hippocampal excitability after KA treatment.

REFERENCES

1. C. Gall, Seizures induce dramatic and distinctly different changes in enkephalin, dynorphin, and CCK immunoreactivities in mouse hippocampal mossy fibers. *J. Neurosci.* 8:1852-62 (1988).
2. J. S. Hong, J. F. McGinty, L. Grimes, T. Kanamatsu, J. Obie, and C. L. Mitchell, Seizure-induced alterations in the metabolism of hippocampal opioid peptides suggest opioid modulation of seizure-related behaviors, *NIDA Res. Monogr.* 82:48-66 (1988).

3. J. F. McGinty, T. Kanamatsu, J. Obie, and J. S. Hong, Modulation of opioid peptide metabolism by seizures: differentiation of opioid subclasses. *NIDA Res. Monogr.* 71:89-101 (1986).
4. T. Kanamatsu, J. Obie, L. Grimes, J. F. McGinty, K. Yoshikawa, S. Sabol, and J. S. Hong, Kainic acid alters the metabolism of Met5-enkephalin and the level of dynorphin A in the rat hippocampus. *J. Neurosci.* 6:3094-102 (1986).
5. K. R. Pennypacker, D. Walczak, L. Thai, R. Fannin, E. Mason, J. Douglass, and J. S. Hong, Kainate-induced changes in opioid peptide genes and AP-1 protein expression in the rat hippocampus. *J. Neurochem.* 60:204-11 (1993).
6. C. Gall, J. Lauterborn, P. Isackson, and J. White, Seizures, neuropeptide regulation, and mRNA expression in the hippocampus. *Prog. Brain Res.* 83:371-90 (1990).
7. J. S. Hong, L. Grimes, T. Kanamatsu, and J. F. McGinty, Kainic acid as a tool to study the regulation and function of opioid peptides in the hippocampus. *Toxicology* 46:141-57 (1987).
8. J. A. Angulo and B. S. McEwen, Molecular aspects of neuropeptide regulation and function in the corpus striatum and nucleus accumbens. *Brain Res.* 9:1-28 (1994).
9. I. S. Zagon and P. J. McLaughlin, Endogenous opioid systems regulate cell proliferation in the developing rat brain. *Brain Res.* 412:68-72 (1987).
10. I. S. Zagon and P. J. McLaughlin, Identification of opioid peptides regulating proliferation of neurons and glia in the developing nervous system, *Brain Res.* 542:318-23 (1991).
11. R. Racine, V. Okujava, and S. Chipashvili, Modification of seizure activity by electrical stimulation. 3. Mechanisms. *Electroencephalogr. Clin. Neurophysiol.* 32:295-9 (1972).
12. S. T. Young, L. J. Porrino, and M. J. Iadarola, Cocaine induces striatal c-fos-immunoreactive proteins via dopaminergic D1 receptors. *Proc. Natl. Acad. Sci. U.S. A.* 88:1291-5 (1991).
13. T. Popovici, A. Represa, V. Crepel, G. Barbin, M. Beaudoin, and Y. Ben Ari, Effects of kainic acid-induced seizures and ischemia on c-fos-like proteins in rat brain. *Brain Res.*. 536:183-94 (1990).
14 K. R. Pennypacker, L. Thai, J. S. Hong, and M. K. McMillian, Prolonged expression of AP-1 transcription factors in the rat hippocampus after systemic kainate treatment. *J. Neurosci.* 14:3998-4006 (1994).
15. J. I. Morgan, D. R. Cohen, J. L. Hempstead, and T. Curran, Mapping patterns of c-fos expression in the central nervous system after seizure. *J. Neurosci.* 237:192-7 (1987).
16. J. I. Morgan and T. Curran, Stimulus-transcription coupling in the nervous system: involvement of the inducible proto-oncogenes fos and jun. *Annu. Rev. Neurosci.* 14: 421-51 (1991).
17. M. Frotscher and J. Zimmer, Lesion-induced mossy fibers to the molecular layer of the rat fascia dentata: identification of postsynaptic granule cells by the Golgi-EM technique. *J. Comp. Neurol.* 215:299-311 (1983).

18. D. L. Tauck and J. V. Nadler, Evidence of functional mossy fiber sprouting in hippocampal formation of kainic acid-treated rats. *J. Neurosci.*. 5:1016-22 (1985).
19. M. M. Okazaki, P. G. Aitken, and J. V. Nadler, Mossy fiber lesion reduces the probability that kainic acid will provoke CA3 hippocampal pyramidal cell bursting. *Brain Res.* 440: 352-6 (1988).
20. J. S. Hong, Hippocampal opioid peptides and seizures, in: "The Dentate Gyrus and Its Role in Seizures" C.E. Ribak, C.M. Gall and I. Mody, eds., Elsevier Science Publishers, New York (1992).

MODULATION OF VESICULAR GLUTAMATE RELEASE DURING ANOXIA

Norman Hershkowitz and Alexander N. Katchman

Department of Neurology
Georgetown University Medical School
Washington, DC 20007

INTRODUCTION

Despite the fact that stroke is a leading cause of morbidity and mortality, there is very little the clinician has to offer for its acute management. Clarification of the mechanisms that contribute to cell death during hypometabolic states such as stroke and anoxia may lead to novel strategies for therapy. It appears that excitatory amino acids (glutamate and aspartate) released during stroke contribute to the resulting neuronal death through the mechanism of excitotoxicity[1,2]. However, the mechanism of this release and its regulation are poorly understood.

Within minutes of the onset of various hypometabolic states, glutamate is released, but there is also a paradoxical suppression of synaptic activity [3,4]. This depression results from alterations in synaptic physiology[5]. These apparently contradictory results (glutamate release and synaptic suppression) have sometimes been reconciled by hypothesizing that glutamate is released from metabolic and not from vesicular pools (the pools released with synaptic activity)[6,7]. Metabolic glutamate is proposed to accumulate extracellularly due to a suppression or a reversal of the Na^+/glutamate cotransporter (caused by alterations in ionic gradients and resting membrane potential)[8]. Indeed, hypoxia-induced glutamate release is not dependent on extracellular calcium, consistent with a metabolic source (but see Bosley et al.[9]), and voltage-gated Ca^{++} currents, which trigger synaptically elicited transmitter release, are rapidly suppressed by hypoxia[10]. Direct evidence that hypoxia-induced glutamate comes from metabolic stores, however, is lacking. In this chapter we present evidence that anoxia-induced glutamate may in fact be derived from a vesicular source and describe mechanisms of its regulation. These data help to resolve the apparent paradox of simultaneous synaptic depression and increased glutamate release during anoxia.

METHODS

Experiments were performed with 400-μm thick *in vitro* hippocampal slices obtained from 21- to 25-day-old rats. Slices were perfused with oxygenated artificial cerebrospinal fluid (ACSF) in an interface chamber at a temperature of 34ºC. Hypoxia was

Neurodegenerative Diseases
Edited by Gary Fiskum, Plenum Press, New York, 1996

produced by changing the perfusion medium to ACSF saturated with 95%N_2/5%CO_2 and simultaneously switching the humidified gas to humidified 95%N_2/5%CO_2. For a complete description of the experimental methods see Hershkowitz et al.[11].

CA1 Recording

Whole-cell patch recordings were made from CA1 neurons with K-gluconate electrodes voltage clamped at V_H=-60 mV. Spontaneous excitatory postsynaptic currents (EPSCs) were analyzed by computer software (SCAN) that allowed the construction of time/amplitude raster plots (see Fig 1B) and cumulative spontaneous current curves (see Fig. 1C). The mean frequency of these currents was calculated by dividing the total number of EPSCs by the total time of measurement (4-5 min). Evoked synaptic currents were elicited by stimulation of the Schaffer collateral-commissural projection with a bipolar electrode placed in the stratum radiatum.

CA3 Recordings

Whole-cell patch recordings were made from CA3 neurons by using the patch electrodes but in current clamp mode. In some cells input resistance was monitored by hyperpolarizing current steps (0.1-0.5 nA; 0.5 - 1.0 Hz).

RESULTS AND DISCUSSION

In contrast to the prevailing view, glutamate from synaptic vesicles may contribute to at least a part of the glutamate released during hypoxia. Using whole-cell patch clamp in CA1 neurons of hippocampal slices, we have demonstrated that exposure to hypoxia results in an increase in the frequency of spontaneous transient inward currents (Fig. 1A)[11], as evidenced by the increase in slope of the cumulative plot (Fig. 1C). These spontaneous events chiefly represent glutamatergic miniature excitatory postsynaptic currents (mEPSCs), as indicated by their resistance to tetrodotoxin and sensitivity to glutamate receptor antagonists[11]. In spite of the changes in frequency, no obvious changes in the distribution of the peak amplitudes were apparent (see Fig. 1B and Hershkowitz et al.[11])

These results were unexpected because the increase in mEPSC frequency occurred at a time when the orthodromically elicited synaptic response is nearly completely inhibited. The lack of effect on the inhibition of the mEPSCs by hypoxia (Fig. 1B and see Hershkowitz et al.[11]) indicates that the locus of excitatory synaptic suppression is presynaptic.

Spontaneous vesicular transmitter release is thought to be directly proportional to presynaptic Ca^{++} concentration[12]. The elicited orthodromic response appears to depend upon increases in presynaptic Ca^{++} resulting from the activation of voltage-gated channels. The fact that the elicited response was suppressed at a time when the mEPSC frequency increased suggested to us that the frequency increase may result from an elevation in terminal calcium concentration from a source other than influx through voltage-gated channels. We therefore tested the effect of inhibiting various possible sources of the Ca^{++} on mEPSCs under anoxic conditions. The hypoxia-induced increase in mEPSC frequency was unaffected by removal of extracellular Ca^{++}, by the addition of high concentrations of Cd^{++} (a nonspecific inhibitor of voltage-gated calcium channels)[13], or by a combination of these two treatments[14]. However, drugs that inhibit calcium mobilization from intracellular pools (endoplasmic reticulum) (e.g., dantrolene) caused nearly complete suppression of the anoxia-induced increase in mEPSC frequency[14]. Thus some of the arguments against the importance of anoxia-induced vesicular release are flawed. Anoxia

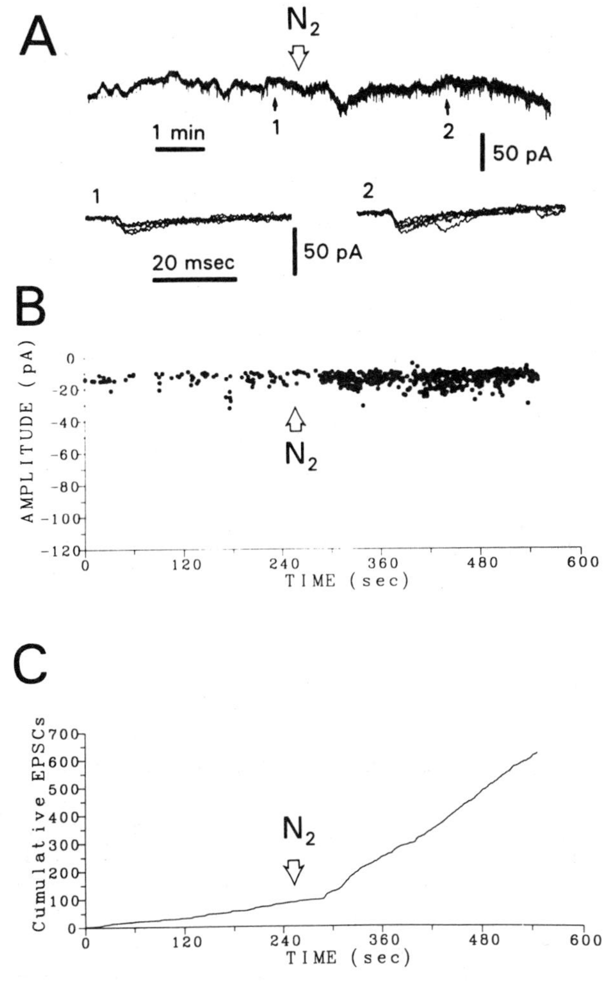

Figure 1. Anoxia-induced alterations in spontaneous activity in a single CA1 cell under control conditions (voltage-clamp mode, V_H=-60 mV).

A. The upper trace is a continuous chart recording. Lower insets are inward currents at an expanded time scale corresponding to the times indicated by small arrows. The open arrow labeled N_2 represents the onset of the anoxic condition in this and all subsequent figures.

B. A time/amplitude raster plot where the peak amplitude of each spontaneous event is plotted as a function of the time of its occurrence.

C. A cumulative plot of the occurrence of transient inward currents. The slope of this plot represents the frequency of events

may induce vesicular release that is not dependent on action potential-triggered, synaptically mediated transmission or transmembrane calcium flux.

Adenosine is a ubiquitous neuromodulator that inhibits excitatory glutamate synaptic transmission. Stimulation of the presynaptic A_1 subtype adenosine receptor reduces glutamate release[15]. Extracellular adenosine concentrations are increased by anoxia and ischemia[16]. As excitotoxicity resulting from ischemia- or anoxia-induced glutamate release appears to cause cell death, alterations in adenosine metabolism may function as an important protective homeostatic mechanism. Thus, adenosine antagonists generally enhance ischemia- and anoxia-induced cell injury, and agonists have proven to be neuroprotective[17].

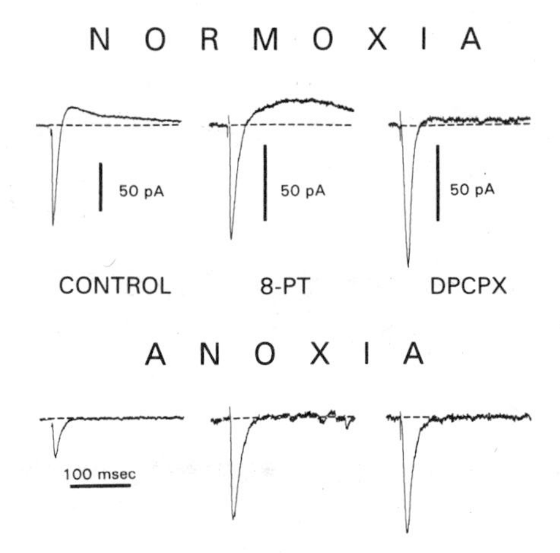

Figure 2. Effects of hypoxia on postsynaptic currents elicited by stimulation of the Schaffer collateral pathway in CA1 neurons in control artificial cerebral spinal fluid (ACSF), and ACSF containing 200 nM of DPCPX and 20μM of 8-PT. Each column represents a different cell with the upper trace obtained during normoxia and the lower trace obtained 3 min. after the initiation of anoxic conditions. Synaptic responses were elicited at a holding potential of -60 mV. The initial inward (downward) current represents the excitatory EPSC, and the later outward (upward) current represents the inhibitory postsynaptic current (IPSC). Current calibration is noted to the right of each record, time calibration is the same in each record.

During hypoxic and ischemic conditions excitatory synaptic activity is rapidly suppressed, as a result of a presynaptic failure. It has been shown by Fowler and ourselves[18,19] that the adenosine A_1 receptor antagonists, 8-phenyltheophilline (8-PT) and 8-cyclopentyl-1,3-dipropylxanthine (DPCPX) protect orthodromically elicited synaptic activity from hypoxia-induced suppression. Thus the anoxia-induced inhibition of the EPSC is markedly reduced in the presence of adenosine antagonists (Fig. 2).

Although adenosine elicits a postsynaptic inward potassium current, which may contribute to synaptic inhibition in the hippocampus[20], the predominant locus of adenosine's inhibitory effect is likely to be presynaptic[15]. This observation is consistent with the presynaptic locus of hypoxia-induced inhibition described above. Interestingly, adenosine antagonists do not protect the inhibitory postsynaptic current (IPSC) from anoxia-induced suppression (Fig. 2 and Katchman et al.[19]). This is consistent with our finding that suppression of the $GABA_A$ IPSC results from a change in the postsynaptic chloride gradient[21].

These data suggest that an important role of the anoxia-induced increase in adenosine may be the suppression of action potential-dependent synaptic glutamate release. However, as noted previously, two forms of vesicular glutamate release occur [spontaneous vesicular (mEPSCs) and action potential-dependent (EPSC)]. Under normoxic conditions adenosine may decrease the frequency of spontaneous mEPSCs[22,23]. This being the case, we hypothesized that the increase in adenosine during anoxia would reduce spontaneous mEPSC frequency. Thus, two opposing processes would modulate spontaneous vesicular release--the increase in terminal free calcium secondary to intracellular calcium mobilization, which will increase mEPSC frequency, and suppression of mEPSC frequency by increases in the extracellular concentration of the neuromodulator adenosine.

To test this hypothesis, we examined the effect of anoxia on spontaneous synaptic activity in cells preincubated in the specific A_1 receptor antagonist DPCPX (200 nM). It was expected that the increase in mEPSC frequency would be greater under such conditions because of the elimination of adenosine as a factor in vesicular release. Figure 3 presents one such experiment. An increase in mEPSC frequency is apparent from the chart

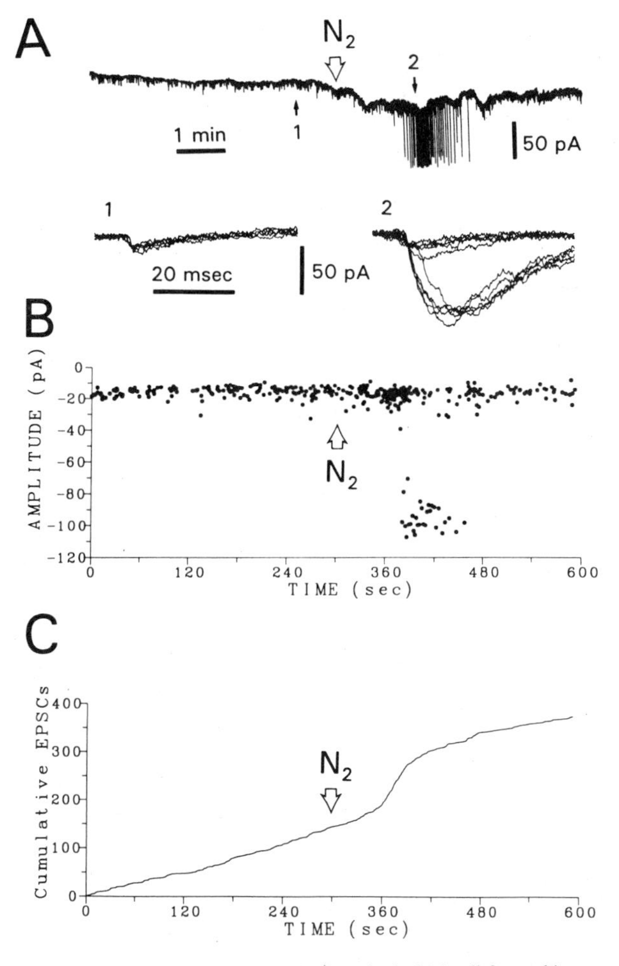

Figure 3. Anoxia-induced alterations in spontaneous activity in a single CA1 cell from a hippocampal slice preincubated in 0.2 μM of DPCPX (V_H=-60 mV).
A. The upper trace is a continuous chart recording. Lower insets are inward currents at an expanded time scale corresponding to the points indicated by small arrows.
B. A time/amplitude raster plot.
C. A cumulative plot of the occurrence of transient inward currents.

recording (Fig 3A) and the increase in slope of the cumulative plot (Fig. 3B). However these data also reveal a new, larger amplitude population of inward currents (see Fig. 3A and B). 17 of 20 cells examined showed a similar large amplitude currents.

There was a slightly greater anoxia-induced increase in the mean frequency of inward currents in the presence of DPCPX compared to control (Table 1). However, a simple comparison of frequency is not justified as the new larger events may represent a different phenomenon. The presence of these larger events also resulted in a significant increase in the mean amplitude of inward currents following anoxia in DPCPX, which was not seen in the control experiments (Table 1).

A careful examination of the nondrug control experiments revealed a similar occurrence of large synaptic events in a very small percentage of cells (2 of 17). This number was sufficiently small as to result in no significant increase in the mean amplitude of spontaneous events (Table 1). In an attempt to identify these currents, we performed a number of experimental manipulations. Thus in a series of experiments (n=12), both large and small amplitude inward currents were absent during the normoxic and anoxic

Table 1. Anoxia-induced alterations in spontaneous synaptic events in control conditions and in the presence of DPCPX.

	CONTROL		DPCPCX	
	O_2	N_2	O_2	N_2
Amplitude (pA)	17.0±1.0	18.0±1.0	16.0±1.0	22.0±1.0
Frequency (Hz)	0.6±0.2*	1.0±0.2	0.4±0.2*	1.6±0.3
n	17		20	

Values are mean ± SE.
*Normoxia (O_2) significantly different from anoxia (N_2) control (paired t test: $P<0.05$).

period when slices were preincubated in DPCPX (200 nM) and the glutamate antagonists 6-cyano-7-nitroquinoxaline-2,3-dione (CNQX, 50 μM) and 3-(2-carboxypiperazine-4-yl)propyl-l-phosphonic acid (CPP, 20 μM). This result shows that all of the measured inward currents are glutamatergic. In another series of experiments (n=6), the CA1 region of slices preincubated in DPCPX (0.2 μM) was completely isolated from its major excitatory input by removal of the CA3 region (minislices). No large amplitude inward transient currents were observed during anoxia in these experiments. This can be appreciated from Table 2, which demonstrates no difference in the mean amplitude of events following anoxia. Furthermore 20 cells recorded in slices preincubated in tetrodotoxin (TTX) as well as DPCPX were studied. As in the surgically manipulated slices, no large inward currents were observed. This can be appreciated from the lack of significant alteration in mean amplitude of the inward currents following anoxia (Table 2). These data suggest that the larger currents represent unitary, action potential-dependent, glutamatergic excitatory postsynaptic currents (EPSCs).

It is also apparent from Table 2 that after deafferentation the anoxia-induced increase in frequency is equal to or slightly reduced from that seen in control studies (compare to Table 1 and our previous studies[11,14]). While frequency was increased in CA1 minislices , the increase was not significant, likely because a small number of cells were examined. It therefore appears that adenosine functions to inhibit action potential-dependent glutamatergic vesicular release (EPSCs), but not mEPSCs, during anoxia and in doing so reduces excitotoxic damage.

Table 2. Anoxia-induced alterations in spontaneous synaptic events in the presence of an adenosine antagonist following deafferentation.

	DPCPX + TTX		DPCPCX Minislices	
	O_2	N_2	O_2	N_2
Amplitude (pA)	16.0±0.4	16.3±0.5	16.0±1.0	15.0±1.0
Frequency (Hz)	0.7±0.1*	1.3±0.2	0.3±0.2	0.5±0.2
n	20		6	

Values are mean ± SE.
*Normoxia (O_2) significantly different from anoxia (N_2) control (paired t test: $P<0.05$).

As the spontaneous CA1 EPSCs appeared to originate from the Schaffer collateral projection of CA3 pyramidal cells, activity in CA3 neurons was studied with current clamp. Figure 4 represents one such experiment. Note that a depolarization followed by a repolarization (a "nub") occurs approximately 1.5 minutes after the onset of anoxia. The nub appears to elicit repetitive bursts of action potentials as well as single action potentials. Many of the single action potentials observed during this time appeared to arise from spontaneous (presumably mEPSCs) synaptic activity. In other cells, only a single burst was apparent. In a number of studies, we monitored membrane input resistance by an 0.1-0.5 nA hyperpolarizing current injection. Anoxia lead to an initial reduction in input resistance (64 % of normoxic value) that was followed by a transient recovery (97 % of normoxic value) during the nub. This suggests that the depolarizing nub resulted from a turning off of an ionic conductance (perhaps a K+ conductance). A depolarization associated with a increase in resistance would act as an ideal stimulus for the initiation of action potentials. This is not only because the depolarization brings the membrane potential closer to the action potential threshold but also because the increase in transmembrane resistance increases the space constant. The latter allows the axon hillock to sense voltage changes at greater distances on the dendritic tree (for example, those due to mEPSCs).

The depolarizing nub has generally not been observed in other electrophysiologic studies of anoxia. Thus, most investigators have reported a monophasic reduction in membrane input resistance[4]. This raises the issue as to whether the nub may represent an artifact of the patch technique. We do not think it is artifactual, because there was a remarkable overlap between the increase in action potential-dependent EPSCs in the CA1 region in the presence of DPCPX and nub- associated CA3 action potentials. Table 3 compares the time course of these two phenomena. The time of the largest unitary EPSC was used as an index to measure the period of action potential-dependent EPSCs in CA1. The mean time to the largest EPSC in CA1 cells lies well within the time interval between the first and last nub-induced action potential in CA3. This correlation suggests that the

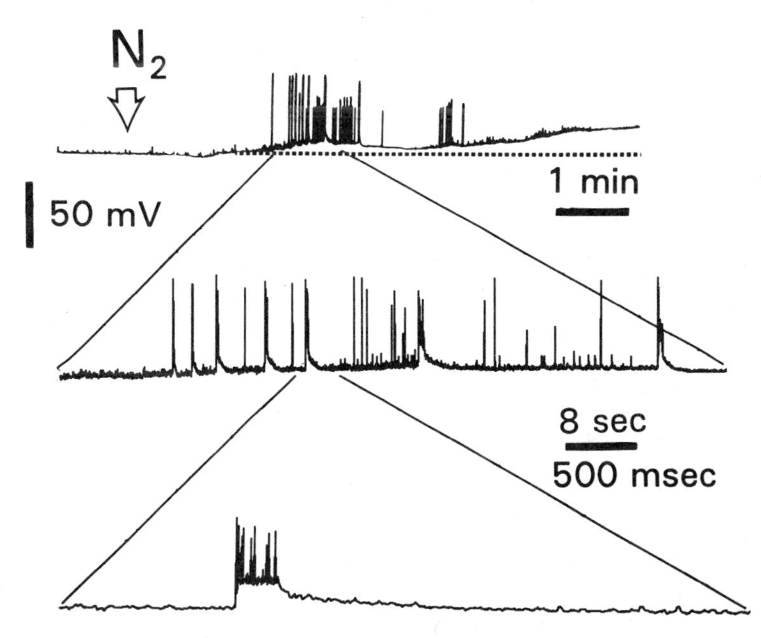

Figure 4. Anoxia-induced activity in CA3 cell measured as voltage changes in current-clamp mode [resting membrane potential (RMP)=-50 mV]. Lower traces represent an expanded time scale in the intervals indicated by the lines.

CA3 phenomenon is real and probably causally related to the CA1 EPSCs. We feel the difficulty in identifying the nub in other studies results from an artifact of the use of sharp electrodes. Penetration with such electrodes would cause a tear around the point of entry and result in an artifactually decrease in leak resistance. Anoxia-induced swelling would tend to enlarge this tear causing a further decrease in this nonspecific leak resistance eclipsing the nub related changes in cell membrane resistance.

Table 3. Comparison of the time of onset of the nub-induced action potential in CA3 cells with EPSC activity in CA1 cells.

	Nub-induced CA3 action potential		CA1 EPSC (in DPCPX)
	First	Last	Largest
Time following anoxia (sec.)	104±9	157±8	134±11
n	19		14

Values are mean ± SE.

In conclusion, there is an increase in vesicular glutamate release during the early stages of anoxia. This increase is predominantly represented by action potential-independent vesicular release measured as spontaneous mEPSCs. This glutamate release results from intracellular calcium mobilization. Although adenosine does not modulate this anoxia-induced, TTX-resistant vesicular release (mEPSCs), it does suppress action potential-propagated glutamate release (EPSCs). Such activity generally is not seen unless tissue is exposed to an adenosine A_1 antagonist. This is likely and important contributing factor to adenosine's natural protective effect. The EPSCs observed in CA1 (in the presence of the antagonist) appears to be triggered by a depolarization in CA3 resulting from a decrease in conductance. Although a K^+ channel is a logical candidate, the actual current underlying this depolarization has not yet been identified.

Acknowledgments

The authors would like to extend their appreciation to Dr. Katrina L. Kelner for editorial assistance and Laura Grubb for technical assistance. This work was supported by NIH grant NS32426.

REFERENCES

1. S.M. Rothman, J.H.Thurston and R.E. Hauhart, Delayed neurotoxicity of excitatory amino acids in vitro, *Neuroscience* 22:471 (1987).
2. D.W. Choi, J. Koh and S. Peters, Pharmacology of glutamate toxicity in cortical cell cultures: Attenuation by NMDA antagonists, *J. Neurosci.* 8:185 (1988).
3. A. Mayevsky and B. Chance, Metabolic responses of the awake cerebral cortex to anoxia hypoxia spreading depression and epileptiform activity, *Brain Res.* 98:149 (1975).
4. N. Fujiwara, H. Higashi, K. Shimoji and M. Yoshimura, Effects of hypoxia on rat hippocampal neurones *in vitro*. *J. Physiol.* 384:131, (1987).
5. H. Collewijn and A. Van Harreveld. Intracellular recording from cat spinal motoneurone during acute asphyxia, *J. Physiol.* 185:1 (1966).
6. M. Ikeda, T. Nakazawa, K. Abe, T Kaneko and K. Yamatsu, Extracellular accumulation of glutamate in the hippocampus induced by ischemia is not calcium dependent - *in vitro* and *in vivo* evidence, *Neurosci. Lett.* 96:202 (1989).

7. J. Sanchez-Prieto and P. Gonzalez. Occurrence of a large Ca^{2+}-independent release of glutamate during anoxia in isolated nerve terminals (synaptosomes), *J. Neurochem.* 50:1322 (1988).
8. I. Torgner and E. Kvamme, Interrelationship between glutamate and membrane-bound ATPases in nerve cells, *Mol. Chem. Neuropathol.* 12:19 (1990).
9. T.M. Bosley, P.L. Woodhams, R.D. Gordon and R. Balozs, Effects of anoxia on the stimulated release of amino acid neurotransmitters in the cerebellum *in vitro*. *J. Neurochem.* 40:189 (1983).
10. K. Krnjevic and J. Leblond, Changes in membrane currents of hippocampal neurons by brief anoxia, *J. Neurophys.* 62:15 (1989).
11. N. Hershkowitz N, A.N. Katchman and S. Veregge, Site of synaptic depression during hypoxia: A patch clamp analysis, *J. Neurophysiol.* 69:432 (1993).
12. E. Alnaes and R. Rahamimoff, On the role of mitochondria in transmitter release from motor nerve terminals, *J.Physiol.* 248:285 (1975).
13. A.P. Fox, M.C. Nowycky and R.W. Tsien. Kinetic and pharmacological properties distinguishing three types of calcium currents in chick sensory neurones, *J. Physiol.* 394:149 (1987).
14. A.N. Katchman and N. Hershkowitz, Early anoxia-induced vesicular glutamate release results from mobilization of calcium from intracellular stores, *J. Neurophysiol.* 70:1 (1993).
15. C.R. Lupica, W.R. Proctor, and T.V. Dunwiddie, Presynaptic inhibition of excitatory synaptic transmission by adenosine in rat hippocampus: analysis of unitary EPSP variance measured by whole-cell recording, *J. Neurosci.* 12:3753 (1992).
16. H. Hagberg , P. Anderson, J. Lacarewicz, I. Jacobson, S. Butcher, M. Sandberg. A. Lehmann, and A.X. Hambereger, Extracellular adenosine, inosine, hypoxanthine, and xanthine in relation to tissue nucleotides and purines in rat striatum during transient ischemia, *J. Neurochem.* 49:227 (1985).
17. K.A. Rudolphi, P. Schubert, F.E. Parkinson and B.B. Fredholm, Neuroprotective role of adenosine in cerebral ischemia, *Trends Pharmacol. Sci.* 13:434 (1992).
18. J.C. Fowler, Adenosine antagonists delay hypoxia-induced depression of neuronal activity in hippocampal brain slice, *Brain Res.* 490:378 (1989).
19. A.N. Katchman and N. Hershkowitz, Adenosine antagonists prevent hypoxia-induced suppression of excitatory but not inhibitory synaptic currents, *Neurosci. Lett.* 159:123 (1993).
20. U. Gerber, R.W. Greene, H.L. Haas, and D.R. Stevens, Characterization of inhibition mediated by adenosine in the hippocampus of the rat *in vitro*. *J. Physiol.* 417:567 (1989).
21. A.N. Katchman, S. Vicini and N. Hershkowitz, Mechanism of early anoxia-induced suppression of the $GABA_A$-mediated inhibitory postsynaptic current, *J. Neurophysiol.* 71:1128 (1993).
22. K.P. Sholz and R.J. Miller, Inhibition of quantal transmitter release in the absence of calcium influx by G protein-linked adenosine receptor at hippocampal synapses, *Neuron.* 8:1139 (1992).
23. S.M. Thompson, M. Capogna and M. Scanziani, Presynaptic inhibition in the hippocampus, *Trends Neurosci.* 16:222 (1993).

A MODEL FOR THE EXPRESSION OF DIFFERENT GLUTAMATE TRANSPORTER PROTEINS FROM A RAT ASTROCYTE-TYPE GLUTAMATE TRANSPORTER GENE

Raymond S. Roginski, MD, PhD

Assistant Professor, Department of Anesthesia
UMDNJ-Robert Wood Johnson Medical School
Clinical Academic Building, Suite 3100
125 Paterson Street/New Brunswick, NJ 08901-1977

INTRODUCTION

The mammalian brain uses the largest variety of mechanisms of gene expression of any organ to generate many different proteins from a single gene. Nowhere is this phenomenon more obvious and relevant than in the families of genes encoding the ionotropic glutamate (Glu) receptors (iGluRs)[1]. The differences introduced usually affect the physiologic role of the variant protein. For example, the gene encoding the NMDAR1 iGluR makes nine splice variants[2] that display distinct properties when expressed in exogenous systems[2,3]. A second example involves RNA editing of the GluR2 primary transcript that results in the incorporation of the amino acid (AA) arginine (R) instead of glutamine (Q) at the R/Q site in the putative intramembranous[4] domain of the iGluRs. This alteration abolishes the calcium permeability of the edited GluR2 subunit.

Glu transporters (GluTs) are another multigene family involved in Glu neurotransmission[5]. My laboratory has cloned and is studying the functional expression of an astrocyte GluT, GluT-R[6], whose sequence is nearly identical to GLT-1 cDNA[7,8] (see below for differences). GluT-R cDNA lacked the putative GLT-1 translation initiation site[8] (TIS; termed TISα1 below), but it contained an excellent[9] TIS (designated TISβ in this model), located in the same reading frame but 120 AA downstream from TISα1. The TISα1 and β proteins' predicted lengths are 573 AA and 453 AA, respectively. Since Glu receptors are known to employ multiple mechanisms of gene expression, I reasoned that GluT genes might behave this way as well. This chapter will review previous work from other laboratories and my own on the rat astrocyte-type GluT, represented by the cDNA clones GLT-1 and GluT-R. Data presented at the Neurodegenerative Diseases '95 symposium that reveal another TIS will then be described. The last section will discuss a unifying hypothesis to explain how at least three proteins can be produced from the GluT gene.

DATA WITH WHICH THE MODEL MUST BE CONSISTENT: GluT GENE PRODUCTS

Protein Species Detected with Anti-GluT Antibodies

Danbolt and colleagues[10] developed a purification protocol to isolate the putative glial GluT (later termed GLT-1 by this group) from rat brain homogenates. Subsequently these authors raised a polyclonal antiserum against the purified transporter[11]. In the former publication, the major purified protein band migrated with M_r 80,000; in the latter, 73,000. In both references the authors commented that the transporter showed a very broad band. The apparent size decreased by about M_r 10,000 with deglycosylation[11] but the band remained broad. We have raised and affinity-purified an antiserum against the C-terminal 18 AA of GluT-R/GLT-1[12]. This antiserum also detects a broad band (M_r 75,000) in rat brain homogenates. Upon deglycosylation, the size decreased by M_r 10,000 but the band width was maintained. Another group[13] raised an antiserum against the C-terminal 15 AA of the GluT sequence and saw a 73,000 M_r broad band. The reason for the broadness of the band has not been elucidated, but glycosyl residues do not appear to be the cause. Perhaps multiple, closely-migrating species exist and/or novel covalent modifications are present. Both phenomena could contribute to the unusual width of the larger band.

Another aspect of the purification protocol[10] was that several smaller bands co-purified with the 80,000 M_r band. One of these showed an M_r of ~50,000 and migrated broadly, although not as broadly as the larger band. This band did not react with the polyclonal anti-protein antiserum[11] under the conditions described. However, our anti-carboxyl antiserum detected a specific, immunoreactive M_r 50,000 band (in addition to the larger band) in total brain homogenates that appears to be about the same size as the co-purifying band. This band most likely represents a specific TISβ translation product (rather than a degradation product) because it co-migrates with the predominant *in vitro* translation (IVT) product detected with rat brain poly(A)$^+$ RNA and TIS$\alpha\beta$ cRNA templates[14]. The 50,000 M_r species is much less abundant than the M_r 75,000 protein.

Although the GluT-R and GLT-1 cDNAs most probably originated from the same chromosomal gene, there are four differences within the region we have deposited in the GenBANK (accession number U15098). Two single-base "insertions" occur 5′ to TISα1. Another "insertion" in the 3′ untranslated region generates a polyadenylation signal (AATAAA) in GluT-R that is absent in GLT-1. This sequence may be involved in the processing of the 2 kb mRNA species. The fourth difference is a "substitution," an A in codon 401 (isoleucine) of GluT-R that corresponds to a G in codon 521 (valine) of GLT-1. Interestingly, an example of RNA editing resulting in either isoleucine or valine has been described in an iGluR subunit[1].

Northern Blots and 5′ Rapid Amplification of cDNA Ends (RACE)

In total brain RNA, the 1.6 kb GluT-R TISβ cDNA detects three bands, an intense 11 kb species and two less abundant, smaller RNAs, 3.5 kb and 2 kb in size[14]. We have isolated three putative mRNA 5′ ends by 5′ RACE[15], located at -550, -300 and +41 relative to the 5′ end reported for GLT-1 cDNA[8]. Further experiments reported at this meeting[16] revealed that some transcripts corresponding to the -550 and/or -300 5′ ends included an alternative exon 156 bases long between TISα1 and TISβ. The DNA sequence showed a termination codon that closed the TISα1 reading frame. However, another site, termed TISα2, was found upstream from the main GluT reading frame. Remarkably, TISα2 reads into the main GluT reading frame and predicts an amino terminus of 12 residues that are completely different from the initial six AA of the TISα1 translation product. The TISα2 product is predicted to be 579 AA long. An entirely analogous situation has been described for two glycine transporter variants, GLYT-1 and GLYT-2[17], that differ at their amino termini and have distinct locations in the CNS and peripheral tissues.

THE MODEL: ALTERNATE SPLICING AND DOWNSTREAM (INTERNAL) INITIATION

A possible evolutional justification for deriving several proteins from a gene is conservation of genetic material. However, the opposite strategy, gene duplication followed by mutational divergence, has also been observed. In order to account for different 5′ ends, the most likely explanation would be to have several promoters. The 5′ RACE data are consistent with this part of the model, although specific RNA cleavage sites are an alternative hypothesis.

To explain the different mRNA sizes observed in Northerns, differential 3′ end cleavage must also be postulated. Preliminary 3′ RACE experiments performed in my laboratory with a cDNA synthesis primer that contains a 3′ terminal oligo(dT) segment have provided evidence for a 3′ end very close to the carboxyl terminus of the GluT reading frame. Combined with 5′ RACE data for a 5′ terminus at +41, it appears likely that the 2 kb mRNA species can be explained. Since all 5′ ends are less than 700 bases from the reading frame, the 11 kb mRNA must have an extremely large (~ 8.5 kb) 3′ untranslated region, the function of which is currently unclear.

Since no mRNA species lacking TISα1 and α2 were revealed by RACE experiments and since *in vitro* translation data showed that cRNA corresponding to the +41 terminus produced predominantly the smaller protein, my expectation is that the 2 kb mRNA species is responsible for the TISβ protein *in vivo*. The shorter protein is present in smaller amounts than the larger forms, consistent with the low abundance of its presumed mRNA.

The 573 AA protein probably derives from the 11 kb mRNA species, which probably incorporates the -550 and/or -300 5′ ends. Perhaps the longer 5′ end allows translation to proceed more efficiently from TISα1 than in the +41 transcripts. Although TISα1 has the advantage of being the most upstream initiation site, it is also weaker than α2 and β by virtue of pyrimidines at the key -6 and -9 positions[9].

The 579 AA protein (see Diagram) is probably less abundant than the 573 AA, based upon the PCR amplification experiment that resulted in its detection[16]. However, PCR is not rigorously

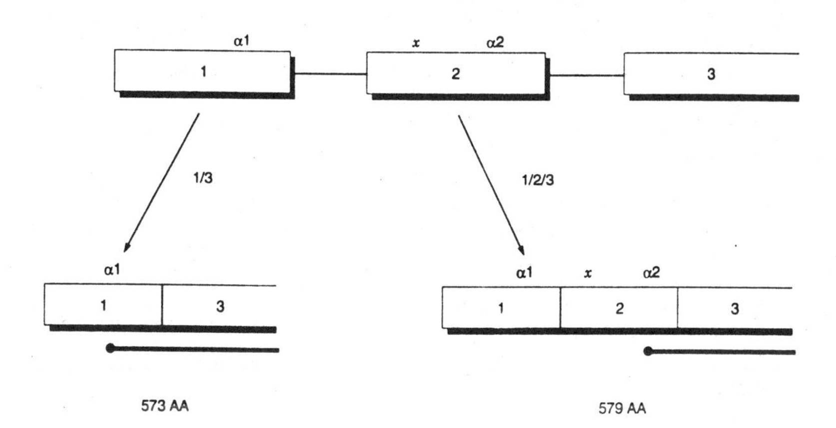

quantitative and protein levels are not always proportional to mRNA amounts. Therefore, quantitation of the levels of the 579 AA transporter will be performed with a specific antiserum generated against the unique amino terminal residues (in progress).

CONCLUSIONS

My colleagues and I have begun to categorize different gene products of the astrocyte-type GluT gene. We have found evidence for three mRNA species and three proteins, although we currently have data for the 579 AA protein only from the deduced AA sequence of the alternatively spliced exon (see Diagram). To my knowledge, the genomic structure of this gene has not been reported, but the model predicts the locations of three exons and two introns.

Downstream initiation has been postulated to account for the production of the 453 AA protein. To my knowledge, this is the first example of downstream initiation in the mammalian brain. The 579 AA protein is predicted to be translated from an alternatively spliced transcript. This mRNA includes an exon that terminates the α1 reading frame and initiates a new amino terminus (at the α2 site) that reads into the main translation frame. This situation appears to be similar to two recently described glycine transporter variants[17].

This model of GluT gene expression may explain certain extant data about this transporter subtype and should be helpful for interpreting future experiments designed to elucidate the presumed distinct physiological roles of the different transporter proteins. For example, the predicted transmembrane topology of the 453 AA protein differs markedly from the 573 AA protein[14,15]. In fact, the smaller protein is predicted to resemble the iGluRs in structure. Therefore, it would not be surprising if it were expressed in other cells, e.g., neurons, in addition to astrocytes. Many interesting observations, some of which may be relevant to human neurodegenerative diseases, are likely to sprout from the GluT field in the near future.

Acknowledgments

The author wishes to acknowledge support from the Department of Anesthesia, UMDNJ-Robert Wood Johnson Medical School and the Foundation for Anesthesia Education and Research (FAER). The author is the recipient of a Burroughs-Wellcome/FAER Young Investigator Award.

REFERENCES

1. M. Hollmann and S. Heinemann, Cloned glutamate receptors, *Annu. Rev. Neurosci.* 17:31(1994).
2. M. Hollmann, J. Boulter, C. Maron, L. Beasley, J. Sullivan, G. Pecht, and S. Heinemann, Zinc potentiates agonist-induced currents at certain splice variants of the NMDA receptor, *Neuron* 10:943(1993).
3. G.M. Durand, P. Gregor, X. Zheng, M.V. Bennett, G.R. Uhl, and R.S. Zukin, Cloning of an apparent splice variant of the rat N-methyl-D-aspartate receptor NMDAR1 with altered sensitivity to polyamines and activators of protein kinase C, *Proc. Natl. Acad. Sci. USA* 89:9359(1992).
4. M. Hollmann, C. Maron, and S. Heinemann, N-Glycosylation site tagging suggests a three transmembrane domain topology for the glutamate receptor GluR1, *Neuron* 13:1331(1994).
5. Y. Kanai, C.P. Smith, and M.A. Hediger, A new family of neurotransmitter transporters: the high-affinity glutamate transporters, *FASEB Journal* 7:1450(1993).
6. R. Roginski, K. Choudhury, M. Marone, and H. Geller, Molecular characterization and expression of a rat brain glutamate transporter cDNA, *Anesthesiology* 79(3A): A787(1993).
7. B.I. Kanner, Glutamate transporters from brain. A novel neurotransmitter transporter family, *FEBS Lett.* 325:95(1993).
8. G. Pines, N.C. Danbolt, M. Bjoras, Y. Zhang, A. Bendahan, L. Eide, H. Koepsell, J. Storm-Mathisen, E. Seeberg, and B.I. Kanner, Cloning and expression of a rat brain L-glutamate transporter, *Nature* 360:464;768(1992).
9. M. Kozak, An analysis of 5′-noncoding sequences from 699 vertebrate mRNAs, *Nuc. Acids Res.* 15:8125(1987).
10. N.C. Danbolt, G. Pines, and B.I. Kanner, Purification and reconstitution of the sodium- and potassium-coupled glutamate transport glycoprotein from rat brain, *Biochem.* 29:6734(1990).
11. N.C. Danbolt, J. Storm-Mathisen, and B.I. Kanner, An [Na^+ + K^+]coupled L-glutamate transporter purified from rat brain is located in glial cell processes, *Neurosci.* 51:295(1992).
12. H. Lee and R.S. Roginski, Characterization of glutamate transporters from rat striatum with affinity-purified antibodies, *Anesthesiology*, manuscript submitted.
13. J.D. Rothstein, L. Martin, A.I. Levey, M. Dykes-Hoberg, L. Jin, D. Wu, N. Nash, and R.W. Kuncl, Localization of neuronal and glial glutamate transporters, *Neuron* 13:713(1994).
14. K. Choudhury, R.S. Roginski, S. Meiners, M. Marone, and H.M. Geller, Cloning and expression of GluT-R, a novel, high-affinity rat brain glutamate transporter, manuscript in preparation.
15. R. Roginski, K. Choudhury, A. Basma, M. Marone, and H. Geller, Rat brain cDNA GluT-R encodes a functional glutamate transporter, *Anesthesiology* 81(3A): A870(1994).
16. R.S. Roginski, K. Choudhury, and J. Yospin, Rat brain glutamate transporter mRNA variant predicts an alternate amino terminus for the astrocyte-type transporter, *Proceedings of the XVth Washington International Spring Symposium. Neurodegenerative Diseases '95: Molecular and Cellular Mechanisms and Therapeutic Advances.*
17. B. Borowsky, E. Mezey, and B.J. Hoffman, Two glycine transporter variants with distinct localization in the CNS and peripheral tissues are encoded by a common gene, *Neuron* 10:851(1993).

GLUTAMATE TRANSPORTERS: MOLECULAR MECHANISMS OF FUNCTIONAL ALTERATION AND ROLE IN THE DEVELOPMENT OF EXCITOTOXIC NEURONAL INJURY

Davide Trotti[1], Niels C. Danbolt[2], Barbara Lodi Rizzini[1], Paola Bezzi[1], Daniela Rossi[1], Giorgio Racagni[1] and Andrea Volterra[1]

[1]Center of Neuropharmacology, Institute of Pharmacological Sciences, University of Milan, 20133, Italy and [2]Department of Anatomy, Institute of Basic Medical Sciences, University of Oslo, N-0317, Oslo, Norway.

GLUTAMATE UPTAKE AND TRANSPORTER SUBTYPES

Glutamate uptake is essential to maintain resting extracellular glutamate ($[Glu]_O$) $\leq$ 1 μM, a concentration not producing significant activation of excitatory aminoacid (EAA) receptors[1]. Thereby, sharp synaptic signalling takes place upon glutamate release, while receptor overstimulation leading to excitotoxic neuronal damage is avoided. Modulatory changes of glutamate uptake may affect fast excitatory transmission[2], while impaired or reversed transport likely participates to the neuropathology of ischemia/reperfusion injury[3] and amyotrophic lateral sclerosis[4] (ALS).

Glutamate uptake process is electrogenic and involves inward cotransport of 2 Na^+ ions and outward countertransport of K^+ and OH^-, with net inward movement of one positive charge[5]. 4 different transporter subtypes have been cloned so far: EAAC1, located in nerve cells of several brain regions and the spinal cord; GLT-1, similarly widespread in CNS, but confined to astrocytic processes; EAAT4, concentrated in cerebellar neurons; and GLAST, seen at highest level in cerebellar Bergmann's glia, but also present in the glial cells of other brain areas[6,7]

ARACHIDONIC ACID INHIBITION OF GLUTAMATE TRANSPORT

Arachidonic acid (AA) rapidly inhibits glutamate uptake[8,9]. To study its mode of action we have utilized a highly purified brain glutamate transporter (shown to be GLT-1) functionally reconstituted in artificial liposomes[10]. AA potently and reversibly inhibits uptake also by this preparation, suggesting that it directly interacts with the transport process. The action of AA is mimicked by fatty acids with similar degree of cis-unsaturation, but not by saturated, trans-unsaturated ones and by arachidonate esters. Therefore, both a folded carbon chain and a free carboxyl group are required for molecular interaction. AA effect is reversed by rapid dilution of the incubation medium. This

Neurodegenerative Diseases
Edited by Gary Fiskum, Plenum Press, New York, 1996

procedure does not remove the AA bound to liposomal lipids (≈95% of the added AA), suggesting that only the AA partitioning to the aqueous phase affects transport by binding to the transporter protein or to the protein-lipid boundary. Pre-saturation of lipidic binding sites with AA itself or with its inactive ethyl ester results in a 50 to 100-fold increase of inhibitory potency. Therefore, by binding AA tightly, the membrane lipids extract it from the water phase, significantly reducing its active concentration.

OXYGEN RADICAL INHIBITION OF GLUTAMATE TRANSPORT

We have recently shown that glutamate uptake in cortical glial cultures is persistently inhibited upon brief exposure to reactive oxygen species (ROS) such as superoxide anion (O_2^-) and hydrogen peroxide (H_2O_2)[11]. This effect is antagonized and partly reversed by disulfide-reducing agents (e.g. dithiothreitol, DTT), suggesting that protein SH groups are target for inhibition. H_2O_2 attenuates electrogenic glutamate uptake current in voltage-clamped astrocytes with minor or no effect on resting membrane conductance: therefore, its action on glutamate transport could be direct. Indeed, we have now evidence that H_2O_2 reduces uptake by highly purified GLT-1 reconstituted into liposomes. This data demonstrates that at least one glutamate transporter subtype is molecular target for oxidants.

Although both AA and ROS directly inhibit glutamate transport, their mechanisms of action are different and independent[12]. Thus, full additivity exists when AA and ROS are applied together. Moreover, distinct pharmacological agents selectively block either effect, while AA+ROS are antagonized only by a combination of them.

ALTERED GLUTAMATE TRANSPORT AND SELECTIVE VULNERABILITY

Alteration of glutamate transport is neurotoxic

Experiments in cell cultures suggest that alteration of glutamate transport with competitive inhibitors such as threo-ßOH-aspartate (THA) or t-pyrrolidin-dicarboxylate (PDC) results in excitotoxic neuronal death. Thus, chronic application of THA to organotypic spinal cord cultures produces slow neuronal degeneration mediated by AMPA/kainate receptors[13]. On the other hand, we find that acute application of PDC to cortical neuroglial co-cultures causes NMDA-dependent neuronal cell loss within 24 h. This effect of PDC correlates with an induced rise of ambient $[Glu]_o$, due to 2 distinct mechanisms: (a) reduced uptake of synaptically released glutamate and (b) transporter-mediated efflux from both neurons and glia[14].

Arachidonic acid in ischemia

Early during ischemia AA and other fatty acids are released from phospholipids via PLC and PLA_2 and accumulate in the brain due to failure of reacylation, an energy-dependent process[15]. AA is specifically released at EAA synapses via NMDA-dependent PLA_2 activation[16] and its accumulation is highest at those rich in NMDA receptors (e.g. in area CA1 of hippocampus). Most of this AA is not metabolized because (1) oxidative reactions are abolished by anoxia; (2) neurons and glia even in normoxia have very low metabolic capacity compared to extra-blood-brain-barrier tissues. Therefore, glutamate transport in ischemia is likely targeted by excess free AA. Indeed, synaptosomes from ischemic and post-ischemic (1h) tissue display reduced uptake capacity and this defect can be counteracted by treatment with albumin, a free fatty acid-binding protein[17]. Interstingly, in the post-ischemic period, NMDA responses of CA1 pyramidal cells are found potentiated, a phenomenon called "anoxic LTP"[18]. In spite of normal glutamate release, reduced uptake with AA might prolong $[Glu]_0$ in the synaptic space and contribute to enhanced NMDA activation.

Oxygen radicals in ALS

ALS is a lethal disease of unknown pathogenesis causing selective degeneration of cortical and spinal motor neurons. Recently, defects in Cu/Zn superoxide dismutase (SOD-1) gene have been reported in the familial form of ALS[19]. SOD-1 is a key enzyme in the scavenging of cytosolic O_2^-: its alteration might lead to O_2^- accumulation or to the formation of related noxious species (e.g. peroxynitrite and hydroxyl radical). Another defect specific to ALS is functional impairment of glutamate uptake in the motor cortex and spinal cord[4]. Rothstein and co-workers, by use of transporter subtype-specific antibodies, find selective loss of GLT-1 protein[20]. GLT-1 in the spinal cord is located in the astrocytic processes surrounding motor neurons[7]: its loss might cause local $[Glu]_0$ rise, with excitotoxic consequences[13]. The reasons why GLT-1 is lost in ALS are unknown. However, according to our data, one could hypothesize that pathological rise of ROS (or related species) in the motor neuron environment results in GLT-1 inactivation.

REFERENCES

1. S.A. Lipton and P.A. Rosenberg, Excitatory amino acids as a final common pathway for neurologic disorders. *N. Engl. J. Med.* 330:613 (1994)
2. G. Tong and C.E. Jahr, Block of glutamate transporters potentiates postsynaptic excitation. *Neuron* 13:1195 (1994)
3. M. Szatkowski and D. Attwell, Triggering and execution of neuronal death in brain ischemia: two phases of glutamate release by different mechanisms. *Trends Neurosci.* 17:359 (1994)
4. J.D. Rothstein, L.J. Martin and R.W. Kunkl, Decreased glutamate transport by the brain and spinal cord in amyotrophic lateral sclerosis. *N. Engl. J. Med.* 326:1464 (1992)
5. N.C. Danbolt, The high affinity uptake system for excitatory amino acids in the brain. *Progr. Neurobiol.* 44:377 (1994)
6. J.D. Rothstein, L. Martin, A.I. Levey, M. Dykes-Hoberg, L. Jin, D. Wu, N. Nash and RW Kunkl RW, Localization of neuronal and glial glutamate transporters. *Neuron* 13:713 (1994)
7. K.P. Lehre, L.M. Levy, O.P. Ottersen, J. Storm-Mathisen and N.C. Danbolt, Differential expression of two glial glutamate transporters in the rat brain: quantitative and immunocytochemical observations. *J. Neurosci.* 15:1835 (1995)
8. B. Barbour, M. Szatkowski, N. Ingledew and D. Attwell, Arachidonic acid induces a prolonged inhibition of glutamate uptake into glial cells. *Nature* 342:918 (1989)
9. A. Volterra, D. Trotti, P. Cassutti, C. Tromba, A. Salvaggio, R.C. Melcangi and G Racagni, High sensitivity of glutamate uptake to extracellular free arachidonic acid levels in rat cortical synaptosomes and astrocytes. *J. Neurochem.* 59:600 (1992)
10. D. Trotti, A. Volterra, K.P. Lehre, D. Rossi, O. Gjesdal, G. Racagni and N.C. Danbolt, Arachidonic acid inhibits a purified and reconstituted glutamate transporter directly from the water phase and not via the phospholipid membrane. *J. Biol. Chem.* (1995) in press.
11. A. Volterra, D. Trotti, C. Tromba, S. Floridi and G. Racagni, Glutamate uptake inhibition by oxygen free radicals in rat cortical astrocytes. *J. Neurosci.* 14:2924 (1994)
12. A. Volterra, D. Trotti and G. Racagni, Glutamate uptake is inhibited by arachidonic acid and oxygen radicals via two distinct and additive mechanisms. *Mol. Pharmacol.* 46:986 (1994)
13. J.D. Rothstein, L. Jin, M. Dykes-Hoberg, and R.W. Kunkl, Chronic inhibition of glutamate uptake produces a model of slow neurotoxicity. *Proc. Natl. Acad. Sci. USA* 90:6591 (1993)
14. A. Volterra, P. Bezzi, B. Lodi Rizzini, D. Trotti and G. Racagni, Excitotoxicity by transporter-mediated glutamate efflux from neurons and glia in non-ischemic conditions. *Soc. Neurosci. Abs.* 21: (1995) in press

15. S.Renchrona, E. Westerberg, B. Åkesson and B.K. Siesjö, Brain cortical fatty acids during and following complete and severe incomplete ischemia. *J. Neurochem.* 38:84 (1982)
16. A. Dumuis, M. Sebben, L. Haynes, J.P. Pin and J. Bockaert, NMDA receptors activate the arachidonic acid cascade system in striatal neurons. *Nature* 336:68 (1988)
17. F.S. Silverstein, K. Buchanan and M.V. Johnston, Perinatal hypoxia-ischemia disrupts striatal high affinity [^{3}H]glutamate uptake into synaptosomes. *J. Neurochem.* 47:1614 (1986)
18. V. Crépel, C. Hammond, P. Chinestra, D. Diabira and Y. Ben-Ari, A selective LTP of NMDA receptor-mediated currents induced by anoxia in CA1 hippocampal neurons. *J. Neurophysiol.* 70:2045 (1993)
19. D.R. Rosen et al., Mutations of Cu/Zn superoxide dismutase gene are associated with familial amyotrophic lateral sclerosis. *Nature* 362:59 (1993)
20. J.D. Rothstein, M. Van Kammen, A.I. Levey, L. Martin and R.W. Kunkl, An analysis of glutamate transporter subtypes in amyotrophic lateral sclerosis: a selective loss of glial glutamate transporter GLT-1. *Ann. Neurol.* (1995) in press.

NEURODEGENERATIVE DISEASE AND OXIDATIVE STRESS: INSIGHTS FROM AN ANIMAL MODEL OF PARKINSONISM

Teresa G. Hastings[1] and Michael J. Zigmond[2]

Departments of Neurology[1] and Neuroscience[2]
University of Pittsburgh
Pittsburgh, PA 15260

INTRODUCTION

Parkinson's disease (PD), Alzheimer's disease (AD), and amyotrophic lateral sclerosis (ALS) are severe neurological diseases which collectively affect one of every five individuals. At first glance these disorders seem to have little in common. However, we will briefly review some of the reasons to believe that the diseases are interrelated and, thus, that insights from one can lead to a better understanding of the others. We then will explore a particular hypothesis regarding the mechanisms underlying the neuropathology of the disorders — that they involve an excess of reactive metabolites, including free radicals, which promote specific patterns of degeneration in different groups of individuals. Our discussion will focus on data that we have collected in our attempt to understand the basis of the selective vulnerability of dopamine (DA)-containing neurons in PD.

PD, AD, AND ALS: WHAT DO THEY HAVE IN COMMON?

Guam Neurodegeneration Complex

One of the first indications that there might be links among PD, AD, and ALS came from an epidemic of neurodegeneration that occurred in the Western Pacific between the late 1940s and the 1970s[1,2]. Centered in Guam, the epidemic ultimately accounted for 20% of all deaths on the island. Its cause remains a matter of conjecture. However, considerable evidence points to an endogenous toxin present in the cycad plant, whose use for the production of flour had been revived during World War II in response to the lack of food.

Patients affected by the epidemic did not exhibit a single syndrome. Instead, a wide variety of clinical symptoms were seen, including those of PD, AD, and ALS, as well as combinations of individual symptoms that normally characterize these diseases. For example, whereas ALS was common, many individuals exhibited the bradykinesia, tremor, and rigidity of PD together with AD-like memory deficits and personality disturbances.

Clinical and Neuropathological Similarities Among the Disorders

The Guam epidemic suggests that most if not all of the symptoms commonly associated with PD, AD, and ALS can be initiated by a single agent. In retrospect, there have long been reasons for considering this possibility (Table 1). First, there are the parallels in the natural history of the three diseases: They are usually sporadic and progressive in nature, and do not emerge as clinical entities in most affected individuals until age 50-70.

Neurodegenerative Diseases
Edited by Gary Fiskum, Plenum Press, New York, 1996

A second reason for considering a common underlying mechanism is that combinations of symptoms of PD, AD, and ALS are not limited to victims of the Guam epidemic. For example, signs of dementia are frequently associated with the later phases of parkinsonism[3], whereas "parkinsonian" motor disturbances are often seen in AD[4] and in those cases of ALS that survive more than a few years[5].

Perhaps the most telling reason for suspecting a common mechanism comes from neuropathological examinations. In each of the three diseases a particular subset of neurons is more affected than others: DA neurons in PD, cholinergic neurons in AD, and motor neurons in ALS. This itself indicates a parallel. However, more telling is the observation that overlapping neuropathology is common. For example, whereas the loss of neurons in the nucleus basalis and of acetylcholine in the forebrain is associated with AD, it also occurs in a significant number of patients diagnosed with PD[6,7]. Likewise, Lewy bodies as well as the loss of substantia nigra neurons, the diagnostic criteria for PD, can be seen in AD[8,9] as well as ALS[10]. And both neurofibrillary tangles and deposits of ubiquitin-like material can be detected in many cases of each of the three disorders[11].

Table 1
Clinical Features Often Common to
Parkinson's disease, Alzheimer's disease, and ALS

	PD	AD	ALS
Natural History			
progressive	x	x	x
onset at 50-70 yrs	x	x	x
Clinical symptoms			
bradykinesia	xx	x	x
rigidity	xx	x	x
tremor	xx	x	x
dementia	x	xx	x
Neuropathology			
neuronal loss			
mesencephalic DA	xx	x	x
forebrain ACh	x	xx	
motoneuron			xx
Lewy bodies	xx	x	x
ubiquitin	xx	x	x
neurofibrillary tangles	x	xx	x

Possible Common Etiologies

Appel may have been the first to suggest that PD, AD, and ALS were different manifestations of a common deficit[12]. His hypothesis was that each of the disorders was a result of an impairment in the availability of a *neurotrophic factor*, an idea that has received support from subsequent research[13,14]. A second suggestion of a common mechanism was that of Calne and coworkers, who emphasized *environmental causes*[15]. With few exceptions, such as that of the Guam epidemic, there has been little evidence for environmental agents that might underlie any of the disorders, let alone all three. Yet, since we do not really know much about what to be looking for, this may well be a better reflection of our ignorance than the weakness of the hypothesis. *Impaired mitochondrial function*[16] and *excitotoxicity*[17] also have been proposed as causal factors.

Reactive Oxygen Species

Our own attention has been drawn to *oxidative stress* caused by an excess of free radicals, or more generally, reactive oxygen species (ROS). These compounds are products of normal metabolism as well as neuropathological states. As suggested by the term, ROS are characterized by a high degree of reactivity caused by the presence of unpaired electrons, such as occur in the hydroxyl radical (•OH) or at least a strongly electrophilic site. As a result, ROS have the potential to cause a variety of cellular insults, including lipid peroxidation, DNA fragmentation, and modification of proteins. This potential for cytotoxicity may help to explain the evolution of a plethora of mechanisms normally available to regulate the availability of ROS[18]. Recently, it has been suggested that oxidative stress is a common feature in many neurodegenerative disorders, including PD, AD, and ALS[19,20].

An excess of ROS has most commonly been invoked in discussions of aging[21]. This is noteworthy since there are many parallels between aging and the disorders under consideration[22], and it has often been suggested that some of the symptoms of PD, AD, and ALS represent an acceleration of the normal aging process. Moreover, evidence has accumulated suggesting ROS involvement in specific neurodegenerative diseases.

Evidence linking oxidative stress to neurodegeneration is particularly in the case of PD[23-27]. First, substantia nigra neurons appear to sustain an active level of oxidation even under normal conditions. This is indicated by the accumulation within DA neurons of neuromelanin, a polymerization product of oxidized DA. Indeed, it is the neurons of the mesencephalon highest in neuromelanin that are most susceptible to degradation in PD[28,29]. In addition, post-mortem analyses of substantia nigra from PD patients indicates abnormalities in the formation and/or handling of ROS in PD. Observations include decreased glutathione (GSH) and GSH peroxidase[30,31], increased iron[32-34] and decreased ferritin[35] (but see also[36,37]), increased superoxide dismutase[38], and increased lipid peroxidation[39].

More recently, ROS also have been implicated in AD[40]. For example, free radicals can cause the formation of ß-amyloid protein[41] and, indeed, ß-amyloid can itself form free radicals[42]. In addition, increased iron has been detected in senile plaques[43], and signs of oxidative stress have been detected in post-mortem brain samples from AD patients[44-46]. The possible connection between ALS and free radicals has been strongly suggested by the recent observation of a mutation at the locus responsible for the genetic expression of Cu,Zn-superoxide dismutase in patients with familial ALS[47]. Additional support is provided by the observation of increased protein carbonyls in CNS tissue from patients who had died with ALS[48] and of free radical-induced toxicity in cultred spinal cord neurons[49].

Our interest in the oxidative stress hypothesis began from a consideration of the mechanism of action of 6-hydroxydopamine (6-OHDA), a drug that is cytotoxic by virtue of its rapid oxidation to form such reactive species as superoxide anions (O_2^-), hydrogen peroxide (H_2O_2), •OH, and 6-OHDA quinones (Figure 1). In the remainder of this review we will outline our work with 6-OHDA, DA, and related reactive metabolites in relation to PD, returning in the end to comment on the possible relevance of our work for neurode nerative disease more generally.

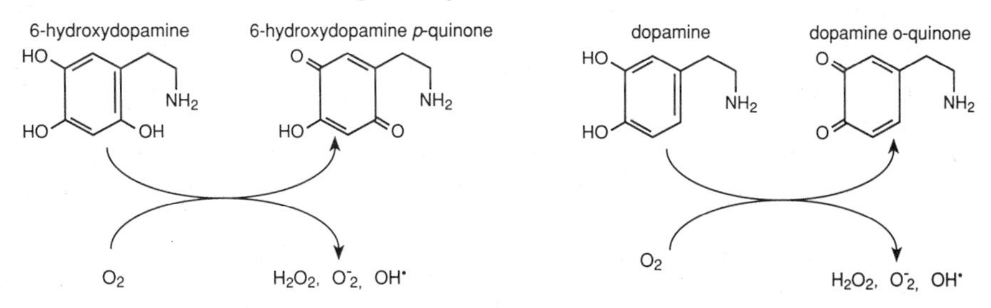

Figure 1: Oxidation of 6-hydroxydopamine and dopamine

AN ANIMAL MODEL OF PARKINSONISM

In low concentrations, 6-OHDA is selectively accumulated in catecholamine neurons where it autoxidizes to form the ROS noted above. As has been reviewed elsewhere, several parallels exist between 6-OHDA-treated rats and patients with PD[50,51]. These include an apparent causal relation between the loss of nigrostriatal DA neurons and neurological symptoms, a preclinical phase that persists until the loss of DA in the caudate-putamen is almost complete, and the temporary reversal of symptoms with L-DOPA.

The Basis of the Preclinical Stage of PD

The absence of gross neurological dysfunction despite extensive loss of DA neurons is associated with an increase in the capacity of the remaining DA neurons to deliver transmitter to denervated sites. Our group has examined this phenomenon in a variety of ways. Studies using neostriatal slices prepared from control and 6-OHDA-lesioned rats demonstrated that partial destruction of neostriatal DA terminals increased the fractional overflow of DA. The increase in overflow was related to lesion size, reaching 7-fold with losses of DA greater than 90%. Further studies have suggested that this apparent increase in DA efflux per terminal is due to a reduction in DA reuptake as well as an increase in DA release from residual terminals[52,53], and is accompanied by an increase in DA synthesis[54]. This increase in DA synthesis and release, coupled with the decrease in DA reuptake, presumably explained the observation that the extracellular concentration of DA is not significantly different from normal unless the loss of DA terminals exceeds 80%[55-57].

The Mechanism of Action of DOPA

The principal treatment for PD is the systemic administration of L-DOPA. A precursor to DA, L-DOPA appears to increase the release of transmitter from residual DA neurons as well as other elements within the striatum[57-59]. One consequence of this increase in the synthesis and release of DA is a very marked rise in the extracellular concentration of DA and DOPAC in the neostriatum of rats lesioned with 6-OHDA. For example, in the presence of an inhibitor of peripheral decarboxylation, L-DOPA (100 mg/kg) caused a 2-fold increase in extracellular DA in intact animals as assessed by in vivo microdialysis. However, in animals with large bilateral neostriatal DA depletions, L-DOPA increased extracellular DA by 14-fold, to a level that was several times higher than that seen in control animals[57]. This increase in DA availability is presumably the basis for the therapeutic effect of L-DOPA.

DA AS A NEUROTOXIN

The Oxidation of DA

The hypothesis that DA contributes to the neuropathology associated with PD was first suggested because DA, like 6-OHDA, is a reactive molecule that will oxidize to form free radicals and quinones (Figure 1)[60]. The formation of these reactive metabolites can occur by one of three routes. First, the catechol ring of DA can undergo oxidation by 1 or 2 electron transfers, forming various reactive compounds. Second, brain can oxidize DA enzymatically[61], including via the peroxidase activity contained in prostaglandin synthase[62]. And third, H_2O_2 is formed as a normal product of DA metabolism by monoamine oxidase, and without adequate reduction by GSH peroxidase, the H_2O_2 can be broken down through interaction with metal ions into •OH[63]. In short, conditions that increase the concentration and/or turnover of DA should increase the potential for formation of reactive metabolites.

A Role for DA in Neurotoxic Events

There is now a large body of evidence indicating that DA can, indeed, participate in neurotoxic events. For example, pharmacological or lesion-induced depletion of DA can attenuate the neurotoxic effects of a variety of challenges, including systemic administration of methamphetamine[64,65], intracerebral administration of excitatory amino acids[66], and ischemia[67]. Moreover, the exposure of cells to DA has been shown to have neurotoxic consequence[68,69]. However, these studies provide little or no evidence for *selective* effects of DA on dopaminergic neurons, and thus fall short of providing an explanation for the neurodegenerative profile of PD.

The cytotoxicity of DA and related catechol analogues has been attributed in part to the reactive quinone form of the oxidized molecule, at least under in vitro conditions[70]. The electron-deficient DA quinone appears to react with nucleophilic sulfhydryl groups on proteins, inactivating enzymes and receptors. The DA quinone may also form conjugates with endogenous reducing agents, GSH or cysteine, by reacting with thiol groups of these molecules. Indices of in vivo DA oxidation have been obtained by measuring the amount of free cysteinyl-catechol derivatives formed in dopaminergic brain regions of several species, including humans[71]. Cysteinyl-catechol conjugates also increase with age[72] and ascorbic acid deficiency[73].

DA Toxicity: Impact of DA Concentration and Antioxidants

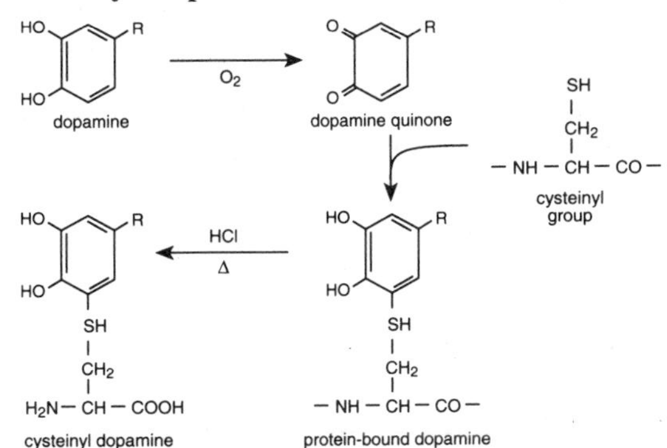

Figure 2: Formation of cysteinyl-dopamine. This reaction scheme depicts the stages of dopamine oxidation and subsequent binding to cysteinyl residues on protein followed by acid hydrolysis to isolate modified amino acid residues.

The electron deficient quinones formed during the oxidation of DA will react with nucleophilic sulfhydryl groups, a major source of which is provided by the cysteinyl residues of proteins (Figure 2). Thus, binding of DA to cysteinyl residues may serve as an index of DA oxidation; it may also be a cytotoxic event. To examine conditions under which such binding might occur, we incubated neostriatal slices with ^{3}H-DA under various buffer conditions and then measured the amount of protein-bound radioactivity.Results showed that DA oxidized in vitro to form DA quinone, which then bound to the cysteinyl groups of protein. This could be prevented by the addition of ascorbate or GSH (Figure 3).

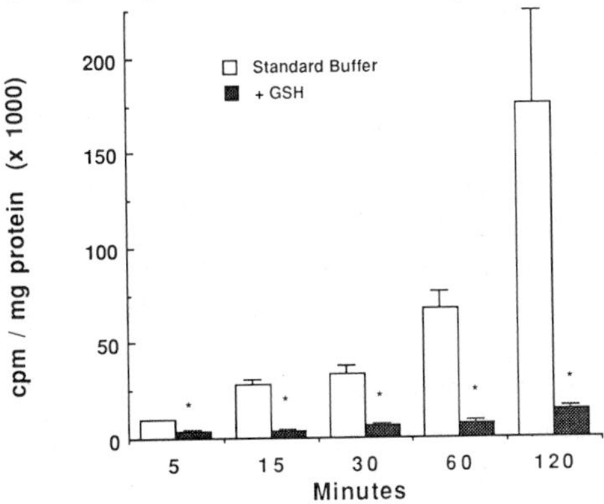

Figure 3: In vitro binding of dopamine (DA) to protein in striatal tissue. Slices were incubated with 60 nM ^{3}H-DA for the indicated times at 37°C in standard Krebs buffer or with buffer containing 0.01, 0.10, or 1.00 mM GSH. Shown is mean $\pm$ S.E. M. for total bound tritium. All GSH concentrations gave the same results and the data are combined. Additional experiments indicated that DA and its metabolite DOPAC bound to cysteinyl groups on protein[74].

Next, we sought to determine whether the oxidation of DA and its binding to protein would also occur in vivo. Together with our colleague David Lewis, high concentrations of DA (0.05-1.0 μmol in 2 μl) were injected into neostriatum, and tissue surrounding the injection site was assayed 24 h later. Protein-bound cysteinyl-DA and cysteinyl-DOPAC rose in proportion to the concentration of DA injected, increasing 35- to 55-fold above baseline at the highest dose. In addition, 7 days after DA administration cellular damage and gliosis were observed at the injection site. This centralized area of nonspecific damage was surrounded by a region of specific loss of tyrosine hydroxylase immunoreactivity with no detectable loss of synaptophysin immunoreactivity or alterations in cellular architecture. As in the case of the formation of cysteinyl-catechols, the extent of damage was dependent upon the DA concentration. Moreover, co-administration of DA with an equimolar concentration of ascorbate or GSH greatly reduced both protein cysteinyl-catechol formation and the loss of tyrosine hydroxylase immunoreactivity (Figure 4)[75]. These findings suggest that DA can cause specific toxicity to dopaminergic neurons, as had previously been shown by Filloux and Townsend[76], that the binding of DA to protein is correlated with this toxicity, and that both result from the oxidation of DA.

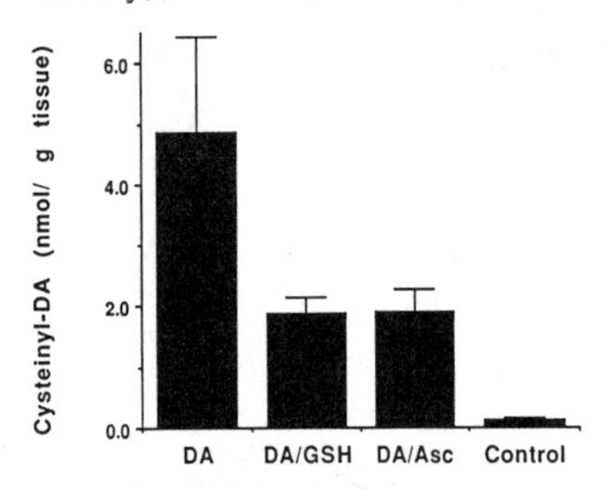

Figure 4: Neurotoxic effects of dopamine after intrastriatal injection. Protein-bound cysteinyl-DA was measured 24 h after intrastriatal injections of 1 μmol dopamine alone (DA) or in combination with equimolar glutathione (DA/GSH) or ascorbate (DA/Asc). Control animals received no treatment. Shown are mean ± S.E.M.

Hypothesis: Reactive Metabolites of DA Play a Role in PD

We hypothesize that there are at least two conditions under which toxic products of DA oxidation might lead to degeneration of DA neurons: an increase in the rate of formation of the DA oxidation products, and a decrease in the capacity of brain to inactivate these reactive compounds. There are several reason to believe that these events might take place in the PD patient. First, increased DA turnover, and thus the potential for increased oxidative stress, is indicated by the increased ratio of DA metabolites to DA that has been detected in post-mortem brain tissue[77]. This may be further exacerbated by chronic treatment with L-DOPA, although there is no conclusive evidence for L-DOPA-induced toxicity in the treatment of PD[78]. Second, as has already been summarized above, there is considerable evidence for deficiencies in the ability of PD patients to protect themselves against the neurotoxic effects of ROS.

SUMMARY AND CONCLUSIONS

Accumulating evidence suggests a role for oxidative stress in the neuropathology of PD. Our data and those of others suggests that the stress may derive at least in part from an accumulation of reactive metabolites of DA, possibly including the DA quinone. This accumulation may result from an elevation in the rate at which the metabolites are formed or a decrease in the capacity for buffering against the ROS once produced. The increase in DA turnover that occurs during both the preclinical phase of the disease and the clinical management with L-DOPA may play a role in these events.

Similar transmitter-specific alterations in the formation or inactivation of ROS may be involved in the neuropathological basis of other neurodegenerative diseases, as well. Indeed, it is possible that a common defect in the handling of ROS underlies each of the disorders, with the precise nature of the neuropathology and resulting symptoms resulting from additional genetic or environmental factors that convey differential susceptibility to specific groups of neurons.

The oxidative stress hypothesis of neurodegenerative disease is only one of many that have been proposed in recent years. However, it is noteworthy that many of the other theories may themselves involve the formation of ROS as an intermediary step. Thus, defects in mitochondrial function could result from or cause the formation of excess free radicals. Moreover, oxidative stress has been implicated in the toxicity associated with excitatory amino acids[79,80] and the withdrawal of neurotrophic factors[81,82].

Attempts to treat PD as well as other neurodegenerative diseases with antioxidants have not led to any striking indications of efficacy[83] (but see also[84]). However, in light of the accumulating evidence in support of a role for oxidative stress in the neuropathology of PD, AD, and ALS, we believe that a much greater effort to test the impact of such an approach is warranted.

Acknowledgment

Many colleagues have contributed to the research and concepts summarized in this review, including Elizabeth D. Abercrombie, Andrew Giovanni, David A. Lewis, Li Ping Liang, and Ariel Rabinovic. The authors also thank Susan D. Giegel and Deborah A. Malanosky for help in preparing this manuscript. This work was supported in part by U.S. Public Health Service Grants NS19608, MH43947, MH29670, MH45156, MH00058, AG05133, and NS09076.

REFERENCES

1. R.M. Garruto and Y. Yase, Neurodegenerative disorders of the western Pacific: The search for mechanisms of pathogenesis, *Trends in Neurosci.* 9:368-374 (1986).
2. D.E. Lilienfeld, D.P. Perl, and C.W. Olanow, Guam neurodegeneration, in: "Neurodegenerative Diseases", D.B. Calne, W.B. Saunders, Philadelphia (1994).
3. R.G. Brown, C.D. Marsden, How common is dementia in Parkinson's disease? *Lancet* 1:1262-1265 (1984).
4. P.K. Molsa, R.J. Marttila, V.K. Rinne, Extrapyramidal signs in Alzheimer's disease, *Neurology* 34:1114-1116, (1984).
5. L.P. Rowland, Natural history and clinical features of amyotrophic lateral sclerosis and related motor neuron diseases, in: "Neurodegenerative Diseases", D.B. Calne, W.B. Saunders, Philadelphia (1994).
6. M. Ruberg, A. Ploska, F. Javoy-Agid, and Y. Agid, Muscarinic binding and choline acetyltransferase activity in Parkinsonian subjects with reference to dementia, *Brain Res.* 232:129-139 (1982).
7. J.M. Candy, R.H. Perry, E.K. Perry, D. Irving, G. Blessed, A.F. Fairbairn, and B.E. Tomlinson, Pathological changes in the nucleus of Meynert in Alzheimer's and Parkinson's Diseases, *J. Neurol. Sci.* 54:277-289 (1983).
8. M. Tabaton, A. Schenone, P. Romagnoli, and G.L. Mancardi, A quantitative and ultrastructural study of substantia nigra and nucleus centralis superior in Alzheimer's disease, *Acta Neuropathol.* 68:218-223 (1985).
9. T. Uchihara, H. Kondo, K. Kosaka, and H. Tsukagoshi, Selective loss of nigral neurons in Alzheimer's disease: A morphometric study, *Acta Neuropathol.* 83:271-276 (1992).
10. S. Kato, M. Oda, and H. Tanabe, Diminution of dopaminergic neurons in the substantia nigra of sporadic amyotrophic lateral sclerosis, *Neuropathol. App. Neurolog.* 19:300-304 (1993).
11. K. Iqbal and I. Grundke-Iqbal, Neurofibrillary tangles, in: "Neurodegenerative Diseases", D.B. Calne, W.B. Saunders, Philadelphia (1994).
12. S.H. Appel, A unifying hypothesis for the cause of amyotrophic lateral sclerosis, Parkinsonism, and Alzheimer disease, *Ann. Neurol.* 10:499-505 (1981).
13. R.M. Lindsay, C.A. Altar, J.M. Cedarbaum, C. Hyman, and S.J. Wiegand, The therapeutic potential of neurotrophic factors in the treatment of Parkinson's disease, *Exp. Neurol.* 124:103-118 (1993).

14. F. Hefti, Growth factors and neurodegeneration, in: "Neurodegenerative Diseases", D.B. Calne, W.B. Saunders, Philadelphia (1994).
15. D.B. Calne, E. McGeer, A. Eisen, P. Spencer, Alzheimer's disease, Parkinson's disease, and motoneurone disease: Abiotrophic interaction between ageing and environment? *Lancet* 2:1067 (1986).
16. H. Reichmann and P. Riederer, Mitochondrial disturbances in neurodegeneration, in: "Neurodegenerative Diseases", D.B. Calne, W.B. Saunders, Philadelphia (1994).
17. R. Horowski, H. Wachtel, L. Turski, and P.-A. Löschmann, Glutamate excitotoxicity as a possible pathogenetic mechanism in chronic neurodegeneration, in: "Neurodegenerative Diseases", D.B. Calne, W.B. Saunders, Philadelphia (1994).
18. B. Halliwell, and J.M.C. Gutteridge, Oxygen radicals and the nervous system, *Trends Neurosci.* 8:22-26 (1985).
19. J.T. Coyle and P. Puttfarcken, Oxidative stress, glutamate, and neurodegenerative disorders, *Science* 262:689-694 (1993).
20. P. Jenner, Oxidative damage in neurodegenerative disease, *Lancet* 344:796-798 (1994).
21. D. Harman, The aging process, *Proc. Natl. Acad. Sci.* 78:7124-7128 (1981).
22. D.M.A. Mann, Vulnerability of specific neurons to aging, in: "Neurodegenerative Diseases", D.B. Calne, W.B. Saunders, Philadelphia (1994).
23. G. Cohen and P. Werner, Free radicals, oxidative stress, and neurodegeneration, in: "Neurodegenerative Diseases", D.B. Calne, W.B. Saunders, Philadelphia (1994).
24. P. Jenner, D.T. Dexter, J. Sian, A.H.V. Schapira, and C.D. Marsden, Oxidative stress as a cause of nigral cell death in Parkinson's disease and incidental Lewy body disease, *Ann. Neurol.* 32:S82-S87 (1992).
25. C.W. Olanow, An introduction to the free radical hypothesis in Parkinson's disease, *Ann. Neurol.* 32:S2-S9 (1992).
26. E.C. Hirsch, Why are nigral catecholaminergic neurons more vulnerable than other cells in Parkinson's disease? *Ann. Neurol.* 32:S88-S93 (1992).
27. M.B.H. Youdim, Inorganic neurotoxins in neurodegenerative disorders without primary dementia, in: "Neurodegenerative Diseases", D.B. Calne, W.B. Saunders, Philadelphia (1994).
28. E. Hirsch, A.M. Graybiel, and Y.A. Agid, Melanized dopaminergic neurons are differentially susceptible to degeneration in Parkinson's disease, *Nature* 334:345-348 (1988).
29. A. Kastner, E.C. Hirsch, 0. Lejeune, F. Javoy-Agid, and Y. Agid, Is the vulnerability of neurons in substantia nigra of patients with Parkinson's disease related to their neuromelanin content? *J. Neurochem.* 59:1080-1089 (1992).
30. T.L. Perry, D.V. Godin, and S. Hansen, Parkinson's disease: A disorder due to nigral glutathione deficiency? *Neurosci. Lett.* 33:305-310 (1982).
31. S.J. Kish, C. Morito, and 0. Hornykiewicz, Glutathione peroxidase activity in Parkinson's disease brain, *Neurosci. Lett.* 58:343-346 (1985).
32. K.M. Earle, Studies on Parkinson's disease including x-ray fluorescent spectroscopy of formalin fixed brain tissue, *J Neuropathol. Exp. Neurol.* 27:1-14 (1968).
33. E. Sofic, W. Paulus, K. Jellinger, P. Riederer, and M.B.H. Youdim, Selective increase of iron in substantia nigra zona compacta of Parkinsonian brains, *J. Neurochem.* 56:978-982 (1991).
34. D.T. Dexter, F.R. Wells, F. Agid, Y. Agid, A.J. Lees, P. Jenner, and C.D. Marsden, Increased nigral iron content in postmortem parkinsonian brain, *Lancet* 2:1219-1220 (1987).
35. D.T. Dexter, A. Carayon, M. Vidailhet, M. Ruberg, F. Agid, Y. Agid, A.J. Lees, F.R. Wells, P. Jenner, and C.D. Marsden, Decreased ferritin level in brain in Parkinson's disease, *J. Neurochem.* 55:16-20 (1990).
36. K. Jellinger, W. Paulus, I. Grundke-Iqbal, P. Riederer, and M.B. Youdim, Brain iron and ferritin in Parkinson's and Alzheimer's diseases. *J. Neural Trans.* 2:327-340 (1990).
37. P. Riederer, E. Sofic, W.-D. Rausch, B. Schmidt, G.P. Reynolds, K. Jellinger, and M.B.H. Youdim, Transition metals, ferritin, glutathione and ascorbic acid in parkinsonian brain. *J. Neurochem.* 52:515-520 (1989).
38. R.J. Marttila, H. Lorentz, and U.K. Rinne, Oxygen toxicity protecting enzymes in Parkinson's disease: increase of superoxide dismutase-like activity in the substantia nigra and basal nucleus, *J. Neurol. Sci.* 86:321-331 (1988).
39. D. Dexter, C. Carter, F. Agid, Y. Agid, A.J. Lees, P. Jenner, C.D. Marsden, Lipid peroxidation as cause of nigral cell death in Parkinson's disease, *Lancet* 2:639-640 (1986).
40. S. Whyte, K. Beyreuther, and C.L. Masters, Rational therapeutic strategies for alzheimer's disease, in: "Neurodegenerative Diseases", D.B. Calne, W.B. Saunders, Philadelphia (1994).
41. T. Dyrks, E. Dyrks, T. Hartmann, C.L. Masters, K. Beyreuther, Amyloidogenicity of ßA4 and ßA4-bearing APP fragments by metal catalysed oxidation, *J. Biol. Chem.* 267:18210-18217 (1992).

42. K. Hensley, J.M. Carney, M.P. Mattson, M. Aksenova, M. Harris, J.F. Wu, R.A. Floyd, and D.A. Butterfield, A model for ß-amyloid aggregation and neurotoxicity based on free radical generation by the peptide: Relevance to Alzheimer's disease, *Proc. Natl. Acad. Sci.* 91:3270-3274 (1994).
43. P.F. Good, D.P. Perl, L.M. Bierer, and J. Schmeidler, Selective accumulation of aluminum and iron in the neurofibrillary tangles of Alzheimer's disease: A laser microprobe (LAMMA) study, *Ann. Neurol.* 31:286-292 (1992).
44. R.N. Martins, C.G. Harper, G.B. Stokes, C.L. Masters, Increased cerebral glucose-6-phosphate dehydrogenase activity in Alzheimer's disease may reflect oxidative stress, *J. Neurochem.* 46:1042-1045 (1986).
45. J.S. Richardson, K.V. Subbarao, and L.C. Ang, Autopsy of Alzheimer's brains show increased peroxidation to an in vitro iron challenge, in: "Alzheimer's Disease: Basic mechanisms, Diagnosis and Therapeutic Strategies", K. Iqbal, ed., Wiley, New York (1991).
46. C.D. Smith, J.M. Carney, P.E. Starke-Reed, C.N. Oliver, E.R. Stadtman, R.A. Floyd, and W.R. Markesbery, Excess brain protein oxidation and enzyme dysfunction in normal aging and in Alzheimer diseases, *Proc. Natl. Acad. Sci.* 88:10540-10543 (1991).
47. D. Rosen, T. Siddique, D. Patterson, Mutations in Cu/Zn superoxide dismutase gene are associated with familial amyotrophic lateral sclerosis, *Nature* 362:59-62 (1993).
48. A.C. Bowling, J.B. Schulz, R. H. Brown, M.F. Beal, Superoxide dismutase activity, oxidative damage, and mitochondrial energy metabolism in familial and sporadic amyotrophic lateral sclerosis, *J. Neurochem.* 61:2322-2325 (1993).
49. S.U. Kim, Tissue culture models of neurodegeneration, in: "Neurodegenerative Diseases", D.B. Calne, W.B. Saunders, Philadelphia (1994).
50. M.J. Zigmond, and E.M. Stricker, Animal models of parkinsonism using selective neurotoxins: clinical and basic implications, *Int. Rev. Neurobiol.* 31:1-79 (1989).
51. M.J. Zigmond, T.G. Hastings, and E.D. Abercrombie, Neurochemical responses to 6-hydroxydopamine and L-DOPA therapy: implications for Parkinson's disease, *Ann. N.Y. Acad. Sci.* 648:71-86 (1992).
52. M.K. Stachowiak, R.W. Keller Jr, E.M. Stricker, and M.J. Zigmond, Increased dopamine efflux from striatal slices during development and after nigrostriatal bundle damage, *J. Neurosci.* 7:1648-1654 (1987).
53. G.L. Snyder, R.W. Keller, and M.J. Zigmond, Dopamine efflux from striatal slices after intracerebral 6-hydroxydopamine: evidence for compensatory hyperactivity of residual terminals, *J. Pharmacol. Ther.* 253:867-876 (1990).
54. L.P. Liang, and M.J. Zigmond, Dopamine synthesis in neostriatal slices after intraventricular 6-hydroxy-dopamine, *Soc. Neurosci. Abstr.* 19:401 (1993).
55. T.E. Robinson, and I.Q. Wishaw, Normalization of extracellular dopamine in striatum following recovery from a partial unilateral 6-OHDA lesion of the substantia nigra: a microdialysis study in freely moving rats, *Brain Res.* 450:209-224 (1988).
56. W.Q. Zhang, H.A. Tilson, K.P. Nanry, P.M. Hudson, J.S Hong, and M.K. Stachowiak, Increased dopamine release from striata of rats after unilateral nigrostriatal bundle damage, *Brain Res.* 461:335-342 (1988).
57. E.D. Abercrombie, A.E. Bonatz, and M.J. Zigmond, Effects of L-DOPA on extracellular dopamine in striatum of normal and 6-hydroxydopamine-treated rats, *Brain. Res.* 525:36-44 (1990).
58. R.W. Keller Jr., W.G. Kuhr, R.M. Wightman, and M.J. Zigmond, The effect of L-dopa on in vivo dopamine release from nigrostriatal bundle neurons, *Brain Res.* 447:191-194 (1988).
59. G.L. Snyder, and M.J. Zigmond, The effects of L-DOPA on *in vitro* dopamine release from striatum, *Brain Res.* 508:181-187 (1990).
60. D.G. Graham, Oxidative pathways for catecholamines in the genesis of neuromelanin and cytotoxic quinones, *Mol. Pharmacol.* 14:633-643 (1978).
61. M.B. Grisham, V.J. Perez, and J. Everse, Neuromelanogenic and cytotoxic properties of canine brainstem peroxidase, *J. Neurochem.* 48:876-882 (1987).
62. T.G. Hastings, Enzymatic oxidation of dopamine: The role of prostaglandin H synthase, *J. Neurochem.* 64:919-924 (1995).
63. G. Cohen, The pathobiology of Parkinson's disease: biochemical aspects of dopamine neuron senescence, *J. Neural Trans.* 19:89-103 (1983).
64. C.J. Schmidt, J.K. Ritter, P.K. Sonsalla, G.R. Hanson, and J.W. Gibb, Role of dopamine in the neurotoxic effects of methamphetamine, *J. Pharmacol. Exp. Ther.* 233:539-544 (1985).
65. M. Johnson, D.M. Stone, G.R. Hanson, and J.W. Gibb, Role of the dopaminergic nigrostriatal pathway in methamphetamine-induced depression of the neostriatal serotonergic system, Eur. J. Pharm. 135:231-234 (1987).
66. A.G. Chapman, N. Durmuller, G.J. Lees, and B.S. Meldrum, Excitoxicity of NMDA and kainic acid is modulated by nigrostriatal dopaminergic fibers, *Neurosci. Letts.* 107:256-260 (1989).

67. J. Weinberger, J. Nieves-Rosa, and G. Cohen, Nerve terminal damage in cerebral ischemia: protective effect of alpha-methyl-para-tyrosine, *Stroke* 16:864-870 (1985).
68. P.A. Rosenberg, Catecholamine toxicity in cerebral cortex in dissociated cell culture *J. Neurosci.*8:2887-2894 (1988).
69. P.P. Michel, and F. Hefti, Toxicity of 6-hydroxydopamine and dopamine for dopaminergic neurons in culture, *J. Neurosci. Res.* 26:428-435 (1990).
70. D.G. Graham, S.M. Tiffany, W.R. Bell Jr., and W.R. Gutknecht, Autoxidation versus covalent binding of quinones as the mechanism of toxicity of dopamine 6-hydroxydopamine and related compounds toward C1300 neuroblastoma cells *in vitro, Mol. Pharmacol.* 14:644-653 (1978).
71. B. Fornstedt, E. Rosengren, and A. Carlsson, Occurrence and distribution of 5-S-cysteinyl derivatives of dopamine dopa and dopac in the brains of eight mammalian species, *Neuropharmacol.* 25:451-454 (1986).
72. B. Fornstedt, A. Brun, E. Rosengren, and A. Carlsson, The apparent autoxidation rate of catechols in dopamine-rich regions of human brains increases with the degree of depigmentation of substantia nigra, J. Neural Trans. 1:279-295 (1989).
73. B. Fornstedt, and A. Carlsson, Vitamin C deficiency facilitates 5-S-cysteinyldopamine formation in guinea pig striatum, *J. Neurochem.* 56:407-414 (1991).
74. T.G. Hastings, and M.J. Zigmond, Identification of catechol-protein conjugates in neostriatal slices incubated with ^{3}H-dopamine: impact of ascorbic acid and glutathione, *J. Neurochem.* 63:1126-1132 (1994).
75. T.G. Hastings, D.A. Lewis, and M.J. Zigmond, Intrastriatally administered dopamine: evidence of selective neurotoxicity associated with dopamine oxidation, *Soc. Neurosci. Abst,.* 20:413 (1994).
76. F. Filloux, and J.J. Townsend, Pre- and post-synaptic neurotoxic effects of dopamine demonstrated by intrastriatal injection, *Exp. Neurol.* 119:79-88 (1993).
77. H. Bernheimer, W. Birkmayer, 0. Hornykiewicz, K. Jellinger, and F. Seitelbeger, Brain dopamine and the syndromes of Parkinson and Huntington: clinical morphological and neurochemical correlations, *J. Neurol. Sci.* 20:415-455 (1973).
78. J. Blin, A.-M. Bonnet, and Y. Agid, Does levodopa aggravate Parkinson's disease? *Neurology* 38:1410-1416 (1988).
79. I.J. Reynolds, and T.G. Hastings, Glutamate induces the production of reactive oxygen species in cultured forebrain neurons following NMDA receptor activation, *J. Neurosci.* 15:3318-3327 (1995).
80. J.B. Schulz, D.R. Henshaw, D. Siwek, B.G. Jenkins, R.J. Ferrante, P.B. Cipolloni, N.W. Kowall, B.R. Rosen, and M.F. Beal, Invovlement of free radicals in excitotoxicity in vivo, *J. Neurochem.* 64:2239-2247 (1995).
81. Z. Pan, and R. Perez-Polo, Role of nerve growth factor in oxidant homeostasis: glutathione metabolism, *J. Neurochem.* 61:1713-1721 (1993).
82. M. Mayer, and M. Noble, N-Acetyl-L-cysteine is a pluripotent protector against cell death and enhancer of trophic factor-mediated cell survival *in vitro, Proc. Natl. Acad. Sci.* 91:7496-7500 (1994).
83. Parkinson Sutdy Group, DATATOP: Effects of tocopherol and deprenyl on the profression of disability in early Parkinson's disease. *N. Eng. J. Med.* 328:176-183.
84. S. Fahn, A pilot trial of high-dose alpha-tocopherol and ascorbate in early Parkinson's disease, *Ann. Neurol.* 32:S128-S132 (1992).

A PROGRESSIVE BALLOONING OF MYELINATED, TH-POSITIVE AXONS IS PRODUCED BY MPTP IN THE DOG

James S. Wilson, Blair H. Turner, and James H. Baker

Department of Anatomy
College of Medicine
Howard University
Washington, D.C. 20059

INTRODUCTION

Systemic injections of 1-methyl-4-phenyl-1,2,3,6-tetrahydropyridine (MPTP) produce selective degeneration of the dopaminergic neurons of the nigrostriatal pathway in some mammals such as dogs, monkeys, and humans but not others such as rats. The mechanism of MPTP's actions has been studied by biochemical and pharmacological techniques which have found that MPTP's toxicity requires its oxidation by glial MAO-B to 1-methyl-4-phenylpyridinium (MPP+)[1-3]. Surprisingly, MPP+ is not selectively toxic to dopaminergic neurons but damages or kills most cells with which it comes into contact including hepatocytes and granule cells of the cerebellum[4,5].

Considering the non-specific nature of the toxicity of MPP+, it is paradoxical that the distribution of MPP+, following a systemic injection, does not coincide with the site of lesion. For instance, the region of the brain with the highest levels of MPP+ is the *locus coeruleus* which is not lesioned by the drug; whereas the *substantia nigra*, which contains moderate amounts of MPP+, is lesioned by the drug[6,7]. To explain the specificity of MPTP's actions, Snyder and colleagues have suggested that MPP+ is concentrated to toxic levels in nigral neurons by the dopamine uptake carrier for which it is a substrate[8]. In support of this hypothesis, our studies have shown that blockage of the dopamine transporter with GBR 12909 reduced the size of the lesion induced by MPTP in dog[9] (unpublished observations). Although these biochemical and pharmacological studies have explained why neurons without a dopamine uptake carrier are not damaged by MPP+, they have not explained why MPTP produces nigrostriatal degeneration in humans, monkeys and dogs but not in other mammals, such as rats, which have a dopamine uptake carrier.

For the last several years, we have been documenting the spatiotemporal sequence of degenerative changes in the cell bodies, axons and terminals of the nigrostriatal pathway induced by MPTP[10,11]. Our morphologic approach assumes that inferences can be drawn about the mechanisms of toxicity from the timing and location of pathological changes. We have chosen the dog model because a single, systemic injection of MPTP consistently produces not only the acute release of dopamine, as it does in the rodent[12-14], but also

Neurodegenerative Diseases
Edited by Gary Fiskum, Plenum Press, New York, 1996

selectively lesions the nigrostriatal pathway. In this study, we describe the effects of MPTP on nigrostriatal axons and suggest how axonal morphology might explain the preferential vulnerability of some dopaminergic neurons to MPTP.

METHODS

Adult male beagles received a single intravenous injection of MPTP at a dose (3 mg/kg) that produces a permanent and extensive loss of neurons staining for tyrosine hydroxylase (TH) in the substantia nigra and depletion of striatal dopamine[3,11].

In the light microscopic studies, dogs were sacrificed one day to three years after an overdose injection of pentobarbital and perfused transcardially with a formalin fixative. Brains were stereotaxically blocked, removed from the skull and placed in a 30% sucrose solution until sinking. Coronal or sagittal sections (30um) were cut on a freezing microtome. Adjacent sections were stained by the Fink-Heimer, cupric silver, Nissl, fiber and hematoxylin-eosin (H-E) methods as previously described[11,15]. Immunohistochemical staining for TH, myelin-basic protein, and neurofilaments were carried out using the ABC-DAB technique (Vector Laboratories, Burlingame, CA) as previously described[15].

In the electron microscopic study, one dog was sacrificed 8 days after an injection of MPTP. Vibrotome sections were cut through the substantia nigra and processed for TH-immunocytochemistry as previously described[16].

RESULTS

Examination of the nigrostriatal pathway with the light microscope revealed swollen TH-positive axons one day after an MPTP injection. This was the earliest pathological change observed with the light microscope and preceded terminal degeneration or degenerative changes in TH-positive soma in the substantia nigra. After one day there was a progressive increase in the number and size of the TH-positive swollen axons seen in the nigrostriatal pathway. The ballooning of axons reached a plateau at 8 days when many swollen axons had obtained a diameter of 30um or more. At post-injection day 14, large, swollen axons were still evident along the nigrostriatal pathway, but their numbers were declining. These changes are illustrated in Figure 1 A-E.

By eight days after the injection, two distinct regions were identified in the swollen axons cut coronally: 1) a central core surrounded by 2) a peripheral ring. The central core was basophilic in H-E stains and was argyrophilic in Fink-Heimer stains (Fig. 1F). It also reacted most intensely to monoclonal antibodies against phosphorylated H-neurofilaments (Fig. 1G). In contrast, monoclonal antibodies against myelin-basic protein labelled most intensely the peripheral ring which was acidophilic in H-E stains. The differential staining of two regions within the ballooning axons suggests that the axons damaged by MPTP are myelinated.

In parasagittal sections from animals surviving 8 days, the swellings were seen to be restricted to only an axonal segment and did not involve the entire length of the axon. This gave the damaged axon a 'club-like' appearance (Fig. 1H) with the club-head pointing rostrally towards the striatum and the tail pointing caudally towards the substantia nigra. Axons could not be followed rostrally past the 'club-like' swelling. Although these club-like swellings could be seen throughout the extent of the nigrostriatal pathway, their numbers were densest at two locations: a proximal location at the rostral tip of the substantia nigra and a midway location within the hypothalamus halfway between the substantia nigra and striatum. In freshly stained tissue a halo of TH-reactivity was seen immediately outside of the 'club-like' swellings suggesting that TH had leaked out of the damaged axon.

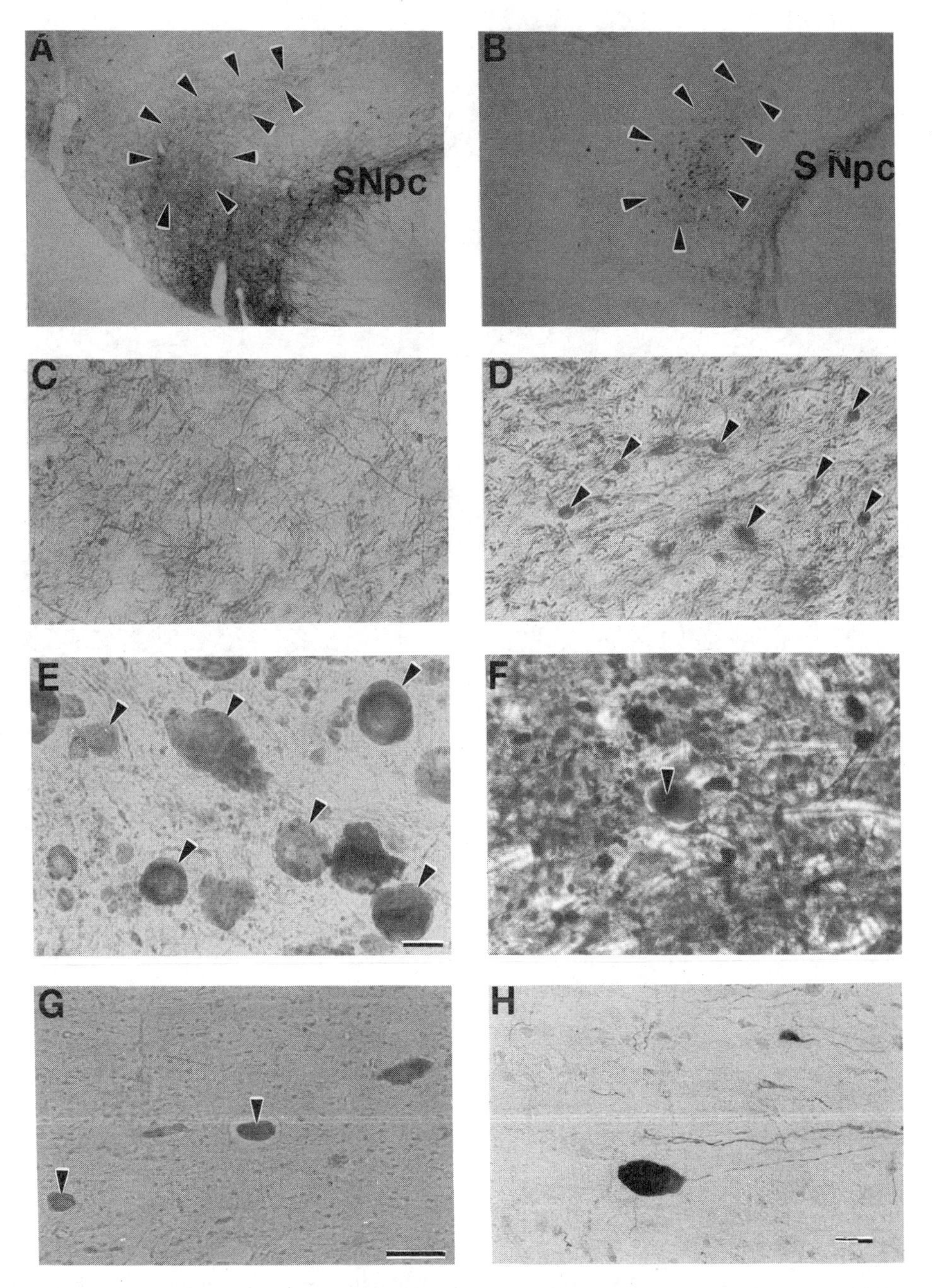

FIGURE 1 Effects of MPTP observed in the nigrostriatal pathway with the light microscope. A&B: A photograph taken at low magnification through the nigrostriatal pathway before and 14-days after an MPTP injection, respectively. Arrows indicate location of nigrostriatal pathway. C-E: A higher magnification of the nigrostriatal pathway before, 1-day after and 14-days after MPTP, respectively. Diagonal arrows indicate some of the swollen axons. Calibration bar for C-E in E equals 20uM. F&G: Two regions are seen within the ballooning axons cut coronally 8-days after an injection. Vertical arrows indicate core region of swollen axon. Calibration bar for F&G in G equals 50uM. H: The ballooning axons appear 'club-like' in sagittal sections 8-days after an injection. Calibration bar in H equals 20uM. Sections A-E & H were stained with antibodies against TH. F&G were stained for degeneration and phosphorylated H-neurofilaments, respectively. SNpc: substantia nigra pars compacta.

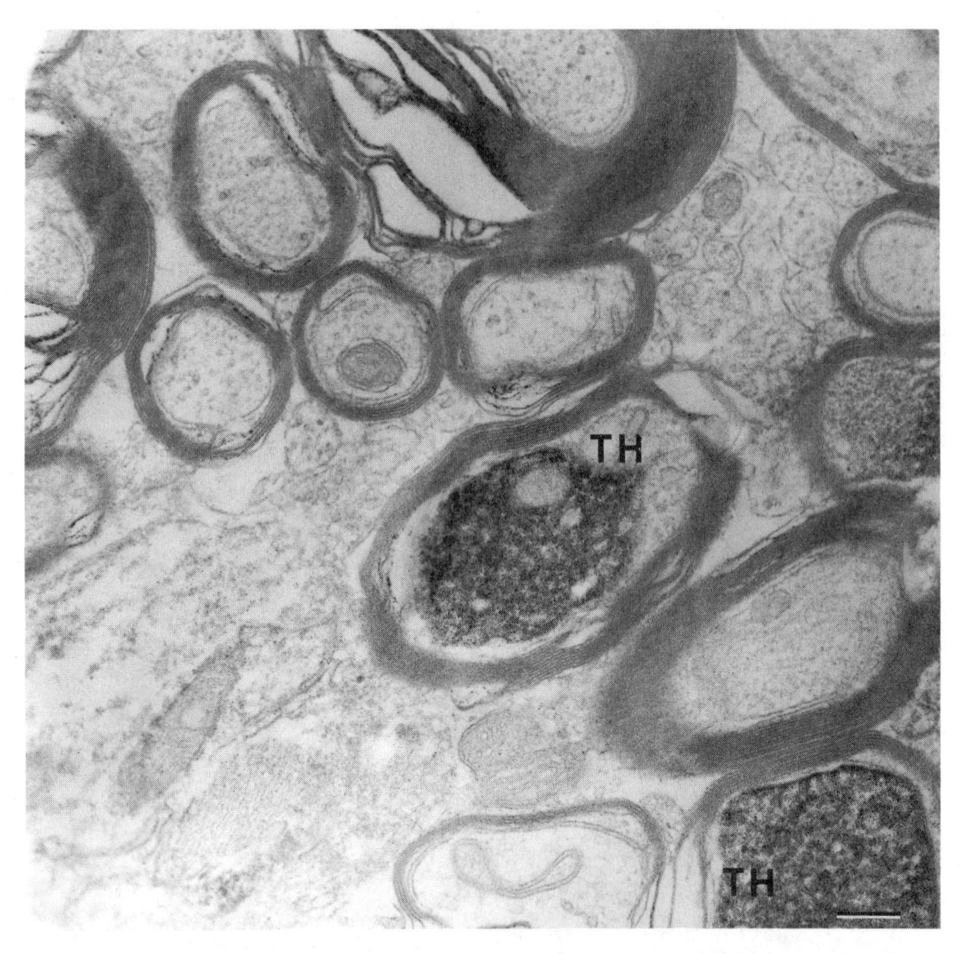

FIGURE 2 Electron micrograph through the nigrostriatal pathway 8-days after an injection of MPTP. Two myelinated axons are labelled by antibodies against TH; however, they are not swollen in this figure. Calibration bar in lower left-hand corner equals 2.3um.

Examination of the substantia nigra with the electron microscope revealed that TH-positive axons are myelinated. Examples of this is seen in figure 2.

DISCUSSION

Two of our principal findings were 1) that swelling of TH-positive axons in the nigrostriatal pathway was the earliest pathological change we observed with the light microscope and 2) that axons continued to balloon progressively for the next week until some reached a diameter of 30um or more. It is difficult to imagine how neurons could survive the proximal portion of their axons swelling to the extent found in this and other studies[17]. These data suggest that the ballooning of axons might be the final pathway leading to death. However, our experiments were not designed to determine if swelling axons resulted in cell death or if they were a manifestation of a cell dying.

Direct evidence demonstrating that MPTP-induced axonal damage can produce cell death is provided by experiments involving intracarotid injects of MPTP in monkeys. By directly injecting MPTP into the internal carotid artery, brain regions supplied by this artery have a "first pass" exposure to MPTP about 15-fold greater than the remainder of the body. Because the internal carotid artery supplies the striatum and the basilar artery supplies the substantia nigra, intracarotid infusions of MPTP expose only the terminals of the nigrostriatal system to a high concentration of the toxin. Once taken up by the terminals, labeled-MPP+ was retrogradely transported along the nigrostriatal pathway; however, the toxin never reached the cell bodies of the substantia nigra[18]. In fact, the label was transported no further than the proximal portion of the nigrostriatal pathway at one of two locations where we found severely swollen axons. Even though MPP+ remained in the axons and never reached the cell bodies of dopamine neurons, it still produced cell death [19]. Therefore, axotomy is one of the pathways by which MPTP can kill neurons.

Based on our current understanding of the mechanism of MPTP's action, the axon (including its terminals) is the most likely cellular compartment where MPP+ would initially produce its greatest injury following a systemic injection. As discussed earlier, MPP+ is sequestered in neurons by the dopamine uptake carrier. Because ligands for the dopamine transporter label the striatum significantly more intensely than the substantia nigra[20-22], these data suggest that a dopaminergic axon should be exposed to considerably higher concentration of MPP+ than its soma or dendrites. This has been confirmed experimentally[6]. Once in the axon, MPP+ is taken up into the mitochondria where it uncouples the electron transport chain at complex I[4]. The resulting depletion of ATP should then most affect the synapse because the neuron's energy metabolism is highest there[23]. Therefore, these data indirectly suggest that axons and their terminals might be the primary site of cellular insult that leads to neural death, especially at low doses of MPTP.

If axotomy is one mechanism by which MPTP can kill neurons, are there morphological differences in the axons of rodents and carnivores that might explain why MPTP is toxic in dogs and not in rats? Based on studies in rodents, the nigrostriatal pathway is generally described as being composed of fine, unmyelinated axons[24]. However, in this study, we found that many of the TH-positive axons within the substantia nigra of dog are relatively large and well myelinated. Similarly, dopaminergic axons within the striatum are myelinated in monkey[25]. Given these morphological differences, this raises the question of whether or not large, myelinated axons are preferentially vulnerable to the toxic effect of MPTP or other chemicals.

It is well known that large myelinated axons are preferentially vulnerable to certain injuries including head trauma, amyotrophic lateral sclerosis, and exposure to many chemical toxins[26,27]. The mechanism of this preferential vulnerability is not known; however, it may be related to the unique morphology of myelinated axons. A number of laboratories have demonstrated that the larger the axon, the more relative constriction is found at the nodes of Ranvier[28,29]. For instance, unmyelinated axons have no constrictions along their length because they lack nodes of Ranvier. In contrast, cross-sectional area at the node of Ranvier in axons more than 4um in diameter is reduced to less than 20% of its internodal value. These constrictions provide potential sites for the mechanical blockage of axonal transport following injury[30]. Thus, when exposed to certain toxins or pathological conditions, the nodes of Ranvier are the sites of a rapid buildup of neurofilaments and organelles as they are transported down the axon from the cell body[31,32]. This buildup produces a blockage of axonal transport resulting in swelling of the fibers and possibly the mechanical breakup of the axon. Deprived of essential nutrients from the soma, the terminals then degenerate; deprived of trophic factors from the terminals, the soma degenerates by apoptosis[33].

MPTP might produce axotomy leading to cell death through a similar mechanism. We suggest that it is difficult to produce a lesion in rats with MPTP because their nigrostriatal pathway is composed of small fibers with few axonal constrictions and potential sites for congestion of axonal transport. In contrast, the nigrostriatal pathway of dog is preferentially vulnerable to MPTP because its axons are large and myelinated.

Therefore, there is a greater probably in dog than in rat that axonal transport will become congested and blocked at the nodes of Ranvier especially if the axon becomes disorganized because of a loss of energy.

LITERATURE CITED

1. R.E. Heikkila, F.S. Manzino, F.S. Cabbat, and R.C. Duvoisin, Protection against the dopaminergic neurotoxicity of 1-methyl-4-phenyl-1,2,3,6-tetrahydropyridine by monamine oxidase inhibitors, Nature (London), 311:467(1984).
2. J.W. Langston, I. Irwin, E.B. Langston, and L.S. Forno, Pargyline prevents MPTP-induced parkinsonism in primates, Sci., 225:1480(1984).
3. J.N. Johannessen, C.C. Chiueh, J.P. Bacon, N.A. Garrick, R.S. Burns, V.K. Weise, I.J. Kopin, J.E. Parisi, and S.P. Markey, Effects of 1-methyl-4-phenyl-1,2,3,6-tetrahydropyridine in the dog: Effect of pargyline pretreatment, J. Neurochem ., 53:582(1989).
4. T.P. Singer, R.R. Ramsay, K. McKeown, A. Trevor, and N.E. Castagnoli Jr., Mechanism of the neurotoxicity of 1-methyl-4-phenylpyridinium (MPP+), the toxic bioactivation product of 1-methyl-4-phenyl-1,2,3,6-tetrahydropyridine (MPTP), Toxicol., 49:17(1988).
5. A.M. Marini, J.P. Schwartz, and I.J. Kopin, The neurotoxicity of 1-methyl-4-phenylpyridinium in cultured cerebellar granule cells, J. Neurosci., 9:3665(1989).
6. S.P. Markey, J.N. Johannesen, C.C. Chieuh, R.S. Burns, and M.A. Herkenham, Intraneuronal generation of a pyridinium metabolite may cause drug-induced parkinsonism, Nature (London), 311:464(1984).
7. J.A. Javitch, G.R. Uhl, and S.H. Snyder, Parkinsonism-inducing neurotoxin, N-methyl-4-phenyl-1,2,3,6-tetrahydropyridine: Characterization and localization of receptor binding sites in rat and human brain, Proc. Natl. Acad. Sci. USA, 81:4591(1984).
8. S.H. Snyder, R.J. D'Amato, J.S. Nye, and J.A. Javitch, Selective uptake of MPP+ by dopamine neurons is required for MPTP toxicity: Studies in brain synaptosomes and PC-12 cells, in: "MPTP: A Neurotoxin Producing a Parkinsonian Syndrome," S.P. Markey, N. Castagnoli Jr., A.J. Trevor, and I.J. Kopin, eds., Academic Press, Inc., San Diego (1986).
9. J.S. Wilson, Effects of GBR 12909, a specific dopamine uptake blocker, on MPTP's toxicity in dog, Soc. Neurosci. Abstr., 15:138(1989).
10. S.C. Rapisardi, V.O.P. Warrington, and J.S. Wilson, Effects of MPTP on the fine structure of neurons in substantia nigra of dogs, Brain Res., 512:147(1990).
11. J.S. Wilson, B.H. Turner, G.D. Morrow, and P.J. Hartman, MPTP produces a mosaic-like pattern of terminal degeneration in the caudate nucleus of dog, Brain Res., 423:329(1987).
12. J.S. Wilson, D.T. Shearer, A.K. Adelakun, and R.G. Carpentier, Mechanisms of the inotropic actions of MPTP and MPP+ on isolated atria of rat, TAP, 111:49(1991).
13. C.J. Schmidt, L.A. Matsuda, and J.W. Gibb, In vitro release of tritiated monoamines from rat CNS tissue by the neurotoxic compound 1-methyl-4-phenyl-1,2,3,6-tetrahydropyridine, Europ. J. Pharmacol., 103:255(1984).
14. D.J.S. Sirinathsinghji, R.P. Heavens, and C.S. McBride, Dopamine-releasing action of 1-methyl-4-phenyl-1,2,3,6-tetrahydropyridine(MPTP) and 1-methyl-4-phenylpyridinium (MPP+) in the neostriatum of the rat as demonstrated in vivo by the push-pull perfusion technique: dependence on sodium but not calcium ions, Brain Res., 443:101(1988).
15. B.H. Turner, J.S. Wilson, J.C. McKenzie, and N. Richtand, MPTP produces a pattern of nigrostriatal degeneration which coincides with the mosaic organization of the caudate nucleus, Brain Res., 473:60(1988).

16. J.K. Young, J.C. McKenzie, and J.H. Baker, Association of iron-containing astrocytes with dopaminergic neurons of the arcuate nucleus, J. Neurosci. Res., 25:204(1990).
17. D.M. Jacobowitz, R.S. Burns, C. Chiueh, and I.J. Kopin, N-methyl-4-phenyl-1,2,3,6-tetrahydropyridine (MPTP) causes destruction of the nigrostriatal but not the mesolimbic dopamine system in the monkey, Psychopharmacol. Bull., 20:416(1984).
18. M. Herkenham, M.D. Little, K.S. Bankiewicz, S.C. Yang, and Others, Selective retention of MPP+ within the monoaminergic systems of the primate brain following MPTP administration: an in vivo autoradiographic study, Neurosci., 40:133(1991).
19. K.S. Bankiewicz, E.H. Oldfield, C.C. Chiueh, D.M. Doppman, D.M. Jacobowitz, and I.J. Kopin, Hemiparkinsonism in monkeys after unilateral internal carotid artery infusion of 1-methyl-4-phenyl-1,2,3,6-tetrahydropyridine (MPTP), Life Sci., 39:7(1986).
20. B.J. Ciliax, M.R. Kilbourn, M.S. Haka, and J.B.J. Penney, Imagining the dopamine uptake sites with ex vivo [^{18}F] GBR 13119 binding autoradiography in rat brain, J. Neurochem ., 55:619(1990).
21. B. Scatton, A. Dubois, M.C. Dubocovich, N.R. Zahniser, and D. Fage, Quantitative autoradioraphy of ^{3}H-nomifensine binding sites in rat brain, Life Sci., 36:815(1985).
22. J.A. Javitch, S.M. Strittmatter, and S.H. Synder, Differential visualization of dopamine and norepinephrine uptake sites in rat brain using [^{3}H]Mazindol Autoradiography, J. Neurosci., 5:1513(1885).
23. L. Sokoloff and S. Takahashi, Functional activation of energy metabolism in nervous tissue: where and why, This Volume, (1995).
24. V.M. Pickel, E. Johnson, M. Carson, and J. Chan, Ultrastructure of spared dopamine terminals in caudate-putamen nuclei of adult rats neonatally treated with intranigral 6-hydroxydopamine, Brain Res. Dev. Brain Res, 70:75(1992).
25. M.Y. Arsenault, A. Parent, P. Seguela, and L. Descarries, Distribution and morphological characteristics of dopamine-immunoreactive neurons in the midbrain of the squirrel monkey (Saimiri sciureus), J. Comp. Neurol, 267:489(1988).
26. G. Krinke, H. Schaumberg, P. Spencer, P. Thomann, and R. Hess, Clioquinol and 2,5-hexanedione induce different types of distal axonopathy in the dog, Neuropath. (Berl.), 47:213(1979).
27. J.T. Povlishock, The morphopathologic response to experimental head injuries of varying severity, in: "Central Nervous System Status Report," DP Becker and JT Povlishock, eds., William Byrd Press, Richmond, Va (1985).
28. C.S. Raine, Differences between the nodes of Ranvier of large and small diameter firs in the P.N.S, J. Neurocyto., 11:935(1982).
29. M. Rydmark, Nodal axon diameter correlates linearly with internodal axon diameter in spinal roots of the cat, Neurosci. Lett., 24:247(1981).
30. C.H. Berthold and A. Mellstrom, Peroxidase activity at consecutive nodes of Ranvier in the nerve to the medial gastrocnemius muscle after intramuscular administration of horseradish peroxidase, Neurosci., 19:1349(1986).
31. J.B. Cavanagh, The 'dying back' process, Arch. Pathol. Lab. Med., 103:659(1979).
32. J.B. Cavanagh, The pattern of recovery of axons in the nervous system of rats following 1,5,-hexanediol intoxication: a question of rheology?, Neuropath. App. Neurobiol., 8:19(1982).
33. H. Mochizuki, N. Nakamura, K. Nishi, and Y. Mizuno, Apoptosis is induced by 1-methyl-4-phenylpyridinium ion (MPP+) in ventral mesencephalic-striatal co-culture in rat, Neurosci. Lett., 170:191(1994).

HIV-1 COAT PROTEIN GP120 INDUCES NEURONAL INJURY TO CULTURED DOPAMINE CELLS

Barbara A. Bennett, Daniel E. Rusyniak, and Charlotte K. Hollingsworth

Department of Physiology and Pharmacology
Bowman Gray School of Medicine
Winston-Salem, NC 27157

INTRODUCTION

Human immunodeficiency virus (HIV)-1 infection is commonly associated with neurological manifestations that can produce devastating cognitive and motor impairments.[1-4] Concomitant with this are CNS disturbances which can include various organic mental disorders.[5] These CNS disturbances may occur in the absence of opportunistic infections as well as in the absence of direct infection of neurons by HIV-1.[4] In fact, HIV-1 has only been demonstrated in macrophage or microglial cells and not in neurons.[6-9] Previous studies have shown that infection with HIV-1 results in neurotoxicity which can be elicited by viral particles, the most likely of which is the glycoprotein, gp120.[10-12] This viral protein has been shown to produce neurotoxicity at very low concentrations (pM) in various neuronal culture systems[10-19] as well as modify astrocyte function.[20-22] Thc mechanism(s) by which gp120 induces neurotoxicity are not understood, but there is evidence that NMDA receptors, Ca^{2+} channels, and nitric oxide (NO) are important mediators.[11,13-15,18,19]

Autopsy findings of HIV(+) brains reveal a consistent morphological abberation in which there is a spongiform appearance (small rounded vacuoles) in certain brain areas. One of the areas in which these spongiform changes are seen is the substantia nigra,[23] an area in the ventral mesencephalon closely allied with cognitive and motor function. This largely dopaminergic area may play a significant role in the development and progression of AIDS dementia complex (ADC). In fact, there is substantial clinical evidence suggesting the involvement of dopaminergic neuronal dysfunction in ADC.[1,24] The aim of the current study is to demonstrate that recombinant gp120 (rgp120) induces dopaminergic neuronal damage in primary midbrain cultures by examining various neurochemical and immunochemical parameters.

EXPERIMENTAL PROCEDURES

Culture Preparation

Neurodegenerative Diseases
Edited by Gary Fiskum, Plenum Press, New York, 1996

Fetal rats (E15) were obtained from timed-pregnant Sprague Dawley rats and dissociated as described previoiusly.[25] The mesencephalon was identified as extending from the caudal border of the diencephalon to the caudal border of the mesencephalic flexure. This area includes the dopamine (DA) cells of the SN (A9) and the VTA (A10), but excludes the noradrenergic cells of the locus coeruleus. Tissues were rinsed, mechanically dispersed and plated. Culture media consisted of DMEM (4.5 gm glucose/L, no pyruvate) supplemented with 10% fetal calf serum, 2 mM glutamine, 100 μg/ml streptomycin and 100 Units/ml penicillin. This media was removed after 24 hrs, replaced with media containing 2% serum for 24 hrs and this was then exchanged for serum free media (N2 media) which was utilized throughout the remainder of the study. Cultures were maintained in an atmosphere of 5% CO_2/95% air at 37°C. After 8-10 days of culture, cells were exposed for **72 hrs to 10^{-13} to 10^{-8} M rgp 120** (recombinant gp120 from HIV-1_{SF2}.[26-29] In separate experiments, MK-801 (1 μM) or 7-nitroindazole (7-NI; 10μM) were co-administered with rgp120 for 72 hrs. Anti-gp120 (1:500)[28,30] was preincubated with rgp120 for 30 min at room temperature and this mixture was added to the cultures for 72 hrs.

High Affinity [^{3}H]Dopamine Uptake

After treatment of the cultures with rgp120, media was removed and replaced with fresh media for an additional 18-24 hrs. Cultures were then rinsed and [^{3}H]DA uptake was analyzed as previously described[31]. Non-specific uptake was determined by the addition of mazindol (10 μM) and specific uptake was determined by subtracting non-specific counts from the total. Data are expressed as percent of control after correction for non-specific uptake. Significance was determined by analysis of variance (ANOVA) followed by Tukey's post-hoc analysis or by Student's t-test where appropriate.

Immunocytochemical Procedures

Lab Tek slide chambers were prepared for the immunocytochemical studies utilizing 3 x 10^5 cells per well. The cultures were fixed with 4% paraformaldehyde (30 min) followed by a 24 hr incubation with the primary antisera. Tissues were processed by the Vectastain protocol (Vector Laboratories) using diaminobenzidine (DAB) as the chromogen. The antisera used were tyrosine hydroxylase (TH; 1:1,000, PelFreez) and neuron specific enolase (NSE; 1:15,000, Polysciences). Results represent TH(+) or NSE(+) cells/well. The morphometric data (process length and cell number) were analyzed blind to the experimental condition using a morphometric software package. After a 3 day exposure to rgp120 (10^{-12} to 10^{-8} M), morphometric data were obtained from 50 cells per treatment group that were extracted from 4 experiments. Significance level for the immunocytochemical data were determined by ANOVA followed by Tukey's post-hoc analysis.

DNA fragmentation

Cells were removed from the plates by rinsing with ice-cold PBS and pelleted by centrifugation at 200g for 10 min. The pellet was lysed with 0.5 ml of hypotonic lysing buffer, pH 7.5 (10 mM Tris, 1 mM EDTA, 0.2% Triton X-100). The lysate was immediately centrifuged at 13,000g for 10 min and the supernatant, containing fragmented DNA, was collected. DNA from the 13,000g supernatants was precipitated overnight at -20°C in 50% isopropanol and 0.5 M NaCl. The precipitates were then collected after centrifugation at 13,000g, air dried, and resuspended in 10 mM Tris, 1 mM EDTA, pH 7.4. Loading buffer contained 15 mM EDTA, 2% SDS, 50% glycerol and 0.5% bromphenol blue and was added to the samples at 1:5 (v/v), and samples heated to 65°C for 10 min. Electrophoresis was in

0.75% agarose for 2 hrs at 70 V. DNA was visualized by staining with ethidium bromide and then photographed.

RESULTS

Exposure of midbrain cultures to rgp120 for 3 days resulted in a concentration-dependent decrease in the ability of the cells to transport [^{3}H]DA (Figure 1). Uptake was maximally reduced after exposure to 1 nM rgp120 (approximately 35%), and there were significant reductions with 100 pM rgp120, also.

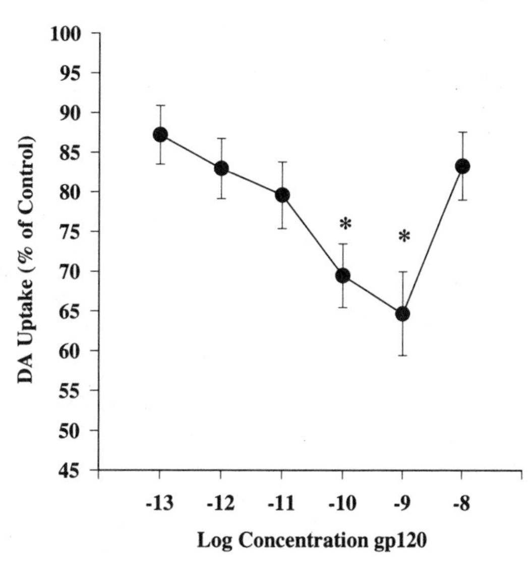

Figure 1. The concentration-dependent effects of rgp120 on midbrain culture [3H]DA uptake. The exposure period was 3 days. Values are means ± SEM.

Addition of known inhibitors of rgp120 during the exposure period resulted in a reduction in the effects of the viral protein. As shown in Figure 2, addition of 1 μM MK-801, 10 μM 7-NI or anti-gp120 antibodies attenuated the effects of rgp120 on [^{3}H]DA uptake. Others have shown that these agents can prevent the effects of rgp120 and our findings suggest that the detrimental effects of rgp120 on dopaminergic cells are probably mediated by similar mechanisms which are thought to include both NMDA- and NO-facilitated events.[11-15]

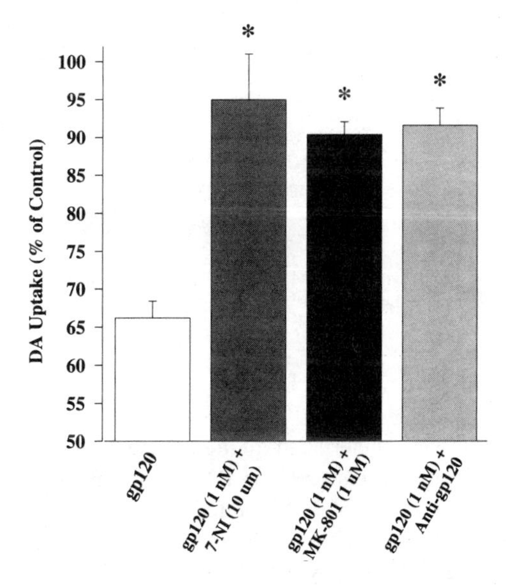

Figure 2. Attenuation of damage due to rgp120 (1 nM) exposure by 7-NI, MK-801 and anti-gp120 antibodies. The exposure period was 3 days. Values are means ± SEM.

Sister cultures were examined for DA cell survival after treatment with the glycoprotein. After a 3 day exposure period to rgp120, the cultures were fixed and stained with antisera to tyrosine-3-monooxygenase (TH) and neuron-specific enolase (NSE). The number of DA cells per well was approximately 250-300 and the density of neurons averaged 180 neurons/mm^2. There was no change in either the number of DA cells surviving or the density of neurons remaining after a 3 day treatment with rgp120 (10^{-12} to 10^{-8} M).

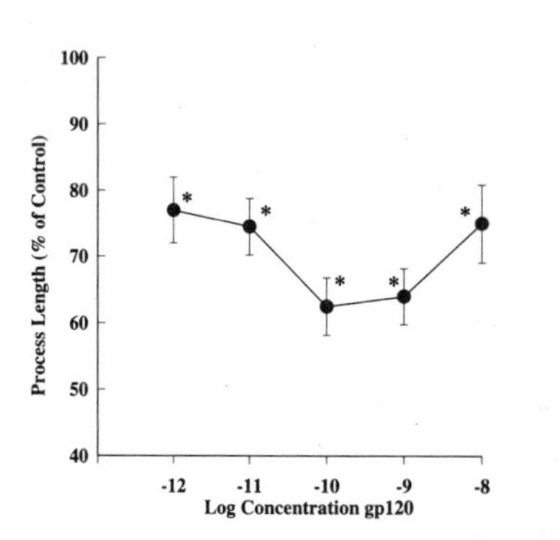

Figure 3. Reduction in DA process length after exposure to rgp120 for 3 days. There is a concentration-dependent decrease in process length. Values are means ± SEM.

Morphological assessment of these same cultures was performed by examining process length of DA cells. Measurement of DA process length revealed that rgp120 caused a concentration-dependent reduction in length, suggesting that arborization is affected by the presence of rgp120 (Figure 3). Average process length for controls was 200 μm and this was reduced to approximately 120 μm (a 40% reduction) in cultures treated with 100 pM rgp120. Illustrating this effect are photomicrographs showing neuronal process length in control and rgp120-treated cultures (Figure 4). Panel A shows control (untreated) cells and B is representative of treated (100 pM) cultures, illustrating a pronounced reduction in process length.

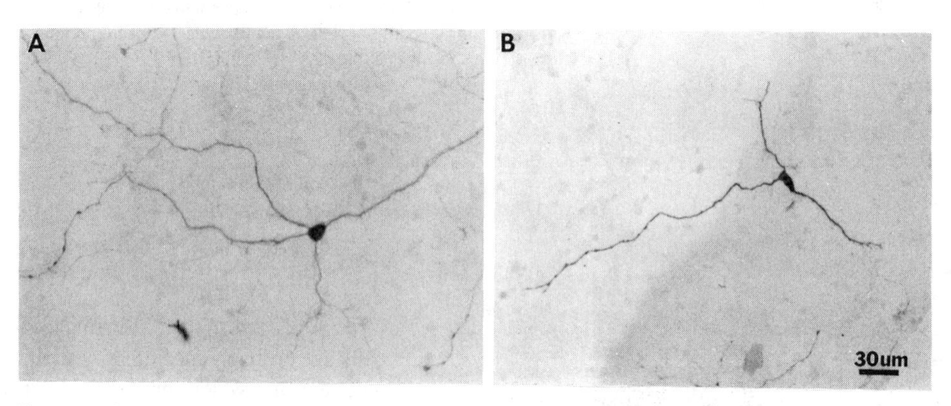

Figure 4. Photomicrograph illlustrating DA cells after exposure to rgp120 (1 nM) for 3 days. A) non-treated culture and B) cells treated with rgp120 displaying reduced process length.

A more prolonged exposure to rgp120 was undertaken to determine if further deleterious effects could be induced by examining the cultures for DNA fragmentation. Midbrain cultures were grown for 10 days and then rgp120 (500 pM) added. There was a 15 min pre-exposure period, consisting of Mg^{2+}-free media and 25 μM glutamate in the presence and absence of rgp120. This pre-exposure period was followed by the addition of rgp120 (500 pM) to the cultures for 10 days. This media was then removed and fresh media added for 24 hrs. The cells (approximately 20 million per treatment group) were then collected and processed for DNA fragmentation. The results are shown in Figure 5, with lane 1 showing the migration pattern of a 123 bp DNA ladder. Without the addition of rgp120, no signs of DNA fragmentation could be detected (lane 2). However, after incubation of the cultures with 500 pM rgp120 for 10 days, the DNA was clearly fragmented, as visualized by the DNA migration in the agarose gels (lane 3). The visible bands mark the degradation products (oligonucleosomal fragments) in multiples of approximately 180 base pairs.

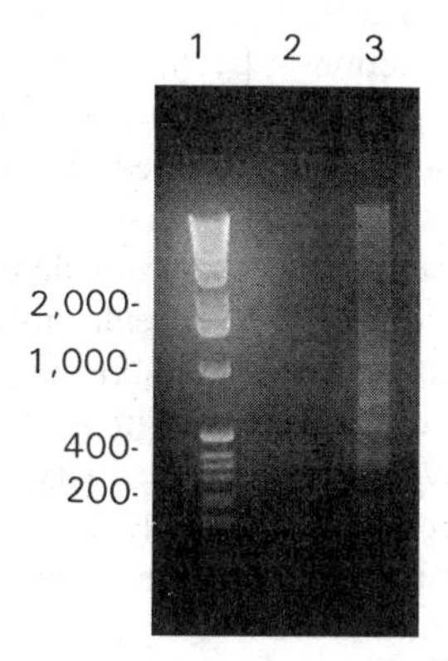

Figure 5. Agarose gel electrophoresis of DNA extracted from midbrain cultures. The DNA was isolated from: lane 1, 123 bp DNA ladder; lane 2, control cultures; lane3, rgp120-treated cultures.

DISCUSSION

In the present study, we have demonstrated that the HIV-1 coat protein, rgp120, in extremely low concentrations induces damage to a specific neuronal subpopulation, midbrain DA cells. This was evidenced by a reduction in DA uptake, an index of neuronal function, as well as a reduction in the neuronal cytoarchitecture. In addition, with a prolonged exposure period, rgp120 promoted the typical pattern of DNA fragmentation that is suggestive of apoptosis. These experiments indicate that rgp120 is capable of inducing dopaminergic damage *in vitro* suggesting that this type of damage may contribute to the deleterious effects observed *in vivo*.

The dopaminergic dysfunction observed after rgp120 was prevented by co-administration of an NMDA receptor blocker, MK-801. Previous studies have shown that NMDA antagonists were capable of preventing the toxicity associated with rgp120 administration.[11,18] Administration of Ca^{2+} channel antagonists has also been shown to attenuate the glycoprotein-induced toxicity. [13,18] These and other studies suggest that calcium entry into the cells via either the ligand-operated or the voltage-activated channels may be important triggers for a large release of intracellular Ca^{2+} and the subsequent damage induced by low concentrations of rgp120.

The present study also addressed the involvement of nitric oxide (NO) in producing the deleterious effects of rgp120 by adding an inhibitor of nitric oxide synthetase (NOS), 7-nitroindazole (7-NI). This compound inhibits both neuronal and macrophage NOS, being differentially selective for the cNOS (K_i=0.16 μM) compared to the iNOS (K_i=1.6 μM)[32]. The concentration used in these studies (10 μM) was not selective for either isoform. Co-administration of 7-NI with 1 nM rgp120 restored DA uptake to 95% of control values, reinforcing the possibility that NO synthesis is important in the cascade of events that ultimately results in neuronal damage. The specificity of the effect of rgp120 in our midbrain culture system is supported by the protective action of anti-gp120 antibodies.

Previous studies have shown that exposure to the viral protein gp120 can induce apoptosis in either T cells[33-35] or in rat cortical cell cultures.[17,36] Apoptosis is a physiological control mechanism that is characterized by degradation of DNA to oligonucleosomal fragments, nuclear condensation and cell death.[37] There are many unanswered questions regarding the trigger for an apoptic event and, as such, there are no known mechanisms by which the HIV-1 viral protein initiates apoptosis in cultured neurons. Although rgp120 did not produce significant cell death utilizing a shorter exposure period (3 days), either the length of the exposure (10 days) or the addition of 25 μM glutamate (which by itself had no effect on controls) was responsible for the enhanced neural damage that was observed.

The results from the present study suggest that rgp120 is capable of producing neuronal damage to dopaminergic cells and cause a dysfunction in this subset of aminergic nerves. The reduction in [^{3}H]DA uptake that was observed most likely reflects a change in V_{max} (or the number of uptake sites) since there was a decrease in process length and transporter sites are known to be located along processes as well as at terminals.[38] Other neurodegenerative diseases are believed to be accompanied by presynaptic terminal loss,[39,40] such as in Alzheimer's disease, or decreases in dendritic length,[41] such as in Parkinson's disease. Indeed, pathology studies of HIV-encephalitis has found that there is cortical neuron loss as well as decreased dendritic length and branching.[42] The fact that antiviral agents can improve the cognitive and motor dysfunction in AIDS patients suggests that some of the neuronal damage may be reversible in nature, while irreversible damage would more likely occur with cell death. Treatment of the early signs of AIDS dementia, before nerve cell loss is extensive, may prevent or reverse the progression of the severe neurological complications associated with the later stages of ADC.

ACKNOWLEDGEMENTS

The following reagents were obtained through the AIDS Research and Reference Reagent Program, AIDS Program, NIAID, NIH: rgp120 from HIV-1_{SF2} and antiserum to HIV-1_{SF2} gp120 from Dr. Kathy Steimer, Chiron Corporation. This work was supported by grants from the National Institue on Drug Abuse and (DA05073 to B.A.B.).

Correspondence should be sent to Dr. Barbara Bennett, at Department of Physiology and Pharmacology, Bowman Gray School of Medicine, Medical Center Blvd, Winston-Salem, North Carolina 27157.

REFERENCES

1. J.C. Ameisen and A. Capron, Cell dysfunction and depletion in AIDS: the programmed cell death hypothesis, *Immunol. Today* 12:102-104 (1991).
2. J. Artigas, F. Niedobitek, G. Grosse, W. Heise, and G. Gosztonyi, Spongiform encephalopathy in AIDS dementia complex: report of five cases., *J. Acquir. Immune Defic. Syndromes* 2:374-381 (1989).
3. B.A. Bennett, C.E. Hyde, J.R. Pecora, and J.E. Clodfelter, Long-term cocaine administration is not neurotoxic to cultured fetal mesencephalic dopamine neurons, *Neurosci. Lett.* 153:210-214 (1993).
4. B.A. Bennett, C.E. Hyde, J.R. Pecora, and J.E. Clodfelter, Differing neurotoxic potencies of methamphetamine, mazindol, and cocaine in mesencephalic cultures, *J. Neurochem.* 60:1444-1452 (1993).
5. D.J. Benos, B.H. Hahn, J.K. Bubien, S. Ghosh, N.A. Mashburn, M.A. Chaikin, G.M. Shaw, and E.N. Benveniste, Envelope glycoprotein gp120 of human immunodeficiency virus type 1 alters ion transport in astrocytes: implications for AIDS dementia complex, *Proc. Natl. Acad. Sci. USA* 91:494-498 (1994).
6. L. Brenneman, G.L. Westbrook, S.P. Fitzgerald, D.L. Ennist, K.L. Elkins, M.R. Ruff, and C.B. Pert, Neuronal cell killing by the envelope protein of HIV and its prevention by vasoactive intestinal peptide., *Nature* 335:639-642 (1988).
7. C. Cerruti, M. Drian, J. Kamenka, and A. Privat, Localization of dopamine carriers by BTCP, a dopamine uptake inhibitor, on nigral cells cultured *in vitro.* , *Brain Res.* 555:346-355 (1991).
8. A. Ciardo and J. Meldolesi, Effects of the HIV-1 envelope glycoprotein gp120 in cerebellar cultures. $[Ca^{2+}]_i$ increases in a glial cell subpopulation., *Euro. J. Neurosci.* 5:1711-1718 (1993).

9. V.L. Dawson, T.M. Dawson, G.R. Uhl, and S.H. Snyder, Human immunodefiency virus type 1 coat protein neurotoxicity mediated by nitric oxide in primary cortical cultures, *Proc. Natl. Acad. Sci. USA* 90:3256-3259 (1993).
10. E.B. Dreyer, P.K. Kaiser, J.T. Offerman, and S.A. Lipton, HIV-1 coat protein neurotoxicity prevented by calcium channel antagonists., *Science* 248:364-367 (1990).
11. D.H. Gabuzda, D.D. Ho, S.M. de la Monte, M.S. Hirsch, T.R. Rota, and R.A. Sobel, imunohistochemical identification of HTLV-III antigen in brains of patients with AIDS., *Ann. Neurol.* 20:289-295 (1986).
12. D.H. Gabzuda and M.S. Hirsch, Neurologic manifestations of infection with human immunodeficiency virus., *Ann. Int. Med.* 107:383-391 (1987).
13. D. Giulian, E. Wendt, K. Vaca, and C.A. Noonan, The envelope glycoprotein of human immunodeficiency virus type-1 stimulates release of neurotoxins from monocytes., *Proc. Natl. Acad. Sci. USA* 90:2769-2773 (1993).
14. D. Giulian, K. Vaca, and C.A. Noonan, Secretion of neurotoxins by mononuclear phagocytes infected with HIV-1., *Science* 250:1593-1596 (1990).
15. M.-L. Gougeon and L. Montagnier, Apoptosis in AIDS, *Science* 260:1269-1270 (1993).
16. H. Groux, G. Torpier, D. Monte, Y. Mouton, A. Capron, and J.C. Ameisen, Activation-induced death by apoptosis in $CD4^+$ T cells from human immunodeficiency virus-infected asymptomatic individuals, *J. Exp. Med.* 175:331-340 (1992).
17. N.L. Haigwood, C.B. Barker, K.W. Higgins, P.V. Skiles, G.K. Moore, K.A. Mann, D.R. Lee, J.W. Eichberg, and K.S. Steimer, Evidence for Neutralizing Antibodies Directed Against Conformational Epitopes of HIV-1 gp120., in: "Vaccines 90", Cold Spring Harbor Laboratories, New York, pp. 313-320 (1990).
18. K.D. Kieburtz, L.G. Epstein, H.A. Gelbard, and J.T. Greenamyre, Excitotoxicity and dopaminergic dysfunction in the acquired immunodeficiency syndrome dementia complex: therapeutic implications, *Arch Neurol* 48:1281-1284 (1991).
19. S. Koenig, H.E. Gendelman, J.M. Orenstein, M.C. Dal Canto, G.H. Pezeshkpour, M. Yungbluth, F. Janotta, A. Aksamit, M.A. Martin, and A.S. Fauci, Detection of AIDS virus in macrophages in brain tissue from AIDS patients with encephalopathy., *Science* 233:1089-1093 (1986).
20. J.A. Levy, A.D. Hoffman, S.M. Kramer, J.M. Shimabukuro, and L.S. Oshiro, Isolation of lymphocytopathic retroviruses from San Francisco patients with AIDS., *Science* 225:840-842 (1984).
21. S.A. Lipton, Models of neuronal injury in AIDS: another role for the NMDA receptor?, *Trends Neurosci.* 15:75-79 (1992).
22. S.A. Lipton, N.J. Sucher, P.K. Kaiser, and E.B. Dreyer, Synergistic effects of HIV coat protein and NMDA receptor-mediated neurotoxicity., *Neuron* 7:111-118 (1991).
23. T.M Lo, C.J. Fallert, T.M. Piser, and S.A. Thayer, HIV-1 envelope protein evokes intracellular calcium oscillations in rat hippocampal neurons., *Brain Res.* 594:189-196 (1992).
24. E. Masliah, N. Ge, M. Morey, R. DeTeresa, R.D. Terry, and C.A. Wiley, Cortical dendritic pathology in human immunideficiency virus encephalitis, *Lab. Invest.* 66:285-291 (1992).
25. E. Masliah, R.D. Terry, R.M. DeTeresa, and L.A. Hansen, Immunohistochemical quantification of the synapse-related protein synaptophysin in Alzheimer's disease, *Neurosci. Lett.* 103:234-239 (1989).
26. W.E.G. Muller, H.C. Schroder, H. Ushijima, J. Dapper, and J. Bormann, gp120 of HIV-1 induces apoptosis in rat cortical cell cultures: prevention by memantine, *Eur. J. Pharmacol.* 226:209-214 (1992).
27. B.A. Navia, B.D. Jordan, and R.W. Price, The AIDS dementia complex. I. Clinical features, *Ann. Neurol.* 19:517-524 (1986).

28. S. Patt, H.J. Gertz, L. Gerhard, and J. Cervos-Navarro, Pathological changes in dendrites of substantia nigra neurons in Parkinson's disease: a Golgi study, *Histol. Histopathol.* 6:373-380 (1991).
29. S. Perovic, C. Schleger, G. Pergande, S. Iskric, H. Ushijima, P. Rytik, and W.E.G. Muller, The triaminopyridine flupirtine prevents cell death in rat cortical cells induced by N-methyl-D-aspartate and gp120 of HIV-1, *Eur. J. Pharmacol.* in press:(1995).
30. S.W. Perry, Organic mental disorders caused by HIV: update on early diagnosis and treatment, *Am J. Psychiatry* 147:696-710 (1990).
31. R.J. Pomerantz, D.R. Kuritzkes, S.M. De LaMonte, T.R. Rota, A.S. Baker, D. Albert, D.H. Bor, E.L. Feldman, R.T. Schooley, and M.S. Hirsch, Infection of the retina by human immunodeficiency virus type I., *New England J. Med.* 317:1643-1647 (1987).
32. R.W. Price, B. Brew, J. Sidtis, M. Rosenblum, A.C. Scheck, and P. Cleary, The brain in AIDS: central nervous system HIV-1 infection and AIDS dementia complex, *Science* 239:586-592 (1988).
33. R. Sanchez-Pescador, M.D. Power, P.J. Barr, K.S. Steimer, M.M. Stempien, S.L. Brown-Shimer, W.W. Gee, A. Renard, A. Randolph, J.A. Levy, D. Dina, and P.A. Luciw, Nucleotide sequence and expression of an AIDS-associated retrovirus (ARV-2)., *Science* 227:484-492 (1985).
34. T. Savio and G. Levi, Neurotoxicity of HIV coat protein gp120, NMDA receptors, and protein kinase C: a study with rat cerebellar granule cells cultures., *J. Neurosci. Res.* 34:265-272 (1993).
35. C.J. Scandella, J. Kilpatrick, W. Lidster, C. Parker, J.P. Moore, G.K. Moore, K.A. Mann, P. Brown, S. Coates, B. Chapman, F.R. Masiarz, and K.S. Steimer, Nonaffinity purification of recombinant gp120 for use in AIDS vaccine development., *AIDS Res. Hum. Retroviruses* 9:1233-1244 (1993).
36. K.S. Steimer, G.A. Van Nest, N.L. Haigwood, E.M. Tillson, C. George-Nascimento, P.J. Barr, and D. Dina, Recombinant *env* and *gag* Polypeptides in Characterizing HIV-1 Neutralizing Antibidies., in: "Vaccines 88", Cold Spring Harbor Laboratories, Cold Spring Harbor, NY, pp. 347-355 (1988).
37. B.A. Watkins, H.H. Dorn, W.B. Kelly, R.C. Armstrong, B.J. Potts, F. Michaels, C.V. Kufta, and M. Dubois-Dalcq, Specific tropism of HIV-1 for microglial cells in primary human brain cultures, *Science* 249:549-553 (1990).
38. G.T. Williams, Programmed cell death: apoptosis and oncogenesis, *Cell* 65:1097-1098 (1991).
39. D.J. Wolff, G.A. Datto, R.A. Smatovicz, and R.A. Tempsick, Identification and characterization of a calmodulin-dependent nitric oxide synthase from GH3 pituitary cells, *J. Biol. Chem.* 268:9425-9429 (1993).
40. S.S. Zhan, K. Beyreuther, and H.P. Schmitt, Synaptophysin immunoreactivity of the cortical neuropil in vascular dementia of Binzwanger type compared with the dementia of Alzheimer type and nondemented controls, *Dementia* 5:79-87 (1994).

PLATELETS AS PERIPHERAL MODEL OF GLUTAMATE-RELATED EXCITOTOXICITY IN PARKINSON'S DISEASE

Carlo Ferrarese, Graziella Bianchi, Marianna Bugiani, Tiziana Cogliati, Maura Frigo, Davide Passoni, Nicoletta Pecora, Roberto Piolti, Clara Pozzi, Rachele Tortorella and Lodovico Frattola

Department of Neurology, University of Milan - Ospedale San Gerardo, Monza ITALY

INTRODUCTION

Over-stimulation of glutamate receptors and oxidative stress are two possible mechanisms that cooperate to selective degeneration of dopaminergic cells in Parkinson's Disease (PD)[1,2,3]. Defects in mitochondrial enzymes, which make dopaminergic cells more susceptible to oxidative stress, have been found not only in substantia nigra of PD patients[4], but also in their peripheral tissues[5] and blood platelets[6,7,8,9]. Recent studies also indicated that impaired mitochondrial function in platelets of PD patients is a characteristic of the disease and is not a consequence of pharmacological treatment[10].

Platelets have been used as a model of biochemical processes that occur in the brain; for example, abnormalities of dopamine uptake into platelets[11] or platelet granules[12] have been shown in PD. Since the energy-dependent glutamate uptake in the platelets is similar to that observed in neurons[13], it is reasonable to hypothesize that it could be affected by the modifications of mitochondrial enzymes described in PD. In the present study we investigated glutamate concentration and uptake in platelets of PD patients, as possible peripheral markers of a central glutamatergic dysfunction in this disorder.

PATIENTS AND METHODS

Patients

For the present study, 34 PD patients were selected, of which 29 had degenerative PD and 5 suffered from vascular parkinsonian syndrome. All patients were between stage 2 and 4 of the Hohen and Yahr scale[14], with a severity of disease between 21 and 70, according to the Unified Parkinson's Disease Rating Scale (UPDRS)[15]. Duration of disease was between 2 and 14 years. All patients underwent clinical and neuropsychological (Mini Mental Status Examination)[16] evaluation, to assess the degree of cognitive impairment, and CT scan.

Neurodegenerative Diseases
Edited by Gary Fiskum, Plenum Press, New York, 1996

L-dopa was the standard treatment adopted in all cases.

10 age- and sex-matched controls, without any evidence of neuropsychiatric disorder or cognitive dysfunction, were also selected for this study.

After informed consent, 30 ml of fasting blood were collected and immediately processed for platelet preparation. Platelets were stored at -80 °C until use for biochemical studies.

Glutamate Assay by HPLC

Platelets were homogenized in H_2O, aminoacids were extracted with perchloric acid, and the mixture was neutralized with potassium carbonate, to avoid hydrolysis of glutamine[17]. Platelet extracts were filtered with Millipore filters and aminoacids were derivatized with o-phthaldialdehyde before HPLC injection. Glutamate levels were analyzed by reverse-phase HPLC with a multistep gradient of acetate buffer and metanol/tetrahydrofuran, followed by fluorimetric detection, as previously described[17]. Glutamate peaks were easily detected in platelets extract (Figure 1).

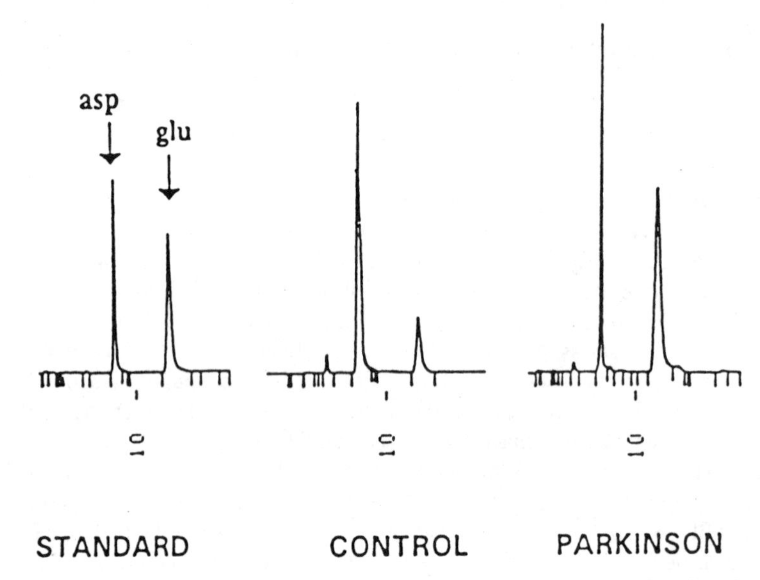

Figure 1. Chromatographic profile of platelets extracts from a control and a PD patients.

Glutamate Uptake Studies

For glutamate uptake, the method of Mangano and Schwarcz[13] was followed, with minor modifications. Briefly, platelets were resuspended in 0.32 M sucrose, and 50 μl aliquots were added to 425 μl of 25mM Tris-citrate buffer, pH 7.0, containing NaCl or choline chloride to study sodium dependent uptake. Samples were preincubated for 15 min at 37°C. Uptake was started by addition of 25 μl [^{3}H]Glu (5 μCi; 30 μM final concentration) and stopped after 30 min by addition of 2 ml ice cold Tris-citrate buffer containing 1 mM cold Glu. Platelets were separated by vacuum filtration and filters were extracted by Formula-989 (Du Pont, MA, USA) and counted in a β-counter (60% efficiency). Net uptake was determined by subtracting no-sodium blanks from the total counts and expressed as fmol Glu/mg protein/30 min.

Statistical Analysis

All results are expressed as mean ± Standard Deviation (S.D.). Student's two-tail t-test was used to assess the significance of differences between means. One-way analysis of variance (ANOVA) was used to compare controls and patient groups.

Statistic analysis of correlations was performed with the Pearson's correlation test.

RESULTS

Table 1 shows glutamate levels in platelets of PD patients and controls. A significant 80% increase of glutamate levels was observed in PD patients, compared to controls.

Table 1. Glutamate levels in platelets of controls and PD patients.

Patients (number)	Mean Age	Age Range	Sex M/F	Glutamate (mean ± SD) $nmol/10^{10}$ platelets
Controls (10)	76	40 - 90	4/6	18 ± 5.7
Parkinson (34)	68	50 - 83	25/9	33 ± 14 [1]

[1] $p < 0.05$

PD patients were then divided into subgroups according to their clinical-neuroradiological characteristics: idiopathic PD, patients with dementia and brain atrophy (demented PD), patients with other neurological signs and ischemic areas at the CT (ischemic PD). The increase in glutamate content was more evident in idiopatic PD, while it was variable and not significant in demented or ischemic patients (Table 2).

Table 2. Glutamate levels in platelets of controls and subgroups of PD patients.

Patients	No. of subjects	Glutamate (mean ± SD) $nmol/10^{10}$ platelets
Controls	10	18 ± 5.7
Idiopathic PD	13	36 ± 21 [1]
Demented PD	16	32 ± 20
Ischemic PD	5	29 ± 29

[1] $p = 0.02$

Possible correlations between platelet glutamate levels and duration or severity of the disease were also investigated, but no significant correlation was observed.

Glutamate uptake studies were then carried out in platelets of 17 PD patients and in 10 age-matched controls. A significant 35% reduction of glutamate uptake ($p < 0.05$) was observed in PD patients, with respect to controls (Table 3). The decrease was more consistent (50% reduction) and significant ($p < 0.01$) in the subgroup of idiopathic PD; no significant change was observed in demented PD (Table 3).

Table 3. Glutamate uptake in platelets of controls and PD patients.

Patients	No. of subjects	Glutamate uptake (mean ± SD) fmol/ mg prot/30 min
Controls	10	2.1 ± 0.95
Total PD	17	1.3 ± 0.82 [1]
Idiopathic PD	12	0.9 ± 0.43 [2]
Demented PD	5	1.9 ± 1.00

[1] $p < 0.05$
[2] $p < 0.01$

A significant correlation was observed between the severity of PD, measured by the UPDRS, and the decrease of glutamate uptake in platelets ($r= -0.61$; $p < 0.05$) (Figure 2).

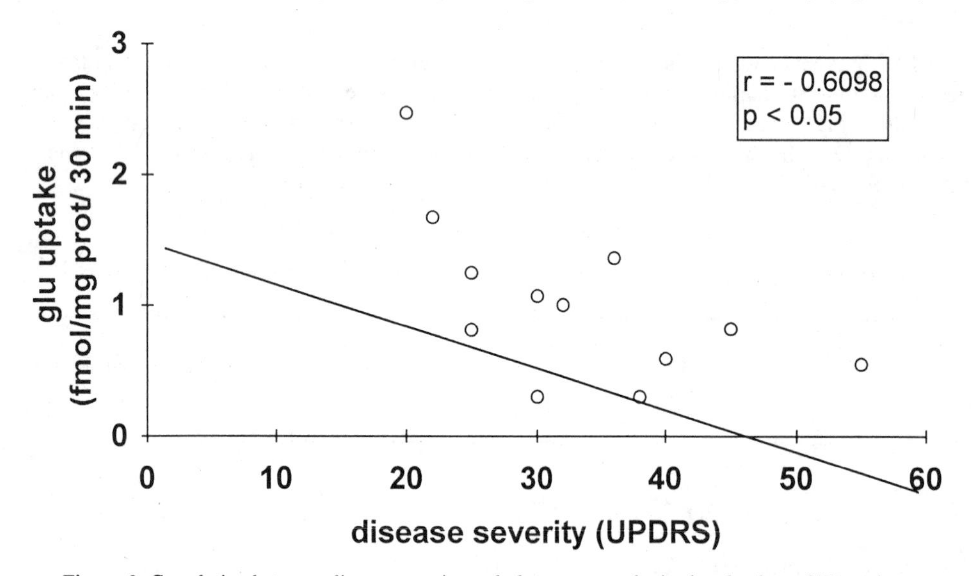

Figure 2. Correlation between disease severity and glutamate uptake in the platelets of PD patients.

DISCUSSION

In the present study we described the increase of glutamate levels and the decrease of glutamate uptake in the platelets of PD patients. Both modifications seem to be specific for the idiopathic PD patients, while they are not significant in demented or ischemic patients.

The decrease in sodium- and energy-dependent glutamate uptake observed in this study could be related to the impairment of mitochondrial function, previously reported in platelets of PD patients[6,7,8,9,10]. Since similar deficiencies of mitochondrial enzymes have been described also in the brain[4,5], it is possible that a decreased glutamate uptake in substantia nigra may lead to the excitotoxic damage of dopaminergic neurons. Alternatevely, platelets *per se* might contribute to neuronal damage by releasing glutamate[18].

The finding of increased levels of glutamate in platelets of PD patients is intriguing, but of more difficult explanation and apparently in contrast with the decreased uptake. Several mechanisms might be responsible for the observed increase, and some of them could be related to the reported mitochondrial dysfunction. The energy-dependent synthesis of glutamine could be impaired or, alternatively, the rate of glutamine deamination, an energy producing reaction, could be enhanced. In both cases, the result would be the increase in the cellular content of glutamate. In agreement with this interpretation, it has been reported that, in glucose-deprived platelets *in vitro*, glutamate concentration rises in parallel with glutamine decrease. The latter is therefore metabolized no further than glutamate[19].

Taken together, our data on increased intracellular level and decreased uptake of glutamate in the platelets of PD patients might suggest the possibility to use platelets as peripheral model of exitotoxic phenomena in PD. This gains interest if we consider that these modifications are particularly significant in idiopathic PD, where neuro-degeneration is specific for dopaminergic neurons. Our hypothesis is further supported by previous reports on mitochondrial abnormalities, that are specific for idiopathic PD and are absent in Parkinson-plus syndromes[9].

However, further experiments are necessary to investigate the mechanisms of glutamate alteration in platelets and its relationship with what might happen in glutamatergic terminals in the brain.

REFERENCES

1. M.F. Beal, Does impairment of energy metabolism result in excitotoxic neuronal death in neurodegenerative illnesses?, *Ann Neurol.* 31:119 (1992).
2. S.Fahn and G. Cohen, The oxidant stress hypothesis in Parkinson's disease: evidence supporting it, *Ann Neurol.* 32:804 (1992).
3. J.T.Coyle and P. Puttfarcken, Oxidative stress, glutamate, and neurodegenerative disorders, *Science.* 262:689 (1993).
4. N. Hattori, M. Tanaka,T. Ozawa and Y. Mizuno, Immunohistochemical studies on Complexes I, II, III and IV of mithocondria in Parkinson's disease, *Ann Neurol.* 30:563 (1991).
5. J.M. Shoffner, R.L. Watts, J.L. Juncos, A. Torroni and D.C. Wallace, Mitochondrial oxidative phosphorylation defects in Parkinson's disease, *Neurology.* 41 suppl 1:152 (1991).
6. W.D. Parker, S.J. Boyson and J.K. Parks, Abnormalities of the electron transport chain in idiopathic Parkinson's disease, *Ann Neurol.* 26:719 (1989).
7. D. Krige, M.T. Carroll, J.M. Cooper, C.D. Marsden and A.H.V. Schapira, Platelet mithocondrial function in Parkinson's disease, *Ann Neurol.* 32:782 (1992).
8. H. Yoshino, Y Nakagawa-Hattori, T Kondo and Y Mizuno, Mitochondrial complex I and II activities and platelets in Parkinson's disease, *J Neural Transm.* 4:27 (1992).
9. R. Benecke, P. Strumper and H Weiss, Electron transfer complexes I and IV of platelets are abnormal in Parkinson's disease but normal in Parkinson-plus syndromes, *Brain.* 116:1451 (1993).
10. C.W. Shults, F. Nasirian, D.M. Ward, K. Nakano, M. Pay, L.R. Hill and R.H. Haas, Carbidopa/levodopa and selegiline do not affect platelet mitochondrial function in early parkinsonism, *Neurology.* 45:344 (1995).
11. A. Barbeau, G. Campanella, R.F. Butterworth and K. Yamada, Uptake and efflux of

^{14}C-dopamine in platelets: evidence for a generalized defect in Parkinsons's disease, *Neurology.* 25:1 (1975).
12. J. M. Rabey, H. Shabtai, E. Graff and Z. Oberman, [^{3}H] dopamine uptake by platelet transport chain in idiopathic Parkinson's disease, *Life Sci.* 3:1753 (1993).
13. R.M. Mangano and R. Schwarcz, The human platelet as a model for the glutamatergic neuron: platelet uptake of L-glutamate, *J Neurochem.* 36:1067 (1981).
14. M.M. Hoehn and M.D. Yahr, Parkinsonism: onset, progression and mortality, *Neurology.* 17:427 (1967).
15. S. Fahn, R.L. Elton and The Members of the UPDRS development committee, Unified Parkinson's disease rating scale, *in*: "Recent Develpments in Parkinson's disease," S. Fhan, D. Marsden and D. Calne ed., MacMillan, London (1987).
16. M.F. Folstein, S.E. Folstein and P.R. McHugh, "Mini-Mental State": a pratical method for grading the cognitive state of patients for the clinician, *J Psychiatr Res.* 12:189 (1975).
17. C. Ferrarese, N. Pecora, M. Frigo, I. Appollonio and L. Frattola, Assessment of reliability and biological significance of glutamate levels in cerebrospinal fluid, *Ann Neurol.* 33:316 (1993).
18. R. Joseph, C. Tsering, S. Grunfeld, and K.M.A. Welch, Platelet secretory products may contribute to neuronal injury, *Stroke.* 22:1448 (1991)
19. S. Murphy, S. Munoz, M. Parry-Billings, and E.Newsholme, Amino acid metabolism during platelet storage for transfusion, *Br J Haematol.* 81:585 (1992).

BRAINSTEM MOTONEURON CELL GROUPS THAT DIE IN AMYOTROPHIC LATERAL SCLEROSIS ARE RICH IN THE GLT-1 GLUTAMATE TRANSPORTER

Loreta Medina[1], Griselle Figueredo-Cardenas[1],
J.D. Rothstein[2], and Anton Reiner[1]

[1]Department of Anatomy and Neurobiology, University of Tennessee, Memphis, TN 38163
[2]Department of Neurology, Johns Hopkins University, Baltimore, MD 21205

INTRODUCTION

Several studies have suggested that excessive Ca^{2+} influx into cells may underlie the selective destruction of motoneurons observed in the sporadic form of Amyotrophic Lateral Sclerosis (ALS)[1-4]. Consistent with this, in the cranial motor nuclei that survive better in sporadic ALS (i.e. oculomotor, trochlear and abducens motor nuclei), a high percentage of the motoneurons are enriched in the Ca^{2+} buffering protein parvalbumin in normal monkey and normal human brainstem[5,6]. In contrast, in the cranial motor nuclei that are dramatically affected in sporadic ALS (i.e. trigeminal, facial, and hypoglossal motor nuclei), only a low percentage of the motoneurons contain parvalbumin[5,6]. One possible clue as to why excessive Ca^{2+} influx into cells may occur in sporadic ALS has been provided by Rothstein and colleagues[3]. In brief, they have found a significant decrease of high-affinity glutamate transport in the motor cortex and spinal cord of sporadic ALS victims[3], which seems to be due mainly to a profound loss of a specific glutamate transporter subtype, GLT-1, which is localized to astroglia[7]. A minor loss of another glutamate transporter subtype, EAAC1 (localized to neurons), also occurs in the motor cortex of ALS victims which may be secondary to the death of cortical neurons[7]. A defect in either glutamate transporter subtype would lead to an excess of extracellular glutamate in the vicinity of neurons depending on that transporter for glutamate clearance from the extracellular space. Increased extracellular glutamate could lead to excess Ca^{2+} entry into motoneurons via glutamate-gated or voltage-activated Ca^{2+} channels. If a defect in glutamate transport underlies the pathogenesis of sporadic ALS, either one or both of the glutamate transporter

Neurodegenerative Diseases
Edited by Gary Fiskum, Plenum Press, New York, 1996

subtypes found to be lost in sporadic ALS should be present in abundance in the affected motor nuclei in normal conditions. To investigate this, in this study we have used immunohistochemical methods to analyze the localization of GLT-1 and EAAC1 glutamate transporters in the different cranial motor nuclei of normal monkey brainstem.

MATERIAL AND METHODS

Formaldehyde-fixed brainstems from three rhesus monkeys (*Macaca mulatta*) were used in the present study. Forty microns sections at the level of the various brainstem motor nuclei (oculomotor, trochlear, trigeminal, facial, and hypoglossal motor nuclei) were obtained using a freezing microtome. Some sections were stained immunohistochemically using specific polyclonal antibodies against either the GLT-1 glutamate transporter or the EAAC1 glutamate transporter as described previously[8]. In brief, sections were rinsed three times in Tris-buffered saline (TBS, 0.05 M, pH 7.4), preincubated for 30 min in 4% normal goat serum in TBS containing 0.1% Triton X-100, and then incubated in the primary antibody (rabbit anti-GLT-1, 0.17 μg/ml; or rabbit anti-EAAC1, 2μg/ml) with 2% normal goat serum, diluted in TBS containing 0.1% Triton X-100 for 48-72 hours at 4°C. Following primary antibody incubation, sections were rinsed in TBS and incubated sequentially in a donkey anti-rabbit IgG (Jackson; 1:50; 1 hour at room temperature), and then in a rabbit peroxidase anti-peroxidase complex (Sternberger; 1:200; 1 hour at room temperature). Finally, sections were rinsed in TBS and stained using a standard diaminobenzidine reaction.

To localize the brainstem motor cell groups, adjacent sections were either stained immunohistochemically for choline acetyltransferase[6], or were cresyl violet stained.

LOCALIZATION OF GLT-1 AND EAAC1 IN MONKEY BRAINSTEM MOTOR NUCLEI

In agreement with a previous report on GLT-1 localization in rat brain[8], our results in monkey indicated that GLT-1 immunoreactivity was generally intense in most brainstem regions, and was largely absent from white matter tracts. Ultrastructural studies in rat brain have shown that GLT-1 is localized to astroglial cells and processes[8]. The morphological appearance at light microscopic level of the GLT-1 immunoreactive elements in monkey brainstem revealed that GLT-1 was also restricted to astroglial cells and processes which surrounded and in many instances covered or enveloped neurons in the different monkey brainstem regions. For example, GLT-1 immunoreactivity was observed to be dense to very dense in the superior colliculus, the central gray, the substantia nigra pars compacta, the molecular layer and Purkinje cell layer of the cerebellum, the pontine nuclei and the inferior olive. All cranial motor nuclei studied (i.e. oculomotor, trochlear, trigeminal, facial, and hypoglossal motor nuclei) also contained dense to extremely dense GLT-1 immunoreactivity. The immunoreactivity appeared to be very dense or extremely dense in both the trigeminal and facial motor nuclei, in which most motoneuron cell bodies and proximal dendrites were enveloped by very thick sheaths of GLT-1

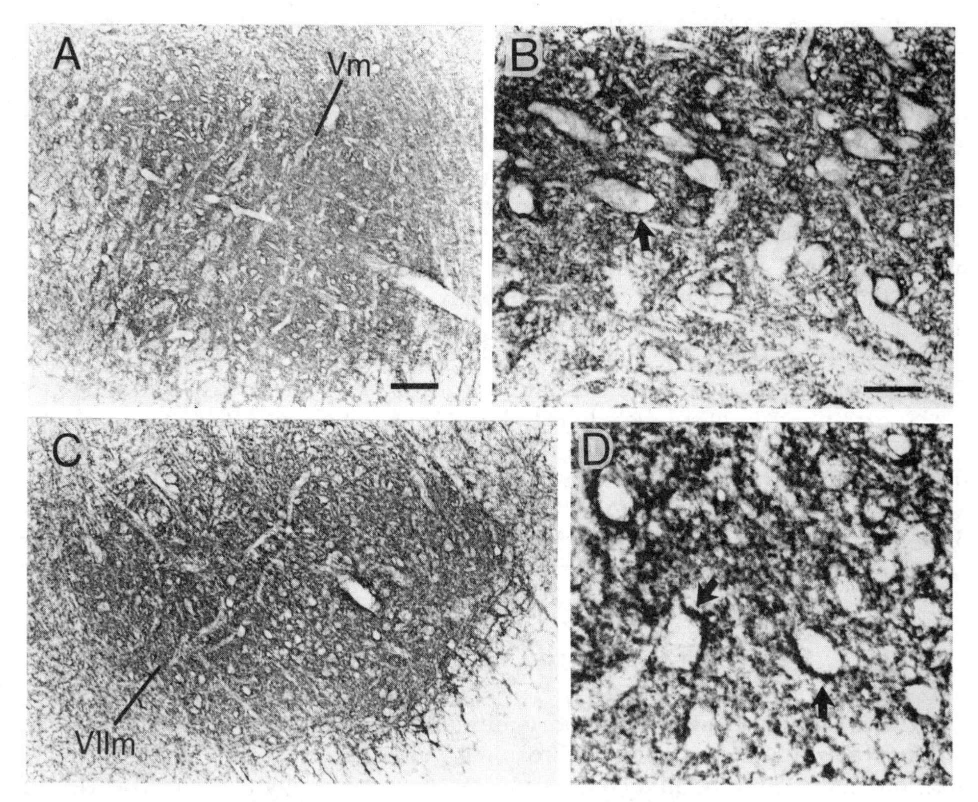

Figure 1. A, C: Low magnification photomicrographs of the trigeminal (A) and facial (C) motor nuclei stained immunohistochemically for GLT-1. B, D: High magnification photomicrographs showing details of GLT-1 immunoreactivity in the trigeminal (B) and facial (D) motor nuclei. Note the very dense GLT-1 immunoreactivity of astroglial cells and processes in both motor cell groups (extremely dense in the facial nucleus). Most motoneuron cell bodies and proximal dendrites are enveloped by very thick sheaths of GLT-1 immunoreactive astroglial processes (arrows). Abbreviations: Vm, trigeminal motor nucleus; VIIm, facial motor nucleus. Scale bars: A, C = 200 μm (scale in A); B, D = 50 μm (scale in B).

immunoreactive glial processes (Fig. 1). The immunoreactivity in both the oculomotor and trochlear motor nuclei appeared to be less intense than in the facial and trigeminal motor nuclei (Fig. 2). Finally, the intensity of GLT-1 immunoreactivity in the hypoglossal motor nucleus appeared to be intermediate between that in the facial/trigeminal motor nuclei and that in the oculomotor/trochlear motor nuclei. Computer-assisted image analysis of the optical density of GLT-1 immunoreactive elements observed in each motor cell group was carried out as described previously[9]. The optical density of immunoreactive elements was also measured in other brain regions, such as the molecular layer of the cerebellum, the central gray, the superior colliculus, the pontine nuclei, as well as in a representative brainstem fiber tract, the superior cerebellar peduncle. Image analysis revealed that the density of GLT-1 immunoreactive elements was highest in the molecular layer of the cerebellum, and lowest (nearly equal to background levels) in the fiber tract. Among brainstem motor cell groups, image analysis revealed that the density of GLT-1 immunoreactive elements was highest in the facial motor nucleus, followed by the trigeminal motor nucleus, and the density was lowest in the trochlear motor nucleus.

Our results in monkey indicated the presence of numerous EAAC1 immunoreactive neurons and processes in several brainstem regions, such as the central gray, the dorsal raphe nucleus, and the medial and superior vestibular nuclei. The Purkinje cells of the cerebellum were also intensely immunoreactive for EAAC1. A dense net of EAAC1 immunoreactive processes was present in the nucleus of the descending trigeminal tract, and numerous fibers and dispersed cells were also observed in the brainstem reticular formation. The majority of the motoneurons of two representative brainstem motor cell groups studied, the trochlear nucleus (an ALS-resistant motor cell group) and the facial nucleus (an ALS-vulnerable motor cell group), were not labeled for EAAC1, with the exception of a few motoneurons which appeared to be extremely lightly labeled (Fig. 3). A few small labeled neurons, which probably represent interneurons, were observed in the trochlear motor nucleus. Finally, although most motoneuron cell bodies and proximal dendrites were not labeled for EAAC1, a low to moderate number of EAAC1 immunoreactive processes of unknown origin were present in the trochlear and facial motor nuclei.

RELATION TO A GLUTAMATE TRANSPORT DEFECT IN SPORADIC ALS

Increased levels of the excitatory amino acid glutamate have been observed in the brain, spinal cord, and cerebrospinal fluid of ALS victims, suggesting that excitotoxicity may be involved in the motoneuron degeneration observed in this disease[2,10]. Supporting the excitotoxic hypothesis in ALS, long-term ingestion of the excitotoxin β-N-methylamino-L-alanine (BMAA) has been implicated in the pathology of the Guamián form of ALS[11]. A recent study has revealed that the abnormal glutamate levels observed in sporadic ALS may be related to a defect in the glutamate re-uptake systems[3]. Concretely, a significant decrease in high-affinity glutamate transport that is disease-specific has been observed in the motor cortex and spinal cord of sporadic ALS victims[3]. Defective glutamate transport could lead to motoneuron death in the following manner: (1) defective clearance of glutamate from the extracellular space by glutamate transporters could lead to exposure of surrounding neurons to abnormally high concentrations of glutamate[3]; and (2) increased extracellular glutamate could lead to excess Ca^{2+} influx into cells via glutamate-gated or voltage-activated Ca^{2+} channels, which would produce lethal damage to cells[1,12-14]. Supporting the hypothesis that a defect in glutamate transport may underlie the motoneuron loss observed in sporadic ALS, selective inhibition of glutamate transport in cultured organotypic spinal cord slices produces slow motoneuron degeneration[15]. The neurotoxicity produced by chronic inhibition of glutamate transport in vitro seems to be mediated by non-N-methyl-D-aspartate (non-NMDA) receptors, but not by NMDA receptors[15]. The decreased glutamate transport observed in motor cortex and spinal cord of sporadic ALS victims seems to be mainly related to a dramatic loss of one specific subtype of high-affinity glutamate transporter: GLT-1, localized to astroglia[7]. The other two subtypes of high-affinity glutamate transporters, EAAC1 (localized to neurons), and GLAST (localized to glial cells and some neurons), are either only modestly lost (EAAC1) or remain unchanged (GLAST) in ALS motor cortex[7]. Since in the cortex EAAC1 is specifically found in pyramidal neurons,

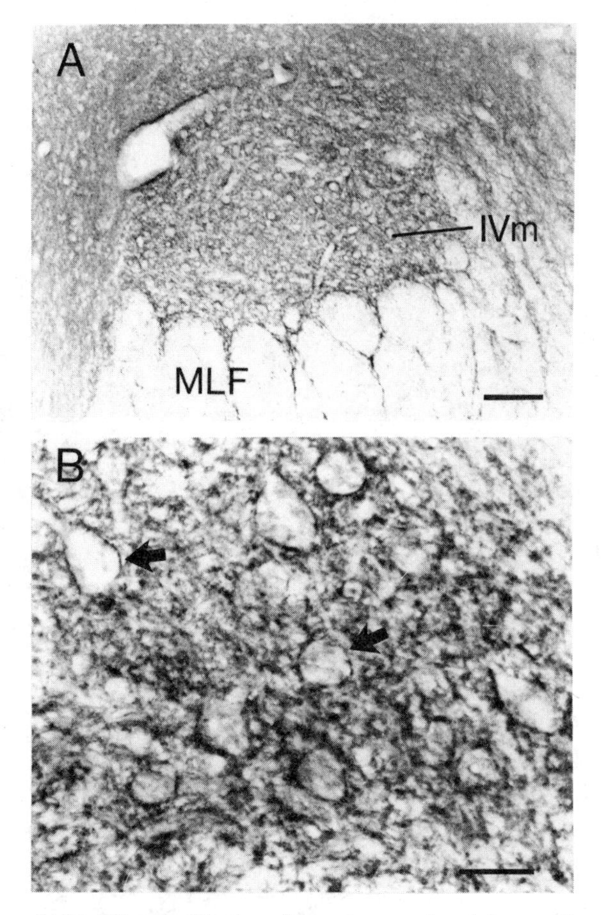

Figure 2. Low (A) and high (B) magnification photomicrographs of the trochlear motor nucleus stained immunohistochemically for GLT-1. Note the numerous immunoreactive astroglial processes that surround the unlabeled motoneurons. Many of them cover or envelop the motoneuron cell bodies (arrows). Abbreviations: IVm, trochlear motor nucleus; MLF, medial longitudinal fasciculus. Scale bars: A = 200 μm; B = 50 μm.

the minor loss of EAAC1 in ALS motor cortex has been associated to the death of cortical neurons[7]. Our results in monkey have indicated that all brainstem motor cell groups are rich in GLT-1, and relatively poor in EAAC1. Therefore, if a loss of GLT-1 glutamate transporter occurs in sporadic ALS, all brainstem motoneurons should be affected. However, not all brainstem motoneurons are equally affected in ALS: brainstem motor cell groups controlling the extraocular muscles (i.e., oculomotor, trochlear, and abducens motor nuclei) survive well in ALS, whereas other brainstem motor cell groups (i.e. trigeminal, facial, and hypoglossal motor nuclei) are dramatically affected[6]. Our results indicate that both ALS-resistant (i.e. oculomotor and trochlear motor nuclei) and ALS-vulnerable (i.e. trigeminal, facial and hypoglossal motor nuclei) motor cell groups of the brainstem are rich in GLT-1. Since defective glutamate transport can lead to excess Ca^{2+} influx into cells, those neurons that are enriched in Ca^{2+}-buffering proteins, such as parvalbumin, would be better able to withstand the damage. This is the case of ALS-resistant motor cell groups of the brainstem, in which the majority of the motoneurons contain parvalbumin[5,6]. In contrast,

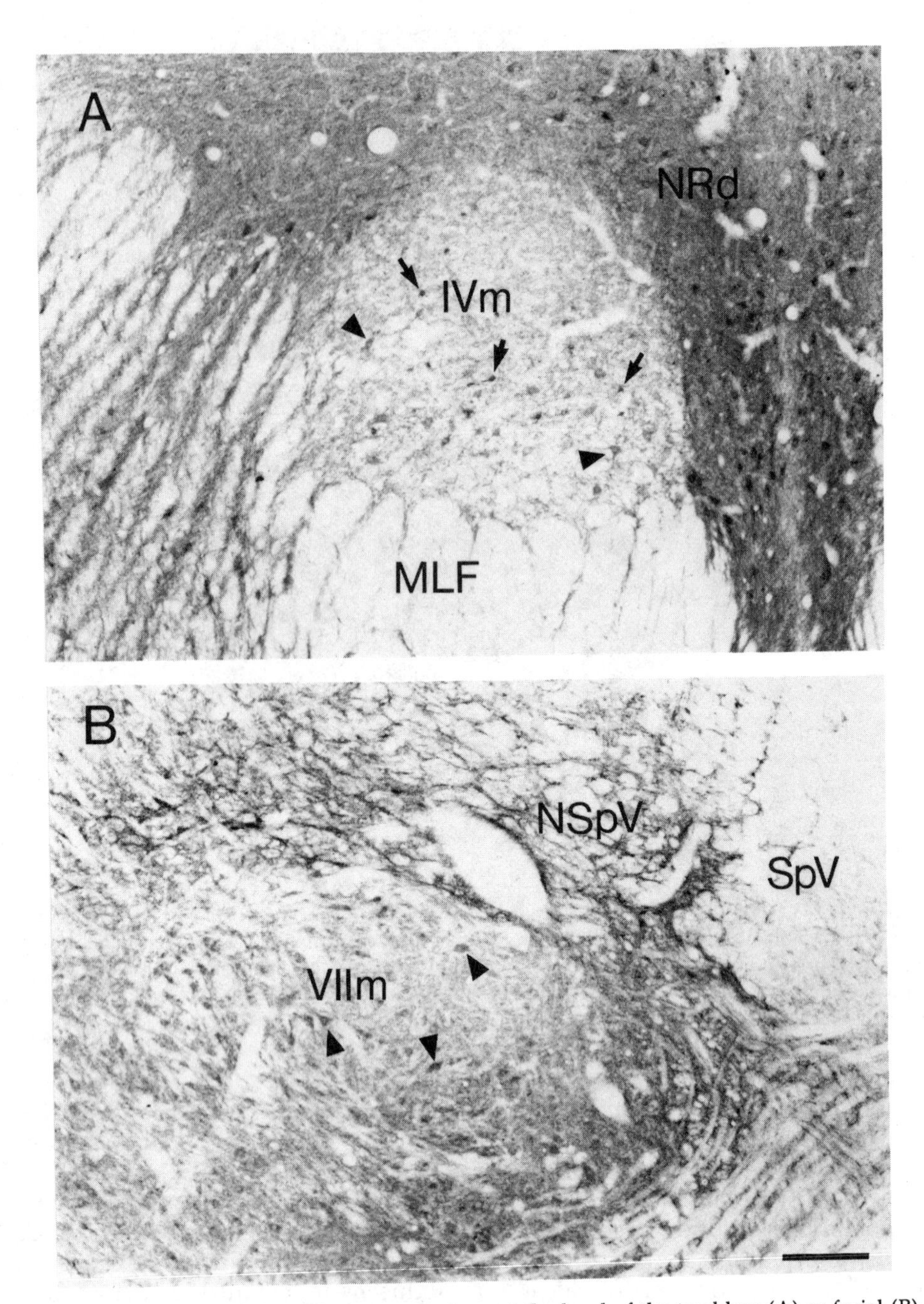

Figure 3. Photomicrographs of brainstem sections at the level of the trochlear (A) or facial (B) motor nuclei stained immunohistochemically for EAAC1. The majority of the motoneurons in the trochlear (IVm) and facial (VIIm) motor nuclei are not labeled for EAAC1, with the exception of a few motoneurons that seem to be lightly labeled (arrowheads). Further, in the trochlear nucleus a few small neurons, which are probably interneurons, are strongly immunoreactive for EAAC1 (arrows). Numerous strongly immunoreactive neurons and a dense immunoreactive neuropil are present in the dorsal raphe nucleus and the nucleus of the spinal trigeminal tract. Abbreviations: IVm, trochlear motor nucleus; MLF, medial longitudinal fasciculus; NRd, dorsal raphe nucleus; NSpV, nucleus of the spinal trigeminal tract; SpV, spinal trigeminal tract; VIIm, facial motor nucleus. Scale bar: A, B = 200 μm (scale in B).

the majority of the motoneurons in the ALS-vulnerable brainstem motor cell groups lack parvalbumin[5,6], and would be unable to resist an excess of Ca^{2+} influx into cells in the case of a defective glutamate transport.

Our results also indicated that non-motor cell groups, such as the superior colliculus, the central gray, the substantia nigra, or the cerebellum, are also rich in GLT-1 glutamate transporter. However, these non-motor cell groups are not affected in ALS. The selectivity of sporadic ALS for motoneurons may be related to the fact that the decreased glutamate transport observed in the brain of ALS victims seems to be highly region-specific. For example, glutamate transport levels are significantly decreased in the motor cortex and spinal cord of sporadic ALS victims, but not in the striatum, the visual cortex or the hippocampus[3]. The latter three regions are rich in GLT-1 glutamate transporter in both normal humans and rats and do not show any morphological change in ALS[8,16]. Similarly to the striatum, the visual cortex or the hippocampus, it is possible that glutamate transporter levels remain unchanged in other non-motor cell groups of the brain in sporadic ALS. Another possibility is that in some regions the loss of a specific glutamate transporter subtype may be compensated by unaffected or less affected subtypes of glutamate transporters. In this respect, while brainstem and spinal cord motoneurons seem to be only rich in GLT-1, many regions of the brain, including the cerebral cortex, the striatum, and the cerebellum, are rich not only in GLT-1, but also in EAAC1 and/or GLAST glutamate transporters[8,16]. Finally, other factors that need to be considered for analyzing different vulnerability among brain and spinal cord regions in sporadic ALS are the presence of calcium chelators in cells, or the cellular expression of non-NMDA receptors.

ACKNOWLEDGMENTS

We thank Dr. Howard T. Chang (University of Tennessee at Memphis), Dr. Randall J. Nelson (University of Tennessee at Memphis), and Dr. Paul Gamlin (University of Alabama at Birgmingham) for generously providing us with the fixed monkey brainstem that we used for this study. We also thank Mr. Marl Bush for his excellent technical assistance. This research was supported by NS-19620 (A.R.), NS-28721 (A.R.), NS-33958 (J.D.R.), the Hereditary Disease Foundation (A.R.), the Muscular Dystrophy Association (J.D.R.), and the Neuroscience Center of Excellence of the University of Tennessee (L.M.).

REFERENCES

1. D.W. Choi, Glutamate neurotoxicity and diseases in the nervous system, *Neuron* 1:623 (1988).
2. J.D. Rothstein, G. Tsai, R.W. Kuncl, L. Clawson, D.R. Cornblath, D.B. Drachman, A. Pestronk, B.L. Stauch, and J.T. Coyle, Abnormal excitatory amino acid metabolism in amyotrophic lateral sclerosis, *Ann. Neurol.* 28:18 (1990).
3. J.D. Rothstein, L.J. Martin, and R.W. Kuncl, Decreased glutamate transport by the brain and spinal cord in amyotrophic lateral sclerosis, *N. Engl. J. Med.* 326:1464 (1992).
4. S.H. Appel, Excitotoxic neuronal death in amyotrophic lateral sclerosis, *Trends Neurosci.* 16:3 (1993).
5. A. Reiner, S. Anfinson, and G. Figueredo-Cardenas, Motoneurons that are resistant to ALS are preferentially enriched in the calcium binding protein parvalbumin, *Neurosci. Abstr.* 19:197 (1993).
6. A. Reiner, L. Medina, G. Figueredo-Cardenas, and S. Anfinson, Brainstem motoneuron pools that are selectively resistant in amyotrophic lateral sclerosis are preferentially

enriched in parvalbumin: evidence from monkey brainstem for a calcium-mediated mechanism in sporadic ALS, *Exp. Neurol.* 131:239 (1995).
7. J.D. Rothstein, M. Van Kammen, A.I. Levey, L. Martin, and R.W. Kuncl, Selective loss of glial glutamate transporter GLT-1 in amyotrophic lateral sclerosis, *Ann. Neurol.* 37 (1995) in press
8. J.D. Rothstein, L. Martin, A.I. Levey, M. Dykes-Hoberg, L. Jin, D. Wu, N. Nash, and R.W. Kuncl, Localization of neuronal and glial glutamate transporters, *Neuron* 13:713 (1994).
9. G. Figueredo-Cardenas, K.D. Anderson, Q. Chen, C.L. Veenman, and A. Reiner, Relative survival of striatal projections neurons and interneurons after intrastriatal injection of quinolinic acid in rats, *Exp. Neurol.* 129:37 (1994).
10. A. Plaitakis, E. Constatakakis, and T. Smith, The neuroexcitotoxic amino acids glutamate and aspartate are altered in the spinal cord and brain in amyotrophic lateral sclerosis, *Ann. Neurol.* 24:446 (1988).
11. P.S. Spencer, P.B. Nunn, S. Hugon, A.C. Ludolph, S.M. Ross, D.N. Roy, and R.C. Robertson, Guam amyotrophic lateral sclerosis-parkinsoniam dementia linked to a plant excitant neurotoxin, *Science* 239:517 (1987).
12. M.F. Beal, Mechanisms of excitotoxicity in neurologic diseases, *FASEB J.* 6:3338 (1992).
13. M.F. Beal, Does impairment of energy metabolism result in excitotoxic neuronal death in neurodegenerative illnesses?, *Ann. Neurol.* 31:119 (1992).
14. S. Orrenius, M.J. Burkitt, G.E.N. Kass, J.M. Dypbukt, and P. Nicotera, Calcium ions and oxidative cell injury, *Ann. Neurol.* 32:S33 (1992).
15. J.D. Rothstein, L. Jin, M. Dykes-Hoberg, and R.W. Kuncl, Chronic inhibition of glutamate uptake produces a model of slow neurotoxicity, *Proc. Natl. Acad. Sci. USA* 90:6591 (1993).
16. M. Dykes-Hoberg, L.J. Martin, A.I. Levey, D. Rye, N. Nash, L. Jin, R.W. Kuncl, and J.D. Rothstein, Cellular and ultrastructural localization of glutamate transporter subtypes in rat and human brain, *Neurosci. Abstr.* 20:927 (1994).

PROGRESSIVE NEURODEGENERATION IN RAT BRAIN AFTER CHRONIC 3-VO OR 2-VO

J.C. de la Torre[1], B.A. Pappas[2], T. Fortin[2], M. Keyes[2], and C. Davidson[2]

[1]University of New Mexico
Division of Neurosurgery
2211 Lomas Blvd., N.E.
Albuquerque, New Mexico 87131

[2] Carleton University
Department of Psychology
Colonel By Drive
Ottawa, Ontario K15 5B6

INTRODUCTION

Progressive neurodegeneration in the aging population is commonly associated with human dementia, including Alzheimer's disease (AD) (Rose and Hennebery, 1994)

Although cerebral location and patterns of degeneration vary according to the specific dementia, it is generally agreed that in Alzheimer's disease, neurodegenerative changes begin in the CA1 hippocampal region and in the entorhinal cortex. (Ameral and Insausti, 1990; West et al, 1994)

Memory loss is a cardinal symptom of degenerative dementia and this loss is expressed early in AD when visuo-spatial orientation becomes impaired. (Mendis and Mohr, 1993).

We have reported that chronic 3-vessel occlusion (3-VO) involving both carotid arteries and the left subclavian artery in 10 month middle aged rats results in a clinico-pathologic picture resembling AD (de la Torre and Fortin 1994; de la Torre et al, 1992a). Over a 12 week period, the 3-VO model is characterized by progressive visuo-spatial impairment, selective CA1 neuronal damage confined to the hippocampus, membrane phospholipid synthesis increase in the hippocampal area, reduced cerebral blood flow and an absence of focal neurological deficits (de la Torre et al, 1992b;de la Torre et al, 1993a; de la Torre et al, in press)

These pathologic changes have been described in AD (Grubb 1977;Mckhann et al, 1981; Nitsch 1991; Pettegrew 1988; Prohovnik et al 1988; Sulkava et al, 1983; Wade et al, 1987)

In the present study, we compared 2-vessel occlusion (2-vo) consisting of bilateral carotid artery ligation in 10 month old Sprague-Dawley rats with age-matched 3-VO rats, particularly with regard to the relative onset of morphologic and memory changes resulting from this vascular insult. One of the objectives of comparing 3-VO and 2-VO was to determine whether the latter approach (being also the simpler surgical procedure) could induce the same pathologic picture as 3-VO since one of the drawbacks of 3-VO is that aged rats (>18 months old) can not be used due to the high mortality associated with this procedure, a problem not seen in 19 month rats subjected to 2-VO (de la Torre et al 1993b)

METHODS

Male Sprague-Dawley rats 10 months old weighing 550-650 grams (obtained from Charles River Canada, St. Constance, Quebec) were subjected to 3-VO or 2-VO using a surgical approach previously described (de la Torre et al, 1992a; de la Torre and Fortin, 1994). For 3-VO, permanent occlusion of both carotid arteries and the left subclavian artery was done following intubation of the trachea with a PE-190 catheter and placing the animal on a Harvard rodent respirator set at 50 strokes/minute and 4.5 ml volume. A ventral, midline incision from mid-larynx to about 3 cm above the xiphoid process was made using an electrosurgical scalpel and the chest was opened at the 5th rib interspace and cut cephalad as we have described (de la Torre and Fortin, 1994). The left subclavian artery was identified and tied off with 5-0 silk suture thus removing the left vertebral artery flow to the brain. The chest muscles were sutured and the left and right carotid arteries were carefully separated from the vagus nerve and carotid sheath, then doubly ligated with 5-0 silk or other non-absorbable suture. A second group of rats underwent 2-VO using the same procedure as with 3-VO but without intubation or left subclavian artery occlusion. Mortality from 2-VO/3-VO was 0.8% and 5% respectively.

Rats were divided into 3 groups: Group I: 3-VO (N=48); Group II: 2-VO (N=40); Group III: control, non-occlusion, sham surgery (N=76).

After 2-VO, rats were tested in a modified Morris water maze test from 1 to 25 weeks at periodic intervals. Rats subjected to 3-VO were tested from 1-12 weeks. The water maze test consisted of the rats ability to find an invisible underwater platform within 120 seconds (platform location latency) following 4 trials/day (de la Torre et al, 1993a). Total swim path lengths and percentage of total path spent in the platform quadrant were video recorded and analyzed using one-way ANOVAS for each test period.

Eight rats from each group underwent ^{31}P-magnetic high resolution imaging at 12 weeks post-op using a Bruker 4.7 T/30 cm internal bore Medspec system equipped with a custom-built activity-shielded gradient system as we have previously described (de la Torre et al, 1992a).

At the end of 2,9, and 25 weeks, rats from each group were killed by fixation perfusion (de la Torre et al,1992a). The tissue was then sectioned at 4 um at the level of the CA1 using plates 20-23 from Paxinos and Watson (1982) rat brain atlas.

Neuronal damage was identified by the presence of a pyknotic nucleus and/or hyperchromatic cytoplasm using the Palmgren silver stain (de la Torre et al, 1992a)

Serial sections were stained with glial fibrillary acidic protein (GFAP) immunoreaction (1:400, Sigma). The number of damaged neurons/total number of neurons and the density of GFAP in the CA1 and oriens layer respectively was evaluated using a Jandel Sigma

Scan software program connected to a PC via a digital tablet (de la Torre et al, 1993a). Group means, standard deviations and statistical analyses were computed by the Jandel program. A p value <0.05 was considered significant.

RESULTS

Table 1 summarizes the results of the water maze test evaluation for 3-VO, 2-VO, and their respective control groups at 7,14 and 63 days following surgery. The data indicate that following 3-VO, visuo-spatial memory impairment was evident 7 days post-op and continued for at least 63 days. It appears from analysis of the platform location latency that 3-VO rats became more dysfunctional

Table 1

Mean platform location latencies in rats subjected to 2-VO or 3-VO as compared to non-occluded controls. After 7 days post-op, only 3-VO rats show impairment of visuo-spatial tasks in the water maze test. Which worsens from day 14 to day 63 post-op. 2-VO rats show moderate memory function impairment 14 days post-op but some learning ability is regained by day 63 post-op despite continuation of visuo-spatial task impairment as compared to control.

POST-OP DAY OF TEST	2 - VO	CONTROL	3 - VO	CONTROL
	(Mean platform location latency in seconds)			
7	18 ± 4	15 ± 3	29 ± 8*	16 ± 2
14	47 ± 11*	8 ± 4	38 ± 9*	9 ± 2
63	22 ± 6*	7 ± 1	62 ± 15*	8 ± 4

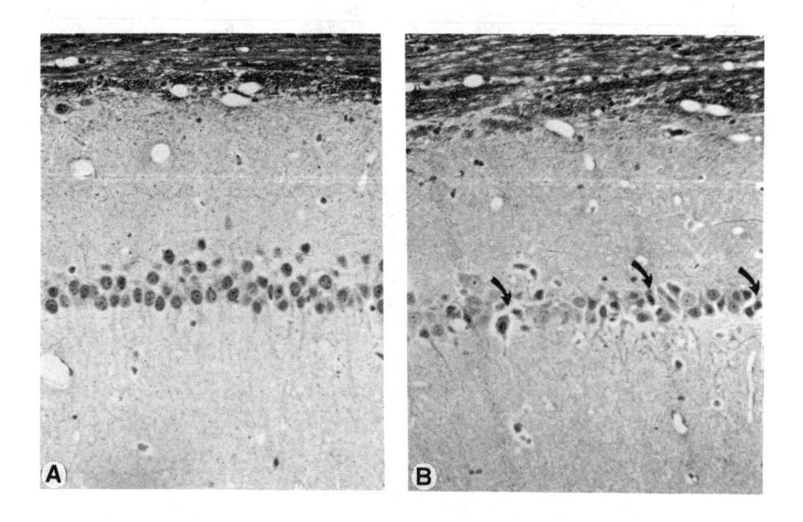

Figure 1

Normal appearance of CA1 hippocampal neurons 9 weeks after 2-vessel occlusion (2-VO) is observed (A). Following 3-VO at the same time period, moderate damage to CA1 neurons is seen in association with perineuronal spaces (B, arrows). Palmgren silver stain, 240X.

with increasing time so that their ability to locate the hidden submerged platform deteriorated over the 9 week observation period. By contrast, 2-VO rats began to show memory impairment in the water maze test 2 weeks post-op and by day 63, initiated a <u>mild</u> recovery (Table 1) which remained relatively unchanged to the end of the observation period 175 days post-op (not shown on Table 1). The 2-VO rats despite their mild recovery from day 14 post-op still showed memory impairment to visuo-spatial tasks relative to the controls.

Histological examination of 3-VO rats revealed that selective neuronal damage in CA1 hippocampus was generally present 3 weeks post-op and that the number of damaged neurons grew progressively higher by week 9 post-op (Fig. 1, B). The neuronal damage in 3-VO rats was accompanied by a relative increase in GFAP density in the oriens layer of the hippocampus but not elsewhere. GFAP increase was observed both in the degree of hypertrophy and the total number of astrocytes present (hyperplasia). The GFAP density increase but not CA1 neuronal damage appeared to correlate positively with the severity of memory impairment in 3-VO rats. In 2-VO rats, CA1 neuronal damage was not observed in weeks 1-9 (Fig. 1, A) but neuronal <u>loss</u> was recorded in week 25 (175 days post-op). Since 2-VO rats were not killed for histology between week 9 and 25, it is not known whether neuronal damage during that interval occurred prior to the total reduction of CA1 neurons seen on week 25.
^{31}P-magnetic resonance imaging on 2-VO and 3-VO at week 12 post-op, showed a hyperintense signal localized to hippocampus and a smaller, localized signal in the fronto-parietal cortex of 3-VO rats indicating possible tissue damage of this region (Fig. 2B).

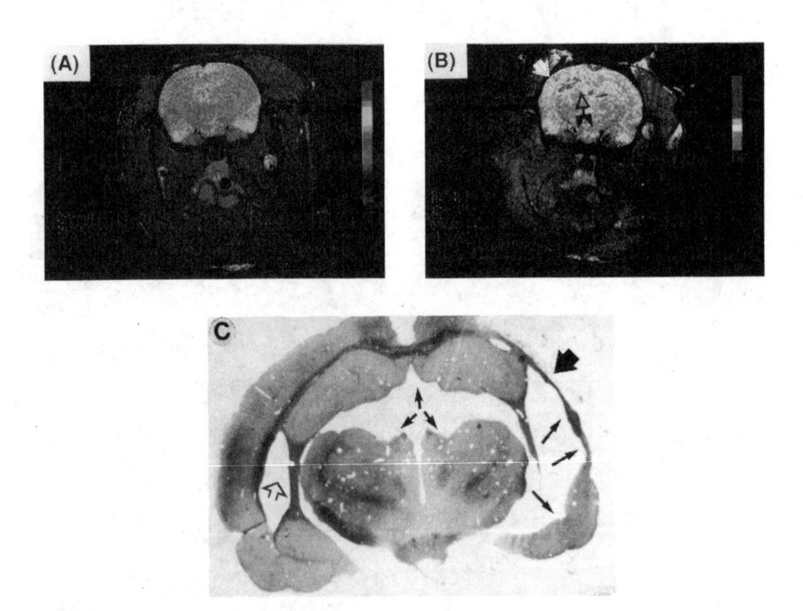

Figure 2

^{31}P-magnetic resonance imaging of rat brain at the level of the dorsal hippocampus 12 weeks after 2-VO (A) and 3-VO (B). 2-VO brain appears normal while 3-VO shows a hyperintense signal in the hippocampus (B, open arrow) and a smaller signal in the fronto-parietal cortex (B, white arrow). After 25 weeks of 2-VO, only mild neuronal loss of CA1 is observed (not shown); in 3-VO rats severe parietal and entorhinal cortex atrophy with hippocampal neuron damage (C, thin black arrows and large black arrow) and ventriculomegaly is noted (C, open arrow,).

2-VO rats imaged 12 weeks post-op did not show such changes (Fig. 2A). The damage to CA1 observed histologically and by neuroimaging of 3-VO rats examined 12 weeks post-op was noted to progress to atrophy and degeneration of the parietal cortex as well as one side of the CA1 region when rats were examined post-mortem 25 weeks post-op (Fig. 2C). These atrophic-degenerative changes were not observed 25 weeks post-op in 2-VO rats.

Comparison of CA1 neuronal damage between 3-VO and 2-VO showed that damage proceeded progressively in 3-VO rats from week 3-25 post-op but no damage was observed from week 1-9 in 2-VO rats and only mild CA1 cell loss was seen on week 25 post-op.

DISCUSSION

Chronic cerebrovascular hypoperfusion for 1-25 weeks in middle aged 10 month old rats using 3-VO or 2-VO results in neurodegeneration of CA1 neurons which is associated with an increase density of reactive astrocytosis.

The differences between 3-VO and 2-VO appear to be in the damaging impact on CA1 neurons. 3-VO rats show a definite progression in the number of CA1 neurons affected by this vascular insult which ostensibly continues beyond 9 weeks post-op extending to the fronto-parietal and entorhinal cortex by week 25 (de la Torre et al, in press). CA1 neuronal damage after 3-VO correlates positively with a GFAP density increase, a suggestive finding that may implicate the reactive astrocytes in the pathogenic onset of the observed neuronal damage. In 2-VO rats by contrast, no damage to CA1 was observed even at 9 weeks post-op. Unfortunately, our study did not determine whether progressive neurodegeneration occurred between week 9 and 25 in 2-VO rats since no histology was available for this time period. We did observe that on week 25 post-op, 2-VO showed a loss in the total number of neurons in CA1 which was associated with a significant increase in the GFAP density of the oriens layer which borders the CA1 subfield.

This finding in 2-VO rats is similar to the neuronal damage and increased GFAP seen in 3-VO animals and lends further support to the notion that reactive astrocytosis is responsible not only for the CA1 damage but also for the resulting memory impairment. If this statement has some validity, one then might ask, what triggers the reactive astrocytosis in the hippocampus?

In this regard, Mossakowski and his group (1994) have reported that the selective vulnerability of CA1 neurons in the gerbil may lie in their reduced capillary density and consequently lower blood supply of this hippocampal sector as compared to neighboring brain regions. Although it is not certain whether this finding also applies to rat CA1 vascular density (Schmidt-Kastner and Freund, 1991),it is generally accepted that cerebral ischemia or hypoxia can lead to the formation of reactive astrocytes, as exemplified in Alzheimer's disease (Delacourte, 1990; Duffy et al, 1980; Haarpin et al, 1990).

A more likely explanation for the development of visuo-spatial memory impairment after 2-V0 or 3-VO may be the effect of persistent lower hippocampal blood flow (de la Torre et al, 1992a) on neurotrophic factors or proteins localized in the CA1 neurons or glia.

The reason for this line of thinking is that there is an absence of CA1 damage and reactive astrocytosis in the hippocampus of 2-VO rats, yet there is significant impairment of memory acquisition 14 days post-op. Pilot data on 19 month rats subjected to 2-VO for 14 days revealed no CA1 damage or increased GFAP density, but when in situ hybridization for APP-770 and APP-695 were performed on the hippocampus, 2-VO rats showed a lower expression of these proteins when compared to non-occluded controls (de la Torre et al, unpublished observations).

It is of interest to note that Wistar rats subjected to 2-VO undergo a more violent reaction to this vascular insult than Sprague-Dawley rats. Following 2-VO, Wistar rats are reported to develop a 71% mortality when 7 weeks old and a 3.6 % mortality when 5 months old (Ni et al, 1994). In another study on adult Wistar rats, 2-VO induced a 21% mortality within 48 hours associated with extensive neuronal damage outside the CA1 (Tsuchiya et al, 1992). Our experience with over 250 Sprague-Dawley rats 10-19 months old shows that 2-VO results in less than 1% mortality while neuronal damage is absent for the first 12 weeks post-op and eventually results in selective, mild loss of CA1 neurons after 25 weeks post-op. However, it should be stated that even Sprague-Dawley rats may show variable vulnerability to 2-VO/3-VO depending on the individual supplier. We therefore recommend that the name of the supplier and the mortality rate be stated in articles dealing with this cerebrovascular insult.

Although the reason for these interspecies, intercolony differences is not known, we feel it is an important consideration when choosing a rat model for 3-VO or 2-VO investigations.

In conclusion, we have presented evidence that mild or moderate (2-VO/3-VO) chronic cerebral hypoperfusion in middle aged rats results in a syndrome consisting of morphological, memory and metabolic abnormalities consistent with progressive neurodegeneration. The model may be useful in dementia research and as a screen for therapy aimed at age-related memory disorders.

CONCLUSIONS

1. **2-VO AND 3-VO INDUCE SELECTIVE NEURODEGENERATION OF CA1**

2. **IN 3-VO, NEURODEGENERATION EXTENDS TO CORTEX AND RESULTS IN ATROPHIC CHANGES WITH VENTRICULOMEGALY 25 WEEKS POST-OP; NO SUCH CHANGES OBSERVED IN 2-VO**

3. **INCREASED GFAP DENSITY 3-25 WEEKS POST-OP MAY BE LINKED TO CA1 DAMAGE AFTER 2-VO OR 3-VO**

4. **VISUO-SPATIAL MEMORY IMPAIRMENT BECOMES PROGRESIVELY WORSE 1-9 WEEKS POST-OP AFTER 3-VO; SOME "RELEARNING" AFTER 2-VO IS SEEN 1-25 WEEKS POST-OP BUT NOT ENOUGH FOR TOTAL RECOVERY**

5. **ChAT ACTIVITY IS UNCHANGED IN HIPPOCAMPUS/CAUDATE N. AFTER 2-VO**

6. 2-VO OR 3-VO MAY BE GOOD MODELS TO SCREEN POTENTIAL THERAPY IN PROGRESSIVE NEURODEGENERATION AFFECTING MEMORY FUNCTION IN AGING RATS

Acknowledgements

This study was supported in part by the Ontario Heart and Stroke Foundation and by Dedicated Health Research Funds of the University of New Mexico School of Medicine.

REFERENCES

Amaral, D.G., and Insausti, R, 1990, Hippocampal formation. In: The Human Nervous System, Academic Press, New York, pp. 711-755

Ball, M.,1977, Neuronal loss, neurofibrillary tangles and granulovacuolar degeneration in the hippocampus with aging and dementia. Acta Neuropathol. 37:111-117

de la Torre, J.C., Fortin, T., Park, G., Butler, K., Kozlowski, P., Pappas, B., de Socarraz, H., Saunders, J. And Richard, M., 1992a, Chronic cerebrovascular insufficiency induces dementia-like deficits in aged rats. Brain Res 582:186-195

de la Torre, J.C., Fortin, T., Park, G., Saunders, J., Kozlowski, P., Butler, K., de Socarraz, H., Pappas, B. and Richard, M., 1992b, Aged but not young rats develop metabolic, memory deficits after chronic brain ischemia. Neurol Res 14(Suppl.), 177-180

de la Torre, J.C., Fortin, T., Park, G. and Pappas, B., 1993b, Spatial memory loss without morphological damage to CA1 neurons. Soc Neurosci Abstr. 19:1045

de la Torre, J.C. and Fortin, T., A chronic physiological rat model of dementia, 1994, Behavior Brain Res 63:35-40

de la Torre, J.C., Fortin, T. and Saunders, J., Correlates between NMR spectroscopy, diffusion weighted imaging and CA1 morphometry following chronic brain ischemia. J Neurol Sci Res, in press

de la Torre, J.C., Fortin, T., Park, G., Pappas, B., Saunders, J. and Richard, M., 1993a, Brain blood-flow restoration "rescues" chronically damaged rat CA1 neurons. Brain Res 623:6-15

Delacourte, A., 1993, General and dramatic glial reaction in Alzheimer brains. Neurology 40:33-37

Duara, R., 1994, Neuroimaging with CT and MRI in Alzheimer's disease. In: Alzheimer's Disease. R. D. Terry, R. Katzman, K. Bick (eds) Raven Press, New York, pp. 75-86

Duffy, P.E., Rapport, M. and Graf, L., 1980, Glial fibrillary acidic protein and Alzheimer-type dementia. Neurology 30:778-782

Grubb, R., Raichle, M., Gado, M.,Eichling, J. and Hughes, C., 1977, Cerebral blood flow, oxygen utilization and blood volume in dementia. Neurology 27:905-910

Harpin, L.L., Delaere, P. and Javoy-Agid, F., 1990, Glial fibrillary acidic protein and BA4 protein deposits in temporal lobe of aging brain and senile dementia of Alzheimer-type. J Neurosci Res 27:587-594

Jack, C.R., Petersen, R., O'Brien, P., and Tangalos, E., 1992, MR - based hippocampal volumetry in the diagnosis of Alzheimer's disease. Neurology 42:183-188

McKhann, G., Drachman, D. And Folstein, M., 1981, Clinical diagnosis of Alzheimer's disease. Neurology 34:939-944

Mendis, T., and Mohr, E., 1993, Dementia: A clinical approach. Canad J Diag 2:91-105

Mossakowski, M.J., Wrzolkown, T., Tukaj, C., and Gadamski, R., 1991, Comparative morphometric analysis of terminal vascularization of hippocampal CA1 and CA3 sectors in mongolian gerbils. Folia Neuropathol 32:1-7

Nitsch, R., Blustajn, J., Wurtman, R. and Growdon, J., 1991, Membrane phospholipid metabolites are abnormal in Alzheimer's disease. Neurology 41 (Suppl.) 1:269

Pettegrew, J.W., Kanagasabai, P., Moossy, J., Martinez, J., 1988, Correlation of phosphorous-31 magnetic resonance spectroscopy and morphologic findings in Alzheimer's disease. Arch Neurol 45:1093-1096

Prohovnik, I., Mayeux, R., Sackheim, H., Smith, G., Stern, Y. and Alderson, P.Y., 1988, Cerebral perfusion as a diagnostic marker of early Alzheimer's disease. Neurology 38:931-937

Rose, C.D., and Henneberry, R.C., 1994, Etiology of the neurodegenerative disorders: A critical analysis. Neurobiol Aging 15:233-234

Schmidt-Kastner, R., and Freund, T.G., 1991, Selective vulnerability of the hippocampus in brain ischemia. Neuroscience 40:599-636

Sulkava, R., Haltia, M., Paetau, A., Wikstrom. J. and Palo, J., 1983, Accuracy of clinical diagnosis in primary degenerative dementia: correlation with neuropathological findings. J Neurol Neursurg Psychiatry 46:9-13

Wade, J.P., Mirsen, T. and Hachinski, V., The clinical diagnosis of Alzheimer's disease. Arch Neurol 44:24-27

West, M.J., Coleman, P., Flood, D., and Troncoso, J., 1994, Differences in the pattern of hippocampal neuronal loss in normal ageing and Alzheimer's disease. Lancet 344:769-772

IN VIVO BIOLOGY OF APP AND ITS HOMOLOGUES

Sangram S. Sisodia, Ph.D.,[1,4] Gopal Thinakaran, Ph.D.,[1,4] Cornelia S. von Koch, B.A.,[3,4] Hilda H. Slunt, B.S.,[4] A. Jane I. Roskams, Ph.D.,[2,3] Cheryl A. Kitt, Ph.D.,[1,4] Eliezer Masliah, M.D.,[6] Vassilis E. Koliatsos, M.D.,[1-4] Peter R. Mouton, Ph.D.,[1,4] Lee J. Martin, Ph.D.,[1,3,4] Randall R. Reed, Ph.D.,[2,3,5] Gabrielle V. Ronnett, M.D., Ph.D.,[2,3] Hui Zheng, Ph.D.,[7] Lex H.T. Van der Ploeg, Ph.D.,[7] and Donald L. Price, M.D.[1-4]

[1]Department of Pathology
[2]Department of Neurology
[3]Department of Neuroscience
[4]Neuropathology Laboratory
[5]Howard Hughes Medical Institute, Department of Molecular Biology & Genetics
The Johns Hopkins University School of Medicine
Baltimore, Maryland
[6]Departments of Neuroscience and Pathology
University of California
San Diego, California
[7]Merck Research Laboratories
Rahway, New Jersey

INTRODUCTION

A principal pathological feature of Alzheimer's disease is the deposition of ß-amyloid protein (Aß) in brain parenchyma.[1-3] Aß, an ~4-kD polypeptide, is derived from larger type I integral membrane glycoproteins, termed the amyloid precursor proteins (APP).[4-9] APP is a member of a family of homologous amyloid precursor-like proteins (APLPs), including APLP1[10] and APLP2.[11-13] APP, APLP1, and APLP2 show considerable homology in the N-terminal cysteine-rich domain, the cytoplasmic tail and transmembrane sequences, but the APLPs differ from APP by the conspicuous absence of the Aß region. APP homologues have also been identified in *C. elegans* (*apl*-1)[14] and *Drosophila* (*appl*),[15] and these molecules lack the Aß region as well. Flies with a deleted *APPL* gene are viable and have a defective fast phototaxis response, a behavior that is partially rescued by the introduction of human APP cDNA. The biological functions of APP and the APLPs in the mammalian nervous system are far from clear, although several studies have suggested that APP play roles in cell-cell or cell-matrix interactions,[16-18] calcium hemostasis,[19] growth promoting activities,[20] and the formation/maintenance of synapses *in vivo*.[21] More

Neurodegenerative Diseases
Edited by Gary Fiskum, Plenum Press, New York, 1996

recently, investigations in cultured cells[22] and in transgenic animals[23] have provided support to the notion that APP is neuroprotective.

Gene Targeting of *APP*

In an attempt to assess APP function *in vivo*, we used a gene targeting strategy to generate mice with functionally inactivated *APP* genes. [24] We demonstrate that mice with a homozygous mutation in the *APP* gene are viable and do not express *APP* gene-derived transcripts or polypeptides. We have examined the expression of mRNAs encoding each of the APP homologues in mice with inactivated APP genes and have failed to demonstrate any compensatory changes in the levels of transcripts encoding the APLPs. Moreover, we were unable to detect any alterations either at the level of cellular morphology or cortical organization in the central nervous system (CNS) of *APP*-targeted mice relative to wild-type littermates. Using stereological methods, we have not detected any quantitative differences in synaptic number in the hippocampus of *APP*-targeted mice relative to wild-type littermates. Finally, in an attempt to ascertain a potential functional role for APP in neuronal sprouting/synaptogenesis, we lesioned the perforant pathway by aspiration of the entorhinal cortex. Two weeks following the lesion, we failed to observe any differences between *APP*-targeted and wild-type mice in the level of reactive synaptogenesis in the outer molecular layers of the dentate gyrus, as assayed by acetylcholinesterase histochemistry.

APLP1 Biology

APLP1, the first described APP homologue, was shown to be expressed specifically and at high levels in adult brain.[10] Furthermore, an antibody directed against an epitope in the APLP1 cytoplasmic domain recognized polypeptides of 65 kD and 33 kD in extracts prepared from mouse brain and neuroblastoma cells. We have raised polyclonal antibodies against a synthetic peptide corresponding to the C-terminal 11 amino acids of APLP1 and now document that APLP1 expressed in mouse brain migrates at ~95 kD in SDS-PAGE, indistinguishable to APLP1 generated in cells transiently transfected with full-length APLP1 cDNA. We confirm that APLP1 is expressed at high levels in the CNS and that there is little selectivity between brain regions. We have analyzed the developmental expression of *APLP1* in mouse by *in situ* hybridization and document that the first detectable expression of APLP1 transcripts is coincident with the onset of neurogenesis (days 9-10 p.c.) with newly born, postmitotic neurons in the developing CNS and in the spinal cord expressing abundant APLP1 mRNA. The pattern of expression of APLP1 mRNA is distinct from the generally ubiquitous distributions of APP and APLP2 transcripts at these early developmental stages. Furthermore, we have used RT-PCR to examine the expression of APLP1 mRNA in a pluripotent mouse cell line, P19, stimulated to differentiate by treatment with retinoic acid. We show that the appearance of APLP1 mRNA parallels the expression of neurofilament mRNA. Interestingly, we have not detected any APLP1 transcripts in differentiating P19 cells or in mouse development that encode the KPI domain. The biological function(s) of APLP1 at early stages of neuronal differentiation are unclear and await the generation and characterization of mice with targeted *APLP1* alleles.

APLP2 Expression and Processing

At present, APLP2 is the best characterized member of APLPs. Like APP, APLP2 is encoded by several mRNAs derived by alternative splicing.[12,13,25] Both APP and APLP2 are expressed in brain and peripheral tissues,[12,13] and *in situ* hybridization studies disclose similarities in the distributions of APP and APLP2 mRNAs in mouse[13] and human[12] CNS. Finally, APLP2 matures through the same unusual secretory/cleavage pathway as APP.[13] However, in contrast to APP-695 and APP-751/770, APLP2-751 is a substrate for modification by a single chondroitin sulfate glycosaminoglycan (CS GAG).[26] We and others have reported that alternative splicing of APLP2 pre-mRNA gives rise to several APLP2 transcripts. Although the predominant APP transcripts expressed by neurons encode the KPI-forms of APP, neurons primarily express APLP2 mRNAs that encode KPI-containing APLP2. Additional alternatively spliced APLP2 transcripts that contain or lack sequences encoding a 12 amino acid miniexon are differentially expressed in specific neuronal populations. Interestingly, encoded polypeptides that lack sequences encoded by the miniexon are substrates for modification by CS GAG chains at a single serine residue. Remarkably, insertion of sequences encoded by the miniexon fully abrogates CS GAG attachment. To assess the generality of the novel mechanism of regulation of CS GAG modification in members of the APP/APLP family, we examined the levels of CS GAG modification of APP isoforms encoded by alternatively spliced APP mRNAs. Indeed, earlier reports indicated that APP is a substrate for modification by CS GAG in C6 glioma cells[27,28] despite our failure to demonstrate CS GAG modification of transiently expressed APP-695 or -770 isoforms CHO and COS-1 cells.[26] On the other hand, L-APP isoforms, which lack 18 amino acids encoded by exon 15, contain the sequence ENEGSGL; this sequence is generated by a fusion of sequences encoded by exons 14 and 16 and is surprisingly similar to sequences flanking the CS GAG attachment site of APLP2-751. We demonstrate that, like APLP2-751, L-APP isoforms are also substrates for CS GAG addition. The molecular mechanisms involved in regulating the levels of alternatively spliced transcripts that encode APP/APLP2 isoforms are unknown. Furthermore, the biochemical mechanism(s) by which exon 15 of APP-770/751/695 and the 12 amino acid insert in APLP2-763/706 disrupt CS GAG modification of respective core proteins have not been fully clarified. Nevertheless, we suggest that the CS GAG addition represents a cell- or developmental-specific mechanism to generate additional functional diversity for each polypeptide.

APLP2 Biology *In Vivo*

To define the distribution of APLP2 in the rodent nervous system, we generated antibodies against full-length mouse APLP2-751 or the C-terminal 12 amino acids of APLP2. We established the specificity of these reagents by Western blot analysis of extracts prepared from mammalian cells transiently transfected with cDNAs encoding APP and each of the described homologues. Consistent with previous *in situ* hybridization studies,[13] our investigations with the APLP2-specific antibodies revealed that APLP2 is expressed in a variety of regions, including cerebral cortex, hippocampus, and thalamus. Confocal microscopy revealed prominent colocalization of APLP2 with MAP2, a postsynaptic marker, in cortex and hippocampus. Moreover, APLP2 mRNA is highly expressed in sensory neurons in the olfactory epithelium, and the encoded polypeptide is abundant in olfactory sensory axons and axon terminals in glomeruli. In contrast to its subcellular distribution in cortex and hippocampus, APLP2 is clearly present in both pre- and postsynaptic compartments in the olfactory bulb. Notably, mRNA encoding CS GAG-modified forms of APLP2 are highly represented in the olfactory epithelium, relative to alternatively spliced mRNA that encode CS GAG-free forms of APLP2. As would be anticipated, high levels of sensory neuron-synthesized CS GAG-modified APLP2 accumulates in the olfactory bulb. Because sensory neurons in the olfactory epithelium are in a state of continual turnover, axons of newly generated neurons must establish synaptic connections with neurons in the olfactory bulb in adult life. In view of the evidence in support for CS proteoglycans in aspects of cell migration and neuronal outgrowth, we suggest that CS GAG-modified APLP2 plays an important role in axonal pathfinding and/or synaptic interactions of newly born olfactory sensory neurons with respective targets in the olfactory bulb. To assess the potential biological functions of APLP2 *in vivo,* we have used a gene targeting strategy to generate mice with inactivated *APLP2* genes. We demonstrate that APLP2 is absent in these animals, and our current efforts are aimed at assessing the levels of homologous proteins. Mice with inactivated *APLP2* genes will be valuable reagents for assessing the role of APLP2 in experimental paradigms of degeneration and regeneration in the CNS and peripheral nervous system.

ACKNOWLEDGEMENTS

This work was supported by grants from the U.S. Public Health Service (NIH NS 20471, AG 05146, NS 07179), the American Health Assistance Foundation, and the Adler Foundation. Drs. Price, Koliatsos, and Martin are recipients of a Leadership and Excellence in Alzheimer's Disease (LEAD) award, and Drs. Price and Koliatsos are recipients of a Javits Neuroscience Investigator Award (NIH NS 10580). Dr. Sisodia is the recipient of a Zenith Award from the Alzheimer's Association.

REFERENCES

1. G.G. Glenner and C.W. Wong, Alzheimer's disease: initial report of the purification and characterization of a novel cerebrovascular amyloid protein, *Biochem. Biophys. Res. Commun.* 120:885-890 (1984).

2. G.G. Glenner and C.W. Wong, Alzheimer's disease and Down's syndrome: sharing of a unique cerebrovascular amyloid fibril protein, *Biochem. Biophys. Res. Commun.* 122:1131-1135 (1984).

3. C.L. Masters, G. Simms, N.A. Weinman, G. Multhaup, B.L. McDonald, and K. Beyreuther, Amyloid plaque core protein in Alzheimer disease and Down syndrome, *Proc. Natl. Acad. Sci. USA* 82:4245-4249 (1985).

4. J. Kang, H.-G. Lemaire, A. Unterbeck, J.M. Salbaum, C.L. Masters, K.-H. Grzeschik, G. Multhaup, K. Beyreuther, and B. Müller-Hill, The precursor of Alzheimer's disease amyloid A4 protein resembles a cell-surface receptor, *Nature* 325:733-736 (1987).

5. N. Kitaguchi, Y. Takahashi, Y. Tokushima, S. Shiojiri, and H. Ito, Novel precursor of Alzheimer's disease amyloid protein shows protease inhibitory activity, *Nature* 331:530-532 (1988).

6. P. Ponte, P. Gonzalez-DeWhitt, J. Schilling, J. Miller, D. Hsu, B. Greenberg, K. Davis, W. Wallace, I. Lieberburg, F. Fuller, and B. Cordell, A new A4 amyloid mRNA contains a domain homologous to serine proteinase inhibitors, *Nature* 331:525-532 (1988).

7. R.E. Tanzi, A.I. McClatchey, E.D. Lampert, L. Villa-Komaroff, J.F. Gusella, and R.L. Neve, Protease inhibitor domain encoded by an amyloid protein precursor mRNA associated with Alzheimer's disease, *Nature* 331:528-530 (1988).

8. T.E. Golde, S. Estus, M. Usiak, L.H. Younkin, and S.G. Younkin, Expression of β amyloid protein precursor mRNAs: recofnition of a novel alternatively spliced form and quantitation in Alzheimer's disease using PCR, *Neuron* 4:253-267 (1990).

9. G. König, U. Mönning, C. Czech, R. Prior, R. Banati, U. Schreiter-Gasser, J. Bauer, C.L. Masters, and K. Beyreuther, Identification and differential expression of a novel alternative splice isoform of the βA4 amyloid precursor protein (APP) mRNA in leukocytes and brain microglial cells, *J. Biol. Chem.* 267:10804-10809 (1992).

10. W. Wasco, K. Bupp, M. Magendantz, J.F. Gusella, R.E. Tanzi, and F. Solomon, Identification of a mouse brain cDNA that encodes a protein related to the Alzheimer disease-associated amyloid-beta-protein precursor, *Proc. Natl. Acad. Sci. USA* 89:10758-10762 (1992).

11. C.A. Sprecher, F.J. Grant, G. Grimm, P.J. O'Hara, F. Norris, K. Norris, and D.C. Foster, Molecular cloning of the cDNA for a human amyloid precursor protein homolog: evidence for a multigene family, *Biochemistry* 32:4481-4486 (1993).

12. W. Wasco, S. Gurubhagavatula, M.d. Paradis, D.M. Romano, S.S. Sisodia, B.T. Hyman, R.L. Neve, and R.E. Tanzi, Isolation and characterization of *APLP2* encoding a homologue of the Alzheimer's associated amyloid β protein precursor, *Nature Genetics* 5:95-99 (1993).

13. H.H. Slunt, G. Thinakaran, C. von Koch, A.C.Y. Lo, R.E. Tanzi, and S.S. Sisodia, Expression of a ubiquitous, cross-reactive homologue of the mouse β-amyloid precursor protein (APP), *J. Biol. Chem.* 269:2637-2644 (1994).

14. I. Daigle and C. Li, *Apl-1* a *Caenorhabditis elegans* gene encoding a protein related to the human β-amyloid protein precursor, *Proc. Natl. Acad. Sci. USA* 90:12045-12049 (1993).

15. D.R. Rosen, L. Martin-Morris, L. Luo, and K. White, A *Drosophila* gene encoding a protein resembling the human β-amyloid protein precursor, *Proc. Natl. Acad. Sci. USA* 86:2478-2482 (1989).

16. B.D. Shivers, C. Hilbich, G. Multhaup, M. Salbaum, K. Beyreuther, and P.H. Seeburg, Alzheimer's disease amyloidogenic glycoprotein: expression pattern in rat brain suggests a role in cell contact, *EMBO J.* 7:1365-1370 (1988).

17. D. Schubert, L.-W. Jin, T. Saitoh, and G. Cole, The regulation of amyloid β protein precursor secretion and its modulatory role in cell adhesion, *Neuron* 3:689-694 (1989).

18. F.G. Klier, G. Cole, W. Stalleup, and D. Schubert, Amyloid β-protein precursor is associated with extracellular matrix, *Brain Res.* 515:336-342 (1990).

19. M.P. Mattson, S.W. Barger, B. Cheng, I. Lieberburg, V.L. Smith-Swintosky, and R.E. Rydel, β-amyloid precursor protein metabolites and loss of neuronal Ca^{2+} homeostasis in Alzheimer's disease, *Trends Neurosci.* 16:409-414 (1993).

20. T. Saitoh, M. Sundsmo, J.-M. Roch, T. Kimura, G. Cole, D. Schubert, T. Oltersdorf, and D.B. Schenk, Secreted form of amyloid β protein precursor is involved in the growth regulation of fibroblasts, *Cell* 58:615-622 (1989).

21. L. Mucke, E. Masliah, W.B. Johnson, M.D. Ruppe, M. Alford, E.M. Rockenstein, S. Forss-Petter, M. Pietropaolo, M. Mallory, and C.R. Abraham, Synaptotrophic effects of human amyloid β protein precursors in the cortex of transgenic mice, *Brain Res.* 666:151-167 (1994).

22. M.P. Mattson, B. Cheng, A.R. Culwell, F.S. Esch, I. Lieberburg, and R.E. Rydel, Evidence for excitoprotective and intraneuronal calcium-regulating roles for secreted forms of the β-amyloid precursor protein, *Neuron* 10:243-254 (1993).

23. L. Mucke, C.R. Abraham, M.D. Ruppe, E.M. Rockenstein, S.M. Toggas, M. Mallory, M. Alford, and E. Masliah, Protection against HIV-1 gp120-induced brain damage by neuronal expression of human amyloid precursor protein (hAPP), *J. Exp. Med.,* in press.

24. H. Zheng, M.-H. Jiang, M.E. Trumbauer, D.J.S. Sirinathsinghji, R. Hopkins, D.W. Smith, R.P. Heavens, G.R. Dawson, S. Boyce, M.W. Conner, K.A. Stevens, H.H. Slunt, S.S. Sisodia, H.Y. Chen, and L.H.T. Van der Ploeg, β-amyloid precursor protein-deficient mice show reactive gliosis and decreased locomotor activity, *Cell*, in press.

25. R. Sandbrink, C.L. Masters, and K. Beyreuther, Similar alternative splicing of a non-homologous domain in βA4-amyloid protein precursor-like proteins, *J. Biol. Chem.* 269:14227-14234 (1994).

26. G. Thinakaran and S.S. Sisodia, Amyloid precursor-like protein 2 (APLP2) is modified by the addition of chondroitin sulfate glycosaminoglycan at a single site, *J. Biol. Chem.* 269:22099-22104 (1994).

27. J. Shioi, J.P. Anderson, J.A. Ripellino, and N.K. Robakis, Chondroitin sulfate proteoglycan form of the Alzheimer's β-amyloid precursor, *J. Biol. Chem.* 267:13819-13822 (1992).

28. J. Shioi, L.M. Refolo, S. Efthimiopoulos, and N.K. Robakis, Chondroitin sulfate proteoglycan form of cellular and cell-surface Alzheimer amyloid precursor, *Neurosci. Lett.* 154:121-124 (1993).

LYMPHOCYTE AMYLOID PRECURSOR PROTEIN mRNA ISOFORMS IN NORMAL AGING AND ALZHEIMER'S DISEASE

Richard P. Ebstein,[1] Lubov Nemanov,[1] Gregory Lubarski,[1] Marina Dano,[2] Teres Trevis,[2] and Amos Korczyn,[2]

[1]Shapiro Molecular Neurobiology Laboratory, S. Herzog Memorial Hospital, P.O.Box 35300, Jerusalem 91351, Israel & [2]Neurology Department, Tel-Aviv University Medical School, Israel

INTRODUCTION

The amyloid precursor protein (APP) molecule is expressed on the surface of most cells including neurons.[1] The three major isoforms of APP in brain contain 770, 751 and 695 amino acids, respectively. The two longer forms of the molecule contain a Kunitz-type serine protease inhibitor domain (KPI). These longer isoforms (APP_{770} and APP_{751}) are the principal species found in non-neuronal cells whereas the shorter APP_{695} is the major form in differentiated neurons.[2] Isoform changes have also been observed in AD and following astrocyte, glial and microglial, cell activation.[3] It is still unknown which isoforms of APP and which cell types contribute to β/A4 peptide deposition in AD although a number of different cell types are potential contributors to the amyloid load.

The expression of the APP mRNA isoforms in easily-accessible peripheral cells offers the possibility of examining, using benign clinical procedures, possible changes in APP mRNA as a result of aging and disease. Since post-mortem studies of APP mRNA isoforms in AD brains are sometimes difficult to interpret due to artifactual changes associated with such variables as postmortem course, peripheral cells offer a complementary strategy for studying possible age or disease-associated alterations in amount of APP mRNA that may parallel changes occurring in less accessible brain tissue. In the current investigation we quantified the amount of lymphocyte APP mRNA isoforms, using a reverse transcriptase - polymerase chain reaction (RT-PCR) procedure, obtained from a group of 64 cognitively-intact subjects ranging in age from 20-91 years and in 19 sporadic Alzheimer's disease patients.

METHODS

Biochemical Methods

Lymphocytes were isolated from whole fresh blood by density gradient centrifugation over Ficol-Paque (Pharmacia). RNA was purified using Trizol reagent (Gibco). Quantitative RT-PCR was carried out using the procedure of Horikoshi et al.[4] The forward primer (complementary to APP bases 958-977) is: CACCACAgAgTCTgTggAAg and the back primer

Neurodegenerative Diseases
Edited by Gary Fiskum, Plenum Press, New York, 1996

(complementary to APP bases 1213-1194) is: AggTgTCTCgAgATACTTgT. Human β-actin served as the control standard. The primers for β-actin amplification were: (i) forward primer: gAgAAgATgACCCAgATCATgT (ii) back primer: ACTCCATgCCCAggAAggAAg g. First strand cDNA synthesis was performed using APP-1213 as a primer and Moloney murine leukemia virus (MLV) reverse transcriptase (BioLab, New England Nuclear). The actin mixture was diluted 1:800 and the APP mixture was diluted 1:80 and 4 μl from the first strand synthesis from each reaction mixture were used in separate PCR reactions.

The PCR reaction mixture (50 μl final volume) consisted of the following components: 0.56 U Taq polymerase and supplied buffer (MBI, BioLabs), 40 pmoles forward and back primers (either actin or APP), and 200 μM dATP, dGTP, dTTP and 20 μM dCTP. 1-2 μCi α-$dCTP^{32}$ was added to the reaction mixture to monitor the course of the reaction and to quantify the subsequent PCR fragments. The PCR conditions were: 94°C for 1 min, 55°C for 1 min, 72°C for 2 min. A final extension was carried out at 70°C for 5 min. A Hybaid thermal cycler was used to carry out the PCR.

Aliquots of the reaction mixture were then electrophoresed on a 6% acrylamide gel (1mm) in a Hoefer vertical gel apparatus at 150 V constant voltage for approximately 2.5 hours. Actin and APP were from the same subject were loaded on the same lane. In initial experiments a γ-ATP^{32} labeled molecular weight ladder was run simultaneously to subsequently ascertain the size of the amplified fragments. Following electrophoresis, the gel was dried using a BioRad gel dryer and then exposed to X-ray film (Kodak X OMAT AR) for appropriate periods. The autoradiogram was analyzed by initially scanning with a Hewlit Packard laser scanner and then intensity of the bands quantified using Quantiscan desitometry software (Biosoft).

Patient Selection

Cognitively-intact subjects ranged in age from 20-90 years (n=64). The younger subjects (20-55 years) were examined by routine medical history taking. All subjects underwent a complete medical history and routine laboratory tests. The cognitively-intact elderly subjects (>55 years) were additionally examined by administration of the Minimental examination (mean=27.64±0.22) and were selected for the study only in the absence of any indications of dementia. Patients (n=19) classified as having probable Alzheimer's disease met DSM III (R) criteria. The average score on the Minimental test was 14.51±1.92 and the average duration of dementia was 55.9±6.2 months.

RESULTS

The amount of the two principal isoforms of APP mRNA present in lymphocytes (APP_{751} and APP_{770}) was determined by the method of Horikoshi and his colleagues[4] who developed a revised approach to competitive quantitative PCR. The key features of this method are that (i) the relative, rather than the absolute, levels of gene expression are determined by comparing the ratio of PCR products generated by amplification of the target DNA segment and an endogenous internal standard gene (β-actin) in separate reactions and that (ii) linear amplification regions are determined simply by serial dilution of the cDNA (first strand) synthesis reaction sample, without the need for quantifying input RNA. The shorter isoform APP_{695} mRNA is not detectable by the relatively short exposure times employed in the present study. This species represents only a small fraction of the total APP mRNA in this lymphocyte preparation and as reported previously,[5] the 751 and 770 transcripts are ~10-fold more abundant than the 695 transcript in lymphocytes.

We observed a moderate but significant (r=0.293, p=0.019, Pearson) age-associated increase in the relative (compared to β-actin) amount of APP_{751} mRNA in lymphocytes

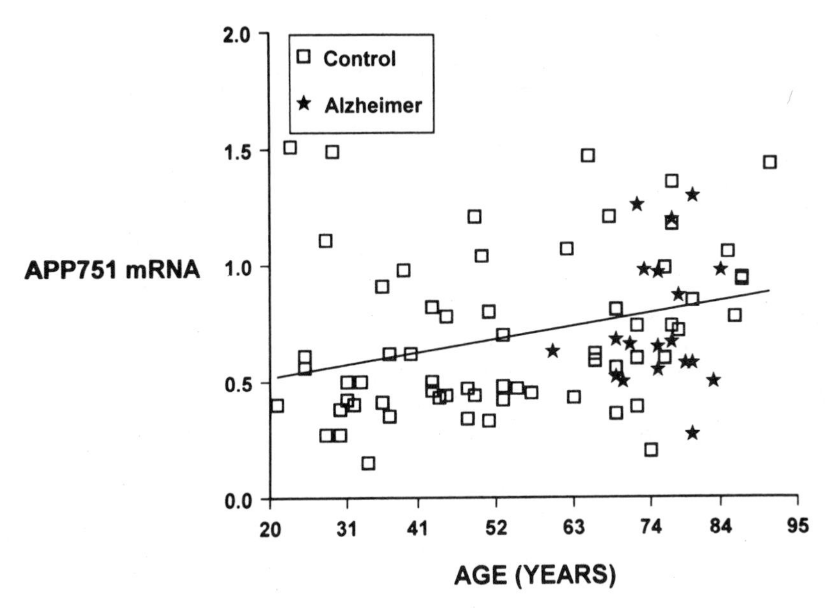

FIGURE 1. Relative changes (compared to β-actin) in lymphocyte APP_{751} mRNA levels during normal aging and in Alzheimer's disease.

obtained from 64 cognitively-intact subjects ranging in age from 20 to 91 years of age (Fig 1). No significant difference was observed between the relative APP_{751} mRNA levels in 19 Alzheimer's disease patients and the 27 cognitively-intact older (> 55 years) subjects.

DISCUSSION

The current study demonstrates an age-associated increase in lymphocyte APP_{751} mRNA obtained from cognitively-normal subjects. No difference was observed between the relative amounts of this mRNA isoform in lymphocytes from AD patients and control subjects > 55 years of age. The AD subjects show an increase in the amount of the major lymphocyte isoform which is consistent with their biological age.

It is of interest to compare the findings in the present study that demonstrate an increase in lymphocyte APP_{751} mRNA in normal aging to previous investigations of APP mRNA in human brain. Although a number of studies of APP mRNA isoform levels in postmortem brain tissue from sporadic AD patients (in comparison to normal brain) have been carried out, these investigations have yielded conflicting results.[6-9] Absence of consensus among these various reported postmortem measurements of APP mRNA isoforms may be due to variable changes in APP mRNA following brain death as well as methodological differences.

Several recent investigations[10-12] have analyzed APP mRNA and APP protein synthesis in fibroblasts isolated from patients with familial (autosomal dominant) Alzheimer's disease (FAD). These studies are useful in assessing the relevancy of the current findings in lymphocytes to accumulations of amyloid in normal aging and sporadic AD. The results reported by Querfurth and his colleagues[10] demonstrate that changes in APP mRNA levels are also reflected in changes in APP protein levels. In the current study, only mRNA levels were examined and the observed changes are meaningful only in so far as they reflect changes in APP protein. The results reported by Querfurth and his colleagues[10] suggest that this is not an unreasonable assumption. Moreover, the results reported by Querfurth and his colleagues[11] demonstrate that changes in fibroblast APP mRNA levels are most likely reflected in similar changes in brain mRNA levels for the APP protein since cosegregation of fibroblast APP mRNA synthesis is oberved with haplotypes in the 14q24.3 region which has been demonstrated by linkage methods to be associated with AD in the FAD-1 pedigree. Similarly, it may not be an undue speculation to suggest that the currently reported changes in lymphocyte APP mRNA levels also occur during in the brain during the course of normal aging and may contribute to the overall amyloid load in some individuals. Further studies are required to determine how ubiquitous are age-associated changes in APP mRNA and protein accumulation.

Querfurth and his colleagues[10] also examined fibroblasts obtained from Down's syndrome patients and found an approximately two-fold increase in APP mRNA attributed to the duplication of the APP gene in trisomic patients. Increased APP mRNA and protein synthesis in DS is a critical antecedent in the development of senile plaques in these patients and strengthens the argument that even relatively moderate variations (two-fold) in APP transcription rates can have serious consequences in contributing to the amyloid load and subsequent disease development.

In addition to the role of amyloidosis in neuronal degeneration, it is worth considering that changes in lymphoycte APP levels associated with normal aging and disease state may also influence the immunological competence of these cells. In both normal aging and AD there is some evidence for immunological dysfunction similar to those found in Down's syndrome.[13] These immune abnormalities may be due to abnormal expression or processing of APP in lymphocytes perhaps due to excessive accumulation of APP occurring during the normal course of human aging.

Although the average changes for each of the three cohort groups (<55, >55 years of age and AD) in APP_{751} mRNA isoform levels observed in the current study are not large (<55 years 0.61±0.054, n=37 vs >55 years 0.79±0.07, n=27 Student's t=2.22, p=0.03; AD 0.76±0.07, n=19) much greater individual differences between subjects are observed for the relative amounts of lymphocyte APP_{751} mRNA. For example in the <55 year old cohort group, a 10 fold variation in relative APP_{751} mRNA levels was observed (min=0.15, max=1.51). The marked individual differences in amount of APP_{751} mRNA isoforms observed among the subjects in the cognitively-intact group suggests the possibility that for some individuals increased accumulation of APP associated with normal aging might be an additional risk factor in developing AD.

In summary, various factors including mutation events, aging, and stress have now been shown to alter the amounts of APP mRNA isoforms as well as APP protein levels in peripheral cells and may reflect similar events occurring in brain. Studies using human peripheral cells offer a convenient model system for evaluating the possible role played by changes in APP mRNA and protein levels in contributing to the development of sporadic AD and are deserving of further investigation.

Acknowledgements

The authors would like to thank the Joseph Levy Charitable Foundation (London), the Heinz and Anna Kroch Foundation (London) and the Shapiro-Grover-Leff-Simon Families (Montreal) for generous financial support. Dr. Lubov Nemanov is a recipient of a fellowship from the Israeli Ministry of Absorption (Scientific Section).

REFERERENCES

1. T.E. Golde, T.E., S. Estus, M. Usiak, L.H. Younkin, and S.G. Younkin, Expression of beta amyloid protein precursor mRNAs: Recognition of a novel alteratively spliced form and quantitation in Alzheimer's disease using PCR. Neuron 4: 253 (1990).

2. J.P. Anderson, F.S. Esch, P.S. Keim, K. Sambamurti, I. Lieberburg, and N.K. Robakis, Exact cleavage site of Alzheimer amyloid precursor in neuronal PC-12 cells. Neurosc Lett 128: 126 (1991)
3. S. Itagaki, P.L. McGeer, H. Akiyama, S. Zhu, and D.J. Selkoe, Relationship of microglia and astrocytes to amyloid deposits of Alzheimer's disease. J Neuroimmunol 24: 173 (1989)
4. T. Horikoshi, K.D. Danenberg, T.H.W. Stadlbauer, M. Volkenandt, L.C.C. Shea, K. Aigner, B. Gustavsson, L. Leichman, R. Frosing, M. Ray, N.W. Gibson, C.P. Spears, and P.V. Danenberg, Quantitation of thymidylate synthetase, dihydrofolate reductase and DT-diaphorase gene expression in human tumors using the polymerase chain reaction. Cancer Research 52: 108 (1992).
5. S. Ledoux, N. Rebai, A. Dagenais, I.T. Shaw, J. Nalbantoglu, R.P. Sekaly, and N.R. Cashman, Amyloid precursor protein in peripheral mononuclear cells is upregulated with cell activation. J Immunol 150: 5566 (1993).
6. S.A. Johnson, T. Mcneill, B. Cordell, and C.E. Finch, Relation of neuronal APP-751/APP-695 mRNA ratio and neuritic plaque density in Alzheimer's disease. Science 248: 854 (1990).
7. G. Konig, J.M. Salbaum, O. Wiestler, W. Lang, H.P. Schmitt, C.L. Masters, and K. Beyreuther, Alternative splicing of the βA4 amyloid gene of Alzheimer's disease in cortex of control and Alzheimer's disease patients. Mol Brain Res 9: 259 (1991).
8. S.Tanaka, L. Liu, J. Kimura, S. Shiojiri, Y. Takahashi, N. Kitaguchi, S. Nakamura, and K. Ueda , Age-related changes in the proportion of amyloid precursor protein mRNAs in Alzheimer's disease and other neurological disorders. Mol Brain Res 15: 303 (1992).
9. S.Tanaka, S. Nakamura, K. Ueda, M. Kameyama, S. Shiojiri, Y. Takahashi, N. Kitaguchi, and H. Ito, Three types of amyloid protein precursor mRNA in the human brain: their differential expression in Alzheimer's disease. Biochem Biophys Res Comm 157: 472 (1988).
10. H.W. Querfurth, E.M. Wijsman, P.H. St George-Hyslop, and D.J. Selkoe βAPP mRNA transcription is increased in cultured fibroblasts from the familial Alzheimer's disease-1 family. Mol Brain Res 28: 319 (1995).
11. M. Citron, C. Vigo-Pelfrey, D.B. Teplow, C. Miller, D. Schenk, J. Johnston, B. Winblad, N. Venizelos, L. Lannfelt, and D. Selkoe, Excessive production of amyloid beta-amyloid protein by peripheral cells of symptomatic and presymptomatic patients carrying the Swedish familial Alzheimer disease mutaion. Proc Natl Acad Sci (USA) 91: 11993 (1994).
12. J.A. Johnston, R.F. Cowburn, S. Norgren, B. Wiehager, N. Venizelos, B. Winblad, C. Vigo-Pelfrey, D. Schenk, L. Lannfelt, and C. O'Neill Increased β-amyloid release and levels of amyloid precursor protein (APP) in fibroblast cell lines from family members with the Swedish Alzheimer's disease APP670/671 mutation FEBS Lett 354: 274 (1994).
13. V.K.Singh, H.H. Fudenberg, and F.R. Brown III, Immunologic dysfunction: simultaneous study of Alzheimer's and older Down's syndrome patients. Mech Ageing Dev 37: 257 (1987).

DIFFERENTIAL REGULATION OF APP SECRETION BY APOLIPOPROTEIN E3 AND E4

Benjamin Wolozin[1], Jasna Basaric-Keys[1], Robert Canter[1], Yunhua Li[1], Dudley Strickland[2] and Trey Sunderland[1]

[1]Section on Geriatric Psychiatry, NIMH, Bethesda MD 20892 and [2]American Red Cross Labs, Rockville, MD 20855

INTRODUCTION

Apolipoprotein E (apo E) is a cholesterol transport protein that is present both in blood as well as in central nervous system, where its levels increase following nerve injury. Presumably the function of apo E is to facilitates lipid uptake [1]. In the periphery, apo E is synthesized by the liver and taken up throughout the body via low density lipoprotein receptors [1]. In the central nervous system, apo E is synthesized by astrocytes and microglial cells following nerve injury [1]. Apo E avidly binds lipids, in particular cholesterol, and it is taken up by the axon during regeneration [2]. Thus, the action of apo E plays an important role in the ability of the brain to cope with neuronal injury. This role may also be critical in the ability of the brain to cope with a neurodegenerative illness, such as Alzheimer's disease.

There are three different forms of apo E: apo E2, apo E3 and apo E4. These differ by only two amino acids, residues 112 and 158. Apo E2 has cys - cys at positions 112 and 158 respectively; apo E3 is cys - arg, and apo E4 is arg - arg. Recent studies indicate that the allele coding for apo E4, e4, is an important risk factor for Alzheimer's disease [3]. In some families, the risk for Alzheimer's Disease (AD) increases from 20% to 90% for individuals with increasing numbers of e4 alleles [4]. The increased risk of AD appears to stem from abnormal association of the aß peptide with apo E4. Studies of Alzheimer pathology show that Apo E4 accumulates with aß in neuritic plaques [5, 6]. In addition, *in vitro* studies show that, in high micromolar range, apo E and aß can bind to each other [3, 5, 7]. This association appears to promote the aggregation of aß, which has been shown to be toxic at high levels [8, 9]. Hence the term 'pathologic chaperone' has been used to describe one possible role of apo E in Alzheimer's pathophysiology [8].

Apo E interacts with cells through specific receptors that recognize the apo E/lipid complex. The principle apo E receptor in peripheral cells, such as hepatocytes, is the LDL receptor [1]. In contrast, the predominant apo E receptor on neurons is the LDL receptor related protein, LRP, which is abundant in the brain [6]. Upon binding to LRP, the apo E/LRP complex is internalized where the lipids are used by the neuron. The signal transduction capabilities of LRP are unknown, although it appears that apo E does have the capacity to modulated signal transduction systems such as cAMP-related transcriptional activity [10]. The action of LRP, in turn, is also regulated. The LRP receptor associated protein, RAP regulates LRP function by controlling its internalization; in the presence of RAP, LRP is not internalized [11]. Thus, the apo E/ receptor complex not only regulates other processes, but is itself tightly regulated.

The importance of apo E in neuronal function and the complex receptor biology for apo E suggests that the biology of apo E is indeed complicated. The association between aß and

Neurodegenerative Diseases
Edited by Gary Fiskum, Plenum Press, New York, 1996

apo E might be part of a physiologic pathway by which apo E controls neuronal. Similarly, the mechanism by which the association of apo E and aß leads to an increased risk for Alzheimer's disease may be more detailed than the simple formation of a toxic aggregated aß product. As a preliminary step in understanding the relationship between apo E, aß and neuronal function, we have studied the regulation of aß and its precursor, amyloid precursor protein (APP) by apo E in order to determine whether there is a regulatory connection. We find that apo E increases the secretion of both APP. Moreover, the apo E3 and E4 isozymes show significant differences in their profiles of action.

METHODS

Cell culture: PC12 cells were grown on Matrigel (Collaborative Research) coated dishes with RPMI containing 10% bovine fetal serum and 5% horse serum. All studies were done on subconfluent cells - confluent cells did not respond to apo E. Prior to treatment with apo E, the cells were transferred to DMEM and incubated overnight. Then the apo E was added at the concentrations and times as described. Apolipoprotein E3 and E4 were produced recombinantly (Pan Vera, Madison, WI). The lipids used in the study were obtained from a lipid concentrate (Gibco/BRL) solution containing cholesterol, cod liver oil, pluronic F68 and DL-a-tocopherol-acetate.
Analysis of secreted APP: Following treatment of the cells, the medium was removed and spun at 14,000 rpm for 15 min. APP was then isolated from the supernatant using agarose-coupled heparin (BioRad). For the heparin isolations, 200 μl heparin/agarose was added to each sample of medium and incubated 1 hr at 4°C; the amount of heparin used was double that required to bind all the APP in the presence or absence of apo E as judged by titration experiments (data not shown). The heparin/APP complexes were then spun down at 10,000 rpm for 5 min and washed three times with PBS/0.05%Tween. Following the final spin, 40 μl of gel loading buffer containing 5% ß-mercaptoethanol was added, the samples were boiled 8 min and run electrophoresed on 10% tris/glycine gels. Immunoblots were performed with antibodies to the extracellular domain of APP, 22C11 (Boehringer-Mannheim).
Analysis of apo E binding: Following incubation of PC12 cells with apo E, the cells were washed twice with PBS, scraped into a buffer containing protease inhibitors, ultrasonicated and frozen. Following harvesting, protein concentrations were determined by a Bradford assay (BioRad) and the samples were electrophoresed on 10% tris/glycine gels using 30 μg protein per lane transferred to nitrocellulose. Immunoblots were performed using anti-apolipoprotein E antibody (Calbiochem).

RESULTS

Apo E Increases APP_S Levels

In order to study the action of apo E, PC12 cells were grown at subconfluent levels and then incubated overnight in serum free medium in order to lipid starve the cells and remove any residual apo E or lipid present in the serum. Apo E2, 3 or 4 with or without lipid concentrate was then added to the cells and incubated for 1 - 24 hrs. The secreted APP (APP_S) was then isolated with heparin-agarose and immunoblotted using 22C11 as described in the Methods section. While the cells responded to apo E in both lipid rich and lipid free conditions, the response of the cells in the presence of lipids was both much greater and more consistent than the response in the absence of lipids (fig. 1). Densitometric quantitation of immunoblots showed that, in the presence of lipids, 10 μg/ml apo E 2-4 increased APP_S levels by 110 ± 5%, with all isoforms being equally active. In contrast, in absence of lipids, apo E increased APP_S levels by only 43%. Thus, apo E exerts a powerful regulatory influence over APP secretion.

Interestingly, while the three forms of apo E were equally active at the highest doses examined, apo E3 and E4 showed distinctly different dose response curves (fig. 1). While the increase in APP secretion in response to apo E3 is monophasic, the action of apo E4 is biphasic, decreasing APP_S levels at low doses of apo E4 and increasing APP_S levels at higher doses of apo E4 (fig. 1). Thus, while apo E3 and E4 are equally active at high doses, the two isoforms differ greatly in action at lower doses.

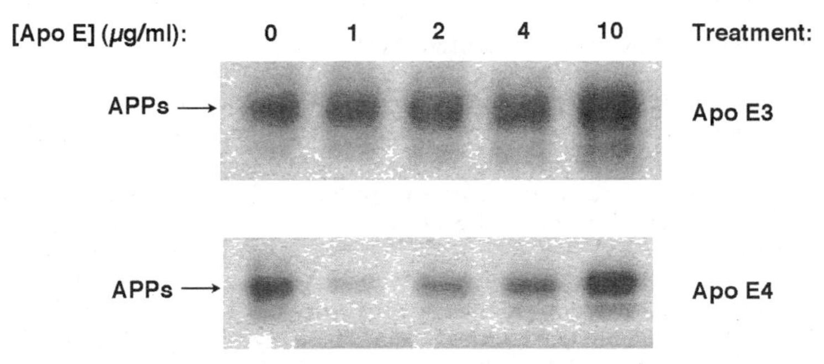

Figure 1. Apo E3 and E4 modulate APP secretion. Increasing doses of apo E3 increased APP secretion, while increasing doses of apo E4 had a biphasic effect on APPs. At low doses (1-2 μg/ml), apo E4 decreased APPs, while higher doses of apo E4 increased APPs.

Differential Kinetics of Apo E Binding

In order to begin to understand the mechanism underlying the differences in activity among the apo E isozymes, the cellular binding of apo E was analyzed. Apo E was added to serum starved PC12 cells and incubated at varying doses or for varying lengths of time. The cells were then washed, lysed and the cell lysates were immunoblotted with anti-apo E antibody. Dose response curves showed increases in apo E binding between 1 - 10 μg/ml, however the dose response profile for apo E3 and E4 were similar. In contrast, the kinetics of apo E binding were significantly different. Apo E3 achieved maximal binding within 30 min of addition and the apo E was gone eliminated 6 hrs after addition (fig 2 A). Apo E4, on the other hand, did not achieve maximal

binding until 1-3 hrs after binding and significant amounts of apo E4 were still detectable 6 hrs after addition of apo E4 (fig 2B). For both species, though, the apo E was almost completely eliminated 24 hrs after addition of apo E. Given that elimination of apo E occurs via internalization of the apo E and subsequent degradation, these data suggest that apo E4 is internalized at a slower rate the apo E3.

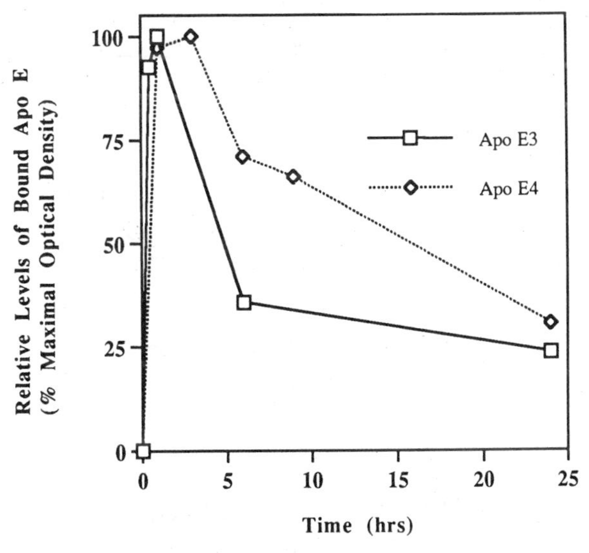

Figure 2. Time course of apo E3 and E4 binding to PC12 cells. Apo E3 or E4 (5 μg/ml) was incubated with PC12 cells for varying periods of time. The washed lysates were then immunoblotted with anti-apo E antibody and quantitated using video desitometry. While both apo E3 and E4 bound rapidly to the cells, apo E4 was eliminated more slowly, presumably due to decreased internalization.

Cellular Binding of Apo E is Mediated by LRP

Studies using RAP showed that the binding of apo E on the PC12 cells is mediated by the LRP. Previous studies have shown that under basal conditions PC12 cells have very little LDL receptor [12]. If cellular binding of apo E is mediated via the LRP, internalization of the apo E/LRP complex can be prevented by RAP. In order to investigate this, we incubated cells with apo E3 or E4 for 1.5 hrs with or without 1 μM RAP. The cells were then washed with either PBS, to observe total binding, or 0.1 M glycine, pH 2.5 in order to observe internalized apo E. As shown in figure 3A, RAP increased the total binding of both apo E3 and E4. This increase in binding results from the prevention of LRP-mediated internalization and subsequent degradation of apo E in the lysosome. Thus, apo E binding is stabilized by RAP. The prevention of internalization could be directly observed after using the pH 2.5 glycine wash, which strips away any surface bound apo E leaving only internalized apo E. Under these conditions, application of RAP greatly reduced the amount of APP detected indicating that RAP was preventing the internalization of the apo E (data not shown). Thus, cellular apo E binding is mediated by the LRP and internalization of the apo E/LRP complex can be prevented by concurrent incubation with RAP.

Further studies with RAP showed that the apo E-related increases in APP_S were mediated by LRP. Incubation of serum starved PC12 cells with apo E3 increased APP_S secretion, however the extent of this increases could be reduced by concurrent treatment with RAP (fig 3B). This indicates that the increase in APP_S elicited by apo E is mediated by the LRP.

DISCUSSION

The apo E4 isozyme is an important risk factor for Alzheimer's disease. At least part of the risk appears to arise from the ability of apo E4 to enhance aß aggregation [8]. The ability of apo E to bind aß raises the question of whether there is a physiologic link between the two proteins. We have now shown that apo E is a potent regulator of APP secretion. Previously, we have shown that apo E modulates APP secretion when added without addition of exogenous lipids, however the present studies were done in the presence of lipids because this mimics physiologic conditions better [13]. Addition of exogenous lipids improved the response of Apo E, both increasing the size of the response and increasing the consistency of the results. We have now observed that additIn the presence of lipids, both apo E3 and E4 were capable of increasing APP secretion. Using RAP to inhibit internalization of bound apo E, we were able to show that apo E exerts its effect on APPs via the LDL receptor related protein (LRP). Co-incubation of RAP with apo E blocked any

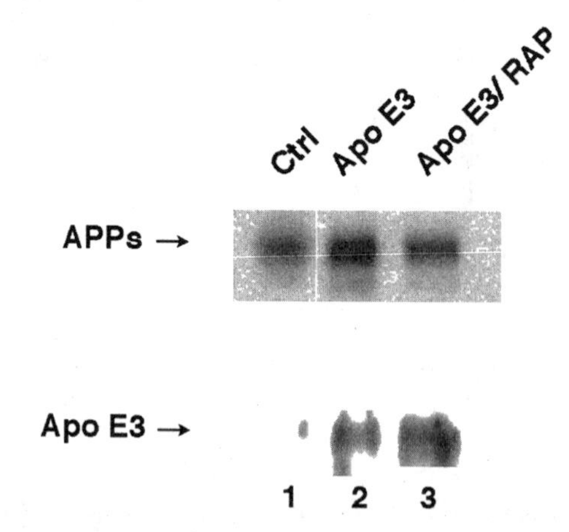

Figure 3. RAP inhibits the internalization of apo E and the apo E-induced increases in APP secretion. PC12 cells were treated with apo E3 (10 μg/ml) or apo E3 + RAP (1 μg/ml). RAP prevented any changes in APPs levels induced by apo E3 and it increased the levels of bound apo E3, presumably by preventing internalization of apo E.

changes in APP secretion. Thus, the mechanism of action of apo E in undifferentiated PC12 cells proceeds via binding to LRP.

The ability of apo E to regulate APPs suggests that there is a physiological link between apo E and APP. This finding supplements previous work showing that both apo E and APP are important in neurite outgrowth. One possible mechanism by which apo E contributes to neurite outgrowth is by providing lipids to facilitate the rapid increase in membrane content as the neurite grows. The mechanism by which APP facilitates neurite outgrowth is currently unknown, but it may involve stimulation cellular adhesion perhaps via a tyrosine phosphorylation-linked mechanism [14, 15]. By stimulating of APP secretion, apo E is therefore promoting neurite outgrowth consistent with its previously proposed function. The ability of apo E to stimulate APP secretion, though, suggests that apo E is acting as more than just a lipid transport protein; it appears to be modulating the levels of other proteins that are involved in neurite outgrowth. Thus, apo E may be both a lipid transport protein and a growth factor.

Interestingly, the actions of apo E3 and E4 differ markedly. For instance, while apo E3 increases APP secretion in a monophasic manner, with increasing apo E levels producing increased APP secretion, apo E4 displays a biphasic dose response pattern. At low doses (1-2 µg/ml) apo E4 decreases APP secretion, while higher doses of apo E4 (10 µg/ml) increase APP secretion. The proteins also display different binding kinetics. Following binding, apo E3 is rapidly internalized, so that bound apo E3 has largely been removed 3 hrs after addition of the protein. In contrast, apo E4 is internalized slowly and is readily detectable on the PC12 cells for more than 6 hrs after initial application. We are only now beginning to explore the effects of these differences in behavior however they could affect cellular functioning. Recent studies have shown that apo E3 and E4 have opposing actions on dorsal root ganglion cells, with apo E3 promoting neurite extension and apo E4 inhibiting neurite extension [16]. While the prevailing view is that the increased risk for Alzheimer's disease associated with the apo E4 isozyme derives from the increased tendency of apo E4 to promote aggregation of aß, the differential actions of apo E3 and E4 on neurite extension provide an alternative model accounting for the association of apo E4 with Alzheimer's disease [8]. Perhaps these opposing actions derive from the differential actions of apo E3 and E4 on PC12 cells. The inhibition of APP secretion by low doses of apo E4 may also reflect the inhibition of particular molecular pathways important to neurite extension. The identification of LRP as the mediator of the apo E signals in PC12 cells provides a useful system for investigating these questions.

REFERENCES

1. R. Mahley, (1988): Apolipoprotein E: Cholesterol transport protein with expanding role in cell biology. Science 240:622-30.

2. J. Boyles, C. Zoellner, L. Anderson, L. Kosik, R. Pitas, K. Weisgraber, D. Hui, R. Mahley, P. Gebicke-Haerter, M. Ignatius and E. Shooter, (1989): A role for apolipoprotein E, apolipoprotein A-1 and low density lipoprotein receptors in cholesterol transport during regeneration and remyelination of the rat sciatic nerve. J. Clin. Invest. 83:1015-31.

3. W. Strittmatter, A. Saunders, D. Schmechel, M. Pericak-Vance, J. Enghild, G. Salvesen and A. Roses, (1993): Apolipoprotein E: high-avidity binding to ß-amyloid and increased frequency of type 4 allele in late-onset familial Alzheimer disease. PNAS 90:1977-81.

4. E. Corder, A. Saunders, W. Strittmatter, D. Schmechel, P. Gaskell, G. Small, A. Roses, J. Haines and M. Pericak-Vance, (1993): Gene dose of apolipoprotein E type 4 allele and the risk of Alzheimer's disease in late onset families. Science 261:921-3.

5. D. Schmechel, A. Saunders, W. Strittmatter, B. Crain, C. Hulette, S. Joo, M. Pericak-Vance, D. Goldgaber and A. Roses, (1993): Increased amyloid ß-peptide deposition in cerebral cortex as a consequence of apolipoprotein E genotype in late-onset Alzheimer disease. PNAS 90:9649-53.

6. G. Rebeck, J. Reiter, D. Strickland and B. Hyman, (1993): Apolipoprotein E in sporadic Alzheimer's Disease: allelic variation and receptor interactions. Neuron 11:575-80.

7. W. Strittmatter, K. Weisgraber, D. Huang, L. Dong, G. Salvesen, M. Pericak-Vance, D. Schmechel, A. Saunders, D. Goldgaber and A. Roses, (1993): Binding of human apolipoprotein E to synthetic amyloid beta peptide: isoform-specific effects and implications for late-onset Alzheimer disease. PNAS 90:8098-102.

8. J. Ma, A. Yee, H. Brewer, S. Das and H. Potter, (1994): Amyloid-associated proteins alpha(1)-antichymotrypsin and apolipoprotein E promote assembly of Alzheimer beta-protein into filaments. Nature 372:92-94.

9. D. Loo, A. Copani, C. Pike, E. Whittemore, A. Walencewicz and C. Cotman, (1993): Apoptosis is induced by ß-amyloid in cultured central nervous system neurons. PNAS 90:7951-5.

10. M. Reyland and D. Williams, (1991): Suppression of cAMP-mediated signal transduction in mouse adrenocortical cells which express apolipoprotein E. J. Biol. Chem. 266:21099-104.

11. T. Willnow, Z. Sheng, S. Ishibashi and J. Herz, (1994): Inhibition of hepatic chylomicron remnant uptake by gene transfer of a receptor antagonist. Science 264:1471-4.

12. M. Ignatius, E. Shooter, R. Pitas and R. Mahley, (1987): Lipoprotein uptake by neuronal growth cones in vitro. Science 236:959-62.

13. B. Wolozin, J. Basaric-Keys, R. Canter, Y. Li, D. VanderPutten and T. Sunderland, (1995): Differential regulation of APP secretion by apolipoprotein E3 and E4. Ann. New York Acad Sci. in press:

14. D. Schubert, L. W. Jin, T. Saitoh and G. Cole, (1989): The regulation of amyloid ß protein precursor secretion and its modulatory role in cell adhesion. Neuron 3:689-694.

15. S. Greenberg, E. Koo, D. Selkoe, W. Qiu and K. Kosik, (1994): Secreted ß-amyloid precursor protein stimulates mitogen-activated protein kinase and enhances tau phosphorylation. Proc. Natl. Acad. Sci. (USA) 91:7104-8.

16. B. Nathan, S. Bellosta, D. Sanan, K. Weisgraber, R. Mahley and R. Pitas, (1994): Differential effects of apolipoprotein E3 and E4 on neuronal growth in vitro. Science 264:850-2.

APOLIPOPROTEIN E UPTAKE IS INCREASED BY BETA-AMYLOID PEPTIDES AND REDUCED BY BLOCKADE OF THE LDL RECEPTOR

Uwe Beffert[1], Nicole Aumont[1], Doris Dea[1], Jean Davignon[2] and Judes Poirier[1]

[1]Douglas Hospital Research Centre
McGill Centre for Studies in Aging
Department of Neurology and Neurosurgery
McGill University
[2]Clinical Research Institute of Montreal
Montreal, Quebec, Canada H4H 1R3

INTRODUCTION

Apolipoprotein E (apoE) is a lipid-binding, 37 kDa, 299 amino acid glycoprotein involved in cholesterol and phospholipid transport and metabolism. ApoE mediates the uptake of lipid complexes through interaction with the apoB/ApoE (LDL) receptor and other receptors[1]. The LDL receptor pathway consists of cell surface binding of apoE-containing lipoproteins followed by internalization and degradation of the lipoprotein by a lysosomal pathway[2]. In the peripheral nervous system, apoE has been proposed to be involved in the transport of cholesterol in repair, growth and maintenance of membranes during development or following injury[3,4]. In the central nervous system, ApoE released from astrocytes in response to injury such as entorhinal cortex lesions in the rat plays a pivotal role in the redistribution of cholesterol and phospholipids during synaptic remodeling and compensatory synaptogenesis[5,6].

The structural gene for apoE is polymorphic, leading to three major apoE isoforms, designated apoE2, apoE3 and apoE4, which differ by amino acid substitutions at one or both of two sites, residues 112 and 158[7]. The E2 isoform has Cys residues at site 112 and 158, E3 has a Cys residue at 112 and Arg at site 158 whereas E4 has Arg at both sites. Individuals may therefore be homozygotes (ϵ2/ϵ2, ϵ3/ϵ3 or ϵ4/ϵ4) or heterozygotes (ϵ2/ϵ3, ϵ2/ϵ4 or ϵ3/ϵ4). Recently, the frequency of the apoϵ4 allele was shown to be significantly increased in sporadic and late-onset familial cases of Alzheimer's disease (AD)[8-11]. ApoE has also been immunohistochemically localized to extracellular senile plaques, intracellular neurofibrillary tangles, and vascular amyloid, the neuropathological lesions characterisitic of AD[12,13]. Furthermore, a correlation has also been demonstrated between senile plaque density and apoϵ4 allele copy number in the hippocampus[14] and cortical areas[15,16] of AD subjects, with ϵ4 homozygotes having the greatest density of plaques. *In vitro*, the beta-amyloid (Aβ) peptide,

Neurodegenerative Diseases
Edited by Gary Fiskum, Plenum Press, New York, 1996

originally isolated from AD neuritic plaques, has been shown to avidly bind apoE[17]. To further understand the mechanism by which apoE may be linked to amyloid and AD, we have investigated apoE and amyloid interactions in a primary cell culture model.

MATERIALS AND METHODS

Primary hippocampal neuronal cultures were obtained from the hippocampus of embryonic (E17-E18) Long-Evans rats as described by Mitchell[18]. Briefly, the brains of the embreyos were removed, the hippocampus dissected and placed in Hanks Balanced Salt Solution (HBSS) containing 15 mM HEPES. The tissue was incubated with 0.25% trypsin for 10 min. @ 37°C and then rinsed several times in Dulbecco's Modified Eagle Medium (DMEM) containing 10% fetal calf serum, penicillin/streptomycin and fungizone. Neurons were plated in poly-D-lysine coated culture wells at a density of 10^6 cells per 16 mm^2 well. After 24 hours, the media was changed to serum-free and supplemented with Gibco neuronal growth supplement B-27. The neurons were treated with exogenous apoE/amyloid on the sixth day after plating.

Human apoE3 was purified from the plasma of pregenotyped apoE-homozygous individuals as previously described by Brissette[19]. Briefly, 10 μM phenylmethylsulfonyl fluoride, 0.3 mM EDTA and 3 mM sodium azide were added to all sera and then ultracentrifuged at 8.5×10^4 *g* for 30 min. at 16°C and then twice at 1.8×10^6 *g*-h to isolate and wash the VLDL. An equal volume of isopropanol was added to a 1.5 mg/ml VLDL solution while agitating vigorously and then centrifuged at 10,000 *g* for 10 min. The isopropanol was evaporated under a jet of nitrogen and acetone (1.2 times the volume) was added to the remaining solution and centrifuged at 20,000 *g* for 20 min. The pellet containing apoE was resuspended in NaCl (150 mM) and the acetone extraction was repeated. Solvent-extracted apolipoproteins were delipidated by extracting two times with chloroform/methanol (2:1), each time followed by centrifugation at 20,000 *g* for 20 min. Delipidated proteins were dissolved in Tris-HCl buffer (10 mM; pH 7.4) containing NaCl (150 mM), EDTA (0.3 mM) and urea (5.0 M). Solutions were then centrifuged at 10,000 g for 10 min. and then dialyzed against the same buffer containing no urea.

Apolipoprotein E3-containing liposomes were prepared according to the method of Matz and Jonas[20] with the following conditions: 100 mol of phosphatidylcholine (PC)/100 mol cholate/25 mol cholesterol/mol of apolipoprotein E. The mixtures were incubated for 12 hours @ 4°C and were dialyzed for 48 hours with four changes of HCl (10 mM) buffer pH 7.4 containing NaCl (150 mM) and EDTA (0.3 mM). ApoE3-liposomes were isolated via centrifugation at 3×10^6 *g*-h at a density of 1.125 g/ml and the isolated fraction was dialyzed against a NaCl (150 mM) solution pH 7.5 containing 0.3 mM EDTA and 3 mM sodium azide.

To block uptake of apoE into the neurons, anti-LDL receptor antibody (1 μg/ml; Amersham Int'l, UK) was diluted into the media of the neurons for 2 hours prior to apoE and amyloid treatment. Human apoE3 (1 μg/ml final) was incubated in the presence and absence of amyloid peptide fragments 1-28, 25-35 and 1-40 (1 μg/ml final; Bachem, CA) for two hours at 37°C. The apoE and amyloid was then diluted to final concentration in the neuronal media for 24 hours. The next day, medium from each condition was collected and the cells were washed three times with ice-cold PBS. Cell viability was assessed by measuring lactate dehydrogenase activity (LDH kit; Sigma) from the medium. The cells were removed from the petries, placed into eppendorf tubes and sonicated briefly to disrupt the membranes. Samples were spun down, aliquoted and stored at –80°C for protein assay and Western experiments.

Using the method of Laemmli[21], neuronal homogenates (25 μg) were mixed with 5X reducing sample buffer, boiled for one minute and run on Novex 12% tris-glycine precast gels

(Helixx Technologies, Scarborough, Canada) for SDS-PAGE. Proteins were electrophoretically transferred onto nitrocellulose membranes (Amersham, Oakville, Canada), dried and then blocked for one hour at room temperature (RT) with 5% dried milk in TBS-T (Tris buffered saline with 0.1% Tween 20 pH 7.6). Membranes were washed briefly in TBS-T and then probed with goat anti-human apoE antibody (International Immunology Corporation; Murrieta, CA) diluted 1:2000 in TBS-T for one hour at RT. Following three rinses in TBS-T, membranes were incubated with rabbit anti-goat biotinylated secondary antibody (Calbiochem) 1:10,000 in TBS-T for 20 min., rinsed again, and incubated for another 20 min. with streptavidin-horseradish peroxidase conjugate (Amersham) 1:5000 in TBS-T. After washing for one hour in TBS-T, membranes were detected via Amersham's Enhanced Chemiluminescence (ECL) kit and exposed to Kodak XAR5 film for 2 to 10 minutes.

RESULTS AND DISCUSSION

Primary neuronal cultures from rat were treated with human apoE3 in the presence and absence of amyloid peptides for a period of 24 hours. Figure 1 demonstrates the levels of apoE as detected by Western blotting in the homogenates collected from these cultures. As expected, no endogenous apoE was detected from untreated control cultures (fig. 1; Control) or from cells treated with micelles only (fig. 1; Micelles: liposomes consisting of phosphatidylcholine, cholate and cholesterol). Most of the apoE synthesized in the central nervous system has been localized to astrocytes and not neurons.

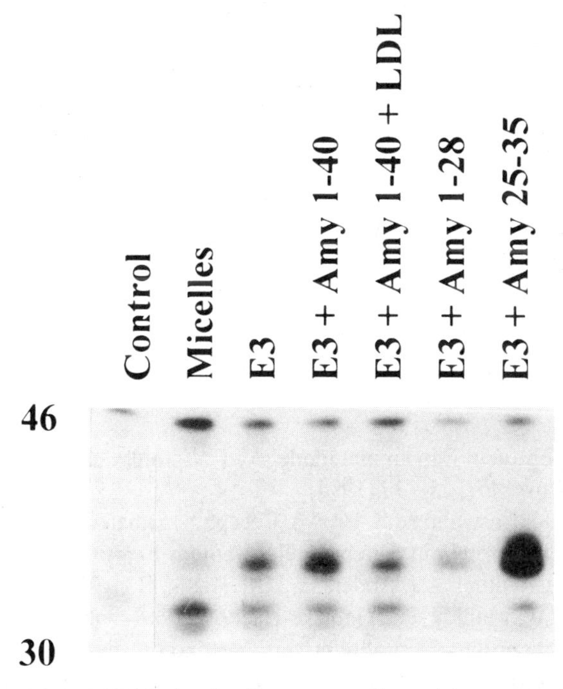

Figure 1. Western blot for apoE (37 kDa band) of homogenates from primary neuronal cultures, representing uptake of apoE following 24 hour treatment. Conditions include **Control**: no treatment, **Micelles**: equal volume of liposomes as for human apoE3 preparation, **E3**: 1 μg/ml human apoE3 micelles in media, **E3 + Amy 1-40**: 1 μg/ml apoE3 micelles and amyloid 1-40 which have been preincubated for 2 hrs at 37°C, **E3+Amy 1-40+LDL**: same as previous except cultures pretreated for 2 hrs with anti-LDL receptor antibody prior to apoE/amyloid incubation, **E3+Amy 1-28**: 1 μg/ml apoE3 micelles and amyloid 1-28, **E3+Amy 25-35**: 1 μg/ml apoE3 micelles and amlyoid 25-35. Numbers at left represent molecular weight markers at 30 and 46 kDa.

In the presence of human apoE3-containing liposomes (fig. 1; E3), Western detection revealed a significant uptake of apoE by the neurons as compared to control conditions. This indicates that rat neurons are indeed capable of uptake of the human apoE protein. Previous studies from our laboratory have demonstrated that free, purified human apoE3 is not internalized into neurons and requires the presence of a lipid partner. In the presence of amyloid 1-40 (fig. 1; E3 + Amy 1-40), the 37 kD band for apoE was significantly increased compared to apoE3 alone, indicating an enhanced uptake of apoE in the presence of amyloid 1-40. Amyloid 1-40 alone did not enhance uptake beyond control levels (data not shown).

Shorter amyloid peptide fragments 1-28 and 25-35 were also examined for alterations to apoE uptake. Amyloid 1-28 (fig. 1; E3 + Amy 1-28), in contrast to the 1-40 peptide showed no enhanced apoE uptake. Amyloid 25-35 (fig. 1; E3 + Amy 25-35) on the other hand demonstrated a marked increase of apoE signal compared to control apoE3 alone. These results demonstrate a sequence specific interaction of the amyloid peptide with human apoE3 and also demonstrates significant differences of interaction between these two peptides with regards to uptake by central nervous system neurons.

To verify whether the apoE uptake observed by Western analysis in this experiment was indeed occurring through the established LDL receptor pathway, cultures were pretreated with anti-LDL receptor antibody for 2 hours prior to apoE3 and amyloid treatment (fig. 1; E3 + Amy 1-40 + LDL). The blockade of the receptor by the LDL receptor antibody significantly reduced the amount of apoE taken up by the neurons. This is consistant with the fact that LDL receptors are known to be present on rat hippocampal neurons and that apoE is internalized through an LDL receptor pathway. Treatment with anti-LDL receptor and apoE3 alone also significanlty reduced the uptake of apoE (data not shown). Futhermore, the antibody by itself had no effect on endogenous rat apoE levels (data not shown).

Neurotoxicity of amyloid peptides is a major concern when using primary neurons as a model. LDH levels obtained from the media of the amyloid and apoE3 treated conditions revealed no significant increases compared to controls (data not shown). Protein assay results of the neuronal cell homogenates also displayed no significant variations from well to well, indicating no variations between cell number in each well.

The results presented here demonstrate the usefulness of primary neuronal cultures in examining the potential interactions of apolipoprotein E and beta-amyloid, two proteins with strong links to Alzheimer's disease. The data also suggest that amyloid could act as a regulator of apoE metabolism in neuronal cells.

REFERENCES

1. J. Poirier. Apolipoprotein E in animal models of CNS injury and in Alzheimer's disease, *Trends Neurosci* 17:525-530.(1994)
2. J.L. Goldstein, S.K. Basu and M.S. Brown. Receptor-mediated endocytosis of low-density lipoprotein in cultured cells, *Methods Enzymol* 98:241-260.(1983)
3. J.K. Boyles, C.D. Zoellner, L.J. Anderson, L.M. Kosik, R.E. Pitas, K.H. Weisgraber, D.Y. Hui, R.W. Mahley, P.J. Gebicke-Haerter, M.J. Ignatius and E.M. Shooter. A role for apolipoprotein E, apolipoprotein A-I, and low density lipoprotein receptors in cholesterol transport during regeneration and remyelination of the rat sciatic nerve, *J Clin Invest* 83:1015-1031.(1989)
4. J.K. Boyles, L.M. Notterpek and L.J. Anderson. Accumulation of apolipoproteins in the regenerating and remyelinating mammalian peripheral nerve. Identification of apolipoprotein D, apolipoprotein A-IV, apolipoprotein E, and apolipoprotein A-I, *J Biol Chem* 265:17805-17815.(1990)

5. J. Poirier, M. Hess, P.C. May and C.E. Finch. Astrocytic apolipoprotein E mRNA and GFAP mRNA in hippocampus after entorhinal cortex lesioning, *Brain Res Mol Brain Res* 11:97-106.(1991)
6. J. Poirier, A. Baccichet, D. Dea and S. Gauthier. Cholesterol synthesis and lipoprotein reuptake during synaptic remodelling in hippocampus in adult rats, *Neuroscience* 55:81-90.(1993)
7. J. Davignon, R.E. Gregg and C.F. Sing. Apolipoprotein E polymorphism and atherosclerosis, *Arteriosclerosis* 8:1-21.(1988)
8. J. Poirier, J. Davignon, D. Bouthillier, S. Kogan, P. Bertrand and S. Gauthier. Apolipoprotein E polymorphism and Alzheimer's disease, *Lancet* 342:697-699.(1993)
9. A.M. Saunders, W.J. Strittmatter, D. Schmechel, P.H. George-Hyslop, M.A. Pericak-Vance, S.H. Joo, B.L. Rosi, J.F. Gusella, D.R. Crapper-MacLachlan, M.J. Alberts, C. Hulette, B. Crain, D. Goldgaber and A.D. Roses. Association of apolipoprotein E allele epsilon 4 with late-onset familial and sporadic Alzheimer's disease, *Neurology* 43:1467-1472.(1993)
10. E.H. Corder, A.M. Saunders, W.J. Strittmatter, D.E. Schmechel, P.C. Gaskell, G.W. Small, A.D. Roses, J.L. Haines and M.A. Pericak-Vance. Gene dose of apolipoprotein E type 4 allele and the risk of Alzheimer's disease in late onset families [see comments], *Science* 261:921-923.(1993)
11. J. Nalbantoglu, B.M. Gilfix, P. Bertrand, Y. Robitaille, S. Gauthier, D.S. Rosenblatt and J. Poirier. Predictive value of apolipoprotein E genotyping in Alzheimer's disease: results of an autopsy series and an analysis of several combined studies, *Ann Neurol* 36:889-895.(1994)
12. Y. Namba, M. Tomonaga, H. Kawasaki, E. Otomo and K. Ikeda. Apolipoprotein E immunoreactivity in cerebral amyloid deposits and neurofibrillary tangles in Alzheimer's disease and kuru plaque amyloid in Creutzfeldt-Jakob disease, *Brain Res* 541:163-166.(1991)
13. T. Wisniewski, B. Frangione. Apolipoprotein E: a pathological chaperone protein in patients with cerebral and systemic amyloid, *Neurosci Lett* 135:235-238.(1992)
14. J. Poirier, I. Aubert, P. Bertrand, R. Quirion, S. Gauthier, J. Nalbantoglu. Apolipoprotein E4 and cholinergic dysfunction in Alzheimer's disease. In: Giacobini E, Becker R, eds. Advances in Alzheimer Therapy. Boston: Birkhauser, (1994)
15. D.E. Schmechel, A.M. Saunders, W.J. Strittmatter, B.J. Crain, C.M. Hulette, S.H. Joo, M.A. Pericak-Vance, D. Goldgaber and A.D. Roses. Increased amyloid beta-peptide deposition in cerebral cortex as a consequence of apolipoprotein E genotype in late-onset Alzheimer disease, *Proc Natl Acad Sci USA* 90:9649-9653.(1993)
16. U. Beffert, J. Poirier. Apolipoprotein E, plaques, tangles and cholinergic dysfunction in Alzheimer's disease. *Ann N Y Acad Sci* (In Press)
17. W.J. Strittmatter, K.H. Weisgraber, D.Y. Huang, L.M. Dong, G.S. Salvesen, M. Pericak-Vance, D. Schmechel, A.M. Saunders, D. Goldgaber and A.D. Roses. Binding of human apolipoprotein E to synthetic amyloid beta peptide: isoform-specific effects and implications for late-onset Alzheimer disease, *Proc Natl Acad Sci USA* 90:8098-8102.(1993)
18. J.B. Mitchell, K. Betito, W. Rowe, P. Boksa and M.J. Meaney. Serotonergic regulation of type II corticosteroid receptor binding in hippocampal cell cultures: evidence for the importance of serotonin-induced changes in cAMP levels, *Neuroscience* 48:631-639.(1992)
19. L. Brissette, P.D. Roach and S.P. Noel. The effects of liposome-reconstituted

apolipoproteins on the binding of rat intermediate density lipoproteins to rat liver membranes, *J Biol Chem* 261:11631-11638.(1986)
20. C.E. Matz, A. Jonas. Micellar complexes of human apolipoprotein A-I with phosphatidylcholines and cholesterol prepared from cholate-lipid dispersions, *J Biol Chem* 257:4535-4540.(1982)
21. U.K. Laemmli. Cleavage of structural proteins during the assembly of the head of bacteriophage T4, *Nature* 227:680-685.(1970)

THE NON-AMYLOID-β COMPONENT OF ALZHEIMER'S DISEASE PLAQUE AMYLOID: COMPARATIVE ANALYSIS SUGGESTS A NORMAL FUNCTION AS A SYNAPTIC PLASTICIZER.

Julia M. George and David F. Clayton

Department of Cell & Structural Biology
and The Beckman Institute
University of Illinois
Urbana, IL 61801

A second intrinsic peptide component of Alzheimer's Disease amyloid has recently been identified and called the "Non-Amyloid-β Component" (NAC)[1]. The peptide's precursor (NACP) is the homologue of a brain-specific presynaptic protein of unknown function, identified independently in at least four other laboratories. The first report of this protein was in 1988 by Maroteaux et al.[2], who identified it as a component enriched in a biochemical preparation of synaptic vesicles from the electric organ of Torpedo. Later, they cloned related sequences from the rat, and showed them to be enriched in telencephalic regions of the rat brain[3]. They originally termed this protein "synuclein," although early evidence for nuclear localization[2] has yet to be corroborated. Nakajo and colleagues identified a closely-related protein, first in bovine brain[4] and then from rats[5]. They showed this protein to be phosphorylated on serine residues, and have suggested it is also phosphorylated on tyrosine[6]; they named the protein Phosphoneuroprotein-14-kDa (PNP-14). More recently, Jakes et al.[7], using an antibody originally raised against paired helical filaments, detected two forms of the same human protein, one more closely related to PNP-14 and the other more related to the original synuclein. Consistent among all these reports has been the observation that the protein(s) is especially abundant in the telencephalon (especially the hippocampus) where it is distinctly enriched in presynaptic elements[2, 7-10].

In our own laboratory, we identified a member of the same family in experiments designed to clone genes that regulate neural plasticity in a favorable animal model, the songbird[11-14]. We have shown that the RNA is regulated in the circuit that controls song learning, so that it is most abundant during song memorization and then abruptly drops, in the nucleus most clearly associated with song memorization[11, 12]. The encoded protein is virtually identical to rat synuclein 1 protein and human NACP; here we refer to the canary sequence as ***synelfin***, to capture the two most significant features of the protein that are invariant among all these isolates from diverse species: 1) the protein is enriched in synaptic terminals; 2) the protein sequence is organized around an 11 amino-acid periodicity. As we describe elsewhere[11], this 11-residue periodicity appears to reflect the specific conservation of an amphipathic α-helical structure that bears signature features of apolipoproteins.

We do not suggest that synelfin necessarily functions as an apolipoprotein, however, as it is different in at least three important aspects. First, it is primarily intracellular, whereas apolipoprotein is primarily extracellular (although there are reports of detection of apolipoprotein E in the cytosol of some neurons[15, 16]). Second, unlike the apolipoprotein family, the degree of evolutionary sequence conservation extends beyond the maintenance of an amphipathic periodicity, and involves the precise conservation of an explicit amino acid sequence throughout the domain of 11-mers. Finally, the synelfin/NACP/synuclein/PNP-14 proteins share a highly acidic C-terminus, a feature not found in apolipoproteins (Fig. 1).

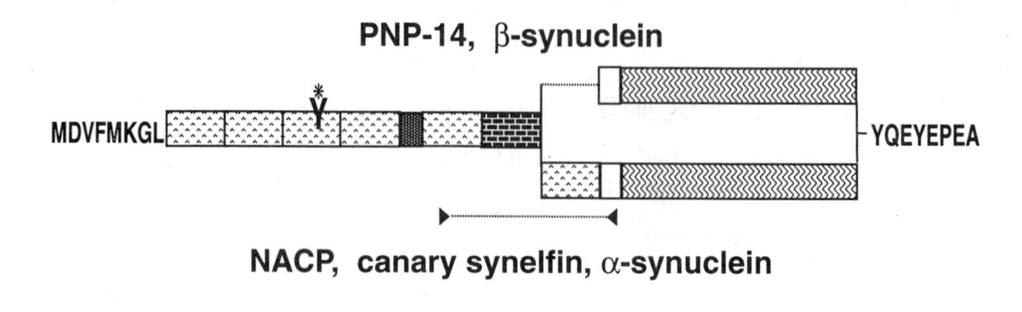

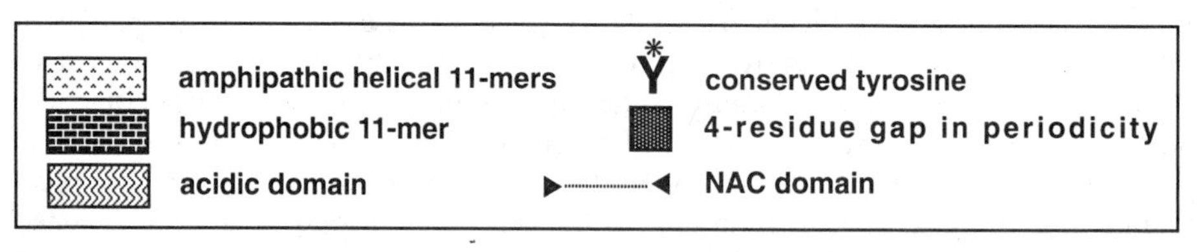

Figure 1. Conserved structural features of the synelfin family proteins. The features shown are common to sequences isolated from humans[1, 7], rodents[3, 5, 6], cows[4] and birds[11]. The sequences share a common size of 134-143 amino acids. All sequences are organized around the presence of either 6 (PNP-14, β-synuclein) or 7 (NACP, canary synelfin, α-synuclein) repetitions of a consensus motif that has a predicted amphipathic α-helical structure similar to apolipoprotein[11]. Each repeating unit has distinctive conserved sequence features; notably, each repeating unit is more similar to the homologous unit in other species than it is to its neighboring repeats in the same molecule[11]. The NAC domain that forms an intrinsic component of plaque amyloid[1] and nucleates amyloid fibril formation[17] is centered about repeat unit 6; this repeat is also unusual in its lack of polar residues and in its potential to participate in β-pleated sheet structures. In both humans and rats, two forms of the protein have been found that differ primarily in the presence or absence of repeat 7, and in the precise sequence of their carboxy-terminal acidic domains.

A physical relationship to apolipoproteins is obviously suggestive in the context of Alzheimer's Disease (AD), since the ApoE4 allele is a predisposing factor in the development of late onset AD[18]. Initial genetic studies have not yet provided further insight into the relationship to AD, however. The human NACP gene maps to chromosome 4[19, 20], and no differences have been detected in the protein-coding region in AD vs. controls[19]. Furthermore, no disease-associated differences in NACP expression have yet been detected [1, 21], apart from the presence of the NAC peptide within plaque amyloid. As the protein is abundant in the telencephalon and highly conserved among vertebrates, these observations collectively suggest that synelfin/NACP must serve an important function in the normal forebrain, and this function may somehow be perturbed in the development of AD-related pathology.

The focus of our current research, therefore, is to define the normal function of this protein in the brain. Here we suggest the specific hypothesis that synelfin (et al.) has a normal function in activity-dependent synaptic plasticity. The evidence for this is as follows:

1) the protein is enriched in presynaptic terminals, but not tightly associated with synaptic vesicles[4, 8-10, 22, 23].
2) the protein is a kinase substrate, suggesting a mechanism for modulation of its function by synaptic activity[6]. A particularly intriguing site of potential phosphorylation is the tyrosine in the middle of repeat unit 3 (Fig. 1) -- this is the only tyrosine present within the domain of 11-mers, and it is specifically conserved in all isolates of the protein to date.
3) the protein is most abundant in neural structures which are generally considered to display high levels of functional and anatomical plasticity in adulthood. These structures include the hippocampus, cortex, and the projections of cerebellar granule neurons onto Purkinje cell dendrites[3, 8, 10, 11]
4) in the avian song control system, expression of the protein declines with the end of the "sensitive period" for song acquisition[11, 12], although it remains high throughout the rest of the telencephalon in adult birds.
5) the domain of 11-mers is structurally related to the lipid-binding domains of apolipoprotein, suggesting a possible molecular interaction with free or membrane-bound phospholipids[11, 12].

REFERENCES

1. K. Ueda, *et al.*, Molecular cloning of cDNA encoding an unrecognized component of amyloid in Alzheimer disease, *Proc. Natl. Acad. Sci. (USA)* 90:11282 (1993).
2. L. Maroteaux, J. Campanelli, and R. Scheller, Synuclein: a neuron-specific protein localized to the nucleus and presynaptic terminal, *J. Neurosci.* 8:2804 (1988).
3. L. Maroteaux, and R. Scheller, The rat brain synucleins; family of proteins transiently associated with neuronal membrane, *Mol. Brain Res.* 11:335 (1991).
4. S. Nakajo, K. Omato, T. Aiuchi, T. Shibayama, I. Okahashi, H. Ochiai, Y. Nakai, K. Nakaya, and Y. Nakamura, Purification and characterization of a novel brain-specific 14-kDa protein, *J. Neurochem.* 55:2031 (1990).
5. T. Tobe, S. Nakajo, A. Tanaka, A. Mitoya, K. Omata, K. Nakaya, M. Tomita, and Y. Nakamura, Cloning and characterization of the cDNA encoding a novel brain-specific 14-kDA protein, *J. Neurochem.* 59:1624 (1992).
6. S. Nakajo, K. Tsukada, K. Omata, Y. Nakamura, and K. Nakaya, A new brain-specific 14-kDa protein is a phosphoprotein: its complete amino acid sequence and evidence for phosphorylation, *Eur. J. Biochem.* 217:1057 (1993).
7. R. Jakes, M. Spillantini, and M. Goedert, Identification of two distinct synucleins from human brain, *FEBS Let.* 345:27 (1994).
8. A. Iwai, E. Masliah, M. Yoshimoto, N. Ge, L. Flanagan, H. Rohan de Silva, A. Kittel, and T. Saitoh, The precursor protein of Non-Aβ Component of Alzheimer's Disease Amyloid is a presynaptic protein of the central nervous system, *Neuron* 14:467 (1995).
9. T. Shibayama-Imazu, I. Okahashi, K. Omata, S. Nakajo, H. Ochiai, Y. Nakai, T. Hama, Y. Nakamura, and K. Nakaya, Cell and tissue distribution and developmental change of neuron specific 14 kDa protein (phosphoneuroprotein 14), *Brain Res.* 622:17 (1993).
10. S. Nakajo, S. Shioda, Y. Nakai, and K. Nakaya, Localization of phosphoneuroprotein 14 (PNP 14) and its mRNA expression in rat brain determined by immunocytochemistry and in situ hybridization, *Mol. Brain Res.* 27:81 (1994).

11. J.M. George, H. Jin, W.S. Woods, and D.F. Clayton, Characterization of a novel protein regulated during the critical period for song learning in the zebra finch, (submitted).
12. J.M. George, H. Jin, W. Woods, and D.F. Clayton, A novel RNA is transiently induced in the song circuit of zebra finches during the critical period for song acquisition, *Soc. Neurosci. Abstr.* 19:807 (1993).
13. D.F. Clayton, M. Huecas, and K.L. Nastiuk, Analysis of preferential gene expression in the song control nucleus HVc of canaries, *Soc. Neurosci. Abstr.* 14:605 (1988).
14. D.F. Clayton, M.E. Huecas, E.Y. Sinclair-Thompson, K.L. Nastiuk, and F. Nottebohm, Probes for rare mRNAs reveal distributed cell subsets in canary brain, *Neuron* 1:249 (1988).
15. S.-H. Han, C. Hulette, A.M. Saunders, G. Einstein, M. Pericak-Vance, W.J. Strittmatter, A.D. Roses, and D.E. Schmechel, Apolipoprotein E is present in hippocampal neurons without neurofibrillary tangles in Alzheimer's Disease and in age-matched controls, *Exp. Neurol.* 128:13 (1994).
16. S.-H. Han, G. Einstein, K.H. Weisgraber, W.J. Strittmatter, A.M. Saunders, M. Pericak-Vance, A.D. Roses, and D.E. Schmechel, Apolipoprotein E is localized to the cytoplasm of human cortical neurons: a light and electron microscopic study, *Journal of Neuropathol Exp Neurol* 53:535 (1994).
17. H.Y. Han, P.H. Weinreb, and P.T. Lansbury, The core Alzheimers peptide NAC forms amyloid fibrils which seed and are seeded by beta-amyloid - is NAC a common trigger or target in neurodegenerative disease, *Chemistry & Biology* 2:163 (1995).
18. E. Corder, A. Saunders, W. Strittmatter, D. Schmechel, P. Gaskell, G. Small, A. Roses, J. Haines, and M. Pericak-Vance, Gene dose of apolipoprotein E type 4 allele and the risk of Alzheimer's disease in late onset families, *Science* 261:921 (1993).
19. D. Campion, *et al.*, The NACP/Synuclein gene - chromosomal assignment and screening for alterations in Alzheimer disease, *Genomics* 26:254 (1995).
20. X.H. Chen, *et al.*, The Human NACP/Alpha-Synuclein gene - chromosome assignment to 4q21.3-Q22 and TaqI RFLP analysis, *Genomics* 26:425 (1995).
21. M.C. Irizarry, T.W. Kim, R.E. Tanzi, D. Rosene, J.M. George, D.F. Clayton, and B.T. Hyman, Characterization of Synelfin/Synuclein/NACP in the central nervous system of Alzheimer's Disease patients, (in prep.).
22. J.M. George, and D.F. Clayton, Differential regulation in the avian song control circuit of an mRNA predicting a highly conserved protein related to protein kinase C and the *bcr* oncogene, *Mol. Brain Res.* 12:323 (1992).
23. G. Withers, J. George, G. Banker, and D.F. Clayton, Antibodies to HAT-3, a novel songbird protein, recognize synaptic elements in cultured rat hippocampal neurons., *Soc. Neurosci. Abstr.* 20:1437 (1994).

MUTATIONS OF HUMAN CU, ZN SUPEROXIDE DISMUTASE EXPRESSED IN TRANSGENIC MICE CAUSE MOTOR NEURON DISEASE

Mark E. Gurney*, Arlene Y. Chiu†, Mauro C. Dal Canto*, John Q. Trojanowski‡, and Virginia M.-Y. Lee‡;

*Northwestern University, Chicago USA; †City of Hope Medical Center, Duarte USA; ‡University of Pennsylvania, Philadelphia USA.

INTRODUCTION

ALS causes the degeneration of motor neurons in cortex, brainstem and spinal cord with consequent paralysis and death.[1] Most cases of ALS are sporadic and have an unknown etiology.[2] However, about 10-15% of all ALS cases are inherited. An adult-onset, autosomal dominantly inherited trait is the predominant form[3], although there is a rare, recessively inherited childhood-onset form of ALS in which survival can be quite long[4]. In 1991, a fraction of families with the familial form of ALS (FALS) showed linkage to a disease locus on human chromosome 21q.[5] Shortly thereafter, in 1993, the target of mutation on chromosome 21q was shown to be the gene (*SOD1*) encoding Cu,Zn superoxide dismutase (Cu,Zn SOD). At least 22 different missense mutations causing the substitution of one amino acid for another have now been found in FALS kindreds. Cu,Zn SOD is a metalloenzyme that catalyzes the dismutation of superoxide ($O_2 \cdot^-$) to hydrogen peroxide (H_2O_2). The copper ion provides the redox center for the dismutation of superoxide, while the zinc ion plays a structural role. Three different genes encoding superoxide dismutases are present in the human genome. All three enzymes contain a transition metal in their active site, but differ in their subcellular localization. Only Cu,Zn SOD is mutated in FALS. Cu,Zn SOD is primarily cytosolic and is expressed in every cell within the body.[6,7] Why only motor neurons are affected by the mutations found in FALS is unknown. The mutations of Cu,Zn SOD found in affected families are primarily amino acid substitutions in structural regions of the polypeptide.[8] No deletions of the human *SOD1* gene have been described which suggests that expression of the mutant polypeptide is required for pathogenesis.

To test the hypothesis that FALS results from a direct action of the mutant polypeptide, we and others have generated transgenic mice that express mutant forms of human Cu,Zn SOD found in affected families. Our results suggest that FALS is due to gain-of-function mutations in the enzyme. We tested the Cu,Zn SOD amino acid substitution $gly^{93} \rightarrow$ ala in mice[9], while other groups have tested the $gly^{85} \rightarrow$ arg and $gly^{37} \rightarrow$ ala mutations.[10,11] Only those mouse lines that express the mutant polypeptide at the highest levels develop motor neuron disease. In each case, the mice developed progressive paralysis of their limbs and died by 4-6 months of age. Conversely, mice that express wild-type human Cu,Zn SOD do not develop clinical motor neuron disease.[7,12] Expression of wild-type human Cu,Zn SOD in mice causes mild changes at the neuromuscular junction[10,13] and slight vacuolar changes in the proximal axons of ventral horn motor neurons.[14] However, it

Table 1. Comparison of survival, transgene copy number, human Cu,Zn SOD protein in brain and total brain Cu,Zn SOD activity between transgenic lines

Line	G1L	G1H	N29
mutation	G93A	G93A	wild-type
survival (days)	170 ± 16	139 ± 12	>440
human *SOD1* copy number (per diploid genome)	18 ± 1.3	25 ± 1.5†	8 ± 1.5
human Cu,Zn SOD protein‡ (ng/μg protein)	2.3 ± 0.5	3.1 ± 0.5†	6.8 ± 0.7
total SOD activity‡ (U/μg protein)	0.9 ± 0.2	1.2 ± 0.3	0.9 ± 0.1
Increase in brain SOD activity (transgenic / non-transgenic)&	4.1	5.5	4.1

† Student's t-test, p<0.001 vs. G1L transgenic line, & SOD activity in nontransgenic brain was 0.22 ± 0.03 U/mg protein; ‡ values for human Cu,Zn SOD protein and total SOD acitvity differ from Gurney et al. due to the addition of 0.1% Triton X-100 to the homogenization buffer.

also protects mice from a range of oxidative insults including MPTP+ toxicity[15] and ischemic damage due to stroke.[16]

EXPRESSION OF HUMAN CU,ZN SOD IN TRANSGENIC MICE

To produce transgenic mice expressing wild-type or mutant forms of human Cu,Zn SOD, we utilized an 11.7 kB genomic EcoR1 x BamH1 fragment that contained the entire human *SOD1* coding sequence together with promoter and enhancer sequences needed for expression of the gene in mice.[17] We derived multiple lines of transgenic mice expressing wild-type human Cu,Zn SOD as well as the gly[93]→ ala (G93A) mutation of human SOD found in FALS kindreds. These differ in transgene copy number and in the amount of human SOD expressed in brain. A description of three of these lines can be found in Table 1. The two lines designated G1L and G1H are sublines that derive from the same founder (designated G1). In the F2 generation, we identified a male who stably transmitted an increased number of transgene copies to his progeny. Segregation of the transgene locus is Mendelian, suggesting that an in-register, unequal cross-over event which increased transgene copy number from 18 to 25 copies per diploid genome must have occurred during meiotic recombination. With the increase in transgene copy number, there was an increase in the amount of human Cu,Zn SOD protein expressed in brain. The amount of human SOD expressed in brain when normalized to transgene copy number was consistently higher in the transgenic lines expressing wild-type as compared to mutant human SOD as shown for the N29 line in Table 1. The difference in levels of expression is consistent with the finding *in vitro* that the *SOD1* mutations found in FALS kindreds decrease the stability of the protein.[18] The G93A mutation chosen for analysis in these experiments has little effect on enzymatic activity. The G1L, G1H and N29 lines all had elevated total brain Cu,Zn SOD enzymatic activity in proportion to the amount of human Cu,Zn SOD expressed in brain.

CLINICAL DISEASE IN MUTANT *SOD1* MICE

Clinical disease is fully penetrant in both the G1H and G1L sublines of mice expressing the G93A mutant form of Cu,Zn SOD, while N29 mice that express comparable

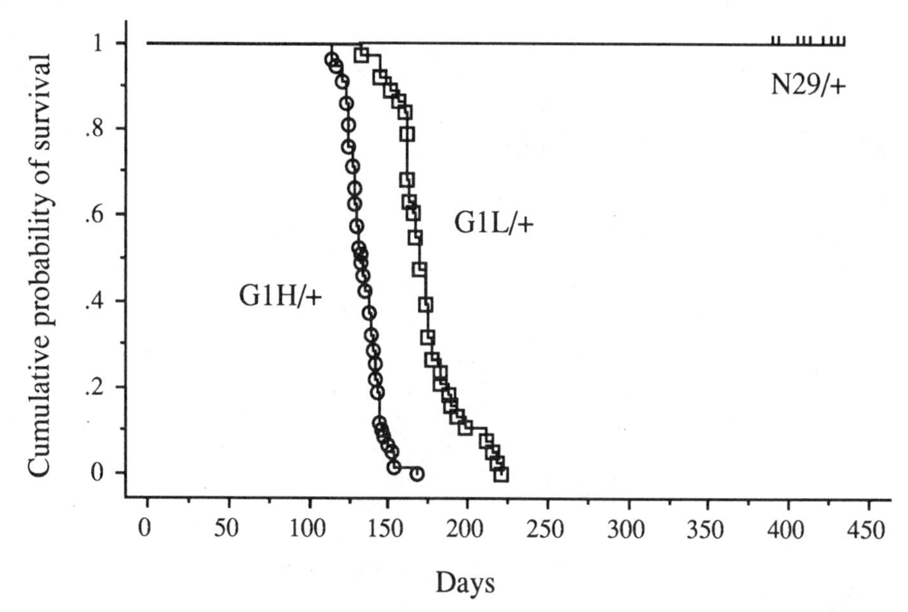

Figure 1. Kaplan-Meier plot comparing cumulative probability of survival of hemizygous transgenic mice in the G1H, G1L and N29 lines. The G1H and G1L mice express a mutant form of human Cu,Zn SOD, while N29 mice express the wild-type human enzyme. Censored data for healthy animals are indicated by the upward ticks.

amounts of wild-type human Cu,Zn SOD in brain remain healthy for as long as 440 days of age (Figure 1). Thus, these results argue that the genetic mechanism of disease is a toxic gain-of-function in the mutant enzyme.

The timing and severity of disease differs between the G1H and G1L sublines of transgenic mice. The G1H mice have an earlier disease onset, i.e. 91 ± 14 days as compared to 144 ± 4 days for G1L mice, and survival also is shorter. G1H mice reach end-stage disease at 134 ± 11 days of age, while G1L mice survive slightly longer to 173 ± 20 days. Since the amount of mutant human Cu,Zn SOD expressed in brain is greater in G1H mice as compared to G1L, this implies that the timing and severity of motor neuron disease is controlled by the expression level of a toxic protein. Clinical disease has not been observed in N29 mice expressing comparable levels of wild-type human Cu,Zn SOD for up to 440 days.

Many of the earliest signs of clinical disease in G1H and G1L mice parallel the clinical findings in human ALS as shown in Table 2. In human ALS, degeneration of the

Table 2. Comparison of clinical findings in human ALS and in transgenic G1H and N29 mice.

Finding	ALS	G1H	N29
hyperreflexia	+	+	-
crossed spread of spinal reflexes	+	+	-
fasiculations	+	-	-
clonus	+	+	-
shaking of the limbs	-	+	-
progressive paralysis	+	+	-
death due to clinical disease	+	+	-

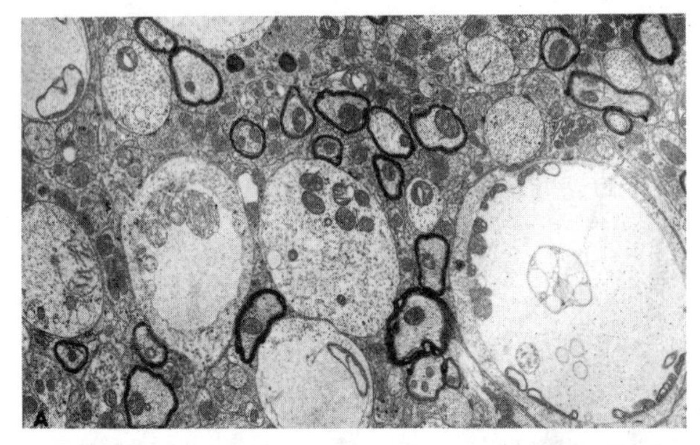

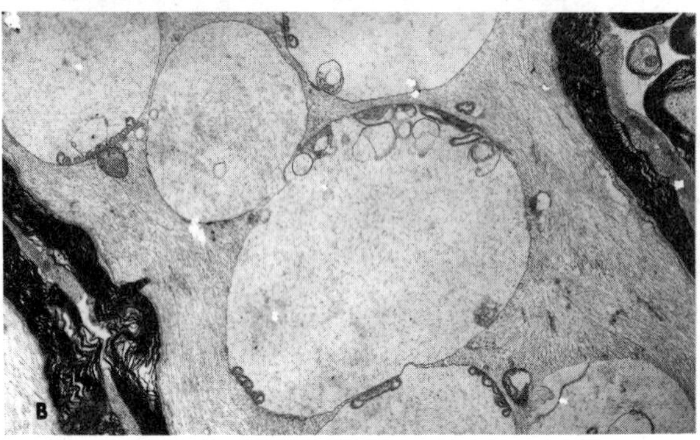

Figure 2. (A) Transgenic G1H mouse expressing mutant human Cu,Zn SOD. Ultrastructural view of the anterior horn neuropil showing swollen dendritic processes. The swollen process in the upper left corner displays a peculiar alteration of the mitochondria with linearization of their structure. (B) Transgenic N29 mouse expressing wild-type human Cu,Zn SOD. Ultrastructural view of a motor axon containing numerous vacuoles that is passing through the anterior column of the spinal cord . These show the same peculiar mitochondrial alteration as observed in mice expressing mutant human Cu,Zn SOD.

cortical spinal tract leads to disinhibition of spinal reflexes. Patients become hyperreflexive and show crossed spread of spinal reflexes to the opposite limb. Spasticity (resistance to passive movement of the limbs) becomes evident and spontaneous contractions of denervated muscle fibers (fasiculations) can be seen beneath the skin or in the tongue. Many of these clinical signs also are present in G1H and G1L mice. Both types of mice show hyperreflexia, crossed spread of spinal reflexes, clonus and spasticity. Fasiculation is not present, but may be difficult to detect beneath the furry skin of the mouse. Shaking of the limbs when suspended in the air is the most consistent diagnostic indication of the onset of disease in G1H and G1L mice, but this sign has no clinical counterpart in human ALS.

As the disease progresses in G1H and G1L mice, proximal muscle weakness with marked atrophy develops. Weakness and atrophy usually are more evident in the hindlimbs than in the forelimbs. As weakness becomes more pronounced, spasticity and hyperreflexia become less so, perhaps because of the increasing weakness. The shaking of the limbs becomes less apparent and another, unrelated tremor of the distal joints of the toes develops which occurs in the absence of movement in the hind foot. At end-stage disease, G1H mice are severely paralyzed and lie on their side. They generally are alert, but do not move in response to tapping on their cage or when gently prodded. When removed from their cage and placed on their side, the mice are unable to right themselves. Mice were euthanized when no longer able to right themselves within 30 seconds of being placed on their side.

PATHOLOGY IN MUTANT *SOD1* MICE IS RESTRICTED TO BRAINSTEM, SPINAL CORD, NERVE AND MUSCLE

Despite the widespread expression of mutant human Cu,Zn SOD in somatic tissues of G1H mice, disease is restricted to brainstem, spinal cord, peripheral nerve and muscle . No consistent pathological changes were detected in somatic tissues with the exception of mild

Table 3. Timing of pathological changes in the spinal cords of G1H mice expressing the G93A mutant form of human Cu,Zn SOD and in N29 mice expressing wild-type human Cu,Zn SOD.

G1H Mice						
Age (days)	35	45	82	111	130	140
Vacuolation	+	+	++	++	++	++
Motor Neuron Loss	-	-	+	++	+++	+++
Spheroids	-	-	+	+	++	++
Ubiquitination	-	-	+	+	++	++
Astrocytosis	-	-	+	+	++	++

N29 Mice					
Age (days)	51	82	111	132	199
Vacuolation	-	-	-	-	+/-
Motor Neuron Loss	-	-	-	-	-
Spheroids	-	-	-	+	++
Ubiquitination	-	-	-	+/-	+/-
Astrocytosis	-	-	-	-	-

biliary hyperplasia in liver and mild glomerular vacuolation in kidney which were noted in both G1H and N29 mice. At end-stage disease, mice were severely paralyzed and most showed loss of weight, signs of dehydration, and signs of respiratory insufficiency. These clinical signs are all consistent with the pathology noted in the spinal cord. Motor neurons innervating the limbs were reduced by 50% and myelinated phrenic nerve axons innervating the diaphragm also showed a severe reduction in number. Denervation of skeletal muscle is detected with the same time course as the loss of motor neurons in the spinal cord, although at early stages in the disease, reinnervation kept pace with the death of motor neurons and few denervated end plates were visible in skeletal muscle. Reinnervation is primarily by nodal sprouts arising from the intramuscular nerve. As the loss of motor neurons progresses with age, however, compensatory processes fail and increasing numbers of muscle fibers become denervated. Thus, clinical impairment is a consequence of motor neuron disease.

MITOCHONDRIAL VACUOLATION IN MUTANT *SOD1* MICE

The earliest signs of pathology develop in the spinal cords of G1H and G1L mice well in advance of clinical signs of disease. By 35 days of age in G1H mice, an unusualvacuolation of mitochondria and smooth endoplasmic reticulum is noted in the proximal axons of spinal cord motor neurons. As the animals age, the vacuolation spreads to the cell body and dendrites of ventral horn neurons. Vacuolation also occurs in motor-related areas of the brainstem, although the timing of vacuolation in the brainstem is delayed by about one month in comparison to the spinal cord. Vacuolation remains prominent in the spinal cords of G1H mice at end-stage disease, although it becomes less so in G1L mice.

Initially, most of the changes in mitochondria are limited to the dilation of the internal cristae.[12] These then separate and fragment as the intermembrane space becomes swollen. This causes an unusual splitting of the inner and outer mitochondrial membranes with linearization of the remaining cristae (Figure 2). The permeability properties of the outer mitochondrial membrane are relatively nonspecific,[19] so it seems unlikely that the basis of the swelling is due to ionic disequilibrium. Small amounts of Cu,Zn SOD may be present within the mitochondrial intermembrane space[20,21] and could contribute to this pathology. The vacuolar changes in endoplasmic reticulum and mitochondria might contribute to disease by impairing either energy metabolism or calcium homeostasis within vulnerable neurons.

PATHOLOGY AND MOTOR NEURON LOSS IN SPINAL CORD

Cytoplasmic ubiquitination of protein, accumulation of neurofilaments in axonal spheroids, gliosis and motor neuron loss become evident in the ventral horn at roughly the onset of clinical disease (Table 3). In addition, there was degeneration of the ascending sensory tracts in an intermediate zone of the dorsal columns of the spinal cord as seen in familial ALS, as well as phosphorylation of neurofilaments in the motor neuron cell body and occasional Lewy bodies. Mice at end-stage disease show up to 50% loss of cervical and lumbar motor neurons, however, neither thoracic nor cranial motor neurons show appreciable loss despite the development of extensive vacuolar changes. Autonomic motor neurons are spared in patients and that also was observed in this transgenic model. Counts of brainstem, preganglionic parasympathetic neurons of the vagus and preganglionic sympathetic neurons in the thoracic cord were normal. Thus, these results indicate that somatic motor neurons innervating the limbs were selectively vulnerable to loss, whereas other populations of somatic and autonomic cholinergic neurons were spared. Loss of somatic motor neurons was restricted to G1H mice expressing mutant human Cu,Zn SOD. N29 mice expressing wild-type human Cu,Zn SOD had no loss of cholinergic motor neurons from the spinal cord at 170 days of age.

CURRENT ISSUES

For ALS and other neurodegenerative disease we need to address the following questions:

What is the mechanism of neuronal damage and degeneration?
Why are particular types of neurons selectively vulnerable?
Do neurons die, by apoptosis, necrosis or some other pathway?
What determines the age-dependence of disease penetrance?

We are closest to providing an answer to the first question. Our data argue for a gain-of-function mechanism of damage in the transgenic mouse model. Expression of human Cu,Zn SOD containing a G93A mutation at high levels in transgenic mouse brain causes fully penetrant motor neuron disease, while expression of comparable amounts of wild-type human Cu,Zn SOD does not. Since the G93A mutant form of Cu,Zn SOD retains enzymatic activity, there is an increase in total brain Cu,Zn SOD activity. Therefore motor neuron disease in this model is not due to a decrease in Cu,Zn SOD activity. Instead, the data argue that mutation has introduced a novel toxic property into the enzyme. Since over 22 different missense mutations of *SOD1* have been identified in FALS kindreds, the alteration in Cu,Zn SOD conformation or function introduced by mutation is likely to be global rather than specific.

Two hypotheses regarding the pathogenesis of familial ALS and the nature of the gain-of-function in Cu,Zn SOD seem reasonable in light of current data. First, disease may be caused by oxidative damage. Mutation of Cu,Zn SOD may enhance catalysis of a normally unfavorable side reaction such as the decomposition of hydrogen peroxide to hydroxyl radical and hydroxyl anion[22] or the catalysis of protein nitration on tyrosine residues by peroxynitrite with release of hydroxyl radical.[23] Peroxynitrite forms spontaneously by the reaction of superoxide with nitric oxide. Both of these reactions are catalyzed inefficiently by wild-type Cu,Zn SOD which may account for the limited vacuolar pathology noted in N29 mice. These mechanisms of pathogenesis require that the copper redox center be retained in the active site of the enzyme. Mutation could also cause the release of free copper from the enzyme at inappropriate sites within neurons or at concentrations sufficient to catalyze the generation of reactive oxygen species.[24] If this were so, then mutations that cause release of copper from the enzyme should cause more severe disease. That does not occur, however, as a mild form of ALS with prolonged survival is seen in a Japanese kindred with a $his^{46} \rightarrow arg$ mutation of one of the histidines that coordinates copper in the active site.[25,26] Alternatively, free-radical mechanisms may not play a role in disease. Instead, the mutated protein (or a fragment thereof) may be toxic in and of itself, much as is hypothesized for the βA4 peptide deposited as amyloid in Alzheimer's disease.[27] For example, mutant Cu,Zn SOD may alter the metabolism of neurofilament proteins leading to the accumulation of neurofilament rich inclusions or aggregates in the perikarya and/or processes of affected motor neurons as demonstrated for the G1H mice.[28] While the mechanism whereby this might occur is unclear at this time, the hypothesis that the neurofilament rich inclusions could disrupt or physically "block" axonal transport in ALS is supported by the recent demonstration of defective axonal transport in a transgenic model of motor neuron disease that was generated by the over expression of neurofilament proteins and the accumulation of neurofilament aggregates in affected spinal cord neurons.[29] By preventing the orthograde transport of trophic or other factors from the periphery that promote the survival of motor neurons, or by depriving synapses and the distal portions of axons of orthogradely transported proteins required for normal synaptic function or homeostasis, neurofilament-rich inclusions may play a central role in the pathogenesis of motor neuron dysfunction and death in ALS. Thus, to identify potential targets for therapeutic intervention in ALS, it will be important to explore these issues further in the transgenic mice described here.

However, it remains to be seen whether or not the insights provided by molecular genetics into the pathogenesis of familial ALS will lead to an understanding of the sporadic disease. Sporadic and familial ALS have a similar clinical course and findings, with only minor differences in age of onset, sex ratio, survival, and site of onset.[3] This suggests that there may be common elements to the pathogenesis of the two diseases. One common element may be damage to mitochondria as suggested by recent findings with respect to glutamate excitotoxicity and sporadic ALS (Figure 3).

Glutamate is the major excitatory amino acid neurotransmitter in the CNS. It interacts with two classes of receptors that differ in ligand specificity and ionic conductances. These two receptor classes are distinguished by their responses to kainate (KA) and *N*-methyl-D-aspartate (NMDA) which are both chemical analogs of glutamate. Both receptor subtypes are present on motor neurons.[30] NMDA-receptors can be opened by glutamate only if the motor neuron membrane is depolarized. When opened, the NMDA-receptor allows the influx of Ca^{++} as does a subtype of KA-receptor. Increases in intracellular Ca^{++} due to influx through glutamate receptors have been postulated to be responsible for neuron damage in ischemia, epilepsy, spinal cord trauma and neurodegenerative disease.[31] Motor neurons may be particularly vulnerable to damage caused by elevated intracellular Ca^{++} due to low levels of expression of Ca^{++} binding proteins such as calbindin D-28K and parvalbumin.[32]

Compelling evidence for the involvement of glutamate excitotoxic damage in sporadic ALS comes from the recent clinical trial conducted with riluzole in which a 12% improvement in survival was observed.[33] Riluzole is thought to decrease glutaminergic neurotransmission by inhibiting the release of glutamate from synaptic terminals. Consistent with the clinical benefit of riluzole is the observation in sporadic ALS that glutamate is elevated in fasting plasma[34] and in cerebrospinal fluid.[35] Increased CSF levels of glutamate in sporadic ALS may be due to decreased reuptake of glutamate into synaptic terminals.[36]

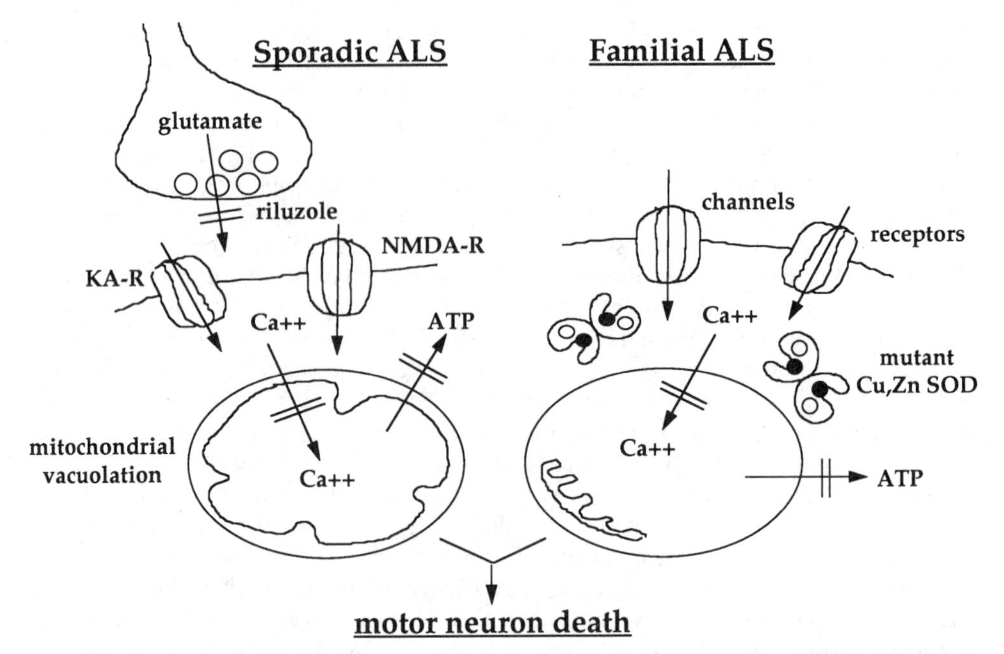

Figure 3. A common endpoint of damage in sporadic and familial ALS? Both excitotoxic lesions and expression of mutant human Cu,Zn SOD in mice cause characteristic mitochondrial cytopathology. In each instance this could lead to impaired mitochondrial sequestration of Ca^{++} or bioenergetics. Both have been implicated in pathways leading to neuron death.

Whatever the mechanism of damage caused by mutated Cu,Zn SOD, the mitochondria in vulnerable neurons are an early and selective target of pathogenesis. Mitochondrial vacuolation also is caused by excitotoxic lesion with glutamate and by the NMDA receptor antagonist MK801.[37] Glutamate excitotoxicity causes swelling of both the inner and outer mitochondrial membranes, while in mutant SOD1 mice, there is a swelling of primarily the outer mitochondrial membrane and the intermembrane space (Figure 3). The difference in cytopathology probably indicates that the cause of mitochondrial damage also is different; however, its end result may be the same. Mitochondria are crucially important for the buffering of intracellular calcium concentrations and for the biogenesis of ATP. The prominent mitochondrial cytopathology could indicate that both functions are impaired. Currently, we do not know if vacuolation is only a feature of the transgenic mouse model or if it also occurs in human ALS.

These hypotheses suggest directions for future research. First, if oxidative mechanisms play a role in pathogenesis, then it should be possible to detect their biochemical signature. One expects to find oxidation of lipids, proteins or nucleic acids. Second, there should be biochemical correlates to the mitochondrial cytopathology. Motor neurons may buffer Ca^{++} less well. For example, an increase in steady-state levels of intracellular Ca^{++} may increase the frequency of miniature end plate potentials at the neuromuscular junction, while impaired buffering of transient Ca^{++} increases may cause an increase in quantal content. Impairment of mitochondrial bioenergetics is implied in FALS by the elevation of complex I activity.[38] Does a similar elevation of complex I occur in the mouse model? Findings in the mouse model need to be referred back to the FALS patient. The mitochondrial cytopathology is an early feature of disease in the transgenic mice. It may not be possible to biopsy the spinal cord in early FALS, but it certainly is possible to biopsy muscle in order to assess the condition of mitochondria in the motor nerve terminal.

ACKNOWLEDGMENTS

This work was supported by grants from the Muscular Dystrophy Association and from the National Institutes of Health.

REFERENCES

1 Mulder, D.W. (1982) Clinical limits of amyotrophic lateral sclerosis. In: Rowland, L.P., (Ed.) Human Motor Neuron Diseases, pp. 15-22. New York: Raven Press.
2 Tandan, R. and Bradley, W.G. (1985) Amyotrophic lateral sclerosis: Part 1. Clinical features, pathology and ethical issues in management. *Ann. Neurol. 18*, 271-280.
3 Mulder, D.W., Kurland, L.T., Offord, K.P., and Beard, C.M. (1986) Familial adult motor neuron disease: Amyotrophic lateral sclerosis. *Neurol. 36*, 511-517.
4 Ben Hamida, M., Hentati, F., and Ben Hamida, C. (1990) Hereditary motor system diseases (chronic juvenile amyorophic lateral sclerosis): conditions combining a bilateral pyramidal syndrome with limb and bulbar atrophy. *Brain 113*, 347-363.
5 Siddique, T et al. (1994) Linkage of a gene causing familial amyotropic lateral sclerosis to chromosome 21 and evidence of locus heterogeneity. *New Englnd. J. Med. 324*, 1381-1384.
6 Crapo, J.D., Oury, T., Rabouille, C., Slot, J.W., and Chang, L-Y. (1992) Copper, zinc superoxide dismutase is primarily a cytosolic protein in human cells. *Proc. Natl. Acad. Sci. USA 89*, 10405-10409.
7 Beckman, J.S., Carson, M., Smith, C.D. and Koppenol, W.H. (1993) ALS, SOD and peroxynitrite. *Nature 364*, 584.
8 Deng, H.X. et al. (1993) Amyotrophic lateral sclerosis and structural defects in Cu,Zn Superoxide dismutase. *Science 261*, 1047-1051.
9 Gurney, M.E., Pu, H., Chiu, A.Y., Dal Canto, M.C., Polchow, C.Y., Alexander, D.D., Caliendo, J., Hentati, A., Kwon, Y.W., Deng, H.-X., et al. (1994) Motor neuron degeneration in mice expressing a human Cu, Zn superoxide dismutase mutation. *Science 264*, 1772-1775.
10 Ripps, M.E., Huntley, G.W., Hof, P.R., Morrison, J.H., and Gordon, J.W. (1995) Transgenic mice expressing an altered murine superoxide dismutase gene provide an animal model of amyotrophic lateral sclerosis. *Proc. Natl. Acad. Sci. USA 92*, 689-693.
11 Price, D. et al., see chapter from this symposium
12 Avraham, K.B., Sugarman, H., Rotshenker, S. and Groner, Y. (1991) Down's syndrome: morphological remodelling and increased complexity in the neuromuscular junction of transgenic CuZn-superoxide dismutase mice. *J. Neurocytol. 20*, 208-215.
13 Avraham, K.B., Schickler, M., Sapoznikov, D., Yarom, R. and Groner, Y. (1988) Down's syndrome: Abnormal neuromuscular junction in tongue of transgenic mice with elevated levels of human Cu,Zn-superoxide dismutase. *Cell 54*, 823-829.
14 Dal Canto, M.C. and Gurney, M.E. (1994) The development of CNS pathology in a murine transgenic model of human ALS. *Am. J. Pathol. 145*, 1271-1280.
15 Przedborski, S., Kostic, V., Jackson-Lewis, V., Naini, A.B., Simonetti, S., Fahn, S., Carlson, E., Epstein, C.J., and Cadet, J.L. (1992) Transgenic mice with increased Cu/Zn-superoxide dismutase activity are resistant to *N*-methyl-4-phenyl-1,2,3,6-tetrahydropyridine-induced neurotoxicity. *J. Neurosci. 12*, 1658-1667.
16 Yang, G., Chan, P.H., Chen, J., Carlson, E., Chen, S.F., Weinstein, P., Epstein, C.J. and Kamii, H. (1994) Human copper-zinc superoxide dismutase transgenic mice are highly resistant to reperfusion injury after focal cerebral ischemia. *Stroke 25*, 165-70.
17 Epstein, C.J., Avraham, K.B., Lovett, M., Smith, S., Elroy-Stein, O., Rotman, G., Bry, C. and Groner, Y. (1987) Transgenic mice with increased Cu/Zn-superoxide dismutase activity: Animal model of dosage effects in Down syndrome. *Proc. Natl. Acad. Sci. USA 84*, 8044-8048.
18 Borchelt, D.R. et al. (1994) Superoxide dismutase 1 with mutations linked to familial amyotrophic lateral sclerosis possesses significant activity. *Proc. Natl. Acad. Sci. USA 91*, 8292-8296.
19 Benz, R. (1990) Biphysical properties of porin pores from mitochondrial outer membrane of eukaryotic cells. *Experientia 46*, 131-137.
20 Weisiger, R.A. and Fridovich, I. (1973) Mitochondrial superoxide dismutase. Site of synthesis and intramitochondrial localization. *J. Biol. Chem. 248*, 4793-4796.
21 Liou, W. Chang, L.Y. Geuze, H.J. Strous, G.J. Crapo, J.D. and Slot, J.W. (1993) Distribution of CuZn superoxide dismutase in rat liver. *Free Radical Biol. Med. 14*, 201-207.
22 Yim, M.B., Chock, P.B. and Stadtman, E.R. (1993) Enzyme function of copper,zinc superoxide dismutase as a free radical generator. *J. Biol. Chem. 268*, 4099-4105.
23 Beckman, J.S., Carson, M., Smith, C.D. and Koppenol, W.H. (1993) ALS, SOD and peroxynitrite. *Nature 364*, 584.

24 Ribarov, S.R. and Bochev, P.G. (1984) The interaction of copper chloride with the erythrocyte membrane as a source of activated oxygen species. *Gen. Physiol. Biophys. 3*, 431-435.

25 Ogasawara, M. et al. (1993) Mild ALS in Japan associated with novel SOD mutation. *Nature Genetics 5*, 323-324.

26 Caffi, M.T., Battistoni, A., Polizio, F., Desideri, A., and Rotilio, G. (1994) Impaired copper binding by the H46R mutant of human Cu,Zn superoxide dismutase, invovled in amyotrophic lateral sclerosis. *FEBS Letters 356*,314-316.

---

27 Selkoe, D. J. (1994) Alzheimer's disease: a central role for amyloid *J. Neuropathol. Exp. Neurol. 53*, 427-428.

28 Tu, P.-H., Raju, P., Robinson, K.A., Gurney, M.E., Trojanowski, J.Q., and Lee, V. M.-Y. (1995) Transgenic mice carrying a human mutant superoxide dismutase transgene develop neuronal cytoskeletal pathology resembling human amyotrophic lateral sclerosis. *Proc. Natl. Acad. Sci. USA. in press.*

29 Collard, J.-F., Cote, F. and Julien, J.-P. (1995) Defective axonal transport in a transgenic mouse model of amyotrophic lateral sclerosis. *Nature 375*, 61-64.

30 Shaw, P.J. (1994) Excitotoxicity and motor neurone disease. *J. Neurol. Sci. 124(Suppl.)*, 6-13.

31 Choi, D.W. (1988) Glutamate neurotoxicity and disease of the nervous system. *Neuron 1*, 623-634.

32 Ince, P.G. et al. (1992) Parvalbumin and calbindin D-28K in the human motor system and in motor neuron disease. *Neuropathol. Appl. Neurobiol.*

33 Bensimon G. Lacomblez L. Meininger V. (1994) A controlled trial of riluzole in amyotrophic lateral sclerosis. *New Englnd. J. Med. 330*, 585-591.

34 Plaitakis, A. Mandeli, J., Fesdjian, C. and Sivak, M.A. (1991) Dysregulation of glutamate metabolism in ALS: correlation with gender and disease type. *Neurol. 41*, 392-393.

35 Rothstein, J.D. et al. (1990) Abnormal excitatory amino acid metabolism in amyotrophic lateral sclerosis. *Ann. Neurol. 28*, 18-25.

36 Rothstein, J.D., Martin, L.J., and Kuncl, R.W. (1992) Decreased glutamate transport by the brain and spinal cord in amyotrophic lateral sclerosis. *New Englnd. J. Med. 326*, 1464-1468.

37 Fix, A.S. et al. (1993) Neuronal vacuolization and necrosis induced by the noncompetitive N-metyl-D-aspartate (NMDA) antagonist MK(+)801 (dizocilpine maleate): A light and electron microscopic evaluation of rat retrosplenial cortex. *Exp. Neurol. 123*, 204-215.

38 Bowling, A.C., Schulz, J.B., Brown, R.H. Jr, and Beal, F.M. (1993) Superoxide dismutase activity, oxidative damage, and mitochondrial energy metabolism in familial and sporadic amyotrophic lateral sclerosis. *J. Neurochem. 61*, 2322-2325.

EFFECT OF THE KENNEDY MUTATION OF THE ANDROGEN RECEPTOR ON GENE EXPRESSION IN NEUROBLASTOMA CELLS.

P.A. Yerramilli-Rao, O. Garofalo, P.N. Leigh, and J.-M. Gallo.

Departments of Neurology,

Institute of Psychiatry,

De Crespigny Park,

London, SE5 8AF.

United Kingdom

INTRODUCTION

Kennedy's syndrome (X-linked bulbar and spinal muscular atrophy) is an X-linked disorder characterised by sensory and lower motor neurone degeneration.[1,2,3] The discovery of a mutation in the androgen receptor (AR) gene in Kennedy's syndrome reinforced the hypothesis that an X-linked factor might be involved in the disease process of Amyotrophic Lateral Sclerosis (ALS), because of the prevalence of the disease in males, (ratio 1.6:1). This mutation, which is tightly linked with the disease phenotype, consists of an increased number of CAG repeats present in the first exon of the AR gene, encoding a polyglutamine chain.

The AR gene is expressed in neurones of the motor cortex and spinal cord,[4,5] and thus molecular action of this gene in motor neurones may be linked to motor neuronal degeneration and may be associated to the pathogenesis of ALS. The study of genes which are regulated by the AR therefore represents a way in which gene products important to neuronal survival may be identified.

The human neuroblastoma cell line SH-SY5Y[6] is a good model for motor neurones since these cells differentiate into a neuronal phenotype, exhibit mainly cholinergic characteristics,[7] and respond to many of the same trophic factors as motor neurones.[8,9] We have previously shown, by Northern blotting, that SH-SY5Y cells expressed the AR.[10] By using a modification of the differential display technique, initially described by Liang and Pardee[11], we have shown that specific appear to be expressed in response to 5αdihyrotestosterone (DHT) in SH-SY5Y cells.[10]

SH-SY5Y cells transfected with Kennedy's mutations of the AR (i.e containing increased CAG repeat sequences), can be studied to determine whether expression of androgen-regulated genes, as identified previously, may be modified.

MATERIALS AND METHODS

Cell Culture

SH-SY5Y cells were maintained in Dulbecco's Minimal Essential Medium (DMEM)/Ham's F-12 (1:1, v/v) supplemented with 15%(v/v) heat inactivated foetal bovine serum (HIFBS), 2mM glutamine and, penicillin [100IU/ml]/streptomycin [100mg/ml] in a humidified atmosphere of 5% CO_2, 95% air at 37° C. The cells were subcultured at an approximate ratio of 1:5 in 75cm^2 flasks (approximately 5x10^{4}cells/cm^2).

For androgen-treatment, cells were grown in phenol-red free medium supplemented with 15%(v/v) charcoal-stripped HIFBS, for 48h before addition of 0.5nM of 5αDHT.

Differential Display of RNA

Total cellular RNA was isolated from control and androgen-treated cells and analysed by differential display essentially as described previously.[8,9]

cDNA Cloning

PCR amplified cDNAs were analysed on a standard 6% polyacrylamide/8M urea sequencing gel in Tris-borate buffer. Gels were run at 1800V constant voltage and 55° C. [γ-^{32}P]ATP labelled ΦX174DNA/*Hinf*I (Promega), was used as a size estimation marker.

To recover radioactive cDNA bands, polyacrylamide gels were wrapped in Saranwrap without fixation or drying, and exposed directly to X-ray film overnight. Bands of interest were detected by aligning the gel with the developed autoradiogram. cDNA bands demonstrating reproducible differences were excised and recovered by electroelution, using the S&S Biotrap™ system (Schleicher and Schuell).

Recovered cDNAs were ethanol precipitated and the pellet resuspended in sterile water after a 95% (v/v) ethanol wash. Re-amplification by PCR was performed using the same reaction mixture and conditions as for differential display except for dNTP concentrations which were 20mM and that the reactions were performed in the absence of radioisotopes. Re-amplified DNA was ligated into the pCRII cloning vector (Invitrogen Corp., San Diego.) according to manufacturer's instructions. Plasmid DNA was then purified using a modified alkaline lysis procedure.

Sequence analysis

DNA sequencing was performed directly from the pCRII cloning vector. Both strands of the cDNA insert were sequenced using the commercially available Cyclist™ Exo$^-$ *Pfu* DNA sequencing kit (Stratagene) and the T7 and M13 Reverse primers. The nucleotide sequences obtained were compared with known sequences by searching the GenBank and EMBL data bases using the BLAST network service.

RESULTS

Identification of genes specifically expressed in response to 5αDHT in SH-SY5Y cells

We previously showed that the primer combination of T_{11}CA (5'-TTT TTT TTT TTC A-3') and LtK3 (5'-CTT GAT TGC C-3') displayed 60-70 cDNA bands upon polyacrylamide gel electrophoresis of the amplification products of reverse transcribed RNA, ranging in size from 100bp to 500bp. In SH-SY5Y cells the pattern of amplified cDNAs obtained from control cells and cells treated with 0.5nM, 50nM or 5000nM 5αDHT (17B-hydroxy-5αandrostan-3-one) for 48h were, for the most part similar. PCR differential display is capable of detecting qualitative differences in relative mRNA levels, but may also be semiquantitative as the intensity of certain cDNA bands were found to be reproducibly increased or decreased. However there were a number of specific apparent differences (i.e cDNA bands that were either present or absent from treated cells). In total nine candidate clones exhibited specific induction upon androgen treatment. These clones were of the following sizes:

Clone 1= 480bp, Clone 2= 400bp, Clone 3= 310bp, Clone 4= 300bp, Clone 5= 260bp, Clone 6= 225bp, Clone 7= 210bp, Clone 8= 195bp, Clone 9= 180bp.

These changes were confirmed by repeating the PCR and differential display procedures. All of these nine were up-regulated by 5αDHT, and maximal effects were reached at 0.5nM, close to the dissociation constant (K_d) of the AR.[11] No further changes were seen at increasing concentrations of DHT.

Analysis of differentially expressed genes

The cDNAs of identified mRNA species were successfully recovered from an unfixed polyacrylamide gel, reamplified and, cloned into the pCRII vector using the TA cloning kit.

All nine clones have been sequenced in both directions using the T7 and M13 Reverse primers, and these partial nucleotide sequences have been compared with gene data base sequences. To date we have found no significant homology with any published genes for the sequences obtained from the cDNA fragments of bands one to eight, suggesting that they may represent previously unknown genes associated with androgen induction. The cDNA fragment from clone nine was found to be highly homologous to *Xenopus leavis* calmodulin dependent protein kinase II beta subunit (Fig. 1). The 180bp fragment was 96% identical to bases 2569-2597 (3' untranslated region) of the 2609bp published sequence.

```
Clone 9:          AGCTAGCTAGCTAGCTAGCTAGCTAAGCCAATTCCAGCACACTGG

                  CGGCCTTACTAGTGGATCCAGCTCGGTACCAAGCTT GCATGCATA
                                                       ||  | |||||||
Xenopus laevis calmodulin dependent                    TT GGATGCATA  2579
protein kinase II ß subunit

                  GCTTGAGTA TTCTATAGT TCACACCTAAATAGCATTCTATCATAT
                  ||||||||| |||||||||
                  GCTTGAGTA TTCTATAGT                                2597

                  GTCATAGCTTTCGATTTAATTTACCTCCATCCCCTCACACTATTC

                  ACTCTAAAA
```

Figure 1. Comparison of nucleotide sequence of clone 9 with *Xenopus leavis* calmodulin dependent protein kinase II beta subunit.

DISCUSSION

Weiner[12], originally proposed that in some cases, the mechanism of motor neurone death in sporadic ALS related to the loss of androgen receptors. Indeed, in typical sporadic ALS more men are affected than women, with a male to female ratio of 1.5:1.[13] Thus a sex-linked factor may be involved in the disease process. The AR gene has been mapped to Xq 11-12,[14] and some have discussed the possibility of AR abnormality as the cause of ALS.[15]

Kennedy's disease is an X-linked disorder characterised by sensory and lower motor neurone degeneration.[16] The discovery of a mutation in the AR gene in Kennedy's disease[1] supports the notion that AR dysfunction is important in motor neurone degeneration. This mutation consists of an increased number of CAG repeats, encoding for a polyglutamine chain, present in the first exon of the AR gene, and is tightly linked with the disease phenotype. This is the only type of described mutation in the hAR gene associated with motor neurone degeneration.[17]

Expansion of triplet repeats constitutes a now common form of "dynamic" mutation,[18] which have been found to cause the diseases of Fragile X syndrome, myotonic dystrophy, Huntingdon's disease and Kennedy's syndrome. In the case of the gene for the AR, which has the repeat in a coding region, the expanded repeat region results in decreased transactivation of the androgen response element (ARE).[19,20] There is now some evidence to show that the efficiency of nucleosome assembly formation increases with expanded triplet blocks, suggesting that such blocks may repress transcription through the creation of stable nucleosomes.[21]

There are many examples suggesting that androgens may exert a trophic activity towards motor neuronal survival. Torand-Allerand and co-workers, have demonstrated that gonadal steroids can regulate neurite outgrowth in both *in vivo* and *in vitro* systems.[22] It has recently been demonstrated that testosterone differentially regulates tubulin mRNA isoforms after injury in hamster facial motor neurones, in that β_{II}-tubulin mRNA levels were selectively altered by the steroid, whereas b_{III}- or α_{I}-tubulin were not.[23] In some regions of the mouse brain, the production of β-NGF appears to be regulated by testosterone.[24] Thus, androgens, as other steroids, control gene expression, and it is therefore important to identify androgen-controlled genes in neurones, as this could lead to an understanding of the mechanisms of motor neurone cell death.

Using the methodology of differential display of mRNA, between 60-70 amplified cDNAs have been studied; there were specific reproducible differences between mRNAs extracted from androgen-treated and untreated SH-SY5Y cells although the patterns of mRNA species between treated and control cells were very similar. Nine separate cDNA bands were identified as being putatively regulated by 5αDHT. All nine of these cDNAs have been sequenced, but only clone nine corresponded to a known gene sequence. The remaining eight clones may therefore represent unknown genes that may not have previously been identified. Since the differential display method provides clones of only 150-400 base pairs in length from the extreme 3'-region of mRNA, it is frequently difficult to identify the coding region if the cDNA has a long 3'-untranslated region. The lack of homology for clones one to eight with known sequences does not eliminate the possibility that they may be homologous to known proteins or protein families, since non-coding 3' region of mRNA species show great species to species variation.

Kennedy's mutants of the AR have a decreased transactivation function[19,20]; the genes identified by differential display may therefore be downregulated in Kennedy's disease. This can be studied by analysing the level of expression of such genes in SH-SY5Y cells transfected with Kennedy's mutant forms of the AR (i.e containing an increased CAG repeat sequence).

Acknowledgments

This work was supported by the Motor Neurone Disease Association of Great Britain and the British Medical Research Council..

REFERENCES

1. La Spada A.R., Wilson E.M., Lubahn D.M., Harding A.E., and Fischbeck K.H. Androgen receptor gene mutations in X-linked spinal and bulbar muscular atrophy. *Nature.* **352:**77 (1991)

2. Harding A.E., Thomas P.K., Baraitser M., Bradbury P.G., Morgan-Hughes J.A., Ponsford J.R. X-linked recessive bulbospinal neuropathy: a report of ten cases. *J.Neurol.Neurosurg. Psychiatry* **45:**1012 (1982).

3 Sobue G., Hashizume Y., Mukai E., Hirayama M., Mitsuma T., Takahashi A. X-linked recessive bulbospinal neuropathy. *Brain.* **112:**209 (1989).

4 Sar, M., and Stumpf, W.E. Androgen concentration in motor neurones of cranial nerves and spinal cord. *Science.* **197:**77 (1977).

5 Simerly, R.B., Chang, C., Muramatsu, M., and Swanson, L.W. Distribution of androgen and oestrogen receptor mRNA-containing cells in the rat brain: an *in situ* hybridisation study. *J.Comp.Neurol.* **294:**76 (1990).

6 Biedler J.L., Helson L., and Spengler B.A. Multiple neurotransmitter synthesis by human neuroblastoma cell lines and clones. *Cancer Res.* **38:**3751 (1973).

7 Biedler J.L., Helson L., and Spengler B.A. Morphology and growth, tumorigenicity, and cytogenetics of human neuroblastoma cells in culture. *Cancer Res.* **33:**2643 (1973).

8 Squinto S.P., Aldrich T.H., Lindsay R.M., Morrissey D.M., Panayotatos N., Bainco S., Furth M.E., and Yancopoulos G. Identification of functional receptoors for ciliary neurotrophic factor on neuronal cell lines and primary culture. *Neuron.* **5:**757 (1990).

9 Kaplan D.R., Matsumoto K., Lucarelli E., and Thiele C.J. Induction of TrkB by retinoic acid mediates biologic responsiveness to BDNF and differentiation of human neuroblastoma cells. *Neuron.* **11:**321 (1993).

10 Yerramilli-Rao P., Garofalo O., Whatley S., Leigh P.N., and Gallo J.-M. Androgen-controlled specific gene expression in neuroblastoma cells. *J. Neurol. Sci.* (In Press).

11 Tilley W.D., Marcelli M., Wilson J.D., and McPhaul M.J. Characterisation and expression of a cDNA encoding the hAR. *Proc. natl. Acad. Sci. U.S.A.* **86:**327 (1989).

12 Weiner L.P. Possible role of androgen receptors in Amyotrophic Lateral Sclerosis: a hypothesis. *Arch. Neurol.* **37:**129 (1989).

13 Kurtzke J.F. Risk factors in Amyotrophic Lateral Sclerosis. In: Rowland L.P., ed. *Advances in Neurology.* New York: Raven Press, **56:**245 (1991).

14 Brown C.J., Goss S.J., and Lubahn D.B., *etal.* Induction of the TRPM-2 gene in cells undergoing programmed death. *Mol, Cell. Biol.* **9:**3473 (1989).

15 Garofalo O., Figlewicz D.A., Leigh P.N., Powell J.F., Meninger V., Dib M., and Rouleau G.A. Androgen receptor gene polymorphisms in amyotrophic lateral sclerosis. *Neuromusc. Disord.* **3(3):**195 (1993).

16 Kennedy W.R., Alter M., and Sung J.H. Progressive spinal and bulbar muscular atrophy of late onset: a sex-linked recessive trait. *Neurology.* **18:**671 (1968).

17 McPhaul M.J., Marcelli M., Tilley W., Griffin J.E., Isidro-Guitierrez R.F., and wilson J.D. Molecular basis of androgen resistance in a family with a qualitative abnormality of the androgen receptor and responsive to high-dose androgen therapy. *J. Clin.Invest.* **87:**1413 (1991).

18 Richards R.I., Holman K., Friend K., Kremer E., Hillen D., Staples A., Brown W.T., Goonwardena P., Tarleton J., Schwartz C., *et al., Nature Genet.* **1:**257 (1992).

19 Mhatre A.N., Trifiro M.A., Kaufman M., Kazemi-Esfarjani P., Figlewicz D., Rouleau G., and Pinsky L. Reduced transcriptional regulatory competence of the androgen receptor in X-linked spinal and bulbar muscular atrophy. *Nature Genetics.* **5:**184 (1993).

20 Chamberlain N.L., Driver E.D., and Miedfeld R. The length and location of CAG trinucleotide repeats in the androgen receptor N-terminal domain affect transactivation function. *Nucl. Acids. Res.* **22(15):**3181 (1994).

21 Wang Y.-H., Amirhaeri S., Kang S., Wells R.D., and Griffith J.D. Preferential nucleosome assembly at DNA triplet repeats from the Myotonic Dystophy gene. *Science.* **265:**669 (1994).

22 Toran-Allerand C.D. Sex steroids and the development of the newborn mouse hypothalamus and preoptic area *in vitro:* implications for sexual differentiation. *Brain Res.* **106:**407 (1976).

23 Jones K.J., and Oblinger M.M. Androgenic regulation of tubulin gene expression in axotomised hamster facial motor neurones. *J.Neurosci.* **14**(6): 3620 (1994).

24 Katoh-Semba R., Semba R., Kato H., Ueno M., Arakawa Y., and Kato K. Regulation by androgen of levels of the b subunit of NGF and its mRNA in selected regions of the mouse brain. *J. Neurochem.* **62:** 2141 (1994).

GLUTAMATE-PROMOTED SURVIVAL IN HIPPOCAMPAL NEURONS: A DEFECT IN MOUSE TRISOMY 16

Linda L. Bambrick[1], Paul J. Yarowsky[2], and Bruce K. Krueger[1]

[1]Department of Physiology
[2]Department of Pharmacology
University of Maryland School of Medicine
660 West Redwood Street
Baltimore, MD 21201

ABSTRACT

Micromolar concentrations of glutamate, acting at non-NMDA, kainate-preferring receptors, increased the survival of cultured mouse hippocampal neurons maintained in serum-free, chemically-defined medium. Glutamate in excess of 20 μM was excitotoxic. Thus, the survival versus glutamate dose response relation was bell-shaped with an optimal glutamate concentration near 1 μM. Hippocampal neurons from mice with the genetic defect, trisomy 16 (Ts16), died 2–3 times faster than normal (euploid) neurons. Moreover, glutamate, at all concentrations tested, failed to increase survival of Ts16 neurons. Ts16 is a naturally-occurring mouse genetic abnormality, the human analog of which (Down syndrome) leads to altered brain development and Alzheimer's disease. These results demonstrate that the Ts16 genotype confers a defect in the glutamate-mediated survival response of hippocampal neurons and that this defect can account for their accelerated death.

INTRODUCTION

Since virtually no neurons are generated in the adult mammalian brain, mechanisms must exist to ensure the survival of neurons over a lifetime. It has been suggested that all cells require a constant trophic signal to block cell death[1] and the loss of neurons associated with normal aging and the accelerated neuronal death characteristic of neurodegenerative disorders could result from the failure of one or more intrinsic survival mechanisms. In this communication, we describe a novel survival mechanism for mouse hippocampal neurons which is absent in neurons from the trisomy 16 mouse, an animal model for neurodegeneration.

Alzheimer's disease (AD) is a neurodegenerative disorder characterized by death of neurons in the basal forebrain, cerebral cortex and hippocampus. Down syndrome

Neurodegenerative Diseases
Edited by Gary Fiskum, Plenum Press, New York, 1996

(DS,trisomy 21) is of interest for the study of neurodegeneration because all DS individuals develop dementia and the neuropathological symptoms of AD[2,3]. A substantial part of human chromosome 21, including genes coding for β-amyloid precursor protein, the free radical scavenging enzyme superoxide dismutase, and the type 5 ionotropic GluR (GluR5), together with most of the region associated with DS, can be found on mouse chromosome 16 and the genes in this region are triplicated in the trisomy 16 (Ts16) mouse[4,5], suggesting that the Ts16 mouse may be an animal model for DS. Although Ts16 mice do not survive beyond about 18 days gestation, Ts16 fetuses have been shown to exhibit abnormal brain development together with other abnormalities that parallel the symptoms of DS[6-10]. Furthermore, since Ts16 cholinergic basal forebrain neurons appear to be less likely to survive when grown *in vitro*[11] or transplanted into normal mouse brain[12], neuron survival as well as development may be affected in Ts16.

Since neurodegeneration in the hippocampus is a characteristic finding in AD, we focused our studies of Ts16 on the hippocampus and found that Ts16 hippocampal neurons *in vitro* die at an accelerated rate. We found that micromolar concentrations of glutamate promote the survival of euploid neurons, but that this response is lacking in Ts16 cells.

EXPERIMENTAL PROCEDURES

Generation of trisomic mice and karyotyping. Male mice doubly heterozygous for appropriate Robertsonian chromosome translocations were mated with C57Bl/6J female mice. The next day was designated as day 1 of gestation (E1). Normal and trisomic fetuses were easily distinguished, however, karyotypes were routinely generated from livers to confirm the genotype. Mice were obtained from Jackson Laboratories (Bar Harbor, ME).

Neuron cell culture. Cultures of embryonic hippocampus were prepared from euploid and Ts16 mouse fetuses at E16. For each Ts16 fetus, a euploid fetus from the same litter was used. The hippocampi were freed of meninges, digested with trypsin and dissociated by trituration in MEM10/10 (minimum essential medium with Earle's salts, 2mM glutamine, 10% fetal bovine serum, 10% horse serum with penicillin and streptomycin). Cells were plated at 50,000 cells per coverslip on polylysine-treated 12mm glass coverslips photoetched with a lettered grid of 175 μm X 175 μm squares (Eppendorf, Hamburg, Germany). At 1 div the MEM 10/10 was replaced with MEM supplemented with B27[13] (obtained from Gibco-BRL, Long Island, NY) and containing 10 μM cytosine-B-D-arabinofuranoside (Sigma) to kill proliferating cells including astrocytes and neuroblasts. B27 supplement contains optimized concentrations of neuron survival factors including putrescine, triiodothyronine, cortisol, progesterone, transferrin, superoxide dismutase, biotin, antioxidants and insulin, but no other polypeptide survival factors. At 2 div the medium was changed to MEM with B27. The cultures were maintained at 37°C, 95% air, 5% CO_2. Each coverslip was kept in a separate well, 2-4 coverslips were used for each condition in each experiment.

Materials and reagents. Glutamate and D-2-Amino-5-phosphonovaleric acid (APV) were obtained from Sigma. 6-Cyano-7-nitroquinoxaline-2,3-dion (CNQX) was obtained from Research Biochemicals International (Natick, MA). CNQX, APV and glutamate were prepared as stock solutions in MEM. Unless otherwise stated, all tissue culture media and supplements were obtained from Gibco-BRL.

RESULTS

Accelerated death of Ts16 neurons. Cultures of hippocampal neurons were prepared from euploid and Ts16 mouse fetuses at embryonic day 16 (E16), plated on coverslips etched with a lettered grid, and maintained in serum-free medium containing B27 supplement[13]. Astrocytes comprised less than 10% of the cells throughout these experiments. In order to quantitatively analyze neuron survival, the total number of cells identifiable as live neurons by phase contrast microscopy in each grid square was counted daily, over a six-day period. Cell counts at each day are expressed as a percentage of the starting number of cells at 2 div. The results of five experiments are summarized in Figure 1. About half of the euploid neurons survived to 5 div (filled circles). In contrast, the Ts16 neurons died more than twice as fast as euploid neurons with fewer than 10% of the Ts16 neurons remaining at 5 div (open squares).

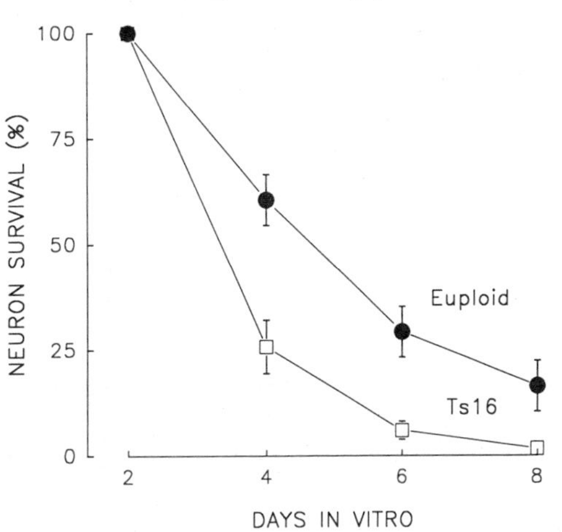

Figure 1. Survival of euploid (filled circles) and Ts16 (open squares) hippocampal neurons maintained in B27-supplemented MEM expressed as a percentage of the cells present at 2 div. Data are the means ± SEM of 5 experiments.

Normal morphology and bFGF response of Ts16 neurons. Ts16 and euploid cell body size and neuron shape were not obviously different at 2 div. An analysis of the number of neurites per neuron (Fig. 2) revealed that both Ts16 and euploid neurons had an average of 3–4 neurites per cell with similar distributions around the mean

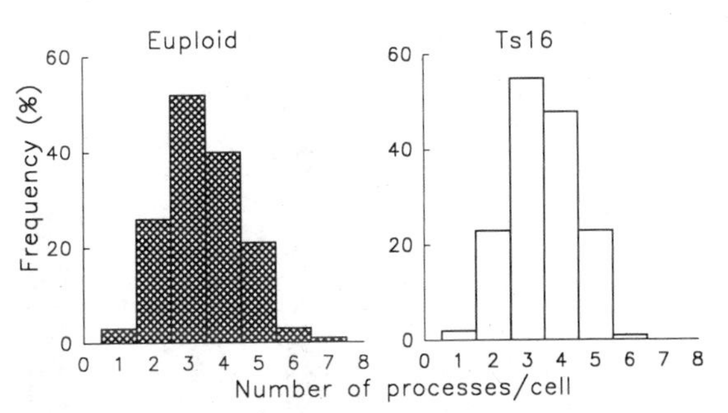

Figure 2. The number of processes on 146 euploid (left) and 152 Ts16 (right) hippocampal neurons at 4 div. Histograms show the frequency distributions of neurons with 1, 2, 3, etc. neurites. The two distributions are not significantly different.

value. Thus, despite the increased death of Ts16 neurons, there was no evidence for systematic morphological differences between live euploid and Ts16 neurons at 2 div.

Rodent hippocampal neuron survival is promoted by basic fibroblast growth factor (bFGF), which is normally produced by astrocytes[14]. The ability of Ts16 neurons to respond to applied growth factors was studied by treatment with 0.5–50 ng/ml bFGF.

Addition of bFGF enhanced the survival of both euploid (filled bars) and Ts16 (open bars) neurons (Fig. 3). Basic FGF dose response curves revealed that euploid and Ts16 neurons are equally sensitive to bFGF ($K_{bFGF} \approx 3$ ng/ml) with maximal stimulation at 10 ng/ml. Therefore, Ts16 neurons have a normal response to bFGF. Further, the ability of bFGF to rescue Ts16 cells shows that Ts16 neurons are not irrevocably destined to die at the time of bFGF addition (2 div). Indeed, in the presence of 3 ng/ml bFGF, the survival of Ts16 neurons is indistinguishable from the survival of euploid neurons in the absence of bFGF. However, bFGF does not cure the Ts16 deficit since Ts16 neurons are still less viable than similarly treated euploid neurons even at maximally effective concentrations of bFGF.

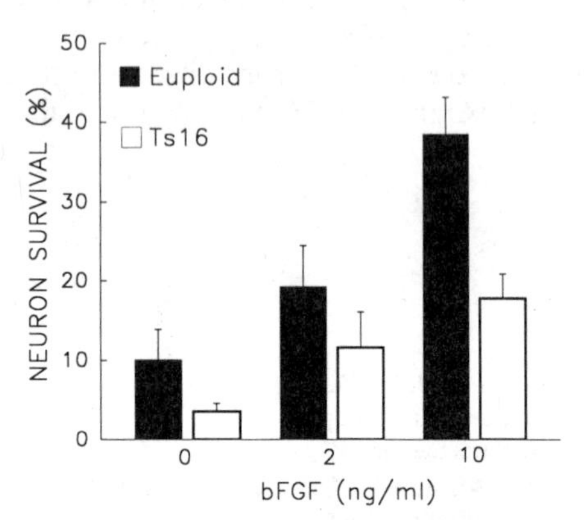

Figure 3. The effect of 2 and 10 ng/ml bFGF on euploid (filled bars) and Ts16 (open bars) hippocampal neuron survival. Survival was determined as in Figure 1. Data are the means ± SEM of 4-5 experiments.

Abnormal glutamate response of Ts16 neurons. Since both excitotoxic death and neuron survival can be mediated through glutamate receptors[15–18], the presence of the gene coding for a kainate-preferring GluR on murine chromosome 16[5] raised the possibility that the Ts16 neurons might have an altered response to glutamate which could lead to their accelerated death. Elevated GluR expression could result in an increased sensitivity to endogenous glutamate in the cultures, which could cause neuron death by an excitotoxic mechanism. In order to test this hypothesis, euploid and Ts16 hippocampal neurons were treated with either the NMDA-type GluR blocker, AP-V, or the kainate/AMPA-type (K/A) ionotropic GluR blocker, CNQX. Neither GluR blocker improved the survival of Ts16 neurons indicating that the increased rate of Ts16 neuron death was not due to increased excitotoxicity as hypothesized. However, CNQX (but not AP-V; data not shown) *decreased* the survival of euploid neurons (Figure 4). This suggests that activation of GluRs by endogenous glutamate contributes

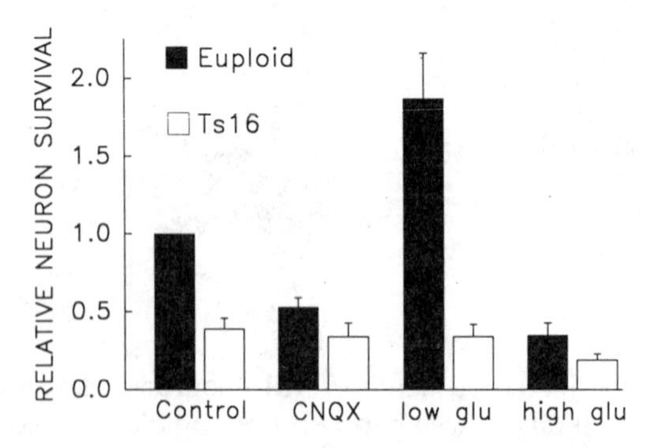

Figure 4. Effect glutamate (1 μM or 20 μM) and of the kainate-preferring GluR blocker, CNQX (50 μM), on euploid and Ts16 hippocampal neuron survival. Survival was determined as described in Figure 1. Filled bars: euploid; open bars: Ts16 neurons.

to euploid neuron survival and that either this response is lacking in Ts16 neurons or insufficient glutamate is present in Ts16 cultures to promote neuron survival.

The effects of GluR activation on euploid and Ts16 neuron survival were further investigated by determining survival at 5 div in the absence or presence of varying concentrations of glutamate added at 2 div. The addition of micromolar concentrations of glutamate enhanced euploid neuron survival with a maximally effective glutamate concentration of about 0.5 to 1 μM. In contrast, glutamate concentrations in the same range had no effect on Ts16 survival. Concentrations of glutamate 20 μM or greater deceased the survival of both euploid and Ts16 neurons to levels below those observed in the absence of added glutamate, probably reflecting excitotoxicity. These results are summarized in Figure 4. Overall, the mean maximal effect of glutamate on euploid neurons was a 2.5 ± 0.5-fold increase in survival as compared to the survival in the absence of added glutamate and presence of CNQX. Glutamate failed to rescue Ts16 neurons at any concentration tested. Therefore, the lack of effect of CNQX on Ts16 neurons resulted from their inability to respond to glutamate rather that from a lack of glutamate in the medium.

DISCUSSION

Glutamate is required for optimal survival of normal hippocampal neurons. The decreased survival of euploid neurons in the presence of CNQX and the increase in their survival induced by addition of micromolar glutamate (Fig. 4, filled bars) shows that glutamate acting through K/A GluRs promotes the survival of normal hippocampal neurons.

There is considerable evidence demonstrating that survival of cerebellar granule cell neurons *in vitro* is decreased by block of NMDA or non-NMDA GluRs while NMDA and glutamate themselves promote granule cell survival[16–18]. *In vivo*, NMDA receptor antagonists also decrease neuron survival in the dentate gyrus[19]. These results, together with studies showing a trophic role for metabotropic GluR activation in cerebellar Purkinje and granule cell survival[20,21] and our finding of a trophic role for K/A GluR activation in the hippocampus, suggest that glutamate may act throughout the brain as a trophic factor as well as a neurotransmitter and that its trophic actions may be mediated by distinct receptor subtypes in different neuronal populations. Moreover, since micromolar glutamate is known to be present in the adult brain[22], glutamate may contribute to neuron survival throughout life.

Glutamate-promoted hippocampal neuron survival is absent in the Ts16 mouse. The failure of CNQX or added glutamate to affect Ts16 neuron survival (Fig. 4), suggested that the accelerated Ts16 neuron death was due to the inability of the Ts16 neurons to respond to endogenous glutamate. Since the survival of euploid neurons in the presence of K/A GluR blockers was virtually indistinguishable from that of Ts16 neurons, this single defect is sufficient to account for the accelerated death of Ts16 neurons. The decreased survival of Ts16 neurons illustrates the profound consequences of the absence of one survival mechanism.

A "glutamate window" for optimal neuron survival. Exposure to very high concentrations of glutamate is toxic to neurons. However, while elevated glutamate is neurotoxic, previous studies in the cerebellum[16–18] and our results in the hippocampus imply that glutamate levels that are too low are also detrimental to neuron survival, suggesting the existence of a "glutamate window" — a range of glutamate concentrations optimal for neuronal survival. While we have not directly measured the concentration of endogenous glutamate in our cultures, the substantial increase in euploid neuron survival observed with 0.2−1 μM added glutamate (Fig. 5) indicates that endogenous levels are in a similar range. Thus we estimate that the optimal "window" for glutamate is in the 1-3 μM range. Interestingly, it has been reported that in the adult hippocampus, extracellular levels of glutamate *in vivo* are about 3 μM[22], and thus would be expected to provide maximal trophic support based on our *in vitro* experiments.

A dynamic, neuroprotective role for astrocytes. Maintaining glutamate concentrations within this trophic "window" may be an important *in vivo* role for astrocytes. One of the commonly accepted physiological roles for astrocytes in the brain is to accumulate glutamate and convert it to the inactive amino acid glutamine, thus protecting neurons from the excitotoxic effects of high glutamate[23−25]. Recently, cultured rat hippocampal astrocytes have also been shown to be capable of releasing glutamate by a Ca^{2+}-dependent mechanism[26]. Taken together with our findings, this observation suggests that astrocytes may play a dual role in ensuring neuron survival not only by protecting neurons from high, excitotoxic levels of glutamate, but also by supplying glutamate when endogenous levels fall below those that are optimal for neuron survival.

Redundant survival mechanisms may mask the absence of glutamate-promoted survival. The defect in glutamate-induced survival is not necessarily fatal, as the accelerated neuron death can be overcome, at least in part, by addition of another survival factor (e.g., bFGF, Fig. 3). The presence of such multiple trophic inputs may help to explain one of the features of many neurodegenerative diseases including AD, which is that neurodegeneration ensues only after several decades of normal function, even in individuals genetically fated to develop the disease. Neurons with a defect in one survival mechanism may live, provided adequate amounts of other survival factors are available. The decreased production of neurotrophic factors with aging[27] could unmask such a defect, leading to neuron death. Our experiments in the Ts16 mouse suggest the hypothesis that a defect in the glutamate survival pathway could contribute to the pathophysiology of neurodegenerative diseases such as AD.

ACKNOWLEDGEMENTS

The authors' research has been supported by NIH grants NS16285 and AG10686 (to BKK), by an Alzheimer's Association/Helen and Philip Brody Pilot Research Grant (to BKK), by an Alzheimer's Association, Northern Virginia Chapter pilot grant (to LLB), and by NSF grant BNS89-198941 (to PJY).

REFERENCES

1. M.C. Raff, B.A. Barres, J.F. Burne, H.S. Coles, Y. Ishizaki, and M.D. Jacobson, Programmed cell death and the control of cell survival: Lessons from the nervous system. *Science* 269:695-700 (1993).
2. H. Potter, Review and hypothesis: Alzheimer disease and Down syndrome — chromosome 21 nondisjunction may underlie both disorders. *Am. J. Hum. Genet.* 48:1192-1200 (1991).
3. D.M.A. Mann, Down's syndrome and Alzheimer's disease: Towards an understanding of the pathogenesis. in "Neurodegeneration", A.J. Hunter and M. Clark, eds. Academic Press, San Diego, CA (1992).
4. J.T. Coyle, M.L. Oster-Granite, R.H. Reeves, and J.D. Gearhart, Down syndrome, Alzheimer's disease and the trisomy 16 mouse. *Trends Neurosci.* 11:390-394 (1988).
5. P. Gregor, R.H. Reeves, E.W. Jabs, X. Yang, W. Dackowski, J.M. Rochelle, R.H. Brown, Jr., J.L. Haines, B.F. O'Hara, G.R. Uhl, and M.F. Seldin, Chromosomal localization of glutamate receptor genes: relationship to familial amyotrophic lateral sclerosis and other neurological disorders of mice and humans. *Proc. Natl. Acad. Sci. USA* 90:3053-3057 (1993).
6. H.S. Singer, M. Tiemeyer, J.C. Hedreen, J. Gearhart, and J.T. Coyle, Morphologic and neurochemical studies of embryonic brain development in murine trisomy 16. *Dev. Brain Res.* 15:155-166 (1984).

7. J.D. Gearhart, M.L. Oster-Granite, R.H. Reeves, and J.T. Coyle, Developmental consequences of autosomal aneuploidy in mammals. *Dev. Genet.* 8:249-265 (1987).
8. J. Kiss, M. Schlumpf, and R. Balazs, Selective retardation of the development of the basal forebrain cholinergic and pontine catecholaminergic nuclei in the brain of trisomy 16 mouse, an animal model of Down's syndrome. *Dev. Brain Res.* 50:251-264 (1989).
9. J.E. Sweeney, C.F. Hohmann, M.L. Oster-Granite, and J.T. Coyle, Neurogenesis of the basal forebrain in euploid and trisomy 16 mice: an animal model for developmental disorders in Down syndrome. *Neuroscience* 31:413-425 (1989).
10. M.L. Oster-Granite, The trisomy 16 mouse as an animal model relevant to studies of growth retardation in Down's syndrome. in "Growth Hormone Treatment in Down's Syndrome", S. Castells and K.E. Wisniewski, eds. John Wiley and Sons Ltd, New York (1993).
11. P. Corsi and J.T. Coyle, Nerve growth factor corrects developmental impairments of basal forebrain cholinergic neurons in the trisomy 16 mouse. *Proc. Natl. Acad. Sci. USA* 88:1793-1797 (1991).
12. D.M. Holtzman, Y. Li, S.J. DeArmond, M.P. McKinley, F.H. Gage, C.J. Epstein, and W.C. Mobley, Mouse model of neurodegeneration: atrophy of basal forebrain cholinergic neurons in trisomy 16 transplants. *Proc. Natl. Acad. Sci. USA* 89:1383-1387 (1992).
13. G.J. Brewer, J.R. Torricelli, E.K. Evege, and P.J. Price, Optimized survival of hippocampal neurons in B27-supplemented Neurobasal, a new serum-free medium combination. *J. Neurosci. Res.* 35:567-576 (1993).
14. K. Unsicker, H. Reichert-Preibsch, R. Schmidt, B. Pettmann, G. Labourdette, and M. Sensenbrenner, Astroglial and fibroblast growth factors have neurotrophic functions for cultured peripheral and central nervous system neurons. *Proc. Natl. Acad. Sci. USA* 84:5459-5463 (1987).
15. D.W. Choi, Excitotoxic cell death. *J. Neurobiol.* 23:1261-1276 (1992).
16. R. Balazs, O.S. Jorgensen, and N. Hack, N-Methyl-D-aspartate promotes the survival of cerebellar granule cells in culture. *Neuroscience* 27:437-451 (1988).
17. R.D. Burgoyne, M.E. Graham, and M. Cambray-Deakin, Neurotrophic effects of NMDA receptor activation on developing cerebellar granule cells. *J. Neurocyt.* 22:689-695 (1993).
18. G.-M. Yan, B. Ni, M. Weller, K.A. Wood, and S.M. Paul, Depolarization or glutamate receptor activation blocks apoptotic cell death of cultured cerebellar granule neurons. *Brain Res.* 656:43-51 (1994).
19. E. Gould, H.A. Cameron, and B.S. McEwen, Blockade of NMDA receptors increases cell death and birth in the developing rat dentate gyrus. *J. Comp. Neurol.* 340:551-565 (1994).
20. H.T.J. Mount, C.F. Dreyfus, and I.B. Black, Purkinje cell survival is differentially regulated by metabotropic and ionotropic excitatory amino acid receptors. *J. Neurosci.* 13:3173-3179 (1993).
21. A. Copani, V.M.G. Bruno, V. Barresi, G. Battaglia, D.F. Condorelli, and F. Nicoletti, Activation of metabotropic glutamate receptors prevents neuronal apoptosis in culture. *J. Neurosci.* 64:101-108 (1995).
22. A. Lehman, H. Isacsson, and A. Hamberger, Effects of *in vivo* administration of kainic acid on the extracellular amino acid pool in the rabbit hippocampus. *J. Neurochem.* 40:1314-1320 (1983).
23. K. Sugiyama, A. Brunori, and M.L. Mayer, Glial uptake of excitatory amino acids influences neuronal survival in cultures of mouse hippocampus. *Neuroscience* 32:779-791 (1989).
24. M.P. Mattson and B. Rychlik, Glia protect hippocampal neurons against excitatory amino acid-induced degeneration: involvement of fibroblast growth factor. *Int. J. Devl. Neurosci.* 8:399-415 (1990).
25. J.D. Rothstein, L. Jin, M. Dykes-Hoberg, and R.W. Kuncl, Chronic inhibition of glutamate uptake produces a model of slow neurotoxicity. *Proc. Natl. Acad. Sci. USA* 90:6591-6595 (1993).
26. V. Parpura, T.A. Basarsky, F. Liu, K. Jeftinija, S. Jeftinija, and P.G. Haydon, Glutamate-mediated astrocyte-neuron signalling. *Nature* 369:744-747 (1994).
27. G. Pepeu, L. Ballerini, and A.M. Pugliese, Neurotrophic factors in the aging brain: a review. *Arch. Gerontol. Geriatr.* Suppl.2:151-158 (1991).

RNA MESSAGE LEVELS IN NORMALLY AGING AND IN ALZHEIMER'S DISEASE(AD)-AFFECTED HUMAN TEMPORAL LOBE NEOCORTEX

Walter J. Lukiw[1,2], Donald R. McLachlan[1] and Nicolas G. Bazan[2],

[1]Centre for Research in Neurodegenerative Disease, Tanz Neuroscience Building, University of Toronto, Toronto Canada M5S IA8
[2]LSU Neuroscience Center, Louisiana State University School of Medicine, New Orleans, Louisiana, USA 70112-2234

INTRODUCTION

The expression of genetic information in neurons and glia, as in all eukaryotic cells, is dependent upon a hierarchy of control, from the acquisition of a transcriptionally competent chromatin conformation to the final assembly and compartmentalization of the relevant gene products. As outlined in Table 1, this genetic signal transduction pathway from DNA to RNA to protein can be further categorized into a series of control gates involving DNA transcription and other post-transcriptional aspects of gene regulation (1-6) and (b) a series of translational and post-translational control points (8-12); the shuttling of RNA messages out of the nucleus and into the cytoplasm represents one prominent intermediary control point (7). Amongst the eukaryotic cells, neurons of the mammalian brain appear to be unique in that the rate of genetic information flow through these control gates appears to be particularly rapid[1], and both quantitatively and qualitatively large amounts of brain-specific DNA transcription products are generated[1,2].

Clearly, *the commitment to transcribe brain DNA templates into RNA messages represents one, if not the most important, control point in the regulation of brain gene expression.* In the normally aging human brain, patterns of DNA transcription and the quantity, quality and distribution of RNA message molecules typically mirrors the biochemical and physiological status of the brain cells. Importantly, the expression of brain genes may become altered upon exposure to novel environmental factors, or as the result of, or in response to, disease.

In order to evaluate the contribution of specific DNA transcription products to the functioning of neurologically normal- control and age-matched Alzheimer's disease (AD) affected human neocortical cells, in this ongoing study, we have to date quantitated total RNA* message levels for 37 genes of neurobiological interest in 54, 0.7 to 25 hour post-

*combined messenger RNA (mRNA) and heterogeneous nuclear RNA (hnRNA).

Neurodegenerative Diseases
Edited by Gary Fiskum, Plenum Press, New York, 1996

Table 1. Points of Control in Eukaryotic Gene Expression

1. formation and maintenance of a transcriptionally competent chromatin structure
2. epigenetic mechanisms (i.e. DNA methylation)
3. initiation of DNA transcription / RNA message synthesis
4. rate of DNA transcription / RNA message synthesis
5. RNA message stability / longevity
6. RNA message processing (capping, intron removal, polyadenylation)
7. shuttling / transport of RNA message from the nucleus to the cytoplasm
8. intracellular RNA message targeting
9. RNA message - ribosome interactions / efficiency and rate of translation
10. post-translational modification of primary gene products
11. selective activation / inactivation / splicing of primary gene products
12. assembly / compartmentalization of processed gene products

mortem human temporal lobe neocortices, isolated from 23 neurologically normal control and 31 age-matched sporadic AD brains (Table 2) . Because many brain genes are expressed to different extents in different neocortical areas[2,] we have elected to study the abundance of DNA transcription products only in the superior temporal lobe of the human neocortex (Brodmann Areas 22, 41 and 42), a neocortical association area moderately to severly affected by the AD process.

EXPERIMENTAL DESIGN FOR RNA ANALYSIS FROM HUMAN BRAIN TISSUE

The selection of control and AD affected human brain tissues was based strictly on CERAD guidelines[3]. Brain tissues were critically matched for age and post-mortem interval (PMI; Table 2). The experimental protocol for tissue dissection and manipulation, total RNA isolation, characterization and quantitation from control and AD-affected human brain tissues have been described elsewhere[2,4]. Briefly, RNA was isolated from temporal lobe grey matter employing methods modified from Chomzynski[5] utilizing TRIzol™ reagent (Bethesda Research Laboratories). This single-step procedure uses a mono-phasic solution of phenol and guanidine isothiocyanate to generate DNA, RNA and protein from as little as 0.5 gram of the same tissue sample. Total RNA message levels were normalized using an internal RNA standard, such as β-actin, an RNA message whose level is not

Table 2. Summary of Human Neocortical Tissues Used for RNA Message Isolation

Neurological Status	N	Age (mean+/-SD; yrs)	Post-Mortem Interval[1] (mean+/- SD; hrs)
normal + controls[2]	23	69.5 +/- 6.9	6.8 +/- 11.1
Alzheimer's disease[3]	31	67.9 +/- 6.6	6.5 +/- 18.5

[1]Interval of post-mortem brain removal to freezer storage at -81°C.
[2]normals comprised cognitively intact, normally aged humans; control brains included dialysis encephalopathy, Parkinsons disease, amyotrophic lateral sclerosis; all of these latter brains failed to meet the current diagnosis for AD using CERAD criteria; there were none to sparse senile plaques (SP) and neurofibrillary tangles (NFT) in the superior temporal lobe neocortex.
[3]All brains passed CERAD criteria; all patients were severly demented and numerous SP and NFT were a consistent feature of the superior temporal lobe neocortex.

Table 3. Seven selected RNA message abundances, expressed as a ratio of Alzheimer's disease (AD) over normal + control, in the human temporal lobe neocortex, and the human chromosomal assignments for these DNA transcripts.

RNA Message	AD/control (%)	Chromosomal Locus
BC200 (brain cytoplasmic, 200 nucleotide)	30	unknown
NF-L (neurofilament light chain protein)	33	8p21
growth inhibitory factor (GIF; MT-related)	39	16p
synaptophysin (synapse protein)	51	Xp11.23-Xp11.22
α-tubulin (microtubule protein)	58	13q11
cytochrome oxidase (mitochondrial)	60	mitochondrial
NF-M (neurofilament medium chain protein)	64	8p21

significantly changed in AD[4], or by comparing against a constant amount of DNA isolated from the same tissue sample. Some of the dot blotting procedures have been automated using a Beckman Biomek 1000 Laboratory Automation Workstation. Total and relative RNA signal levels were quantitated via electronic autoradiography using a Packard Instant Imager, a BIO-RAD GS-250 Molecular Imager or by using conventional autoradiography using Fuji RX or Kodak Biomax MR X-ray film. Out of the 37 RNA messages quantitated, the seven greatest reductions in AD/normal + control RNA signal strength are shown in Table 3, along with the chromosomal assignments for these genes.

ALZHEIMER'S DISEASE (AD) INVOLVES ALTERATIONS IN GENE EXPRESSION

AD remains a devastating neurological dysfunction of uncertain etiology. While features of the familial or hereditary form of AD appear to be linked to genetic loci on human chromosomes 14, 19 and 21, the genetic mechanisms associated with the more common idiopathic or sporadic form of AD remain incompletely understood.

In sporadic AD cases, brain cells of the temporal lobe neocortex show significant reductions in specific RNA message populations. RNA message levels of 60 to 64 percent AD/normal + control included the neurofilament medium chain (NF-M) protein and cytochrome oxidase (ANOVA ~ 0.02 to 0.04[4]; Table 3). RNA message levels detected at less than 60 percent AD/control included those coding for the cytoskeletal and synaptic forming elements α-tubulin, synaptophysin, the growth inhibitory factor (GIF) metallothionein III, neurofilament light chain (NF-L) protein and and the putative RNA message carrier BC200, to 58, 33, 51, 39 and 30 percent, respectively, of both age and post-mortem interval matched control levels (ANOVA $\leq$0.01[4]; Table 3). Interestingly, the most dramatic reduction in RNA message abundance was found to be for BC200[2], a 200 nucleotide brain cytoplasmic RNA that may be involved in RNA transport[6], the intermediate control point in the flow of brain genetic information as depicted in Table 1.

SUMMARY

1. In this study, 18SrRNA, 28SrRNA and the Alu repetitive sequence RNA (~7SL) were by far the most abundant RNA species detected in RNA prepared from the neurons and glia of normally aging human temporal lobe neocortex.

2. Amongst the 37 RNA messages analyzed in this ongoing study, the RNA message coding for synaptophysin was found to be the most abundant protein-coding RNA message. RNA species at levels of 50%, or greater than the level of this synaptic vesicle-specific membrane protein included only the microtubule protein α-tubulin at 55.3%, glial fibrillary acidic protein, GFAP at 61.7%, and the neuron-specific neurofilament light chain protein NF-L at 95.1%.
3. RNA signals essential to brain cell structures and functions including (a) the synapse of the neuron (NF-L and synaptophysin), (b) the axonal and dendritic branching of the neuron (NF-L and α-tubulin) and (c) the cytoskeleton of the glia (GFAP and α-tubulin) are amongst the most abundant RNA messages detected in the normal human temporal lobe neocortex in this study.
4. There is a marked variability in the *individual* RNA message abundance within each normal human temporal lobe neocortex. This represents a kind of *RNA signature* for a moderately large group of RNA messages. The RNA profile for each individual may influence both aspects of normal genetic function and/or predisposition to disease.
5. Within the superior temporal lobe neocortex of patients afflicted with AD, statistically significant reductions in RNA message levels were measured for the neurofilament proteins NF-L and NF-M, mitochondrial cytochrome oxidase, α-tubulin, synaptophysin, growth inhibitory factor (GIF) and BC200 RNA to 33, 64, 60, 58, 51, 39 and 30 percent, respectively, of age and post-mortem interval matched control levels (ANOVA≤0.01[4]; Table 3). Reductions in these RNA messages may play some role in the altered neuronal morphology and growth, energy metabolism, RNA message translocation, signal processing and other neuronal functions associated with AD affected brain cells.
6. The human chromosomal loci for the genes encoding these significantly depressed RNA messages (Table 3) may be a good place to look for genetic abberations associated with the AD process.

ACKNOWLEDGEMENTS

We would like to thank Drs. John Deck and Catherine Bergeron of the Canadian Brain Tissue Bank, Toronto, Canada, Drs. Larry Carver and Hector LeBlanc of the Louisiana State University Brain Tissue Bank, New Orleans, Louisiana, USA, and Dr. Evgeny Rogaev of the National Research Center of Mental Health, Academy of Medical Sciences, Moscow, Russia, for their generous supply of human brain tissues and helpful discussions on RNA extractions and quantitation.

REFERENCES

1. R.J. Thompson, Studies on RNA synthesis in two populations of nuclei from the mammalian cerebral cortex, J. Neurochem. 21:19 (1973).
2. W.J. Lukiw, P. Handley, L.Wong, and D.R.C. McLachlan, BC200 RNA in normal human neocortex, non-Alzheimer dementia (NAD) and senile dementia of the Alzheimer type (AD), Neurochem Res. 17:591 (1992).
3.. S.S. Mirra, A. Heyman, D. McKeel, S.M. Sumi, B.J. Crain, L.M. Brownlee, F.S. Vogel, J.P. Hughes, G. van Belle, L.Berg, and participating CERAD neuropathologists, The consortium to establish a registry for Alzheimer's disease (CERAD) Part II. Standardization of the neuropathologic assessment of Alzheimer's disease, Neurology 41:479 (1991).

4. W.J. Lukiw, N.G. Bazan, and D.R.C. McLachlan, Relative RNA message abundance in control and Alzheimer's disease-affected human neocortex, *in:* "Research Advances in Alzheimer's Disease and Related Disorders" K.Iqbal, J.A. Mortimer, B.Winblad and H.M.Wisniewski, eds., John Wiley & Sons Ltd, New York (1995).
5. P. Chomzynski, A reagent for the single step-simultaneous isolation of RNA, DNA and proteins from cell and tissue samples, Biotechniques 15:532 (1993).
6. H. Tiedge, W. Chen, and J. Brosius, Primary structure, neural specific expression, and dendritic location of human BC200 RNA, J. Neurosci. 13:2382 (1993) .

FUNCTIONAL ACTIVATION OF ENERGY METABOLISM IN NERVOUS TISSUE: WHERE AND WHY

Louis Sokoloff and Shinichi Takahashi

Laboratory of Cerebral Metabolism
National Institute of Mental Health
Bethesda, Maryland 20892, U.S.A.

INTRODUCTION

In tissues that do physical work, such as heart, skeletal muscle, and kidney, rates of energy metabolism under physiological conditions vary more or less in proportion to the amount of work being done by the tissue. To demonstrate such a relationship in brain has been difficult because, first of all, the exact nature of the physical work done by brain tissue is not obvious, and, secondly, the brain mediates a variety of functions, each of which is localized in discrete regions specific for the function and not in the tissue as a whole. It is only recently that it has become possible to measure the rates of energy metabolism in such discrete structural and functional components of the nervous system in conscious, behaving animals and man and to relate these rates to the levels of functional activity within them.

DETERMINATION OF LOCAL RATES OF ENERGY METABOLISM IN NERVOUS TISSUES

Under normal physiological conditions the energy metabolism of neural tissue is supported almost exclusively by the oxidative metabolism of glucose.[1] In ketotic states, such as those associated with starvation, high fat diets, diabetes, etc., when the production and, therefore, also the blood levels of ketone bodies are elevated, D-β-Hydroxybutyrate and acetoacetate can substitute, at least in part, for glucose and be oxidized by the brain in significant amounts.[2] Therefore, except in such ketotic states and also, of course, when the tissue is hypoxic, the rate of glucose utilization is almost stoichiometrically equivalent to the rate of oxygen consumption needed for the complete oxidation of glucose to CO_2 and H_2O. In such normal physiological conditions glucose is, therefore, as good a measure of the total energy metabolism in neural tissues as oxygen consumption.[1]

The development of the autoradiographic 2-deoxy-D-glucose (DG) method[3] made it possible to measure rates of local glucose utilization ($lCMR_{glc}$) simultaneously in all local regions of the nervous system in normal, conscious, behaving animals in various functional states. The method was designed to take advantage of the spatial localization provided by

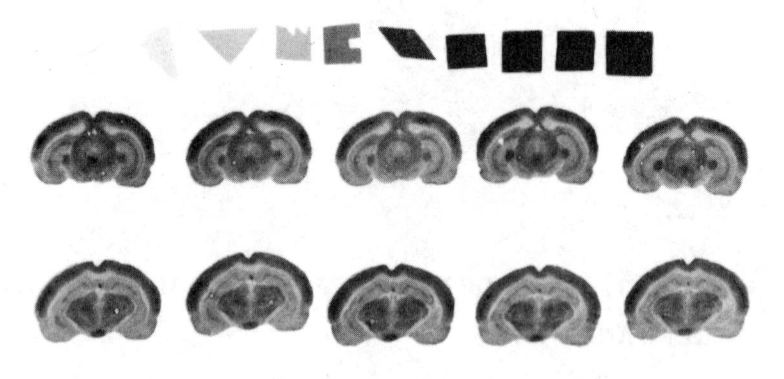

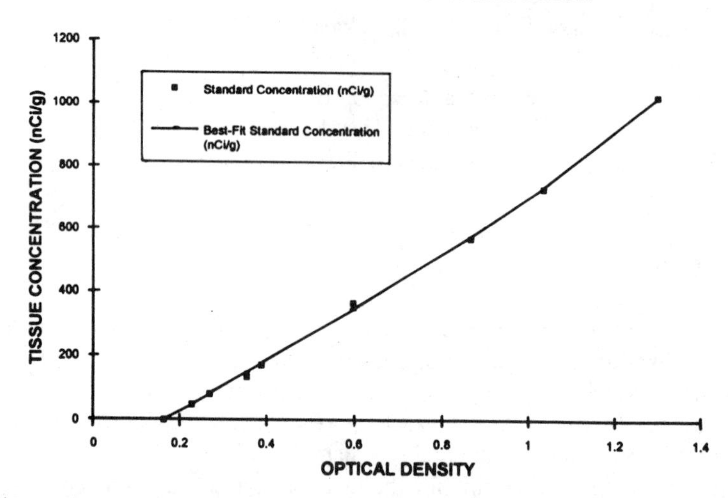

Figure 1. [^{14}C]Deoxyglucose autoradiograms of 20 μm coronal sections of brain from conscious rat and of [^{14}C]methylmethacrylate standards used to quantify ^{14}C concentrations in tissues by quantitative densitometry. The line drawing, which illustrates the relationship between the optical density (darkness) in the autoradiograms and the ^{14}C concentration in the tissue, is determined by densitometric measurements of the autoradiographic images of the calibrated ^{14}C-labeled methylmethacryalate standards (geometric images at top).

quantitative autoradiography. With quantitative autoradiography it is possible to determine the local concentrations of radioactive isotopes in tissues by densitometric analysis of autoradiograms of tissue sections and appropriately calibrated radioactive standards (Fig. 1). The theoretical basis of the DG method has been described in detail previously.[3] Briefly, it is based on the kinetic analysis of a model of the uptake of 2-deoxyglucose and glucose from the blood and their phosphorylation by hexokinase in neural tissues (Fig. 2). The method was originally designed for use with autoradiography of sections of nervous tissue prepared from animals that had been labeled *in vivo* with tracer doses of 2-[^{14}C]deoxyglucose. The spatial resolution within the tissue achieved with the quantitative autoradiographic technique is about 200 μm.[4] The method was subsequently adapted for use with the positron-emitter 2-[^{18}F]fluoro-2-deoxy-D-glucose ([^{18}F]FDG) and positron emission tomography (PET) and applied to man.[5,6] Rates of glucose utilization in some representative cerebral structures of the normal conscious albino rat and monkey are listed in Table 1.[3,7]

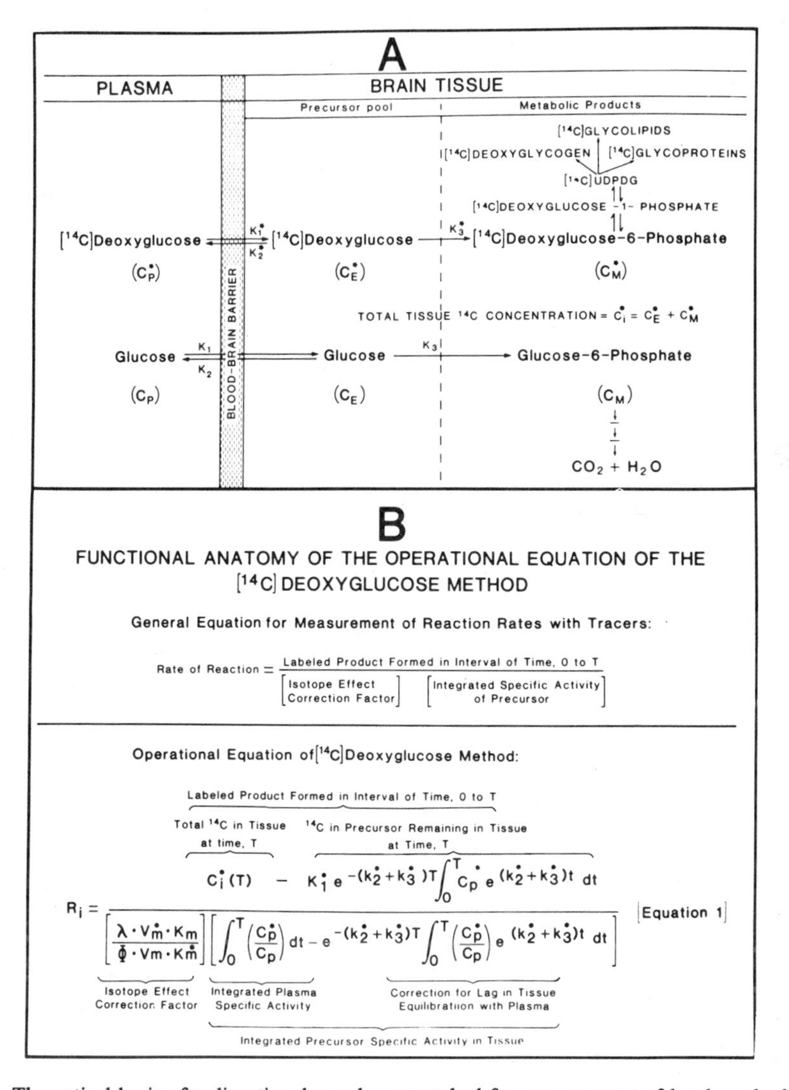

$$R_i = \frac{C_i^*(T) - k_1^* e^{-(k_2^*+k_3^*)T} \int_0^T C_p^* e^{(k_2^*+k_3^*)t}\,dt}{\left[\frac{\lambda \cdot V_m^* \cdot K_m}{\Phi \cdot V_m \cdot K_m^*}\right]\left[\int_0^T \left(\frac{C_p^*}{C_p}\right) dt - e^{-(k_2^*+k_3^*)T} \int_0^T \left(\frac{C_p^*}{C_p}\right) e^{(k_2^*+k_3^*)t}\,dt\right]}$$

Figure 2. Theoretical basis of radioactive deoxyglucose method for measurement of local cerebral glucose utilization. **A.** Theoretical model. C_i^* represents total ^{14}C concentration in a single homogeneous tissue of the brain; C_P^* and C_P represent the concentrations of [^{14}C]deoxyglucose and glucose in arterial plasma, respectively; C_E^* and C_E are their respective concentrations in the tissue pools that serve as substrates for hexokinase; C_M^* is the total concentration of primary and secondary products of [^{14}C]deoxyglucose phosphorylation in the tissue. K_1^* and k_2^* represent the rate constants for carrier-mediated transport of [^{14}C]deoxyglucose from plasma to tissue and back from tissue to plasma, respectively; k_3^* is the rate constant for phosphorylation of [^{14}C]deoxyglucose by hexokinase; k_1, k_2, and k_3 are the equivalent rate constants for glucose. [^{14}C]Deoxyglucose and glucose share and compete for the carrier that transports them both between plasma and tissue and for hexokinase which phosphorylates them to their respective hexose-6-phosphates. **B.** Functional anatomy of the operational equation of the radioactive deoxyglucose method. T represents the time of termination of the experimental period, and t is the variable time. λ equals the ratio of the distribution space of deoxyglucose in the tissue to that of glucose; Φ equals the fraction of glucose which, once phosphorylated, continues down the glycolytic pathway; V_m and K_m represent the maximal velocity and Michaelis-Menten constant of hexokinase for glucose; and V_m^* and K_m^* are the equivalent constants for 2-deoxyglucose. The other symbols are the same as those defined in **A.**[3]

Table 1. Representative values for local crebral gucose utilization in normal conscious albino rat and monkey (μmols/100 g/min)*

Structure	Albino Rat[a] (n=10)	Monkey[b] (n=7)
	Gray Matter	
Visual cortex	107 ± 6	59 ± 2
Auditory cortex	162 ± 5	79 ± 4
Parietal cortex	112 ± 5	47 ± 4
Sensory-motor cortex	120 ± 5	44 ± 3
Thalamus: lateral nucleus	116 ± 5	54 ± 2
Thalamus: ventral nucleus	109 ± 5	43 ± 2
Medial geniculate body	131 ± 5	65 ± 3
Lateral geniculate body	96 ± 5	39 ± 1
Hypothalamus	54 ± 2	25 ± 1
Mammillary body	121 ± 5	57 ± 3
Hippocampus	79 ± 3	39 ± 2
Amygdala	52 ± 2	25 ± 2
Caudate-putamen	110 ± 4	52 ± 3
Nucleus accumbens	82 ± 3	36 ± 2
Globus pallidus	58 ± 2	26 ± 2
Substantia nigra	58 ± 3	29 ± 2
Vestibular nucleus	28 ± 5	66 ± 3
Cochlear nucleus	113 ± 7	51 ± 3
Superior olivary nucleus	133 ± 7	63 ± 4
Inferior colliculus	197 ± 10	103 ± 6
Superior colliculus	95 ± 5	55 ± 4
Pontine gray matter	62 ± 3	28 ± 1
Cerebellar cortex	57 ± 2	31 ± 2
Cerebellar nuclei	100 ± 4	45 ± 2
	White Matter	
Corpus callosum	40 ± 2	11 ± 1
Internal capsule	33 ± 2	13 ± 1
Cerebellar white matter	37 ± 2	12 ± 1
	Weighted Average for Whole Brain	
	68 ± 3	36 ± 1

* The values are the means ± S.E.M. of the number of animals indicated in parentheses. [a] From Sokoloff *et al.*[3] [b] From Kennedy *et al.*[7]

RELATIONSHIP BETWEEN LOCAL FUNCTIONAL ACTIVITY AND LOCAL ENERGY METABOLISM IN NEURAL TISSUES

The results of applications of the autoradiographic 2-[^{14}C]deoxyglucose method in animals and the PET [^{18}F]FDG technique in man to conditions with altered functional activities in discrete anatomically defined regions of the nervous system have clearly established that functional and metabolic activities in animals and man are as closely linked in nervous tissues as in other tissues.[8,9] Functional activation of a variety of sensory, motor, and neuroendocrine systems have been found to be associated with increased $lCMR_{glc}$ in the specific structures known to subserve those functions, and depression of those functions result in decreased metabolism in the same regions.[8,9] It was previously known that local cerebral blood flow was similarly altered in association with changes of local cerebral functional activity,[10,11] and it was assumed that the changes in blood flow were secondary adjustments of the circulation to meet the altered local energy demands. Proof that the rate of energy metabolism was also altered awaited, however, the development of the methods for measuring local energy metabolism in the nervous system *in vivo*.

Metabolic Effects of Decreased Functional Activity

Evidence that $lCMR_{glc}$ declines when functional activity is reduced can be directly visualized in the [^{14}C]deoxyglucose autoradiograms in Fig. 3. Fig. 3A is an autoradiogram of striate cortex of a normal conscious monkey with both eyes open and stimulated by a moving pattern of black and white lines. Laminar heterogeneity in layers of cortical area 17 can be seen. Bilateral visual occlusion lowers $lCMR_{glc}$ in all components of the primary visual pathway; in the striate cortex $lCMR_{glc}$ is reduced to an almost uniform low level with very little laminar heterogeneity (Fig. 3B). The monkey's visual system is binocular with approximately 50 percent crossing of pathways from the two eyes at the optic chiasma. The inputs from the two eyes remain segregated all the way to the striate cortex where they terminate in adjacent columns, one each for each eye, i.e., the ocular dominance columns of Hubel and Wiesel.[12] Occlusion of one eye results in reduced input, lower functional activity, and $lCMR_{glc}$ in the columns in the visual cortex normally served by that eye. In the columns served by the intact "seeing" eye, $lCMR_{glc}$ remains normal, leading to visualization of the ocular dominance columns in the deoxyglucose autoradiograms (Fig. 3C).[13] Results such as these in a variety of neural systems have confirmed that local energy metabolism is indeed reduced when local functional activity is decreased.

Metabolic Effects of Increased Functional Activity

The converse is also true; increased local functional activity raises $lCMR_{glc}$. For example, tweaking the whiskers of rodents increases $lCMR_{glc}$ in the "whisker-barrel" regions of the contralateral sensory cortex.[14] Moving an arm and hand increases $lCMR_{glc}$ in the contralateral motor cortex and, in fact, in all of the components of the motor and sensory pathways that are involved in the task.[15] The metabolic effects of increased functional activity in a neuroendocrine pathway in the rat is illustrated in the [^{14}C]deoxyglucose autoradiograms in Fig. 4, which were prepared from the brains of rats that were either salt-loaded or made hypotensive by administration of the α-adrenergic blocking agent, phenoxybenzamine. Both osmotic stimulation and hypotension are known to activate the hypothalamo-neurohypophysial pathway from the supraoptic and paraventricular nuclei in the hypothalamus to the neural lobe of the pituitary and to stimulate secretion of vasopressin. It can be seen that both conditions markedly increase $lCMR_{glc}$ selectively in the posterior pituitary.[16] It should be noted that hypotension also stimulates $lCMR_{glc}$ in the supraoptic and paraventricular nuclei while the osmotic stimulation does not; the reasons for this apparent discrepancy are discussed below.

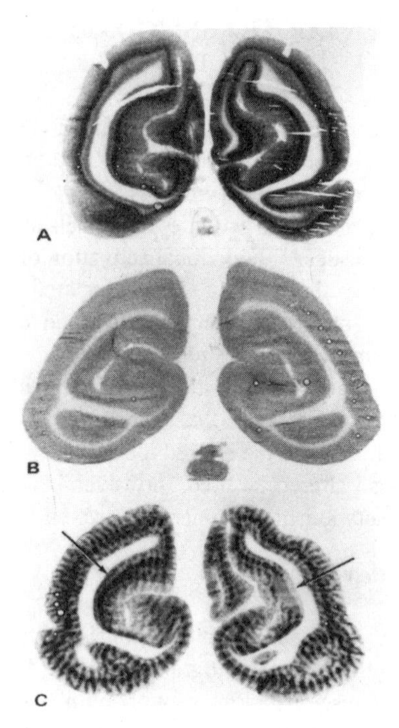

Figure 3. [^{14}C]Deoxyglucose autoradiograms demonstrating effects of acute visual occlusion on local glucose utilization in striate cortex of monkey. The greater the density (i.e., darkness) the greater is the rate of glucose utilization. **(A)** Striate cortex from monkey with both eyes open. Note heterogeneity in the laminae; the darkest lamina corresponds to Layer IV. **(B)** Striate cortex from monkey with both eyes patched. Note general reduction in density and almost complete disappearance of the laminar heterogeneity. **(C)** Striate cortex from monkey with only right eye patched. Left side of autoradiogram corresponds to left hemisphere contralateral to occluded eye. Note alternating dark and light columns traversing full thickness of the striate cortex; these are the ocular dominance columns. The dark bands represent the columns for the open eye; the light bands represent the columns for the patched eye and demonstrate the reduced glucose utilization resulting from the reduced visual input. The arrows point to regions of bilateral asymmetry; these are the loci of representation of the "blind spots" of the visual fields. From Kennedy *et al.*[13]

Quantitative Aspects of Relationship between Functional and Metabolic Activities

The changes in rates of glucose utilization that occur with functional activation or depression are quantitatively related to the level of the functional activity. For example, in dark-adapted albino and Norway brown rats retinal stimulation with light flashes of calibrated intensity and random frequencies between 5 and 10 Hz results in selective increases of $lCMR_{glc}$ in the primary projection zones from the retina (e.g., superficial layer of superior colliculus and dorsal nucleus of lateral geniculate) that are approximately proportional to the logarithm of the intensity of illumination up to a maximum of 70 lux in albino rats and 700 lux in pigmented rats; $lCMR_{glc}$ is unchanged in divisions of these and other structures that do not receive direct projections from the retina (Fig. 5).[17] In this range of illumination the metabolic response to the stimulus in affected structures appears, therefore, to obey the Weber-Fechner Law. The metabolic rates rise more steeply in albino than in pigmented rats, reach a peak at about 70 lux, and then rapidly decline with increasing stimulus intensity; the more rapid rise in albino rats probably reflects more light reaching the retina, and the decline at higher light intensities probably results from retinal overload and damage from excessive light. In pigmented rats, in which the retina is less vulnerable to excessive light intensity, $lCMR_{glc}$ increases progressively with stimulus intensities up to 700 lux, but between 700 and 7000 lux it rises less steeply in the lateral geniculate body and remains more or less constant in the superficial layer of the superior colliculus (Fig. 5).[17]

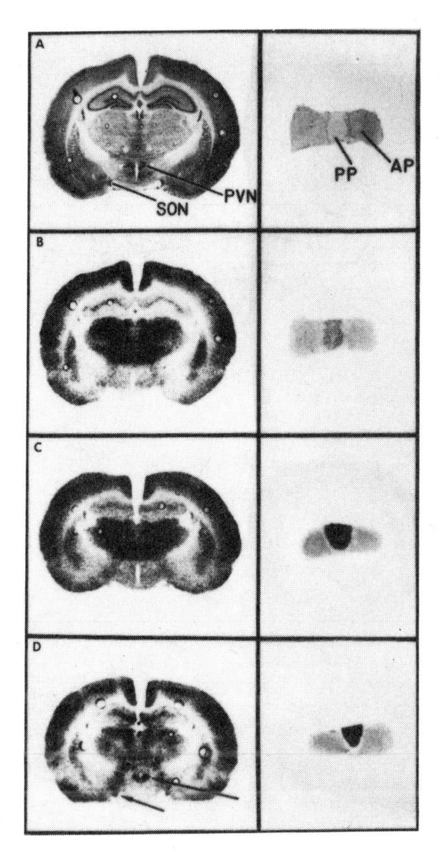

Figure 4. Effects of activation of hypothalamo-neurohypophysial pathway by salt-loading or hypotension on local cerebral glucose utilization in the rat. **(A)** Histological sections of brain stained with cresyl violet (Nissl) and pituitary stained with toluidine blue demonstrating positions of supraoptic nucleus (SON), paraventricular nucleus (PVN), posterior pituitary (PP), and anterior pituitary (AP). **(B)** [^{14}C]Deoxyglucose autoradiograms of brain and pituitary from normal control rat drinking only water. **(C)** [^{14}C]Deoxyglucose autoradiograms from rat given 2% (w/v) NaCl to drink for 5 days. Note selective marked increase in density in posterior hypophysis, indicating increased glucose utilization. **(D)** [^{14}C]Deoxyglucose autoradiograms from rat made hypotensive by administration of 20 mg/kg of phenoxybenzamine 45-60 minutes prior to administration of the [^{14}C]deoxyglucose. Note selective increases in labeling of supraoptic and paraventricular nuclei and posterior pituitary. From Schwartz *et al.*[16]

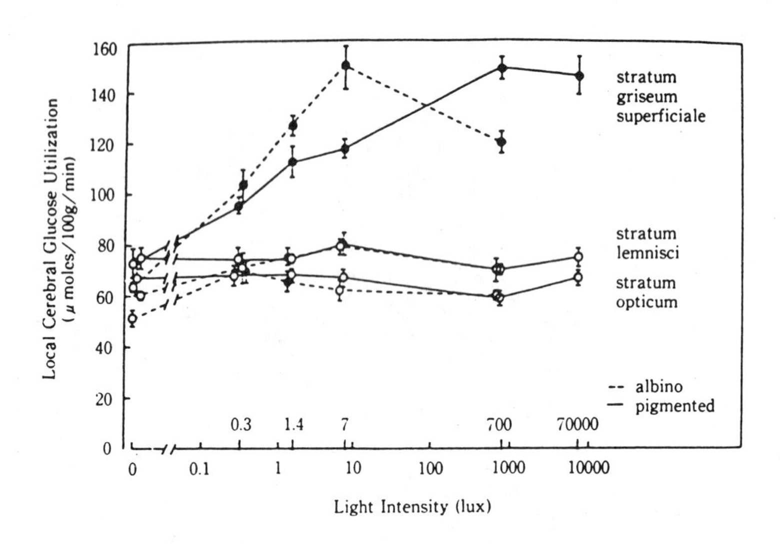

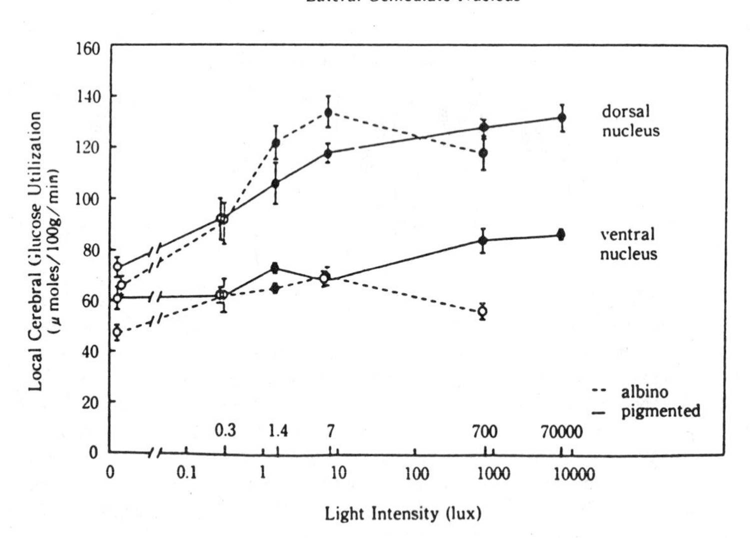

Figure 5. Rates of glucose utilization as function of intensity of retinal illumination with randomly timed light flashes in various layers of superior colliculus and lateral geniculate nucleus of dark adapted albino and pigmented rats. From Miyaoka *et al.*[17]

SITES OF FUNCTION-LINKED CHANGES IN ENERGY METABOLISM IN NERVOUS TISSUES

Traditional electrophysiological techniques generally focus on recording of spikes from neuronal cell bodies and relate these to functional activity. We, therefore, assumed at first that the highest rates of $lCMR_{glc}$ and the increases in energy metabolism observed during functional activation were in perikarya. Some of the observations were, however, inconsistent with this assumption. For example, in the autoradiograms of rat brains in Fig. 1 a dark band, reflecting a high rate of glucose utilization, can be seen to run through the hippocampus. We first thought that this band represented the pyramidal cell layer, but careful comparison of the autoradiograms and histologically stained sections revealed that it represented the molecular layer, a layer rich in neuropil and not cell bodies. Similarly, in Fig. 3A it can be seen that $lCMR_{glc}$ in the striate cortex of the monkey with both eyes open is highest in a sub-layer of Layer IV (e.g., the darkest band parallel to cortical surface) that is not particularly rich in cell bodies but is the layer in which axonal terminals of the afferent geniculocalcarine pathway form synapses with dendrites of neurons situated in other laminae of the visual cortex.[13] It is this neuropil-rich layer that shows the greatest reduction in glucose utilization when visual input is interrupted (Fig. 3B). Also, in Figs. 4B and 4C it can be seen that osmotic stimulation with salt-loading markedly increases $lCMR_{glc}$ in the posterior pituitary, 40% of which consists of axonal terminals of the afferent hypothalamo-neurohypophysial tract[18] while the supraoptic and paraventricular nuclei in the hypothalamus, the sites of the cell bodies from which this tract originates, are essentially unaffected. In contrast, hypotension induced by α-adrenergic blockade markedly stimulates $lCMR_{glc}$ in these nuclei (Fig. 4D); hypotension induced by hemorrhage or other means has the same effect. Instead of direct action on the cell bodies, as occurs with salt-loading, in the case of hypotension the cell bodies in these nuclei receive afferent inputs from the brain stem (e.g., from the nucleus tractus solitarius) that are part of the central neural pathways of the baroreceptor reflexes. Observations like these have established that it is mainly the energy metabolism in regions rich in synapses and neuropil that is linked to functional activity rather than the metabolism of the perikarya.

To compare the effects of functional activation on $lCMR_{glc}$ simultaneously in the perikarya and nerve terminals of the same pathway, Kadekaro *et al.*[19] stimulated the sciatic nerve of the anesthetized rat electrically at different frequencies and measured glucose utilization by the [^{14}C]deoxyglucose method in both the body of the dorsal root ganglia and the dorsal horn of corresponding segments of the lumbar spinal cord. The body of the dorsal root ganglion contains perikarya devoid of neural processes; cell bodies free of axonal terminals and dendritic processes could then be examined at the same time as the nerve terminals of the same pathway in the dorsal horn of the spinal cord. The results confirmed that it is the glucose utilization in the region of the nerve terminals and not the cell bodies that is linked to the functional activity. Glucose utilization in the dorsal horn of the lumbar cord increased linearly with the frequency of stimulation while the region of the cell bodies of the same pathway in the dorsal root ganglia showed no metabolic changes in response to the electrical stimulation (Fig. 6). The failure of the perikarya in the dorsal root ganglion to respond metabolically to electrical stimulation is at first glance surprising. There is evidence, however, that soma membrane of neurons is not very excitable or productive of action potentials. Studies with patch clamps have shown that voltage-dependent Na^+ channels are relatively sparse in soma membranes; Smith[20] has obtained electrophysiological evidence with patch clamp techniques that soma and dendrites of spinal cord neurons and soma of dorsal root ganglion cells cultured *in vitro* do not generate action potentials. Freygang[21] and Freygang and Frank[22] concluded from analyses of extracellular potentials recorded from single spinal motor neurons and single neurons in the lateral geniculate nucleus that the

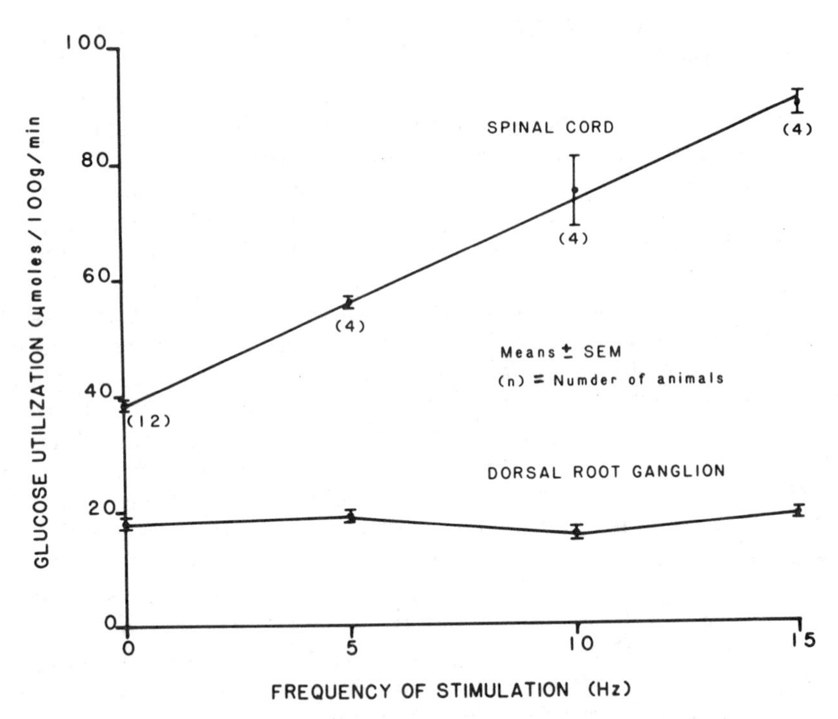

Figure. 6. Frequency-dependent effects of electrical stimulation of sciatic nerve on glucose utilization in dorsal root ganglion and dorsal horn of lumbar spinal cord. Error bars represent S.E.M. of number of animals in prentheses. From Kadekaro *et al.*[19]

soma-dendritic membrane can be driven synaptically to produce post-synaptic potentials but not propagating action potentials. Inasmuch as action potentials may mediate coupling of energy metabolism and functional activity, if they are absent or sparse in perikarya, then functional activation of glucose metabolism would not be expected (see below). Some energy metabolism does, of course, proceed in cell bodies even at rest; it is probably mainly used to support vegetative and biosynthetic processes needed to maintain cellular structural and functional integrity rather than those directly related to functional activity.

The finding that the rate of energy metabolism is linked to spike frequency in neuropil and not in cell bodies also resolves the question of the metabolic changes that occur with inhibition and excitation. There have been experiments in which increased glucose utilization was seen in structures in which electrophysiological evidence indicated that there must have been inhibition of neuronal activity. This raised the question of whether active inhibition required energy just like excitation. It appears now that it is the electrical activity in the afferent nerve terminals that is associated with the energy consumption, and this activity is the same whether excitatory or inhibitory neurotransmitters are released at the terminals. The energy metabolism of the post-synaptic cell bodies, whether activated or inhibited, is not significantly altered, and to determine which has occurred, one can look downstream at the next synapses to which those neurons project. Glucose utilization would be reduced in their projection zones if they were inhibited and increased if they were excited.

CHEMICAL MECHANISMS LINKING ENERGY METABOLISM TO FUNCTIONAL ACTIVITY

Muscles move and support masses against gravitational forces; the heart pumps blood against a pressure head; kidneys transport water and solutes against osmotic and concentration gradients. The need for energy to support such work is obvious. The nature of work done by nervous tissue that requires energy is less obvious. Clearly, the work of nervous tissue must be related to the generation, propagation, and conduction of action potentials. Indeed, glucose utilization does increase almost linearly with the frequency of action potentials in the dorsal horn of the spinal cord (Fig. 6), and similar results have been obtained by Yarowsky *et al.*[23] in the superior cervical ganglion. Action potentials do not themselves consume energy; they are produced by the movements of K^+ ions from inside the cell to the extracellular space and of Na^+ ions from the extracellular space into the cell. In the regions with increased frequency of action potentials extracellular K^+ concentrations can be expected to rise. Neurochemists have known for decades that increasing K^+ concentration in the external medium stimulates respiration of brain slices *in vitro*. Shinohara *et al.*[24] has found that increasing extracellular K^+ concentration in the cerebral cortex *in vivo*, which depolarizes the membranes and causes spreading depression, stimulates cortical glucose utilization *in vivo* as well (Fig. 7). Increased extracellular K^+ and/or intracellular Na^+ concentrations stimulate the activity of Na^+,K^+-ATPase, an enzyme that uses the energy of

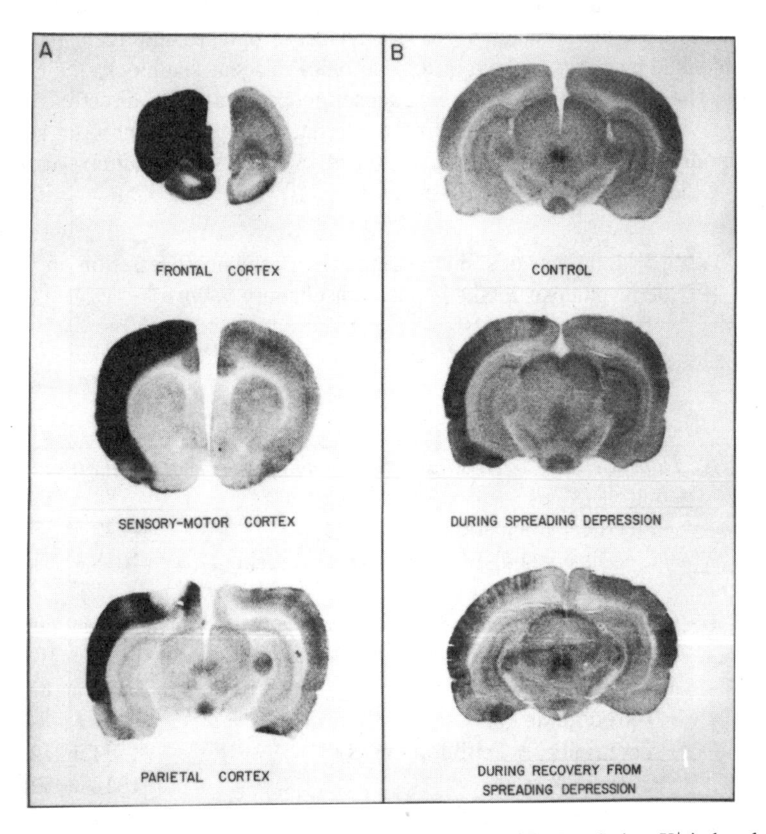

Figure 7. [^{14}C]Deoxyglucose autoradiograms showing changes in $lCMR_{glc}$ during K^+-induced spreading cortical depression and during recovery in the rat; the greater the density, the higher is the $lCMR_{glc}$. In all cases the experimental side is on the left and was treated with KCl; the control side on the right was treated with equivalent concentrations of NaCl. **(A)** Autoradiograms of brain sections at various levels of cerebral cortex from conscious rat in which spreading cortical depression was produced and sustained on left side by 5 M KCl applied to surface of intact dura over parietal cortex every 15-20 minutes; right side was treated comparably with NaCl. **(B)** Autoradiograms of brain sections at level of parietal cortex from three rats under barbiturate anesthesia. Top section is from normal, anesthetized rat; middle section is from similarly anesthetized rat in which 80 mM KCl in artificial cerebrospinal fluid was repeatedly applied directly to the surface of the left parietal cortex; bottom section is from similarly anesthetized rat studied immediately after return of cortical D.C. potential to normal after a single wave of spreading depression induced by a single application of 80 mM KCl to the parieto-occipital cortex of the left side. From Shinohara *et al*[24]

ATP to transport Na^+ back out of the cell and K^+ back into the cell to restore the ionic gradients to normal. ATPase activity lowers the ATP concentration and ATP/ADP ratio and increases the ADP, phosphate acceptor, and inorganic phosphate concentrations within the cell; these are all intracellular changes that can stimulate both glycolysis and electron transport in the respiratory chain.

The possibility that the activation of Na^+,K^+-ATPase activity by membrane depolarization might provide a mechanism for coupling of energy metabolism to functional activity was examined in rat posterior pituitary glands incubated *in vitro* under conditions in which they could be stimulated electrically to secrete vasopressin.[25] The posterior pituitary gland is highly enriched with axon terminals of the hypothalamo-hypophysial tract which account for greater than 42 per cent of the gland's total volume.[18] Electrical stimulation of the gland increased [^{14}C]deoxyglucose uptake, indicating increased glucose utilization, and this increase in uptake was completely blocked by addition of ouabain, a specific inhibitor of Na^+,K^+-ATPase that did not inhibit spike generation or hormone secretion under the conditions of the experiment (Table 2A). The membranes were also depolarized by the addition of veratridine, an agent that opens Na^+ channels and allows Na^+ entry into the cells. Veratridine also markedly stimulated [^{14}C]deoxyglucose uptake, and again the increase in uptake was blocked by either ouabain or tetrodotoxin, an agent that blocks the Na^+ channels (Table 2B). The increased [^{14}C]deoxyglucose uptake was not directly related to a stimulation of vasopressin secretion; the posterior pituitary glands cannot be stimulated to secrete hormone in Ca^{++}-free medium, and yet even in such medium veratridine still stimulated [^{14}C]deoxyglucose uptake (Table 2C).

Table 2. Influence of sodium pump activity and neurosecretion on [^{14}C]deoxyglucose uptake in posterior pituitary *in vitro* [a]

Condition	[^{14}C]Deoxyglucose Uptake (Cpm/100 µg protein/15 min)
A. *Dependence on activation of sodium pump --- Electrical Stimulation*	
Controls (4)	988 ± 19
+ Electrical stimulation at 10 Hz (4)	1272 ± 57 [b]
+ Electrical stimulation at 10 Hz + ouabain (4)	1018 ± 51 [c]
B. *Dependence on activation of sodium pump --- Opening Na+ channels*	
Controls (14)	1381 ± 50
+ Veratridine (14)	1891 ± 85 [b]
+ Tetrodotoxin (9)	1209 ± 84 [c]
+ Veratridine, + Tetrodotoxin (8)	1551 ± 72 [c]
+ Ouabain (4)	1318 ± 57 [c]
+ Veratridine, + Ouabain (4)	1218 ± 120 [c]
C. *Independence from activation of secretion ---- (Ca^{2+}-free medium)*	
Controls (in Ca^{++} free medium) (6)	1142 ± 38
+ Veratridine (in Ca^{++}- free medium) (6)	1681 ± 78 [b]

[a] The values represent means ± S.E.M. of results obtained in number of experiments indicated in parentheses. From Mata *et al.*[25]

[b] Indicates statistically significant difference from controls ($p<0.001$).

[c] Indicates no statistical significant difference from controls.

It appears then that the energy metabolism associated with the electrical and functional activities of nervous tissue is not used directly in the generation and propagation of action potentials. The energy is used to restore the ionic gradients and resting membrane potentials that were partly degraded during the excitation phase. In this respect, it is equivalent to what used to be called "the heat of recovery."

ROLE OF ASTROGLIA IN FUNCTIONAL ACTIVATION OF ENERGY METABOLISM

The increases in glucose utilization provoked by functional activation occur mainly in synapse-rich neuropil and are directly proportional to the spike frequency in the nerve terminals. Action potentials result from the movements of K^+ from inside to outside and of Na^+ from outside to inside the cell. Presumably, therefore, the higher the spike frequency, the greater are the increases in intracellular Na^+ concentration ($[Na^+]_i$) and extracellular K^+ concentration ($[K^+]_o$), and, consequently, the magnitude of Na^+,K^+-ATPase activity and energy metabolism needed to restore the ion gradients back to resting levels. Such restoration must obviously occur in neuronal elements from which the action potentials are derived, but astrocytes might also be involved. Astrocytes have been reported to regulate $[K^+]_o$ by passive diffusion[26,27] or active transport[28] following increases in $[K^+]_o$ resulting from neuronal excitation,[29,30] and there have been reports based on studies with tissue slices[31-34] or cultured cells[35-39] that energy is consumed in the process. The spatial resolution of the autoradiographic deoxyglucose method is presently inadequate to identify the specific cellular or subcellular elements in neuropil that share in the increases in energy metabolism *in vivo*. It remains, therefore, uncertain whether the function-driven increases in glucose metabolism are limited to axonal and/or dendritic processes or include astrocytic processes that envelop the synapses. To approach this question we have been attempting to simulate *in vitro* some of the changes in the extracellular environment that might result from increased spike activity *in vivo* and to examine their effects on glucose metabolism of neurons and astroglia in culture.[40] Thus far, we have examined the effects of increased extracellular K^+ and intracellular Na^+ concentrations and of the excitatory neurotransmitter, L-glutamate.

Culture Conditions

Neuronal and mixed neuronal-glial cultures were prepared from fetal rat striatum on embryonic day 16. The striatal tissue was excised and mechanically disrupted by passage through a 22-gauge needle, and the dissociated cells were cultured in high-glucose (25 mM) Dulbecco's modified Eagle medium containing 10% (v/v) fetal bovine serum, streptomycin (100 µg/ml), and penicillin (100 U/ml) at 37 °C in humidified air containing 7% CO_2. For neuronal cultures viable cells that excluded trypan-blue were placed in either six-well or in the eight center wells of 24-well poly(L-lysine)-coated culture plates, and cytosine arabinoside (10 µM) was added 2-3 days later. The cultures were used for assays after 6-8 days *in vitro*. For mixed neuronal-astroglial cultures, cells were placed similarly in poly(L-lysine)-coated plates, but no cytosine arabinoside was added.

Astroglial cultures were prepared from cerebral cortex of newborn rats or striatum removed on embryonic day 16. Results were similar with cultures from both sources. Meninges and blood vessels were removed from brain tissue obtained from newborn rats, and the fronto-parietal cortices were dissected out and mechanically disrupted. The dissociated cells (2.5 X 10^5 cells/ml) were cultured in medium like that used for neuronal and mixed neuronal-astroglial cultures (see above) in uncoated 75 cm^2 -culture flasks at 37°C in humidified air containing 7% CO_2. Culture medium was changed every two days until the cultures reached confluence. The flasks were then shaken overnight at room temperature to eliminate loosely adherent process-bearing cells. The adherent cells were treated with trypsin-EDTA solution diluted 1:5 with Dulbecco's phosphate-buffered saline for 1-2 minutes at 37 °C, washed with fresh culture medium, and placed in wells of uncoated culture plates to obtain secondary cultures of astroglia. Culture medium was changed every 3 days, and the cultures were used when confluent (day 19-22) or one week later (day 26-29). Some of the cultures were treated with dibutyryl cAMP (0.5 mM) for 24 or 72 hours prior to the experiment to induce morphological differentiation of the astroglia.

Assay Conditions

The standard reaction mixture for assay of [^{14}C]deoxyglucose phosphorylation contained the following: 110 mM NaCl, 5.4 mM KCl, 1.8 mM $CaCl_2$, 0.8 mM $MgSO_4$, 0.9 mM NaH_2PO_4, 44 mM $NaHCO_3$, 2.0 mM D-glucose, and 50 µM [^{14}C]deoxyglucose (containing 2.5 µCi/ml) in a final volume of 0.2 ml per well in 24-well plates or 1.0 ml per well in six-well plates. The pH was adjusted with CO_2 to approximately 7.2 just before use. When the KCl content of the medium was raised, the NaCl concentration was decreased by an equivalent amount to maintain constant osmolality. Immediately before the assay the culture medium was replaced with the reaction mixture described above, except for the lack of [^{14}C]DG, and preincubated at 37 °C in humidified air with 7% CO_2 for 15 minutes. The preincubation reaction mixture was then replaced with reaction mixture containing [^{14}C]deoxyglucose and one of four different concentrations of K^+ (2.7, 5.4, 28, and 56 mM) and incubated for 10, 15, or 30 minutes.

The effects of veratridine (75 µM) and monensin (10 µM), both of which enhance Na^+ entry, were examined with a K^+ concentration of 5.4 mM and 15 minutes of incubation. The effects of 500 µM L-glutamate were examined under similar conditions. When choline was substituted for Na^+ to obtain a Na^+-free medium, or inhibitors (e.g. 1 mM ouabain to inhibit Na^+,K^+-ATPase activity; 10 µM tetrodotoxin to block voltage-dependent Na^+ channels; 1 mM DL-2-amino-5-phosphonovaleric acid to block NMDA receptors; 100 µM CNQX to inhibit non-NMDA receptors; 1 mM DL-threo-β–hydroxyaspartatic acid to inhibit sodium-dependent glutamate transport) were added, the cultures were first preincubated with the appropriate medium but without [^{14}C]deoxyglucose for 15 minute before final incubation with [^{14}C]deoxyglucose.

At the end of the incubation standard reaction mixture without [^{14}C]deoxyglucose was added to the cultures and incubated for 5 minutes to allow efflux of remaining free unphosphorylated [^{14}C]deoxyglucose from the cells. After such treatment 96-98% of the total ^{14}C recovered was in [^{14}C]deoxyglucose-6-phosphate and its secondary acidic metabolites. The cell carpets were then washed quickly 3 times with ice-cold Dulbecco's phosphate-buffered saline and dissolved by digestion in 0.2 ml of 1.0 M NaOH for 2 hours at room temperature. Fifty µl of the dissolved cells were removed for protein determination;[41] another 50 µl were used for assay of ^{14}C by liquid scintillation counting with external standardization. The value for disintegrations per minute (DPM) was divided by the measured protein content for each well, and the results are expressed as the mean DPM ± SEM of quadruplicate wells.

Effects of Increased $[K^+]_o$ on Rates of [^{14}C]Deoxyglucose Phosphorylation.

Increasing $[K^+]_o$ from 5.4 to 28 and 56 mM statistically significantly stimulated [^{14}C]DG phosphorylation in both neuronal and mixed neuronal-astroglial cultures (day 6) at 15 and 30 minutes ($p<0.05$); there was a lag in the stimulation because it was not seen at 10 minutes of incubation (Fig. 8). In pure secondary astroglial cultures base-line rates of [^{14}C]DG phosphorylation varied considerably among preparations, but no stimulation by increased $[K^+]_o$ in the range of 2.7 to 56 mM was found in any of them (Fig. 8, Table 3A). The age of the astroglial cultures was not a factor; neither 22 day nor 29 day astroglial cultures showed elevation of [^{14}C]DG phosphorylation in response to increased $[K^+]_o$ (Table 3B). Also, addition of dibutyryl cAMP (dbcAMP) to the culture medium, which transforms astrocytes from protoplasmic to fibrous types, raised the baseline rate of [^{14}C]DG phosphorylation in 22 day old astroglial cultures but did not lead to a stimulation of [^{14}C]DG phosphorylation by elevated $[K^+]_o$ (Table 3C).

These results appear to be in conflict with reports that increased $[K^+]_o$ stimulates [^{14}C]deoxyglucose uptake in cultured astroglia.[35-39] Differences in the cell cultures or incubation conditions may be involved. The astroglial cultures in our studies were derived from cortex of newborn rats; they were secondary cultures and were very pure containing no neurons and virtually no cells without glial fibrillary acidic protein (GFAP). The astroglia in these other studies were derived from mouse brain and may not have been secondary cultures and possibly, therefore, not as pure. Also, there may be some influence of buffer conditions used in the incubation. We have also occasionally observed enhancement of [^{14}C]deoxyglucose phosphorylation with increased $[K^+]_o$ when HEPES or phosphate buffers were used instead of bicarbonate buffers. This phenomenon is currently under study.

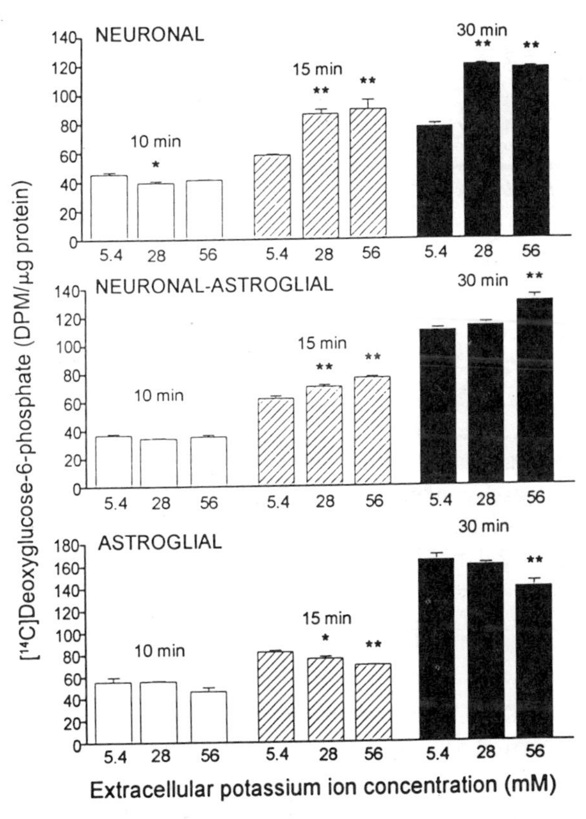

Figure 8. Effects of $[K^+]_o$ on rates of [^{14}C]deoxyglucose phosphorylation in neuronal (day 6), mixed neuronal-astroglial (day 6), and astroglial (day 19, no dbcAMP) cultures. Values are means ± SEM obtained from quadruplicate wells. Numbers above bars indicate duration of incubation. *$p<0.05$, **$p<0.01$ compared to the 5.4 mM $[K^+]_o$ (Dunnett's test for multiple comparison). Representative of a minimum of three such experiments for each condition. From Takahashi *et al.*[40]

Table 3. Effects of increasing extracellular potassium ion concentration on rates of [^{14}C]deoxyglucose phosphorylation in cultured astroglia (DPM/μg of protein)

Culture Conditions	Incubation Time (min.)	Extracellular K^+ Conccntration			
		2.7 mM	5.4 mM	28 mM	56 mM
A. *Effects of expanded range of extracellular K^+ concentration*					
20 Day Culture	10	39 ± 1	40 ± 1	39 ± 1	39 ± 0
No dbcAMP	15	65 ± 2	64 ± 0	73 ± 1[a]	69 ± 2
	30	144 ± 3	142 ± 2	133 ± 3	125 ± 4[a]
B. *Effects of age of culture*					
22-Day Culture	10	--	112 ± 2	108 ± 3	74 ± 3[b]
No dbcAMP	15	--	143 ± 4	131 ± 1[c]	127 ± 3[b]
	30	--	273 ± 8	251 ± 3[c]	215 ± 3[b]
29-Day Culture	10	--	99 ± 2	87 ± 4[c]	60 ± 1[b]
No dbcAMP	15	--	147 ± 2	144 ± 3	124 ± 3[b]
	30	--	328 ± 3	283 ± 6[b]	235 ± 3[b]
C. *Effects of treatment of cultures with dbcAMP*					
22-Day Culture					
No dbcAMP	15	73 ± 1	75 ± 3	81 ± 3	81 ± 8
+dbcAMP-1 day	15	96 ± 3[d]	90 ± 3[d]	100 ± 2[d]	95 ± 2
+dbcAMP-3 days	15	116 ± 1[d]	106 ± 1[d]	125 ± 3[d]	116 ± 7[d]

NOTE: Values are means ± SEM of quadruplicate wells and are representative of at least three experiments for each condition. The culture ages were 19-22 days for **A**, 19-22 and 26-29 days for **B**, and 19-22 days for **C**. From Takahashi *et al.*[40]

[a] $p<0.01$ compared to 2.7 mM $[K^+]_o$; [b] $p<0.01$, [c] $p<0.05$ compared to 5.4 mM $[K^+]_o$;
[d] $p<0.01$ compared to values in absence of dbcAMP (Dunnett's ***t*** test).

Effects of Veratridine and Monensin on [^{14}C]Deoxyglucose Phosphorylation.

Facilitating Na^+ entry into cells with veratridine (75 μM), which opens membrane voltage-dependent Na^+ channels, or with the Na^+ ionophore, monensin (10 μM), stimulated phosphorylation of [^{14}C]deoxyglucose in astroglia by 20% and 171%, respectively (Fig. 9). Ouabain (1 mM) reduced the basal rate of phosphorylation to about 20% of control values and completely eliminated the stimulation by veratridine, but only partially suppressed the stimulation by monensin (Fig. 9). Tetrodotoxin (10 μM) did not alter basal rates of [^{14}C]deoxyglucose phosphorylation in astroglia; it did, however, eliminate the effects of veratridine but had no effect on the stimulation of [^{14}C]deoxyglucose phosphorylation by monensin (Fig. 9). Replacement of Na^+ with choline eliminated the stimulations by both veratridine and monensin (Fig. 9).

Effects of Glutamate on Phosphorylation of [^{14}C]Deoxyglucose in Astroglia.

Glutamate stimulated [^{14}C]deoxyglucose phosphorylation in astroglia (Fig. 10), and the stimulation was concentration-dependent in the range of 0-500 μM (data not shown). The stimulation by glutamate was eliminated by 1 mM ouabain or incubation in Na^+-free medium but was unaffected by tetrodotoxin or NMDA and non-NMDA receptor antagonists (Fig. 10). The addition of DL-threo-β-hydroxyaspartate, an inhibitor of glutamate-Na^+ co-transport,[42] itself markedly stimulated [^{14}C]deoxyglucose phosphorylation, and glutamate

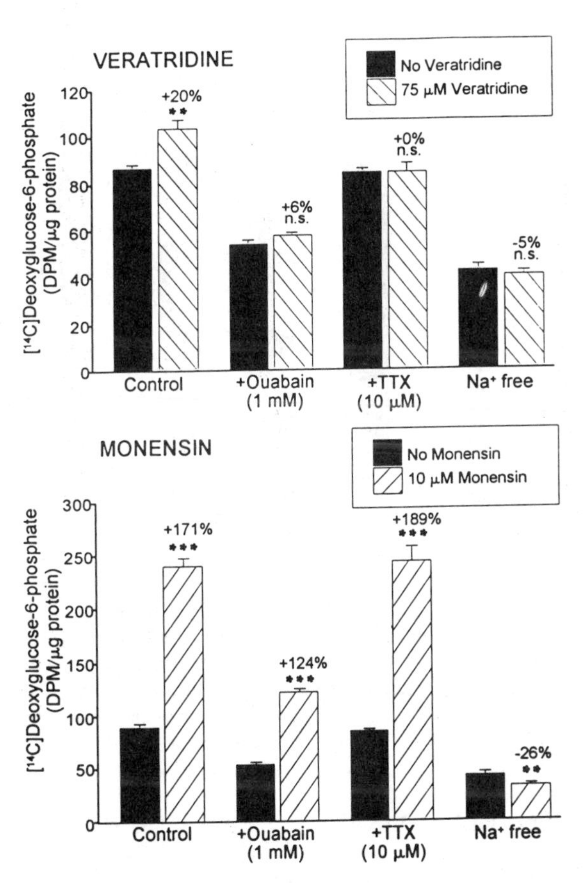

Figure 9. Effects of veratridine (upper) and monensin (lower) on rates of [^{14}C]deoxyglucose phosphorylation in astroglia (day 21, no dbcAMP), and the effects of ouabain, tetrodotoxin (TTX), and Na^+-free medium on the veratridine and monensin effects. Values are means ± SEM obtained from quadruplicate wells. Numbers above bars indicate percent difference from appropriate controls. Representative of three such experiments in three separate astroglial preparations. **$p<0.01$; ***$p<0.001$; n.s., not statistically significant, compared to appropriate controls (grouped *t* test). From Takahashi *et al.*[40]

had no additional stimulating effect in its presence (Fig. 10). The stimulation by hydroxyaspartate was blocked by ouabain, completely absent during incubation in Na^+-free medium, and unaffected by tetrodotoxin and inhibitors of NMDA and non-NMDA receptors (data not shown). These results are consistent with the possibility that hydroxyaspartate inhibits glutamate transport competitively but is itself co-transported with Na^+ into the cells. These results agree closely with those recently reported by Pellerin and Magistretti,[43] although that they did not observe a stimulation of deoxyglucose phosphorylation by hydroxyaspartate. It appears then that glutamate can stimulate glucose utilization in astroglia, not by acting on known glutamate receptors but by increasing $[Na^+]_i$ by its co-transport into the cells with Na^+.

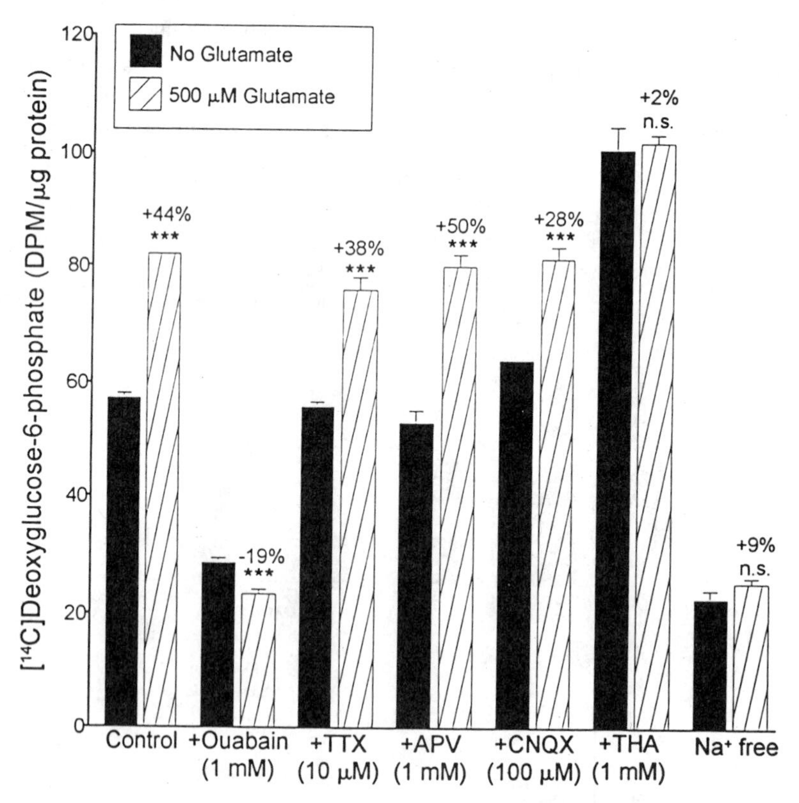

Figure. 10. Effects of glutamate on [^{14}C]deoxyglucose phosphorylation in astroglia (day 22, no dbcAMP), and effects of ouabain, tetrodotoxin (TTX), DL-2-amino-5-phosphonovaleric acid (APV), CNQX, DL-threo-β-hydroxyaspartic acid (THA), and Na^+-free medium on the glutamate effects. Values are means ± SEM from quadruplicate wells. Numbers above bars indicate percent differences from appropriate controls. Representative of three such experiments with three separate astroglial preparations. ***$p<0.001$, n.s., not statistically significant, compared to each control (grouped *t* test). From Takahashi *et al.*[40]

SUMMARY AND CONCLUSIONS

Numerous studies have established that functional activation in the nervous system is associated with increased glucose utilization in the terminal projection zones of the activated pathways. The increases in glucose utilization occur mainly in neuropil and little if at all in perikarya and are more or less proportional to the increased spike frequency in the metabolically activated regions. Action potentials result from depolarization-induced rapid inward Na^+ current that leads to increased $[Na^+]_i$ and an efflux of K^+ increasing the $[K^+]_o$. Increased $[K^+]_o$ was used in the present studies to depolarize neurons in culture; the increases in $[K^+]_o$ and $[Na^+]_i$ could then activate neuronal Na^+,K^+-ATPase and thus stimulate energy consumption to restore ionic gradients to resting levels. The same processes must certainly operate *in vivo* in the neuronal processes in the neuropil during neuronal excitation. Astroglial processes in the neuropil surround the synapses between neuronal elements. In assays with bicarbonate buffers, which are commonly used in studies with astrocytic cultures, raising $[K^+]_o$ failed to stimulate astroglial glucose utilization. Astroglia, however, do not produce action potentials, and depolarization by increased $[K^+]_o$ or other means should then not directly raise $[Na^+]_i$. It was recently reported, however, that depolarization of mouse astroglial membranes by increased $[K^+]_o$ results in reversal of electrogenic Na^+/HCO_3^- co-transport, normally operating from inside to outside the cell, thus leading not only to inward transport of Na^+ but also intracellular alkalinization.[44] Raising pH should stimulate phosphofructokinase activity and increase glucose utilization. Possibly it was this phenomenon that was responsible for reported observations of K^+- induced stimulations of glucose utilization in cultured astrocytes.[42-46] That buffer conditions may play a part is suggested by our occasional observation of K^+-induced increases in glucose utilization when HEPES or phosphate was used for buffering instead of bicarbonate. Even if increased $[K^+]_o$ does not directly stimulate energy metabolism in astroglia, it is clear that increased $[Na^+]_i$ does, as shown by the effects of veratridine and monensin. Veratridine causes persistent activation of voltage-dependent sodium channels[45,46] allowing $[Na^+]_i$ to rise in accordance with the sodium ion gradient across the cell membrane. Astroglial Na^+,K^+-ATPase activity should respond to an elevation in $[Na^+]_i$ inasmuch as $[Na^+]_i$ in astroglia is about 10-20 mM and the K_m of Na^+,K^+-ATPase for Na^+ is about 10 mM.[47] Veratridine stimulated [^{14}C]deoxyglucose phosphorylation not only in neurons but in astroglia as well. The veratridine stimulation was completely inhibited by 1 mM ouabain, indicating that the stimulation was dependent on Na^+,K^+-ATPase activity, and, furthermore, it was also blocked by tetrodotoxin, indicating that the stimulation was mediated by Na^+ entry into the cells through voltage-dependent sodium channels. Monensin, a carboxylic ionophore that mediates exchange of external Na^+ for internal H^+,[48] also stimulated [^{14}C]deoxyglucose phosphorylation, partly by activation of Na^+,K^+-ATPase activity due to increased $[Na^+]_i$ and probably also by increased phosphofructokinase activity and glycolysis due to the decreased intracellular H^+ concentration.[49] The results of the present studies are consistent with these effects; the monensin stimulation was only partially blocked by ouabain, presumably only the part due to stimulation of Na^+,K^+-ATPase activity. These results agree well with those of Erecinska *et al.*[50] in rat synaptosomes but only partly with those of Yarowsky *et al.*,[33] who found the monensin stimulation of [^{14}C]deoxyglucose phosphorylation in astroglia to be completely inhibited by ouabain.

It is clear that [^{14}C]deoxyglucose phosphorylation in astroglia is robustly stimulated by increased $[Na^+]_i$ through a ouabain-sensitive mechanism. It is less clear whether it also responds to elevated $[K^+]_o$. Astroglial membranes are depolarized by increased $[K^+]_o$, but the depolarization is not associated with action potentials[56] that would directly raise Na^+ entry into the cells. This may be due to either low density or different properties of the voltage-dependent Na^+ channels in astroglia.[52] Barres *et al.*[53] observed that type 1 astrocytes prepared from optic nerve do have voltage-dependent Na^+ channels, but they open at more negative potentials and more slowly than Na^+ channels in neurons in response to depolarization. Also, Hisanaga *et al.*[54] have reported that exposure to high $[K^+]_o$ elicits c-*fos* expression in neurons but not in astroglia. These studies indicate that neurons and astroglia respond differently to increased $[K^+]_o$.

The failure of elevated $[K^+]_o$ to increase glucose utilization in astroglia, at least when examined with bicarbonate buffering, indicates that K^+ uptake by astrocytes does not require energy. Astroglia do take up K^+ as manifested by the ability of an astroglial syncytium to move $[K^+]_o$ from areas of intense neuronal stimulation to areas of lower $[K^+]_o$, e.g., spatial buffering of potassium.[52] Also, astroglia in primary cultures have been reported to take up K^+ avidly,[51,55] and the uptake rate is higher than that of neurons.[56] Astroglial membranes, however, have a much higher K^+ conductance than that of neurons, and the K^+ uptake need not, therefore, be energy-dependent.

The question remains whether astroglia contribute to the increases in local energy metabolism that accompany increased function-related spike activity in neuropil. The present results indicate that neuronal elements do respond with increased glucose utilization to changes in the extracellular ionic environment associated with spike activity (i.e., increased $[K^+]_o$), and astroglial energy metabolism is stimulated by increased Na^+ entry. Spike activity in axonal processes has other consequences in addition to those on ion fluxes; it leads to release of neurotransmitters that could activate Na^+ entry into astrocytes and thus stimulate astroglial energy metabolism. Glutamate is the most common and almost ubiquitously distributed excitatory neurotransmitter in the central nervous system. Astroglia actively take up glutamate immediately after its release to keep its concentration in extracellular space below neurotoxic levels. At least three different Na^+-dependent glutamate transporters have been cloned,[57-59] and one of them is expressed mainly in astrocytes.[60] When glutamate is taken up by astroglia, Na^+ is co-transported, resulting in increased $[Na^+]_i$ and depolarization,[61] and in stimulation of [^{14}C]eoxyglucose phosphorylation as well. This stimulation of glucose metabolism is eliminated when Na^+ in the external medium is replaced by choline. It would appear then that astroglia can participate in the increased consumption of glucose associated with functional activity.

REFERENCES

1. D.D. Clarke and L. Sokoloff, Circulation and energy metabolism of the brain; in: *Basic Neurochemistry, Fifth Edition*, G. Siegel, B.W.Agranoff, R.W. Albers, and P. Molinoff, eds, Raven Press, New York (1994), pp. 645-680.
2. O.E. Owen, A.P. Morgan, H.G. Kemp, J.M. Sullivan, M.G. Herrera, G.F. Cahill, Jr: Brain metabolism during fasting. *J. Clin. Invest.* 46:1589-1595 (1967).
3. L. Sokoloff, M, Reivich, C. Kennedy, M.H. Des Rosiers, C.S. Patlak, K.D. Pettigrew, O. Sakurada, M. Shinohara, The [^{14}C]deoxyglucose method for the measurement of local cerebral glucose utilization: heory, procedure, and normal values in the conscious and anesthetized albino rat. *J. Neurochem*. 28:897-916 (1977).
4. C. B. Smith, Localization of actvity-associated changes in metabolism of the central nervous system with the deoxyglucose method: Prospects for cellular resolution, in: *Current Methods in Cellular Neurobiology, Vol. I, Anatomical Techniques*, J.L. Barker and J.F. McKelvy, eds,John Wiley, New York (1983), pp. 269-317.
5. M. Reivich, D. Kuhl, A. Wolf, J. Greenberg, M. Phelps, T. Ido, V. Cassella, J. Fowler, E. Hoffman, A. Alavi, P. Som, and L. Sokoloff, The [^{18}F]fluoro-deoxyglucose method for the measurement of local cerebral glucose utilization in man. *Circ. Res*. 44:127-137 (1979).
6. M. E. Phelps, S.C. Huang, E.J. Hoffman, C. Selin, L. Sokoloff, and D.E. Kuhl, Tomographic measurement of local cerebral glucose metabolic rate in humans with (F-18)2-fluoro-2-deoxy-D-glucose: validation of method, *Ann. Neurol.*; 6:371-388 (1979).
7. C. Kennedy, O. Sakurada, M. Shinohara, J.W. Jehle, and L. Sokoloff, Local cerebral glucose utilization in the normal conscious Macaque monkey. *Ann. Neurol.* 4: 293-301 (1978).

8. L. Sokoloff, Localization of functional activity in the central nervous system by measurement of glucose utilization with radioactive deoxyglucose. *J. Cereb. Blood Flow Metab.* 1: 7-36 (1981).
9. M.E. Phelps, J.C. Mazziotta, and S.C. Huang, Study of cerebral function with positron computed tomography. *J. Cereb. Blood Flow Metab.* 2:113-162 (1982).
10. L. Sokoloff, Local cerebral circulation at rest and during altered cerebral activity induced by anesthesia or visual stimulation, in: *The Regional Chemistry, Physiology and Pharmacology of the Nervous System*, S.S. Kety and J. Elkes, eds., Pergamon Press, Oxford (1961), pp. 107-117.
11. N.A. Lassen, D. Ingvar, and E. Skinhøj, Brain function and blood flow. *Sci. Am.* 239: 62-71 (1978).
12. D.H. Hubel and T.N. Wiesel, Receptive fields and functional architecture of monkey striate cortex. *J. Physiol. (London),* 195: 215-243, (1968).
13. C. Kennedy, M.H. Des Rosiers, O. Sakurada, M. Shinohara, M. Reivich, J.W. Jehle, and L. Sokoloff, Metabolic mapping of the primary visual system of the monkey by means of the autoradiographic [^{14}C]deoxyglucose technique. *Proc. Natl. Acad. Sci., U.S.A.* 73:4230-4234 (1976).
14. P.J. Hand, The 2-deoxyglucose method, in: *Neuroanatomical Tracing Methods*, L. Heimer and M.J. Robards, eds., Plenum Press, New York (1981), pp. 511-538.
15. H.E. Savaki, C. Kennedy,, L. Sokoloff, and M. Mishkin, Visually guided reaching with the forelimb contralateral to a "blind" hemisphere: a metabolic study in monkeys. *J. Neurosci.* 13:2772-2789 (1993).
16. W. Schwartz, C.B. Smith, L. Davidsen, H. Savaki, L. Sokoloff, M. Mata, D. J. Fink, and H. Gainer, Metabolic mapping of functional activity in the hypothalamo-neurohypophysial system of the rat. *Science* 205:723-725 (1979).
17. M. Miyaoka, M. Shinohara, M. Batipps, K.D. Pettigrew, C. Kennedy, and L. Sokoloff, The relationship between the intensity of the stimulus and the metabolic response in the visual system of the rat. *Acta Neurol. Scand.*60 [Suppl 70]):16-17 (1979).
18. J.J. Nordmann, Ultrastructural morphometry of the rat neurohypophysis. *J. Anat.* 123:213-218 (1977).
19. M. Kadekaro, A.M. Crane, and L. Sokoloff, Differential effects of electrical stimulation of sciatic nerve on metabolic activity in spinal cord and dorsal root ganglion in the rat. *Proc. Natl. Acad. Sci., U.S.A.* 82:6010-6013 (1985).
20. T.G. Smith, Jr., Sites of action potential generation in cultured neurons. *Brain Res.* 288:381-383 (1983).
21. W.H. Freygang, Jr., An analysis of extracellular potentials from single neurons in the lateral geniculate nucleus of the cat. *J. Gen. Physiol.* 41:543-564 (1958).
22. W.H. Freygang, Jr. and K. Frank, Extracellular potentials from single spinal motoneurones. *J. Gen. Physiol.* 42:749-760 (1959).
23. P. Yarowsky, M. Kadekaro, and L. Sokoloff, Frequency-dependent activation of glucose utilization in the superior cervical ganglion by electrical stimulation of cervical sympathetic trunk. *Proc. Natl. Acad. Sci., U.S.A.* 80:4179-4183 (1983).
24. M. Shinohara, B. Dollinger, G. Brown, S. Rapoport, L. Sokoloff, Cerebral glucose utilization: Local changes during and after recovery from spreading cortical depression. *Science* 203:188-190 (1979).
25. M. Mata, D.J. Fink, H. Gainer, C.B. Smith, L. Davidsen, H. Savaki, W.J. Schwartz, and L. Sokoloff, Activity-dependent energy metabolism in rat posterior pituitary primarily reflects sodium pump activity. *J. Neurochem.* 34: 213-215 (1980).
26. R.K. Orkand, J.G. Nicholls, and S.W. Kuffler, Effect of nerve impulses on the membrane potential of glial cells in the central nervous system of amphibia. *J. Neurophysiol.* 29:788-806 (1966).

27. F. Medzihradsky, P.S. Nandhasri, V. Idoyaga-Vargas, and O.Z. Sellinger, A comparison of ATPase activity of the glial cell fraction and the neuronal perikaryal fraction isolated in bulk from rat cerebral cortex. *J. Neurochem.* 18:1599-1603 (1971).
28 F.A. Henn, H. Haljamäe, and A. Hamberger, Glial cell function: active control of extracellular K^+ concentration. *Brain Res.* 43: 437-443 (1972).
29. L. Hertz, Drug-induced alterations of ion distribution at the cellular level of the central nervous system. *Pharmacol. Rev.* **29,** 35-65 (1977).
30. M. Erecinska and I.A. Silver, Metabolism and role of glutamate in mammalian brain. *Progress in Neurobiol.* **43,** 37-71 (1994).
31. M.A. Kai-Kai and V.W. Pentreath, High resolution analysis of [3H]2-deoxyglucose incorporation into neurons and glial cells in invertebrate ganglia: histological processing of nervous tissue for selective marking of glycogen. *J. Neurocytol.* 10:693-708 (1981).
32 V.W. Pentreath and M.A. Kai-Kai, Significance of the potassium signal from neurons to glial cells. *Nature* **295,** 59-61 (1982).
33. P. Yarowsky, A.F.Boyne, R. Wierwille, and N. Brookes, Effect of monensin on deoxyglucose uptake in cultured astrocytes: energy metabolism is coupled to sodium entry. *J. Neurosci.* 6:859-866 (1986).
34. R.S. Badar-Goffer, O.Ben-Yoseph, H.S. Bachelard, and P.G. Morris, Neuronal-glial metabolism under depolarizing conditions. A ^{13}C-n.m.r. study. *Biochem. J.* **282,** 225-230 (1992).
35. C. J. Cummins, R.A Glover, and O.Z. Sellinger, Neuronal cues regulate uptake in cultured astrocytes. *Brain Res.* 170: 190-193 (1979).
36. C.J. Cummins, R.A. Glover, and O.Z. Sellinger, Astroglial uptake is modulated by extracellular K^+. *J. Neurochem.* 33:779-785 (1979).
37 N. Brookes and P. J. Yarowsky, Determinants of deoxyglucose uptake in cultured astrocytes: the role of the sodium pump. *J. Neurochem.* 44:473-479 (1985).
38. L. Hertz and L. Peng, Energy metabolism at the cellur level of the CNS. *Can. J. Physiol. Pharmacol.* 70 (Suppl.):S145-S157 (1992).
39. L. Peng, X. Zhang, and L. Hertz, High extracellular potassium concentrations stimulate oxidative metabolism in a glutamatergic neuronal culture and glycolysis in cultured astrocytes but have no stimulatory efect in a GABAergic neuronal culture. *Brain Res.* 663:168-172 (1994).
40. S.Takahashi, B.F. Driscoll, M.J. Law, and L. Sokoloff, Role of sodium and potassium in regulation of glucose metabolism in cultured astroglia. *Proc. Natl. Acad. Sci., U.S.A.* 92: 4616-4620 (1995).
41. P.R. Smith, R.I. Krohn, G.T Hermanson, A.K. Mallia, F.H. Gartner, M.D. Provenzano, E. K. Fujimoto, N.M. Goeke, B.J. Olson, and D.C. Klenk, Measurement of protein using bicinchoninic acid. *Anal. Biochem.* 15:76-85 (1985).
42. B. Flott and W. Seifert, Characterization of glutamate uptake in astrocyte primary cultures from rat brain. *Glia* 4: 293-304 (1991).
43. L. Pellerin and P.J. Magistretti,.Glutamate uptake into astrocytes stimulates aerobic glycolysis: A mechanism coupling neuronal activity to glucose utilization. *Proc. Natl. Acad. Sci. U.S.A.* 91:10625-10629 (1994).
44. N. Brookes and R.J. Turner, K^+-induced alkalinization in mouse cerebral astrocytes mediated by reversal of electrogenic Na^+-HCO_3 cotransport. *Am. J. Physiol.* 267 (*Cell Physiol.* 36):C1633-C1640 (1994).
45. P.P. Li and T.D. White, Rapid effects of veratridine, tetrodotoxin, gramicidin D, valinomycin and NaCN on the Na^+, K^+ and ATP contents of synaptosomes. *J. Neurochem.* 28: 967-975 (1977).

46. W.A. Catterall, Cellular and molecular biology of voltage-gated sodium channels. *Physiol. Rev.* 72 (suppl.):S15-S48 (1992).
47. H.K. Kimelberg, S. Biddlecome, S. Narumi, and R.S. Bourke, ATPase and carbonic anhydrase activities of bulk-isolated neuron, astroglia and synaptosome fractions from rat brain , *Brain Res.* 141:305-323 (1978).
48. B.C Pressman and M. Fahim, Pharmacology and toxicology of the monovalent carboxylic ionophores. *Annual. Rev. Pharmacol. Toxicol.* 22:465-490 (1982).
49. B. Trivedi and W.H. Danforth, Effect of pH on the kinetics of frog muscle phosphofructokinase *J. Biol. Chem.* 241:4110-4112 (1966).
50. M. Erecinska, F. Dagani, D. Nelson, J. Deas, and I.A. Silver, Relations between intracellular ions and energy metabolism: A study with monensin in synaptosomes, neurons, and C6 glioma cells. *J. Neurosci.* 11:2410-2421 (1991).
51. G. Moonen, G. Frank, and E. Schoffeniels, Glial control of neuronal excitability in mammals: I. Electrophysiological and isotopic evidence in culture. *Neurochem. Int.* **2,** 299-310 (1980).
52. B.A. Barres, New roles for glia. *J. Neurosci.* 11:3685-3694 (1991).
53. B.A. Barres, L.L.Y. Chun, and D.P. Corey, Glial and neuronal forms of the voltage-dependent sodium channel: characteristics and cell-type distribution. *Neuron* 2:1375-1388 (1989).
54. K. Hisanaga, S.M. Sagar, K.J. Hicks, R.A. Swanson, and F.R. Sharp, c-fos proto-oncogene expression in astrocytes assocxiated wuth differentiatyion or proliferation but not depolarization. *Mol. Brain Res.* 8:69-75 (1990).
55. L. Hertz, An intense potassium uptake into astrocytes, its enhancement by high concentrations of potassium, and its possible involvement in potassium homeostasis at the cellular level. *Brain Res.* 145: 202-208 (1978).
56. L. Hertz, Features of astrocytic function apparently involved in the response of central nervous tissue to ischemia-hypoxia. *J. Cereb. Blood Flow Metab.* 1:143-153 (1981).
57. Y. Kanai and M.A. Hediger, Primary structure and functional characterization of a high-affinity glutamate transporter. *Nature (London)* 360: 467-471 (1992).
58. G. Pines, N.C. Danbolt, M. Bjørås, Y. Zhang, A. Bendahan, L. Eide, H. Koepsell, J. Storm-Mathisen, E. Seeberg , and B.I. Kanner, Cloning and expression of a rat brain L-glutamate transporter. *Nature (London)* 360:464-467 (1992).
59. T. Storck, S. Schulte, K. Hofmann, and W. Stoffel, Structure, expression, and functional analysis of a Na^+-dependent glutamate/aspartate transporter from rat brain. *Proc. Natl. Acad. Sci. USA* 89: 10955-10959. (1992)
60. J.D. Rothstein, L. Martin, A.I. Levey, M. Dykes-Hoberg, L. Jin, D. Wu, N. Nash, and R.W. Kuncl, *Neuron* 13:713-725 (1994).
61. C.A. Bowman and H.K. Kimelberg, Excitatory amino acids depolarize rat brain astrocytes in primary culture. *Nature (London)* 311: 656-659. (1984).

BIOENERGETICS IN OXIDATIVE DAMAGE IN NEURODEGENERATIVE DISEASES

M. Flint Beal

Neurology Service
Massachusetts General Hospital
32 Fruit Street
Boston, MA 02114

ABSTRACT

A major theory of neuronal damage in neurodegenerative diseases is that activation of excitatory amino acid receptors may play a role. In the case of neurodegenerative diseases activation of excitatory amino acid receptors may occur as a consequence of a defect in energy metabolism, which then leads to partial neuronal depolarization and activation of voltage-dependent NMDA receptors. This leads to calcium influx into the cell which is followed by activation of several deleterious processes. Amongst these are the generation of free radicals by mitochondria and the activation of nitric oxide synthase (NOS), leading to production of nitric oxide. Nitric oxide can react with superoxide to produce peroxynitrite, which can then damage lipids, proteins and DNA. Consistent with these proposed mechanisms we found that 3-nitropropionic acid, an irreversible succinate dehydrogenase inhibitor, produces selective striatal lesions in both rats and primates by a secondary excitotoxic mechanism. This compound produces striatal lesions and delayed chorea and dystonia following accidental ingestion in man. We found that it produces energy depletion in the striatum in vivo, and secondary excitotoxic lesions. Chronic low grade systemic administration of 3-nitropropionic acid produces striatal lesions which show age-dependence, spiny neuron dendritic changes, and selective neuronal vulnerability similar to that seen in Huntington's disease. Administration of 3-nitropropionic acid to primates resulted in apomorphine inducible chorea. Histologic examination with GFAP and calbindin staining showed a dorsal-vental gradient of cell loss, and sparing of NADPH-diaphorase neuron. These lesions therefore closely resemble Huntington's disease, and in primates there is an apomorphine inducible chorea. The lesions are accompanied by increased hydroxyl radical generation and they are attenu ated by both free radical spin traps and a selective inhibitor of neuronal NOS. The free radical spin

Neurodegenerative Diseases
Edited by Gary Fiskum, Plenum Press, New York, 1996

traps attenuate hydroxyl radical production, yet they have no effects on striatal energy metabolism or striatal electrophysiologic activity in vivo, arguing against a direct effect on excitatory amino acid receptors. The NOS inhibitor also attenuates MPTP neurotoxicity. Free radical and NOS inhibitors may therefore be useful in the treatment of Huntington's disease and Parkinson's disease.

INTRODUCTION

During the past several years there have been notable advances in the understanding of the molecular defects underlying the pathogenesis of chronic neurodegenerative diseases such as Huntington's disease (HD), Alzheimer's disease (AD), Parkinson's disease (PD), and Amyotrophic Lateral Sclerosis (ALS). Two major advances occurred in 1993. These included the demonstration of the existence of point mutations in the gene encoding the enzyme copper/zinc superoxide dismutase in some familial ALS patients[1] and the finding that the genetic abnormality in HD was an increase in CAG repeats in a gene encoding an unknown protein (HD Study Group).[2] Despite these advances the biochemical mechanisms whereby molecular defects can lead to slowly progressive neuronal degeneration remain an enigma. Some evidence however supported the proposition that defects in mitochondrial energy metabolism may play an intrinsic role.[3] Furthermore strong evidence is emerging that defects in mitochondrial energy production may result in oxidative damage to both mitochondrial DNA as well as other neuronal macromolecules.[4] Defects in energy metabolism may lead to neuronal damage by secondary excitotoxic mechanisms.[3, 5] There appears to be a complex interaction between defects and energy metabolism, excitotoxicity, and oxidative damage to cells.

Excitotoxicity refers to neuronal cell death caused by activation of excitatory amino acid receptors. The concept of slow or weak excitotoxicity was proposed to explain the slow evolution of neurodegenerative diseases which characteristically have a delayed onset and then slowly progress over many years.[3, 5] One mechanism to account for slow excitotoxicity that this may occur as a consequence of a defect in energy metabolism. Under this scenario an inability to maintain cellular ATP levels may lead to partial neuronal depolarization, relief from the voltage dependent magnesium block of the N-methyl-D-aspartate (NMDA) receptor and persistent receptor activation by ambient glutamate levels. There is substantial experimental evidence in support of such a mechanism in vitro. The initial report in support of this was provided by the work of Novelli and coworkers.[6] They demonstrated that glutamate was much more toxic in the setting of energy impairment induced by either hypoglycemia or impairment of oxidative phosphorylation caused by cyanide. Subsequently the work of Zeevalk and Nicklas also showed that inhibition of oxidative metabolism or of glycolysis could lead to excitotoxic cell injury in the chick retina in the absence of any increase in extracellular glutamate levels.[7] Furthermore they showed there graded titration of membrane potential with potassium could mimic this type of excitotoxic neuronal injury.

The key mediators of cellular toxicity following activation of excitatory amino acid receptors have been debated. There is substantial consensus that an influx of intracellular calcium is a key event. Calcium accumulation in cultured cortical neurons correlates with subsequent neuronal degeneration.[8] The direct measurement of intracellular concentrations is a less accurate predictor. The degree of accumulation from the extracellular space correlates better with ensueing cell death. This may be because of intracompartmental buffering of calcium particularly in the mitochondria.[9] Calcium influx via the NMDA receptor is much more effective in mediating cell death in that occuring through non-NMDA receptors or voltage dependent calcium channels.[10] This suggests that there is compartmentalization of calcium dependent neurotoxic processes within neurons with a preferential localization in the submembrane space adjacent to the NMDA receptors. Further evidence favoring calcium as being a key mediator is the observation that cell permeable calcium chelators reduce excitotoxic neuronal damage in vitro and ischemic neuronal injury in vivo.[11]

Increased intracellular calcium levels can initiate a number of deleterious processes. Two of these which we have recently focused on are the generation of free radicals and the activation of nitric oxide synthase. The initial work suggesting that nitric oxide synthase plays a role in excitotoxic cell injury was that of Dawson and colleagues.[12] They demonstrated that nitric oxide synthase inhibitors reduce glutamate neurotoxicity in cultured cortical and striatal neurons. They also showed that hemoglobin which scavenges nitric oxide was neuroprotective. Subsequent work showed that pretreatment of cultures with low doses of quisqualate, which preferentially killed nitric oxide synthase containing neurons, blocked glutamate neurotoxicity in the cultures.[13] Further work in vitro however was controversial. A more direct way of examining whether nitric oxide plays a role in neuronal injury comes from study in transgenic mice which have knockouts of the neuronal isoform of nitric oxide synthase. These animals show reduce influx in size in focal models of cerebral ischemia.[14] We have recently demonstrated that these mice are also resistant to striatal lesions induced by the mitochondrial toxin malonate (unpublished findings).

Direct evidence linking glutamate to excitotoxic cell death was recently obtained using electron paramagnetic resonance in cultured cerebellar neurons.[15] These authors demonstrated that NMDA receptor activation leads to superoxide generation which could be attenuated by either superoxide dismutase or by blocking the NMDA receptor with an NMDA antagonist. Furthermore they showed that the generation of free radicals was attenuated by chelating extracellular calcium. Dykens also showed that calcium leads to free-radical generation by isolated mitochondria.[16] Recent studies have directly demonstrated that activation of excitatory amino acid receptors leads to an influx of calcium into the cell, which is then buffered by mitochondria, and which leads to the generation of free radicals within cells as demonstrated by measurements with fluorescent dyes.[17] These data provide direct evidence that glutamate induced influx of calcium into mitochondria leads to free radical generation.[18] We have obtained evidence in vivo demonstrating excitotoxicity can be linked to free radical generation. We found that free radical spin traps can attenuate excitotoxicity produced by intrastriatal injections of an NMDA, kainate and AMPA. In a similar manner transgenic mice overexpressing superoxide dismutase show attenuation of NMDA neurotoxicity in vivo.[19]

If mitochondrial dysfunction leads to secondary excitotoxic neuronal damage which is similar to that which occurs in neurodegenerative diseases then one ought to be able to produce lesions in animals with mitochondrial toxins which will then replicate some of the characteristic neurochemical and neuropathologic features of neurodegenerative diseases. This was initially shown to be the case with MPTP which produces and excellent model PD. MPTP is converted into 1 methyl-4-phenylpyridinium (MPP^+), an inhibitor of mitochondrial complex I. Administration of MPTP to mice leads to a depletion of striatal ATP concentrations, as does intrastriatal administration of MPP^+.[20, 21]

We have recently investigated the effects of 3-nitroproprionic acid (3-NP), a naturally occuring plant and micotoxin that is an irreversible inhibitor of complex II (succinate dehydrogenase). This compound is known to produce illness in livestock. It is of particular interest since it produces disease in humans which have occurred in outbreaks in China after ingestion of mildewed sugar cane by China.[22] These large outbreaks have occurred in as many 1,000 patients. The illness is characterized by development of an encephalopathy associated with stupor. Following recovery of consciousness there is delayed onset of a movement disorder. The movement disorder has characterized by both choreiform movements as well as dystonia and torticollis. The CT scans of the patients show putaminal necrosis. We studied 3-NP neurotoxicity in both rats and nonhuman primates.[23-26] Intrastriatal administration in rats produces dose-dependent lesions, as well as ATP depletions and age-dependent increases in lactate.[23] These increases parallel age dependent increases in lesion size. Systemic administration of 3-NP to rats produces selective striatal lesions.[25, 26] MRI demonstrated that these lesions were accompanied by focal increases in lactate in the basal ganglia which preceded the onset of degeneration as shown by T2 weighted imaging. Three lines of evidence suggest that these lesions are caused by an excitotoxic mechanism. The first is that the lesions are attenuated by prior decortication which removes the cortico-striatal glutamatergic input.[25] The second piece of evidence is that the lesions are attenuated by administration of lamotrigine, an inhibitor of glutamate release. A third line of evidence is that the lesions are also attenuated by the non-NMDA antagonist NBQX.[27] We carried out microdialysis studies which showed that at a time point of metabolic inhibition at which lactacte concentrations were increased 2-fold there was no increase in extracellular glutamate concentrations.[25] These studies therefore suggest that 3-NP produces lesions by a secondary excitotoxic mechanism. Chronic low grade administration of 3-NP to rats by Alzet pump for 1 month produced selective lesions in the dorso-lateral striatum which showed sparing of NADPH-diaphorase neurons. In addition Golgi studies showed proliferative changes in the dendrites of spiny neurons which are characteristic of HD.

We recently extended our studies to nonhuman primates.[24] The neuroanatomy of the primate basal ganglia much more closely resembles that of man than does that of rodents and is therefore possible to produce chorea, the cardinal clinical feature of HD. Initial studies of MRI scans show that there were selective lesions as shown by T2-weighted imaging in the basal ganglia of primates. Subsequent studies examined the effects of chronic administration of 3-NP over 4-6 weeks in both Macacca species and baboons. Chronic administration resulted in an apomorphine inducible movement disorder that closely resembles that seen in HD. The animals show facial dyskinesia, dystonia, dyskinesia of the extremities, and choreiform movements. Both a clinical rating scale and a quantitative analysis of tangential velocity of individual movements confirm that 3-NP treated animals had a significant increase in choreiform and dystonic movements. Furthermore with chronic administration of 3-NP 2 of the baboons have now developed spontaneous dyskinetic and dystonic posturing of the lower extremities. This is the first time that one has been able to produce spontaneous persistent dyskinesias in primates.

Histological evaluation of the lesions show that they are strikingly reminiscent of those that occur in HD. There is a depletion of both nissl stained and calbindin stained neurons. In contrast medium sized aspiny neurons staining with NADPH diaphorase and large striatal neurons were spared. In addition Golgi studies showed proliferative changes in the dendrites of spiny neurons that were similar to changes which occur in HD. Furthermore the patch matrix compartmentalization of the striatum which is preserved in HD, was spared by the lesions. Lastly there was sparing of the nucleus accumbens which is spared in HD. These findings therefore demonstrate that a mitochondrial toxin administered chronically to primates can reproduce both the histologic features as well as the movement disorder which is characteristic of HD.

We have recently carried out a series of studies to investigate the mechanisms of lesions produced by both MPTP and 3-NP. We initially examined the effects of free radical spin traps.[18] Examined the effects of the free radial spin trap 2-sulfo-tert-phenylbutylnitrone. Administration of this compound significantly attenuated the dopamine depletion produced by a mild course of administration of MPTP.[28] It however was much less or ineffective against more marked dopaminergic depletions. We also examined the effects of energy repletion with coenzyme Q_{10} and nicotinamide.[29] We found that administration of the combination of these 2 compounds produced significant neuroprotective effects against moderate and mild dopaminergic depletion caused by MPTP. Lastly we investigated whether the nitric oxide synthase inhibitor, 7-nitroindazole, could attenuate dopaminergic depletion produced by MPTP.[30] This compound is of interest because it appears to be a relatively selective inhibitor of the neuronal isoform of nitric oxide synthase in vivo. It does not show any effects on blood pressure or on cetylcholine induced vasorelaxation. We administered this compound following induction of both moderate and severe dopaminergic depletion with MPTP. We found that 7-nitroindazole produced dose-dependent virtually complete neuroprotection. To examine the mechanism of the lesions we measured 3-nitrotyrosine. 3-Nitrotyrosine is a product of a nitration reaction which occurs when there is an increased production of the compound peroxynitrite. Peroxynitrite is generated by the reaction of nitric oxide with superoxide. This is an exceptionally fast chemical reaction which occurs 3-fold faster than reaction of superoxide with superoxide dismutase. It occurs at a reaction rate of 6.7×10^9 moles per second. It is therefore essentially diffusion limited. We found that MPTP induced a significant increase in 3-nitrotyrosine concentrations in the striatum of mice. These increases were completely blocked by pretreatment with 7-nitroindazole consistent with its ability to block generation of nitric oxide and therefore to block the subsequent generation of peroxynitrite.

We have also carried out studies to determine whether oxidative damage plays a role in 3-nitroproprionic acid neurotoxicity.[31] We initially administered salicylate and showed that 3-nitroproprionic acid induced a significance increase in 2,3 and 2,5 dihydroxybenzoic acid/salicylate. These 2 compounds are products of hydroxylation of salicylate and have been found to be useful markers for measuring oxidative stress in vivo. We also demonstrated that there is a significant increase in 8-hydroxy-2-deoxyguanosine in nuclear DNA of 3-NP treated rats. This compound is a useful marker for oxidative damage to DNA. Furthermore we found that the free radical spin trap DMPO significantly attenuates 3-NP induced neurotoxicity in rats. This evidence therefore suggests that oxidative damage plays a direct role in 3-NP neurotoxicity.

We also investigated whether nitric oxide and peroxynitrite may play a role in 3-NP neurotoxicity. We examined whether 7-nitroindazole could attenuate lesions produced by 3-NP in rats. Administration of 7-nitroindazole with 3-NP over 5 days completely blocked the lesions in the treated animals as compared to saline treated controls. The controls showed a significant increase in both DHBA/salicylate and 3-nitrotyrosine which were blocked by pretreatment with 7-nitroindazole. The fact that 7-nitroindazole block the increase in DHBA/salicylate and 3-nitrotyrosine suggests that peroxynitrite plays a role in mediating oxidative damage associated with 3-NP neurotoxicity.

As reviewed above substantial evidence suggest that there is an interplay between defects and energy metabolism, excitotoxicity and oxidative damage and the pathogenesis of neurodegenerative diseases. These mechanisms may be downstream consequences of the primary genetic defects in these illnesses. They are likely to interact with processes involved in normal aging which may account for the delayed onset of these diseases. There is substantial evidence that there age-dependent decreases in mitochondrial function. It is therefore possible that with normal aging, and a genetic defect affecting energy metabolism one may pass a critical threshold leading to the onset of illness. The present studies have shown that and 3-NP models of neurodegenerative disease are associated with both oxidative stress and nitric oxide generation.

The present observations have implications for the therapy of neurodegenerative diseases. They suggest that therapeutic possibilities would include excitatory amino acid antagonists, energy enhancing therapies such as coenzyme Q_{10}, free radical scavengers and nitric oxide synthase inhibitors. It is possible that these therapies either alone or in combination may prove useful in attempting to treat a variety of neurodegenerative diseases.

ACKNOWLEDGEMENTS

This work was supported by NIH grants NS16367, NS31579, NS32365, PO1 AG11337, PO1 AG12992 and P50 AG05134. The secretarial assistance of Sharon Melanson is gratefully acknowledged.

REFERENCES

1. D.R. Rosen, T. Siddique, D. Patterson et al., Mutations in Cu/Zn superoxide dismutase gene are associated with familial amyotrophic lateral sclerosis, *Nature* 362:59 (1993).

2. The Huntington's Disease Collaborative Research Group, A novel gene containing a trinucleotide repeat that is expanded and unstable on Huntington's disease chromosomes, *Cell* 72:971 (1993).

3. M.F. Beal, Does impairment of energy metabolism result in excitotoxic neuronal death in neurodegenerative illnesses?, *Ann Neurol.* 31:119 (1992).

4. M.F. Beal, E. Brouillet, B.G. Jenkins, R. Henshaw, B. Rosen and B.T. Hyman, Age-dependent striatal excitotoxic lesions produced by the endogenous mitochondrial inhibitor malonate, *J Neurochem.* 61:1147 (1993).

5. R.L. Albin and J.T. Greenamyre, Alternative excitotoxic hypotheses, *Neurology* 42:733 (1992).

6. A. Novelli, J.A. Reilly, P.G. Lysko et al., Glutamate becomes neurotoxic via the N-methyl-D-aspartate receptor when intracellular energy levels are reduced, *Brain Res.* 451:205 (1988).

7. G.D. Zeevalk and W.J. Nicklas, Chemically induced hypoglycemia and anoxia: relationship to glutamate receptor-mediated toxicity in retina, *J Pharmacol Exp Ther.* 253:1285 (1990).

8. D.M. Hartley, M.C. Kurth, L. Bjerkness et al., Glutamate receptor-induced $^{45}Ca^{2+}$ accumulation in cortical cell culture correlates with subsequent neuronal degeneration, *J Neurosci.* 13:1993 (1993).

9. J.L. Werth and S.A. Thayer, Mitochondria buffer physiological calcium loads in cultured rat dorsal root ganglion neurons, *J Neurosci.* 14:348 (1994).

10. M. Tymianski, M.P. Charlton, P.L. Carlen et al., Source specificity of early calcium neurotoxicity in cultured embryonic spinal neurons, *J Neurosci.* 13:2085 (1993).

11. M. Tymianski, M.C. Wallace, I. Spigelman et al., Cell-permeant Ca^{2+} chelators reduce early excitotoxic and ischemic neuronal injury in vitro and in vivo, *Neuron* 11:221 (1993).

12. V.L. Dawson, T.M. Dawson, E.D. London et al., Nitric oxide mediates glutamate neurotoxicity in primary cortical cultures, *Proc Natl Acad Sci USA* 88:6368 (1991).

13. V.L. Dawson, T.M. Dawson, D.A. Bartley et al., Mechanisms of nitric oxide mediated neurotoxicity in primary brain cultures, *J Neurosci.* 13:2651 (1993).

14. Z. Huang, P.L. Huang, N. Panahian et al., Effects of cerebral ischemia in mice deficient in neuronal nitric oxide synthase, *Science*, 265:1883 (1994).

15. M. Lafon-Cazal, S. Pietri, M. Culcasi et al., NMDA-dependent superoxide production and neurotoxicity, *Nature* 364:535 (1993).

16. J.A. Dykens, Isolated cerebral and cerebellar mitochondria produce free radicals when exposed to elevated Ca^{2+} and Na^{+}: implications for neurodegeneration, *J Neurochem.* 63:584 (1994).

17. L.L. Dugan, S.L. Sensi, L.M.T. Canzoniero et al., Imaging of mitochondrial oxygen radical production in cortical neurons exposed to NMDA, *Soc Neurosci Abst.* 20:1532 (1994).

18. J.B. Schulz, D.R. Henshaw, D. Siwek, B.G. Jenkins, R.J. Ferrante, P.B. Cipolloni, N.W. Kowall, B.R. Rosen and M.F. Beal, Involvement of free radicals in excitotoxicity in vivo, *J Neurochem.* 64:2239 (1995).

19. P.H. Chan, J. Chen, J. Gafni et al., N-methyl-D-aspartate-mediated neurotoxicity is associated with oxygen-derived free radicals, *Princeton Conf Cerebrovasc Dis.* in press:(1995).

20. P. Chan, L.E. DeLanney, I. Irwin et al., Rapid ATP loss caused by 1-methyl-4-phenyl-1,2,3,6-tetrahydropyridine in mouse brain, *J Neurochem.* 57:348 (1991).

21. E. Storey, B.T. Hyman, B. Jenkins, E. Brouillet, J.M. Miller, B.R. Rosen and M.F. Beal, MPP^{+} produces excitotoxic lesions in rat striatum due to impairment of oxidative metabolism, *J Neurochem.* 58:1975 (1992).

22. A.C. Ludolph, F. He, P.S. Spencer et al., 3-Nitropropioinic acid - exogenous animal neurotoxin and possible human striatal toxin, *Can J Neurol Sci.* 18:492 (1991).

23. E.P. Brouillet, B.G. Jenkins, B.T. Hyman, R.J. Ferrante, N.W. Kowall, R. Srivastava, D. Samanta Roy, B.R. Rosen and M.F. Beal, Age-dependent vulnerability of the striatum to the mitochondrial toxin 3-nitropropionic acid, *J Neurochem.* 60:356 (1993).

24. E. Brouillet, P. Hantraye, R. Dolan, A. Leroy-Willig, M. Bottlaender, I. Isacson, M. Maziere, R.J. Ferrante and M.F. Beal, Chronic administration of 3-nitropropionic acid induced selective striatal degeneration and abnormal choreiform movements in monkeys, *Soc Neurosci Abst.* 19:409 (1993).

25. M.F. Beal, E. Brouillet, B.G. Jenkins, R.J. Ferrante, N.W. Kowall, J. M. Miller, E. Storey, R. Srivastava, B. R. Rosen and B. T. Hyman, Neurochemical and histologic characterization of excitotoxic lesions produced by the mitochondrial toxin 3-nitropropionic acid, *J Neurosci.* 13:4181 (1993).

26. U. Wullner, A.B. Young, J.B. Penney and M.F. Beal, 3-Nitropropionic acid toxicity in the striatum, *J Neurochem.* 63:1772 (1994).

27. L. Turski and H. Ikonomidou, Striatal toxicity of 3-nitropropionic acid prevented by the AMPA antagonist NBQX, *Soc Neurosci Abst.* 20:1677 (1994).

28. J.B. Schulz, D.R. Henshaw, R.T. Matthews and M.F. Beal, Coenzyme Q_{10} and nicotinamide and a free radical spin trap protect against MPTP neurotoxicity, *Exp Neurol.* 132:1 (1995).

29. M.F. Beal, D.R. Henshaw, B.G. Jenkins, B.R. Rosen and J.B. Schulz, Coenzyme Q_{10} and nicotinamide block striatal lesions produced by mitochondrial toxin malonate, *Ann Neurol.* 36:882 (1994).

30. J.B. Schulz, R.T. Matthews, M.M.K. Muqit, S.E. Browne and M.F. Beal, Inhibition of neuronal nitric oxide synthase by 7-nitroindazole protects against MPTP-induced neurotoxicity in mice, *J Neurochem.* 64:936 (1995).

31. J.B. Schulz, R.T. Matthews, D.R. Henshaw and M.F. Beal, Inhibition of neuronal nitric oxide synthase (NOS) protects against neurotoxicity produced by 3-nitropropionic acid, malonate, and MPTP, *Soc Neurosci Abst.* 20:1661 (1994).

IS THERE A RELATIONSHIP BETWEEN CONDITIONS ASSOCIATED WITH CHRONIC HYPOXIA, THE MITOCHONDRIA, AND NEURODEGENERATIVE DISEASES, SUCH AS ALZHEIMER'S DISEASE ?

Carl. R. Merril[1], Steve Zullo[1], and Hossein Ghanbari[2]

[1]Laboratory of Biochemical Genetics, NIMH, NIH, NIMH Neuroscience Center at Saint Elizabeths, Washington D.C. 20032
[2]Nymox Corp., Johnson City, TN 37604

The central nervous system (CNS) is extremely sensitive to hypoxia, as neuronal cells require a high rate of energy metabolism to maintain transmembrane potentials. As these cells normally rely on aerobic mitochondrial metabolism to generate the required energy, a lack of oxygen results in a shift toward the anaerobic utilization of glucose. Such a metabolic shift may result in a cascade of potentially deleterious events beginning with an increased production of lactic acid and a decreased rate of ATP production.[1] Increased lactic acid concentrations may cause a decrease in cellular pH, followed by the release of iron and the reaction of ferrous ions with hydrogen peroxide to produce hydroxyl radicals. A decrease in ATP may adversely affect DNA replication, transcription,[2] and mRNA translation. In addition, the levels of inter-cellular calcium will increase as the ATP dependent calcium ion pump mechanism fails. Increased calcium levels can activate: endonucleases, nitric oxide synthetase, phospholipases and proteases. All of the above alterations in cellular metabolism may result in damage to the cellular DNA, particularly the mitochondrial DNA (Figure 1).

Mitochondrial DNA (mtDNA) is especially sensitive to mutations due to free radicals. Such oxidative modifications of mtDNA have been well documented.[3,4] Mitochondrial DNA damaged by free radical modification may be more prone to deletion mutations which may be catalyzed by calcium activated endonucleases and other enzymes involved in homologous recombination, topoisomerase cleavage, and/or in slip-replication.[5]

The lack of oxygen has been found to affect myocardial tissue. An increase of mtDNA deletion mutations, 8 to 2200 fold, has been reported in human cardiac tissue from individuals suffering from ischemic heart disease when compared with age matched controls.[6] While a number of reports have suggested that deletion mutations of mtDNA are primarily an effect of aging, the observation that the occurrence of deleted mtDNA decreases in cardiac tissue after the age of 80 suggests that factors other than aging may be responsible for these deletions.[7] In addition, we found higher levels of the mtDNA deletion, $mtDNA^{4977}$, 12 fold in the putamen and 5 fold in the superior frontal gyrus, from

Neurodegenerative Diseases
Edited by Gary Fiskum, Plenum Press, New York, 1996

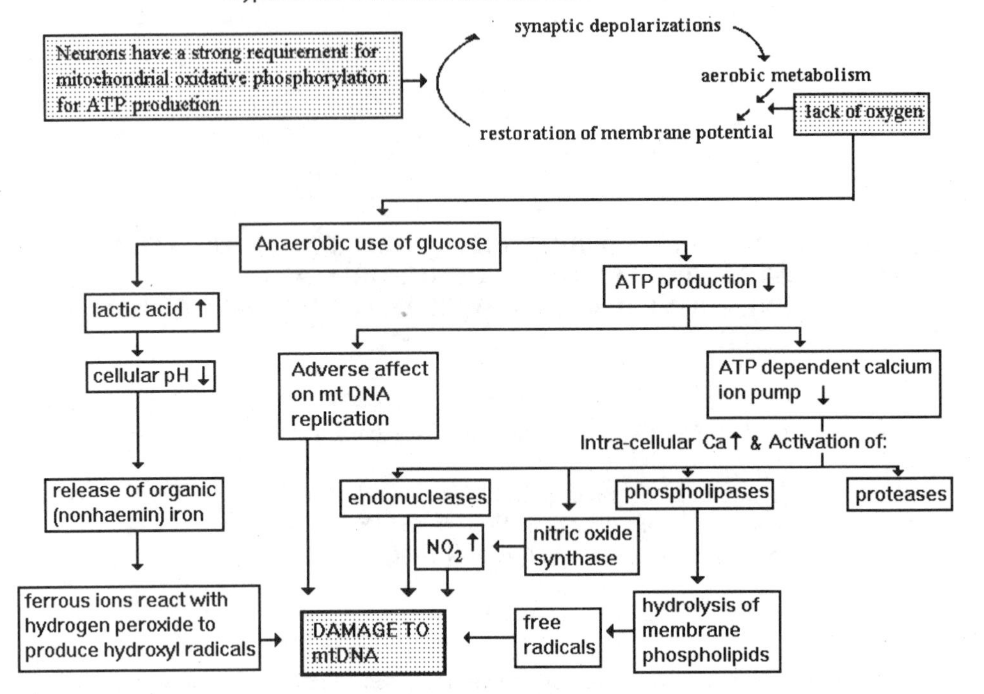

Figure 1. Neuronal cells rely on mitochondrial aerobic oxidative phosphorylation (OXPHOS) to generate energy in the form of ATP, a shift toward anaerobic utilization of glucose results from a lack of oxygen. These events may result in damage to mitochondrial DNA.

individuals who had evidence of conditions associated with chronic hypoxia (such as cardiovascular disease) as compared with age comparable individuals (between 34 and 73 years) who were free of such conditions.[8]

Cardiovascular problem	CNS observation	References
75% or greater obstruction of coronary vessels or occlusion of at least one major coronary artery	Increased incidence of cortical senile plaques	9
History of myocardial infarction	Five fold increase in the occurrence of dementia	10
Rabbits fed high cholesterol diets	Increase in Alzheimer's-like amyloid immunoreactive in brain tissue	11
Apolipoprotein E-4 is associated with higher plasma cholesterol levels and an increase risk of heart disease	The Apolipoprotein E-4 genotype is also associated with Alzheimer's disease	12-16

Figure 2. Is there a relation between cardiovascular disease and neurodegeneration ?

These observations are in agreement with a number of reports in the literature which suggest that cardiovascular problems may be associated with neurodegenerative changes in the CNS (Figure 2). Senile plaques, one of the most consistent features of Alzheimer's disease, have been reported to be increased in patients who had either greater than 75% obstruction of a coronary artery or occlusion of at least one major coronary artery.[9] Similarly, in a prospective study, women who had a history of myocardial infarction had a five-fold increase in the occurrence of dementia.[10] Moreover, rabbits fed a diet high in cholesterol were found to have "Alzheimer-like β-amyloid" immunoreactivity in their brain tissue.[11] In this regard, apolipoprotein E-4 has been clearly demonstrated in population studies to be associated with higher plasma cholesterol levels and an increased risk of heart disease.[12-16] Recently the association of apolipoprotein E-4 with Alzheimer's disease has stimulated a search for a critical role for apolipoprotein E-4 in the CNS.[12] While the direct intercellular CNS effects of apolipoprotein E-4 may be of primary importance, the E-4 polymorphism may also be involved in neurodegenerative processes through its association with heart disease.

If the mitochondrial genome is directly involved with the onset of neurodegenerative diseases such as Alzheimer's disease, as suggested above, there should be some evidence for a genetic association between the mitochondrial genome and this disease. In this regard, Alzheimer's disease has clearly been shown to be multi-factorial (Figure 3) and it has been suggested that some of the genetic complexity of Alzheimer's disease may be due to novel mtDNA variants and the occurrence of OXPHOS defects in patients with this disease.[19-22] Also, in a recent comparison of Alzheimer's disease individuals under the age of 75 years versus age-matched controls, the levels of $mtDNA^{4977}$ deletions in the cortical regions from the Alzheimer's disease individuals were four times as high as the levels found in the controls.[23]

Genetic linkage, chromosome	Gene	References
14	Candidate genes include HSPA2 (a 70kd heat shock protein) and c-*Fos*	17
19	apolipoprotein E-4	12
21	amyloid precursor protein	18
mitochondria	novel mtDNA variants associated with OXPHOS defects	19-22
mitochondria	increased levels of $mtDNA^{4977}$ deletions	23,24

Figure 3. Alzheimer's disease is a multi-factorial disorder.

While there are clearly a number of factors associated with neurodegenerative diseases, such as Alzheimer's disease, the role of chronic hypoxia should be carefully examined as a potential trigger mechanism for initiation of the disease state.

REFERENCES

1. M.F. Beal, Does impairment of energy metabolism result in excitotoxic neuronal death in neurodegenerative illnesses? *Ann. Neurol.* 31:119-130 (1992).
2. T. Kadowaki and Y. Kitagawa, Hypoxic depression of mitochondrial mRNA levels in HeLa cell, *Experimental Cell Res.* 192:243-247 (1991).
3. C. Richter, J.-W. Park, and B.N. Ames, Normal oxidative damage to mitochondrial and nuclear DNA is extensive, *Proc. natn. Acad. Sci. U.S.A.* 85:6465-6467 (1988).
4. K.P. Gupta, K.L. van Golen, E. Randerath, and K. Randerath, Age-dependent covalent DNA alterations (I-compounds) in rat liver mitochondrial DNA, *Mut. Res.* 237:17-27 (1990).

5. D.C. Wallace, Diseases of the mitochondrial DNA, *Annu. Rev. Biochem.* 61:1175-1212 (1992).
6. M. Corral-Debrinski, G. Stepien, J.M. Shoffner, M.T. Lott, K. Kanter, and D.C. Wallace, Hypoxemia is associated with mitochondrial DNA damage and gene induction, *JAMA* 266:1812-1816 (1991).
7. K. Hattori, M. Tanaka, S. Sugiyama, T. Obayashi, I. Takayki, T. Satake, H. Yoshihiro, J. Asai, M. Nagano, and T. Ozawa, Age-dependent increase in deleted mitochondrial DNA in the human heart: possible contributory factor to presbycardia. *Am. heart J.* 121:1735-1742 (1991).
8. C.R. Merril, S. Zullo, H. Ghanbari, M.M. Herman, J.E. Kleinman, L.B. Bigelow, J.J. Bartko, and D.J. Sabourin, Possible relationship between hypoxia and brain mitochondrial DNA deletions, Manuscript submitted for publication.
9. D.L. Sparks, H. Liu, S.W. Scheff, C.M. Coyne, and J.C. Hunsaker, Temporal sequence of plaque formation in the cerebral cortex of non-demented individuals, *J. Neuropath. Exp. Neurol.* 52:135-142 (1993).
10. M.K. Aronson, W.L. Ooi, H. Morgenstern, A. Hafner, D. Masur, H. Crystal, W.H. Frishman, D. Fisher, and R. Katzman, Women, myocardial infarction, and dementia in the very old, *Neurol.* 40:1102-1106 (1990).
11. D.L. Sparks, S.W. Scheff, J.C. Hunsaker, III, H. Liu, T. Landers, and D.R. Gross, Induction of Alzheimer-like β-Amyloid immunoreactivity in the brains of rabbits with dietary cholesterol, *Exp. Neurol.* 126:88-94 (1994).
12. W.J. Strittmatter, A.M. Saunders, D.E. Schmechel, M. Pericak-Vance, I. Enchild, G.S. Salvesen, and A.D. Roses, Apolipoprotein E: high-avidity binding to β-amyloid and increased frequency of type 4 allele in late-onset familial Alzheimer's disease, *Proc. natn. Acad. Sci. U.S.A.* 90:1977-1981 (1993).
13. J. Davignon, R.E. Gregg, and C.F. Sing, Apolipoprotein E polymorphism and atherosclerosis, *Atherosclerosis.* 8:1-21 (1988).
14. D.M. Hallman, E. Boerwinkle, N. Saha, C. Sandhozer, H.J. Menzel, A. Caázár, and G. Utermann, The apolipoprotein E polymorphism: A comparison of allele frequencies and effects in nine populations, *Am. J. Hum. Genet.* 49:338-349 (1991).
15. A.M. Cumming and F. Robertson, Polymorphism at the apoprotein-E locus in relation to risk of coronary disease, *Clin. Genet.* 25:310-318 (1984).
16. T. Kuusi, M.S. Nieminen, C. Ehnholm, H. Yki-Järvinen, M. Valle,, E.A. Nikkilä, and M.-R. Taskinen, Apoprotein E polymorphism and coronary artery disease: Increased prevalence of apolipoprotein E-4 in angiographically verified coronary patients, *Atherosclerosis.* 9:237-241 (1988).
17. G.D. Schellenberg, T.D. Bird, E.M. Wijsman, H.T. Orr, L. Anderson, E. Nemens, J.A. White, L. Bonnycastle, J.L. Weber, M.E. Alonso, H. Potter, L.L. Heston, and G.M. Martin, Genetic linkage evidence for a familial Alzheimer's disease locus on chromosome 14, *Science.* 258:668-671 (1992).
18. A. Goate, M.-C. Chartier-Harlin, M. Mullan, J. Brown, F. Crawford, L. Fidani, L. Giuffra, A. Haynes, N. Irving, L. James, R. Mant, P. Newton, K. Rooke, P. Roques, C. Talbot, M. Pericak-Vance, A. Roses, R. Williamson, M. Rossor, M. Owen, and J. Hardy, Segregation of a missense mutation in the amyloid precursor protein gene with familial Alzheimer's disease. *Nature.* 349:704-706 (1991).
19. F.-H. Lin, R. Lin, H.M. Wisniewski, Y.-W. Hwang, I. Grundke-Iqbal, G. Healy-Louie, and K. Iqbal, Detection of point mutations in codon 331 of mitochondrial NADH dehydrogenase subunit 2 in Alzheimer's brains, *Biochem. Biophys. Res. Comm.* 182:238-246 (1992).
20. V. Petruzzella, X. Chen, and E.A. Schon, Is a point mutation in the mitochondrial ND2 gene associated with Alzheimer's disease? *Biochem. Biophys. Res. Comm.* 186:490-497 (1992).
21. F.-H. Lin, and R. Lin, A comparison of single nucleotide primer extension with mispairing PCR-RFLP in detecting a point mutation, *Biochem. Biophys. Res. Comm.* 189:1202-1206 (1992).
22. J.M. Shoffner, M.D. Brown, A. Torroni, M.T. Lott, M.F Cabell, S.S. Mirra, M.F. Beal, C.-C. Yang, M. Gearing, R. Salvo, R.L. Watts, J.L. Juncos, L.A. Hansen, B.J. Crain, M. Fayad, C.L. Reckord, and D.C. Wallace, Mitochondrial DNA variants observed in Alzheimer's disease and Parkinson disease patients, *Genomics.* 17:171-184 (1993).
23. M. Corral-Debrinski, T. Horton, M.T. Lott, J.M. Shoffner, A.C. McKee, M.F. Beal, B.H. Graham, and D.C. Wallace, Marked changes in mitochondrial DNA deletion levels in Alzheimer brains, *Genomics.* 23:471-476 (1994).
24. N.-W. Soong, D.R. Hinton, G. Cortopassi, and N. Arnheim, Mosaicism for a specific somatic mitochondrial DNA mutation in adult human brain, *Nature Genet.* 2:318-323 (1992).

α-KETOGLUTARATE DEHYDROGENASE IN ALZHEIMER'S DISEASE

John P. Blass,[1] Kwan-Fu Rex Sheu,[1] and Rudolph E. Tanzi[2]

[1]Altschul Laboratory for Dementia Research and Will Rogers Institute
Burke Medical Research Institute, Cornell University Medical College
785 Mamaroneck Avenue
White Plains, NY 10605

and

[2]Neurogenetics Laboratory and Laboratory of Genetics and Aging
Massachusetts General Hospital
Harvard Medical School
Boston, MA 02114

INTRODUCTION

Since ancient times, a creative tension has existed between two viewpoints on illness. One, associated with the Platonic physicians of the island of Cnos, looks on diseases as specific entities which can be classified like plants. From this viewpoint, the specific patient in the examining room is a more or less close copy of an idealized patient. In modern times, this view was developed with particular strength by Sydenham in London in the 1600s. The second viewpoint, associated with the Hippocratic physicians of the island of Cos, views illness as occurring when an organism can no longer remain in balance with the challenges of its environment. In modern times, this view was put forward with particular clarity by Claude Bernard in Paris in the 1800s. He emphasized that an illness occurs when living organisms cannot maintain their "milieu interne" in the face of the challenges of their environment.

The current chapter is written from the mechanistic viewpoint of the Hippocratic physicians and of Claude Bernard. Specifically, this chapter reexamines the way that impairments of energy metabolism can contribute to the development of neurodegenerative disorders and specifically in Alzheimer's Disease (AD).

ENERGY (OXIDATIVE) METABOLISM IN NEUROLOGICAL DISEASES

Maintenance of the internal milieu requires a constant supply of energy, since living cells are steady-state systems. The bulk of the required energy is provided through synthesis of ATP in mitochondria, with the energy generated from oxidation of glucose and other substrates. Energy requirements increase with functional demands such as secretion of biosynthesized molecules, maintenance of axon potentials, and responses to a variety of internal and external stressors.

Variations in the enzymatic machinery of oxidative metabolism can be clinically silent in the absence of stressors, but become clinically important when they limit the ability of cells to increase oxidative metabolism in response to increased functional demands. This general principle has been documented for a number of stressors, including specifically glutamate and other excitotoxins (Henneberry, 1989; Beal, 1992) and for deleterious amyloid peptides (Mattson et al, 1993).

The concept of genetic predispositions which become clinically important in the face of appropriate stressors was developed in detail by Sir Archibald Garrod in the 1920s. He pointed out that "inborn predispositions" are much more common than "inborn errors" (Garrod, 1923) . Roses (1995) has used the term "molecular gerontology" to refer to the situation where a genetic variation becomes deleterious under the stress of aging. One can restate in modern molecular terms the classical view that the form of an illness, including an illness of aging, is determined in part by the patient's biological "site of least resistance." To mix older and newer terminology, a gene can represent a "molecular site of least resistance," ie a *locus minoris resistentiae molecularis* (Blass et al, 1977; Blass, 1993a).

Deficiencies in energy metabolism have long been known to be associated with disorders of the nervous system (Blass et al, 1988). Such disorders have been described in the major pathways of energy metabolism: glycogenolysis, glycolysis, the pyruvate dehydrogenase complex (PDHC), the Krebs tricarboxylic acid (TCA) cycle, electron transport, gluconeogenesis, carnitine metabolism, and glutamate dehydrogenase (Blass et al, 1988). Depending in part on the site and partly on the magnitude of the deficiency, disorders associated with deficiencies of energy metabolism include lactic acidosis with mental disability, various hereditary ataxias, and such common neuro-degenerative disorders as Motor Neuron Disease, Huntington Disease, Parkinson Disease, and AD. Compromise of the supply of oxygen and glucose to the brain is a very common source of damage to the nervous system, in stroke and in other types of hypoxic/ischemic brain damage. As discussed below, a number of disorders of energy metabolism are associated with selective vulnerability in the nervous system.

Since impaired energy metabolism is so common in neurodegenerative diseases, an important question about any abnormality found is whether it is part of the primary disease process or is instead a late event in the pathophysiology. We use three criteria to conclude that an abnormality probably occurs early in the pathophysiological process. First, the abnormality is found not only in histologically damaged areas of the brain but also in histologically normal areas as well. Second, the abnormality is found in clinically normal non-neural tissues as well as in the brain. Evidence is particularly strong from cultured fibroblasts or other cultured cells. Both these types of evidence indicate that the abnormality is not attributable simply to anatomical damage. Third, demonstration of an abnormality at the gene level is convincing proof of a "primary abnormality." In using these criteria, it is important to remember that deficiencies in transcription can be late events. Indeed, decreases in the transcription of mitochondrial DNA (mtDNA) are an early responses of cells to a variety of types of mitochondrial damage. Therefore, deficiencies in the activities of the electron transport complexes measured under artificial, *in vitro* conditions can not be assumed to be primary changes, particularly if localized to anatomically damaged tissue. Similarly, part of the process of cell death by apoptosis is mediated by free radicals. Therefore, evidence of free radical damage in tissues in which cells are dying by apoptosis can not of itself be taken as evidence that free radical damage is an early, " triggering event" in the pathophysiology of cell damage in that tissue.

A second important consideration is raised by the fact that energy metabolism is abnormal in so many neurodegenerative diseases - namely, the issue of specificity. The following discussion reviews our and others studies of the oxidative lesion in AD, focusing on this issue.

Energy (Oxidative) Metabolism in AD

The extensive evidence for abnormalities in oxidative metabolism in AD has been reviewed in detail elsewhere (Blass & Gibson, 1991). Demonstration of the deficiency in oxidative metabolism in AD has been accumulated in many laboratories over the last half century. Data come from studies of AD brain *in vivo* and *ex vivo*, in cultured AD cells, and at the cell free and enzymatic levels. Since these findings and their implications have been extensively reviewed (Blass, 1993a,b;

Blass & Gibson, 1991; Blass et al, 1988), the discussion below concentrates on the enzymatic and gene levels, and on aspects of the pathophysiology.

Proteins. The activities of at least two mitochondrial enzymes have been demonstrated to be deficient in AD brain, namely the pyruvate dehydrogenase complex (PDHC) and the α-ketoglutarate dehydrogenase complex (KGDHC). The activity of monoamine oxidase (MAO) has been found to be increased (Adolfsson et al, 1980). Activities of fumarase, another mitochondrial enzyme, is normal. The deficiencies in PDHC and KGDHC activity in AD brain are robust, each having been reported from at least four laboratories with no contravening reports (for references see Blass, 1993a,b; Blass & Poirier, 1995). Deficiencies of PDHC and KGDHC have been found in AD brain not only in regions with cell loss but also in regions normally free of histological damage in this disease, such as caudate nucleus.

Reports of deficiencies in the electron transport chain activities in AD brain and platelets have been more controversial, since they have not been robustly replicable across laboratories (Parker et al, 1990; Van Zuylen et al, 1992). Our own studies have not found consistent evidence for a functionally significant deficiency in electron transport (Sims et al, 1987) nor in the generation of oxygen free radicals (Makar et al, 1995), except perhaps in regions where cells are dying.

The deficiency in activity of KGDHC persists in cultured skin fibroblasts from patients with AD (Sheu et al, 1994). In contrast, activity of a number of other mitochondrial enzymes including PDHC is normal in the cultured AD cells. Mixing experiments indicate that the deficiency in KGDHC activity in cultured AD fibroblasts can not be attributed to the lack of an activator or the presence of an inactivator such as a protease. Activity of a cytoplasmic enzyme, namely transketolase (TK), is also replicably reduced in AD brain (Sheu et al, 1988; Butterworth & Besnard, 1990). The activity of this enzyme is also reduced in cultured AD fibroblasts (Gibson et al, 1988), but this reduction may be secondary to excess proteolysis of this enzyme in the cultured AD cells (Paoletti & Mocali, 1991).

KGDHC consists of three proteins arranged in a large (mw 4-7 million), ordered array. Thus, KGDHC can be described accurately both as an enzyme and as a submitochondrial particle. The first enzyme in the reaction sequence is α-ketoglutarate dehydrogenase (E1k), which is a thiamine pyrophosphate dependent enzyme. The second enzyme in the reaction sequence, dihydrolipoyl succinyl transferase (E2k), uses dihydrolipoic acid as a cofactor; E2k forms the core of the KGDHC complex. The third enzyme, dihydrolipoyl dehydrogenase (E3; lipoamide dehydrogenase) is a component of both the PDHC and KGDHC complexes and also of the branched chain α-ketoacid dehydrogenase (BCDHC).

Immunoblots of the KGDHC protein components in AD cells showed no abnormality in either E1k or E3. For E2k, AD cells contained both the expected immunoreactive band at ~42 kDa but also a second band at ~29 kDa (Sheu et al, 1994). The ~29 kDa band was absent from or present in trace amounts in cells from non-AD controls which had been cultured under the identical conditions. These cell extracts were prepared in the presence of a cocktail of protease inhibitors, so that the appearance of the ~29 kDa band immunoreactive with an antiserum which recognizes E2k suggests an intrinsically aberrant E2k protein and/or an aberrant proteolysis in AD cells *in vivo* resulting in an abnormal breakdown of the normal protein.

Since the E2k protein participates in forming a large, ordered array in a submitochondrial particle, it is a good candidate for the phenomenon of "negative dominance." An abnormal form of one of these proteins, and particularly of the core E2k protein, might be expected to disrupt the ordered array of enzyme proteins in the KGDHC complex and thus interfere with the function of normal protein. Therefore, a single dose of an abnormal gene for E2k or for one of the other KGDHC proteins might well be functionally and therefore clinically significant. In other words, a dominant pattern of inheritance of illness due to a single mutation in the gene coding for E2k or for one of the other KGDHC enzymes would appear to be reasonable.

Genes. To characterize the mitochondrial enzyme defects in AD in more detail, we moved to the molecular genetic level. The location of the genes for the components of PDHC, KGDHC, and TK are known. The genes for the components of PDHC are on the X chromosome (αE1p), chromosome 3 (E2p), and chromosome 7 (E3); unpublished data indicate that the βE1p gene is not on chromosome 14. TK is on chromosome 3. Of the components of KGDHC, two (E1k and E3) are on chromosome 7. The gene for the E2k protein is, however, on chromosome 14q24.3, in the middle of the limited, candidate region for early onset, familial AD (FAD) linked to this chromosome (Ali et al, 1994; Nakano et al, 1994). These findings led us to concentrate on this gene, the DLST gene (for **d**ihydro**l**ipoyl **s**uccinyl **t**ransferase).

These studies were complicated by the presence of a pseudogene for E2k located on chromosome 1 (Ali et al, 1994; Cai et al, 1994; Nakano et al, 1994). This pseudogene appears to be a transposon. It is transcribed, and in some tissues more mRNA is present from the pseudogene than from the true gene. However, stop codons early in the coding sequence preclude the mRNA from the pseudogene being translated to produce an E2k protein.

Four informative polymorphisms have been identified in the DLST gene itself. Two of these are base substitutions in introns. The other two are neutral base substitutions in specific exons, which do not change the encoded amino acids of the protein. No convincing evidence has been found that any of these polymorphisms are correlated with the occurrence of clinical AD in chromsome 14-linked familial AD (FAD) kindreds.

In a series of patients who were heterozygous for the ε4 allele of the ApoE gene, the gene frequency of the marker associated with one of the neutral polymorphisms was significantly higher in AD patients than in the general, US Caucasian population ($P < 0.008$, after correction for multiple statistical comparisons). These results suggest that the DLST gene may interact with the ApoE gene in modulating susceptibility to AD. (See also the discussion below.) Further studies of larger populations and of ethnically homogeneous populations are in progress.

As of June, 1995, the available data thus suggest that DLST is an unlikely candidate gene for the early onset, 14q24.3-linked forms of FAD. Stronger evidence suggests that DLST may be a risk factor or "susceptibility factor" for late onset, apparently sporadic AD. Firm conclusions will require more information on more patients from a variety of ethnic groups, and more detailed comparison of the sequences of the FAD and AD compared to the normal DLST genes.

Pathophysiology. As has been reviewed extensively elsewhere (Blass et al, 1993a,b), deficient activities of KGDHC can be easily related to the pathophysiology, by plausible mechanisms for which there is extensive independent evidence. The *cognitive* changes in AD are similar, except for reversibility, to those in metabolic encephalopathies, in which the rate of cerebral oxidative metabolism is secondarily reduced from a variety of different insults (Gibson et al, 1981). Evidence is particularly clear from studies of hypoxic hypoxia in aviation medicine. The *neurotransmitter* changes of AD, including the cholinergic deficiency, can be explained by alterations in oxidative metabolism (Blass, 1993a,b; Blass & Gibson, 1991,1993; Gibson & Blass, 1982; Hirsch & Gibson, 1984). Abnormalities in *signal transduction* in AD cells are also similar to changes induced by experimental impairments of oxidative metabolism (Huang & Gibson, 1993). Expression of *amyloid precursor protein* (APP) in brain is increased by insults to oxidative metabolism, which can also alter the processing of APP in the direction of more *β/A4 amyloid* (Gabuzda et al, 1994). The deleterious effects of the ε4 allele of ApoE relative to the ε3 and ε2 alleles may be related to the more ready oxidation of the ε4 allele and the effects of the oxidized protein on a number of processes including amyloid aggregation (Roses, 1995). *Inflammatory reactions* can results from the effects of hypoxic/ischemic damage, and both the accumulation of amyloid and the presence of ApoE proteins in AD plaques may be associated with localized inflammatory reactions (Fillit et al, 1991; McGeer et al, 1991; Rogers et al, 1992). *Cytoskeletal* abnormalities can be induced in cultured cells by impairments of oxidative metabolism, including the appearance of epitopes which react with antibodies to paired helical filaments (Blass et al, 1991; Cheng & Mattson, 1992). Our current data suggest that the appearance of these epitopes in cultured skin fibroblasts results from cross-reactivity of the antibodies used with phosphorylated sites on the MAP4 protein which are homologous to those on tau proteins. *Neuronal cell loss* is a frequent consequence of impairment of cerebral oxidative metabolism, which is not surprising in view of the high rate of oxidative metabolism in the brain and the brain's second-to-second dependence on oxidative metabolism to maintain function (Gibson et al, 1981). Furthermore, both clinical and experimental forms of impairment of oxidative metabolism can lead to loss of neurons in a pattern of *selective vulnerability* (Blass, 1993a,b). Indeed, a commonly used model for selective vulnerability is the 4-vessel occlusion model of stroke, in which CA1 neurons in the hippocampus are lost but CA2 neurons are spared (Endoh et al, 1994). Ko (1993) has shown that neurons lost in vulnerable areas of AD brain are enriched in KGDHC activity (Blass, 1993b). Thus, inadequate activity of KGDHC could predispose to many of the abnormalities characteristic of AD (Blass, 1993a,b).

Furthermore, limitations on the activity of KGDHC, perhaps due to an anomaly in the DLST gene coding for its core E2k component, may predispose to cell damage in the face of a variety of stressors, including glutamate and other excitotoxins. Current data suggest that the KGDHC-catalyzed reaction is the rate-limiting step in glucose oxidation in the brain of normal humans and experimental animals (Blass, 1993b). Therefore, a limitation on the ability of cells to increase the metabolic flux across the KGDHC-catalyzed reaction in response to stressful events in the life of neurons could lead to the loss of these brain cells during a lifetime (Blass, 1993a,b). This effect might be expected to be particularly important as people age, since limitation on KGDHC activity may potentiate the effects of the decrease in the rate of mitochondrial oxidative metabolism associated with aging itself (Blass, 1993a).

KGDHC is also an enzyme of glutamate metabolism (Blass, 1993a,b). KGDHC is responsible for the oxidation of α-ketoglutarate derived from glutamate by transamination or by the glutamate dehydrogenase reaction. Neurons increase their rate of oxidation dramatically in response to depolarization (Gibson & Blass, 1982). Thus, neurons which have an abnormally limited ability to increase the rate of metabolic flux across KGDHC may be at double jeopardy in the face of excess glutamate release. They will have an abnormal limitation both on the rate of glutamate metabolism and on the mobilization of energy after depolarization and glutamate receptor activation. The limitation of KGDHC activity could be particularly deleterious in the face of glutamatergic excitotoxicity. In support of this hypothesis is direct data in tissue culture indicating that impairment of glucose metabolism potentiates the excitotoxic effect of glutamate (Henneberry, 1989; Beal, 1992) and, indeed, the deleterious effects of a number of other stressors including toxic amyloid peptides (Cheng & Mattson, 1992; Mattson et al, 1993; Mattson & Goodman, 1995).

IMPLICATIONS FOR AD AND OTHER NEURODEGENERATIVE DISORDERS

The observations described above are consistent with the suggestion that dominantly inherited, early onset FAD results primarily from one of a limited number of inborn errors of metabolism with late clinical onset, while the common later-onset form of AD is a *convergence syndrome*(Blass, 1993a). According to this formulation, a variety of mechanisms converge to cause the characteristic types of scarring which define AD at autopsy and which appear to lead to the dementia. These mechanisms may act in part through their effects on mitochondria. They may include one or more genetic predispositions, perhaps in DLST, which may be clinically silent in earlier life but significant in older people. They may include age-associated damage to mtDNA and therefore to electron transport. They may include the consequences of cardiovascular and cerebrovascular disease, with intermittent or constant compromise of mitochondrial function due to impaired supply of substrate and oxygen. They may include a variety of other "ills the flesh is heir to," which increase the demand on cerebral oxidative metabolism. Since mitochondrial damage appears to occur in both early onset FAD (Sheu et al, 1994) and in the much more common later onset AD, it is possible that the portion of the pathophysiology medicated by inadequate mitochondrial function may be the same in both types of AD.

This formulation could also explain in part the susceptibility of patients with relatively mild, even subclinical AD to metabolic encephalopathies. This clinical phenomenon is well recognized (Gibson et al, 1991). Since metabolic encephalopathy (delirium) typically involves impairments of cerebral oxidative metabolism, AD patients whose cerebral metabolism is already limited would have increased susceptibility to the effect of other conditions which further impair their cerebral oxidative metabolism. These suggestions are very similar to views put forward by Alzheimer, Perusini, Binswanger, Bleuler, and their contemporaries (Bick et al, 1987; Blass et al, 1992). These early workers also distinguished between the early onset form of AD and the more common, later onset "senile dementias." They conceptualized the early onset form as being a discrete disease, while the later onset forms had more complex causation.

As noted above, there is abundant evidence that impairments of oxidative metabolism are a frequent cause of neurodegenerative diseases. This evidence continues to mount. Therefore, oxidative metabolism may be a fertile area to investigate, at the biochemical and molecular levels, in searching for causes of neuropsychiatric diseases whose causes are still not well understood.

REFERENCES

Adolfsson, R. Gottfries C.G., Oreland, L., et al,, Increased activity of brain and platelet monoamine oxidase in dementia of the Alzheimer type, *Life Sci.* 27:1029.

Ali, G., Wasco, W., Cai, X. et al, 1994, Isolation, characterization, and mapping of the dihyrodlipoyl succinyltransferase [E2k] of human α-ketoglutarate dehyrogenase complex, *Somat. Cell Mol. Genet.* 20:99.

Beal, M.F., 1992, Does impairment of energy metabolism result in excitotoxic neuronal death in neurodegenerative diseases? *Ann. Neurol.* 31:119.

Bick, K., Amaducci, L., and Pepeu, G., 1987, The early story of Alzheimer's disease. Livinia Press, Padua.

Blass, J.P., 1993a, Pathophysiology of the Alzheimer's syndrome, *Neurology* 43 *(suppl 4)*:S25.

Blass, J.P., 1993b, The cultured fibroblast model, *J. Neural. Trans. (Suppl)* 44:87.

Blass, J.P., Baker, A.C., Ko, L. et al, 1991, Expression of Alzheimer antigens in cultured skin fibroblasts, *Arch. Neurol. 48:709.*

Blass, J.P., and Gibson, G.E., 1991, The role of oxidative abnormalities in the pathophysiology of Alzheimer's Disease, *Rev. Neurol. (Paris)* 147:513.

Blass, J.P., and Gibson, G.E., 1993, Nonneural markers in Alzheimer disease, *Alz. Dis. Assoc. Dis.* 6:205.

Blass, J.P., Hoyer, S., and Nitsch, R., 1992, In reply, *Arch. Neurol.* 49:800.

Blass, J.P., Milne, J.A., and Rodnight, R., 1977, Newer concepts of psychiatric diagnosis and biochemical research on mental illness. *Lancet* 1:738.

Blass, J.P., and Poirier, J., 1995, Pathophysiology of the Alzheimer's Syndrome, In press.

Blass, J.P., Sheu, K.-F.R., and Cederbaum, J.M., 1988, Energy metabolism in disorders of the nervous system, *Rev. Neurol. (Paris)* 144:543.

Butterworth, R.F., and Besnard, A.M., 1990, Thiamine-dependent enzyme changes in temporal cortex of patients with Alzheimer's disease, *Metab. Brain Dis.* 4:179.

Cai, X., Sabo, P., Ali, G. et al, 1994, A pseudogene of dihyrdrolipoyl succinyltransferase (E2k) found by PCR amplification and direct sequencing in rodent-human cell hybrid DNAs. *Somat. Cell Mol. Genet.* 20:339.

Cheng, B., and Mattson, M.P., 1992, Glucose deprivation elicits neurofibrillary tangle-likeantigenic changes in hippocampal neurons: Prevention by NGF and bFGF, *Exp. Neurol.* 117:114.

Endoh, M., Pulsinelli, W.A., and Wagner, J.A., 1994, Transient global ischemia induces dynamic changes in the expression of bFGF and the FGF receptor, *Brain Res.* 22:76.

Fillit, H., Ding, W.H., Buee, L. et al, 1991, Elevated circulating tumor necrosis factor levels in Alzheimer's disease. *Neurosci Lett* 129:318.

Gabuzda, D., Busciglio, J., Chen, L.B. et al, 1994, Inhibition of eneergy metabolism alters the processing of amyloid precursor protein and induces a potentially amyloidogenic derivative, *J. Biol. Chem.* 269:13628.

Garrod, A.E., 1923, *Inborn Errors of Metabolism, Oxford, London.*

Gibson, G.E., and Blass, J.P., 1982, Metabolism and neurotransmission, in: *Handbook of Neurochemistry*, A. Lajtha, ed., vol. 3, 2nd ed., Plenum Press, N.Y.

Gibson, G.E., Blass, J.P., Huang, H.-M. et al, 1991, The cellular basis of delerium and its relevance to age-related disorders including Alzheimer's disease, *Int. Psychogeriatrics* 3:373.

Gibson, G.E., Pulsinelli, W.A., and Blass, J.P., 1981, Brain dysfunction in mild to moderate hypoxia, *Am. J. Med. 70:1247.*

Gibson, G.E., Sheu, K.-F.R., Blass, J.P. et al, 1988, Reduced activities of thiamine-dependent enzymes in the brains and peripheral tissues of patients with Alzheimer's Disease, *Arch. Neurol.* 45:836.

Henneberry, R.A., 1989, The role of energy in the toxicity of excitatory amino acids, *Neurobiol. Aging* 10:611.

Hirsch, J.A., and Gibson, G.E., 1984, Selective alterations of neurotransmitter release by low oxygen in vitro, *Neurochem. Res.* 9:1039.

Huang, H.-M., and Gibson, G.E., 1993, Altered β-receptor stimulated cAMP formation in cultured skin fibroblasts from Alzheimer donors, *J. Biol. Chem.* 268:14616.

Ko, L., Sheu, K.-F.R., and Blass, J.P., 1993, Chemical neuroanatomy of energy metabolism: immunohistochemical studies in relation to selective vulnerability, *J. Neurochem.* 61: S70.

Makar, T.K., Cooper, A.J.L., Tofel-Grehl, B., et al, 1995, Carnitine, carnitine acetyltransferase, and glutathione in Alzheimer brain, *Neurochem. Res.* 20:705.

Mattson, M.P., Cheng, B., Culwell, A.R. et al, 1993, Evidence for excitoprotective and intraneuronal calcium regulating for secreted forms of the β-amyloid precursor protein. *Neuron* 10:246.

Mattson, M.P., and Goodman, Y., 1995, Different amyloidogenic peptides share a similar mechanism of neurotoxicity involving reactive oxygen species and calcium. *Brain Res.* 676:219-224.

McGeer, P.L., McGeer, E., Kawamata, T. et al, 1991, Reactions of the immune system in chronic degenerative neurological diseases, *Can. J. Neurol. Sci.* 18 *(suppl 3)*: 376.

Nakano, K., Takase, C., Sakamoto, T. et al, 1994, Isolation, characterization, and structural organization of the gene and pseudogene for the dihyrolipoylamide succinyltransferase component of the 2-oxoglutarate dehydrogenase complex, *Eur. J. Biochem.* 224:179.

Paoletti, F., and Mocali, A., 1991, Enhanced proteolytic activities in cultured fibroblasts of Alzheimer patients are revealed by peculiar transketolase alterations, *J. Neurol. Sci.* 105:211.

Parker, W.D., Filley, C.M., and Parks, J.K., 1990, Cytochrome oxidase deficiency in Alzheimer's disease, *Neurology*40:1302.

Rogers, J., Cooper, N.R., Webster, S. et al, 1992, Complement activation by β-amyloid in Alzheimer's disease, *Proc. Natl. Acad. Sci. (USA)* 89:10016.

Roses, A.D., 1995, Alzheimer's disease as a model of molecular gerontology, *J. NIH Res. 7:51.*

Sheu, K.-F.R., Clarke, D.D., Kim, Y.T. et al, 1988, Studies of the transketolase abnormality in Alzheimer's disease, *Arch. Neurol.* 45:841.

Sheu, K.-F.R., Cooper, A.J.L., Lindsay, J.G., et al, 1994, Abnormality of the α-ketoglutarate dehydrogenase complex in fibroblasts from familial Alzheimer's Disease, *Ann. Neurol.* 35:312.

Sims, N.R., Finegan, J.M., and Blass, J.P., 1987, Altered metabolic properties of cultured skin fibroblasts in Alzheimer's disease, *Ann. Neurol.* 21:451.

Van Zuylen, A.J., Bosman, G.J.C.G.M., Ruitenbeek, W., et al, 1992, No evidence for reduced thrombocyte cytochrome oxidase activity in Alzheimer's Disease, *Neurology* 42:1246.

THIAMINE DEFICIENCY AS A MODEL OF SELECTIVE NEURODEGENERATION WITH CHRONIC OXIDATIVE DEFICITS

Noel Y. Calingasan,[1] Kwan-Fu Rex Sheu,[1] Harriet Baker,[1]
Samuel E. Gandy,[2] and Gary E. Gibson[1,3]

[1]Cornell University Medical College at Burke Medical Research Institute, White Plains, NY 10605; [2]Cornell University Medical College, New York, NY 10021; [3]Corresponding author

INTRODUCTION

Experimental thiamine deficiency is a classical model of the molecular changes that underlie the clinical syndrome referred to variously as delirium, acute confusional state[1,2,3,4] or metabolic encephalopathy[5]. This syndrome is characterized by decreased attention and cognition, alertness, orientation and grasp, memory, affect and perception. A wide variety of systemic disorders lead to the development of the syndrome including hypoxia, ischemia, hypoglycemia, some diseases of peripheral organs, ionic imbalance, poisoning, dysfunction of temperature regulation, infection or inflammation of the brain and spinal cord, primary neuronal and glial disorders, acute delirious states (sedative drug withdrawal, drug intoxication, postoperative delirium, intensive care unit delirium), and vitamin and nutritional deficiencies (e.g. thiamine deficiency). Despite the varied etiology, the diverse insults that lead to delirium may act by common metabolic and cellular pathways, as suggested by results from studies of aging and hypoxia.[6,7,8]

Thiamine plays an important role in brain glucose metabolism as a coenzyme in the form of thiamine pyrophosphate (TPP). The TPP-dependent enzymes of oxidative metabolism in the brain are: transketolase (TK), α-ketoglutarate dehydrogenase complex (KGDHC), pyruvate dehydrogenase complex (PDHC) and branched-chain α-ketoacid dehydrogenase complex (BCKDC).[9,10] TK is a key pentose phosphate shunt enzyme that plays an important role in the production of reducing equivalents and pentose sugars. KGDHC is a tricarboxylic acid cycle enzyme that is highly regulated in mitochondrial oxidation. PDHC, on the other hand, is involved in oxidative decarboxylation of pyruvate to acetyl-CoA for entry into the tricarboxylic acid cycle. BCKDC mediates the degradation of leucine, isoleucine and valine. Thus, TPP-requiring enzymes play a crucial role in the regulation of energy metabolism.

Similarities exist between Alzheimer's disease and thiamine deficiency in man and animals. As in Alzheimer's disease, thiamine deficiency is characterized by decreased TPP-dependent enzyme activities,[11,12] selective cell loss,[13,14,15,16] cholinergic deficits[17,18,19] and profound memory loss.[20] In humans, thiamine deficiency leads to a discrete neurological and psychiatric syndrome.[20] The studies described in this chapter suggest that the mechanism

Neurodegenerative Diseases
Edited by Gary Fiskum, Plenum Press, New York, 1996

involved in neurodegeneration induced by thiamine deficiency may have direct relevance to the pathophysiology of age- and gene-related neurodegenerative diseases in man.

ROLE OF AGE AND GENETICS IN THE RESPONSE TO THIAMINE DEFICIENCY

Aging predisposes to the development of delirium in man and to the effects of thiamine deficiency and hypoxia in animals. Elderly people, especially the very old, are uniquely prone to delirium as a consequence of almost any physical illness or of intoxication with even therapeutic doses of commonly used drugs.[3] Similarly, aging predisposes to the effects of metabolic encephalopathy in animal models. The residual synthesis of acetylcholine in an hypoxic aged animal is one third of that in an hypoxic young animal.[21] The increase in open field activity (i.e. total distance travelled) at intermediate stages of thiamine deficiency is exaggerated in aged mice.[22] Although the activity of KGDHC is unaffected by aging, KGDHC activities in aged brains are more sensitive to thiamine deficiency.[22] Thus, aging predisposes the animals to the effects of thiamine deprivation.

Genetic factors influence the course of thiamine deficiency-induced delirium in man and animals. Patients with Wernicke-Korsakoff syndrome show signs of clinical delirium. Wernicke-Korsakoff syndrome is generally attributed to thiamine deficiency and is often related to alcoholism. Acutely, recent memory is impaired more severely than other mental tasks. However, over time chronic lesions may cause deficits in mental functions other than memory.[5] Not all alcoholics develop Wernicke-Korsakoff syndrome which suggests that genetic factors may be involved. The selective vulnerability to thiamine deficiency in different populations appears to be related to genetic variations in TK.[23] The apparent Km for TPP is more than ten times greater for TK in fibroblasts from Wernicke-Korsakoff patients than from controls. The abnormality persists through serial passage in tissue culture in cells grown in medium containing excess thiamine. The abnormality in this enzyme appears inconsequential if the diet is adequate. Thus, Wernicke-Korsakoff syndrome represents an inborn error of metabolism that genetically predisposes the individual to the development of a neurological disorder, but requires the proper "environmental" insult before it is expressed. Why different individuals do not respond similarly to thiamine deficiency is unknown, but genetic factors appear to be involved.[1] A genetic predisposition to the development of the symptoms of thiamine deficiency exists among experimental animals. The open field behavior of Balb/c mice and CD-1 mice differs and responds to thiamine deficiency in a qualitatively different manner.[22] Other studies demonstrate that thiamine deficiency reduces motor activity more in C57BL/J mice than in Balb/cByJ or Nylar mice.[24] In addition, TK activity is more than twice as high in Balb/c as in CD-1 mice.

BEHAVIORAL AND CHOLINERGIC DEFICITS IN THIAMINE DEFICIENCY

Thiamine deficiency produces a physiologically important change in the cholinergic system. Acetylcholine turnover declines in thiamine deficiency; incorporation of isotopes of choline and glucose into acetylcholine decreases by 32% and 48%, respectively.[25] The changes induced in the cholinergic system by thiamine deficiency in animals appear to be behaviorally important. Studies with a simple motor task, the tight rope test (measures the ability to traverse an elevated taut string) demonstrate clear deficits with thiamine deficiency. Persistently decreased tight rope test scores occur in 50% of the treated rats. The decrease occurs long before the onset of weight loss or neurological symptoms. The acetylcholinesterase inhibitor physostigmine improves the low tight rope test scores in thiamine-deficient rats, whereas neostigmine, which acts peripherally, is ineffective. The effect of physostigmine is inhibited by centrally acting muscarinic blockers, but not by peripherally acting analogues. Neither nicotine nor nicotinic antagonists have any effects.[17]

Performance of rats on spontaneous open field behavior is altered in early stages of thiamine deficiency. One aspect of open field behavior is staring, a behavior in which the rat has a tense body posture and seems to be looking at something.[26] Increases in staring behavior can be detected with as little as one day of thiamine deficiency. The acetylcholinesterase inhibitor physostigmine is as effective as thiamine in decreasing staring in thiamine-deficient rats, but again its peripherally-acting analogue neostigmine is ineffective. Central muscarinic blockers impair the effects of physostigmine and muscarinic agonists are as effective as physostigmine in decreasing staring. Thus, thiamine deficiency leads to distinctive behavioral deficits that can be manipulated pharmacologically.

SELECTIVE VULNERABILITY AND THE THIAMINE-DEPENDENT ENZYMES

Distribution of Thiamine-Dependent Enzymes

Although reduced thiamine availability presumably affects all cells throughout the brain, the cell damage is restricted to a few regions including the thalamus, mammillary body, medial geniculate nucleus, inferior colliculus, medial vestibular nucleus and some periventricular areas.[13,14,15,16,27,28] An explanation for selective vulnerability has been elusive. Reduced activities of TPP-dependent enzymes in brains from thiamine-deficient animals has been known since Sir Rudolph Peters' classical studies in 1930.[29] How these changes relate to selective vulnerability of neuronal population is still not clear. Until recently, an implicit assumption in the literature has been that the selective vulnerability must be related to the distribution of TPP-dependent enzymes.

The general metabolic enzymes KGDHC,[30] PDHC[31] and TK[32] are enriched in selective perikarya that are heterogeneously distributed in the brain. Parallel enrichment of these enzymes occurs in many regions such as the medial septal nucleus, nucleus of the diagonal band, hippocampus, red nucleus, pontine nuclei, trapezoid nucleus and most cranial nerve nuclei.[30,32] In the trapezoid nucleus, nucleus ambiguus, inferior olive and cranial nerve nuclei, virtually all cholinergic neurons are enriched with KGDHC.[33] However, the distribution of these enzymes does not correlate with the predilection of particular brain regions to pathological and biochemical lesions induced by thiamine deficiency. TPP-dependent enzymes are not particularly abundant nor reduced in any of the regions vulnerable to thiamine deprivation.

Responses of Thiamine-Dependent Enzymes to Thiamine Deficiency

Thiamine deficiency produces a dramatic, general reduction of the TPP-dependent enzymes KGDHC and TK, but not PDHC.[11,12,34,35,36] A 50% reduction of both KGDHC and TK activities occurs throughout the brain when measured in the presence of a saturating level of TPP, in both pathologically spared and vulnerable regions. However, immunocytochemistry shows either a reduced TK, or an unchanged KGDHC level of the protein. To understand the biochemical and enzymatic basis of the immunocytochemical results of TK and KGDHC and their different responses to thiamine deficiency, we have analyzed the relationships of enzyme activities, protein levels and, in the case of TK, mRNA level.[37]

Immunoblotting analysis reveals a reduction of TK protein in thiamine-deficient brains. Thus, the reduction of the TK enzyme activity is associated with a similar reduction of the TK protein. The level of the TK mRNA remains unaltered on day 13 of thiamine deficiency as analyzed on Northern blots, suggesting that the rate of TK transcription is not likely to be affected. The most likely interpretation of our result is that the reduced level of TK is due to an unstable TK protein, as a result of deprivation of its TPP coenzyme in thiamine deficiency.

Immunoblots of KGDHC show that the levels of its components E1k and E2k are unchanged in thiamine deficiency, a finding consistent with the immunocytochemistry, and in contrast to that of TK. We interpret the lack of reduction of the KGDHC protein to represent the presence of inactive enzyme form(s) that is immunoreactive but catalytically incompetent. These aberrant form(s) may occur due to deprivation of the thiamine coenzyme or other insults. Thus, selective susceptibility is not associated with the responses of TPP-dependent enzymes to thiamine deficiency.

BLOOD-BRAIN BARRIER ABNORMALITIES IN VULNERABLE REGIONS

A breakdown of the blood-brain barrier has been implicated in Alzheimer's disease,[38] Wernicke-Korsakoff syndrome,[39,40] multi-infarct dementia,[41] and in animal models of amyotrophic leukospongiosis[42] and focal ischemia.[43] Blood-brain barrier dysfunction leads to extravasation of plasma proteins such as IgG, and subsequent vasogenic edema.[44,45] The late stages of experimental thiamine deficiency are characterized by pinpoint hemorrhages in the brain indicating a disruption of the blood-brain barrier.

We examined the blood-brain barrier at different stages of thiamine deficiency using IgG as an indicator of blood-brain barrier integrity.[46] IgG immunoreactivity increases in the inferior colliculus and inferior olive on day 10 before the onset of cell loss, hemorrhage and loss of righting reflex. After 11 or 12 days, multiple vulnerable regions show an increased IgG leakage, pinpoint hemorrhage and cell loss. On day 13, intense IgG immunoreactivity is found in areas of damage and hemorrhage such as the thalamus, inferior colliculus, mammillary body, medial geniculate nucleus, medial vestibular nucleus and inferior olive. Western blot analysis of different brain regions confirms the presence of IgG in vulnerable areas but not in preserved regions.

Preliminary electron microscopic studies of capillary endothelia in areas where IgG immunoreactivity is increased at day 13 reveal intact interendothelial tight junctions and basement membrane. In addition, numerous hollow inclusions occur in the thiamine-deficient capillary endothelia. However, the significance of these ultrastructural findings cannot be determined without further studies utilizing serial ultrathin sections taken at different stages of thiamine deficiency.

These results are the first to report a dysfunction of the blood-brain barrier during early stages of thiamine deficiency. The localized accumulation of IgG coincides with the selective regions where cells are destined to die, and in some areas, occurs prior to the appearance of other pathological changes. These findings could alter the interpretation of past studies. For example, the interpretation of binding studies with nimodipine[47] and peripheral-type benzodiazepine receptor ligand[48] requires that the blood-brain barrier is not selectively damaged. Thus, the current data may help elucidate the role of the blood-brain barrier in the pathogenesis of thiamine deficiency, and may be relevant to the pathophysiology of other neurodegenerative diseases associated with oxidative stress.

ACCUMULATION OF AMYLOID PRECURSOR PROTEIN-LIKE IMMUNOREACTIVITY AROUND VULNERABLE REGIONS

The similarities between thiamine deficiency and Alzheimer's disease described in the Introduction, and the demonstration that Alzheimer β-amyloid precursor protein (APP) responds to neuronal injury[49,50,51,52,53,54,55,56] stimulated us to examine APP immunoreactivity in thiamine deficiency[57] using antibody 369, an antiserum against human $APP^{645\text{-}694}$(numbering according to APP_{695} isoform).[58,59,60,61,62,63,64] An accumulation of APP accompanies neurodegeneration in thiamine deficiency. Three, 6 and 9 days of thiamine deficiency do not appear to damage any brain region nor alter APP-like immunoreactivity. However, 13 days of thiamine deficiency produce pathological lesions mainly in the thalamus, mammillary body, inferior colliculus and some periventricular areas. APP-like immunoreactivity accumulates in aggregates of swollen, abnormal neurites and perikarya that delimit the lesions in the thalamus (Figure 1) and medial geniculate nucleus. Western blot analysis of the thalamic region around the lesion shows increased APP holoprotein immunoreactivity. APP-like immunoreactive neurites are scattered in the mammillary body and medial vestibular nucleus. In the inferior colliculus, intense perikaryal APP-like immunostaining occurs along the periphery of the damaged area. However, immunocytochemistry with 4G8 antiserum (specific for amyloid protein), thioflavine S histochemistry and electron microscopy fail to show highly compact fibrillar β-amyloid structure. Double immunofluorescence histochemistry reveals colocalization of APP-like protein and neurofilament NF-H in abnormal neurites. This result indicates that the APP-like immunoreactive structures are not glial processes. Brain areas without pathological lesions show no changes in APP-like immunoreactivity.

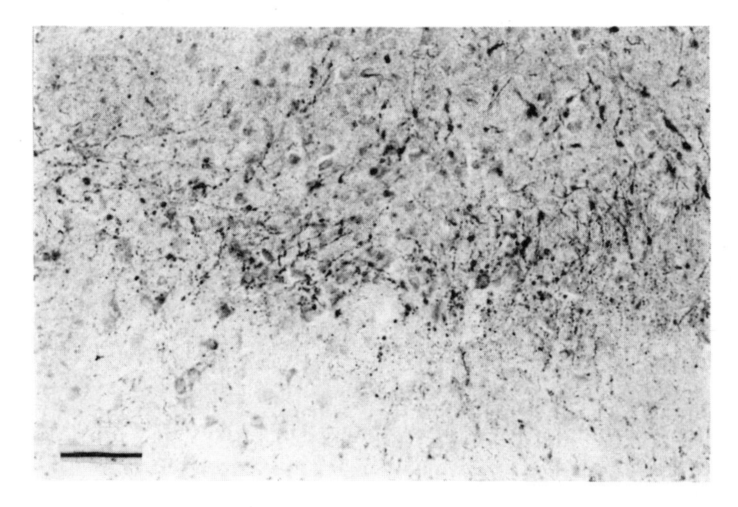

Figure 1. High magnification photomicrograph of a section through the paraventricular nucleus of the thalamus taken from a 13-day thiamine deficient rat showing accumulation of APP-like immunoreactivity in abnormal neurites and perikarya. Note the intensely-stained, beaded, abnormal neurites that delimit the pale area of cell loss (bottom). Scale bar = 100 μm.

These studies show APP-like accumulation in abnormal neurites and perikarya along the periphery of the lesions induced by experimental thiamine deficiency. These results are the first demonstration of a relationship between alterations in oxidative metabolism, APP expression and selective cell loss induced by nutritional/cofactor deficiency.

SUMMARY AND CONCLUSION

Since animals of varying age and genotype (strain) respond differently to thiamine deficiency, this model may have relevance to age- and gene-related neurodegenerative disorders in man. Early stages of thiamine deficiency are associated with behavioral deficits that are clearly cholinergic. Selective vulnerability in thiamine deficiency is associated with a localized increase in IgG and Alzheimer amyloid precursor protein-like immunoreactivities, but is not correlated with the distribution of the thiamine-dependent enzymes or their responses to thiamine deprivation. The sequence of changes in behavior and brain during thiamine deficiency is presented in Figure 2. We hypothesize that thiamine deficiency causes a generalized impairment in oxidative metabolism in the brain which renders select cells or regions more susceptible to a breakdown of the blood-brain barrier. Affected cell populations die and APP accumulates in response to the brain injury. The thiamine deficiency model produces a clear progression of regionally selective pathological changes that begin with abnormalities in the blood-brain barrier and culminate with APP accumulation. Thus, thiamine deficiency may help elucidate the role of the blood-brain barrier and APP metabolism in the pathogenesis of thiamine deficiency and other neurodegenerative diseases associated with oxidative stress.

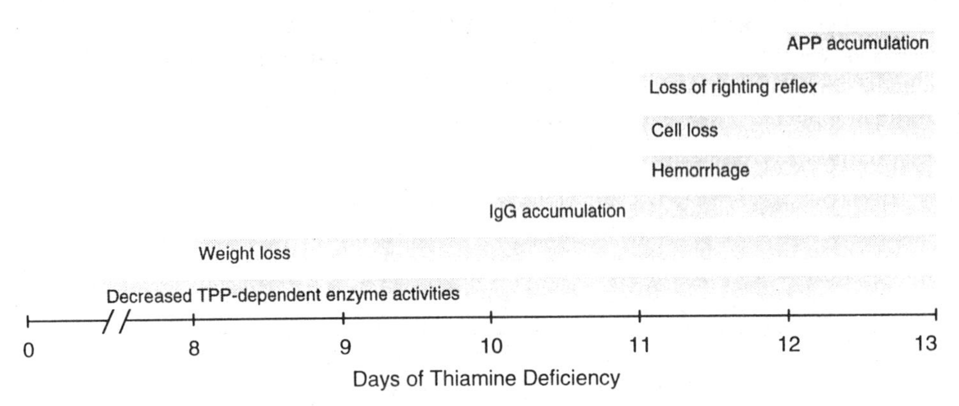

Figure 2. Sequence of changes in behavior and brain during experimental thiamine deficiency

REFERENCES

1. Z.J. Lipowski."Delirium," Charles C. Thomas Publisher (1980).
2. Z.J. Lipowski, Delirium (Acute confusional states), *J. Am. Med. Assoc.* 258:1789-1792 (1987).
3. Z.J. Lipowski, Delirium in the elderly patient, *N. Engl. J. Med.* 320:578-582 (1989).
4. Z.J. Lipowski, Organic brain syndrome: a reformulation, *Comp. Psych.* 19:309-322 (1978).
5. F. Plum and J.B. Posner. "The Diagnosis of Stupor and Coma," 3rd ed., F.A. Davis and Company, Philadelphia, PA (1980).
6. R.A. McFarland and H.N. Evans, Alterations in dark adaptation under reduced oxygen tensions, *Am. J. Physiol.* 7:37 (1939).
7. R.A. McFarland and W.H. Forbes, The effects of variation in the concentration of oxygen and of glucose on dark adaptation, *J. Gen. Physiol.* 24: 69 (1940).
8. R.A. McFarland, F.J.W. Roughton, and M.H. Halperin, The effects of CO_2 and altitude on visual thresholds, *J. Aviation Med.* 15:381-348 (1944).
9. A.L. Lehninger. "Principles of Biochemistry," Worth Publishers, New York, NY (1982).
10. G. Siegel, B. Agranoff, R.W. Albers, and P. Molinoff. "Basic Neurochemistry," Raven Press, New York, NY (1989).
11. G.E. Gibson, H. Ksiezak-Reding, K.-F.R. Sheu, V. Mykytyn, and J.P. Blass, Correlation of enzymatic, metabolic and behavioral deficits in thiamine deficiency and its reversal, *Neurochem. Res.* 9:803-814 (1984).
12. R.F. Butterworth, Cerebral thiamine-dependent enzyme changes in experimental Wernicke's encephalopathy, *Metab. Brain Dis.* 1:165-175 (1986).
13. G. Collins, Glial changes in the brainstem of thiamine-deficient rats, *Am. J. Pathol.* 50:791-802 (1967).
14. P.M. Dreyfus and M. Victor, Effects of thiamine deficiency on the central nervous system, *Am. J. Clin. Nutr.* 9:414-425 (1961).

15. P.J. Langlais and R.G. Mair, Protective effects of glutamate antagonist MK-801 on pyrithiamine-induced lesions and amino acid changes in rat brain, *J. Neurosci.* 10:1664-1674 (1990).
16. E.D. Witt, Neuroanatomical consequences of thiamine deficiency: a comparative analysis, *Alcohol Alcohol.* 2:201-221 (1985).
17. L. Barclay, G.E. Gibson, and J.P. Blass, Impairment of behavior and acetylcholine metabolism in thiamin deficiency, *J. Pharmacol. Exp. Ther.* 217:537-543 (1981).
18. L.L. Barclay, G.E. Gibson, and J.P. Blass, Cholinergic therapy of abnormal open-field behavior in thiamine-deficient rats, *J. Nutr.* 112:1906-1913 (1982).
19. C.V. Vorhees, D.E. Schmidt and R.J. Barrett, Effects of pyrithiamine and oxythiamine on acetylcholine levels and utilization in rat brain, *Brain Res. Bull.* 3:493-496 (1978).
20. M. Victor, R.D. Adams, and G.H. Collins, The Wernicke-Korsakoff syndrome and related neurological disorders due to alcoholism and malnutrition, 2nd ed., *in:* "Contemporary Neurology," F. Plum and F.A. McDowell, eds. F. A. Davis, Philadelphia, PA (1989).
21. G.E. Gibson, C. Peterson, and J. Sansone, Neurotransmitter and carbohydrate metabolism during aging and mild hypoxia, *Neurobiol. Aging* 2:165-172 (1981).
22. G.B. Freeman, P.N. Nielsen, and G.E. Gibson, Effect of age on behavioral and enzymatic changes during thiamin deficiency, *Neurobiol. Aging* 8:429-434 (1987).
23. J.P. Blass and G.E. Gibson, Abnormality of a thiamine-requiring enzyme in patients with Wernicke-Korsakoff Syndrome, *N. Engl. J. Med.* 297:1367-1370 (1977).
24. D.J. McFarland, Mouse phenotype modulates the behavioral effects of acute thiamine deficiency, *Physiol. Behav.* 35:597-601 (1985).
25. L.L. Barclay, G.E. Gibson, and J.P. Blass, The string test: an early behavioral change in thiamine deficiency, *Pharm. Biochem. Behav.* 14:154-157 (1981).
26. L.L. Barclay and G.E. Gibson, Spontaneous open-field behavior in thiamin deficiency, *J. Nutr.* 112:1899-1905 (1982).
27. J.C. Troncoso, N.V. Johnston, K.M. Hess, J.W. Griffin, and D.L. Price, Model of Wernicke's encephalopathy, *Arch. Neurol.* 38:350-354 (1981).
28. I. Watanabe, T. Tomito, K.-G. Hung, and Y. Iwasaki, Edematous necrosis in thiamine-deficient encephalopathy of the mouse, *J. Neuropathol. Exp. Neurol.* 40:454-471 (1981).
29. H.W. Kinnersley and R.A. Peters, Carbohydrate metabolism in birds: brain localization of lactic acidosis in avitaminosis B and its relation to the origin of the symptoms, *Biochem. J.* 24:711-721 (1930).
30. N.Y. Calingasan, H. Baker, K.-F.R. Sheu, and G.E. Gibson, Distribution of the α-ketoglutarate dehydrogenase complex in rat brain, *J. Comp. Neurol.* 346:461-479 (1994).
31. T.A. Milner, C. Aoki, K.-F.R. Sheu, J.P. Blass, and V.M. Pickel, Light microscopic immunocytochemical localization of pyruvate dehydrogenase complex in rat brain: topographical distribution and relation to cholinergic and catecholaminergic nuclei, *J. Neurosci.* 7:3171-3190 (1987).
32. N.Y. Calingasan, K.-F.R. Sheu, H. Baker, E-H. Jung, F. Paoletti, and G.E. Gibson, Heterogeneous expression of transketolase in rat brain, *J. Neurochem.* 64:1034-1044 (1995).
33. N.Y. Calingasan, H. Baker, K.-F.R. Sheu, and G.E. Gibson, Selective enrichment of cholinergic neurons with the α-ketoglutarate dehydrogenase complex in rat brain, *Neurosci. Lett.* 168:209-212 (1994).
34. R.F. Butterworth, J.F. Giguere, and A.M. Besnard, Activities of thiamine-dependent enzymes in two experimental models of thiamine deficiency encephalopathy. 2. α-ketoglutarate dehydrogenase, *Neurochem. Res.* 11:567-577 (1986).

35. G.E. Gibson, P. Nielsen, V. Mykytyn, K. Carlson, and J.P. Blass, Regionally selective alterations in enzymatic activities and metabolic fluxes during thiamin deficiency, *Neurochem. Res.* 14:17-24 (1989).
36. J.F. Giguere and R.F. Butterworth, Activities of thiamine-dependent enzymes in two experimental models of thiamine deficiency encephalopathy: 3. Transketolase, *Neurochem. Res.* 1:305-310 (1987).
37. G.E. Gibson, N.Y. Calingasan, K.-F.R. Sheu and H. Baker, Thiamine deficiency differentially modifies transketolase and α-ketoglutarate dehydrogenase expression in rat brain, *Soc. Neurosci. Abstr.* 20:415 (1994).
38. R.N. Kalaria, The blood-brain barrier and cerebral microcirculation in Alzheimer's disease, *Cereb. Brain Met. Rev.* 4:226-260 (1992).
39. R. Boldorini, L. Vago, A. Lechi, and F. Tedeschi, Wernicke's encephalopathy: occurrence and pathological aspects in a series of 400 AIDS patients, *Acta Bio-Med. Aten. Parm.* 63:43-49 (1992).
40. G. Schroth, W. Wichmann, and A. Valavanis, Blood-brain barrier disruption in acute Wernicke encephalopathy: MR findings, *J. Compu. Asst. Tomog.* 15:1059-1061 (1991).
41. P. Mecocci, L. Parnetti, G.P. Reboldi, C. Santucci, A. Gaiti, C. Ferri, I. Gernini, M. Romagnoli, D. Cadini, and U. Senin, Blood-brain barrier in a geriatric population: barrier function in degenerative and vascular dementias, *Acta Neurol. Scan.* 84:210-213 (1991).
42. N.N. Poleshchuk, V.I. Votyakov, Y.G. Ilkevich, G.P. Duboiskaya, D.G. Grigoriev, and N.D. Kolomiets, Structural and functional changes of blood-brain barrier and indication of prion amyloid filaments in experimental amyotrophic leucospongiosis, *Acta Virol.* 36:293-303 (1992).
43. E.H. Lo, Y. Pan, K. Matsumoto, and N.W. Kowall, Blood-brain barrier disruption in experimental focal ischemia: comparison between in vivo MRI and immunocytochemistry, *Mag. Reson. Imag.* 12:403-411 (1994).
44. M.W.B. Bradbury. "The Concept of a Blood-Brain Barrier," John Wiley, Chichester (1979).
45. I. Klatzo, Presidential address. Neuropathological aspects of brain edema, *J. Neuropathol. Exp. Neurol.* 26:1-14 (1967).
46. R. Schmidt-Kastner, D. Meller, B.-M. Bellander, I. Stromberg, L. Olson, and M. Ingvar, A one-step immunohistochemical method for detection of blood-brain barrier disturbances for immunoglobulins in lesioned rat brain with special reference to false-positive labeling in immunohistochemistry, *J. Neurosci. Methods* 46:121-132 (1993).
47. A.S. Hazell, The pathophysiology of pyrithiamine-induced thiamine deficiency encephalopathy, Thesis, McGill University, pp. 1-176 (1994).
48. D.K. Leong, O. Le, L. Oliva, and R.F. Butterworth, Increased densities of binding sites for the peripheral-type benzodiazepine receptor ligand [h-3]pk11195 in vulnerable regions of the rat brain in thiamine deficiency encephalopathy, *J. Cereb. Blood Flow Metab.* 14:100-105 (1994).
49. N. Otsuka, M. Tomonaga, and K. Ikeda, Rapid appearance of β-amyloid precursor protein immunoreactivity in damaged axons and reactive glial cells in rat brain following needle stab injury, *Brain Res.* 568:335-338 (1991).
50. D.T. Stephenson, K. Rash and J.A. Clemens, Amyloid precursor protein accumulates in regions of neurodegeneration following focal cerebral ischemia in the rat, *Brain Res.* 593:128-135 (1992).
51. K. Iverfeldt, S.I. Walaas, and P. Greengard, Altered processing of Alzheimer amyloid precursor protein in response to neuronal degeneration, *Proc. Natl. Acad. Sci.* 90:4146-4150 (1993).
52. T. Kawarabayashi, M. Shoji, Y. Harigaya, H. Yamaguchi, and S. Hirai, Expression of APP in the early stage of brain damage, *Brain Res.* 563:334-338 (1991).

53. Y. Nakamura, M. Takeda, H. Niigawa, S. Hariguchi, and T. Nishimura, Amyloid β-protein precursor deposition in rat hippocampus lesioned by ibotenic acid injection, *Neurosci. Lett.* 136:95-98 (1992).
54. R. Siman, J.P. Card, R.B. Nelson, and L.G. Davis, Expression of beta-amyloid precursor protein in reactive astrocytes following neuronal damage, *Neuron* 3:275-285 (1989).
55. K. Shigematsu, P.L. McGeer, D.G. Walker, T. Ishii, and E.G. McGeer, Reactive microglia/macrophages phagocytose amyloid precursor protein produced by neurons following neural damage, *J. Neurosci. Res.* 31:443-453 (1992).
56. W.C. Wallace, V. Bragin, N.K. Robakis, K. Sambamurti, D. Van derPutten, C.R. Merril, K.L. Davis, A.C. Santucci, and V. Haroutunian, Increased biosynthesis of Alzheimer amyloid precursor protein in the cerebral cortex of rats with lesions of nucleus basalis of Meynert, *Brain Res.* 10:173-178 (1991).
57. N.Y. Calingasan, S.E. Gandy, H. Baker, K.-F.R. Sheu, K.-S. Kim, H.M. Wisniewski, and G.E. Gibson, Accumulation of amyloid precursor protein-like immunoreactivity in rat brain in response to thiamine deficiency, *Brain Res.* 677:50-60 (1995).
58. J.D. Buxbaum, S.E. Gandy, P. Cicchetti, M.E. Ehrlich, A.J. Czernik, P.J. Fracasso, T. Ramabhadran, A.J. Unterbeck, and P. Greengard, Processing of Alzheimer β/A4-amyloid precursor protein: Modulation by agents that regulate protein phosphorylation, *Proc. Natl. Acad. Sci. USA* 87:6003-6006 (1990).
59. G.L. Caporaso, S.E. Gandy, J.D. Buxbaum, and P. Greengard, Chloroquine inhibits intracellular degradation but not secretion of Alzheimer β/A4 amyloid precursor protein, *Proc. Natl. Acad. Sci. USA* 89:2252-2256 (1992).
60. G.L. Caporaso, S.E. Gandy, J.D. Buxbaum, T.V. Ramabhadran, and P. Greengard, Protein phosporylation regulates secretion of Alzheimer β/A4 amyloid precursor protein, *Proc. Natl. Acad. Sci. USA* 89:3055-3059 (1992).
61. S.E. Gandy, R. Bhasin, T.V. Ramabhadran, E.H. Koo, D.L. Price, D. Goldgaber, and P. Greengard, Alzheimer β/A4-amyloid precursor protein: Evidence for putative amyloidogenic fragment, *J. Neurochem.* 58:383-386 (1989).
62. C. Norstedt, S.E. Gandy, I. Alafuzoff, G.L. Caporaso, K. Iverfeldt, J.A. Grebb, B. Winblad, and P. Greengard, Alzheimer β/A4 amyloid precursor protein in human brain: Aging-associated increases in holoprotein and in a proteolytic fragment, *Proc. Natl. Acad. Sci. USA* 88:8910-8914 (1991).
63. T. Suzuki, A.C. Nairn, S.E. Gandy, and P. Greengard, Phosphorylation of Alzheimer amyloid precursor protein by protein kinase, *Neuroscience* 48:755-761 (1992).
64. R. Bhasin, W. Van Nostrand, T. Saitoh, M. Donets, E. Barnes, W. Quitschke, and D. Goldgaber, Expression of active secreted forms of human amyloid β-protein precursor by recombinant baculovirus-infected cells, *Proc. Natl. Acad. Sci.* 88:10307-10311 (1991).

METABOLIC FAILURE, OXIDATIVE STRESS, AND NEURODEGENERATION FOLLOWING CEREBRAL ISCHEMIA AND REPERFUSION

Gary Fiskum and Robert E. Rosenthal

Departments of Biochemistry and Molecular Biology
and Emergency Medicine
The George Washington University Medical Center
Washington, D.C. 20037

In the United States, 650,000 people die annually from cardiac disease; cardiac arrest accounts for 2/3 (435,000) of these deaths. If an individual suffers a cardiac arrest outside the hospital, he has less than a 15% chance of leaving the hospital alive.[1] Even if one survives initial resuscitative efforts, damage to the brain occurring during ischemia and reperfusion often results in neurologic morbidity or mortality. In one recent study of 262 initially comatose survivors of cardiac arrest, 79% of the patients had died within one year; cerebral failure was the cause of death in 37% of cases. Only 14% of patients were either neurologically normal or slightly impaired at 12 months.[2] Recent advances in preclinical research on brain injury due to transient, global cerebral ischemia have demonstrated that at least four classes of interrelated molecular mechanisms contribute to ischemic neurological impairment. These classes are excitotoxicity, cellular calcium overload, metabolic failure and oxidative stress. This brief review focuses on mechanisms falling within the last two of these classifications.

MITOCHONDRIAL RESPIRATION, CALCIUM TRANSPORT AND FREE-RADICAL GENERATION

It is well established that during even short periods (10-15 min) of complete cerebral ischemia, significant subcellular damage occurs in response to rapid ATP depletion, elevated intracellular Ca^{2+} and H^+ concentrations, and accelerated molecular degradation (e.g. hydrolysis of fatty acyl groups from phospholipids and proteolytic breakdown of proteins).[3] The mitochondrion is a prime subcellular target for ischemic injury and the degree to which it is damaged can limit the rate and extent of cell recovery during reperfusion because of the important roles this organelle plays in ATP formation, anapleurotic metabolism, Ca^{2+}

homeostasis and free radical detoxification.[4] Although impairment of mitochondrial respiratory and Ca^{2+} uptake capacities following cerebral ischemia and the recovery from minor mitochondrial damage following 1-2 hr of reperfusion are documented,[5,6,8] relatively little is known regarding mitochondrial activities following prolonged periods of reperfusion (≥24 hr). Results from our lab[54] and others[6-10] suggest that delayed mitochondrial dysfunction may be a significant event associated with chronic neurological morbidity; however, the mechanisms responsible for latent mitochondrial damage and the intracellular consequences of this event are not yet established. The importance of studying ischemic mitochondrial injury is also emphasized by the likelihood that altered mitochondrial oxidative metabolism has a major impact on important patterns of cerebral metabolism (e.g. glycolytic lactate production and amino acid metabolism[11]). Therefore it will be important in the future to characterize the respiratory and Ca^{2+} uptake activities of both neuronal and glial mitochondria isolated from different regions of the brain and to rigorously establish the temporal distribution of mitochondrial injury during ischemia and during both short and long periods of reperfusion. Correlation of the results of these studies with those obtained from metabolic measurements performed *in vivo* and *ex vivo* will provide insight into the potential metabolic ramifications of mitochondrial injury. Moreover, analysis of the effects of different interventions (e.g. administration of glutamate antagonists or antioxidants) may help elucidate the mechanisms responsible for mitochondrial damage during cerebral ischemia and reperfusion.

Another hypothesis that is being actively pursued is that cerebral ischemia/reperfusion results in either a change in intracellular conditions or an alteration in mitochondrial electron transport that potentiate the ability of these organelles to generate O_2^- and other reactive oxygen species. Although mitochondria from a variety of normal tissues have been shown to be significant producers of O_2- and H_2O_2,[12,13] relatively little evidence actually supports the hypothesis that mitochondrial free radical generation increases following cerebral ischemia/reperfusion. Dykens[14] has shown that a crude brain mitochondrial/synaptosomal preparation will generate increased hydroxyl radicals plus ascorbyl radicals and other carbon-centered radicals when exposed *in vitro* to conditions that are known to exist during ischemia/reperfusion (e.g. high Ca^{2+} concentrations). Hasegawa et al. [15] have also demonstrated with electron spin resonance spectroscopy that an increase in the mitochondrial coenzyme Q_{10} radical occurs *in vivo* in the neonatal mouse brain both during and following inspired N_2 or CO_2-induced hypoxia. In the study by Cino and Del Maestro,[16] mitochondria from a primarily glial origin were isolated following 15 min of decapitation ischemia in the rat and were found to exhibit a subnormal rate of H_2O_2 production. The strongest support for the possibility that mitochondrial free radical generation is abnormally elevated during cerebral ischemia/reperfusion has come from experiments performed *in vitro* with primary cultures of cortical neurons. Both Reynolds and Hastings[17] and Dugan et al.[18] have provided evidence that NMDA receptor-mediated elevations of intracellular Ca^{2+} can promote mitochondrial generation of reactive oxygen species within these cells. However, it still remains to be determined whether this phenomenon occurs *in vivo* and whether it actually contributes to pre-lethal oxidative molecular alterations.

INTERMEDIARY METABOLISM

The strict dependence of mitochondrial oxidative phosphorylation on the presence of O_2 explains the fact that intermediary metabolism is severely perturbed during cerebral ischemia. During hypoxia, glycolysis is transiently activated leading to lactic acidosis[19] and ATP depletion.[20] Elevated levels of NADH result in inhibition of TCA cycle flux and decreases in the level of most TCA cycle intermediates. The role of these ischemic changes in metabolite levels and fluxes during reperfusion is not clear; however, recent studies have suggested significant relationships between levels of TCA cycle metabolites and the extent of ischemic injury.[21]

Delayed tissue damage following ischemia/reperfusion may be linked to free radical-mediated inactivation of key oxidative enzymes as well as enzymes involved in anabolic pathways. Metal containing enzymes, e.g. pyruvate dehydrogenase, succinate dehydrogenase, glyceraldehyde-3-P-dehydrogenase and glutamine synthetase may be especially sensitive to damage.[22] Pyruvate dehydrogenase may be a key enzyme for maintaining normal ATP levels in the brain during reperfusion and recent evidence indicates that this enzyme is particularly sensitive to inactivation following ischemia/reperfusion.[23,24,54] It is also known that hyperglycemia and the associated cerebral lactic acidosis exacerbate neurological impairment associated with cerebral ischemia, but that dichloroacetate, a drug that activates pyruvate dehydrogenase, decreases the severity of ischemic injury.[25] Activation of pyruvate dehydrogenase has the dual effect of increasing metabolic flux into the TCA cycle and decreasing the lactic acidosis generated by a relative shift of energy metabolism toward glycolytic ATP production.

FREE RADICAL-MEDIATED MOLECULAR ALTERATIONS

It is well known that ischemia/reperfusion injury in various tissues has been linked to the generation of free radicals.[26,27] One mechanism which may contribute to cerebral injury following ischemia is the increase in intracellular Ca^{2+} stimulated by excitatory amino acids, and increasing evidence suggests that these two mechanisms are closely associated.[17,18,28,29] There is also increasing evidence that abnormal levels of the free radical nitric oxide can be produced by Ca^{2+} activation of nitric oxide synthetase and that these may contribute to ischemic brain injury, possibly due to generation of other free radicals, e.g. peroxynitrite.[30,31] Most ischemia/reperfusion studies have demonstrated a role for free radical damage to lipids;[32] however, the role of oxidative protein and enzyme damage in reperfusion injury has not been clearly defined.[33-36] Previously it has been determined that many key metabolic enzymes are oxidatively inactivated by metal-catalyzed oxidation systems.[22] The oxidation of some enzymes is also partially inhibited by free radical scavengers and by superoxide dismutase. Based on these findings it has been proposed that the oxidative inactivation of enzymes is a site specific process involving the binding of Fe^{2+} to generate an activated O_2 specie(s), e.g. $OH^{\cdot}$, which oxidizes amino acids near the metal binding site. In addition to the involvement of $OH^{\cdot}$ in oxidative alterations of lipids and proteins during cerebral ischemia/reperfusion, increasing attention is being directed toward reactions of macromolecules with derivatives of nitric oxide. Protein nitration is one such reaction which may play a particularly important role in the inactivation of enzymes and structural proteins. Nitration occurs when peroxynitrite attacks the 3-carbon atom of tyrosine adjacent to the oxygen. It can occur spontaneously but may also be catalyzed by low molecular mass transition metals as well as by superoxide dismutase.[37,38] Significant evidence now indicates that cerebral ischemia and reperfusion results in abnormally elevated levels of nitric oxide[55] which then react extremely rapidly with superoxide ($6.7 \times 10^9\ M^{-1}\ s^{-1}$) to produce peroxynitrite. Peroxynitrite dependent protein nitration could have adverse effects on the structure and function of brain proteins following ischemia/reperfusion since nitration has been shown to inactivate enzymes, e.g. cytochrome P450,[39] and has recently been implicated in the alterations of neurofilament proteins that occur in familial amyotrophic lateral sclerosis[40] as well as in the inflammatory pathogenesis of atherosclerosis[41] and acute lung injury.[42] Further work is need to determine whether such alterations play an important role in ischemic brain injury and whether reperfusion-sensitive enzyme activities, e.g. pyruvate dehydrogenase and glutamine synthetase, are vulnerable to this form of damage.

CEREBRAL RESUSCITATION THROUGH FREE RADICAL ANTAGONISM

A number of post-ischemic interventions aimed toward reducing iron-catalyzed free radical production have been tested.[43] In addition to pharmacological attempts at altering free-radical-mediated brain injury, manipulation of inspired O_2 concentrations during reperfusion may serve as similar experimental and clinically feasible protocols. The proper method for oxygen administration following resuscitation from cardiac arrest remains controversial. Current published guidelines from the American Heart Association suggest that, "The highest possible oxygen concentration (100% is preferable) should be administered as soon as possible to all patients with cardiac or pulmonary arrest".[44] Experimental evidence suggests, though, that indiscriminate usage of 100% oxygen may actually prove damaging following resuscitation,[45] possibly as a result of increased brain lipid peroxidation.[46,47] Technology has now become available which allows *in vivo* monitoring of brain oxygen concentrations.[48,49] When cortical oxygen concentrations were monitored and controlled during reperfusion from 20 min of global ischemia in rats, there was a positive correlation between reoxygenation level and severity of histological neuronal damage.[49] Because of the potential for increased peroxidative membrane damage with hyperoxic reperfusion, it has recently been suggested that post-ischemic reperfusion with relatively hypoxic blood may actually be beneficial following neurologic ischemia.[50]

Amelioration of post-ischemic neurologic injury has also been attempted through the delivery of 100% oxygen at hyperbaric conditions. It has recently been demonstrated that carbon monoxide poisoning in rats results in increased brain lipid peroxidation, but treatment with 100% oxygen at 3 atmospheres absolute (ATA), but not 1 ATA, was found to prevent brain lipid peroxidation, suggesting a direct protective effect for hyperbaric oxygen administration.[51] The effects of hyperbaric O_2 therapy following global cerebral ischemia have been less extensively studied.[52,53] The relative paucity of information concerning the neurological and neurochemical effects of different inspired O_2 concentrations and pressures on ischemic brain injury has led us to embark on a series of studies designed to obtain such information. These studies will improve our understanding of the mechanisms underlying ischemic neurological morbidity and will hopefully lead to the development of improved methods of cerebral resuscitation.

REFERENCES

1. Steuven, H.A., White, E.M., Troiano, P. and Mateer, J.R., Prehospital cardiac arrest - a critical analysis of factors affecting survival. *Resuscitation* 17:251-259 (1989).
2. Da Garavilla, L., Babbs, C.F. and Tacker, W.A., An experimental circulatory arrest model in the rat to evaluate calcium antagonists in cerebral resuscitation. *Am. J. Emerg. Med.* 2:321-326 (1984).
3. Fiskum, G., Mitochondrial damage during cerebral ischemia. *Ann. Emerg. Med.* 14:810-815 (1985).
4. Rosenthal, R.E., and Fiskum, G., Brain mitochondrial function in cerebral ischemia and resuscitation, *in*: "Cerebral Ischemia and Resuscitation", A. Schurr and B.M. Rigor, eds., CRC Press, New York (1990).

5. Rosenthal, R.E., Hamud, F., Fiskum, G., Varghese, P.J., and Sharpe, S., Cerebral ischemia and reperfusion: prevention of brain mitochondrial injury by lidoflazine. *J. Cereb. Blood Flow Metab.* 7:752-758 (1987).
6. Sims, N., and Pulsinelli, W.A., Altered mitochondrial respiration in selectively vulnerable subregions following transient forebrain ischemia in the rat. *J. Neurochem.* 49:1367-1374 (1987).
7. Wagner, K.R., Kleinholz, M., and Myers, R.E., Delayed onset of neurologic deterioration following anoxia/ischemia coincides with appearance of impaired brain mitochondrial respiration and decreased cytochrome oxidase activity. *J. Cereb. Blood Flow Metab.* 10:417-423 (1990).
8. Sciammanna, M.A., Zinkel, J., Fabi, A.Y., Lee, C.P., Ischemic injury to rat forebrain mitochondria and cellular calcium homeostasis. *Biochem Biophys. Acta.* 1134:223-232 (1992).
9. Zaidan, E., and Sims. N.R., The calcium content of mitochondria from brain subregions following short-term forebrain ischemia and recirculation in the rat. *J. Neurochem.* 63:1812-1819 (1994).
10. Sun D., and Gilboe D.D., Ischemia-induced changes in cerebral mitochondrial free fatty acids, phospholipids, and respiration in the rat. *J. Neurochem.* 62:1921-1928 (1994).
11. Peeling, J., Wong, D., and Sutherland, G.R., Nuclear magnetic resonance study of regional metabolism after forebrain ischemia in rats. *Stroke* 20:633-640 (1989).
12. Boveris, A., Oshino, N., and Chance, B., The cellular production of hydrogen peroxide. *Biochem. J.* 128:617-630 (1972).
13. Freeman, B.A., and Crapo, J.D., Hyperoxia increases oxygen radical production in rat lungs and lung mitochondria. *J. Biol. Chem.* 256:10986-10992 (1981).
14. Dykens, J.A., Isolated cerbral and cerbellar mitochondria produce free radicals when exposed to elevated Ca^{2+} and Na^{+}: Implications for neurodegeneration. *J. Neurochem.* 63:584-591 (1994).
15. Hasegawa, K., Yoshioka, H., Sawada, T. and Nishikawa, H., Direct measurement of free radicals in the neonatal mouse brain subjected to hypoxia: An electron spin resonance spectroscopic study. *Brain Res.* 607:161-166 (1993).
16. Cino, M., and Del Maestro, R.F., Generation of hydrogen peroxide by brain mitochondria: the effect of reoxygenation following postdecapitative ischemia. *Arch. Biochem. Biophys.* 269:623-638 (1989).
17.. Reynolds, I.J., and Hastings, T.G., Glutamate induces the production of reactive oxygen species in cultured forebrain neurons following NMDA receptor activation. *J. Neurosci.* 15:3318-3327 (1995).
18. Dugan, L.L., Sensi, S.L., Canzoniero, L.M.T., Handran, S.D., Rothman, S.M., T.-S. Lin, Goldberg, M.P., and Choi, D.W., Mitochondrial production of reactive oxygen species in cortical neurons following exposure to *N*-methyl-D-aspartate. *J. Neurosci.* 15:6377-6388 (1995)..
19. Siesjo, B.K., Elcholm, A., Katsura, K. and Theander, S., Acid-base changes during complete brain ischemia. *Stroke* 21, Suppl. 11:194-199 (1990).
20. Folbergrova, J., Ljunggren, B., Norberg, K., and Siesjo, B.K., Influence of complete ischemia on glycolytic metabolites, citric acid cycle intermediates and associated amino acids in the rat cerebral cortex. *Brain Res.* 80:265-279 (1974).
21. Cannella, D.M., Kapp, J.P., Munschauer, F.E., Markov, A.K., and Shurcard, D.W. , Cerebral resuscitation with succinate and fructose-1,6-diphosphate. *Surg. Neurol.* 31:177-182 (1989).
22. Fucci, L., Oliver, C.N., Coon, M.J., and Stadtman, E.R., Inactivation of key metabolic enzymes by mixed-function oxidation reactions: possible implications in protein

turnover and aging. *Proc. Natl. Acad. Sci. USA* 80:1521-1526 (1983).
23. Katayama, Y., Welsh, F.A., Effect of dichloroacetate on regional energy metabolites and pyruvate dehydrogenase activity during ischemia and reperfusion in gerbil brain. *J. Neurochem.* 52:1817-1822 (1989).
24. Zaidan, E. and Sims, N.R., Selective reductions in the activity of the PDH complex in mitochondria isolated from brain subregions following forebrain ischemia in rats. *J. Cereb. Blood Flow Metab.* 13:98-104 (1993).
25. LeMay, D.R., Zelenock, G.B., and D'Alecy, L.G., Neurological protection by dichloroacetate depending on the severity of injury in the paraplegic rat. *J. Neurosurg.* 73:118-122 (1990).
26. Parks, D.A., Bulkley, G.B., Granger, N., Hamilton, S.R., and McCord, J.M., Ischemic injury in the cat small intestine: Role of superoxide radical. *Gastroenterology* 82:9-15 (1982).
27. Schmidley, J.W., Free radicals in central nervous system ischemia. *Stroke* 21:1086-1090 (1990).
28. Choi, D.W., Methods for antagonizing glutamate neurotoxicity. *Cerebrovasc. Brain Metabol. Rev.* 2:105-147 (1990).
29. Pellegrini-Giampietro, D.E., Cherici, G., Alesiani, M., Carla, V., and Moroni, F., Excitatory amino acid release and free radical formation may cooperate in the genesis of ischemia-induced neuronal damage. *J. Neurosci.* 10:1035-1041 (1990).
30. Bredt, D.S., Synder, S.H., Nitric oxide: A novel neuronal messenger. *Neuron* 8:3-11 (1992).
31. Rordi, R. Beckman, J.S., Bush, K.M. and Freeman, B.A., Peroxynitrite-induced membrane lipid peroxidation: the cytotoxic potential of superoxide and nitric oxide. *Arch Biochem. Biophys.* 228:481-487 (1984).
32. Imaizumi, S. et al. Free radicals and lipid changes in cerebral ischemia, *in*: "Cerebral Ischemia and Resuscitation", A. Schurr and B.M. Rigor, eds.), CRC Press, New York, (1990).
33. Oliver, C.N., Starke-Reed, P.E., Stadtman, E.R. Liu, G.J., Carney, J.M., and Floyd, R.A., Oxidative damage to brain proteins, loss of glutamine synthetase activity, and production of freee radicals during ischemia/reperfusion-induced injury to gerbil brain. *Proc. Natl. Acad. Sci. USA* 87:5144-5147 (1990).
34. Krause, G.S., Degracia, D.J., Skjaerlund, J.M., and O'Neill, B.J., Assessment of free radical-induced damage in brain proteins after ischemia and reperfusion. *Resuscitation* 23:59-69 (1992).
35. Liu, Y., Rosenthal, R., Starke-Reed, P., and Fiskum, G., Inhibition of post-cardiac arrest brain protein oxidation by acetyl-L-carnitine, *Free Rad. Biol. Med.* 15:667-670 (1993).
36. Folbergrova, J., Kiyota, Y., Pahlmark, K., Memezawa, H., Smith, M.-L., and Siesjo, B.K., Eoes ischemia with reperfusion lead to oxidative damage to proteins in the brain? *J. Cereb. Blood Flow Metab.* 13:145-152 (1993).
37. Ischiropoulos, H., Zhu, L., Chen, J., Tsai, M., Martin, J.C., Smith, C.D. and Beckman, J.S., Peroxynitrite-mediated tyrosine nitration catalyzed by superoxide dismutase. *Arch. Biochem. Biophys.* 298:431-437 (1992).
38. Beckman, J.S., Ischiropoulos, H., Zhu, L., van der Woerd, M., Smith, C., Chen, J., Harrison, J., Martin, J.C. and Tsai, M.H., Kinetics of superoxide dismutase and iron catalyzed nitration of phenolics by peroxynitrite. *Arch. Biochem. Biophys.* 298:483-445 (1992).
39. Janing, G.R., Kraft, R., Blank, J., Ristau, O., Rabe, H., and Ruckpaul, K. Chemical modification of cytochrome P-450 LM4. Identification of functionally linked tyrosine residues. *Biochem. Biophys. Acta.* 916:512-523 (1987).

40. Beckman, J.S., Peroxynitrite, superoxide dismutase and tyrosine nitration in neurodegeneration, Abstract of presentation at "Neurodegenerative Diseases '95", the XVth International Washington Spring Symposium, Washington D.C. (1995).
41. Beckmann, J.S., Ye, Y.Z., Anderson, P.G., Chen, J., Accavitti, M.A., Tarpey, M.M. and White, C.R., Extensive nitration of protein tyrosines in human atherosclerosis detected by immunohistochemistry. *Biol. Chem. Hoppe-Seyler* 375:81-88 (1994).
42. Haddad, I.Y., Pataki, G., Hu, P., Galliani, C., Beckman, J.S. and Matalon, S., Quantitation of nitrotyrosine levels in lung sections of patients and animals with acute lung injury. *J. Clin. Invest.* 94:2407-2413 (1994).
43. Rosenthal, R.E., Chanderbhan, R.F., Marshall, Jr., G.H. and Fiskum, G., Prevention of post-ischemic brain lipid conjugated diene production and neurological injury by hydroxyethylstarch-conjugated deferoxamine. *Free Rad. Biol. Med.* 12:29-33 (1992).
44. Standards and guidelines for cardiopulmonary resuscitation and emergency cardiac care. Part III: Adult advanced cardiac life support. *JAMA* 255:2933-2954 (1986).
45. Zwemer, C.F., Whitesall, S.E. and D'Alecy, L.G., Cardiopulmonary-cerebral resuscitation with 100% oxygen exacerbates neurological dysfunction following nine minutes of normothermic cardiac arrest in dogs. *Resuscitation* 27:159-170 (1994).
46. Mickel, H.S., Vaishav, Y.N., Kempski, O., von Lubitz, D., Weiss, J.F., and Feuerstein, G., Breathing 100% oxygen after global brain ischemia in mongolian gerbils results in increased lipid peroxidation and increased mortality. *Stroke* 18:426-430 (1987).
47. Rosenthal, R.E., Miljkovic-Lolic, M., Haywood, Y. and Fiskum, G., Cerebral ischemia/reperfusion: Neurologic effects of hyperoxic resuscitation from experimental cardiac arrest in dogs. *Ann. Emerg. Med.* 25:137 (1995)
48. Wilson D.F., Pastuszko, A., DiGiacomo, J.E., Pawlowski, M., Schneiderman, R., and Delivoria-Papadopoulos, M., Effect of hyperventilation on oxygenation of the brain cortex of newborn piglets. *J. Appl. Physiol.* 70:2691-2696 (1991).
49. Halsey, J.H., Conger, K.A., Garcia, J.H., and Sarvary, E., The contribution of reoxygenation to ischemic brain damage. *J. Cereb. Blood Flow Metab.* 11:994-1000 (1991).
50. Danielisova, V., Marsala, M., Chavko, M., and Marsala, J., Postischemic hypoxia improves metabolic and functional recovery of the spinal cord. *Neurology* 40:1125-1129 (1990).
51. Thom, S.R., Antagonism of carbon monoxide mediated brain lipid peroxidation by hyperbaric oxygen. *Toxicol. Appl. Pharmacol.* 105:340-344 (1990).
52. Kapp, J.P., Phillips, M., Markov, A., and Smith, R.R., Hyperbaric oxygen after circulatory arrest: modification of post-ischemic encephalopathy. *Neurosurg.* 11:496-499 (1982).
53. Iwatsuki, N., Hyperbaric oxygen combined with nicardipine administration accelerates neurologic recovery after cerebral ischemia in a canine model. *Crit. Car Med.* 22:858-863 (1994).
54. Bogaert, Y.E., Rosenthal, R.E. and Fiskum, G., Post-ischemic inhibition of cerebral cortex pyruvate dehydrogenase. *Free Rad. Biol. Med.* 16:811-820 (1994).
55. Nowicki, J.P., Duval, D., Poignet, H., and Scatton, B., Nitric oxide mediates neuronal death after focal cerebral ischemia in the mouse. *Eur. J. Pharmacol.* 204:339-340 (1991).

GLOBAL BRAIN ISCHEMIA AND REPERFUSION: TRANSLATION INITIATION FACTORS

Donald DeGracia, Robert Neumar, Blaine White, and Gary Krause

Department of Emergency Medicine
Wayne State University School of Medicine
Detroit, MI 48201

INTRODUCTION

Cardiopulmonary resuscitation for victims of cardiac arrest, both within and outside of the hospital, succeeds in restoring spontaneous circulation in about 70,000 patients a year in the United States. At least 60% of these patients subsequently die in the hospital as a result of extensive brain damage; only 3-10% of resuscitated patients are finally able to resume their former lifestyles[1]. These statistics may even be optimistic; in a recent report of data for out-of-hospital cardiac arrest, out of a total of 1,445 patients only 10 survived to discharge with the majority of these survivors (6/10) having a spontaneous pulse *en route* to the hospital[2]. Clearly, patients do not often escape cardiac arrest and resuscitation without initiation of brain injury mechanisms, and there are currently no therapeutic approaches that are convincingly effective in ameliorating the brain damage that usually accompanies this clinically important brain insult.

The traditional approach to this problem has been to identify brain injury mechanisms at the cellular level and then attempt to alter them by pharmacologic means. This approach has provided insight into a general causal sequence of brain injury by global ischemia and reperfusion[3]. Rapid energy depletion during complete ischemia leads to neuronal depolarization with release of glutamate from presynaptic terminals and large increases in intracellular Ca^{2+}. Both glutamate-mediated activation of post-synaptic receptors and the opening by depolarization of voltage-dependent Ca^{2+} channels contribute to Ca^{2+} translocation during ischemia. Cytosolic calcium overload during ischemia results in activation of phospholipases with release of free fatty acids from membrane lipids, and may also result in activation of some proteases such as calpain[4]. The reperfusion-induced ultrastructural damage is most prominent in selectively vulnerable neurons in layers 3 and 5 of the cortex and in the CA1 and CA4 zones of the hippocampus and is characterized between 90 minutes and 6 hours reperfusion by depletion of polyribosomes from the rough endoplasmic reticulum, deformation of the Golgi apparatus with the presence of numerous vesicles and membranous whorls, and damage to the plasmalemma[5]. The appearance of substantial ultrastructural damage during reperfusion led to the suggestion that this damage

Neurodegenerative Diseases
Edited by Gary Fiskum, Plenum Press, New York, 1996

was a consequence of oxygen radicals. While the exact identity of these oxygen radicals is unknown, they could be formed by the combination of superoxide, which is generated by enzymatic oxidation of free unsaturated fatty acids that accumulate during ischemia, and iron, which is released during reperfusion from storage proteins[3,4]. Lipid peroxidation during early reperfusion is most prominent in selectively vulnerable neurons[6], and ultrastructural changes similar to those observed in vulnerable brain neurons during reperfusion are induced in cultured neurons specifically by radical-induced damage (unpublished data). In spite of this knowledge gained over the last two decades, attempts to block calcium channels, glutamate receptors, and radical reactions have not led to the development of effective therapy. An alternate approach asks about the neurons' intrinsic abilities to limit damage and initiate early repair. Clearly the neuron's response to ischemia and reperfusion by transcription of mRNA and translation of proteins assumes central importance.

TRANSLATION: EFFECTS OF BRAIN ISCHEMIA AND REPERFUSION

Global brain ischemia results in a transient global suppression of protein synthesis that is especially severe and prolonged in selectively vulnerable neurons[7,8,9]. However early reperfusion is not associated with degradation of nuclear DNA, direct damage to the ribosomes[10,11], or failure of transcription[3,4]. Moreover new mRNAs for proteins likely to be crucial for damage limitation and early initiation of cellular repair are being transcribed[3,4], and kinetic studies of the mRNA translocation system at the nuclear membrane indicate that it remains competent for at least the first 6 hours of post-ischemic reperfusion[12]. We presented studies[13] at this meeting in which we have observed that 90 minutes of reperfusion but not 10 minutes of ischemia is associated with an ~90% inhibition of initiation-dependent protein synthesis during *in vitro* translation of brain homogenates. These results suggest that reperfusion-induced alterations in the process of translation initiation are involved in the the *in vivo* suppression of brain protein synthesis observed during reperfusion[7].

Formation of the translation-initiation complex requires the organized assembly of the small ribosomal subunit, the mRNA to be translated, and the tRNA coding for the first amino acid (always methionine in eukaryotic cells)[14]. This assembly is coordinated by a family of proteins collectively known as eukaryotic initiation factors (eIFs). The eIF-4 and eIF-2 complexes are believed to be the regulatory points for translation initiation. The eIF-4 complex, responsible for introducing the mRNAs into the translation-initiation complex, consists of three subunits, eIF-4A (an ATP-dependent RNA helicase[15]), eIF-4E (which binds to the mRNA's m^7GTP cap), and eIF-4γ (whose exact function is still unknown). The ternary complex containing eIF-2 is responsible for introducing methionyl-tRNA into the translation-initiation complex. Each new initiation cycle requires that eIF-2B exchange GTP for GDP generated on eIF-2 by hydrolysis during the previous initiation cycle. It is known from other model systems that the activity of these eIFs may be altered by proteolysis[16] or modifications of the phosphorylation state[17], both of which can result in suppressed protein synthesis.

We utilized antibodies against eIF-4E, eIF-4γ, eIF-2α, and phosphoserine to examine the effects of global brain ischemia and reperfusion on these important components of the translation initiation system[12]. These experiments identified two potentially important reperfusion-associated alterations of initiation factors. Fragments of eIF-4γ migrating at 90, 69, 54, 43, and 38 kDa consistently appear at 90 minutes reperfusion, and the fragments at 90, 69, and 43 kDA bind eIF-4E. This suggests that these fragments may interfere competitively with the formation of the eIF-4E-eIF-4γ complex, and experiments are now needed to characterize the effects of ischemia and reperfusion on the kinetics of formation of this complex. We also found that post-ischemic reperfusion induces serine phosphorylation of eIF-2α.

Calcium-dependent proteases, such as calpain, are known to degrade eIF-4γ[18], and the fragmentation of eIF-4γ that develops during polio virus infection of cultured HeLa cells is associated with a pronounced loss of translation competence[19]. Furthermore, Roberts-Lewis *et al* observed proteolytic degradation of spectrin, a calpain substrate, during early post-ischemic brain reperfusion[20]. Although our experiments did not directly address the cause of the degradation of eIF-4γ, two of the degradation bands (at 90 and 69 kDa) that appear on our Western blots are consistent in size with the calcium-induced eIF-4γ degradation products observed by Wyckoff *et al.*[17] (we calculate their degradation bands at 90 and 70 kDa), suggesting that eIF-4γ is a a calpain substrate during brain reperfusion.

Fragmentation of eIF-4γ may not be not sufficient to account for the pattern of protein synthesis suppression seen in the selectively vulnerable neurons during reperfusion. Although these neurons generate large amounts of mRNA for HSP-70 during reperfusion, this protein is not efficiently synthesized[21,22]. In elegant experiments with HeLa cells, Joshi-Barve *et al.*[23] showed that introduction of anti-sense nucleotides against mRNA for eIF-4E results in a cellular deficiency of both eIF-4E and eIF-4γ and a dramatic reduction in protein synthesis, although several heat shock proteins, including HSP-70, continued to be translated. This suggests that if inhibition of formation of the eIF-4 complex by fragments of eIF-4γ was the central mechanism in inhibition of translation initiation during reperfusion, then synthesis of heat shock proteins should be able pass by this blockade. This is not the case in selectively vulnerable neurons. The virtual shutdown in protein synthesis in the selectively vulnerable neurons suggests that there are other defects in translation initiation than those that may be expected from inhibited formation of the eIF-4 complex. Serine phosphorylation of eIF-2α during brain reperfusion could be this more general mechanism of inhibition of translation initiation.

Serine phosphorylation of eIF-2α causes a slight shift in its mobility seen with isoelectric focusing, and Burda *et al.*[24] recently reported such a shift associated with brain ischemia and reperfusion. However, they utilized a vessel ligation model that generates severe but incomplete ischemia, and they did not separate the effects of the period of incomplete ischemia from those of the period of reperfusion, nor did they examine the kinetics of initiation-dependent translation. Our results provide direct evidence that serine phosphorylation of eIF-2α is a consequence of post-ischemic brain reperfusion and is associated with a severe inhibition of translation initiation. Serine phosphorylation of eIF-2 α results in a general suppression of translation that is independent of any message specificity. Each initiation cycle requires the regeneration of active eIF-2 by exchange of a eIF-2-bound GDP for a GTP. This process is mediated by eIF-2B and is competitively inhibited by serine-phosphorylation of the α-subunit on eIF-2 in a manner dependent on both the proportion of eIF-2α that is serine phosphorylated and the tissue concentration of eIF-2B[25]. Hu and Wielock[26] found evidence of reduced eIF-2B activity during post-ischemic brain reperfusion, but Burda *et al.*[23] presented kinetic evidence that the observed inhibition of eIF-2B activity could be fully explained by competitive inhibition due to serine-phosphorylation of eIF-2α. This is consistent with our observation that inclusion of a 10-fold excess of eIF-2 with an unphosphorylated α-subunit in a translation reaction with reperfused brain homogenate induces a doubling of the rate of initiation-dependent protein synthesis[12].

It is likely that further studies will find prolonged serine phosphorylation of eIF-2α localized to vulnerable neurons. Phosphorylation of eIF-2α on serine-51 is stimulated by free arachidonate, iron, and lipid hydroperoxides[27,28,29]. Both phospholipase activation with release of free fatty acids during ischemia[30] and lipid peroxidation during reperfusion[6] are more prominent in the vulnerable neurons. Furthermore, the total concentration of lipid peroxidation products in the reperfused brain continues to increase for at least 24 hours[31]. It remains to be determined if the brain kinase responsible for serine phosphorylation of eIF-2 α is similarly stimulated by arachidonate and lipid hydroperoxides. Experiments are now needed to characterize an extended time course of proteolytic degradation of eIF-4γ and serine phosphorylation of eIF-2α, to localize eIF-2α serine phosphorylation, to identify the kinase responsible for reperfusion-induced serine phosphorylation of eIF-2α, and to further examine both *in vitro* and *in vivo* approaches to reversal of the reperfusion-induced inhibition of translation initiation.

This work supported in part by the Emergency Medicine Foundation-Genentech Center for Excellence Award.

REFERENCES

1. G.S. Krause, K. Kumar, B.C. White, S.D. Aust, and J.G. Wiegenstein, Ischemia, resuscitation, and reperfusion: Mechanisms of tissue injury and prospects for protection, *Am Heart J.* 111:768 (1986).
2. A.L. Kellerman, D.R. Staves, and B.B. Hackman, In-hospital resuscitation following unsuccessful prehospital advanced life support "Heroic efforts" or an exercise in futility? *Ann Emerg Med.* 17:589 (1988).
3. B.C. White, L.I. Grossman, and G.S. Krause, Membrane damage and repair in brain injury by ischemia and reperfusion. *Neurology* 43:1656 (1993).
4. B.J. O'Neil, G.S. Krause, L.I. Grossman, G. Grunberger, J.A. Rafols, D.J. DeGracia, R.W. Neumar, B.R. Tiffany, and B.C. White, Global brain ischemia and reperfusion by cardiac arrest and resuscitation: mechanisms leading to death of vulnerable neurons and a fundamental basis for therapeutic approaches, *in:* "Cardiac Arrest: The Pathophysiology and Therapy of Sudden Death," N.A. Paradis, H.R. Halperin, and R.M. Nowak, eds. Williams and Wilkins, Baltimore, Md (1995).
5. J.A. Rafols, B.J. O'Neil, G.S. Krause, R.W. Neumar, and B.C. White, Global brain ischemia and reperfusion: Golgi apparatus ultrastructure in neurons selectively vulnerable to death, *Acta Neuropathol (Berlin).* In press (1995).
6. B.C. White, A Daya, D.J. DeGracia, B.J. O'Neil, J.M. Skjaerlund, G.S. Krause, and J.A. Rafols, Fluorescent histochemical localization of lipid peroxidation during brain reperfusion following cardiac arrest. *Acta Neuropathol (Berlin).* 86:1 (1993).
7. K.A. Hossman, and P. Kleihues, Reversibility of ischemic brain damage. *Arch Neurol.* 29:375 (1973).
8. H.K. Cooper, T. Zalewska, S. Kawakami, and K.A. Hossman, The effect of ischemia and recirculation on protein synthesis in the rat brain. *J Neurochem.* 28:929 (1977).
9. G.A. Dienel, W.A. Pulsinelli, and T.E. Duffy, Regional protein synthesis in rat brain following acute hemispheric ischemia. *J Neurochem.* 35:1216 (1980).
10. G.S. Krause, D.J. DeGracia, J.M. Skjaerlund, and B.J. O'Neil, Assessment of free radical-induced structural damage in brain proteins after ischemia and reperfusion. *Resuscitation* 23:59 (1992).
11. D.J. DeGracia, B.J. O'Neil, G.S. Krause, J.M. Skjaerlund, B.C. White, and L.I. Grossman, Studies of the protein synthesis system in the brain cortex during global ischemia and reperfusion. *Resuscitation* 25:161 (1993).
12. B.R. Tiffany, B.C. White, and G.S. Krause, Nuclear-envelope nucleoside triphosphatase kinetics and mRNA transport following brain ischemia and reperfusion. *Ann Emerg Med.* In press (1995).

13. D.J. DeGracia, R.W. Neumar, B.C. White, and G.S. Krause, Global brain ischemia and reperfusion: Modifications in eukaryotic initiation factors are associated with inhibition of translation initiation. *J Neurochem.* In submission (1995).
14. G.S. Krause, and B.R. Tiffany, Suppression of protein synthesis in the reperfused brain. *Stroke* 24:747 (1993).
15. B.K. Ray, T.G. Lawson, J.C. Kramer, M.H. Cladaras, J.K. Miller, J.W.B. Hershey, J.A. Grifo, W.C. Merrick, and R.A. Thatch, ATP-dependent unwinding of messenger RNA structure by eukaryotic initiation factors. *J Biol Chem.* 260:7651 (1985).
16. B. Buckley, and E. Ehrenfeld, The cap-binding protein complex in uninfected and poliovirus-infected HeLa cells. *J Biol Chem.* 262:13599 (1987).
17. J.W.B. Hershey, Overview: Phosphorylation and translation control. *Enzyme* 44:17 (1990).
18. E.E. Wyckoff, D.E. Croall, and E. Ehrenfeld, The p220 component of eukaryotic initiation factor 4F is a substrate for multiple calcium-dependent enzymes. *Biochem.* 29:10055 (1990).
19. D. Etchison, S.C. Milburn, I. Edery, N. Sonenberg. and J.W.B. Hershey, Inhibition of HeLa cell protein synthesis following poliovirus infection correlates with the proteolysis of a 220,000-dalton polypeptide associated with eucaryotic initiation factor 3 and a cap binding protein complex. *J Biol Chem.* 257:14806 (1982).
20. J.M. Roberts-Lewis, M.J. Savage, V.R. March, L.R. Pinsker, and R. Siman, Immunolocalization of calpain-1-mediated spectrin degradation to vulnerable neurons in the ischemic gerbil brain. *J Neurosci.* 14:3934 (1994)
21. T.S. Nowak Jr, U. Bond, and M.J. Schlesinger, Heat shock RNA levels in brain and other tissues after hyperthermia and transient ischemia. *J Neurochem.* 54:451 (1990).
22. T.S. Nowak Jr, Protein synthesis and the heat shock/stress response after ischemia. *Cerbrovasc Brain Metab Rev.* 2:345 (1990).
23. S. Joshi-Barve, A. DeBenedetti, and R.E. Rhoads, Preferential translation of heat shock mRNAs in HeLa cells deficient in protein synthesis initiation factors eIF-4E and eIF-4γ. *J Biol Chem.* 267:21038 (1992).
24. J. Burda, E.M. Martin, A. Garcia, A. Alcazar, J.L. Fando, and M. Salinas, Phosphorylation of the α subunit of initiation factor 2 correlates with the inhibition of translation following transient cerebral ischaemia in the rat. *Biochem J.* 302:335 (1994).
25. S. Oldfield, B.L. Jones, D. Tanton, and C.G. Proud, Use of monoclonal antibodies to study the structure and function of eukaryotic protein synthesis initiation factor eIF-2B. *Eur J Biochem.* 221:399 (1994).
26. B. Hu, and T. Wieloch, Stress-induced inhibition of protein synthesis initiation: Modulation of initiation factor 2 and guanine nucleotide exchange factor activities following transient ischemia in the rat. *J Neurosci.* 13:1830 (1993).
27. E.I. Rotman, M.A. Brostrom, and C.O. Brostrom, Inhibition of protein syntheis in intact mammalian cells by arachidonic acid. *Biochem J.* 282:487 (1992).
28. R. Hurst, J.R. Schatz, and R.L. Matts, Inhibition of rabbit reticulocyte lysate protein synthesis by heavy metal ions involves the phosphorylation of the α-subunit of the eukaryotic initiation factor 2. *J Biol Chem.* 262:15939 (1987).
29. A.G. De Herreros, C. De Haro, and S. Ochoa, Mechanism of activation of the heme-stabilized translational inhibitor of reticulocyte lysates by calcium ions and phospholipid. *Proc Natl Acad Sci USA.* 82:3119 (1985).
30. A. Umemura, Regional difference in free fatty acids release and the action of phospholipase during ischemia in rat brain. *No To Shinkei.* 42:979 (1990).
31. R.E. Rosenthal, R. Chanderbhan, G. Marshall, and G. Fiskum, Prevention of post-ischemic brain lipid conjugated diene production and neurological injury by hydroxyethyl starch-conjugated deferoxamine. *Free Rad Biol Med.* 12:29 (1992).

BATTEN DISEASE: A TYPICAL NEURONAL STORAGE DISEASE OR A GENETIC NEURODEGENERATIVE DISORDER CHARACTERIZED BY EXCITOTOXICITY?

Steven U. Walkley, Donald A. Siegel, Kostantin Dobrenis

Department of Neuroscience
Rose F. Kennedy Center for Research in Mental Retardation and Human Development
Albert Einstein College of Medicine
Bronx, NY 10461

INTRODUCTION

Batten disease (neuronal ceroid lipofuscinosis) is an inherited neurological disorder of humans and a variety of animal species including dogs, mice, and sheep. Affected individuals appear normal at birth but later exhibit progressive neurological deterioration and death. The spectrum of clinical disease includes retarded mental development and/or dementia, blindness, motor system dysfunction, and seizures, and in late disease the latter can be intractable. The age at which clinical symptoms appear varies and infantile, late infantile, juvenile and adult-onset disease subtypes are recognized. Disease course in individuals with early-onset disease generally is rapid, whereas late-onset disorders exhibit a more protracted course. On postmortem exam, atrophy of cerebral cortex and ballooning of surviving neurons are characteristic features. The latter finding has led to classification of Batten disease as a neuronal storage disorder along with Tay-Sachs, Hurler, and related lysosomal diseases. Although the primary metabolic defect(s) in Batten disease remain unknown, recent research has established that, with the exception of infantile disease variants, a substantial portion of the intracellular storage material is a single protein, subunit *c* of mitochondrial ATP synthase.[1] Current findings suggest that this subunit, which is encoded by nuclear DNA, is synthesized correctly and undergoes normal trafficking to mitochondria; however, its subsequent removal from mitochondria and degradation appear to be delayed.[2] Why this particular mitochondrial component accumulates in cells, and whether its accumulation signals the *primary* metabolic defect in Batten disease, are unknown. Nonetheless, its overwhelming abundance may be tied to cell dysfunction. For example, since many normal cortical GABAergic neurons are believed "mitochondria-rich" relative to other types of neurons,[3] it is possible that these cells could as a consequence be more prone to *c* subunit accumulation and more susceptible to dysfunction in Batten disease.[4] In support of this view, substantial loss of GABAergic neurons has been found to occur in cerebral cortex in Batten disease animal models, and mitochondria in some GABAergic neurons (but not other types of neurons) have been shown to exhibit structural abnormalities.[5,6] In addition, histochemical studies show substantial reduction in cytochrome oxidase activity in all types of cortical and cerebellar neurons and this may be indicative of more widespread mitochondrial dysfunction. The effects of suboptimal mitochondrial function in cortical pyramidal neurons and cerebellar Purkinje cells likely would be exacerbated by loss of inhibitory (GABAergic) inputs to these cells and could lead to heightened vulnerability to chronic, glutamate-driven excitotoxicity and cell death. This type of pathogenic process is characteristically "neurodegenerative" and distinctly different than that known to occur in most storage diseases caused by lysosomal hydrolase deficiencies. Massive death of neurons is not a common feature in the latter diseases. A greater understanding of the mechanisms of neuron death and brain dysfunction in Batten disease can be anticipated to give insight not only into primary metabolic defects, but also into possible ways to treat and/or ameliorate clinical symptoms. Furthermore, understanding the causes and consequences of neuronal cell loss in a genetic model of neurodegeneration in animals may provide insight into similar events in a wide variety of human neurodegenerative diseases.

HISTORICAL PERSPECTIVE

The first disorder recognized as what we refer to today under the umbrella term "Batten disease" was described as a clinical entity in the early 19th century in Norway by Christian Stengel.[7] Nearly a century later, a detailed description of brain pathology in this type of disease was reported by Frederick Batten for cases of "progressive cerebral degeneration" that he observed.[8] Batten made a clear distinction between this disorder and cases of amaurotic family idiocy (later known as Tay-Sachs disease) which had been described by Bernard Sachs in a series of reports beginning in the late 1880's.[9,10] In particular he stressed that even though both types of disorders appeared inherited, other features were distinctly different. For example, Batten's cases did not exhibit race proclivity or the exact same spectrum of clinical manifestations as those described by Sachs. However, there were also similarities, including progressive mental decline and blindness, and surviving neurons appeared swollen and vacuolated in both diseases. These similarities proved sufficiently compelling to lead to a common classification with most Batten disease cases being viewed as clinical variants of Tay-Sachs disease. The problem of disease classification changed in 1963 with the first ultrastructural studies of storage disorders. The characteristic appearance of cytoplasmic storage bodies in Batten disease was found to be unlike that of the membranous cytoplasmic bodies described in Tay-Sachs disease.[11,12] This difference resulted in Batten disease being renamed "neuronal ceroid lipofuscinosis" to indicate the accumulation of ceroid/lipofuscin-type lipopigments as the major storage material.[13] In subsequent years Tay-Sachs disease and most other neuronal storage diseases were found to result from defects in specific lysosomal hydrolases and rapid progress in understanding their disease pathogenesis ensued. But Batten disease exhibited no similar lysosomal enzyme deficiency and the primary molecular defect, or defects, remain undetermined. More recently, the discovery that many forms of Batten disease are characterized by accumulation of subunit *c* of mitochondrial ATP-synthase[1] suggested to some that the term "proteolipid proteinosis" be applied,[14] consistent with the possibility that Batten disease represents a degradative defect of this specific molecule. The induction of a similar state of lipofuscin-like storage with the lysosomal protease inhibitor, leupeptin, is consistent with this view.[15] Thus, whether referred to as ceroid lipofuscinoses or proteolipid proteinosis, Batten disease has remained classified within the neuronal storage disease family.

THE CELLULAR PATHOLOGY OF STORAGE DISORDERS

Batten disease differs from Tay-Sachs and many other typical lysosomal hydrolase deficiencies not only in terms of ultrastructural pathology and evidence of primary lysosomal pathology, but also in terms of secondarily-induced cellular pathology. Tay-Sachs and related diseases are known to be characterized by growth of ectopic dendritic membrane on cortical pyramidal neurons.[16] This new membrane occurs as dendritic spine-covered cellular enlargements ("meganeurites") at the axon hillock to facilitate "storage", and/or as growth of new ectopic dendrites at this same region of the cell. In both cases, the ectopic dendritic membrane exhibits new synapses and the resulting changes in connectivity have been suggested to underlie mental retardation. Recent studies indicate that an increase in intracellular GM2 ganglioside, occurring as a direct or indirect consequence of the primary metabolic defect, is associated with the recapitulation of dendrite growth in many types of storage diseases.[17,18] Golgi studies of Batten disease have revealed meganeurites, but these structures consistently lack spines and synapses[6,19] and thus resemble axon hillock enlargements occurring secondary to lipofuscin accumulation in old age rather than the elaborate regrowth of dendritic membrane described above. Pyramidal neurons in canine Batten disease also lack elevations of GM2 ganglioside.[18]

Another cytopathologic feature of neurons in many storage diseases is neuroaxonal dystrophy or axonal spheroid formation.[16] Spheroids are granular in nature and ultrastructurally are found to contain tubulovesicular profiles, dense bodies, mitochondria, and other organelles. The ultrastructural composition of spheroids is similar in all storage diseases, i.e., they do not exhibit the characteristic pathological cytosomes present in neuronal perikarya (which do tend to be different in different types of storage diseases). Although once viewed as an essentially nonspecific reaction to the storage disease process, a more recent study using animal models has indicated that spheroids predominate in neurons using γ-aminobutyric acid (GABA) as a neurotransmitter.[20] Furthermore, this study showed that the incidence and distribution of axonal spheroids in GABAergic neurons closely correlated with the onset and type of clinical disease displayed by these diseases. In terms of Batten disease, studies in animals, using ovine, canine, and murine models, have revealed an absence of significant numbers of axonal spheroids in GABAergic neurons and in other cell types.[6]

A conspicuous feature of Batten disease that is generally absent in most other neuronal storage diseases is massive neuronal cell death and brain atrophy.[21] Many forms of Batten disease exhibit severe, regionally-selective atrophy, with cerebral cortex commonly being most affected. Indeed, the degree of cortical atrophy can rival that of Alzheimer's disease. Ovine and canine models also exhibit massive cortical atrophy whereas cell death in a putative Batten disease model in mice appears limited to spinal cord.[5] Why neurons are dying is unknown and apart from Batten's original emphasis on cerebral degeneration as a key feature of this disease, it has been intracellular storage rather than cell death that has dominated Batten disease research. Some studies of Batten disease in children have suggested that cell death occurs in a selective fashion, with loss of GABAergic neurons and/or GABAergic synapses predominating.[19,22] Studies of canine Batten disease indicate that subsets of GABAergic neurons in cerebral cortex and cerebellum exhibit differential cell loss but that other types of neurons, notably pyramidal and Purkinje cells, also eventually join this neurodegenerative process.[6]

WHY DO NEURONS DIE IN BATTEN DISEASE?

Neuron death in Batten disease is regionally selective, but intracellular storage is characteristic of all neurons. It is also evident that the storage process begins early and can considerably precede the onset of clinical neurological disease. These findings suggest that the storage material in Batten disease may not be directly cytotoxic, or if it is, that there are specific threshold or other factors that play a critical role. It is possible that characteristics of specific populations of neurons render particular cell types vulnerable to the disease process. Current evidence supports the likelihood of a series of steps in a pathologic cascade that ultimately ends in death of neurons (Fig. 1).

Step 1: Several lines of evidence suggest that mitochondrial function may be suboptimal in many forms of Batten disease. Whether this occurs as a direct result of a primary genetic defect is unknown, but the abnormal accumulation of subunit *c* of mitochondrial ATP-synthase (which makes up approximately 50% of the storage material),[1] and delay in removal of this subunit from mitochondria[2] may be central events leading to mitochondrial compromise. Additional findings which may be indicative of mitochondrial involvement in Batten disease include reduced cytochrome oxidase activity in widespread neuronal populations,[6] structurally abnormal mitochondria in subsets of GABAergic neurons,[5,6] ultrastructural studies suggesting close association between storage material and mitochondria,[6,11] and positron emission tomography studies indicating reduced glucose utilization in brain of Batten disease patients.[23]

Step 2: Intracellular storage in many forms of Batten disease consist not only of *c* subunit accumulation, but also of elevated levels of other proteins, oligosaccharides, phospholipids, neutral lipids, and so forth.[1] The apparent heterogeneity of this material led to the original designation "ceroid lipofuscinosis".[21] There is evidence that suboptimal mitochondrial function may be linked to the accumulation of some of these materials. That is, intraneuronal storage of ceroid-lipopigments has been documented in a disease induced by acetyl ethyl tetramethyl tetralin (AETT),[24] a musk fragrance which is known to uncouple oxidative phosphorylation in mitochondria.[25] Lipopigment storage also accompanies vitamin E (α-tocopherol) deficiency, occurring in association with abnormal lipid peroxidation and the presence of structurally abnormal mitochondria in neurons.[26] Abnormal lipid peroxidation,[27] as well as abnormal mitochondria (see above), have also been reported to occur in Batten disease.

Step 3: Neurons with relatively higher metabolic rates and/or larger numbers of mitochondria may exhibit greater vulnerability to compromise of mitochondrial function than other neurons. In cerebral cortex, GABAergic neurons are believed to be mitochondria-rich[3] and studies of human Batten disease have suggested that the preferential loss of cortical GABAergic neurons or their synapses contributes to disease pathogenesis.[19,22] Studies of Batten disease in animal models also have revealed significant GABAergic cell loss, though some subpopulations (e.g., GABAergic cells immunoreactive for parvalbumin) appear less affected.[6] Loss of inhibitory inputs to other types of cells, such as cortical pyramidal neurons, would likely exacerbate disease impact on these cells since firing rates, and consequently metabolic demands, presumably would increase (see Step 4).

Step 4: Reduced oxidative metabolism in neurons receiving substantial inputs via *N*-methyl-D-aspartate (NMDA) receptors may render these cells vulnerable to a slow form of excitotoxicity. Such a mechanism has been proposed in a variety of human neurodegenerative diseases.[28] According to this scenario, reduced ATP production leads to partial cell depolarization and to a cascade of events secondary to removal of the voltage-dependent Mg^{++} block of NMDA channels, followed by calcium entry and, eventually, excitotoxic cell death. The brain region most susceptible to atrophy in Batten disease, the cerebral cortex, is known to contain abundant NMDA receptors.[29] There is also evidence suggesting that loss of inhibitory inputs to neurons, as described in step 3, could exacerbate this excitotoxic process.[30]

Step 5: Neuron death as a result of the above process could ultimately occur via apoptotic mechanisms or necrosis, or both. Some studies suggest that excitotoxicity-induced neuronal death may occur through apoptosis and a preliminary report on neuron death in Batten disease has suggested that neurons may die by this mechanism.[31] Death of neurons in cerebral cortex, particularly pyramidal neurons which are more abundant and represent greater tissue volume, would lead to the characteristic cortical atrophy.

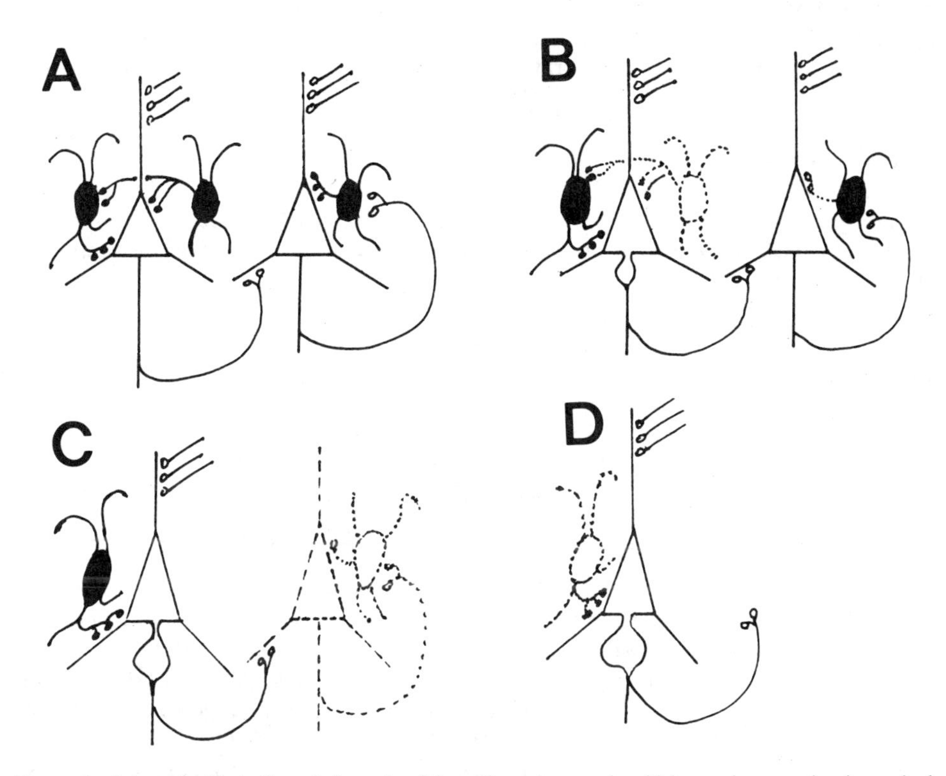

Figure 1. Schematic illustration of elements of the pathogenic cascade which may be occurring in cerebral cortex in Batten disease (see text for details). (A) Neural circuits in the cortex and elsewhere are established in early brain development in normal fashion. (B) Suboptimal mitochondrial function, possibly caused by recycling errors involving subunit *c* of mitochondrial ATP-synthase, may have the greatest initial impact on neurons with more mitochondria (for example, some types of cortical GABAergic cells). This in turn may lead to reduced inhibition of adjacent cells due to loss of GABAergic cells or synapses, and/or to reduced inhibitory cell output. Exact functional consequences may be difficult to predict since loss of inhibition could effect both excitatory and inhibitory cortical neurons. Accumulation of storage material occurs in all neurons, possibly due in part to mitochondrial compromise, and aspiny meganeurites may form at the axon hillock of some pyramidal neurons. (C) Reduced inhibitory drive to pyramidal neurons, coupled with suboptimal mitochondrial function in these cells, could be anticipated to cause increased vulnerability to glutamate-driven excitotoxic mechanisms of cell death. Surviving pyramidal neurons develop larger aspiny meganeurites. (D) Progressive loss of cortical circuits occurs, with cell death possibly involving ensembles of interconnected pyramidal and GABAergic cells. Death of neurons may involve apoptotic mechanisms, or necrosis, or both. (Dark cells/terminals, inhibitory [GABAergic]; open cells/synapses, excitatory; dotted lines indicate dying cells or processes.)

FUTURE DIRECTIONS

Paramount to understanding Batten disease is the determination of the primary genetic defects responsible for this group of disorders. Such a finding could prove immediately insightful in terms of understanding pathogenic mechanisms. But it is also possible that the insights will be more limited, not unlike recent experiences following advances in the molecular genetics of some other inherited neurological disorders. Studies designed to *directly* explore the cellular pathobiology of Batten disease are essential, and given the availability of well documented animal models, such studies are also readily feasible. Knowing the specific disease mechanisms at work in Batten disease may provide insight into events in other types of neurodegenerative diseases and into possible ways to slow or ameliorate clinical symptoms. For example, if mitochondrial compromise and slow excitotoxicity are the basis for cortical cell death in Batten disease, pharmacologically blocking the excitotoxic mechanism,[30], coupled with treatments to minimize other deleterious secondary effects,[32] may slow the loss of neurons, and consequently, the neurological decline.

ACKNOWLEDGEMENTS

We are grateful to Drs. R. Jolly, P. March, and A. Siakotos for discussion and to M. Huang and S. Wurzelmann for expert technical assistance. This work was supported by the NIH (NS 30163).

REFERENCES

1. D.N. Palmer, I.M. Fearnley, S.M. Medd, J.E. Walker, R.D. Martinus, S.L. Bayliss, N.A. Hall, B.D. Lake, L.S. Wolfe, and R.D. Jolly, Lysosomal storage of the DCCD reactive proteolipid subunit of mitochondrial ATP synthase in human and ovine ceroid lipofuscinosis, *in* "Lipofuscin and Ceroid Lipopigments," E.A. Porta, ed., Plenum, New York (1990).
2. J. Ezaki, L.S. Wolfe, T. Higuti, K. Ishidoh, and E. Kominami, Specific delay of degradation of mitochondrial ATP synthase subunit c in late infantile neuronal ceroid lipofuscinosis (Batten disease), *J. Neurochem.* 64:733 (1995).
3. C.R. Houser, J.E. Vaughn, S.H.C. Hendry, E.G. Jones, and A. Peters, GABA neurons in the cerebral cortex, *in* "Cerebral Cortex: Functional Properties of Cortical Cells," E.G. Jones and A. Peter, eds., Plenum, New York (1984).
4. S.U. Walkley and P. March, Biology of neuronal dysfunction in storage disorders, *J. Inher. Metabol. Dis.* 16:284 (1993).
5. S.U. Walkley, P.A. March, C.E. Schroeder, S. Wurzelmann, and R.D. Jolly, Pathogenesis of brain dysfunction in Batten disease, *Am. J. Med. Gen.* 56: (In press).
6. P.A. March, S.U. Walkley, and S. Wurzelmann, Morphological alterations in neocortical and cerebellar GABAergic neurons in a canine model of juvenile Batten's disease. *Am. J. Med. Gen.* 56: (In press).
7. C. Stengel, Account of a singular illness among four siblings in the vicinity of Roraas, *in* "Ceroid Lipofuscinosis (Batten's Disease), D. Armstrong, N. Koppang, and J.A. Rider, eds., Elsevier Biomedical Press, New York (1982).
8. F.E. Batten, Family cerebral degeneration with macular change (so-called juvenile form of family amaurotic idiocy). *Quart. J. Med.* 7:444-453 (1914).

9. B. Sachs, On arrested cerebral development with special reference to its cortical pathology, *J. Nerv. Ment. Dis.* 14:541 (1887).
10. B. Sachs and I. Strauss, The cell changes in amaurotic family idiocy, *J. Exp. Med.* 12:685 (1910).
11. W. Zeman and S. Donahue, Fine structure of the lipid bodies in juvenile amaurotic idiocy. *Acta Neuropath.* 3:144 (1963).
12. R. Terry and M. Weiss, Studies in Tay-Sachs disease II. Ultrastructure of the cerebrum. *J. Neuropathol. Exper. Neurol.* 22:18 (1963).
13. W. Zeman, and P. Dyken, Neuronal ceroid lipofuscinosis (Batten's disease): Relationship to amaurotic family idiocy? *Pediatrics* 44:570 (1969).
14. R.D. Jolly, A. Shimada, I. Dopfmer, P.M. Slack, M.J. Birtles, and D.N. Palmer, Ceroid-lipofuscinosis (Batten's disease): Pathogenesis and sequential neuropathological changes in the ovine model, *Neuropathol. Appl. Neurobiol.* 15:371 (1989).
15. G.O. Ivy, F. Schottler, J. Wenzel, M. Baudry, and G. Lynch, Inhibitors of lysosomal enzymes: Accumulation of lipofuscin-like dense bodies in the brain, *Science* 226:985 (1984).
16. S.U. Walkley, Pathobiology of neuronal storage disease. *Intern. Rev. Neurobiol.* 29:191 (1988).
17. D.A. Siegel and S.U. Walkley, Growth of ectopic dendrites on cortical pyramidal neurons in neuronal storage diseases correlates with abnormal accumulation of GM2 ganglioside. *J. Neurochem.* 62:1852 (1994).
18. S.U. Walkley, Pyramidal neurons with ectopic dendrites in storage diseases exhibit increased GM2 ganglioside-immunoreactivity. *Neuroscience* (In press).
19. R.S. Williams, I.T. Lott, R.J. Ferrante and V.S. Caviness, The cellular pathology of neuronal ceroid lipofuscinosis, *Arch. Neurol.* 34:298 (1977).
20. S.U. Walkley, H.J. Baker, M.C. Rattazzi, M.E. Haskins and J.-Y. Wu, Neuroaxonal dystrophy in neuronal storage disorders: Evidence for major GABAergic neuron involvement, *J. Neurol. Sci.* 104:1 (1991).
21. W. Zeman, The ceroid lipofuscinoses, *in* H.M. Zimmerman, ed., "Progress in Neuropathology," Grune & Straton, New York (1976).
22. H. Braak and H.H. Goebel, Pigmentoarchitectonic pathology of the isocortex in juvenile neuronal ceroid lipofuscinosis: Axonal enlargements in layer IIIab and cell loss in layer V, *Acta Neuropathol.* 46:79 (1979).
23. M. Philippart, C. Messa, and H.T. Chugani, Spielmeyer-Vogt (Batten, Spielmeyer-Sjogren) disease. Distinctive patterns of cerebral glucose utilization, *Brain* 117:1085 (1994).
24. P.S. Spencer, A.B. Sterman, D. Horoupian, and M.M. Foulds, Neurotoxic fragrance produces ceroid and myelin disease, *Science* 204:633 (1979).
25. W. Cammer, Uncoupling of oxidative phosphorylation *in vitro* by the neurotoxic fragrance compound acetyl ethyl tetramethyl tetralin and its putative metabolite, *Biochem. Pharm.* 29:1531.
26. T. Miyagishi, N. Takahata, and R. Iizuka, Electron microscopic studies on the lipo-pigments in the cerebral cortex nerve cells of senile and vitamin E deficient rats. *Acta Neuropathol.* 9:7 (1967).
27. J.A. Rider, G. Dawson, and A. Siakotos, Perspective of biochemical research in the neuronal ceroid-lipofuscinosis, *Am. J. Med. Gen.* 42:519 (1992).
28. M.F. Beal, Does impairment of energy metabolism result in excitotoxic neuronal death in neurodegenerative disease? *Ann. Neurol.* 31:119 (1992).

29. C.W. Cotman, D.T. Monaghan, O.P. Ottersen, and J. Storm-Mathisen, Anatomical organization of excitatory amino acid receptors and their pathways, *TINS* 10:273 (1987).
30. R. Horowski, H. Wachtel, L. Turski, and P.-A. Löschmann, Glutamate excitotoxicity as a possible pathogenic mechanism in chronic neurodegeneration, *in* "Neurodegenerative Diseases," D.B. Calne, ed., W.B. Saunders, Philadelphia (1994).
31. R.N. Boustany, S.C. Lane, and W. Quin, Apoptosis is the mechanism of neuronal cell death and retinal degeneration in Batten disease, *Ann. Neurol.* 36:496 (1994).
32. E.D. Hall, Novel Inhibitors of iron-dependent lipid peroxidation for neurodegenerative disorders. *Ann. Neurol.* 32:S137 (1992).

ENDOGENOUS AMINO ACID PROFILE DURING *IN VITRO* DIFFERENTIATION OF NEURAL STEM CELLS

Eulalia Bazán, Miguel A. López-Toledano, Maria A. Mena, Rafael Martín del Rio, Carlos L. Paíno and Antonio S. Herranz

Dpto. de Investigación-Neurobiología, Hospital Ramón y Cajal, INSALUD, 28034-Madrid, SPAIN

INTRODUCTION

Cell replacement therapy is becoming an alternative approach for the treatment of neurodegenerative diseases. Nevertheless, this approach is limited by the availability of donor tissue and by ethical constraints. Cell culture techniques could provide an unlimited source of cells for transplantation.

Multipotential progenitor cells can be obtained from embryonic and adult murine Central Nervous System (CNS)[1,2]. In culture, these stem cells grow in supension as spheric aggregates (neurospheres) in defined medium supplemented with epidermal growth factor (EGF). Plating them on adhesive substrata induces phenotypic differentiation into either neurons or glial cells that are potentially useful as a source of neural cells for transplantation[3] .

Endogenous amino acids (AAs) in the CNS can play a role as neurotransmitters, metabolic precursors or even as trophic or differentiating agents. The role of γ-aminobutyric acid (GABA) and glutamate (Glu) as trophic factors has been described for several brain regions,[4,5] and the importance of taurine (Tau) in developing brain is well known [6].

The aim of the present study has been to characterize biochemically the process of striatal stem cell differentiation *in vitro*, studying the content of endogenous AAs and its correlation with the presence of the different cell types in the culture.

METHODS

Primary Cultures

Striata from E15 rat embryos were dissected-out and mechanically dissociated. Cell suspensions were grown in a defined medium (that we call DF12), composed of Dulbeco's modified Eagle's medium and Ham's F-12 (1:1), L-glutamine (2 mM), sodium piruvate (1 mM) (all from GIBCO), glucose (0.6%) (from Sigma), N2 components (insulin 25 μg/ml, transferrin 100 μg/ml, progesterone 20 nM, putrescine 60 μM, and sodium selenite 30 nM (all from Sigma except transferrin from Boehringer-Mannheim) and 20 ng/ml EGF (Boehringer-Mannheim). After a minimum of five passages, cells were plated at a density of 38000 cells/cm^2 on poly-l-ornithine coated glass coverslips (Ø 12 mm) or plastic dishes (Ø 35 mm). Cultures were maintained in DF12 for 72 hours and then switched to DF12 without EGF for longer culture periods. Immunocytochemical and biochemical analysis were performed at 0.08 (2 hours), 1, 3, 6, 10 and 20 days post plating (dpp).

Immunocytochemistry

Cells on coverslips were fixed in 4% paraformaldehide for 10 min and post-fixed and permeabilized in ethanol-acetic (19:1) (10 min) at -20 °C. Dual immunolabelling was performed by incubating overnight at 4 °C in a mixture of mouse anti β-tubulin isotype III (1:80 from Sigma) and either anti glial fibrillary protein (GFAP) (1:500 from Serva) or nestin antiserum 130 (1:1000 kindly gifted by Dr. R. McKay and Dr.M. Marvin) made in rabbit. Secondary antibodies (Anti-mouse IgG-FITC 1:40 from Boehringer-Mannheim or Anti-rabbit IgG-LRSC 1:200 from Jackson) were incubated for 30 min at 24°C. Coverslips were then washed three times in PBS and mounted in medium containing p-phenylenediamine and bisBenzimide (Hoechst 33342, Sigma). Fluorescence was detected and photographed with a Zeiss Jena photomicroscope.

Amino Acid Analysis

Cells on plastic dishes were washed three times in 2 ml of PBS and scraped at 4 °C in 250 μl of 0.2 N perchloric acid - 0.2 mM ethylenediaminetetraacetic acid (EDTA) - 0.5 mM sodium meta-bisulfate ($Na_2S_2O_5$). Cell suspensions were sonicated three times for 15 sec and centrifuged at 7200 g x 5 min. The supernatants were frozen at -80 °C until they were analyzed. The pellets were used for protein evaluation following the method described by Stauffer [7].

The samples were analyzed for amino acid content using the reverse-phase high performance liquid chromatography (HPLC) procedure with O-phthaldialdehide (OPA) precolumn derivatization. This method coincides essentially with that described by Jones et al.,[8] except that 3-mercaptopropionic acid was used instead of 2-mercaptoethanol in the derivatization reaction. This change improved the sensitivity of the analysis as much as two-fold (Herranz, A.S., personal communication).

RESULTS AND DISCUSSION

Immunocytochemistry

Plating neurospheres on an adhesive substrata induces phenotypic differentiation (Fig.1). At 2 hours post-plating most of the cells were nestin$^+$ (Fig.2 a, b). Nestin is an intermediate filament abundantly expressed in neuroephitelial stem cells early in embryogenesis, and extinguished in almost all mature CNS cells[9]. Nestin$^+$ cells were still abundant (Fig. 2 c) at 3 dpp, but some became β-tubulin III$^+$ (Fig. 2 d), a specific marker for neurons. At 10 dpp many cells were β-tubulin III$^+$ (Fig.2 e), extending long processes and showing neuronal morphology. At that time, many cells were reactive to GFAP, indicating the presence of astrocytes (Fig.2 f). By 20 dpp most of the cells were GFAP$^+$ (Fig.2 g) and the neurons showed a complex branching pattern (Fig.2, h). Between 10 and 20 dpp there was an evident loss of neurons in the cultures, probably due to their acquired dependence on some trophic factors. Observations in our laboratory suggest that the *in vitro* differentiation pattern is slower after 10 passages (data not shown).

Endogenous Free Amino Acids

The pattern of endogenous free AAs during *in vitro* stem cell differentiation was qualitatively similar for many of the AAs analyzed (Fig. 3). All the AAs, with the exception of methionine (Met) (Fig. 5), were present at 2 h post plating, showing a sharp decrease at 1 dpp. At 3 and 6 dpp many of them recovered their initial values (Fig. 3). The maximal increase was found at 10 dpp. By 20 dpp their levels either remained constant or decreased (Fig. 3), with the exception of glutamine (Gln) (Fig. 4).

It is noteworthy the presence of GABA and Tau, two AAs not supplied in the feeding medium. The pattern of these two AAs was different from the one described above (Fig. 4). GABA was present at 2 h post plating at very low concentration (1.31

nmol/mg prot), and was not detected again until 10 dpp (Fig. 4), when many neurons were present (Fig. 2 e). The role of GABA as a trophic neurotransmitter during brain development is well established,[4] and Reynolds et al[1,2] have shown immunoreactivity for GABA in neurons derived from stem cells after 25 days post plating. Tau was maximal after 10 dpp and decreased to very low values at 20 dpp (Fig. 4). This pattern of expression is similar to that observed during brain development, showing maximal levels in fetuses and decreasing in adult brain[6]. Our results also suggest that neuroepithelial stem cells have the machinery required to synthesize GABA and Tau.

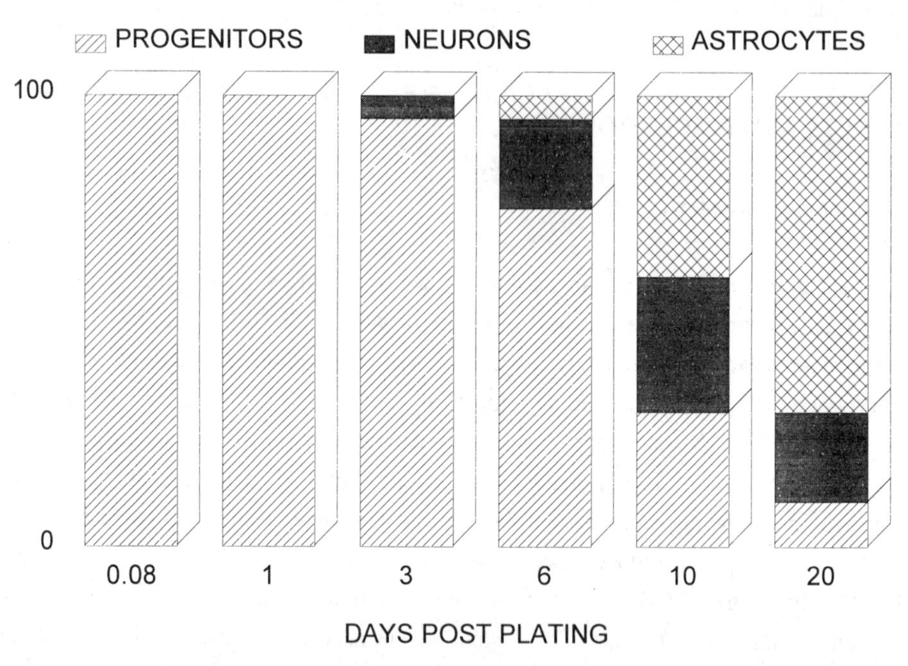

Figure 1. Schematic diagram of the cell composition of cultures at different times post plating. For clarity, only progenitors, neurons and astrocytes are represented (radial glia is included in astrocytes; oligodendrocytes have not been estimated yet; the existence of other CNS derivatives is unknown). Some neurons are already present at 3 dpp. The neuronal population is maximal at the 10 dpp point. Astrocytes constitute the main cell type at 20 dpp.

Glu reached a maximal value after 10 dpp (Fig. 4). At 20 dpp its value decreased to similar levels to those obtained at 2 h, 3 and 6 dpp (Fig. 4). Interestingly, the maximal peak of Glu at 10 dpp correlates with a moment in which Gln shows a sharp decrease (Fig. 4). Since Gln is a key precursor for Glu acting as a neurotransmitter,[10] the reciprocity observed between these two AAs at 10 and 20 dpp might be reflecting the existence of some glutamatergic neurons in our cultures. Immunocytochemical studies in similar preparations,[1,2] have failed to find immunoreactivity for Glu at 25 dpp. Since whave detected a decrease in Glu and an increase in Gln between 10 and 20 dpp the

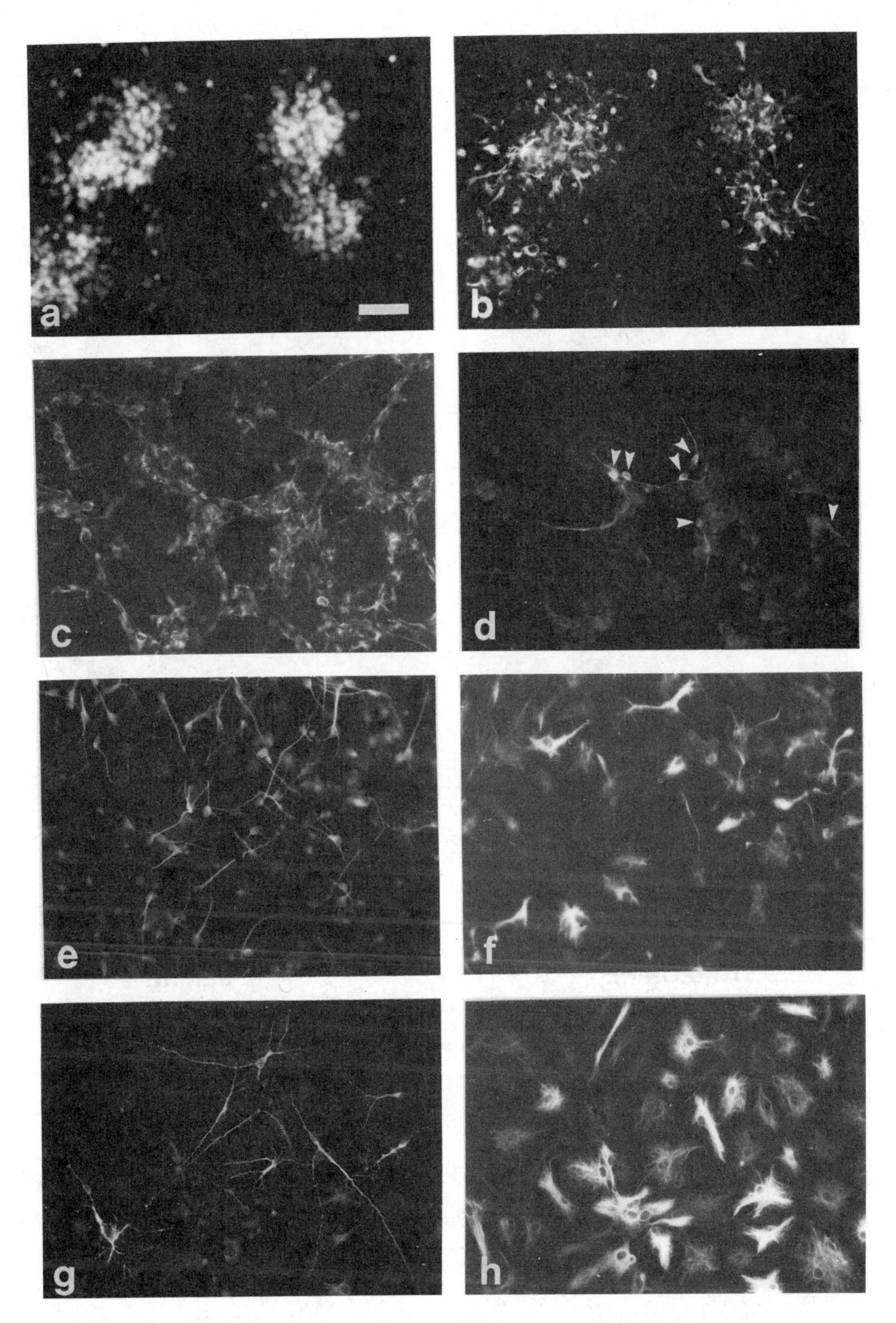

Figure 2. a) Neurospheres 2 h post plating stained with bisBenzimide to show all cell nuclei . b) Same field as *a* stained for nestin. Note that all cells are nestin^{+}. c) Nestin and d) β-tubulin III immunolabelling at 3 dpp; a few cells in the culture start to express neuronal phenotype (arrowheads). e) β-tubulin III and f) GFAP at 10 dpp. g) β-tubulin III and h) GFAP at 20 dpp. Note the developed morphology of both neurons and astrocytes at this moment. Scale bar, 50 μm.

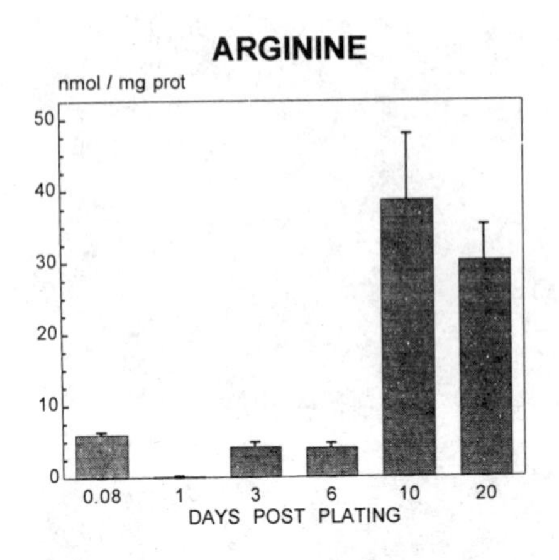

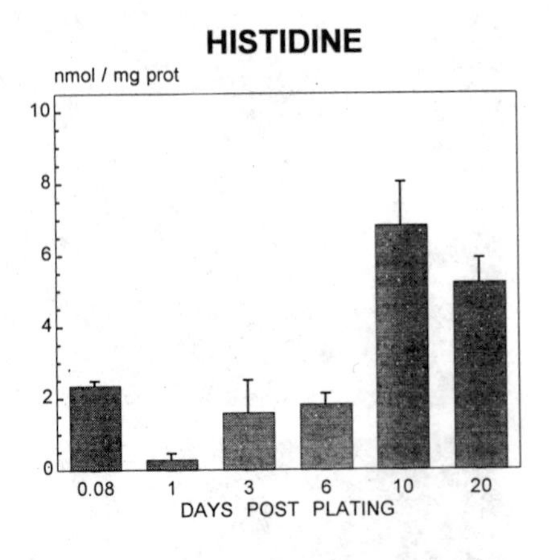

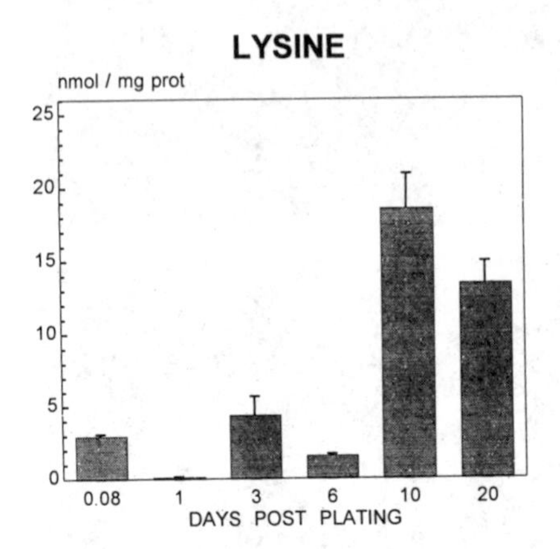

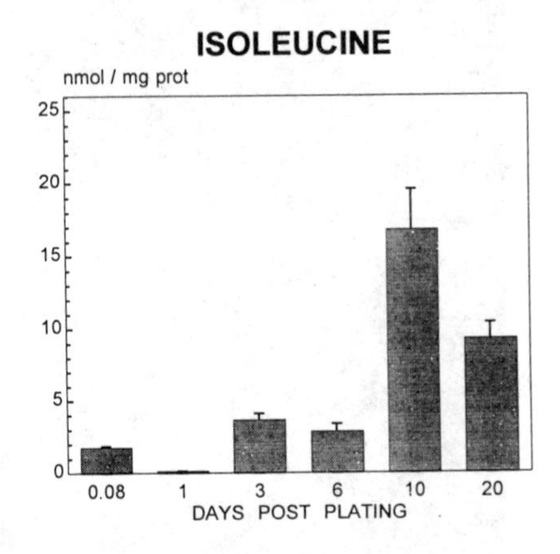

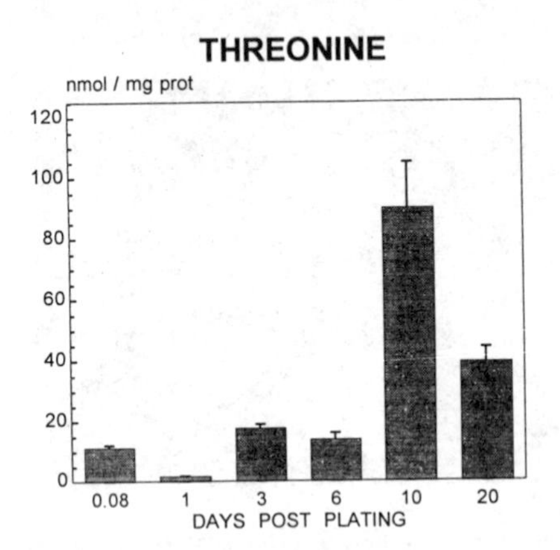

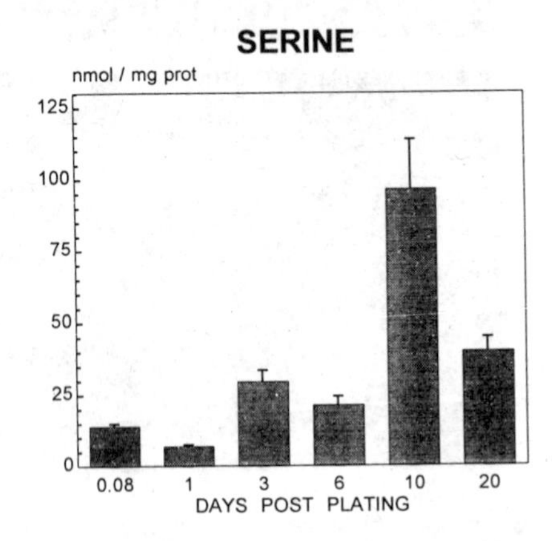

Figure 3. Profile of endogenous amino acids during stem cell differentiation. Histograms represent the mean $\pm$ SEM of 3 different plates obtained from neurospheres after 6 passages. These amino acids follow the general profile described in the text, although their range of concentrations is different.

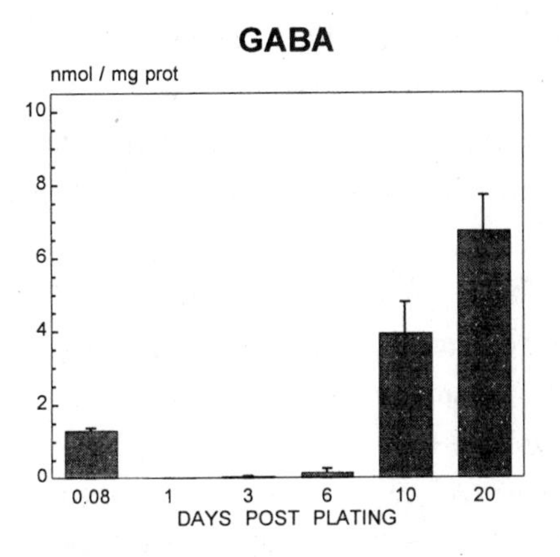

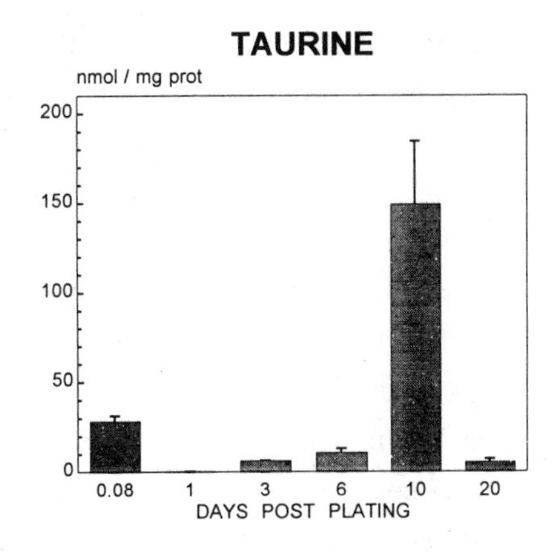

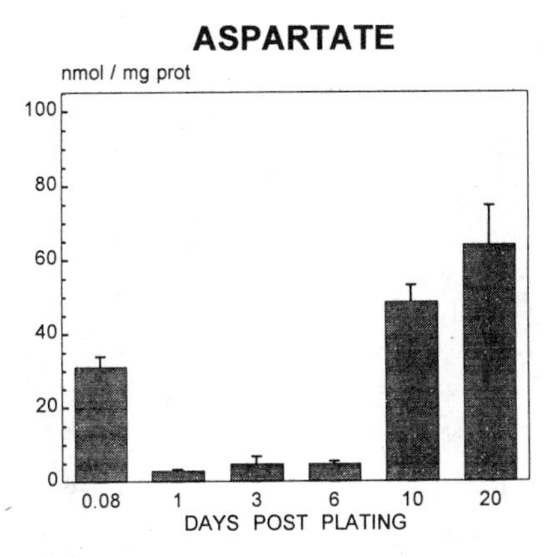

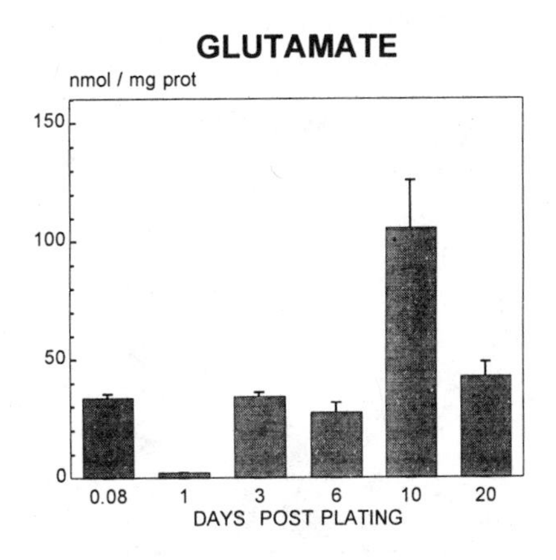

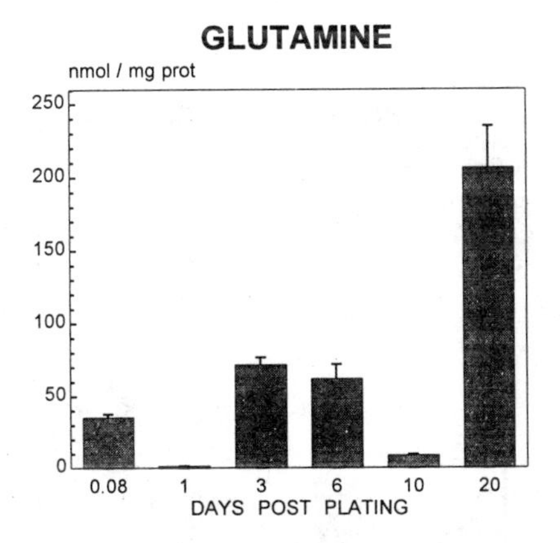

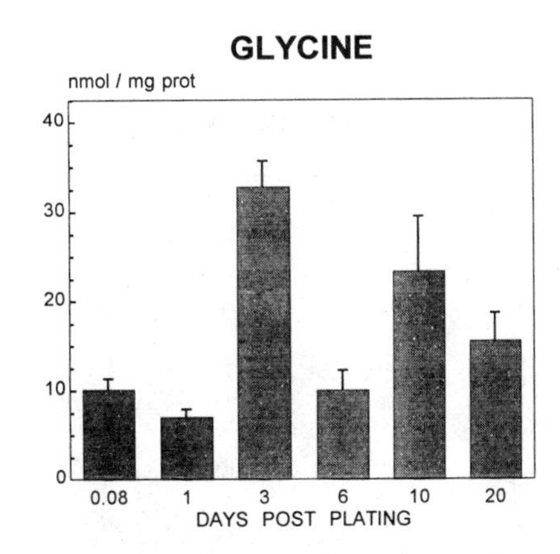

Figure 4. Pattern of endogenous content of 6 amino acids with a putative role as neurotransmitters during stem cell differentiation. Histograms represent the mean ± SEM of 3 different plates obtained from neurospheres after 6 passages.

absence of labeling may be due to the loss of glutamatergic neurons during this period.

Glycine (Gly) showed a particular pattern, presenting two maximal peaks at 3 and 10 dpp and not having any significant decrease at 1 dpp (Fig. 4). The characteristic peak at 3 dpp may indicate an important role for Gly in the initial steps of neuronal differentiation.

Tyrosine (Tyr) presented a pattern similar to that observed for most of the AAs analyzed (Fig. 5). Tyr is the precursor of cathecholamines like dopamine or norepinephrine. However, no levels of these neurotransmitters were detected at 6 or 10 dpp, and no immunoreactivity was observed for tyrosine hydroxilase (TH) at 6, 10 or 20 dpp (data not shown). Trytophan (Trp), an essential AA, was not detectable at any time studied. The absence of Trp (AA supplied in the feeding medium) could be explained by its quick incorporation into proteins[11,12]. A similar mechanism could explain the absence of Met before 10 dpp (Fig. 5).

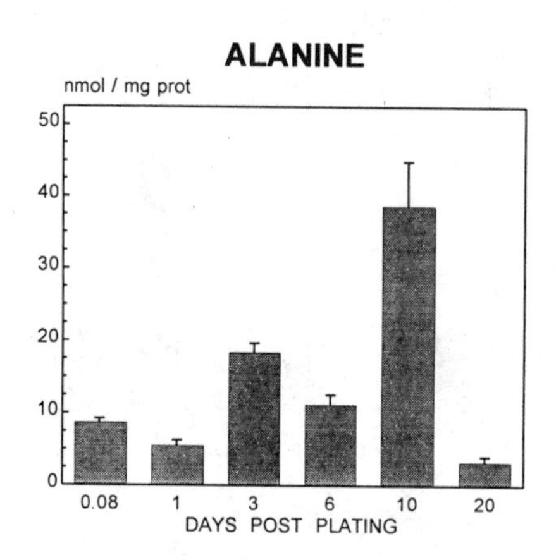

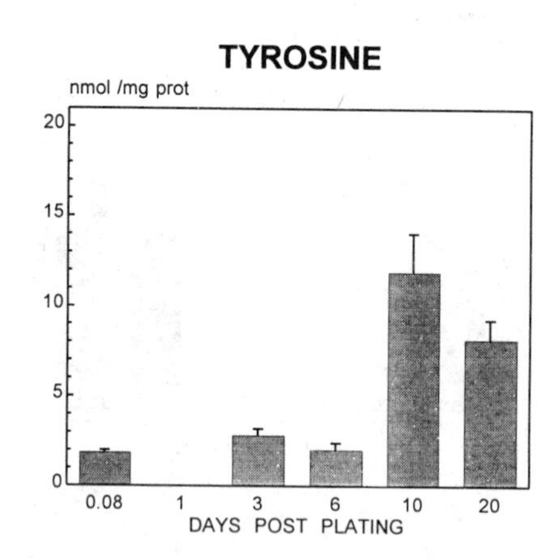

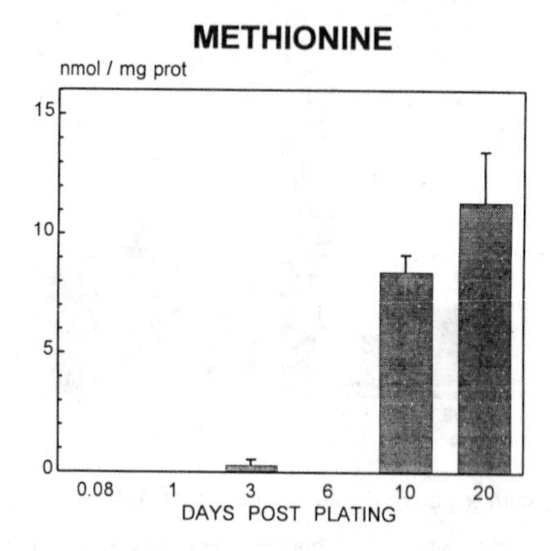

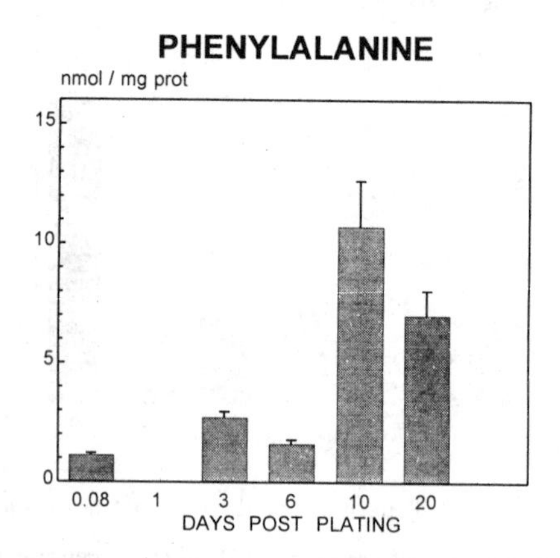

Figure 5. Pattern of endogenous content of 4 amino acids with a putative role as precursors of neurotransmitters (Tyr, Phe) or involved in protein synthesis (Met) during stem cell differentiation. Histograms represent the mean ± SEM of 3 different plates obtained from neurospheres after 6 passages.

CONCLUDING REMARKS

Our results show that *in vitro* differentiation of neural stem cells into neurons or astrocytes is accompanied by important changes in the content of the endogenous free AAs. In general, the pattern of accumulation for most of the AAs analyzed correlates well with their described role as metabolic precursors, trophic factors or neurotransmitters.

Not all AAs show the same profile of endogenous accumulation during the differentiation process. The exceptions to the general pattern may indicate a role for those AAs in development. For instance, Gly is the only AA with a peak of accumulation at 3 dpp, a time when the neuronal phenotype starts to appear.

Stem cells and some of their progeny *in vitro* synthetize Tau and GABA, two AAs needed for the correct development of the CNS.

The finding of molecules, like AAs, involved in the process of differentiation of the CNS provides us with a tool to manipulate this process *in vitro*, and then to obtain cells suitable for transplantation in neurodegenerative diseases.

ACKNOWLEDGMENTS

This research was partially supported by FIS Grants number 93/555 and 94/0494 from Ministerio de Sanidad y Consumo, SPAIN.

REFERENCES

1. B.A. Reynolds, W. Tetzlaff, and S. Weiss, A multipotent EGF responsive striatal embryonic progenitor cell produces neurons and astrocytes, J. Neurosci. 12: 4565-4574 (1992).
2. B.A. Reynolds and S. Weiss, Generation of neurons and astrocytes from isolated cells of the adult mammalian central nervous system, Science 255: 1707-1710 (1992).
3. F.H. Gage, J. Ray, and L.J. Fisher, Isolation characterization, and use of stem cells from CNS, Ann. Rev. Neurosci. 18: 159-92 (1995).
4. G. Barbin, H. Pollard, J.L. Gaïarsa, and Y. Ben-Ari, Involvement of $GABA_A$ receptors in the outgrowth of cultured hippocampal neurons, Neuroscience Letters 152: 150-154 (1993).
5. J.M. Lauder, Neurotransmitters as growth regulatory signals: role of receptors and second messengers, TINS 16 (6): 233-240 (1993).
6. J. A. Sturman, Taurine in development, Physiological Reviews 73 (1) : 119-147 (1993).
7. C.E. Stauffer, A linear standard curve for the Folin Lowry determination of protein, Analytical Biochemistry 69: 646-648 (1975).

8. B.N. Jones, S. Paabo, and S. Stein, Amino acid analysis and enzymatic sequence determination of peptides by an improved O-phthaldialdehide precolumn labeling procedure, J. Liquid Chrmatogr. 4: 565-586 (1981).
9. T. Tohyama, V.M.-Y. Lee, L. B. Rorke, M. Marvin, R.D.G. McKay, and J.Q. Trojanowski, Nestin expression in embryonic human neuroepithelium and in human neuroepithelial tumor cells, Laboratory Investigation 66 (3): 303-313 (1992).
10. A. Hamberger, G. Chiang, E. Nylén, S.W. Scheff, and C.W. Cotman, Stimulus evoked increase in the biosynthesis of the putative neurotransmitter glutamate in the hippocampus, Brain Research 143: 549-555 (1978).
11. J.L. Fando, F. Domínguez and E. Herrera, Tryptophan overload in the pregnant rat: effect on brain amino acid levels and in vitro protein synthesis, J. Neurochem. 37: 824-829 (1981).
12. E.M. Tyobeka and K.L. Manchester, Control of cell-free protein synthesis by amino acids: effects on tRNA charging, Int. J. Biochem. 17: 873-877 (1985).

THE PROTECTIVE ACTION OF NITRONE-BASED FREE RADICAL TRAPS IN NEURODEGENERATIVE DISEASES

Robert A. Floyd

Oklahoma Medical Research Foundation
825 N.E. 13th Street
Oklahoma City, OK 73104

INTRODUCTION

Nitrone-based free radical traps (NRTs) have been shown to protect in several experimental animal models of neurodegenerative diseases and to mitigate other parameters associated with the aging process in brain. Since the prototype NRTs were first utilized in analytical chemistry to trap and characterize the presence of free radicals in chemical and biochemical systems, it is the simplest hypothesis that they protect by trapping crucial free radicals involved in the oxidative damage that occurs in animal models of neurodegenerative diseases. This remains to be rigorously proven. Pertinent data supporting this concept are presented in a framework of the knowledge accrued from studies on oxidative damage in brain.

OXIDATIVE STRESS AND OXIDATIVE DAMAGE

Early studies on oxidative damage to biological systems were focused to a large extent on membrane lipid peroxidation. The discovery of superoxide dismutase in 1969[1] clearly ushered in a more rapid pace of research in this area and fueled the conviction of its underlying importance. There is now overwhelming evidence that oxidative damage is very central to many pathophysiological processes and plays an important role in the etiology of neurodegenerative diseases.[2-5] Figure 1 presents the general concept of oxidative stress. Thus, oxidative damage potential (P_0) is caused by the presence of the continuous flux of semi-reduced oxygen species formed as by-products of aerobic metabolism as well as the presence of the oxidatively damaged products formed. The oxidative damage potential is opposed by and therefore held in a dynamic metastable equilibrium state by antioxidants and antioxidant enzymes acting as the antioxidant defense capacity (A_c) of the system. The semi-reduced oxygen species formed and the reactive oxygen species involved are also noted. Under normal conditions the oxidative damage potential is slightly greater than the antioxidant defense capacity such that a small but significant amount of reactive oxygen species and oxidized products are present at all times thus exerting oxidative stress on the biological system.

Neurodegenerative Diseases
Edited by Gary Fiskum, Plenum Press, New York, 1996

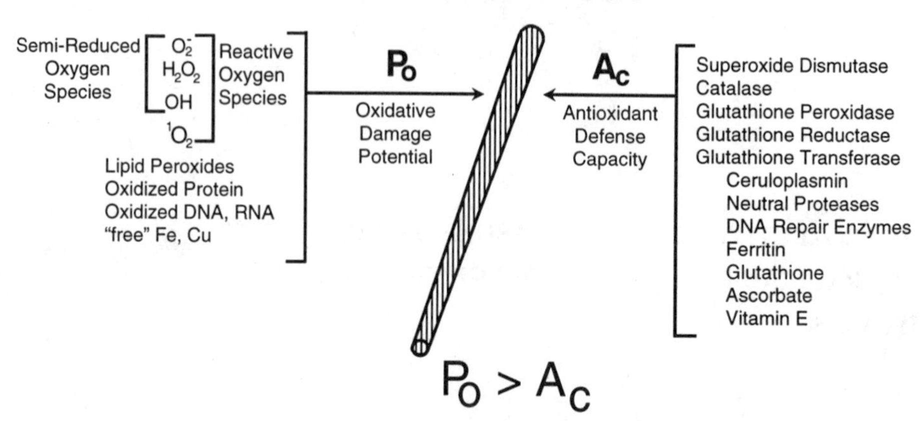

Figure 1. Diagram illustrating the general concept of oxidative stress.

BRAIN IS SUSCEPTIBLE TO OXIDATIVE DAMAGE

Early experiments with brain homogenate showed that it readily peroxidizes when incubated in a phosphate buffer at 25–37°C.[6] In contrast, most other tissue homogenates do not peroxidize under these conditions. It was also noted that brain homogenate did not peroxidize at or near ice temperature and that iron chelators such as desferrioxamine effectively prevented peroxidation[6] thus establishing the importance of "free" iron in brain peroxidation. Ascorbate, which is present at high levels (up to 4 mM) in brain, combined with very low levels (i.e., catalytic levels) of "free" iron readily mediates brain peroxidative damage. As a direct demonstration of the importance of these conditions, we showed that ascorbate plus ADP-ligated Fe, at a Ki of 1.5 μM Fe, caused a dramatic loss in the ability of the striatum P_2 fraction to synthesize dopamine from tyrosine.[7] Ascorbate/Fe also caused peroxidative damage to the striatum P_2 fraction membranes.[7]

Experiments such as those described above plus many other confirmatory observations have established the general notion that brain is highly susceptible to oxidative damage.[3] The established facts supporting this notion are presented in Table 1. Experiments carried out with *in vivo* experimental neurodegenerative disease models also strongly support this concept.

REACTIVE OXYGEN SPECIES AND THEIR QUANTITATION

The reactive oxygen species normally present in biological systems are superoxide ($O_2^{\dot{-}}$), hydrogen peroxide (H_2O_2), and hydroxyl free radicals ($\dot{O}H$). The metal-catalyzed Haber-Weiss reaction, which is the sum of the two reactions shown below, is considered to be the pertinent one that occurs in living systems, i.e.:

$$O_2^{\dot{-}} + Fe^{3+} \longrightarrow Fe^{2+} + O_2$$
$$H_2O_2 + Fe^{2+} \longrightarrow Fe^{3+} + \dot{O}H + OH^-$$

SUM $$H_2O_2 + O_2^{\dot{-}} \longrightarrow \dot{O}H + OH^- + O_2.$$

Table 1. Reasons why brain is sensitive to oxidative damage

- Contains high levels of easily peroxidized unsaturated (20:4, 22:6) membrane fatty acids
- Utilizes large amounts of oxygen per unit weight
- Contains high levels of iron in certain regions and has generally high levels of ascorbate, the combination of iron/ascorbate mediates peroxidation if tissue disorganization occurs
- Contains only modest levels of antioxidants (vitamin E) and antioxidant enzymes and low levels of catalase
- Neurons are post-mitotic cells

The equations as written indicate solution phase reactions; however, it is likely that some, if not most, of the reactive metals, either Fe or Cu, that catalyze the reactions may reside ligated on a biological macromolecule surface or as a low molecular weight ligand. Therefore, the $\dot{O}H$ formed, which reacts at diffusion-limited rates with almost any biological molecule, may attack chemical bonds on the biological surface that complexes with it. This notion of site specificity is the underlying reason which undoubtedly causes much of the oxidative damage that occurs to nucleic acids and proteins.

Expected levels of the reactive oxygen species under normal conditions, depending on the tissue, range from 10 nM for H_2O_2, probably 10^{-11} M for $O_2^{\dot{-}}$, and extremely low levels, perhaps to as low as 10^{-15} M, for $\dot{O}H$. These extremely low levels of the reactive oxygen species is caused in part by the fact that they are so reactive. Thus, they react and hence disappear nearly as fast as they are formed resulting in very low instantaneous levels of the actual reactive species present at any specific time. The extremely low levels of the reactive oxygen species makes it impossible by any present spectroscopic method to directly "visualize" them in biological systems. For this reason, we began, over 20 years ago, two approaches to overcome these limitations. The two approaches are: (A) utilization of exogenous traps and (B) quantitation of distinctive oxidation products formed by oxidative damage to biological molecules. Approach A makes use of the following reaction:

$$\text{trap} + \text{reactive oxygen species} \longrightarrow \text{oxidized trap product}$$

The desirable features needed in a trap include: (i) that it will react rapidly with a specific reactive oxygen species and form a stable oxidized trap product and (ii) that the trap as well as the product does not significantly alter the biological system. In approach B, quantitation of unique oxidized products of proteins, nucleic acids, and lipids have proven very useful such as the extremely sensitive quantitation of 8-hydroxy-2′-deoxyguanosine,[8] an oxidation product of DNA.

Two exogenous traps which we have focused on include salicylate and α-phenyl-*tert*-butyl nitrone (PBN). Salicylate reacts at diffusion-limited rates with $\dot{O}H$ to form two stable products (2,5-dihydroxy benzoic acid and 2,3-dihydroxy benzoic acid) which, along with salicylate, are easily quantitated at very high sensitivity using HPLC and tandem-in line-electrochemical and fluorescence detection methodology.[9] The fact that salicylate rapidly penetrates into all tissues and appears to have very little biological effect, except at high levels, makes this a very effective *in vivo* $\dot{O}H$ trap that is now very widely used. PBN has been utilized as one of a series of nitrones termed spin-traps,[10] but we will refer to this class of compounds as nitrone-based free radical traps or NRTs. The following equation illustrates the reaction of PBN with an unstable (reactive) free radical ($\dot{R}$) to form an NRT-radical adduct. The adduct is a nitroxyl-free radical and long-lived (i.e., relatively stable.)

$$\text{PBN} + R\cdot \longrightarrow \text{NRT-Radical Adduct (Stable)}$$

PBN: $C_6H_5-CH=\overset{+}{N}(O^-)-C(CH_3)_3$ + R• (Free Radical (Reactive)) ⟶ $C_6H_5-CH(R)-N(O\cdot)-C(CH_3)_3$ NRT-Radical Adduct (Stable)

Thus, in effect a reactive-free radical has been trapped by the NRT to form a stable (i.e., unreactive) free radical. In principle, the nitroxyl adduct yields a unique electron spin resonance (ESR) spectrum. The relative stability and the uniqueness of the ESR spectrum has historically made the NRTs useful to confirm, identify, and characterize free radicals in chemical and biochemical reactions.

RESULTS IN EXPERIMENTAL MODELS OF STROKE

Some of the first attempts to evaluate the use of salicylate as an *in vivo* trap for $\dot{O}H$ was done with the Mongolian gerbil global brain ischemia/reperfusion (stroke) model.[11] This animal model has considerable advantages, primarily the fact that ligation of the common carotid arteries prevents blood flow into the forebrain region but does not alter interruption of blood flow in the cerebellum or brain stem. This fact thus provides region specificity to critically assess the validity of the results within the same brain. Table 2 summarizes the facts established utilizing the gerbil model. As noted, an ischemia/reperfusion insult (IRI) to brain caused increased salicylate hydroxylation but only if oxygen was allowed to reenter the anoxic brain. It was also found that hippocampus, the region most severely effected by an IRI, had the highest levels of salicylate hydroxylation as well as protein oxidation. If the ischemia lasts only 2 min before reperfusion, the results showed no increase in salicylate hydroxylation. This protocol also does not cause damage to brain. The activity of glutamine synthetase (GS) drops colinearly with the extent of severity of the IRI protocol. On a related note, it has been shown in the same model that glutamate levels increase after an IRI. This implicates that the loss of GS activity results in a decrease in conversion of glutamate to glutamine thus allowing glutamate, which is toxic to neurons, to accumulate.

PBN was first shown to be neuro-protective in the gerbil stroke model.[2] Thus, if PBN is given either before or a short time (1–2 hrs) after an IRI, it has protective action in both preventing lethality and in preventing death of the CA_1 neurons. It was shown that older gerbils

Table 2. Observations in the gerbil global stroke model

1. Salicylate hydroxylation requires oxygen re-entering ischemic brain.[11]
2. Extent of salicylate hydroxylation is proportional to stroke-mediated injury to brain.[11]
3. Salicylate hydroxylation is brain region specific (i.e., hippocampus extensive, cortex less so, but none observed in brain stem or cerebellum).[12]
4. Protein oxidation is proportional to stroke-mediated injury to brain.[13]
5. Glutamine synthetase activity dropped in proportion to injury to brain.13
6. Stroke induced increase of glutamate in hippocampus[14] as well as retina.[15]
7. Stroke to older gerbils more lethal than to younger gerbils.[2]
8. Lethality prevented by PBN administered either before[2,16,17,18] or shortly after stroke.[16,17,18]
9. PBN prevents CA_1 neuron loss if given before or after stroke.[18]
10. Chronic PBN administration to older gerbils prevents stroke mediated lethality long after ceasing PBN dosing.[19,20]

became much more prone to IRI-mediated lethality than younger gerbils.[2] If PBN is given in a chronic fashion (i.e., lower doses daily) for 14 days, it was shown that the older gerbils became much more resistant to an IRI than the non-treated controls.[20] That is, about 85% of non-treated older gerbils were killed (assessed at 7 days) by 10-min ischemia. In contrast, if the gerbils were treated with PBN at 30 mg PBN twice daily for 14 days, only about 15% died after a 10-min IRI. The resistance to lethal brain IRI was maintained for at least 5 days after ceasing PBN administration. The half-life of PBN loss in rats is about 132 min;[21] therefore, gerbils, which had experienced approximately 60 PBN loss half-lives past the last NRT administration, were still protected from the IRI. This is quite a remarkable fact.

In addition to the gerbil model, the neuro-protective action of NRTs has now been extended to include the rat middle cerebral artery occlusion (MCAO) stroke model. Table 3 summarizes the results obtained when the middle cerebral artery was permanently occluded along with occlusion of the ipsilateral common carotid.[22] Brain edema, infarct volume as well as neurological scores, were assessed in rats which had been given PBN at various times both before and after the MCAO occlusion. The data clearly show that PBN had a protective effect on all three parameters with all four of the PBN treatment regimes. That is, PBN given before as well as after MCAO ligation was protective. Even PBN given starting 12 hrs after the MCAO-ligation showed protective activity. There was a general correlation in the changes that occurred in brain edema, infarct volume, and neurological scores.

In another MCAO model, PBN administered after reperfusion had begun also proved protective. The data are presented in Figure 2. The data were collected in this case utilizing a monofilament transcient MCAO blockage model. Blood flow through the middle cerebral artery was prevented for 2 hrs after which the blockage was removed and PBN administered either after 1 hr of recirculation or after 3 hrs of recirculation. PBN reduced infarct volume in both cases but was more effective if given 1 hr after starting recirculation. Infarct volume was determined 48 hrs after recirculation had started. Thus, it is quite apparent from the MCAO model that PBN is effective delivered after the IRI.

RESULTS IN EXPERIMENTAL MODELS OF BRAIN CONCUSSION

At the present time, no direct evaluation of the neuro-protective action of an NRT in a brain concussion model has been done. But data collected by the Phillis group[24] clearly

Table 3. PBN-mediated protection in focal ischemia-mediated neural cell loss and neurological behavioral scores of rats

Treatment	Infarct volume (mm^3)	Brain edema % change	Neurological behavioral scores	
Ischemia, no PBN	194.9 ± 9.2	6.8 ± 0.8	1.67 ± 0.11	(N=15)
Ischemia, PBN (-) 0.5, +24 hr	80.2 ± 17.9**	3.0 ± 0.7*	1.00 ± 0.20*	(N= 9)
Ischemia, PBN + 0.5, 5, 12 hr, etc.	98.9 ± 18.1**	3.3 ± 0.9*	1.00 ± 0.20*	(N=15)
Ischemia, PBN + 5, 12, 12 hr, etc.	97.9 ± 29.9**	2.2 ± 1.6*	0.40 ± 0.10**	(N=12)
Ischemia, PBN + 12, 12, 12 hr, etc.	130.9 ± 22.8*	2.7 ± 0.8*	1.10 ± 0.20*	(N=13)

*p <0.05, **p<0.01

Stroke administered by permanent middle cerebral artery and ipsilateral common carotid artery occlusion. PBN administered at 100 mg/kg, neurological scores assessed by strength and use of forelimb contralateral to artery occlusion tested for a total of 10 times each deficit was given 0.1 score. Control rats scored 0 deficits. The data are summarized from the results of Cao and Phillis.[22]

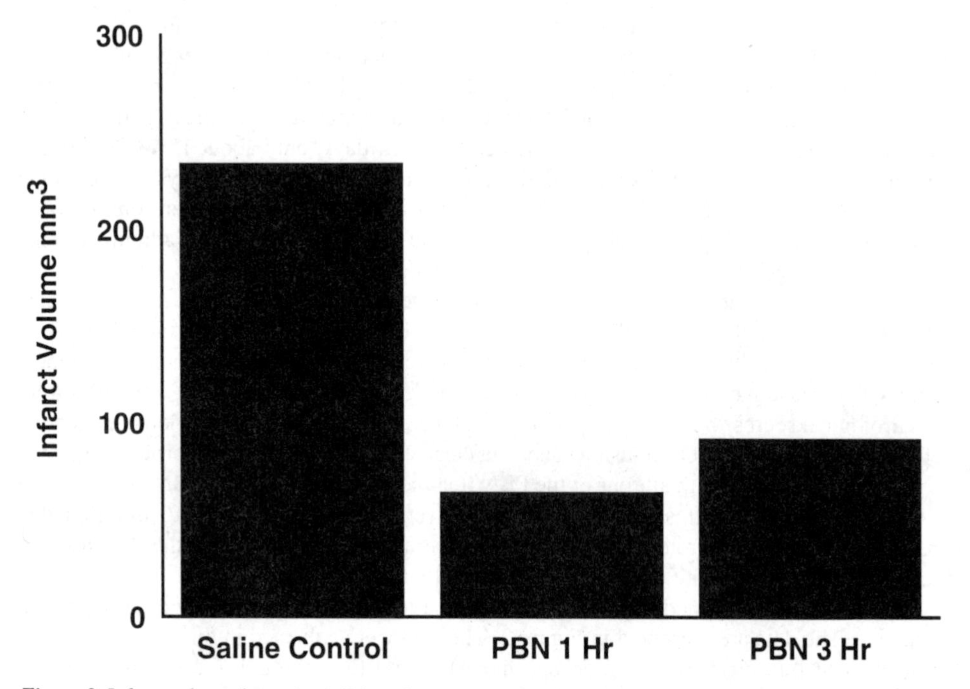

Figure 2. Infarct volume determined 48 hrs after inducing a middle cerebral artery occlusion in rats using a monofilament blockage technique where blockage was maintained for 1 hr after which the blockage was relieved by removing the monofilament. PBN was not administered (saline control) or administered either at 1 hr or 3 hrs after cessation of blockage. This figure is reconstructed from the data of Zhao et al.[23]

indicates that PBN may prove to be protective in a brain concussion model. Their data is summarized in Table 4. They have shown that in a rat brain concussion model, where a cortical cup containing artificial CSF (aCSF) is placed upon the lesioned brain, $\dot{O}H$ is trapped by POBN. There was no $\dot{O}H$ observed in the control non-lesioned brains. PBN pretreatment of the rats prevented the formation of POBN-OH adducts in the lesioned brain. This clearly implicates that PBN could be protective in brain concussion and that its presence acts by suppressing, perhaps by directly trapping, $\dot{O}H$ radicals. This also clearly implicates the involvement of $\dot{O}H$ in the damage that occurs in brain concussion. Interestingly, they observed an ascorbyl free radical in the aCSF of lesioned brain. This observation is pertinent to the proposed action of ascorbate in oxidative damage to brain (Table 1).

Table 4. Observations in the experimental rat brain concussion model*

1. $\dot{O}H$ trapped by POBN in aCSF in cortical cap on lesioned brain
2. No $\dot{O}H$ trapped by POBN in cortical cups of non-lesioned brain
3. PBN administration prevented POBN-OH adducts in cortical cup
4. Ascorbyl radical present in aCSF in cortical cup
5. PBN administration did not alter EEG of normal rat brain

*Observations of Sen, S. et al.[24] aCSF refers to artificial cerebral spinal fluid, POBN to pyridyl-N-oxide-*tert*-butyl nitrone and PBN to α-phenyl-*tert*-butyl nitrone.

Table 5. NRT prevention of NMDA-mediated death of cultured cerebellar granule cells*

Agent	IC_{50}
PBN	10 mM
DMPO	30 mM
TMPO	3 mM

*Mouse cerebellar granule cells were used at 10 days *in vitro* culture. Glycine (3 μM) was used to activate the NMDA receptor. PBN, DMPO, and TMPO refers to α-phenyl-*tert*-butyl nitrone; 5,5-dimethyl pyrroline 1-oxide and 3,3,5,5-tetramethyl pyrroline 1-oxide, respectively. TMPO is a chemical having a permanent nitroxyl-free radical and is not considered an NRT. Data of Lafon-Cazal, M. et al.[25]

RESULTS IN EXPERIMENTAL MODELS OF EXCITOTOXICITY, LOWERED ENERGY PRODUCTION AND DOPAMINE DEPLETION.

At first glance, it would not appear that these three distinct neurotoxicity research areas are related, but Beal's group[26,27] have clearly shown that free radicals are common to all three. These three research areas provide data considered pertinent to the neurodegenerative conditions of stroke, multiple system dystrophy, Huntington's disease, Parkinson's disease, and possibly others. In all of these research areas, NRTs have been shown to be effective in preventing damage in *in vivo* models.[27]

Excitotoxicity refers to neuronal death caused by mechanisms involving excitatory amino acids.[4] Glutamate and related excitatory amino acids account for as much as 40% of all synapses in the CNS.[4] Activation occurs through binding to inotropic receptors mediating depolarization. These receptors have been classified according to their most potent agonists including NMDA (N-methyl-D-aspartate), AMPA (α-amino-3-hydroxy-5-methyl-4-isoxasole-proprionic acid), and KA (kianic acid). Utilizing cultured cerebellar granule cells, Lafon-Cazal et al[25] showed that NMDA receptor stimulation caused superoxide production. The NRTs DMPO (5,5-dimethyl pyrroline 1-oxide) and PBN were able to prevent NMDA-receptor stimulated neuronal death. The data are shown in Table 5. Activation of either the KA

Table 6. NRT-mediated protection in models of excitotoxicity lowered energy production and dopamine depletion

Neurotoxic Agent	Model	Approximate protection by S-PBN (Lesion volume of treated to nontreated x 100)
NMDA	Excitotoxicity	43%
AMPA	Excitotoxicity	69%
KA	Excitotoxicity	48%
3-AP	Multiple system atrophy	37%
MPP^+	Parkinson's disease	62%
Malonate	Huntington's disease	55%

NMDA (N-methyl-D-aspartate), AMPA (α-amino-3-hydroxy-5-methyl-4-isoxasole-proprionic acid), KA (kianic acid), 3-AP (3-acetyl pyridine), MPP^+ (1-methyl-4-phenyl pyridinium); S-PBN (N-*tert*-butyl-α-[2-sulfophenyl]) nitrone was administered (100 mg/kg I.P.) 1 hr before and 2 and 5 hr after striatal injections of neurotoxic agent. Male Sprague-Dawley rats 300–325 gm. Lesion volumes quantitated 1 week after injection (data of Schulz, J.B. et al[26]).

Table 7. Observations pertinent to mechanism of NRT-mediated neuro-protection

Observation	NRT	Reference
1. Prevented malonate-mediated increase in salicylate hydroxylation to form 2,3-dihydroxybenzoate	S-PBN	Schulz, J.B. et al[27]
2. Had no effect on malonate-mediated increase in lactate or loss in ATP levels	S-PBN	Schulz, J.B. et al[27]
3. Slight decrease (15–24%) in malate/glutamate driven respiration in isolated liver mitochondria and very little, if any, effect in oxidative phosphorylation	PBN (4.7 mM) DMPO (24 mM) PBN/DMPO	Wong, P.K. et al[28] Wong, P.W. and Floyd, R.A.[29]

receptor or the voltage-sensitive Ca^{2+} channel did not cause superoxide production in these cells.[25]

Beal's group[27] has used six different neurotoxins including NMDA, AMPA, KA, 3-acetyl pyridine (3-AP), 1 methyl-4-phenylpyridinium (MPP^+) and malonate which have been used as experimental models for various neurodegenerative diseases and have noted that 2-sulfo-phenyl PBN shows considerable protection in all models. The data are summarized in Table 6. The NRT was given as three doses of 100 mg/kg I.P. each, the first 1 hr prior to neurotoxin injection into the brain and the second and third at 2 hrs and 5 hrs post-injection. The data show that the NRT protected in all cases decreasing lesion volume between approximately 30–65%. They noted that in both the malonate and 3-AP model that salicylate hydroxylation was significantly increased due to neurotoxin administration. In addition, in the malonate model they found[26,27] that 2-sulfophenyl-PBN administered 1 hr prior to the neurotoxin caused a significant decrease in salicylate hydroxylation to form 2,3-dihydroxybenzoic acid. It is interesting to also note that pretreatment with the NRT did not have any significant effect on malonate-mediated buildup in lactate or the observed loss in ATP levels.[27] This clearly indicates that the NRT effect is mediating events other than through specific interference with the action of malonate on mitochondria energy production (Table 7). Earlier, we had noted that PBN and DMPO, only at relatively high concentrations, slightly decreased respiration in isolated liver mitochondria[28] and showed very little, if any, effect on oxidative phosphorylation[29] (Table 7).

HOW ARE NRTs ACTING?

In order to understand the mechanistic basis of how the NRTs are protective in models of neurodegenerative diseases, we must keep in mind the large range of time involved of their effectiveness in stroke models as well as the protective role they have in chronic treatment regimes in general brain aging models. Figure 3 brings together some important points that appear to be more pertinent to the stroke model but does help explain the success of NRTs in chronic treatment aging models also. Thus, as tissue oxygen returns to normal levels after brain ischemia, reactive oxygen species rapidly increase within minutes but later decrease to slightly above normal levels even for several hours after the ischemia/reperfusion insult to the brain. It is considered that there are early mediators of oxidative damage increasing rapidly starting within minutes after the brain insult. We consider that if NRTs are present early (i.e., either given prior to or immediately after the insult), they prevent the oxidative damage that occurs early. In addition to the early mediators, there are late mediators, which appear hours after the brain insult and mediate oxidative damage. The induction of the late mediators can be prevented

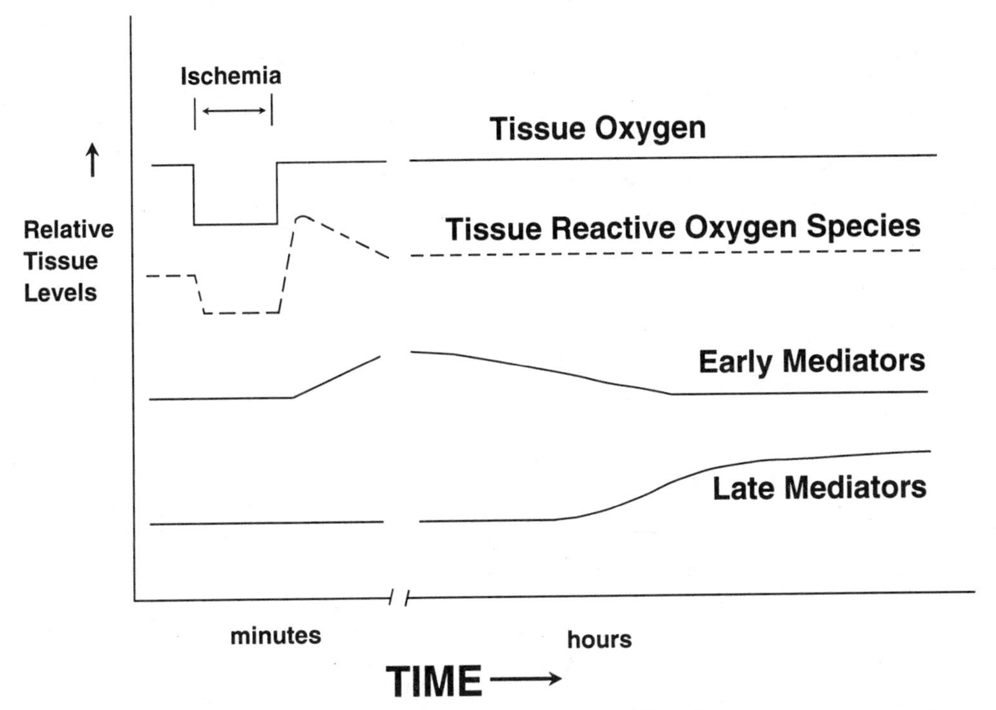

Figure 3. Illustration of the time dependence of oxygen concentration and reactive oxygen species expected to be present in brain that has undergone an ischemia reperfusion insult. There are early, as well as late, mediators of oxidative damage that appears at the expected approximate times illustrated after the brain insult.

by NRTs when these are administered later. It is possible that the early and late mediators are associated with gene induction, cellular influx or remodeling or any number of other possibilities. Little is known of these mediators of oxidative tissue injury, but it is likely that low levels of these factors may be involved in the change that occurs in the aging brain. Thus if the NRTs are given in a chronic fashion, it is considered that they depress the mediators of oxidative damage that are increased in the aging brain. Much more research work needs to be done to elucidate the action of NRTs in understanding their protective action in neurodegenerative diseases.

ACKNOWLEDGMENTS

The author acknowledges the collaborative effort of many colleagues especially Dr. J. M. Carney, who with the author, made several discoveries regarding the neuro-protective action of NRTs. This work was supported in part by NIH grants NS 23307 and AG 09690.

REFERENCES

1. J.M. McCord and I. Fridovich, Superoxide dismutase: An enzymic function for erythrocuperin (hemocuprein), *J. Biol. Chem.* 244(22):6049 (1969).
2. R.A. Floyd, Role of oxygen free radicals in carcinogenesis and brain ischemia, *FASEB J.* 4:2587 (1990).
3. R.A. Floyd and J.M. Carney, Free radical damage to protein and DNA: Mechanisms involved and relevant observations on brain undergoing oxidative stress, *Ann. Neurol.* 32:S22 (1992).

4. J.T. Coyle and P. Puttfarcken, Oxidative stress, glutamate, and neurodegenerative disorders, *Science* 262:689 (1993).
5. E.D. Hall and J.M. Braughler, Central nervous system trauma and stroke. II. Physiological and pharmacological evidence for involvement of oxygen radicals and lipid peroxidation, *Free Radic. Biol. Med.* 6:303 (1989).
6. M.M. Zaleska and R.A. Floyd, Regional lipid peroxidation in rat brain *in vitro*: Possible role of endogenous iron, *Neurochem. Res.* 10:397 (1985).
7. M.M. Zaleska, K. Nagy, and R.A. Floyd, Iron-induced lipid peroxidation and inhibition of dopamine synthesis in striatum synaptosomes, *Neurochem.* 14:597 (1989).
8. R.A. Floyd, J.J. Watson, P.K. Wong, D.H. Altmiller, and R.C. Rickard, Hydroxyl free radical adduct of deoxyguanosine: Sensitive detection and mechanisms of formation, *Free Radic. Res. Commun.* 1:163 (1986).
9. R.A. Floyd, M.S. West, K.L. Eneff, J.E. Schneider, P.K. Wong, D.T. Tingey, and W.E. Hogsett, Conditions influencing yield and analysis of 8-hydroxy-2′-deoxyguanosine in oxidatively damaged DNA, *Anal. Biochem.* 188:155 (1990).
10. E.G. Janzen, Spin trapping, *Acc. Chem. Res.* 4:31 (1971).
11. W. Cao, J.M. Carney, A. Duchon, R.A. Floyd, and M. Chevion, Oxygen free radical involvement in ischemia and reperfusion injury to brain, *Neurosci. Lett.* 88:233 (1988).
12. J.M. Carney, T. Tatsuno, and R.A. Floyd, The role of oxygen radicals in ischemic brain damage: Free radical production, protein oxidation and tissue dysfunction, *in:* "Pharmacology of Cerebral Ischemia," J. Krieglstein and H. Oberpichler-Schwenk, eds., Publisher Wissenschaftliche Verlagsgesellschaft, Stuttgart (1992).
13. C.N. Oliver, P.E. Starke-Reed, E.R. Stadtman, G.J. Liu, J.M. Carney, and R.A. Floyd, Oxidative damage to brain proteins, loss of glutamine synthetase activity, and production of free radicals during ischemia/reperfusion-induced injury to gerbil brain, *Proc. Natl. Acad. Sci. USA* 87:5144 (1990).
14. G. Delbarre, B. Delbarre, F. Calinon, C. Loiret, and A. Ferger, Accumulation of glutamate, asparate and GABA in the hippocampus of gerbil after transient cerebral ischemia, *Soc. Neurosci. Abst.* 15:1174 (1989).
15. G. Delbarre, B. Delbarre, and F. Calinon, Increase of amino acids in gerbil after ischemia, *FASEB J.* 4:A1128 #5008 (1990).
16. J.W. Phillis and C. Clough-Helfman, Protection from cerebral ischemic injury in gerbils with the spin trap agent N-tert-butyl-α-phenylnitrone (PBN), *Neurosci. Lett.* 116:315 (1990).
17. J.W. Phillis and C. Clough-Helfman, Free radicals and ischaemic brain injury: Protection by the spin trap agent PBN, *Med. Sci. Res* 18:403 (1990).
18. C. Clough-Helfman and J.W. Phillis, The free radical trapping agent N-*tert*-butyl-α-phenylnitrone (PBN) attenuates cerebral ischaemic injury in gerbils, *Free Radic. Res. Commun.* 15:177 (1991).
19. J.M. Carney, P.E. Starke-Reed, C.N. Oliver, R.W. Landrum, M.S. Chen, J.F. Wu, and R.A. Floyd, Reversal of age-related increase in brain protein oxidation, decrease in enzyme activity, and loss in temporal and spacial memory by chronic administration of the spin-trapping compound N-*tert*-butyl-α-phenylnitrone, *Proc. Natl. Acad. Sci. USA* 88: 3633 (1991).
20. R.A. Floyd and J.M. Carney, Nitrone radicals traps (NRTs) protect in experimental neurodegenerative diseases, *in:* "Neuroprotective Approaches to the Treatment of Parkinson's Disease and Other Neurodegenerative Disorders," C.A. Chapman, C.W. Olanow, P. Jenner, and M. Youssim, eds., Academic Press Limited, London (1993).
21. G. Chen, T.M. Bray, E.G. Janzen, and P.B. McCay, Excretion, metabolism and tissue distribution of a spin trapping agent, α-phenyl-N-*tert*-butyl-nitrone (PBN) in rats, *Free Radic. Res. Commun.* 9:317 (1990).

22. X. Cao and J.W. Phillis, α-Phenyl-tert-butyl-nitrone reduces cortical infarct and edema in rats subjected to focal ischemia, *Brain Res.* 644:267 (1994).
23. Q. Zhao, K. Pahlmark, M.L. Smith, and B.K. Siesjo, Delayed treatment with the spin trap alpha-phenyl-N-tert-butyl nitrone (PBN) reduces infarct size following transient middle cerebral artery occlusion in rats, *Acta. Physiol. Scand.* 152:349 (1994).
24. S. Sen, H. Goldman, M. Morehead, S. Murphy, and J.W. Phillis, α-Phenyl-*tert*-butyl-nitrone inhibits free radical release in brain concussion, *Free Radic. Biol. Med.* 16:685 (1994).
25. M. Lafon-Cazal, S. Pietri, M. Culcasi, and J. Bockaert, NMDA-dependent superoxide production and neurotoxicity, *Nature* 364:535 (1993).
26. M.F. Beal, Mechanisms of excitotoxicity in neurologic diseases, *FASEB J.* 6:3338 (1992).
27. J.B. Schulz, D.R. Henshaw, D. Siwek, B.G. Jenkins, R.J. Ferrante, P.B. Cipolloni, N.W. Kowall, B.R. Rosen, and M.F. Beal, Involvement of free radicals in excitotoxicity in vivo, *J. Neurochem.* 64:2239 (1995) .
28. P.K. Wong, J.L. Poyer, C.M. DuBose, and R.A. Floyd, Hydralazine-dependent carbon dioxide free radical formation by metabolizing mitochondria, *J. Biol. Chem.* 263:11296 (1988).
29. P.K. Wong and R.A. Floyd, Unpublished observations (1995).

NITRIC OXIDE ACTIONS IN THE NERVOUS SYSTEM

Valina L. Dawson [1,2,3] and Ted M. Dawson[1,3]

Departments of Neurology[1], Physiology[2] and Neuroscience[3]
Johns Hopkins University School of Medicine
600 North Wolfe Street
Pathology 2-210
Baltimore, Maryland 21287

INTRODUCTION

Nitric oxide (NO) for many decades has been known to be a toxic gas, a constituent of air pollution, a component of cigarette smoke, and a by product of microbial metabolism. Only very recently has it been identified as a product of mammalian cells. The unique, although surprising, role for NO as a biological messenger molecule was developed by investigations in the fields of immunology, cardiovascular pharmacology, toxicology, and neurobiology (Dawson and Snyder 1994; Moncada and Higgs, 1993; Nathan, 1992; Feldman et al., 1993). In the nervous system the discovery of NO as a messenger molecule is changing the conventional concepts of how cells in the nervous system communicate. Classical neurotransmitters are enzymatically synthesized, stored in synaptic vesicles, and released by exocytosis from synaptic vesicles during membrane depolarization. These neurotransmitters mediate their biological actions by binding to membrane-associated receptors, which initiates intracellular changes in the postsynaptic cell. The activity of conventional neurotransmitters is terminated by either reuptake mechanisms or enzymatic degradation. There are multiple points at which biological control can be exherted over the production and activity of conventional neurotransmitters. None of these classical biological mechanisms are exploited by the nervous system to regulate the activity of NO. Instead, NO is synthesized on demand by the enzyme NO synthase (NOS) from the essential amino acid, L-arginine. NO is small, diffusible, membrane permeable and reactive. These chemical properties of NO make it a unique neuronal messenger molecule (Feldman et al., 1993). Since the cell can not sequester and regulate the local concentration of NO, the key to regulating NO activity is to control NO synthesis. Putative cellular targets of NO are rapidly being discovered as well as potential physiologic and pathophysiologic roles in the nervous system. NO may regulate neurotransmitter release, it may play a key role in morphogenesis and synaptic plasticity, it may regulate gene expression, and it may mediate inhibitory processes associated with sexual and aggressive behavior. Under conditions of excessive formation, NO is emerging as an important mediator of neurotoxicity in a variety of disorders of the nervous system.

ISOFORMS OF NITRIC OXIDE SYNTHASE

Although as early as 1916 it was noted that mammals excrete more nitrate than is consumed in the diet, it was thought that nitrogen oxide biosynthesis was an exclusive property of microorganisms. It was not until the 1980's that nitrogen oxides were recognized as normal products of mammalian metabolism (Greene et al., 1981a, b). Attempts to understand the vasodilatory effects of organic nitrates (Arnold et al., 1977; Ignarro et al., 1981) and the role of acetylcholine-induced relaxation of vascular smooth muscle (Furchgott and Zawadzki, 1980) led to the discovery that NO, produced in endothelial cells, is an important biological messenger molecule, critical to the regulation of vascular hemodynamics (Ignarro et al., 1987; Palmer et al., 1987). Parallel investigations identified a role for NOS and NO in macrophages (Nathan, 1992; Marletta, 1993) in which there was an L-arginine dependent pathway that produced nitrogen byproducts and contributed to the cytotoxic action of macrophages on tumor cells, protozoa and other microorganisms (Stuehr and Marletta, 1985; Hibbs et al., 1987a, b; Stuehr and Nathan, 1989). The NOS isoform in endothelial cells is constitutively expressed and briefly activated by increases in intracellular calcium while the NOS isoform in macrophages, which normally do not contain detectable levels of NOS protein, requires protein synthesis before the enzyme is expressed. Once NOS is induced in macrophages, it produces large quantities of NO for sustained periods. These initial observations led to the classification of NOS enzymes as constitutive or inducible. However this nomenclature is not entirely accurate. Recent studies have shown that following traumatic or pathologic insult all NOS isoforms can be "induced", in that, new enzyme protein synthesis occurs (Dawson and Snyder, 1994).

Molecular cloning experiments have led to the identification of three NOS genes, neuronal NOS (nNOS), endothelial NOS (eNOS) and immunologic NOS (iNOS) which were named by the tissue from which they were first cloned. All three isoforms have been identified in numerous tissues. nNOS is present not only in neurons but also in skeletal muscle, neutrophils, pancreatic islets, endometrium, and respiratory and gastrointestinal epithelia (Nathan and Xie, 1994). eNOS has also been localized to a small population of neurons in the central nervous system (Dinerman et al., 1994). iNOS can be expressed in most tissues including neurons, astrocytes, and endothelial cells (Oswald et al., 1994; Dawson and Dawson, 1995). In the nervous system, astrocytes and microglia can be induced by a variety of cytokines to express iNOS (Nathan, 1992; Marletta, 1993) which produces large quantities of NO for sustained periods resulting in neuronal damage (Dawson et al., 1994; Chao et al., 1992; Galea et al., 1992). NO produced by astrocytes and microglia may contribute to neurologic damage in several disease states (Murphy et al., 1993).

BIOSYNTHESIS OF NITRIC OXIDE AND REGULATION OF NOS

The biosynthesis of NO has been investigated using both biochemical and molecular biological approaches (Marletta, 1993; Bredt and Snyder, 1994). NO is formed by the stoichiometric conversion of L-arginine to L-citrulline and NO in the presence of oxygen and NADPH through an oxidative-reductive pathway which requires 5 electrons (Bredt and Snyder, 1994). The mechanism and order in which these cofactors and reactants interact with NOS to initiate catalysis differs between the constitutive (nNOS and eNOS) isoforms and the inducible (iNOS) isoform (Figure 1). nNOS and eNOS are expressed as noninteractive monomers which stoichiometrically bind the flavin mononucleotide (FMN), and flavin adenine dinucleotide (FAD) (Feldman et al., 1993; Marletta, 1993; Bredt and Snyder, 1994). On binding heme, tetrahydrobiopterin (BH4) and L-arginine the NOS monomers dimerize (Baek et al., 1993). However, this does not lead immediately to electron flow and catalysis. Increases in intracellular calcium are required which result in calcium/calmodulin complexes that bind to nNOS or eNOS dimers at an exposed, basic, hydrophobic site. The binding of calcium/calmodulin induces a conformational change allowing the flow of electrons from the reductase segment of the enzyme containing the flavoprotein binding domains to the catalytic site containing the heme (Abu-Soud and Stuehr, 1993). The inactive monomers of iNOS, on the other hand, bind calmodulin even at very low concentrations of intracellular calcium (Cho et al., 1992). Dimerization and activation then occurs on the binding of heme, tetrahydrobiopterin (BH_4) and L-arginine. Since calmodulin, heme, tetrahydrobiopterin (BH4) and L-arginine are usually abundant in cells, most iNOS is expressed as functionally active dimers. Therefore, the catalyst for nNOS and eNOS activation is intracellular calcium, while the regulation of iNOS activity is the level of mRNA and protein synthesis and degradation.

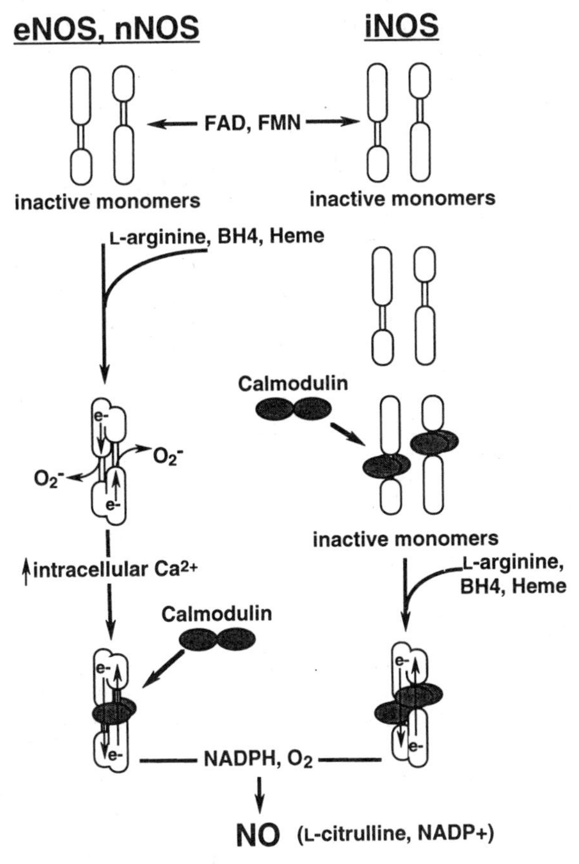

Figure 1. Mechanistic differences between the NOS isoforms. Transcription of NOS results in the translation of inactive monomers. The binding of heme, tetrahydrobiopterin (BH_4), and L-arginine leads to dimerization of nNOS and eNOS. In this configuration, electron flow cannot be completed. The electron may be donated to oxygen which may account for the reports of superoxide anion production by NOS. When intracellular calcium is elevated, calcium binds calmodulin and the complex binds to NOS inducing a conformational change in the dimer and permitting the flow of electrons. In the presence of NADPH and oxygen, NO is produced. In contrast, the inactive monomers of iNOS can bind calmodulin with high affinity without an elevation of intracellular calcium. The necessary conformational change which enables electron flow occurs when heme, BH_4, and L-arginine bind and allow the dimerization of iNOS. Since cells are replete in calmodulin, heme, BH_4, and L-arginine, once iNOS is translated it is in the activated state. Therefore, the activity of nNOS and eNOS is determined by intracellular calcium levels while the activity of iNOS is determined by the levels of mRNA and protein.

The numerous sites for cofactors and substrate binding on nNOS provides multiple opportunities for pharmacologic regulation of catalytic activity. Analogs of L-arginine, such as nitro-L-arginine, compete for binding to the catalytic site and subsequently decrease NO formation. Flavoprotein inhibitors such as diphenyleneiodonium prevent the shuttling of electrons essential for the conversion of L-arginine to L-citrulline and NO. Calmodulin is an essential cofactor for nNOS providing the last link necessary in the shuttling of electrons from the flavoproteins in the reductase end of the enzyme to the heme moiety and catalytic site . Calmodulin antagonists, such as W-7, calmidazolium, and some gangliosides are effective inhibitors of nNOS catalytic activity (Dawson et al., 1993; Dawson et al., 1995). Consensus sites for phosphorylation are evident from the predicted protein sequences of nNOS. *In vitro,* biochemical studies indicate that neuronal NOS can be phosphorylated by calcium/calmodulin-dependent protein kinase, cyclic AMP-dependent protein kinase, cyclic GMP-dependent protein kinase, and by protein kinase C. Phosphorylation of NOS by all these enzymes decreases NOS catalytic activity *in vitro* (Dawson and Snyder, 1994; Bredt et al., 1992; Dinerman et al., 1994). Calcineurin, a protein phosphatase, dephosphorylates NOS and subsequently increase its catalytic activity (Dawson et al., 1993). Multiple levels of regulation of nNOS are thus possible by phosphorylation.

LOCALIZATION OF NITRIC OXIDE SYNTHASE IN THE NERVOUS SYSTEM

NO as a neuronal messenger molecule was first described by Garthwaite and colleagues, who showed that cerebellar granule cells release an endothelium-derived relaxing factor-like substance after exposure to glutamate agonists (Garthwaite et al., 1988). The activation of NOS was linked to stimulation of the N-methyl-D-aspartate (NMDA) class of glutamate receptors (Bredt and Snyder, 1989). Subsequent anatomical localization of nNOS lead to insight into the potential function of NO in the nervous system. Immunohistochemical and *in situ* hybridization studies revealed a unique distribution for nNOS which did not co-localize with any one neurotransmitter (Bredt et al., 1991). In the cerebellum, nNOS is expressed in glutaminergic granule cells and GABAergic basket cells. In the cerebral cortex and corpus striatum, nNOS neurons are colocalized with somatostatin and neuropeptide Y (Dawson et al., 1991). In the pedunculopontine tegmental nucleus of the brainstem, nNOS neurons lack somatostatin and neuropeptide Y, but stain for choline acetyltransferase (Dawson et al., 1991). Although nNOS does not co-localize with a single neurotransmitter, all nNOS neurons identified co-localize with the histochemical stain, NADPH diaphorase (Dawson et al., 1991; Hope et al., 1991). NADPH diaphorase is a histochemical stain originally described by Thomas and Pearse (1964) in which diaphorase enzymes reduce tetrazolium dyes in the presence of NADPH, to a dark blue formazan precipitant. Confirmation that nNOS catalytic activity results in diaphorase staining was obtained when human kidney 293 cells transfected with nNOS cDNA produced cells that stained for both nNOS and NADPH diaphorase (Dawson et al., 1991). The localization of nNOS immunoreactivity with NADPH diaphorase staining in neurons is only observed under appropriate paraformaldehyde fixation. Presumably, paraformaldehyde fixation inactivates virtually all NADPH-dependent oxidative enzymes, with the exception of NOS (Matsumoto et al., 1993). Interestingly, NADPH diaphorase/nNOS neurons are relatively spared from cell death in Huntington's and Alzheimer's Disease, vascular stroke, and NMDA neurotoxicity (Dawson et al., 1992).

With the exception of the cerebellum in which most granule cells express nNOS, nNOS neurons comprise only about 1% to 2% of the total neuronal population in many brain regions, such as the cerebral cortex and corpus striatum. The nNOS neurons are not displayed in any obvious pattern and exhibit properties of medium to large aspiny neurons. Hippocampal pyramidal cells of the CA1 region do not contain detectable amounts of nNOS, however, they express the endothelial isoform. Immunoreactivity for eNOS is highly concentrated within pyramidal cells of the CA1 through CA3 region of the hippocampus as well as granule cells of the dentate gyrus (Dinerman et al., 1994). This markedly contrasts with the staining for nNOS which is absent from CA1 pyramidal neurons and is only concentrated within GABAergic interneurons of the hippocampus. In some brain regions, such as the cerebellum and olfactory bulb, both eNOS and nNOS occur in the same cell populations though in different proportions. NADPH diaphorase staining was also shown to co-localize with eNOS in the CA1 pyramidal cells of the hippocampus by using gluteraldehyde containing fixatives rather than paraformaldehyde fixation (Dinerman et al., 1994).

NITRIC OXIDE AS A NEURONAL MESSENGER

In the peripheral autonomic nervous system depolarization of nNOS expressing myenteric plexus neurons in the gastrointestinal tract causes relaxation of smooth muscle associated with peristalsis (Bult et al., 1990; Desai et al., 1991). The blockade of this process by NOS inhibitors implicates a role for NO as the nonadrenergic-noncholinergic (NANC) neurotransmitter of the gut (Bult et al., 1990; Desai et al., 1991; Nathan, 1992). In the penis, NOS is highly concentrated in the pelvic plexus, the cavernosal nerve and nerve plexus and the adventitia of the deep cavernosal arteries and sinusoids in the periphery of the corpora cavernosa (Burnett et al., 1992). Electrically stimulated penile erections in rats and relaxation of isolated corpus cavernosum strips elicited by nerve stimulation are blocked by NOS inhibitors (Burnett et al., 1992; Rejfer et al., 1992). Thus, NO is the NANC neurotransmitter which regulates penile erection. Neurons derived primarily from cells in the sphenopalatine ganglia that innervate the outer adventitial layers of cerebral blood vessels express nNOS (Nozaki et al., 1993). Thus, cerebral blood flow is regulated by neuronally derived NO in addition to endothelially produced NO. It is possible, in the central nervous system, that activity-dependent activation of nNOS may influence small cerebral arterioles and locally regulate cerebral blood flow (Faraci, 1992; Iadecola, 1993). NO, through increases in intracellular cGMP levels, also mediates increases in regional cerebral blood flow (rCBF) elicited by CO_2 inhalation (Iadecola et al., 1994). In mice lacking the gene for nNOS there is normal augmentation of cerebral blood flow following CO_2 inhalation which is not associated with increases in cGMP levels is not blocked by NOS inhibitors (Irikura et al., 1995). Thus, other compensatory parallel pathways exist which maintain the cerebral circulatory response to CO_2 inhalation that are not mediated through NO or cGMP.

Neurotransmitter release may be regulated by NO. Inhibitors of NOS block NMDA receptor-mediated neurotransmitter release from cortical or striatal synaptosomes (Montague et al., 1994; Hirsch et al., 1993). Similar observations have been made in other model systems in which NOS inhibitors also block the release of neurotransmitters (Dawson and Snyder; 1994). Differentiated PC12 cells release acetylcholine upon depolarization with potassium (Sandberg et al., 1989). Release of both acetylcholine and dopamine from PC12 cells is blocked by NOS inhibitors and it reversed by the addition of excess L-arginine (Hirsch et al., 1993). NO may stimulate the release of neurotransmitters by activation of guanylyl cyclase and initiation of cyclic GMP-dependent protein phosphorylation cascades which augment the phosphorylation of synaptic vesicle proteins associated with neurotransmitter release.

A form of cellular memory, long-term potentiation (LTP) in the hippocampus, has been shown by some investigators to be sensitive to inhibitors of NOS (Zorumski et al., 1993; Schuman and Madison, 1994). Consistent with the idea that NO is involved in LTP is the observation that the inhibition of LTP by NOS inhibitors is reversed by L-arginine. Additionally, hemoglobin, which binds extracellular NO, also blocks LTP. In cultured hippocampal neurons, NO produces an increase in the frequency of spontaneous miniature excitatory post-synaptic potentials and direct application of NO may elicit LTP (Bohme et al. 1991; O'Dell et al., 1991; Haley et al, 1992). Support for the role of NO as a retrograde messenger that travels from pyramidal cells to stimulate release of excitatory transmitter from Schaeffer collaterals is further suggested by the inhibition of LTP by NOS inhibitors directly injected into CA1 pyramidal neurons (Schuman and Madison, 1991). Additionally, LTP which has been thought to be exquisitely specific has been observed to spread to synapses of neighboring neurons by a diffusible signal, potentially NO (Schuman and Madison, 1994). Although there has been disparity in observing an NO component in initiating or maintaining LTP, consistent effects are observed under specific experimental conditions (Zorumski and Izumi, 1993; Schuman and Madison, 1994). Transgenic mice which lack nNOS provide a unique opportunity to explore many of the postulated functions of neuronally derived NO in the nervous system. In the nNOS null mice, LTP induced by weak intensity tetanic stimulation was slightly reduced (O'Dell et al., 1994). This LTP was blocked by NOS inhibitors, just as it is in wild-type mice (O'Dell et al., 1994). Immunostaining with an eNOS specific antibody revealed that eNOS was expressed in CA1 pyramidal cells in both nNOS null and wild-type mice (O'Dell et al., 1994). Thus, eNOS may be more important in LTP than nNOS. Additional retrograde messengers must exist, such as arachidonic acid, carbon monoxide or platelet activating factor, and are likely to be required for LTP as maximal LTP is not blocked with inhibitors of NOS (O'Dell et al., 1994).

NITRIC OXIDE BIOCHEMISTRY

The combination of a nitrogen atom with an oxygen atom yields a species with an unpaired electron. Therefore, NO is by definition a free radical. Most free radicals are chemically very reactive with other chemical species due the free electron. Neurochemists regard NO as highly reactive due to its short half-life ($t_{1/2}$) in comparison to other traditional neuronal messenger molecules. However, chemically, NO reacts with a limited range of chemical species and therefore is not considered a highly reactive free radical (Butler et al., 1995).

NO can react with molecular oxygen (O_2) yielding nitrogen dioxide (NO_2). This reaction occurs rapidly in the gas phase at high concentrations of reactants. However, in solution at concentrations of reactants that exist in cells, the $t_{1/2}$ is several hours, suggesting that under normal physiological conditions, O_2 is not the primary chemical target of NO (Butler et al., 1995). NO rapidly reacts with nitrogen dioxide to yield nitrite ions (NO_2-), the basis of the colormetric Greiss reaction (Ding et al., 1988). The Greiss reaction is a commonly used indirect measure of NO formation from activated cells expressing iNOS such as macrophages and astrocytes in which high concentrations of both nitrogen and oxygen intermediates are rapidly and continuously being formed (Ding et al., 1988). Thiols are commonly assumed to be a major targets for NO. Nitrosothiols with biological relevance have been isolated and characterized including, s-nitrosoglutathione and the nitrosothiol of serum albumin (Stamler et al., 1992). However, the biosynthetic pathway under which this occurs is not understood, in that at physiologic pH, NO does not readily react with thiols (Ignarro and Greutter, 1980).

Recently a plethora of often conflicting biological roles have been assigned to NO. To address this issue investigators have examined the various valence states of NO and their biological function (Lipton et al., 1993; Stamler et al., 1994). In addition to the free radical form there are the oxidized and reduced forms (NO+ and NO-) of NO. All three valence states may exist in the brain and the different valences states may be, in part, responsible for some of the conflicting activities attributed to NO. NO+ is an active nucleophile which is unlikely to be formed at neutral pH or react with any other nucleophile other than water (Butler et al., 1995). However, NO+ can readily be transferred from a nitrosothiol to a second thiol or another nucleophile in a process called transnitrosation (Butler et al., 1995). The work of Stamler and colleagues indicate that NO+ does not exist in the free form, but rather NO+ is donated from complex carrier molecules to exert its biologic activity. The biologic significance of NO- is as yet unknown but it also does not exist in the free form and is donated from complex carrier molecules. NO free radical can exist in the free form in solution and most likely is the form of NO that is produced on stimulation of nNOS.

The biologic role for NO in the S-nitrosylation of many proteins is emerging as an important regulatory system (Stamler, 1994; Stamler et al., 1992). For instance, the NMDA receptor is inactivated by nitrosylation. Thus, NO may physiologically modulate glutaminergic neurotransmission (Lipton et al., 1993). However, recent work suggests that NO inhibits NMDA currents though an interaction with cations rather than the redox modulatory site of the NMDA receptor channel (Fagni et al., 1995). NO also stimulates the apparent auto-ADP ribosylation of glyceraldehyde-3-phosphate dehydrogenase (GAPDH) (Dawson and Snyder, 1994; Bredt and Snyder, 1994). NO may react with an active site cysteine resulting in inhibition of GAPDH catalytic activity and depression of glycolysis. The mechanism of apparent ADP ribosylation is not clear but may reflect direct binding of NAD to the cysteine catalyzed by S-nitrosylation (McDonald and Moss, 1993). Through the formation of intracellular S-nitrosoglutathione, NO can deplete intracellular glutathione levels resulting in a rapid and concomitant activation of the hexose-monophosphate shunt pathway (Clancy et al., 1994). NO also inhibits thioester-linked long chain fatty acylation of neuronal proteins possibly through a direct modification of substrate cysteine thiols in the neuronal growth cone. Thus, NO might reversibly inhibit the growth of rat dorsal root ganglion neurites (Hess et al., 1993).

Probably the most significant biological reaction of NO is with transition metals resulting in NO-metal complexes. A particularly important and well known transition metal target of NO is the iron in the heme moiety of guanylyl cyclase (Ignarro, 1990). The NO-iron interaction in guanylyl cyclase induces a conformational change in the heme moiety which activates the enzyme and results in increased cGMP levels. NO increases prostaglandin production by activating another heme-containing enzyme, cyclooxygenase (Salvemini et al., 1993). NO can also inhibit enzymes which use the heme-moiety is involved in catalysis such as NOS and indoleamine 2,3-dioxygenase. Additional heme-containing proteins that may be targets for NO include catalase, cytochrome c, hemoglobin, and peroxidase.

NO also reacts with non-heme iron, particularly with iron-sulfur clusters, in numerous enzymes including, NADH-ubiquinone oxidoreductase, cis-aconitase, and NADH:succinate oxidoreductase (Nathan, 1992). Recent reports however, suggest that peroxynitrite, a reaction product of NO and superoxide anion, reacts with iron-sulfur clusters rather than NO directly (Hauslanden et al., 1994; Castro et al., 1994). These reactions readily occur after macrophage activation. However, in contrast to the reversible reaction of NO with heme, the reaction of NO with iron-sulfur cluster results in the desolution of the cluster (Henry et al., 1993). Through these interactions NO may inhibit DNA synthesis by binding to the non-heme iron of ribonucleotide reductase. NO can liberate iron by binding to the iron in ferritin, an iron-storage protein. NO can influence iron metabolism at the post-transcriptional level by interacting with cytosolic aconitase. Cytosolic acontitase has dual functions that is regulated by NO. In the absence of NO it functions as cytosolic aconitase. In the presence of NO it functions as the iron responsive binding protein. NO disrupts aconitase activity and exposes its RNA binding site permitting binding of the iron-responsive element binding protein to the iron-responsive element (Drapier et al., 1993; Weiss et al., 1993). Production of NO through stimulation of NMDA receptors in rat brain slices, stimulates the RNA binding function of the iron-responsive element binding protein while diminishing its aconitase activity. The iron-responsive element binding protein also has a discrete neuronal localization in several brain structures, thus it may be an important molecular target for NO action in the brain (Jaffrey et al, 1994).

NO will react with the superoxide anion ($O_2^{\bullet}-$) to produce the potent oxidant, peroxynitrite (ONOO-) (Radi et al., 1991). The rate of this reaction is three times faster than the rate of reaction of the enzyme, superoxide dismutase (SOD), in catalyzing the dismutation of the superoxide anion to hydrogen peroxide. Therefore, when present at appropriate concentrations, NO can effectively compete with SOD for $O_2^{\bullet}-$. Although a simple molecule, ONOO- is chemically complex. It has the activity of the hydroxyl radical and the nitrogen dioxide radical although it does not readily decompose into these entities (Koppenol et al., 1992; Crow et al., 1993). ONOO- is also a potent oxidant which reacts readily with sulfhydryls (Radi et al., 1991) and with zinc-thiolate moieties (Crow et al., 1995). It can also directly nitrate and hydroxylate aromatic rings on amino acid residues (Beckman et al., 1992). Additionally, ONOO- can oxidize lipids (Radi et al., 1991), proteins (Moreno and Pryor, 1992) and DNA (King et al., 1992).

NITRIC OXIDE MEDIATED NEUROTOXICITY

Under normal physiologic conditions NO is predominantly a neuronal messenger molecule. However, when present in higher concentrations, NO, like glutamate, can initiate a neurotoxic cascade (Dawson et al., 1992). Substantial evidence indicates that excess glutamate acting via NMDA receptors mediates cell death in focal cerebral ischemia (Choi, 1988). Glutamate neurotoxicity may also play a part in neurodegenerative diseases such as Alzheimer's disease and Huntington's disease (Meldrum and Garthwaite, 1990; Lipton and Rosenburg, 1994). The activation of NMDA receptors and the subsequent increase in intracellular calcium levels presumably initiates most forms of glutamate neurotoxicity. NMDA applied only for a brief (5 min) period of time sets in motion irreversible processes which result in cell death 12 to 24 hours later. This type of toxicity is exquisitely dependent upon calcium (Figure 2). Since NOS is a calcium-dependent enzyme, we wondered whether activation of NOS could be involved in NMDA neurotoxicity. Treatment of cortical cultures with NOS inhibitors or removal of L-arginine from the media blocks NMDA neurotoxicity (Dawson et al, 1991; 1993). Additionally, reduced hemoglobin, which binds NO and prevents it from reaching its target cells, completely prevents NMDA neurotoxicity. Blockade of NMDA neurotoxicity by NOS inhibitors is also observed in cultures of caudate-putamen and hippocampus. If NO is the mediator of NMDA neurotoxicity, then compounds which release NO directly should also be neurotoxic. Cortical cultures exposed to the NO donors, sodium nitroprusside (SNP), S-nitroso-N-acetylpenicillamine (SNAP) or SIN-1 exhibit a delayed neurotoxicity which follows the same time course as NMDA neurotoxicity (Dawson et al., 1993).

NO has been implicated in glutamate neurotoxicity in a variety of model systems and prolonged application of NOS inhibitors, after the initial exposure to excitatory amino acids, leads to enhanced neuroprotection. Others have failed to confirm NO's role in this phenomenon (Dawson and Snyder, 1994). NO may also protect neurons from glutamate neurotoxicity when cells are exposed to NO prior to exposure to glutamate. Recent studies indicate that NO may possess both neurodestructive and neuroprotective properties depending upon the redox milieu. The NO radical ($NO^{\bullet}$) is neurodestructive, while NO+ (i.e., NO complexed to a carrier molecule) is neuroprotective (Lipton et al., 1993). Cortical cultures from transgenic mice which lack neuronal NOS are relatively resistant to NMDA neurotoxicity (Dawson VL, Huang PL, Fishman MC, Snyder SH and Dawson TM, unpublished observations), thus the potential cytoprotective effects of NO are overwhelmed under conditions of excessive NO formation.

Most NOS inhibitors are not useful in the study of nNOS function in the CNS of intact animals. For instance, experiments with non-selective NOS inhibitors are confounded by inhibition of eNOS which results in hypertension and alterations in cerebral blood flow. Results may also be confounded by a potential interaction of arginine analog NOS inhibitors with arginine dependent cellular processes. L-arginine is an essential amino acid involved in the urea cycle. Disturbances in the urea cycle can affect the TCA cycle and ultimately the energy balance in the cell. Additionally, L-arginine metabolism feeds into the polyamine pathway. Alterations of this pathway can have profound effects on cellular function and survival.

The potential involvement of NO in stroke, the third leading cause of death and morbidity in the United States, has prompted the search for selective nNOS inhibitors. Besides arginine analog NOS inhibitors which compete with L-arginine at the catalytic site, NOS can be inhibited indirectly at several regulatory sites, thus providing alternative strategies for protection against NO-mediated cell death. The flavoproteins, FAD and FMN, are critical for the necessary shuttling of electrons involved in the conversion of L-arginine to NO and L-citrulline. The flavoprotein inhibitor, diphenyleneiodonium, is potently neuroprotective against NMDA neurotoxicity, although it is unlikely to be therapeutically useful. Calmodulin is an essential co-factor for the activation of NOS. Agents which inhibit or bind calmodulin such as calmidazolium or W7 can decrease NOS catalytic activity and provide neuroprotection. Gangliosides, which are neuroprotective in a variety of animal models, may be neuroprotective through inhibition of NOS. Gangliosides inhibit NOS activity and the potency of NOS inhibition closely parallels their affinities for binding calmodulin and providing neuroprotection (Dawson et al., 1995). Phosphorylation of NOS and the attendant reduction of its catalytic activity also provides another potential approach to neuroprotection. The immunosuppressant, FK506 and cyclosporin A, which binds small, soluble receptor proteins designated FK506-binding proteins (FKBP) and cyclophilins respectively, inhibit NOS and are thus neuroprotective. This neuroprotection is due to the interaction of the immunosuppressant with its respective binding protein to inhibit the calcium-activated phosphatase, calcineurin. Inhibition of calcineurin prevents the dephosphorylation and activation of NOS, thus NOS remains in the inactive phosphorylated state (Dawson et al., 1993). The physiologic relevance of this observation was recently confirmed by the report that FK506 is profoundly neuroprotective in a rat model of focal ischemia (Sharkey and Butcher, 1994).

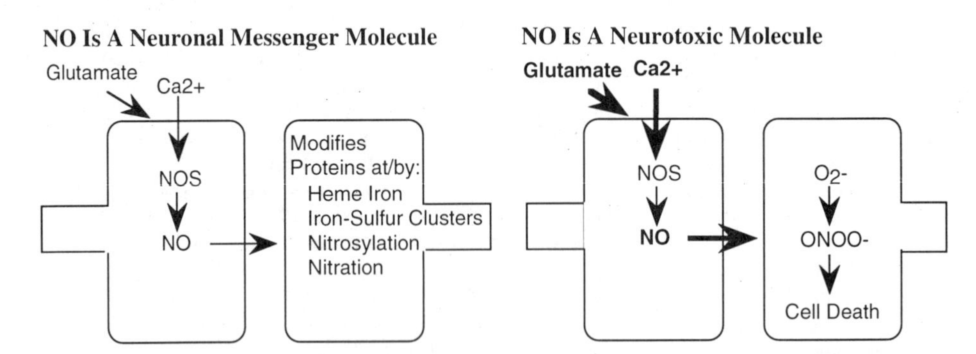

Figure 2. Nitric oxide has dual roles as a neuronal messenger molecule and as a neurotoxin. Under normal physiologic conditions, production of NO results in local modulation of cellular activity through modification of intracellular proteins. Under pathophysiologic conditions, such as ischemic damage in stroke, excessive glutamate is released and NOS is overactive. NO has the opportunity to compete for superoxide anion (O_2-) and peroxynitrite (ONOO-) is formed. It is likely that formation of peroxynitrite is a major pathway for neuronal cell death.

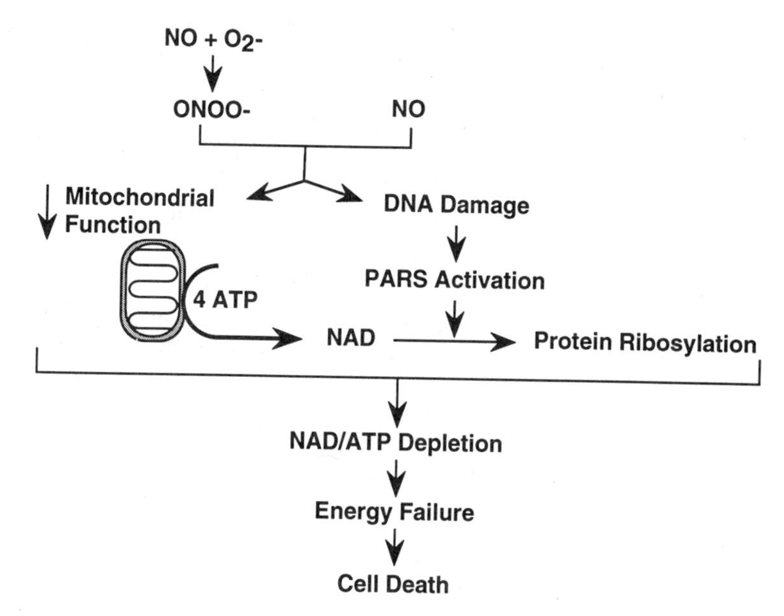

Figure 3. Mechanism of NO-mediated neurotoxicity. Nitric oxide (NO) and or peroxynitrite (ONOO-) may damage DNA and inhibit mitochondrial function which could lead to more free radical formation. The damaged DNA activates poly (ADP-ribose) synthetase (PARS) which transfers ADP-ribose groups to nuclear proteins from NAD. To regenerate NAD from nicotinamide, 4 high energy equivalents of ATP are required. PARS is highly promiscuous and adds numerous ADP-ribose groups to nuclear proteins consuming a very large amount of energy. NMDA-generated free radicals such as NO activate a futile cycle of DNA damage followed by PARS activation which depletes cells of their energy stores, ultimately leading to cell death.

Another potential strategy for developing neuroprotective agents is to identify the targets of NO that participate in neurotoxicity. The best known target of NO is guanylyl cyclase. However, neither inhibitors of guanylyl cyclase nor cell permeable analogs of cGMP have any effect on NMDA or NO neurotoxicity (Lustig et al., 1992; Dawson et al., 1993). Therefore, it is unlikely that activation of guanylyl cyclase and elevation of intracellular cGMP levels participates in the genesis of neurotoxicity. The majority of evidence indicates that NO's toxic effects occur through an interaction with the superoxide anion to form peroxynitrite. The toxic effects of NO or ONOO- may occur through multiple mechanisms (Figure 2). NO can damage DNA with the subsequent activation of the nuclear enzyme poly (ADP ribose) synthetase (PARS). The only known activator of PARS is damaged DNA (Lautier et al., 1993; Zhang et al., 1994). Once activated, PARS catalyzes the attachment of ADP ribose units to nuclear proteins, such as histone and PARS itself. For every mole of ADP ribose transferred, 1 mole of NAD is consumed and four free energy equivalents of ATP are consumed to regenerate NAD (Figure 4). Thus, activation of PARS can rapidly deplete energy stores. Consistent with this possibility is that cortical cell cultures are protected from glutamate and NO neurotoxicity by a series of PARS inhibitors (Zhang et al., 1994). The neuroprotective effects of these agents parallels their potency as PARS inhibitors. Recent studies directly link energy depletion, in that dramatic drops in cellular ATP and NAD levels occur as early as four hours after exposure to cytotoxic concentrations of NMDA. The fall in ATP and NAD levels is completely blocked by PARS inhibitors and glutamate antagonists (Sasaki M., Brahmbhatt H., Kaufmann S. Dawson TM, and Dawson VL., manuscript in preparation).

Following cerebrovascular infarction, release of excitatory amino acids in the extracellular space stimulates NMDA receptors, increasing NOS activity and NO levels. Marked increases in NO production in the brain occur during focal ischemia (Malinski et al., 1993). Once formed, NO can react with the superoxide anion, levels of which are also increased during transient ischemia, to form peroxynitrite. If peroxynitrite is the toxin of physiologic importance, then one would expect that inhibiting accumulation of superoxide anion, or decreasing production of NO, would be associated with decreased brain injury after focal ischemia. Consistent with this notion is the observation that animals treated with SOD before focal ischemia and in transgenic mice which overexpress SOD, the infarct volume following focal ischemia is markedly attenuated (Kinouchi et al., 1991). In a similar manner, NOS inhibitors reduce infarct volume following middle cerebral artery occlusion in mice, rats, and cats (Dawson and Snyder, 1994). Recent studies show that low doses of NOS inhibitors are neuroprotective while high doses are ineffective and suggest that partial inhibition of neuronal NOS is sufficient to obtain an optimal neuroprotective effect (Carreau et al., 1994). Exacerbation of injury occurs at high doses of NOS inhibitors through adverse vascular effects from inhibition of endothelial NOS, resulting in decreased regional cerebral blood flow and increased infarction volume. Further support for this hypothesis is the recent study in transgenic mice which lack nNOS. These mice have reduced infarct volumes compared to age-matched wild-type controls following focal permanent middle cerebral artery occlusion (Huang et al., 1994). When the nNOS null mice are treated with the NOS inhibitor, nitro-L-arginine methyl ester, the dilation of the pial vessels is inhibited and stroke volume is increased. Thus, neuronally derived NO plays an important role in mediating neuronal cell death following focal ischemia and endothelially derived NO plays an important protective role by regulating and maintaining proper cerebral blood flow.

Because glutamate acting via NMDA receptors stimulates NO formation, it would be logical to expect that NOS neurons would be the first cells to succumb to excess NMDA receptor stimulation. However, NOS neurons are resistant to NMDA and NO neurotoxicity (Koh and Choi, 1988; Dawson et al., 1993). It is unknown why NOS neurons are resistant to NMDA and NO neurotoxicity but they probably possess protective "factors" which render them relatively resistant to the toxic NO environment they create. NOS neurons within the striatum are enriched in manganese SOD, and SOD in these neurons may prevent the local formation of toxic peroxynitrite rendering NOS neurons resistant to the toxic actions of NO (Iganaki et al., 1991). Another possibility may be that NO diffuses rapidly away from NOS neurons along a concentration gradient in a manner that prevents the generating neuron from seeing excessive concentrations of NO. More likely, NOS neurons possess other protective mechanisms which have yet to be identified.

Although nNOS is constitutively expressed, under some pathologic insults nNOS can be induced in certain cells through new protein synthesis (Wu et al., 1994). In spinal cord motorneurons, nNOS is expressed after avulsion or lesion of the C6 or C7 root (Wu et al., 1994). The expression of nNOS corresponds with the subsequent death of the motorneurons. Consistent with the concept that NO plays a role in neuronal cell death, treatment with a NOS inhibitor dramatically increases the number of surviving neurons (Wu and Li., 1993).

POTENTIAL ROLES FOR NO IN NEUROPSYCHIATRIC DISORDERS

NO may play a role in neurodegenerative disorders and other forms of neurotoxicity. NO is an important mediator in CNS oxygen toxicity since inhibitors of NOS protect mice against CNS oxygen toxicity (Oury et a., 1992). NO may play a role in the pathogenesis of AIDS dementia as the neurotoxicity in primary cortical cultures induced by the HIV coat protein, gp120, (Lipton et al., 1992), is due in part, to activation of NOS (Dawson et al., 1993). Additionally, gp120 has been shown to induce iNOS in human monocyte-derived macrophages and astrocytoma cell lines (Mollace et al., 1993). Using competitive reverse-transcriptase polymerase chain reaction (RT-PCR) we have observed a 3-fold increase in iNOS in cortical tissue from patients infected with HIV and a 7.5-fold increase in iNOS in patients infected with HIV who developed severe dementia (B Wildemann, M. Sasaki, DC Adamson, J McArthur, J Glass, VL Dawson and TM Dawson manuscript in preparation).

Overproduction of NO may also play a pathologic role in inflammatory disorders of the central nervous system. In primary rat cultures, induction of iNOS results in neuronal cell death 72 hours later (Chao et al., 1992; Dawson et al., 1994). Induction of human iNOS in demyelinating regions of multiple sclerosis brains has been observed (Bo et al., 1994) and NO may be directly toxic to the myelin-producing oligodendrocytes (Merrill et al., 1993). Increased levels of NO production have also been observed in bacterial meningitis (Visser et al., 1994). Recent studies suggest that migraine sufferers have supersensitivity to NO (Thomsen et al., 1993). NO may also contribute to the pathogenesis of Alzheimer's disease. In primary cortical cultures, NOS inhibitors provide neuroprotection against toxicity elicited by fragments of human ß-amyloid (Dawson VL., Brahmbhatt H., Cordell B., Snyder SH., and Dawson TM., manuscript in preparation). NOS neurons are also relatively spared in Alzheimer's disease (Hyman et al., 1992).

The nNOS null transgenic mice provide a unique opportunity to investigate the potential role of NO in behavior. The nNOS null mice exhibit excessive mounting behavior to non-estrous females when compared to the wild type mice implicating a role for NO in sexual behavior. Additionally the nNOS mice are markedly more aggressive than the wild type mice. In a neutral arena, nNOS null mice will initiate more attacks in a shorter amount of time than wild type mice and do not assume a submissive posture like wild type mice (RJ Nelson, GE Demas, PL Huang, MC Fishman, VL Dawson, TM Dawson, and SH Snyder, manuscript submitted.) Thus, neuronally derived NO may play an important inhibitory function in normal sexual and social encounters. Derangements in NO formation or disposition may account for similar disorders in humans.

SUMMARY

NO has clearly revolutionized our thinking about aspects of neurotransmission and neuronal signaling. It has also radically altered our thoughts about how synaptic transmission takes place. NO is emerging as an important regulator of a variety of physiologic processes, however, under conditions of excessive formation, NO is emerging as an important mediator of pathologic nervous tissue damage. Understanding the role of NO in these processes will hopefully lead to the development of selective therapeutic agents and to a better understanding of basic processes underlying normal and pathological neuronal functions.

ACKNOWLEDGMENTS:

VLD is supported by grants from USPHS NS22643, NS33142, the American Foundation for AIDS Research, National Alliance for Research on Schizophrenia and Depression, American Heart Association, the Alzheimer's Association. TMD. is supported by grants from the USPHS NS01578, NS26643, NS33277 and the American Health Assistance Foundation, Paul Beeson Physician Scholars in Aging Research Program and International Life Sciences Institute. The authors own stock in and are entitled to royalty from Guilford Pharmaceuticals, Inc., which is developing technology related to the research described in this article. The stock has been placed in escrow and cannot be sold until a date which has been predetermined by the Johns Hopkins University.

REFERENCES

Abu-Soud HM and Stuehr DJ. Nitric oxide synthases reveal a role for calmodulin in controlling electron transfer. Proc. Natl. Acad. Sci. USA 1993; 90: 10769-10772.

Baek KJ, Thiel BA, Lucas S, and Stuehr DJ. Macrophage nitric oxide subunits. Purification, characterization, and the role of prosthetic groups and substrate in regulating their association into a dimeric enzyme. J. Biol. Chem. 1993; 268: 21120-21129.

Beckman JS, Ischiropoulos H, Zhu L, van der Woerd M, Smith C, Chen J, Harrison J, Martin JC, and Tsai M. Kinetics of superoxide dismutase- and iron catalyzed nitration of phenolics by peroxynitrite. Arch. Biochem. Biophys. 1992; 298: 438-445.

Bo L, Dawson TM, Wesselingh S, et al. Induction of nitric oxide synthase in demyelinating regions of multiple sclerosis brains. Ann. Neurol. 1994; in press.

Bohme GA, Bon C, Stutzmann J-M, Doble A, Blanchard J-C. Possible involvment of nitric oxide in long-term potentiation. Eur. J. Pharm. 1991; 199: 379-381

Bredt DS, Ferris CD, Snyder SH. Nitric oxide synthase regulatory sites. J. Biol. Chem. 1992; 267:10976-10981.

Bredt DS, Glatt CE, Hwang PM, Fotuhi M, Dawson TM, Snyder SH. Nitric oxide synthase protein and mRNA are discretely localized in neuronal populations of the mammalian CNS together with NADPH diaphorase. Neuron. 1991; 7:615-624.

Bredt DS, Hwang PM, Snyder SH. Localization of nitric oxide synthase indicating a neural role for nitric oxide. Nature. 1990; 347:768-770.

Bredt DS, Snyder SH. Nitric oxide, a physiological messenger molecule. Annu. Rev. Biochem. 1994;63:in press.

Bredt DS, Snyder SH. Isolation of nitric oxide synthetase, a calmodulin-requiring enzyme. Proc. Natl. Acad. Sci. U.S.A. 1990; 87:682-685.

Bult H, Boeckxstaens GE, Pelckmans PA, Jordaens FH, Van Maercke YM, Herman AG. Nitric oxide as an inhibitory non-adrenergic non-cholinergic neurotransmitter. Nature. 1990; 345:346-347.

Burnett AL, Lowenstein CJ, Bredt DS, Chang TSK, Snyder SH. Nitric Oxide: a physiologic mediator of penile erection. Science 1992; 257:401-403.

Carreau A, Duval D, Poignet H, Scatton B, Vige X, Nowicki J-P. Neuroprotective efficacy of N^{w}-nitro-L-arginine after focal cerebral ischemia in the mouse and inhibition of cortical nitric oxide synthase. Eur. J. Pharmacol. 1994;256:241-249.

Castro L, Rodriguez M, and Radi R. Aconitase is readily inactivated by peroxynitrite, but not by its precursor, nitric oxide. J. Biol. Chem. 1994; 269: 29409-29415.

Chao CC, Hu S, Molitor TW, Shaskan EG and Peterson PK. Activated microglia mediate neuronal cell injury via a nitric oxide mechanism. J Immunol 1992; 149: 2736-2741.

Cho HJ, Xie QW, Calaycay J, Mumford RA, Swiderek KM, Lee TD, and Nathan C. Calmodulin is a subunit of nitric oxide synthase from macrophages. J. Exp. Med. 1992; 176: 599-604.

Choi DW. Glutamate neurotoxicity and diseases of the nervous system. Neuron. 1988; 1:623-634.

Clancy RM, Levartovsky D, Leszczynska-Piziak J, Yegudin J, Abramson SB. Nitric oxide reacts with intracellular glutathione and activates the hexose monophosphate shunt in human neutrophils: evidence for S-nitrosoglutathione as a bioactive intermediary. Proc. Natl. Acad. Sci. U.S.A. 1994;91:3680-3684.

Crow JP, Beckman JS and McCord JM. Sensitivity of the essential zinc-thiolate moiety of yeast alcohol dehydrogenase to hypochlorite and peroxynitrite. Biochem. 1995; 34: 3544-3552.

Dawson TM, Bredt DS, Fotuhi M, Hwang PM, Snyder SH. Nitric oxide synthase and neuronal NADPH diaphorase are identical in brain and peripheral tissues. Proc. Natl. Acad. Sci. U.S.A. 1991; 88:7797-7801.

Dawson TM, Dawson VL, Snyder SH. A novel neuronal messenger molecule in brain: the free radical, nitric oxide. Ann. Neurol. 1992;32:297-311.

Dawson TM, Hung K, Dawson VL, Steiner JP, Snyder SH. Neuroprotective effects of gangliosides may involve inhibition of nitric oxide synthase. Ann. Neurol. 1995; 37: 115-118.

Dawson TM, Snyder SH. Gases as biological messengers: nitric oxide and carbon monoxide in the brain. J. Neurosci. 1994; 14: 5147-5159.

Dawson TM, Steiner JP, Dawson VL, Dinerman JL, Uhl GR, Snyder SH. Immunosuppressant, FK506, enhances phosphorylation of nitric oxide synthase and protects against glutamate neurotoxicity. Proc. Natl. Acad. Sci. U.S.A. 1993; 90:9808-9812.

Dawson V, Brahmbhatt HP, Mong JA, Dawson TM Expression of inducible nitric oxide synthase causes delayed neurotoxicity in primary mixed neuronal-glial cortical cultures. Neuropharmacol. 19954; 33:1425-1430.

Dawson VL, Dawson TM, Bartley DA, Uhl GR, Snyder SH. Mechanisms of nitric oxide mediated neurotoxicity in primary brain cultures. J. Neurosci. 1993; 13:2651-2661.

Dawson VL, Dawson TM, Uhl GR, Snyder SH. Human immunodeficiency virus type 1 coat protein neurotoxicity mediated by nitric oxide in primary cortical cultures. Proc. Natl. Acad. Sci. U.S.A. 1993; 90:3256-3259.

Desai KM, Sessa WC and Vane JR. Involvement of nitric oxide in the reflex relaxation of the stomach to accomodate food or fluid. Nature 1991; 351: 477-479.

Dinerman JL, Dawson TM, Schell MJ, Snowman A, Snyder SH. Endothelial nitric oxide synthase localized to hippocampal pyramidal cells: implications for synaptic plasticity. Proc. Natl. Acad. Sci. U.S.A. 1994; 91:4214-4218.

Dinerman JL, Steiner JP, Dawson TM, Dawson VL, and Snyder SH. Protein phosphorylation inhibits neuronal nitric oxide synthase. Neuropharmacology. 1994; 33: 1245-1252 .

Ding AH, Nathan CF and Stuehr DJ. Release of reactive nitrogen intermediates and reactive oxygen intermediates from mouse peritoneal macrophages. J. Immunol. 1988; 141: 2407.

Drapier J-C, Hirling H, Wietzerbin J, Kaldy P, Kuhn LC. Biosynthesis of nitric oxide activates iron regulatory factor in macrophages. EMBO J. 1993;12:3643-3649.

Faraci FM. Regulation of the cerebral circulation by endothelium. Pharmac. Ther. 1992; 56:1-22.

Feldman PL, Griffith OW, Stuehr DJ. The surprising life of nitric oxide. Chemical & Engineering News. 1993;12:26-38.

Galea E, Feinstein and Reis DJ. Induction of calcium-dependent nitric oxide synthase activity in primary rat glial cultures. Proc. Natl. Acad. Sci. USA 1992; 89:10945-10949.

Garthwaite J, Charles SL, Chess-Williams R. Endothelium-derived relaxing factor release on activation of NMDA receptors suggests role as intercellular messenger in the brain. Nature. 1988; 336:385-388.

Green LC, Ruiz-de-Luzuriaga K, Wagner DA, Rand W, Istfan N, Young VR, Tannenbaum SR. Nitrate biosynthesis in man. Proc Natl. Acad. Sci. USA 1981a; 78:7764-7768.

Green LC, Tannenbaum SR, Goldman P. Nitrate synthesis in the germfree and conventional rat. Science 1981b; 212:56-68.

Haley JE, Wilcox GL, Chapman PF. The role of nitric oxide in hippocampal long-term potentiation. Neuron 1992; 8: 211-216.

Hausladen A and Fridovich I. Superoxide and peroxynitrite inactivate aconitases, but nitric oxide does not. J. Biol. Chem. 1994; 269: 29405-29408.

Henry Y, Lepoivre M, Drapier JC, Ducrocq C, Boucher JL, and Guissani A. EPR characterization of molecular targets for NO in mammalian cells and organelles. FASEB J 1993; 7: 1124-1134.

Hess DT, Patterson SI, Smith DS, Pate Skene JH. Neuronal growth cone collapse and inhibition of protein fatty acylation by nitric oxide. Nature. 1993;366:562-565.

Hibbs JB Jr., Vavrin Z, Taintor RR. L-arginine is required for expression fo the activated macrophage effector mechanism causing selective metabolic inhibition in target cells. J. Immunol. 1987b: 138: 550-565.

Hibbs, JB Jr., Taintor RR, Vavrin Z. Macrophage cytotoxicity: role for L-arginine deaminase and imino nitrogen oxidation to nitrite. Science 1987a; 235: 473-476.

Hirsch DB, Steiner JP, Dawson TM, Mammen A, Hayek E, Snyder SH. Neurotransmitter release regulated by nitric oxide in PC-12 cells and brain synaptosomes. Cur. Biol. 1993; 3:749-754.

Hope BT, Michael GJ, Knigge KM, Vincent SR. Neuronal NADPH diaphorase is a nitric oxide synthase. Proc. Natl. Acad. Sci. U.S.A. 1991; 88:2811-2814.

Huang PL, Dawson TM, Bredt DS, Snyder SH, Fishman MC. Targeted disruption of the neuronal nitric oxide synthase gene. Cell. 1993;75:1273-1286.

Huang Z, Huang PL, Panahian N, Dalkara T, Fishman MC and Moskowtiz MA. Effects of cerebral ischemia in mice deficient in neuronal nitric oxide synthase. Science 1994; 265:1883-1885.

Hyman BT, Marzloff K, Wenniger JJ, Dawson TM, Bredt DS, Snyder SH. Relative sparing of nitric oxide synthase-containing neurons in the hippocampal formation in Alzheimer's disease. Ann Neurol 1992; 32: 818-820.

Iadecola C, Pelligrino DA, Moskowitz MA and Lassen NA. Nitric oxide synthase inhibition and cerebrovascular regulation. J. Cereb. Blood Flow and Metab. 1994; 14: 175-192.

Iadecola C. Regulation of the cerebral microcirculation during neural activity: is nitric oxide the missing link? Trends Neurosci 1993; 16:206-214.

Ignarro LJ and Gruetter CA. Requirement of thiols for activation of coronary arterial guanylate cyclase by glycerol trinitrate and sodium nitrite: possible involvment of s-nitrosothiols. Biochim. Biophys. Acta. 1980; 631:221-231.

Ignarro LJ, Lippton H, Edwards JC, Baricos WH, Hyman AL, Kadowitz PJ, Gruetter CA Mechanism of vascular smooth muscle relaxation by organic nitrates, nitrites, nitroprusside and nitric oxide: evidence for the involvement of S-nitrosothiols as active intermediates. J. Pharmacol. Exp. Ther. 1981; 218:739-749.

Ignarro LJ. Biosynthesis and metabolism of endothelium-derived relaxing factor. Annu. Rev. Pharmacol. Toxicol. 1990; 30: 535-560.

Inagaki S, Suzuki K, Taniguchi N, Takagi H. Localization of Mn-superoxide dismutase (Mn-SOD) in cholinergic and somatostatin-containing neurons in the rat neostriatum. Brain. Res. 1991; 549:174-177.

Jaffrey SR, Cohen NA, Rouault TA, Klausner RD, Snyder SH. The iron-responsive element binding protein: a novel target for synaptic actions of nitric oxide. Proc. Natl. Acad. Sci, USA. 1994; 91: 12994-12998.

King PA, Adnerson VE, Edwards JO, Gustafson G, Plumb RC, and Suggs JW. A stable solid that generates hydroxyl radical upon dissolution in aqueous solution: Reaction with proteins and nucleic acids. J. Am. Chem. Soc. 1992; 114: 5430-5432.

Kinouchi H, Epstein CJ, Mizue T, et al. Attenuation of focal cerebral ischemic injury in transgenic mice overexpressing CuZn superoxide dismutase. Proc. Natl. Acad. Sci. U.S.A. 1991;88:11158-11162.

Koh J-Y and Choi DW. Vulnerability of cultured cortical neurons to damage by excitotoxins: differential susceptibility of neurons containing NADPH-diaphorase. J. Neurosci. 1988: 8: 2153-2163.

Koppenol WH, Moreno JJ, Pryor WA, Ischiropoulos H, Beckman JS. Peroxynitrite, a cloaked osidant formed by nitric oxide and superoxide. Chem. Res. Toxicol. 1992; 5: 834-842.

Lautier D, Lagueux J, Thibodeau J, Menard L, and Poirier GG. Molecular and biochemical features of poly(ADP-ribose) metabolism. Mol. Cell. Biochem. 1993; 122: 171-193.

Lipton SA, Choi YB, Pan Z-H, et al. A redox-based mechanism for the neuroprotective and neurodestructive effects of nitric oxide and related nitroso-compounds. Nature. 1993;364:626-632.

Lipton SA. Models of neuronal injury in AIDS: another role for the NMDA receptor? Trends Neurosci. 1992;15:75-79.

Lustig HS, von Brauchitsch KL, Chan J, Greenberg DA. Ethanol and excitotoxicity in cultured cortical neurons: differential sensitivity of N-methyl-D-aspartate and sodium nitroprusside toxicity. J. Neurochem. 1992; 577: 343-346.

Malinski T, Bailey F, Zhang ZG, Chopp M. Nitric oxide measured by a porphyrinic microsensor in rat brain after transient middle cerebral artery occlusion. J. Cereb. Blood. Flow. Metab. 1993;13:355-358.

Marletta MA. Nitric oxide synthase structure and mechanism. J. Biol. Chem. 1993;268:12231-12234.

Matsumoto T, Nakane M, Pollock JS, Kuk JE, Forstermann U. A correlation between soluble nitric oxide synthase and NADPH-diaphorase activity is only seen after exposure of the tissue to fixative. Neurosci. Letts. 1993;155:61-64.

McDonald LJ and Moss J. Stimulation by nitric oxide of an NAD linkage to glyceraldehyde-3-phosphate dehydrogenase. Proc. Natl. Acad. Sci. USA 1993; 90: 6238-6241.

Meldrum B, Garthwaite J. Excitatory amino acid neurotoxicity and neurodegenerative disease. Trends. Pharmacol. Sci. 1990; 11:379-387.

Merrill JE, Ignarro LJ, Sherman MP, Melinek J, Lane TE. Microglial cell cytotoxicity of oligodendrocytes is mediated through nitric oxide. J. Immunol. 1993;151:2132-2141.

Mollace V, Colasanti M, Persichini T, Bagetta G, Lauro GM, Nistico G. HIV gp120 glycoprotein stimulates the inducible isoform of NO synthase in human cultured astrocytoma cells. Biochem. Biophys. Res. Comm. 1993;194:439-445.

Moncada S, Higgs A. The L-arginine-nitric oxide pathway. N. Engl. J. Med. 1993;329:2002-2012.

Montague PR, Gancayco CD, Winn MJ, Marchase RB, Friedlander MJ. Role of NO production in NMDA receptor-mediated neurotransmitter release in cerebral cortex. Science. 1994;263:973-977.

Moreno JJ and Pryor QA. Inactivation of α-1-proteinase inhibitior by peroxynitrite. Chem. Res. Toxiocol. 1992; 5: 425-431.

Murphy S, Simmons ML, Agullo L, Garcia A, Feinstein DL, Galea E, Reis DJ, Minc-Golomb D, Schwartz JP. Synthesis of nitric oxide in CNS glial cells. Trends in Neurosci. 1993; 16:323-328.

Nathan C, Xie Q-W. Regulation of biosynthesis of nitric oxide. J. Biol. Chem. 1994;19:13725-13728.

Nathan C. Nitric oxide as a secretory product of mammalian cells. FASEB J. 1992;6:3051-3064.

Nozaki K, Moskowitz MA, Maynard KI, et al. Possible origins and distribution of immunoreactive nitric oxide synthase-containing nerve fibers in rat and human cerebral arteries. J. Cerebral. Blood. Flow. and Metabolism. 1993; 13:70-79.

O'Dell TJ, Hawkins RD, Kandel ER, Arancio O. Tests of the roles of two diffusable substances in long-term potentiation: evidence for nitric oxide as a possible early retrograde messenger. Proc. Natl. Acad. Sci. U.S.A. 1991; 88:11285-11289.

O'Dell TJ, Huang PL, Dawson TM, Dinerman JL, Snyder SH, Kandel ER and Fishman MC. Blockade of long-term potentiation by inhibitors of nitric oxide synthase in mice lacking the neuronal isoform suggests a role for the endothelial isoform. Science. 1994; 265: 542 -546.

Oury TD, Ho Y-S, Piantadosi CA, Crapo JD. Extracellular superoxide dismutase, nitric oxide, and central nervous system O_2 toxicity. Proc. Natl. Acad. Sci. U.S.A. 1992;89:9715-9719.

Radi R, Beckman JS, Bush KM, Freeman BA. Peroxynitrite oxidation of sulfhydryls. The cytotoxic potential of superoxide and nitric oxide. J .Biol. Chem. 1991; 266:4244-4250.

Rajfer J, Aronson WJ, Bush PA, Dorey FJ, Ignarro LJ. Nitric oxide as a mediator of the corpus cavernosum in response to nonadrenergic noncholinergic transmission. New. Eng. J. Med. 1992; 326:90-94.

Salvemini D, Misko TP, Masferrer JL, Seibert K, Currie MG, Needleman P. Nitric oxide activates cyclooxygenase enzymes. Proc. Natl. Acad. Sci. U.S.A. 1993; 90:7240-7244.

Sandberg K, Berry CJ, Eugster E, Rogers TB. A role for cGMP during tetanus toxin blockade of acetylcholine release in the rat pheochromocytoma (PC12) cell lines. J Neurosci 1989; 9:3946-3954.

Schuman EM, Madison DV. Locally distributed synaptic potentiation in the hippocampus. Science. 1994;263:532-536.

Schuman EM, Madison DV. Nitric oxide and synaptic function. Annu. Rev. Neurosci. 1994;17:153-183.

Schuman EM, Madison DV. The intercellular messenger nitric oxide is required for long-term potentiation. Science 1991; 254:1503-1506.

Schumann EM and Madison DV. Nitric oxide and synpatic function. Ann. Rev. Neurosci. 1994; 17: 153-183.

Sharkey J, Butcher SP. Immunophillins mediate the neuroprotective effects of FK506 in focal cerebral ischemia. Nature 1994; 371:336-339.

Stamler JS, Simon DI, Osborne JA, et al. S-nitrosylation of proteins with nitric oxide: synthesis and characterization of biologically active compounds. Proc. Natl. Acad. Sci. U.S.A. 1992; 89:444-448.

Stamler JS, Simon DI, Osborne JA, Mullins ME, Jaraki O, Michel T, Singel DJ and Loscalzo J. S-nitrosylation of proteins with nitric oxide: synthesis and characterization of biolgocially active compounds. Proc. Natl. Acad. Sci. USA 1992: 89: 444-448.

Stamler JS. Redox signalling: Nitrosylation and related target interactions of nitric oxide. Cell 1994; 78: 931-936.

Stuehr DJ and Marletta MA. Mammalian nitrate biosynthesis: mouse macrophages produce nitrite and nitrate in response to *Escherichia coli* lipopolysaccharide. Proc. Natl. Acad. Sci. USA 1985; 82: 7738-7742.

Stuehr DJ, and Nathan CF. Nitric oxide. A macrophage product responsible for cytostasis and respiratory inhibition in tumore target cells. J. Exp. Med. 1989; 169: 1543-1555.

Thomas E, Pearse AGE. The solitary active cells. Histochemical demonstration of damage-resistant nerve cells with a TPN-diaphorase reaction. Acta. Neuropathol. 1964; 3:238-249.

Thomsen LL, Iversen HK, Brinck TA, Olesen J. Arterial supersensitivity to nitric oxide (nitroglycerin) in migraine sufferers. Cephalalgia. 1993;13:395-399.

Visser JJ, Scholten RJPM, Hoekman K. Nitric oxide synthesis in meningococcal meningitis. Ann. Int. Med. 1994;120:345-346.

Weiss G, Goossen B, Doppler W, Fuchs D, Pantopoulos K, Werner-Felmayer G, Wachter H, Hentze MW. Translational regulation via iron-responsive elements by the nitric oxide/NO-synthase pathway. EMBO J. 1993;12:3651-3657.

Wu W and Li L. Inhibition of nitric oxide synthase reduces motoneurons death due to spinal root avulsion. Neurosci. Lett. 1993; 153:121-124.

Wu W, Liuzzi FJ, Schinco FP, Depto AS, Li Y, Mong JA, Dawson TM, Snyder SH. Neuronal nitric oxide synthase is induced in spinal neurons by traumatic injury. Neurosci. 1994; 61: 719-726.

Zhang J, Dawson VL, Dawson TM, Snyder SH. Nitric oxide activation of poly (ADP-ribose) synthetase in neurotoxicity. Science 1994; 263:687-689.

Zorumski CF, Izumi Y. Nitric oxide and hippocampal synaptic plasticity. Biochem. Pharmacol. 1993; 46: 777-785.

AN INFLAMMATORY ROLE FOR NITRIC OXIDE DURING EXPERIMENTAL MENINGITIS IN THE RAT

Kathleen M. K. Boje[1]

[1]Department of Pharmaceutics
School of Pharmacy
University of Buffalo
Buffalo, New York 14260

INTRODUCTION

Nitric oxide (NO) functions as an intracellular messenger and an inflammatory mediator, depending on its concentration and cellular target. An enthusiastic, intensive and multi-disciplinary approach best characterizes the current research efforts focussed on understanding the diverse biological roles of NO. Physiological and pathological effects of NO have been described for the cardiovascular, pulmonary, gastrointestinal, immune, renal, endocrinological and central nervous systems[1,2]. Certainly, knowledge of the pathological effects of NO will facilitate the development potential therapeutic drugs applicable to a broad array of disease states.

Isoforms of NO Synthase

NO is the enzymatic product of NO synthase (NOS) on the substrate, arginine[1,3]. Traditionally, are at least two broadly defined classes of NOS: A constitutively expressed form (cNOS) which requires calcium and calmodulin, and a calcium independent, immunologically induced form (iNOS) which requires de novo synthesis following exposure to bacteria, bacterial products, (e.g. lipopolysaccharides, liptechoic acid) or inflammatory cytokines (e.g. TNF-α, IFN-γ, IL-1, IL-2 and IL-6). It is generally accepted that cNOS activity is a transient response to elevations in intracellular calcium, whereas iNOS activity occurs as prolonged response to an immune challenge. The nomenclature for the cNOS and iNOS isoforms is under refinement as it is apparent that the various isoforms are derived from different gene products, each with its own uniquely distinct molecular, cellular and pharmacological characteristics[1-3].

Depending on its redox state, constitutively produced NO may positively transduce physiological processes or initiate events culminating in cytotoxicity[4]. In contrast, sustained immunoinduced synthesis of NO almost uniformly results in cytotoxicity. NO and other reactive nitrogen oxides (degradation products of NO, e.g. $ONOO^-$, NO_2 and NO_2^-) mediate toxicity through the oxidation of protein sulfhydryls[5] and complexation

Neurodegenerative Diseases
Edited by Gary Fiskum, Plenum Press, New York, 1996

with iron containing respiratory enzymes[6]. Other targets of NO include nitrosylation of protein thiols, tyrosine residues and deoxyribonucleic acids[4].

iNOS Expression in Man

The conclusive demonstration of iNOS expression in human monocytes / macrophages is presently an intractable conundrum[3,7]. Although iNOS was first discovered in rodent macrophages, a similar phenomenon has yet to be reliably observed in human monocytes. Indirect support for iNOS activity in man derives from reports of high concentrations of NO_2^- and NO_3^- (stable degradation products of NO) in septic patients[8,9] and in cancer patients receiving interleukin-2[10,11]. The first direct evidence for an immunoinducible NOS isoform in man was obtained from human hepatocytes cultured in vitro[12]. This was followed by the molecular cloning and expression of iNOS from human hepatocytes[13] and chondrocytes[14], and by the localization of the human iNOS gene to chromosome 17 [15].

Association of NO with CNS Diseases

Throughout the 1990's (a.k.a. "The Decade of the Brain"), there has been an explosive growth in our understanding of the actions of NO in the central nervous system. Under certain conditions, neuronal cNOS activity can either serve as a signal transduction messenger or initiate events leading to neurotoxicity[1]. Deleterious consequences of iNOS activity have been suggested or defined for multiple sclerosis[16,17], experimental autoimmune encephalomyelitis[18,19], cerebral ischemia[20], lymphocytic choriomeningitis[21] and meningitis[22-25].

AN INFLAMMATORY ROLE OF NO DURING MENINGITIS

During meningitis, bacterial endotoxin triggers an inflammatory response which causes serious pathophysiologic alterations; namely, increased blood-brain and blood-CSF barrier permeability, increased CSF albumin concentrations and CSF neutrophilic pleocytosis[26-28]. Increased cerebrovascular permeability promotes disturbances in regional cerebral perfusion, cerebrovascular autoregulation, and cerebral metabolism, in addition to vasogenic cerebral edema, intracranial hypertension, and neuronal dysfunction[29-31]. A variety of inflammatory mediators are postulated to play a critical role in mediating these pathophysiologic alterations. Candidate mediators include cytokines[29-31], arachidonic acid metabolites[29-31], prostaglandins[32-36], leukocyte-endothelial adhesion molecules[31,37], toxic oxygen intermediates[29,32,38] and NO[22-25].

We hypothesized that disruption of the blood-brain and blood-CSF barriers during meningitis may be mediated by pathological production of NO. A number of observations are consistent with this hypothesis. Cultured cerebrovascular endothelial cells and meningeal fibroblasts were immunoinduced to synthesize NO[39,40]. Cultured monolayers of murine cerebral endothelial cells (often used as models of the blood-brain barrier), showed time- and dose-dependent changes in NO production[40,41] and monolayer permeability[42] following lipopolysaccharides (LPS) or cytokine exposure. It was recently reported that NO, a potent vasodilator agent, may mediate hyperemia in meningitis[32,43,44]. Moreover, pathological production of NO increased the vascular permeability of retinal, dermal and pulmonary capillary beds[45-47]. Although the endothelial capillary beds of the brain are morphologically different from peripheral vascular beds, it can be envisaged that pathologic production of NO may similarly increase the permeability of the cerebral vasculature, manifested as alterations in the blood-brain and blood-CSF barriers.

Detection of NO from Meninges of Rats with Meningitis

Using a sensitive chemiluminescent NO detection method, we observed NO production from meningeal tissues obtained from rats previously dosed with intracisternal LPS to induce experimental meningitis[25]. NO was not detected from control rats. The specificity and authenticity of the NO analysis was confirmed by the use of agents known to stabilize NO (e.g. superoxide dismutase), quench NO (e.g. methemoglobin) and inhibit NO synthesis (e.g. N^G-nitroarginine, N^G-methylarginine, N^G-aminoarginine and aminoguanidine)[25]. The time course of NO production by immunoinduced meningeal tissues revealed a 3 hour lag before peak quantities of NO were detected 8 hours after intracisternal LPS dosing. NO production subsequently declined over the next 16 hours, attaining control levels 24 hours after intracisternal treatment[48]. No evidence of a systemic NO inflammatory response to intracisternal LPS was detected in systemic white blood cells[48]. Moreover, inhibition of meningeal NO synthesis by NOS inhibitors followed a pharmacological rank order similar to that observed for the iNOS in rodent macrophage cell lines (N^G-aminoarginine > N^G-methylarginine > aminoguanidine >> N^G-nitroarginine)[48].

Increases in Cerebrovascular Permeability During Meningitis

Previously published studies of meningitis in animals examined the permeability changes of the blood-CSF barrier to ^{125}I-albumin and CSF leukocytosis as measures of disruption of the blood-brain barrier[26-28,30,33,49-53]. Although these reports purport to quantify blood-brain barrier injury, such parameters could be regarded as probes of the integrity of the blood-CSF barrier, where leptomeningeal microvascular injury permits substance influx into the CSF of the subarachnoid spaces and ventricles. Alternatively, a more accurate index of blood-brain barrier permeability would involve measurement of selected markers in brain parenchyma, where injury to the cerebral parenchymal endothelial vasculature would be reflected by substance influx into brain tissues.

The versatility of ^{14}C-sucrose as a quantitative marker of regional blood-brain permeability in hyperosmotic treatments is well established[54-58]. However, this tracer has not been employed for permeability studies in experimental meningitis. By measuring the accumulation of systemically administered ^{14}C-sucrose in regional brain tissues, one can pharmacokinetically assess brain region permeability during meningitis. ^{14}C-sucrose is a suitable marker for determining the integrity of the rat blood-brain and blood-CSF barriers as this tracer is not metabolized, does not bind to plasma proteins or brain capillaries, and is marginally permeable to the blood-brain and blood-CSF barriers under physiologic conditions[54,58].

It was first necessary to pharmacokinetically quantify disruption of the blood-brain barrier during meningitis using ^{14}C-sucrose. We induced experimental meningitis in rats by intracisternal administration of 0, 25 or 200 μg LPS in 10 μl of artificial CSF. Eight hours after intracisternal dosing, groups of rats were either dosed intravenously with ^{14}C-sucrose (5 μCi) for pharmacokinetic determination of regional blood-brain barrier permeability[59] or sacrificed for the determination of meningeal NO production[25,48]. Figure 1 shows the apparent unidirectional (influx) transfer coefficient ($K_{in(app)}$) for a representative brain region (hypothalamus - thalamus) and the meningeal NO production for the corresponding LPS treatments. Significant increases in $K_{in(app)}$ for different brain regions and meningeal NO were observed for the two LPS treatments compared to control (no LPS)[48,59]. There were no statistical differences between the two LPS doses for either $K_{in(app)}$ or meningeal NO production. This intriguing data suggests, but does not prove that there is an association between increased NO synthesis during meningitis and increased permeability of the blood-brain barrier.

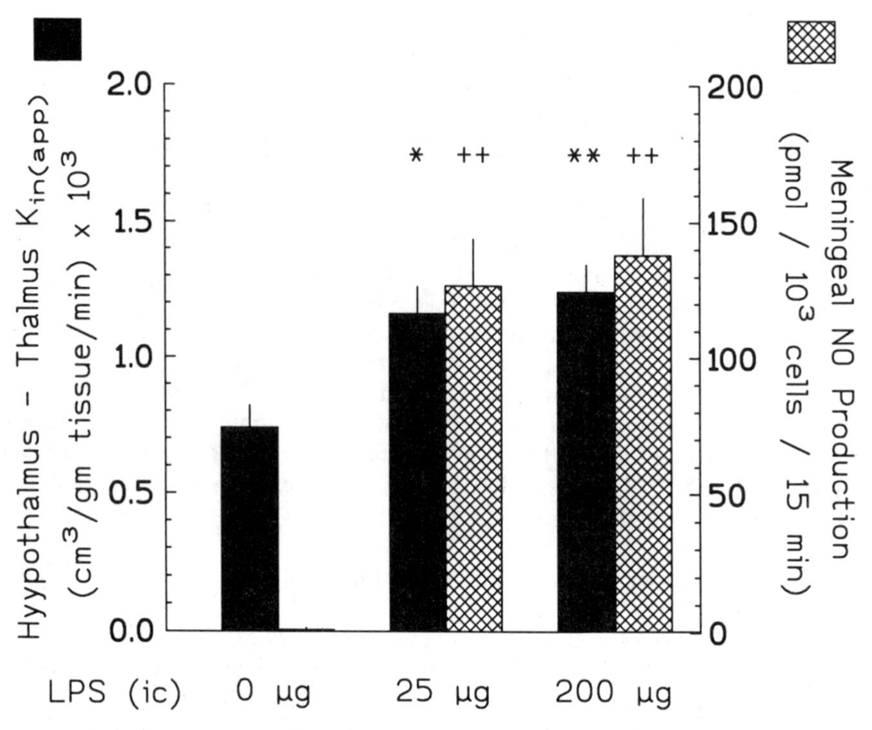

Figure 1. $K_{in(app)}$ and meningeal NO synthesis for intracisternal LPS doses. *P<0.05; **P<0.01 compared to control (0 µg LPS) $K_{in(app)}$ by ANOVA. ++P<0.01 compared to control (0 µg LPS) meningeal NO synthesis by ANOVA. $K_{in(app)}$ for the representative hypothalamus - thalamus brain region was determined from separate groups of 24 rats as described by Boje[59]. Similar significant increases in $K_{in(app)}$ were observed in other brain regions[59]. NO synthesis was determined from separate groups of 3-4 rats using previously published methods[25].

NO as a Mediator of Increases in Cerebrovascular Permeability

One experimental strategy to define a significant role of NO in disruption of the blood-brain barrier required a pharmacologic means to abrogate NO synthesis. This necessitated the rational design of a dosing regimen for in vivo NOS inhibitor administration such that meningeal NO synthesis was negligible during meningitis. The task of defining such a dosing regimen was especially vexing, since it is known that high infusion rates of the cNOS inhibitor, N^G-nitroarginine, evoke significant increases in mean arterial blood pressure, decreases in cortical cerebral blood flow and subsequent disruption of the blood-brain barrier to proteins, presumably due to a hypertensive filtration effect[60-62]. Hence, it was essential to design an NOS inhibitor dosing regimen which effectively suppressed meningeal iNOS activity without provoking hypertensive disruption of the blood-brain barrier[60-62]. There were additional concerns governing the selection of an appropriate NOS inhibitor: (a) the CSF, brain and systemic pharmacokinetics of NOS inhibitors are largely unknown; (b) the timing of inhibitor administration in relationship to LPS dosing should ideally be concurrent with the expression of de novo iNOS; and (c) irreversible iNOS inhibition occurs with prolonged exposure to NOS inhibitor analogues[39,63-66].

Based on published and our own data on the pharmacologic selectivity of NOS inhibitors for meningeal iNOS[25,45,67-69], we chose aminoguanidine as an appropriate iNOS inhibitor. Aminoguanidine was reported to be relatively selective for iNOS[67,68], with an iNOS potency equivalent to N^G-monomethyl-L-arginine (L-NMA). Although aminoguanidine's relative lack of effect on blood pressure is disputed[69], aminoguanidine reportedly was 1.5 - 2 orders of magnitude less effective than L-NMA in causing increases in mean arterial blood pressure[45]. Another important factor in the selection of aminoguanidine was that rudimentary information on the systemic pharmacokinetics of aminoguanidine could be estimated from the literature. Using moment analysis, we estimated an approximate aminoguanidine systemic clearance of 650-730 ml/hr/kg and volume of distribution at steady state of 1.0-2.2 L/kg from the aminoguanidine concentration - time profiles published by Beaven et al.[70]. Assuming linear biexponential pharmacokinetics in rats, knowledge of the clearance and volume of distribution permitted the semi-rational design of an aminoguanidine dosing regimen which would yield predicted aminoguanidine steady stated plasma concentrations of 1 mM. This dosing regimen consisted of a bolus loading dose (1.46 mmoles/kg) followed by a concomitant constant rate infusion (0.505 mmoles/hr/kg)[25].

To directly test the hypothesis that NO contributes to disruption of the blood-brain barrier during meningitis, groups of rats were intracisternally dosed with 0, 25 or 200 µg LPS, followed by intravenous saline or aminoguanidine. Depending on the study protocol, rats were dosed with ^{14}C-sucrose to assess the integrity of the blood-brain barrier[71] or with Evans blue dye for determination of the permeability of the blood-CSF barrier[25]. Meningeal NO production was determined in a separate group of rats were similarly dosed with LPS and intravenous saline or aminoguanidine[25]. Figure 2A shows the percent increase in the blood-brain barrier permeability to ^{14}C-sucrose for the representative brain region, the hypothalamus - thalamus. The percent increase in the blood-CSF barrier to Evans blue dye is presented in Figure 2B, while Figure 2C depicts the corresponding meningeal NO synthesis. Intracisternal LPS clearly provoked significant increases in blood-brain and blood-CSF barrier permeabilities, which was accompanied by a significant increase in meningeal NO production. Administration of aminoguanidine completely attenuated the LPS-induced permeability increases in the blood-brain barrier and partially diminished corresponding increases in the blood-CSF barrier. Aminoguanidine therapy also significantly blunted meningeal NO synthesis. These experiments are persuasive evidence identifying NO as an inflammatory mediator of blood-brain and blood-CSF barrier permeability during meningitis.

SUMMARY

In summary, all available evidence, both published[22-25,43,44] and observational[48,59,71], frame a cogent argument for a fundamental role of NO in contributing to the neuroinflammation of meningitis. Not only is NO produced during meningitis in humans and laboratory rats[22-25,48], but literature data implicate NO as mediator of the early stages of meningeal hyperemia and edema[43,44]. The importance and excitement of the present work is underscored by the identification of NO as a novel, deleterious mediator of blood-brain and blood-CSF barrier disruption during meningitis. Clearly, basic and clinical research throughout the remainder of "The Decade of the Brain" will broaden and better define the role of NO in other neuroinflammatory diseases. This ever expanding knowledge will undoubtedly provide promising, new therapies directed against iNOS in the management of brain diseases.

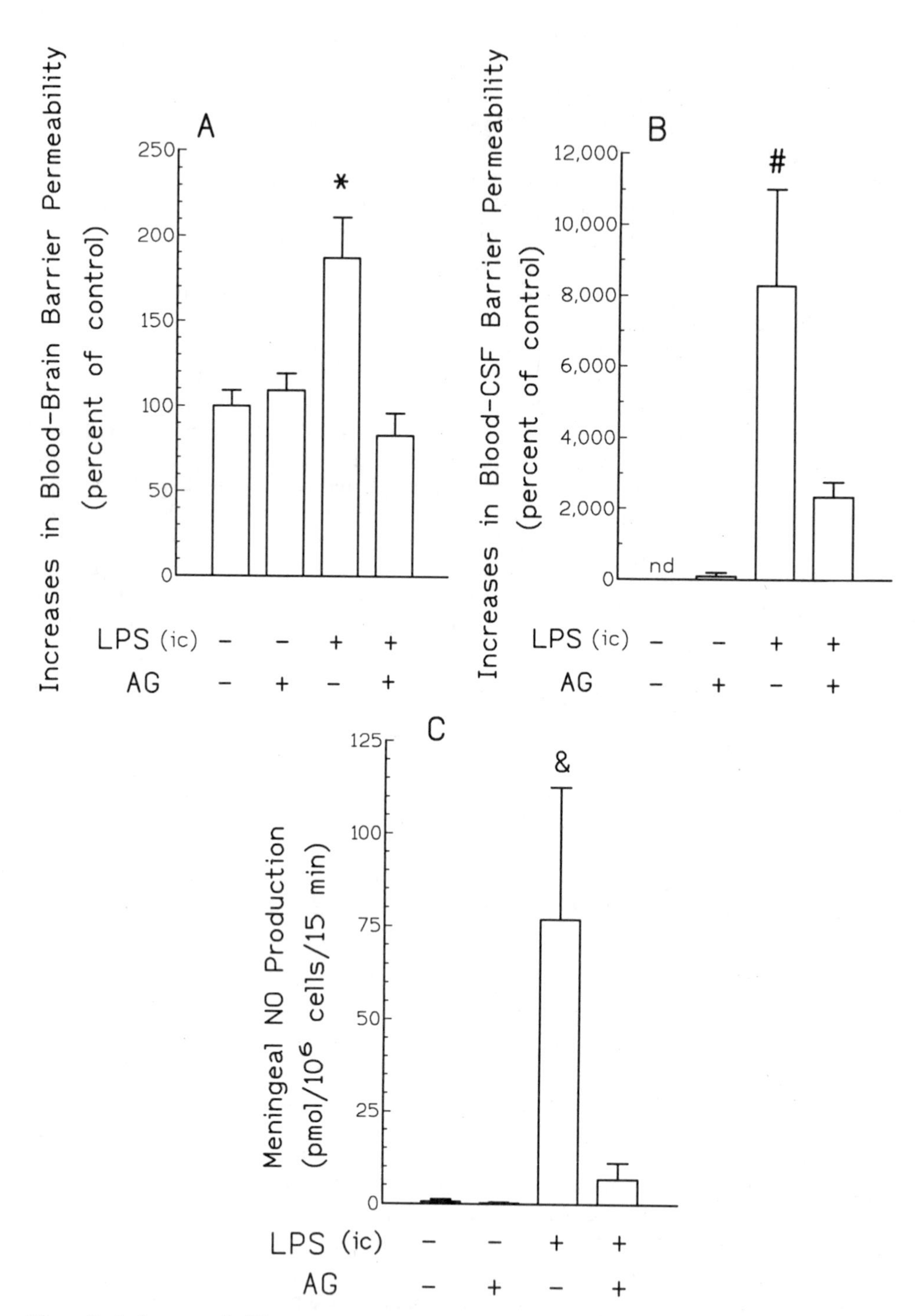

Figure 2. A. Increases in ^{14}C-sucrose blood-brain barrier permeability in the hypothalamus - thalamus brain region. Groups of rats (n = 8) were dosed intracisternally with LPS (+) or vehicle (-) and bolus loading dose and constant infusion of aminoguanidine (AG; +) or saline (-) . Similar results were observed in other brain regions. **B.** Permeability of Evans blue dye across the blood-CSF barrier in groups of rats (n = 6) dosed with LPS and/or AG as described. nd = not done. **C.** NO synthesis from meninges harvested from groups of rats (n = 3 - 4) dosed with LPS and/or AG as described.
*, #, & $P<0.05$ from all other groups by ANOVA.

Acknowledgements

The author thanks Mr. Peter Korytko, Mr. Steven Jodish and Ms. Patricia Neubauer for their excellent technical assistance. Supported in part by the American Association of Colleges of Pharmacy, the Pharmaceutical Research and Manufacturers Foundation, the State of New York and the NIH (NS 31939).

REFERENCES

1. T.M. Dawson and V.L. Dawson, Nitric oxide: Actions and pathological roles, *The Neuroscienctist* 1:7 (1995).
2. H.H.H.W. Schmidt and U. Walter, NO at work, *Cell* 78:919 (1994).
3. C. Nathan and Q.-W. Xie, Nitric oxide synthases: Roles, tolls and controls, *Cell* 78:915 (1994).
4. J.S. Stamler, Redox signaling: Nitrosylation and related target interactions of nitric oxide, *Cell* 78:931 (1994).
5. R. Radi, J.S. Beckman, K.M. Bush, and B.A. Freeman, Peroxynitrite oxidation of sulhydryls: The cytotoxic potential of superoxide and nitric oxide, *J. Biol. Chem.* 266:4244 (1991).
6. J.C. Drapier and J.B. Hibbs, Jr., Differentiation of murine macrophages to express nonspecific cytotoxicity for tumor cells results in L-arginine-dependent inhibition of mitochondrial iron-sulfur enzymes in the macrophage effector cells, *J. Immunol.* 140: 2829 (1988).
7. M. Denis, Human monocytes / macrophages: NO or no NO? *J. Leukocyte Biol.* 55:682 (1994).
8. J.B. Ochoa, A.O. Udekwu, T.R Billiar, R.D. Curran, F.B. Cerra, R.L. Simmons, and A.B. Peitzman, Nitrogen oxide levels in patients after trauma and during sepsis, Ann. Surg. 214: 621 (1991).
9. H.F. Goode, P.D. Howdle, B.E. Walker and N.R. Webster, Nitric oxide synthase activity is increased in patients with sepsis syndrome, *Clinical Sci.* 88:131 (1995).
10. J.B. Hibbs, C. Westenfelder, R. Taintor, Z. Vavrin, C. Kablitz, R.L. Baranowski, J.H. Ward., R.L. Menlove, M.P. McMurry, J.P. Kushner, and W.E. Samlowski, Evidence for cytokine-inducible nitric oxide synthesis from L-arginine in patients receiving interleukin-2 therapy, *J. Clin. Invest.* 89: 867 (1992).
11. J.B. Ochoa, B. Curtis, A.B. Peitzman, R.L Simmons, T.R. Billiar, R. Hoffman, R. Rault, D.L. Longo, W.J. Urba, and A.C. Ochoa, Increased circulating nitrogen oxides after human tumor immunotherapy: correlation with toxic hemodynamic changes, *J. Natl. Cancer Inst.* 84: 864 (1992).
12. A.K. Nussler, M. Di Silvio, T.R. Billiar, R.A. Hoffman, D.A. Geller, R. Selby, J. Madariaga, and R.L. Simmons, Stimulation of the nitric oxide synthase pathway in human hepatocytes by cytokines and endotoxin, *J. Exp. Med.* 176: 261 (1992).
13. D.A. Geller, C.J. Lowenstein, R.A. Shapiro, A.K. Nussler, M. Di Silvio, S.C. Wang, D.K. Nakayama, R.L. Simmons, S.H. Snyder, and T.R. Billiar, Molecular cloning and expression of inducible nitric oxide synthase from human hepatocytes, *Proc. Natl. Acad. Sci. USA* 90:3491 (1993).
14. I.G. Charles, R.J.M. Palmer, M.S. Hickery, M.T. Bayliss, A.P. Chubb, V.S. Hall, D.W. Moss, and S. Moncada, Cloning, characterization, and expression of a cDNA encoding an inducible nitric oxide synthase from the human chondrocyte, *Proc. Natl. Acad. Sci. USA* 90:11419 (1993).
15. N.A. Chartrain, D.A. Geller, P.P. Koty, N.F. Sitrin, A.K. Nussler, E.P. Hoffman,

T.R. Billiar, N. I. Hutchinson, and J.S. Mudgett, Molecular cloning, structure, and chromosomal localization of the human inducible nitric oxide synthase gene, *J. Biol. Chem.* 269: 6765, 1994.
16. A.W. Johnson, J.M. Land, E.J. Thompson, J.P. Bolaños, J.B. Clark, and S.J.R. Heales, Evidence for increased nitric oxide production in multiple sclerosis, *J. Neurol.* 58: 107 (1995).
17. L. Bö, T.M. Dawson, S. Wesselingh, S. Mörk, S. Choi, P.A. Kong, D. Hanley, and B.D. Trapp, Induction of nitric oxide synthase in demyelinating regions of multiple sclerosis brains, *Ann. Neurol.* 36:778 (1994).
18. J.D. MacMicking, D.O. Willenborg, M.J. Wedemann, K.A. Rockett, and W.B. Cowden, Elevated secretion of reactive nitrogen and oxygen intermediates by inflammatory leukocytes in hyperacute experimental autoimmune encephalomyelitis: Enhancement by the soluble products of encephalitogenic T cell, *J. Exp. Med.* 176:303 (1992).
19. A.H. Cross, T.P. Misko, R.F. Lin, W.F. Hickey, J.L. Trotter, and R.G. Tilton, Aminoguanidine, an inhibitor of inducible nitric oxide synthase, ameliorates experimental autoimmune encephalomyelitis in SJL mice, *J. Clin. Invest.* 93:2684 (1994).
20. J.P. Nowicki, D. Duval, H. Poignet, and B. Scatton, Nitric oxide mediates neuronal death after focal cerebral ischemia, *Eur. J. Pharmacol.* 204: 339 (1991).
21. I.L. Campbell, A. Samimi, and C.-S. Chiang, Expression of the inducible nitric oxide synthase: correlation with neuropathology and clinical features in mice with lymphocytic choriomeningitis, *J. Immunol.* 153:3622 (1994).
22. J.J. Visser, R.J.P.M. Scholten, and K. Hoekman, Nitric oxide synthesis in meningococcal meningitis, *Ann. Internal Med.* 120:345 (1994).
23. S. Milstien, N. Sakai, B.J. Brew, C. Krieger, J.H. Vickers, K. Saito, and M.P. Heyes, Cerebrospinal fluid nitrite/nitrate levels in neurologic diseases, *J. Neurochem.* 63:1178 (1994).
24. H.-W. Pfister, A. Bernatowicz, U. Ködel, and M. Wick, Nitric oxide production in bacterial meningitis, *J. Neurol. Neurosurg. Psychiat.* 58:384 (1995).
25. K.M.K. Boje, Inhibition of nitric oxide synthase partially attenuates alterations in the blood-cerebrospinal fluid barrier during experimental meningitis in the rat, *Eur. J. Pharmacol.* 272:297 (1995).
26. V.J. Quagliarello, A. Ma, H. Stukenbrok, and G.E. Palade, Ultrastructural localization of albumin transport across the cerebral microvasculature during experimental meningitis in the rat, *J. Exp. Med.* 174:657 (1991).
27. B. Wispelwey, A.J. Lesse, E.J. Hansen, and W.M. Scheld, Haemophilus influenzae lipopolysaccharide-induced blood brain barrier permeability during experimental meningitis in the rat, *J. Clin. Invest.* 82:1339 (1988).
28. B. Wispelwey, E.J. Hansen, and W.M. Scheld, Haemophilus influenzae outer membrane vesicle-induced blood-brain barrier permeability during experimental meningitis, *Infec. Immun.* 57:2559 (1989).
29. V. Quagliarello and W.M. Scheld, Bacterial meningitis: Pathogenesis, pathophysiology, and progress, *New England J. of Med.* 327:864 (1992).
30. X. Sáez-Llorens, O. Ramilo, M.M. Mustafa, J. Mertsola, and G.H. McCracken, Molecular patho-physiology of bacterial meningitis: Current concepts and therapeutic implications. *J. Ped.* 116:671 (1990).
31. A.R. Tunkel and W.M. Scheld, Pathogenesis and pathophysiology of bacterial meningitis. *Annual Rev. Med.* 44:103 (1993).
32. H.-W. Pfister, K. Frei, B. Ottand, U. Koedel, A. Tomasz, and A. Fontana, Transforming growth factor β_2 inhibits cerebrovascular changes and brain edema

formation in the tumor necrosis factor α independent early phase of experimental pneumococcal meningitis, *J. Exp. Med.* 176:265 (1992).

33. J.L. Kadurugamuwa, B. Hengstler, M.A. Bray, and O. Zak, Inhibition of complement-Factor-5a-induced inflammatory reactions by prostaglandin E_2 in experimental meningitis, *J. Infect. Dis.* 160:715 (1989).
34. X. Sáez-Llorens, O. Ramilo, MM. Mustafa, J. Mertsola, C. de Alba, E. Hansen, and G.H. McCracken, Jr., Pentoxifylline modulates meningeal inflammation in experimental bacterial meningitis, *Antimicrob. Agents Chemother* 34:837 (1990).
35. E. Tuomanen, Breaching the blood-brain barrier, *Scientific American* Feb.:80 (1993).
36. A.-M. Van Dam, M. Brouns, W. Man-A-Hing, and F. Berkenbosch, Immunocytochemical detection of prostaglandin E_2 in microvasculature and in neurons of rat brain after administration of bacterial endotoxin, *Brain Research* 613:331 (1993).
37. E.I. Tuomanen, S.M. Prasad, J.S. George, A.I. Hoepelman, P. Ibsen, I. Heron, and R.M. Starzyk, Reversible opening of the blood-brain barrier by anti-bacterial antibodies, *Proc. Natl. Acad. Sci. USA* 90:7824 (1993).
38. H.-W. Pfister, U. Koedel, R.L. Haberl, U. Dirnagl, W. Feiden, G. Ruckdeschel, and K.M. Einhäupl, Microvascular changes during the early phase of experimental bacterial meningitis, *J. Cerebral Blood Flow and Metab.* 10:914 (1990).
39. K.M. Boje and P. Arora, Microglial-produced nitric oxide and reactive nitrogen oxides mediate neuronal cell death, *Brain Research* 587:250 (1992).
40. R.G. Kilborn and P. Belloni, Endothelial cell production of nitrogen oxides in repone to interferon-γ in combination with tumor necrosis factor, interleukin-1 or endotoxin, *J. Nat. Cancer Inst.* 82:726 (1990).
41. S.S. Gross, E.A. Jaffe, R. Levi, and R.G. Kilborn, Cytokine-activated endothelial cells express an isotype of nitric oxide synthase which is tetrahydrobiopterin-dependent, calmodulin-independent and inhibited by arginine analogs with a rank - order of potency characteristic of activated macrophages, *Biochem. Biophys. Res. Comm.* 178:823 (1991).
42. A.R. Tunkel, S.W. Rosser, E.J. Hansen, and W.M. Scheld, Blood-brain barrier alterations in bacterial meningitis: development of an in vitro model and observations on the effects of lipopolysaccharide, *In Vitro Cell Dev Biol.* 27A: 113 (1991).
43. J.E. Brian, D.D. Heistad, and F.M. Faraci, Dilatation of cerebral arterioles in response to lipopolysaccharide in vivo, *Stroke*, 26:277 (1995).
44. R.L. Haberl, F. Anneser, U. Ködel, and H.-W. Pfister, Is nitric oxide involved as a mediator of cerebrovascular changes in the early phase of experimental pneumococcal meningitis? *Neurol. Res.*16:108 (1994).
45. J.A. Corbet, R.G. Tilton, K. Chang, K.S. Hasan, Y. Ido, J.L. Wang, M.A. Sweetland, J.R. Lancaster, J.R. Williamson, and M.L. McDaniel,Amino-guanidine, a novel inhibitor of nitric oxide formation, prevents diabetic vascular dysfunction, *Diabetes* 41: 552 (1992).
46. M.S. Mulligan, J.M. Hevel, M.A. Marletta, and P.A. Ward, Tissue injury caused by deposition of immune complexes is L-arginine dependent, *Proc. Natl. Acad. Sci.* 88: 6338 (1991).
47. W.G. Mayhan, Role of nitric oxide in modulating permeability of hamster cheek pouch in response to adenosine 5'-diphosphate and bradykinin, *Inflammation* 16: 295 (1992).
48. P.J. Korytko and K.M.K. Boje, submitted for publication.

49. V.J. Quagliarello, W.J. Long, and W.M. Scheld, Morphologic alterations of the blood-brain barrier with experimental meningitis in the rat, *J. Clin. Invest.* 77:1084 (1986).
50. V.J. Quagliarello, B. Wispelwey, W.J. Long, and W.M. Scheld, Recombinant human interleukin-1 induces meningitis and blood-brain barrier injury in the rat, *J. Clin. Invest.* 87:1360 (1991).
51. M.M Mustafa, O. Ramilo, K.D. Olsen, P.S. Franklin, E.J. Hansen, B. Beutler, and G. McCracken, Tumor necrosis factor in mediating experimental haemophilus influenzae type B meningitis, *J. Clin. Invest.* 84:1253 (1989).
52. O. Ramilo, X. Sáez-Llorens, J. Mertsola, H. Jafari, K.D. Olsen, E.J. Hansen, M. Yoshinaga, S. Ohkawara, H. Nariuchi, and G.H. McCracken, Tumor necrosis factor α/cachectin and interleukin 1β initiate meningeal inflammation, *J. Exp. Med.* 172:497 (1990).
53. E. Tuomanen, B. Hengstler, R. Rich, M.A. Bray, O. Zak, and A. Tomasz, Nonsteroidal anti-inflammatory agents in the therapy for experimental pneumococcal meningitis, *J. Infect. Dis.* 155:985 (1987).
54. K. Ohno, K.D. Pettigrew, and S.I. Rapoport, Lower limits of cerebrovascular permeability to nonelectrolytes in the conscious rat, *Am. J. Physiol.* 235:H299 (1978).
55. S.I. Rapoport, W.R. Fredericks, K. Ohno, and K.D. Pettigrew, Quantitative aspects of reversible osmotic opening of the blood-brain barrier, *Am J. Physiol.* 238:R421 (1980).
56. Q.R. Smith, Quantitation of Blood-Brain Barrier Permeability. In Implications of the Blood-Brain Barrier and Its Manipulation, ed. by E.A. Newelt, pp. 85-118, Vol. 1, Plenum Medical Book Co., New York, 1989.
57. Y.Z. Ziylan, P.J. Robinson, and S.I. Rapoport, Differential blood-brain barrier permeabilities to [^{14}C]-sucrose and [^{3}H]inulin after osmotic opening in the rat, *Exp. Neurol.* 79:845 (1983).
58. Y.Z. Ziylan, P.J. Robinson, and S.I. Rapoport, Blood-brain barrier permeability to sucrose and dextran after osmotic opening, *Am. J. Physiol.* 247:R634 (1984).
59. K.M.K. Boje, Cerebrovascular permeability changes during experimental meningitis in the rat, *J. Pharmacol. Exp. Ther.* In press.
60. R. Prado, B.D. Watson, J. Kuluz, and W.D. Dietrich, Endothelium-derived nitric oxide synthase inhibition: effects on cerebral blood flow, pial artery diameter and vascular morphology in rats, *Stroke* 23:1118 (1992).
61. S.I. Rapoport, Opening of the blood-brain barrier by acute hypertension, *Exp. Neurol.* 52: 467 (1976).
62. P. Kubes and N.D. Granger, Editorial comment, *Stroke* 23: 1124 (1992).
63. M.A. Dwyer, D.S. Bredt, and S.H. Snyder, Nitric oxide synthase: Irreversible inhibition by L-N^{G}-nitroarginine in brain in vitro and in vivo, *Biochem. Biophys. Res. Comm.* 176:1136 (1991).
64. N.M. Olken and M.A. Marletta, N^{G}-methyl-L-arginine functions as an alternate substrate and mechanism-based inhibitor of nitric oxide synthase, *Biochem.* 32:9677 (1993).
65. N.M. Olken and M.A. Marletta, N^{G}-allyl- and N^{G}-cyclopropyl-L-arginine: Two novel inhibitors of macrophage nitric oxide synthase, *J. Med. Chem.* 35:1137 (1992).
66. P. Klatt, K. Schmidt, F. Brunner and B. Mayer, Inhibitors of brain nitric oxide synthase: binding kinetics, metabolism and enzyme inactivation, *J. Bio. Chem.* 269:1674 (1994).
67. T.P. Misko, W.M. Moore, T.P. Kasten, G.A. Nickols, J.A. Corbett, R.G. Tilton, M.L. McDaniel, J .R. Williamson, and M.G. Currie, Selective inhibition of

the inducible nitric oxide synthase by aminoguanidine, *Eur. J. Pharmacol.* 233:119 (1993).

68. M.J.D. Griffiths, M. Messent, R.J. MacAllister, and T.W. Evans, Aminoguanidine selectively inhibits inducible nitric oxide synthase, *Br. J. Pharmacol.* 110:963 (1993).
69. F. Laszlo, S.M. Evans, and B.J.R.Whittle, Aminoguanidine inhibits both constitutive and inducible nitric oxide synthase isoforms in rat intestinal microvasculature in vivo, *Eur. J. Pharmacol.* 272:169 (1995).
70. M.A. Beaven, J.W. Gordon, S. Jacobsen and W.B Severs, A specific and sensitive assay for aminoguanidine: Its application to a study of the disposition of aminoguanidine in animal tissues, *J. Pharmacol. Exp. Ther.* 165:14 (1969).
71. K.M.K. Boje, manuscript in preparation.

OXIDATIVE STRESS PLAYS A ROLE IN THE PATHOGENESIS OF FAMILIAL AND SPORADIC AMYOTROPHIC LATERAL SCLEROSIS

Catherine Bergeron[1,2], Connie Petrunka[1], Luitgard Weyer[1]

[1]Centre for Research in Neurodegenerative Diseases and
Department of Pathology
University of Toronto
6 Queen's Park Crescent West
Toronto, Ontario, M5S 1A8, Canada

[2]Department of Pathology
The Toronto Hospital
Toronto, Ontario, Canada

INTRODUCTION

Amyotrophic lateral sclerosis (ALS) is a progressive neurodegenerative disorder characterized by the selective death of upper and lower motorneurons. The disease affects 0.6-2.6:100,000 individuals [1], with a mean age at onset of approximately 55 and a duration of two to three years on average [2]. Motorneuron death results in muscle weakness and paralysis with eventual ventilatory failure and death.

The etiology and pathogenesis ALS remain unclear, and postulated disease mechanisms include altered gene expression or RNA metabolism, excitotoxicity, autoimmunity, altered axoplasmic transport, growth factor deficiency and environmental toxins, either alone or in combination (For review, see Windebank and Williams [3]). More recently oxidative stress has been implicated as an important factor.

OXIDATIVE STRESS AND ANTIOXIDANTS

Oxidative stress is defined as an increase in the generation of oxygen-derived species (superoxide anion, hydroxyl radical, hydrogen peroxide) beyond the ability of antioxidant defenses to cope with them. Such oxygen-derived species are generated as byproducts of normal and aberrant metabolic processes that utilize molecular oxygen. An excess in their production not counterbalanced by a parallel increase in anti-oxidant defenses is detrimental to the cell with consequent oxidative damage to lipids, proteins and nucleic acids (For review see Halliwell [4]).

Neurodegenerative Diseases
Edited by Gary Fiskum, Plenum Press, New York, 1996

Antioxidant defenses include the presence of small molecules such as ascorbic acid, Vitamin E and glutathione as well as free radical scavenging enzymes. These are superoxide dismutase (SOD), glutathione peroxidase (GPX) and catalase. SOD exists in three different forms, cytoplasmic (SOD-1), mitochondrial (SOD-2) and extracellular (SOD-3), and dismutates the superoxide anion by converting it into hydrogen peroxide.

The expression of antioxidative enzymes, including SOD-1, shows marked variation between and within organs and correlates with the level of oxidative metabolism and the cell vulnerability to oxidative damage [5,6]. The brain also shows a differential pattern of expression of antioxidant enzymes along cell types with a predominant neuronal expression of SOD-1 [7,8]. In contrast, glutathione peroxidase and glutathione are exclusively astrocytic [9-12], while SOD-2 is present in both neurons and astrocytes [13].

SOD-1 MUTATIONS IN FAMILIAL ALS

Recent studies have demonstrated mutations in the SOD-1 gene in some cases of familial ALS [14-19]. Variable levels of decreased SOD-1 activity have been associated with several of the mutations [16,20]. Whether or not a loss of enzyme activity of this magnitude is sufficent, by itself, to cause motor neuron death in ALS remains conjectural [21,22]. Furthermore, a significant amount of enzyme activity is retained with some mutations, and a gain of function of the mutant enzyme is now postulated, possibly of a toxic nature [23-26], such as an enhanced affinity for peroxynitrite and subsequent cellular injury through protein nitration [27].

SOD-1 EXPRESSION IN SPORADIC ALS

We have recently documented a 42% increase in SOD-1 mRNA levels in alpha motorneurons of sporadic ALS [28]. Although a primary increase in SOD-1 mRNA levels cannot be entirely ruled out, a more likely explanation is that the expression of the enzyme is induced, possibly in response to oxidative stress. That increased SOD-1 expression can reflect oxidative stress has been demonstrated in rat lung [29], endothelial [30,31] and smooth muscle cells [31] after hyperoxia. SOD-1 induction has also been well documented in gerbil hippocampus after ischemia [32]. The enzyme activity of SOD-1, SOD-2 and catalase was recently measured in the spinal cord of sporadic ALS and found to be normal, while that of glutathione peroxidase was significantly increased [33]. These measurements were performed on homogenized spinal cord tissue, however, and do not take into account the predominantly neuronal localization of SOD-1; one should therefore contemplate the possibility that a normal enzyme activity level in the presence of significant neuronal loss may in fact be associated with increased cellular enzyme levels in the individual residual neurons. Similarly, the increased glutathione peroxidase activity level may reflect in part the astrocytic localization of this enzyme in the presence of reactive gliosis.

Other evidence implicating increased oxidative stress in ALS includes a marked increase in the protein carbonyl content, an index of protein oxidation [34], in Brodmann Area 6 in sporadic ALS [35]. Mitchell et al [36] also demonstrated a decline in whole blood GPX activity which they speculate may reflect increased mobilization and transport of the enzyme as a result of enhanced free radical activity. Increased [^{35}S] glutathione binding sites have been observed in spinal cords from patients with sporadic ALS, possibly as a result of increased concentration of the agonist in the presence of oxidative stress [37]. Finally, SOD-1 immunoreactivity has been demonstrated in intraneuronal neurofilamentous accumulations in both familial [38] and sporadic [39] cases.

SOURCES OF OXIDATIVE STRESS IN ALS

The increased oxidative burden seen in ALS may represent a disease related event. Reactive oxygen species are generated in the course of cell death and neuronal death as a result of various stimuli and can be modulated by the antioxidant activity of the proto-oncogen Bcl-2 [40,41]. Glutamate excitotoxicity has been proposed as a pathophysiological mechanism is ALS [42-44]. In the course of this process superoxide radicals may be generated through several pathways [45] including direct production of oxygen radicals [46-48], calcium mediated activation of xanthine oxidase [49] and the initiation of the arachidonic acid cascade by phospholipase A_2 [50-52]. Finally nitric oxide is also produced [53-55] which may interact with the superoxide anion to form highly reactive peroxynitrites [56]. Importantly, the addition of SOD-1 to several models of excitotoxicity, both NMDA and non-NMDA mediated, has been shown to reduce cell death significantly [49,53,54]. Glutamate excitotoxicity could therefore significantly increase the oxidative burden of motorneurons. Microglial activation also takes place in brain regions involved by ALS [57]. These cells, like other members of the macrophage family, are known to produce both superoxide anions [58,59] and nitric oxide [60] and may possibly contribute to an increased oxidative burden in ALS. The increased iron content detected in ALS spinal cord [33] may also contribute to oxidative stress through several mechanisms [61]. Finally, the motorneurons engaged in the re-innervation process support unusually large motor units with a consequent increase in their oxidative metabolism [62], possibly leading to increased oxidative stress and a compensatory increase in SOD-1 expression.

OXIDATIVE STRESS AND THE PATHOGENESIS OF ALS

While oxidative stress may be generated in the course of ALS, it may also play a primary pathogenetic role in both the familial and sporadic forms of the disease. Altered mitochondrial energy metabolism may occur with aging leading to increased oxidative stress and a predisposition to glutamate excitotoxicity [63]. Environmental factors may also lead to the increased generation of superoxide and other oxygen free radicals in ALS, a possibility raised by the recent appearance of motorneuron disease in horses [64]. Under such circumstances, individuals with altered SOD-1 structure and/or activity would be at special risk of developing ALS. Decreased SOD21 activity could compromise the ability of motorneurons to cope with the additional oxidative burden and lead to motorneuron death [65,66]. Alternatively, the induction of high levels of mutant enzyme with toxic properties may precipitate or accelerate neuronal loss [27].

The unique vulnerability of motor neurons to ALS remains unexplained and is not the result of a low constitutive expression of SOD-1. Using the *in situ* hybridization technique, we have recently demonstrated a high level of SOD-1 expression in human motorneurons, substantia nigra, nucleus basalis, hippocampus and cortex (manuscript in preparation), consistent with the immunohistochemical study of Pardo et al [67]. Furthermore, similar levels of expression are observed in both vulnerable and "resistant" motorneurons such as the oculomotor neurons.

CONCLUSION

Our increasing knowledge of the pathogenesis of ALS reveals a complex disorder, probably heterogeneous and multifactorial like many other late onset disorders such as AD and cancer. Oxidative stress, however, is likely an important pathogenetic event in all forms of ALS. It is therefore important to continue to investigate the possible sources of oxidative stress in ALS, the antioxidant status of motorneurons and the mechanisms whereby oxidative stress may lead to neuronal death, in particular its close interaction with glutamate excitotoxicity and aging.

REFERENCES

1. C. N. Martyn, Epidemiology, *in*: "Motor neuron disease", AC Williams ed., Chapman & Hall Medical, London (1994).
2. R. Tandan, Clinical features and differential diagnosis of classical motor neuron disease, *in*: "Motorneuron disease", AC Williams ed., Chapman & Hall Medical, London (1994).
3. A. J. Windebank, and D. B. Williams, Motor Neuron Disease, *in*: "Peripheral Neuropathy", Peter James Dyck and PK Thomas ed., W.B. Saunders Company, Philadelphia (1993).
4. B. Halliwell, Reactive oxygen species and the central nervous system, *J Neurochem.* 59:1609 (1992).
5. K. E. Muse, T. D. Oberley, J. M. Sempf, and L. W. Oberley, Immunolocalization of antioxidant enzymes in adult hamster kidney, *Histochem J.* 26:734 (1994).
6. T. D. Oberley, L. W. Oberley, A. F. Slattery, L. J. Lauchner, and J. H. Elwell, Immunohistochemical localization of antioxidant enzymes in adult syrian hamster tissues and during kidney development, *Am J Pathol.* 137:199 (1990).
7. P. Zhang, P. Damier, E. Hirsh, Y. Agid, I. Ceballos-Picot, P. Sinet, A. Nicole, M. Laurent, and F. Javoy-Agid, Preferential expression of superoxide dismutase messenger RNA in melanized neurons of human mesencephalon, *Neurosci.* 55:167 (1993).
8. I. Ceballos, F. Javoy-Agid, A. Delacourte, A. Defossez, M. Lafon, E. Hirsch, A. Nicole, P. Sinet, and Y. Agid, Neuronal localization of copper-zinc superoxide dismutase protein and mRNA within the human hippocampus from control and Alzheimer's disease brains, *Free Rad Res Comm.* 12-13:571 (1991).
9. S. P. Raps, J. C. Lai, L. Hertz, and A. J. Cooper, Glutathione is present in high concentrations in cultured astrocytes but not in cultured neurons, *Brain Res.* 493:398 (1989).
10. A. Slivka, C. Mytilineou, and G. Cohen, Histochemical evaluation of glutathione in brain, *Brain Res.* 409:275 (1987).
11. T. K. Makar, M. Nedergaard, A. Preuss, A. Gelbard, A. Perumal, and A. J. Cooper, Vitamin E, ascorbate, glutathione, glutathione disulfide, and enzymes of glutathione metabolism in cultures of chick astrocytes and neurons: evidence that astrocytes play an important role in antioxidative processes in the brain, *J Neurochem.* 62:45 (1994).
12. P. Damier, E. Hirsch, P. Zhang, Y. Agid, and F. Javoy-Agid, Glutathione peroxidase, glial cells and Parkinson's disease, *Neurosci.* 52:1 (1993).
13. P. Zhang, P. Anglade, E. Hirsch, F. Javoy-Agid, and Y. Agid, Distribution of manganese-dependent superoxide dismutase in the human brain, *Neurosci.* 61:317 (1994).
14. D. R. Rosen, T. Siddique, D. Patterson, D. A. Figlewicz, P. Sapp, A. Hentati, D. Donaldson, J. Goto, J. P. O'Regan, H.-X. Deng, Z. Rahmani, A. Krizus, D. McKenna-Yasek, A. Cayabyab, S. M. Gaston, R. Berger, R. E. Tanzi, J. J. Halperin, B. Herzfeldt, R. Van den Bergh, W.-Y. Hung, T. Bird, G. Deng, D. W. Mulder, C. Smyth, N. G. Laing, E. Soriano, M. A. Pericak-Vance, J. Haines, G. A. Rouleau, J. S. Gusella, R. S. Horvitz, and R. H. Brown Jr, Mutations in Cu/Zn superoxide dismutase gene are associated with familial amyotrophic lateral sclerosis, *Nature.* 362:59 (1993).
15. A. Elshafey, G. W. Lanyon, and M. J. Connor, Identification of a new missense point mutation in exon 4 of the Cu/Zn superoxide dismutase (SOD-1) gene in a family with amyotrophic lateral sclerosis, *Hum Mol Genet.* 3:363 (1994).
16. M. Ogasawara, Y. Matsubara, K. Narisawa, M. Aoki, S. Nakamura, and K. Abe, Mild ALS in Japan associated with novel SOD mutation, *Nature Genet.* 5:323 (1993).
17. R. Nakano, S. Sato, T. Inuzuka, K. Sakimura, M. Mishina, H. Takahashi, F. Ikuta, Y. Honma, J. Fujii, N. Taniguchi, and S. Tsuji, A novel mutation in Cu/Zn superoxide dismutase gene in Japanese familial amyotrophic lateral sclerosis, *Biochem Biophys Res Comm.* 200:695 (1994).

18. J. Kawamata, H. Hasegawa, S. Shimohama, J. Kimura, S. Tanaka, and K. Ueda, Leu[106]→Val (CTC→GTC) mutation of superoxide dismutase-1 gene in patient with familial amyotrophic lateral sclerosis in Japan, *Lancet.* 343:1501 (1994).
19. A. Pramatarova, J. Goto, E. Nanma, K. Nakashima, K. Takahashi, A. Takaji, I. Kanazawa, D. A. Figlewicz, and G. A. Rouleau, A two basepair deletion in the SOD 1 gene causes familial amyotrophic lateral sclerosis, *Hum Mol Genet.* 3:2016 (1994).
20. H.-X. Deng, A. Hentati, J. A. Tainer, Z. Iqbal, A. Cayabyab, W.-Y. Hung, E. D. Getzoff, P. Hu, B. Herzfeldt, R. P. Roos, C. Warner, G. Deng, E. Soriano, C. Smyth, H. E. Parge, A. Ahmed, A. D. Roses, R. A. Hallewell, M. A. Pericak-Vance, and T. Siddique, Amyotrophic Lateral Sclerosis and structural defects in Cu,Zn Superoxide dismutase, *Science.* 261:1047 (1993).
21. L. P. Rowland, Amyotrophic lateral sclerosis: human challenge for neuroscience, *Proc Natl Acad Sci USA.* 92:1251 (1995).
22. R. H. J. Brown, Amyotrophic lateral sclerosis: Recent insights from genetics and transgenic mice, *Cell.* 80:688 (1995).
23. M. E. Gurney, H. Pu, A. Y. Chiu, M. C. Dal Canto, C. Y. Polchow, D. D. Alexander, J. Caliendo, A. Hentati, Y. W. Kwon, H.-X. Deng, W. Chen, P. Zhai, R. L. Sufit, and T. Siddique, Motorneuron degeneration in mice that express a human Cu,Zn superoxide dismutase mutation, *Science.* 264:1772 (1994).
24. J. Fujii, T. Myint, H. G. Seo, Y. Kayanoki, Y. Ikeda, and N. Taniguchi, Characterization of wild-type and amyotrophic lateral sclerosis-related mutant Cu,Zn-superoxide dismutases overproduced in baculovirus-infected cells, *J Neurochem.* 64:1456 (1995).
25. M. E. Ripps, G. W. Huntly, P. R. Hof, J. H. Morrison, and J. Gordon, Transgenic mice expressing an altered murine superoxide dimutase gene provide an animal model of amyotrophic lateral sclerosis, *Proc Natl Acad Sci USA.* 92:689 (1995).
26. D. R. Borchelt, M. K. Lee, H. S. Slunt, M. Guarnieri, Z.-S. Xu, P. C. Wong, R. H. J. Brown, D. L. Price, S. S. Sisodia, and D. W. Cleveland, Superoxide dismutase 1 mutations linked to familial amyotrophic lateral sclerosis possesses significant activity, *Proc Natl Acad Sci USA.* 91:8292 (1994).
27. J. S. Beckman, M. Carson, C. D. Smith, and W. H. Koppenol, ALS, SOD and peroxynitrite, *Nature.* 364:584 (1993).
28. C. Bergeron, S. Muntasser, M. J. Somerville, L. Weyer, and M. E. Percy, Copper/Zinc superoxide dismutase mRNA levels are increased in sporadic amyotrophic lateral sclerosis motorneurons, *Brain Res.* 659:272 (1994).
29. M. A. Haas, J. Iqbal, L. B. Clerch, L. Frank, and D. Massaro, Rat lung Cu,Zn superoxide dismutase. Isolation and sequence of a full-length cDNA and studies of enzyme induction, *J Clin Invest.* 83:1241 (1989).
30. L. Jornot, and A. Junod, Response of human endothelial cell antioxidant enzymes to hyperoxia, *Am J Respir Cell Mol Biol.* 6:107 (1992).
31. X.-J. Kong, Cu, Zn superoxide dismutase in vascular cells: changes during cell cycling and exposure to hyperoxia, *Am J Physiol.* 264:L365 (1993).
32. T. Matsuyama, H. Michishita, H. Nakamura, M. Tsuchiyama, S. Shimizu, K. Watanabe, and M. Sugita, Induction of Copper-Zinc superoxide dismutase in gerbil hippocampus after ischemia, *J Cereb Blood Flow Metab.* 13:135 (1993).
33. P. Ince, P. Shaw, J. Candy, D. Mantle, L. Tandon, W. Ehmann, and W. Markesbery, Iron, selenium and glutathione peroxidase activity are elevated in sporadic motor neuron disease, *Neurosci Lett.* 182:87 (1994).
34. E. R. Stadtman, Protein oxidation and aging, *Science.* 257:1220 (1992).
35. A. C. Bowling, J. B. Schulz, R. H. Brown, and M. F. Beal, Superoxide dismutase activity, oxidative damage, and mitochondrial energy metabolism in familial and sporadic amyotrophic lateral sclerosis, *J Neurochem.* 61:2322 (1993).
36. J. Mitchell, J. Gatt, T. Phillips, E. Houghton, G. Roston, and C. Wignall, Cu/Zn superoxide dismutase free radical, and motorneurone disease, *Lancet.* 342:1051 (1993).
37. R. Lanius, C. Krieger, R. Wagey, and C. Shaw, Increased [^{35}S]glutathione binding sites in spinal cords from patients with sporadic amyotrophic lateral, *Neurosci Lett.* 163:89 (1993).
38. A. Shibata Noriuki, M. Hirano, and K. Kobayashi, Immunohistochemical demonstration of Cu/Zn superoxide dismutase in the spinal cord of patients with familial amyotrophic lateral sclerosis, *Acta Histochem Cytochem.* 26:619 (1993).
39. N. Shibata, A. Hirano, M. Kobayashi, S. Sasaki, T. Kato, S. Matsumoto, Z. Shiozawa, T. Komori, A. Ikemoto, T. Umahara, and K. Asayama, Cu/Zn superoxide dismutase-like immunoreactivity in Lewy body-like inclusions of sporadic amyotrophic lateral sclerosis, *Neurosci Lett.* 179:149 (1994).
40. D. J. Kane, T. A. Sarafian, A. Rein, H. Hahn, E. Butler Gralla, J. Selverstone Valentine, T. Örd, and D. E. Bredesen, Bcl-2 inhibition of neural death: decreased generation of reactive oxygen species, *Science.* 262:1274 (1993).

41. D. M. Hockenbery, Z. N. Oltvai, X.-M. Yin, C. L. Milliman, and S. J. Korsmeyer, Bcl-2 functions in an antioxidant pathway to prevent apoptosis, *Cell.* 75:241 (1993).
42. A. Plaitakis, Glutamate dysfunction and selective motor neuron degeneration in Amyotrophic Lateral Sclerosis: A hypothesis, *Ann Neurol.* 28:3 (1990).
43. J. D. Rothstein, L. J. Martin, and R. W. Kuncl, Decrease glutamate transport by the brain and spinal cord in Amyotrophic Lateral Sclerosis, *N Engl J Med.* 326:1464 (1992).
44. J. D. Rothstein, J. Lin, M. Dykes-Hoberg, and R. W. Kuncl, Chronic inhibition of glutamate uptake produces a model of slow neurotoxicity, *Proc Natl Acad Sci USA.* 90:6591 (1993).
45. J. T. Coyle, and P. Puttfarcken, Oxidative stress, glutamate, and neurodegenerative disorders, *Science.* 262:689 (1993).
46. M. Lafon-Cazal, M. Culcasi, F. Gaven, S. Pietri, and J. Bockaert, Nitric oxide, superoxide dismutase and peroxynitrite: putative mediators of NMDA-induced cell death in cerebellar granule cells, *Neuropharmac.* 32:1259 (1993).
47. M. Lafon-Cazal, S. Pietri, M. Culcasi, and J. Bockaert, NMDA-dependent superoxide production and neurotoxicity, *Nature.* 364:535 (1993).
48. A. Y. Sun, Y. Cheng, Q. Bu, and F. Oldfield, The biochemical mechanisms of the excitotoxicity of kainic acid, *Mol Chem Neuropathol.* 17:51 (1992).
49. J. A. Dykens, A. Stern, and E. Trenker, Mechanism of kainate toxicity to cerebellar neurons in vitro is analogous to reperfusion tissue injury, *J Neurochem.* 49:1222 (1987).
50. J. Williams, M. Errington, M. Lynch, and T. Bliss, Arachidonic acid induces a long-term activity-dependent enhancement of synaptic transmission in the hippocampus, *Nature.* 341:739 (1989).
51. J. Lazarewicz, J. Wroblewski, M. Palmer, and E. Costa, Activation of N-methyl-D-aspartate-sensitive glutamate receptors stimulates arachidonic acid release in primary cultures of cerebellar granule cells, *Pharmacology.* 27:765 (1988).
52. A. Dumuis, M. Sebben, L. Haynes, J.-P. Pin, and J. Bockaert, NMDA receptors activate the arachidonic acid cascade system in striatal neurons, *Nature.* 336:68 (1988).
53. S. A. Lipton, Y.-B. Choi, Z.-H. Pan, S. Z. Lei, H.-S. V. Chen, J. Loscalzo, D. J. Singel, and J. S. Stamler, A redox-based mechanism for the neuroprotective and neurodestructive effects of nitric oxide and related nitroso-compounds, *Nature.* 364:626 (1993).
54. V. L. Dawson, T. M. Dawson, D. A. Bartley, G. R. Uhl, and S. H. Snyder, Mechanisms of nitric oxide-mediated neurotoxicity in primary brain cultures, *J Neurosci.* 13:2651 (1993).
55. T. M. Dawson, V. L. Dawson, and S. H. Snyder, A novel neuronal messenger molecule in brain: the free radical nitric oxide, *Ann Neurol.* 32:297 (1992).
56. J. S. Beckman, T. W. Beckman, J. Chen, P. A. Marshall, and B. A. Freeman, Apparent hydroxyl radical production by peroxynitrite: implications for endothelial injury from nitric oxide and 1superoxide, *Proc Natl Acad Sci USA.* 87:1620 (1990).
57. T. Kawamata, H. Akiyama, T. Yamada, and P. L. McGeer, Immunologic reactions in Amyotrophic Lateral Sclerosis brain and spinal cord tissue., *Am J Pathol.* 140:691 (1992).
58. C. A. Colton, and D. L. Gilbert, Production of superoxide anions by a CNS macrophage, the microglia, *FEBS Lett.* 223:284 (1987).
59. M. Tanaka, A. Sotomatsu, T. Yoshida, T. Hirai, and A. Nishida, Detection of superoxide production by activated microglia using a sensitive and specific chemiluminescence assay and microglia-mediated PC12h cell death, *J Neurochem.* 63:266 (1994).
60. S. B. Corradin, J. Mauel, S. Denis Donini, E. Quattrocchi, and P. Ricciardi-Castagloni, Inducible nitric oxide synthase activity of cloned murine microglial cells, *Glia.* 7:255 (1993).
61. B. Halliwell, and J. M. Gutteridge, Free radicals in biology and medicine, *in*: Second ed. Oxford: Clarendon Press, 1989.
62. P. G. Iannuzzelli, X. H. Wang, Y. Wang, and E. H. Murphy, Axotomy-induced changes in cytochrome oxidase activity in the cat trochlear nucleus, *Brain Res.* 637:267 (1994).
63. M. F. Beal, B. T. Hyman, and W. Koroshetz, Do defects in mitochondrial energy metabolism underlie the pathology of neurodegenerative diseases?, *TINS.* 16:125 (1993).
64. H. Mohammed, J. Cummings, T. Divers, B. Valentine, A. de Lahunta, B. Summers, B. Farrow, K. Trembicki-Graves, and A. Mauskopf, Risk factors associated with equine motor neuron disease: a possible model for human MND, *Neurology.* 43:966 (1993).
65. J. D. Rothstein, L. A. Bristol, B. Hosler, R. H. J. Brown, and R. W. Kuncl, Chronic inhibition of superoxide dismutase produces apoptotic death of spinal neurons, *Proc Natl Acad Sci USA.* 91:4155 (1994).
66. C. M. Troy, and M. L. Shelanski, Down-regulation of copper/zinc superoxide dismutase causes apoptotic cell death, *Proc Natl Acad Sci USA.* 91:6384 (1994).
67. C. A. Pardo, Z. Xu, D. R. Borchelt, D. L. Price, S. Sisodia, and D. W. Cleveland, Superoxide dismutase is an abundant component in cell bodies, dendrites, and axons of motor neurons and in a subset of other neurons, *Proc Natl Acad Sci USA.* 92:954 (1995).

MODULATION OF THE MITOCHONDRIAL ANTI-OXYGEN RADICAL DEFENSE OF RAT ASTROGLIAL CELLS IN CULTURE

Emmanuel PINTEAUX, Jean-Christophe COPIN, Marc LEDIG and Georges THOLEY

Centre de Neurochimie du CNRS
Laboratoire de Neurobiologie Ontogénique
5, rue Blaise Pascal
67084- STRASBOURG CEDEX
France.

INTRODUCTION

Oxygen free radicals have been proposed to play a central role in cellular injury that occurs upon postischemic brain tissue reperfusion . Several potential sources of reactive oxygen species (ROS) have been reported: oxidation of accumulated xanthine(X) by xanthine oxidase(XO) [1], uncoupling of the electron transport chain with direct transfer to oxygen [2],etc... In astroglial cells, two forms of superoxide dismutase (SOD) namely Cu,Zn- and Mn-SOD localized respectively in the cytosol and in the mitochondria seem to play a fundamental role in ROS scavenging[3]. Therefore we have investigated biochemically and immunochemically various cytosolic and mitochondrial alterations induced by exposure of rat astroglial cells in primary culture to ROS generated by a X/XO mixture added to the growth medium. In all these experimental conditions we examined the level of Cu,Zn-SOD, glutamine synthetase (GS), both cytosol located enzymes, and of Mn-SOD. In parallel to SOD scavenging enzymes analysis, the expression of stress proteins of the 70 kD family was investigated. Moreover, the effect of almitrine - *(dialylamino 4',6'-triazinyl-2')-1-(bis-parafluoro-benzydryl)-4 piperazine-* vasoactive substance acting on oxygen availability [4] on Mn-SOD and heat shock protein expression has been examined.

RESULTS

Cultures of rat astroglial cells were prepared from cerebral hemispheres of newborn rats[5]. ROS generation was obtained by adding xanthine (1mM final concentration) and 10 mU/ml xanthine oxidase to the cell growth medium during various periods of time. Cu-Zn, Mn-SOD and GS activities were determined as previously reported[5]. Mn-SOD and heat shock protein level were determined immunochemically .

Time effect of X/XO treatment (Fig.1)

Between 0 and 20 hours X/XO treatment a progressive decrease of Cu,Zn-SOD activity is observed: final value corresponds to 55 % of initial value. In parallel, there is also a pronounced decrease of GS activity reaching already 40 % of initial value after 4

Neurodegenerative Diseases
Edited by Gary Fiskum, Plenum Press, New York, 1996

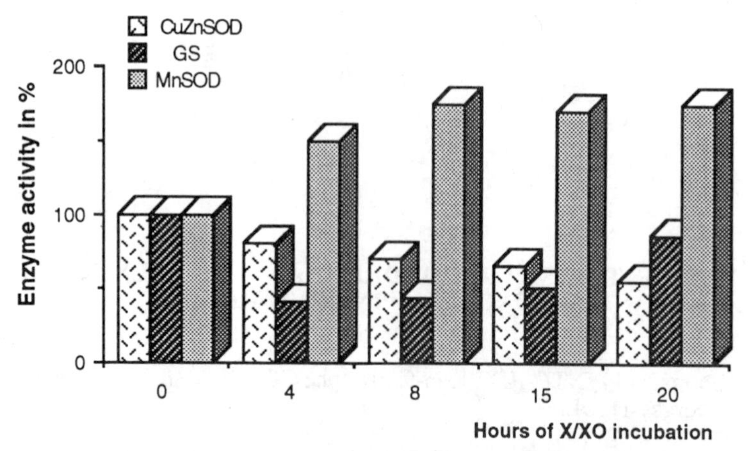

Figure 1. Time effect of X/XO treatment on CuZnSOD,GS and MnSOD activity of rat astroglial cells in primary culture.

hours followed by a recovery period until 20 hours. Concerning Mn-SOD activity, a significant increase by about 50% is already observed after 4 hours treatment and enzyme activity level remains significantly higher than control during the rest of the period tested.

Almitrine effect on MnSOD and 70 kD heat shock protein expression (Fig.2)

Immunochemical analysis of Mn-SOD content of cell extracts underlines a increase of protein expression by more than 100 % after 4 hours X/XO treatment (results not indicated).The increase of Mn-SOD activity by X/XO treatment is significantly accentuated in the presence of 10^{-5} M almitrine. Immunocytochemical analysis of the expression of stress proteins of the 70 kD family reveals that only the constitutive form, Hsc70 is expressed . A more than two fold increase of Hsc 70 cell content is induced by X/XO; moreover its expression is significantly enhanced by addition of 10^{-5} M almitrine.

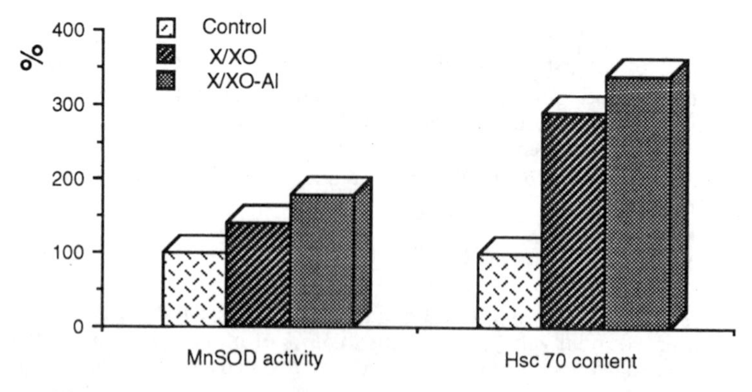

Figure 2. MnSOD activity increase resulting from X/XO treatment is significantly accentuated in the presence of almitrine. In parallel constitutive heat shock (Hsc70)protein synthesis is induced by X/XO treatment; the phenomenon is significantly activated by addition of almitrine.

DISCUSSION

Although several studies have implicated ROS production during postischemic reoxygenation in brain tissue, the specific response of astroglial cells to these radicals is still unclear. In primary culture of rat astrocytes, ROS induce opposite effects on the two forms of oxygen superoxide scavenging enzymes; Cu,Zn- and Mn-SOD. A important decrease of the activity of two cytosolic enzymes, Cu,Zn-SOD and GS is observed. GS is more susceptible than Cu,Zn-SOD to ROS inactivation by considering the fact that the decrease of GS is much more rapid than that of Cu,Zn-SOD. Such decrease could be related to some oxidative inactivation of both enzymes[6]. On the contrary activation of MnSOD corresponding to some induction of the enzyme occurs in these conditions. This observation is in accordance with those of other groups reporting that in epithelial cells, ROS generated by X/XO were associated with specific induction of Mn SOD mRNA[7]. Astroglial cells may respond to oxidant insult by selective induction of antioxidant enzymes like MnSOD; moreover the astroglial MnSOD mitochondrial defense against ROS seems tightly related to Hsc70 synthesis. Almitrine seems to modulate specifically the expression of the mitochondrial form of SOD: the drug could interfere with some second messenger role of oxygen radicals[8] at the mitochondria level.

REFERENCES

1.M.H.O'Regan, M.Smith-Barbour,L.M.Perkins,X.Cao andJ.W.Phillis,The effect of amflutizole, a xanthine oxidase inhibitor, on ischemia evoked purine release and free radical formation in the rat cerebral cortex,*Neuropharmacology*,33,1197(1994)

2.J.F.Turrens,M.Beconi,J.Barilla,V.B.Chavez and J.M.McCord,Mitochondrial generation of oxygen radicals during reoxygenation of ischemic tissues,*Free Rad. Res.Commun.*12-13,681(1991)

3.J.V.Bannister and W.H.Bannister, Aspects of the structure, function and application of superoxide dismutase, CRC Crit.Rev.Biochem.22,111(1987)

4.G.O.Benzi,O.Pastoris,R.F.Villa and A.A.Giuffrida, Influence of aging and exogenous substances on cerebral energy metabolism in post hypoglycemic recovery, Biochem.Pharmacol.34,1477,(1985)

5.J.C.Copin,M.Ledig and G.Tholey,Free radical scavenging systems of rat astroglial cells in primary culture: effects of anoxia and drug treatment,Neurochem.Res.17,677(1992)

6.C.N.Oliver,P.E.Starke-Reed,E.R.Stadtman,G.J.Liv,J.M.Carney and R.A.Floyd,Oxidative damage to brain proteins,loss of glutamine synthetase activity and production of free radicals during ischemia reperfusion induced injury to gerbil brain,Prod.Nat.Acad.Sc.,USA,87,5144(1990)

7.S.Schull,N.H.Heintz,M.Periasamy,M.Monohar,Y.M.Janssen,J.P.Marsh and B.Moosman,Differential regulation of antioxidant enzymes in response to oxidants,J.Biol.Chem. 266,24398(1991)

8 R.Schreck and P.A.Bauerle, A role of oxygen radicals as second messengers, Trends in Cell Biol.1,39(1991)

OXYGEN RADICAL-MEDIATED OXIDATION OF SEROTONIN: POTENTIAL RELATIONSHIP TO NEURODEGENERATIVE DISEASES

Monika Z. Wrona, Zhaoliang Yang, Jolanta Waskiewicz and Glenn Dryhurst

Department of Chemistry and Biochemistry
University of Oklahoma, Norman, OK 73019

INTRODUCTION

Oxygen free radicals have been implicated as a pathoetiological factor in aging[1-5] and in a number of neurodegenerative brain disorders such as Alzheimer's Disease,[1,6-9] Parkinson's Disease,[10-12] transient cerebral ischemia[13] and as a result of methamphetamine[14-16] and ethanol[17] abuse. The brain appears to be particularly vulnerable to oxygen radical-mediated damage, often referred to as *oxidative stress*, because of several biochemical features that include high oxygen consumption, high iron content of some brain regions,[4] relatively low levels of protective enzymes and antioxidants such as the tocopherols,[5] and high content of peroxidizable polyunsaturated fatty acids associated with lipid membranes. It seems to be rather widely accepted that oxygen radicals formed in the central nervous system (CNS) limit their damage to lipids, proteins and nucleic acids.[5,7-13,18] However, it is of relevance to note that in all of the neurodegenerative brain disorders noted previously the serotonergic, noradrenergic and/or dopaminergic systems are seriously damaged. The neurotransmitters employed by these systems, 5-hydroxytryptamine (5-HT; serotonin), norepinephrine (NE), and dopamine (DA) are all very easily oxidized species.[19-22] Thus, it seems likely that these neurotransmitters are also prime targets for oxygen radical-mediated oxidation. Accordingly, a major focus of research in this laboratory is to explore the hypothesis that the aberrant oxidative metabolites of these neurotransmitters might include endotoxins that contribute to the degenerative processes.[23-26]

The neurodegeneration that occurs in AD involves many neurotransmitter systems but only in remarkably selective anatomic regions of the brain. These include long cholinergic,[27-30] noradrenergic,[31-33] serotonergic[29,34,35] and dopaminergic[36] neurons that project from subcortical cell bodies and innervate the cortex, hippocampus and amygdala, regions that sustain severe losses of glutamatergic[37-40] and perhaps other neurons.[41] Thus, neurons that are lost in AD are located either entirely within certain areas of the cortex and hippocampus/amygdala complex or project from subcortical cell bodies and connect to the latter structures. This selective pattern of damage in AD, therefore, argues against a random appearance of brain lesions and points to a progression of the disease from defined starting points along specific connecting neuronal pathways, *i.e.*, a system neurodegeneration.[42] Saper et al.[42] have proposed that such a system degeneration in AD results from the connectivity between affected neuronal pathways. Hardy et al.[43] have taken this idea a step further and suggested that AD is initiated by chronic attack on the axon terminals of long serotonergic and noradrenergic neurons that project from the brain stem at points where they innervate blood capillaries in the association areas of the cortex and hippocampus. These investigators speculate that this chronic attack might be an environmental toxicant, virus or an autoimmune response. This system degeneration concept thus implies that a toxin enters serotonergic and noradrenergic terminals where they innervate blood capillaries which then evokes a retrograde degeneration of these and connected neurons by transneuronal transfer of the toxin. This concept provides an elegant rationale for the anatomic selectivity of the neurodegeneration that occurs in AD. However, evidence for the involvement of a virus, specific environmental toxicants or other such pathogens in AD is lacking. As noted earlier, there is strong evidence for oxygen radical-mediated damage in the neuronal degeneration that occurs in the AD brain.[1,6-9] Indeed, abnormal, but unknown, oxidized forms of 5-HT and its biosynthetic precursor 5-hydroxytryptophan (5-HTPP) have been detected in the cerebrospinal fluid (CSF) of AD patients.[44] Furthermore, hydroxylated forms of 5-HT and DA, *i.e.*, 5,6-dihydroxytryptamine (5,6-DHT)[15] and 6-hydroxydopamine (6-OHDA)[15] have been detected in the brains of rats after methamphetamine administration suggesting that the neurodegenerative effects of this drug might be mediated directly or indirectly by oxygen radicals, particularly the hydroxyl radical (HO·).[14,45] Indeed, 5-HT and catecholamines are known to be readily oxidized by molecular oxygen in the presence of low molecular weight Fe^{2+} to give electrophilic products that bind irreversibly to brain proteins in reactions that are mediated by oxygen radicals.[46] Taken together, these observations prompted us to investigate in more detail the HO·-mediated oxidation of 5-HT and oxidation of this neurotransmitter by molecular oxygen in the presence of catalytic levels of low molecular weight Fe^{2+}.

Hydroxyl radical-mediated oxidation of 5-HT

The Udenfriend et al.[47] system, consisting of ascorbic acid (1.06 mM), $Fe(NH_4)_2 SO_4 \cdot 6 H_2O$ (280 μM) and H_2O_2 (1.0 mM) in pH 7.4 phosphate buffer ($\mu = 0.05$) was employed as the source of HO·. In this system HO· is formed by decomposition of H_2O_2 by Fe^{2+} (Fenton reaction) and the resultant Fe^{3+} is reduced to Fe^{2+} by ascorbic acid (AA; Eqn. 1). 5-HT is very rapidly oxidized by this HO·-generating system.[24] Figure 1A shows an HPLC chromatogram of the product solution formed after oxidation of 5-HT (1.0 mM) with the HO·-generating system for 10 min and clearly shows the presence of four major products: 5,6-DHT, 5-hydroxy-3-ethylamino-2-oxindole (5-HEO), 5,6,7-trihydroxytryptamine (5,6,7-

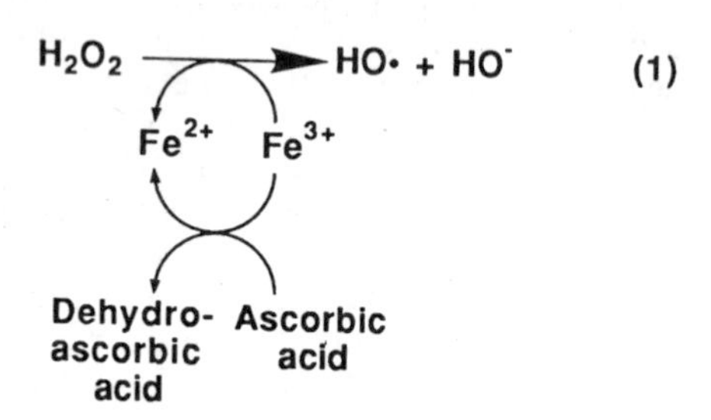

THT) and dimer **1** (structures are shown in the figure and in Scheme 1). The primary step in this reaction involves HO· attack on 5-HT to give 4,5-, 2,5- and 5,6-dihydroxytryptamine radicals which are then subsequently oxidized by a second HO· to give 4,5-dihydroxytryptamine (DHT), 2,5-DHT and 5,6-DHT (Scheme 1).[24] In aqueous solution at physiological pH 2,5-DHT exists predominantly as its 2-keto tautomer 5-HEO, a relatively stable compound. 5,6-DHT, by contrast, is much less stable in the presence of molecular oxygen and could be isolated only if formed in the presence of ascorbic acid which serves to maintain it in its reduced form. Nevertheless, oxidation of 5,6-DHT by molecular oxygen forms a very reactive *o*-quinone, along with H_2O_2 as a byproduct,[48,49] that is further hydroxylated by HO· to give 5,6,7-THT as conceptualized in Scheme 1. The latter is even

Scheme I

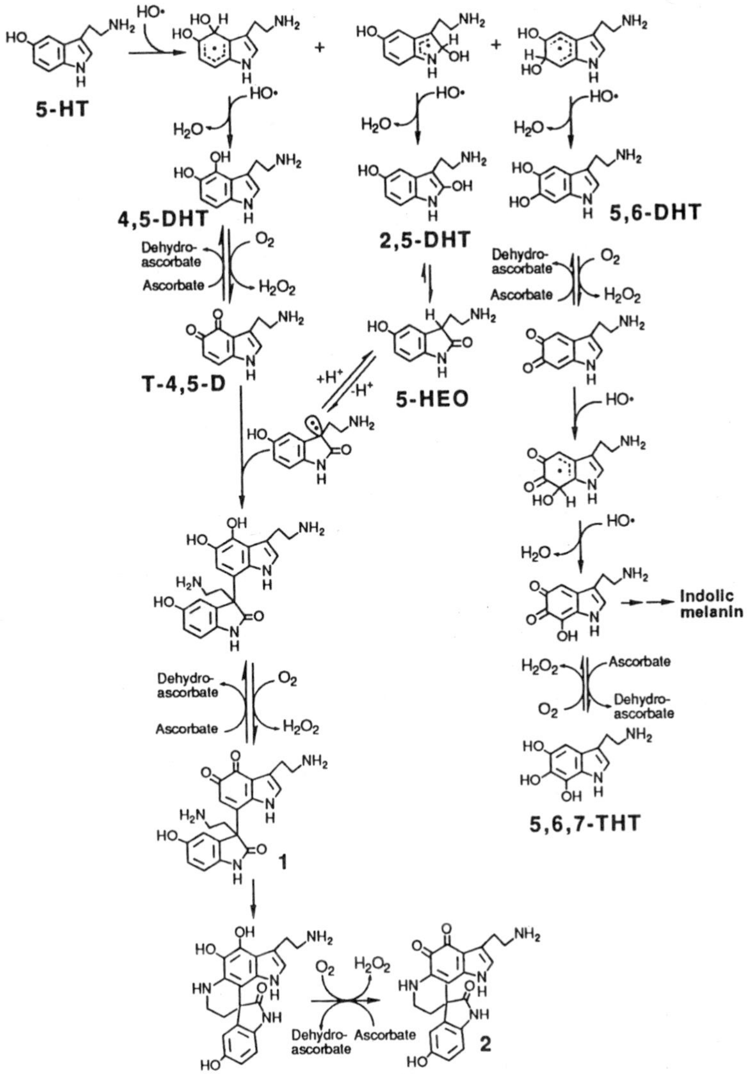

more easily oxidized by molecular oxygen than 5,6-DHT to give an *o*-quinone that subsequently polymerizes to indolic melanin.[49]

4,5-DHT is an extremely unstable compound and is rapidly autoxidized to tryptamine-4,5-dione (T-4,5-D).[50] However, T-4,5-D is not detected as a reaction product (Figure 1A) owing with its facile reaction with 5-HEO to give dimer **1** (Scheme 1). Dimer **1** undergoes a slow intramolecular cyclization and autoxidation to form the pentacyclic compound **2** (Scheme 1) which can be clearly observed as a product after longer reaction times (Figure 1B). The C(7)-position of T-4,5-D is particularly susceptible to attack by nucleophiles[25,50] and the anion of 5-HEO scavenges this dione to give dimer **1** and thence **2** (Scheme 1).[24] However, in the presence of excess free glutathione (GSH), a strong nucleophile, dimers **1** and **2** are not formed, the yield of 5-HEO increases and two new products, **2** and **4**, derived from the reaction of GSH with T-4,5-D appear (Figure 1C). Previous studies[25] have established that T-4,5-D reacts avidly with GSH to give initially 7-*S*-glutathionyl-4,5-dihydroxytryptamine (7-*S*-Glu-4,5-DHT) that is very rapidly autoxidized to give 7-*S*-glutathionyl-tryptamine-4,5-dione (7-*S*-Glu-T-4,5-D) and H_2O_2 as a byproduct (Scheme II).

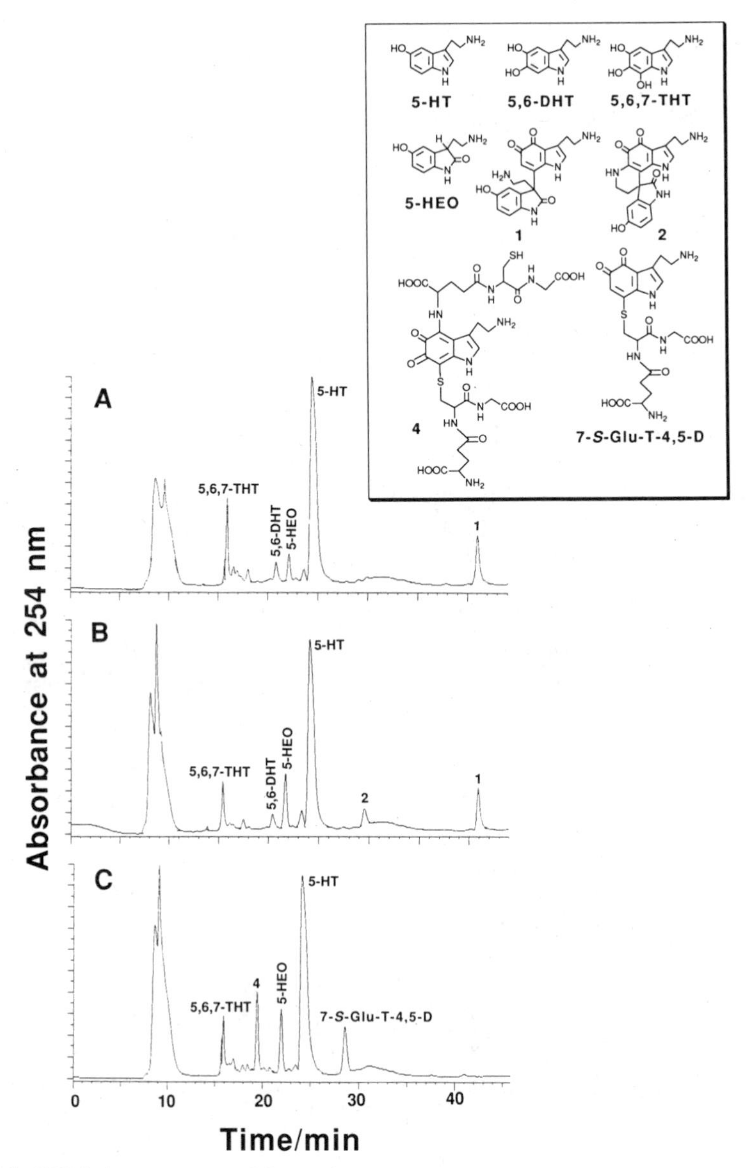

FIGURE 1. HPLC chromatograms of the product solutions obtained following incubation of 5-HT (1 mM) with Fe^{2+} (280 μM), AA (1.06 mM) and H_2O (0.1 mM) in pH 7.4 phosphate buffer at 37°C (A) for 10 min, (B) for 45 min, (C) the same conditions as in (A) except GSH (2 mM) was added to the incubation mixture. Chromatography used HPLCa with Bakerbond reverse phase column (25 x 2.1 cm) and 10 mL injection volume. The mobile phases and gradient profile as in reference[19].

In the presence of HO· and free GSH, 7-*S*-Glu-T-4,5-D undergoes further hydroxylation to **3** which condenses with GSH to give a Schiff base that tautomerizes to **4** (Scheme II). It is of interest to note that the HO·-mediated oxidation of 5-HT generates at least five products (T-4,5-D, 5,6-DHT, 5,6,7-THT, **1** and **2**) that can redox cycle in the presence of ascorbic acid and molecular oxygen to generate H_2O_2 and hence HO· (Eqn. 1) that can potentiate the reaction. In the presence of free GSH **1** and **2** are not formed but 7-*S*-Glu-T-4,5-D and **4** can also redox cycle[25] forming H_2O_2 and, in the presence of Fe^{2+}, HO·.

Scheme II

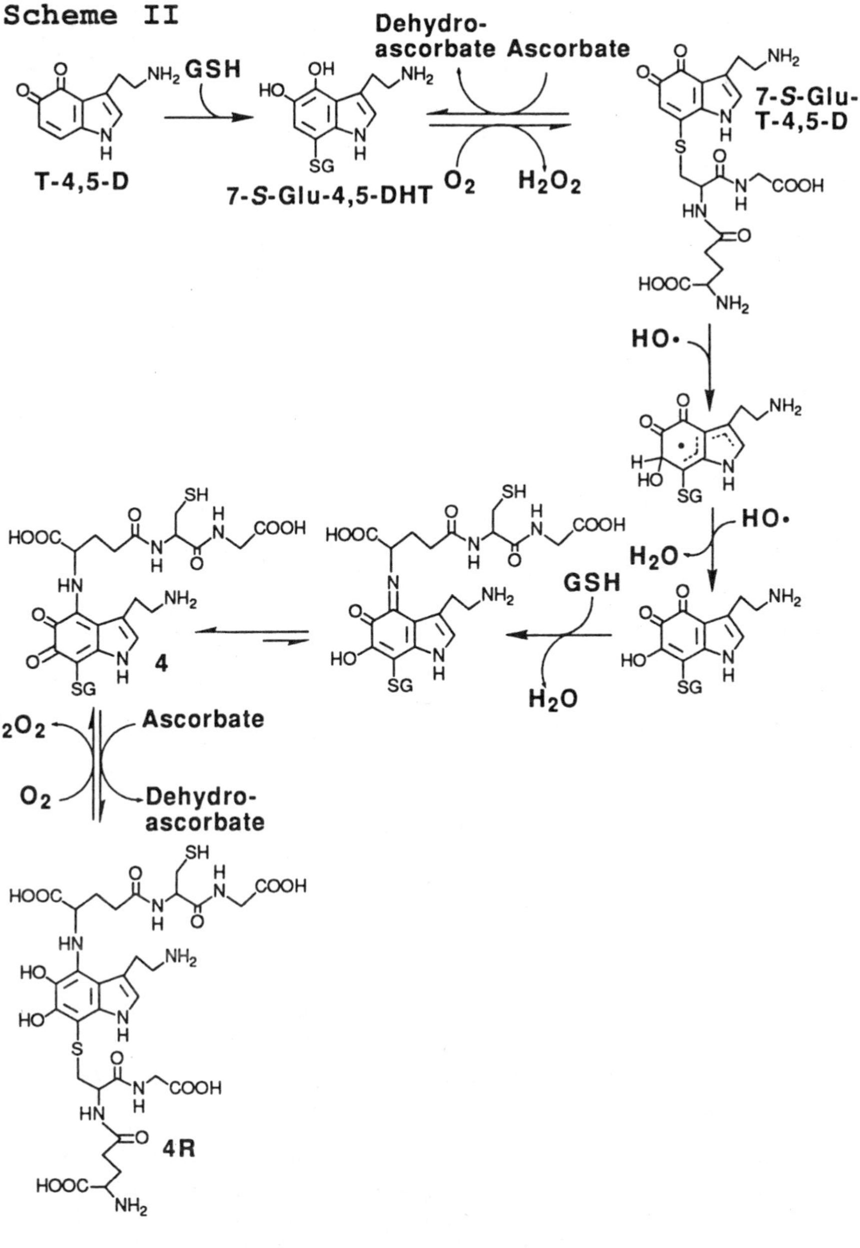

Autoxidation of 5-HT

Using preparative scale HPLC with UV detection the oxidation of 5-HT by molecular oxygen (autoxidation) in buffered aqueous solution at pH 7.4 appears to be very slow and results in formation of 5,5′-dihydroxy-4,4′-bitryptamine (DHBT) and other minor dimeric products after many hours or days.[26] However, by monitoring the autoxidation using HPLC with electrochemical detection (HPLC-EC), a much more sensitive technique, it becomes evident that the reaction is relatively fast.

Figure 2A shows an HPLC-EC chromatogram of the product solution formed when 5-HT (200 μM) was incubated in air-saturated pH 7.4 phosphate buffer ($\mu = 1.0$) for 10 min at 37°C which clearly shows the presence of 5-HEO, 5,6,7-THT and traces of DHBT. After

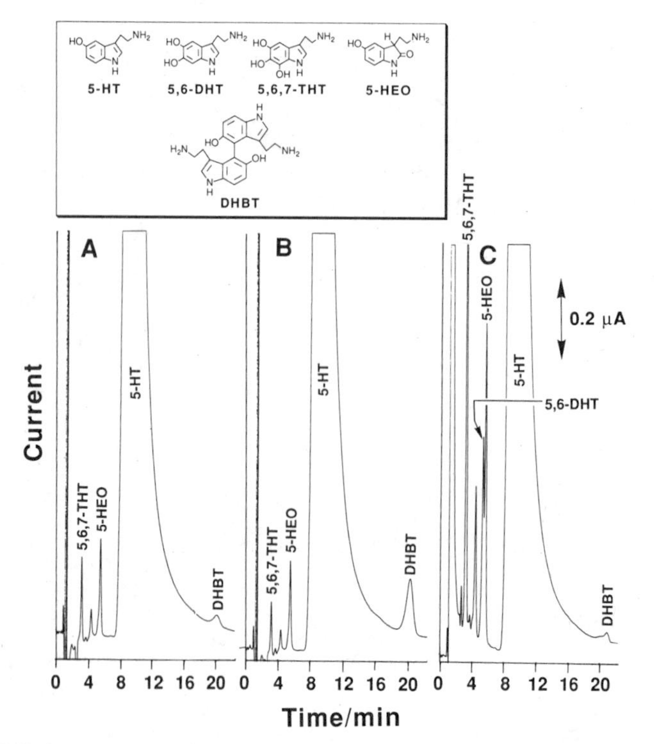

FIGURE 2. HPLC chromatograms of the product mixture formed upon autoxidation of 5-HT (200 μM) in pH 7.4 phosphate buffer (μ = 1.0) at 37°C, (A) after 10 min incubation, (B) after 2h incubation, (C) after 10 min incubation in the presence of AA (200 μM). Chromatography used HPLC with electrochemical detection, BAS reverse phase column (3 μM; 100 x 3.2 mm) and a glassy carbon electrode set at 800 mV vs. Ag/AgCl; 5 μL injection volume. The mobile phase consisted of 0.1 M citric acid, 0.26 mM sodium octyl sulfate, 0.05 mM Na_2EDTA and 0.085% (v/v) diethylamine, pH 2.35; flow rate 0.6 ml min^{-1}.

2h the yields of 5-HEO and 5,6,7-THT declined somewhat while that of DHBT increases significantly (Figure 2B). In such reactions it was not possible to detect 5,6-DHT. However, incubation of 5,6-DHT (50 μM) at pH 7.4 and 37°C resulted in the rapid disappearance of this compound (≤ 30 min) and the appearance of 5,6,7-THT. The latter compound subsequently disappears. These results suggests that autoxidation of 5-HT leads to the initial production of 5,6-DHT that is readily transformed into 5,6,7-THT. Because it is not possible to chemically synthesize pure samples of 5,6,7-THT in order to prepare calibration curves (HPLC-EC peak height *versus* concentration) it was not possible to quantitative the yields of this compound. However, Table 1 shows the yields of 5-HEO and DHBT formed when 5-HT (200 μM) was incubated in air-saturated pH 7.4 phosphate buffer for 10 min at 37°C. The autoxidation of 5-HT was completely inhibited by the transition metal ion complexing agents desferrioxamine and diethylenetriaminetetraacetic acid (DTPA) (Table 1) but was greatly accelerated by addition of Fe^{2+} or Cu^{2+} (50 μM) and, to a lesser extent, by Fe^{3+} (50 μM) (data not shown). These results indicate that trace levels of transition metal ions that always contaminate the buffer systems employed exert a significant catalytic influence on the autoxidation of 5-HT. Ascorbic acid dramatically increases that rate of autoxidation of 5-HT with resultant increases in the yields of 5,6,7-THT and 5-HEO and the appearance of 5,6-DHT (Figure 2C, Table 1). By contrast, in the presence of ascorbic acid the yields of DHBT decrease. These results indicate that autoxidation of ascorbic acid (to dehydroascorbic acid) serves as a primary source of H_2O_2 and hence HO· leading to formation of products characteristic of the HO·-mediated oxidation of 5-HT (Figure 1A, B; Schemes I and II). Using HPLC-EC in the oxidative mode it is not possible to detect T-4,5-D, another expected product of the HO·-mediated oxidation of 5-HT, at nanomolar concentration levels. However, after 2h it is possible to detect **1** and **2** among the products of

Table 1. Concentrations of products formed upon autoxidation of 5-HT (200 μM) in air-saturated pH 7.4 phosphate buffer (μ = 1.0) for 10 min at 37°C and effects of complexing agents, ascorbic acid and glutathione

Added Agent (Concentration)	Concentration[a,b] of 5,6-DHT/ nM	5-HEO/ nM	DHBT/ nM
None	ND[c]	118±4	41±5
Desferrioxamine (200 μM)	ND	ND	ND
DTPA (300 μM)[d]	ND	ND	ND
Ascorbic acid (200 μM)	380±22	323±15	24±3
GSH (50 μM)	ND	129±16	26±3

[a]Product mixture were analyzed by HPLC-EC using a BAS (Bioanalytical Systems, West Lafeyette, IN) 3 μm reversed phase column (100 x 3.2 - mm) and a detector equipped with a glassy carbon electrode set at 800 mV *vs.* Ag/AgCl. The mobile phase consisted of 0.1 M citric acid, 0.26 mM sodium octyl sulfate. 0.05 mM Na_2 EDTA and diethylamine (0.085% v/v). The flow rate was 0.6 mL min^{-1}. Injection volume: 5.0 μL.

[b]Results are reported as the mean of at least 4 replicate determinations ± SEM.

[c]Not detected.

[d]Diethylenetriaminetetraacetic acid.

autoxidation of 5-HT in the presence of ascorbic acid indicating that T-4,5-D is formed as an initial reaction product. Autoxidation of 5-HT in the presence of GSH also results in an increase in the yields of 5-HEO and a decrease in the yield of DHBT (Table 1). However, again, using HPLC-EC in the oxidative mode it is not possible to detect compounds such as 7-*S*-Glu-T-4,5-D and **4** (Scheme II) and hence it is presently not possible to fully evaluate the influence of GSH on the autoxidation of 5-HT.

Further insights into the influence of ascorbic acid on the autoxidation of 5-HT can be derived from the results presented in Table 2. Thus, the HO· scavenger mannitol significantly decreases the formation of 5,6-DHT (and 5,6,7-THT) but has only a minor effect on the yields of 5-HEO and DHBT. Catalase, however, results in a dramatic decrease in the yields of 5,6-DHT (and 5,6,7-THT) and 5-HEO but has only a minor effect on the yields of DHBT. Superoxide dismutase (SOD) also evokes significant decreases in 5,6-DHT (and 5,6,7-THT) and 5-HEO formation and completely blocks formation of DHBT.

Although it has not yet been possible to identify and quantitate all of the products formed as a result of autoxidation of 5-HT and detailed mechanistic pathways remain to be elucidated, the preliminary results presented in Figures 1 and 2 and Tables 1 and 2 permit several important conclusions. First, it is clear that in the presence of molecular oxygen and traces of transition metal ions 5-HT can be oxidized. Second, intraneuronal *antioxidants*, particularly ascorbic acid, greatly potentiate oxidation of 5-HT. Furthermore, under the latter conditions reactive oxygen species (H_2O_2, $O_2^{-\cdot}$, HO·) are formed as well as several hydroxylated tryptamines including the known serotonergic neurotoxin 5,6-DHT.[51]

Although definitive evidence that T-4,5-D is also formed as an initial product of autoxidation of 5-HT, particularly in the presence of ascorbic acid, remains to be obtained, the fact that other products characteristic of the HO·-mediated oxidation of this neurotransmitter are produced strongly suggests that this dione must be a significant product. Similarly, it has not yet been possible to detect 7-*S*-Glu-T-4,5-D and **4** as a product of autoxidation of 5-HT in the presence of GSH, largely as a result of analytical technique limitations. Nevertheless, based upon studies of the HO·-mediated oxidation of 5-HT in the presence of GSH and ascorbic acid and the fact that HO· must be formed when the neurotransmitter is incubated with ascorbic acid it seems inevitable that 7-*S*-Glu-T-4,5-D and **4** must be formed.

Table 2. Effects of antioxidants on the products formed upon autoxidation of 5-HT (200 μM) in air-saturated pH 7.4 phosphate buffer ($\mu = 1.0$) for 10 min at 37°C in the presence of ascorbic acid (200 μM)

Added Agent (Concentration)	Concentration[a] of 5,6-DHT/ nM	5-HEO/ nM	DHBT/ nM
None	380±22[b]	323±15	24±5
Mannitol (2mM)	80±16	281±13	20±5
Catalase (1200 units mL^{-1})	31±5	20±5	15±4
Superoxide dismutase (2,500 units mL^{-1})	170±14	165±11	ND[c]

[a]Conditions for HPLC-EC analysis are shown in Table 1.
[b]Concentrations reported are the mean of at least 4 replicate determinations ± SEM.
[c]Not detected.

Preliminary Biological Studies

Although T-4,5-D cannot be isolated as a solid, it is possible to prepare very pure dilute solutions of this compound.[52] Accordingly, the neurochemical effects of daily infusions of T-4,5-D (1.2 μg in 5 μL of isotonic saline solution pH 7.2) for 19 consecutive days into the right hippocampus of rats (male albino, Sprague-Dawley strain; 330 ± 50 g) via a surgically implanted cannula were studied. Twenty four hours after the last infusion animals were sacrificed by guillotine decapitation, brains removed, dissected, homogenized and the homogenate of the right and left hippocampi of experimentals and controls (infused only with 5 μL of isotonic saline) were analyzed by HPLC-EC[53] with the results shown in Table 3. Thus, T-4,5-D evoked a statistically significant ($p < 0.006$) decrease in 5-HT levels in the right hippocampus of experimentals compared to that measured in the right hippocampus of controls. Furthermore, the decrease of 5-HT levels in the right hippocampus of experimentals were significantly ($p < 0.0001$) reduced compared to those measured in the left hippocampus of these animals.

Table 3. Effect of chronic infusions of T-4,5-D[a] into the right hippocampus of rats on 5-HT levels

	5-HT Levels[b,c] (nmole/g of tissue) Right Hippocampus	Left Hippocampus
Controls (n = 10)	1.57±0.94	1.88±0.13
Experimentals (n = 12)	1.23±0.06[d,e]	1.87±0.10

[a]T-4,5-D (1.2 μg in 5 mL of isotonic saline, pH 7.2) was infused daily for 19 consecutive days.
[b]Determined by HPLC-EC[53]
[c]Values reported are mean ± SEM.
[d] $P < 0.006$ (student t-test) comparing right hippocampus of experimentals to right hippocampus of controls.
[e] $P < 0.0001$ comparing right and left hippocampus of experimentals.

When administered intracerebrally (in 5 μL of isotonic saline) into the brains of mice, 7-*S*-Glu-T-4,5-D is lethal (LD_{50} = 21 μg) and evokes a profound hyperactivity syndrome which persists for approximately 30 - 45 min.[23] Analysis of whole mouse brain for biogenic

Table 4. Effects of intracerebroventricular administration of 7-*S*-Glu-T-4,5-D (21 μg in 5 μL of isotonic saline) to mouse on whole brain levels of neurotransmitters/metabolites

Time	NE	DA	DOPAC % of controls[a]	HVA	5-HT	5-HIAA
30 min	51±6***	80±11*	135±11**	157±14**	81±7*	115±8
1h	119±11	101±7	129±9*	149±12*	98±5	114±6
7 days	117±5	111±4	105±11	96±17	98±5	92±14

[a]Results expressed as mean % of control levels ± SEM; 8 experimentals and 8 controls were used; $^{*}p < 0.05$; $^{**}p < 0.01$; $^{***}p < 0.001$ (student t- test).

amine neurotransmitter/metabolite levels at various times after administration of 7-*S*-Glu-T-4,5-D (Table 4) revealed that during the hyperactivity phase (30 min) NE, DA and 5-HT levels were significantly decreased whereas levels of the metabolites 3,4-dihydroxphenylacetic acid (DOPAC) and homovanillic acid (HVA) were significantly elevated. Although levels of the 5-HT metabolite 5- hydroxyindole-3-acetic acid (5-HIAA) were also elevated at this time they did not reach statistical significance. After 1h, the DA metabolites DOPAC and HVA continued to be statistically increased compared to control levels but at long times all neurotransmitter/metabolite levels returned to control values. These results suggested that an acute dose of 7-*S*-Glu-T-4,5-D evokes a short-lived but significant release and increased turnover of NE, DA and 5-HT.

7-*S*-Glu-T-4,5-D was also tested through the NIMH/NovaScreen® Drug Discovery and Development Program (Contract No. NIMH-2003). Briefly, competitive binding assays were performed in either 250 or 500 μL volumes containing by volume, 80 percent receptor preparation, 10 percent radioligand and 10 percent of the test compound/cold ligand (non-specific binding determinant) in 4 percent dimethylsulfoxide which was diluted to a final concentration of 0.4 percent in the assay. Assays were terminated by rapid vacuum filtration over Whatman glass fiber filters followed by rapid washing with cold buffer. Radioactivity was determined by either liquid scintillation or gamma spectroscopy. The results obtained revealed that 7-*S*-Glu-T-4,5-D at a concentration of 10 μM exhibited 100 percent (mean of two determinations) inhibition of specific binding of [^{3}H] GABA to the $GABA_B$ receptor but had no affinity whatsoever for the $GABA_A$ receptor. The action of the neurotransmitter γ-aminobutyric acid (GABA) and GABA agonists at the $GABA_B$ receptor, located on nerve terminals and postsynaptic sites throughout the brain, decreases the release of biogenic amine, excitatory amino acid and neuropeptide neurotransmitters as well as GABA by blockade of voltage-independent calcium channels.[54] In view of the fact that 7-*S*-Glu-T-4,5-D apparently evokes elevated release and turnover of NE, DA and 5-HT (Table 4) and, perhaps, other neurotransmitters suggests that this compound might be an inverse agonist of the $GABA_B$ receptor.

CONCLUSIONS

The results of this investigation reveal that 5-HT is very readily oxidized by HO· at physiological pH. The initial reaction products are 5-HEO (the more stable keto tautomer of 2,5-DHT), T-4,5-D (formed by the very facile autoxidation of its precursor 4,5-DHT), and 5,6-DHT. Both 5-HEO and 5,6-DHT appears to be dihydroxytryptamine products that are unique to the HO·-mediated oxidation of 5-HT. By contrast, T-4,5-D (*via* 4,5-DHT) is also formed in the electrochemically-driven[19] and various enzyme-mediated oxidation[20] of the neurotransmitter. 5-HEO is a relatively stable compound (*in vitro*) and, hence, might represent a useful analytical marker molecule for intraneuronal formation of HO· and precursor reduced oxygen species ($O_2^{-\cdot}$, H_2O_2) in serotonergic neurons which has been implicated in a variety of neurodegenerative brain disorders including AD.

Both 5,6-DHT[48,49] and 4,5-DHT[50] are very easily oxidized by molecular oxygen to give *o*-quinones with concomitant formation of H_2O_2 as a byproduct. In the presence of ascorbic acid and/or GSH, 5,6-DHT and T-4,5-D undergo redox cycling reactions that further consume molecular oxygen and produce H_2O_2. Secondary products derived from T-4,5-D (*e.g.*, **1**, **2**, 7-*S*-Glu-T-4,5-D, **4**) and 5,6-DHT (*e.g.*, 5,6,7-THT) can also redox cycle in the presence of molecular oxygen and intraneuronal antioxidants such as ascorbic acid[55] and GSH[56] to yield H_2O_2.[24] This, in the presence of trace concentrations of low molecular weight transition metal ions, particularly Fe^{2+}, can provide additional HO· and hence a cascade of reactions in which compounds such as T-4,5-D, 5,6-DHT, 5,6,7-THT, **1**, **2**, 7-*S*-Glu-T-4,5-D and **4** catalyze their own synthesis and produce constantly increasing levels of reactive oxygen species (H_2O_2, $O_2^{-\cdot}$, HO·).

Significantly, oxidation of 5-HT by molecular oxygen can be initiated by trace levels of transition metal ions to give 5-HEO, 5,6,7-THT (presumably via 5,6-DHT) and DHBT. However, the presence of ascorbic acid and GSH, particularly the former, greatly accelerate this reaction. Although the complete details of both the mechanisms and products of the autoxidation of 5-HT in the presence of ascorbic acid and GSH remain to be elucidated, presently available information suggests that 5-HEO, 5,6-DHT, 5,6,7-THT, T-4,5-D, 7-*S*-Glu-T-4,5-D and **4** are probably all formed and, because of their redox cycling properties, contribute to the rapid oxidation of the neurotransmitter and potentiate formation of cytotoxic reduced oxygen species and their own synthesis.

Although 5-HEO does not appear to be neurotoxic or to possess any neuropharmacological properties (work in progress), several other products of the HO·-mediated oxidation and autoxidation of 5-HT are biologically-active substances. For example, 5,6-DHT is a known serotonergic neurotoxin[51] and T-4,5-D appears to be toxic towards serotonergic neurons (Table 3). While an acute dose of 7-*S*-Glu-T-4,5-D does not appear to evoke long-lasting neurodegenerative consequences (Table 4) it does appear to interact quite strongly with the $GABA_B$ receptor as an inverse agonist. In the event that this compound was produced in the cytoplasm of serotonergic axon terminals it is conceivable that its subsequent release and interactions with $GABA_B$ receptors on proximate terminals and postsynaptic sites could potentiate the release of many neurotransmitters including glutamate and evoke anatomically-selective excitotoxic damage.[57]

Oxygen free radicals have been implicated in many degenerative brain disorders including AD,[1,6-9] cerebral ischemia[13] and as a result of methamphetamine abuse.[14-16] Abnormal, but unknown, oxidized forms of 5-HT have been detected in the CSF of AD patients[44] and 5,6-DHT has been detected in rat brain following methamphetamine administration.[16] Furthermore, certain serotonergic and anatomically proximate pathways degenerate in the AD brain.[42,43] Could it be, therefore, that a very early event in the neurodegenerative processes in AD involves an increased level of HO·-mediated oxidation or autoxidation of 5-HT? Such an event, based on the present study, would lead not only to formation of at least one known neurotoxin, 5,6-DHT, but also to a large number of additional aberrant metabolites that as a result of redox cycling reactions could potentiate their own synthesis and greatly elevated levels of cytotoxic reduced oxygen species. These processes, in turn, might contribute to the degeneration of both serotonergic neurons and connected pathways of diverse neurotransmitter contents leading to the remarkably anatomically selective system neurodegeneration that characterizes AD.[42,43]

ACKNOWLEDGMENTS

This work was supported by NIH grant No: GM32367. Additional support was provided by the Research Council and Vice President for Research at the University of Oklahoma. The authors also thank the NIMH/Novascreen® Drug Discovery and Development Program (Contract No: NIMH-2003) for preliminary receptor binding screens.

REFERENCES

1. M.A. Smith, L.M. Sayre, V.M. Monnier and G. Perry, Radical AGEing in Alzheimer's Disease, *Trends Neurosci.*, 18: 172-176 (1995).
2. C.P. LeBel and S.C. Bondy, Oxidative damage and cerebral aging, *Prog. Neurobiol.*, 38: 601-609 (1992).
3. C.W. Olanow, A radical hypothesis for neurodegeneration, *Trends Neurosci.*, 16: 439-444 (1993).
4. M. Gerlach, D. Ben-Shachar, P. Riederer and M.B.H. Yondim, Altered brain metabolism of iron as a cause of neurodegenerative disease? *J. Neurochem.*, 63: 793-807 (1994).
5. B. Halliwell and J.M.C. Gutteridge, Oxygen free radicals and iron in relation to biology and medicine: some problems and concepts, *Arch. Biochem. Biophys.*, 246: 501-514 (1986).
6. R.A. Nixon and A.M. Cataldo, Free radicals, proteolysis and degeneration of neurons in Alzheimer's Disease: how essential is the β-amyloid link? *Neurobiol. Aging* 15: 463-469 (1994).
7. C.D. Smith, J.M. Carney, P.E. Starke-Reed, C.N. Oliver, E.R. Stadtman, R.A. Floyd and W.R. Markesbery, Excess brain protein oxidation and enzyme dysfunction in normal aging and in Alzheimer's Disease, *Proc. Nat. Acad. Sci. U.S.A.*, 88: 10540-10543 (1991).
8. K. Hensley, J.M. Carney, M.P. Mattson, M. Aksenova, M. Harris, J.F. Wu, R.A. Floyd and D.A. Butterfield, A model for β-amyloid aggregation and neurotoxicity based on free radical generation by the peptide: relevance to Alzheimer's Disease, *Proc. Nat. Acad. Sci. U.S.A.*, 91: 3270-3274 (1994).
9. K.V. Subbarao, J.S. Richardson and L.C. Ang, Autopsy samples of Alzheimer's cortex show increased peroxidation in vitro, *J. Neurochem.*, 55: 342-345 (1990).
10. D.T. Dexter, C.J. Carter, F.R. Wells, F. Javoy-Agid, P. Jenner and C.D. Marsden, Basal lipid peroxidation in substantia nigra is increased in Parkinson's Disease, *J. Neurochem.*, 52: 381-389 (1987).
11. P. Jenner, D.T. Dexter, J. Sian, A.H.V. Schapira and D.C. Marsden, Oxidative stress as a cause of nigral cell death in Parkinson's Disease and incidental Lewy Body Disease, *Ann. Neurol.*, (Suppl.), 32: S82-S87 (1992).
12. P. Jenner, A.H.V. Schapira and C.D. Marsden, New insights into the cause of Parkinson's Disease, *Neurology* 42: 2241-2250 (1992).
13. C.N. Oliver, P.E. Starke-Reed, E.R. Stadtman, G.J. Liu, J.M. Carney and R.A. Floyd, Oxidative damage to brain proteins, loss of glutamine synthetase activity and production of free radicals during ischemia/reperfusion-induced injury in gerbil brain, *Proc. Nat. Acad. Sci. U.S.A.*, 87: 5144-5147 (1990).
14. M.J. DeVito and G.C. Wagner, Methamphetamine-induced neuronal damage: a possible role for free radicals, *Neuropharmacology* 28: 1145-1150 (1989).
15. L.S. Seiden and G. Vosmer, Formation of 6-hydroxydopamine in caudate nucleus of the rat after a single large dose of methylamphetamine, *Pharmacol. Biochem. Behav.*, 21: 29-31 (1984).
16. D.L. Commins, K.J. Axt, G. Vosmer and L.S. Seiden, 5,6-Dihydroxytryptamine, a serotonergic neurotoxin is formed endogenously in the rat brain, *Brain Res.*, 403: 7-14 (1987).
17. C. Montoliu, S. Vallés, J. Renau-Piqueras and C. Guerri, Ethanol-induced oxygen radical formation and lipid peroxidation in rat brain: effect of chronic alcohol consumption, *J. Neurochem.*, 63: 1855-1862 (1994).
18. B. Halliwell, Reactive oxygen species and the central nervous system, *J. Neurochem.*, 59: 1609-1623 (1992).
19. M.Z. Wrona and G. Dryhurst, Electrochemical oxidation of 5-hydroxytryptamine in aqueous solution at physiological pH, *Bioorg. Chem.*, 18: 291-317 (1990).
20. M.Z. Wrona and G. Dryhurst, Interactions of 5-hydroxytryptamine with oxidative enzymes, *Biochem. Pharmacol.*, 41: 1145-1162 (1991).

21. F. Zhang and G. Dryhurst, Oxidation chemistry of dopamine: possible insights into the age-dependent loss of dopaminergic nigrostriatal neurons, *Bioorg. Chem.*, 21: 392-410 (1993).
22. M.Z. Wrona, F. Zhang and G. Dryhurst, Electrochemical oxidations of central nervous system indoleamines, catecholamines and alkaloids. Potential significance into neurodegenerative diseases, *J. Chin. Chem. Soc.*, **41**: 231-249 (1994).
23. F. Zhang and G. Dryhurst, Effects of *L*-cysteine on the oxidation chemistry of dopamine: new reaction pathways of potential relevance to idiopathic Parkinson's Disease, *J. Med. Chem.*, 37: 1084-1098 (1994).
24. M.Z. Wrona, Z. Yang, M. McAdams, S. O'Connor-Coates and G. Dryhurst, Hydroxyl radical-mediated oxidation of serotonin: potential insights into the neurotoxicity of methamphetamine, *J. Neurochem.*, 64: 1390-1400 (1995).
25. K-S. Wong, R.N. Goyal, M.Z. Wrona, C.L. Blank and G. Dryhurst, 7-*S*-Glutathionyl-tryptamine-4,5-dione: a possible aberrant metabolite of serotonin, *Biochem. Pharmacol.*, 46: 1637-1652 (1993).
26. M.Z. Wrona, R.N. Goyal, D. Turk, C.L. Blank and G. Dryhurst, 5,5′-Dihydroxy-4,4′-bitryptamine: a potentially aberrant neurotoxic metabolite of serotonin, *J. Neurochem.*, 59: 1392-1398 (1992).
27. P. Davies and A.F.J. Maloney, Selective loss of cerebral cholinergic neurons in Alzheimer's Disease, *Lancet 2*: 1403 (1976).
28. E.K. Perry, R.H. Perry, G. Blessed and B.E. Tomlinson, Necropsy evidence of cerebral cholinergic deficits in senile dementia, *Lancet* 1: 189 (1977).
29. D.M. Bowen, S.J. Allen, J.S. Benton, M.J. Goodhart, E.A. Haan, A.M. Palmer, N.R. Sims, C.C.T. Smith, J.A., Spillane, M.M. Esiri, D. Neary, J.S. Snowden, G.K. Wilcock and A.N. Davison, Biochemical assessment of serotonergic and cholinergic dysfunction in cerebral atrophy in Alzheimer's Disease, *J. Neurochem.*, 41: 266-272 (1983).
30. A.M. Palmer and D.M. Bowen, Neurochemical basis of dementia of the Alzheimer type: contribution of postmortem and antemortem studies, in: *Biological Markers in Dementia of Alzheimer Type*, C. Fowler, L.A. Carlson, C.G. Gottfries and B. Winblad, Eds., Smith-Gordon, London, 1990.
31. A.M. Palmer, P.T. Francis, D.M. Bowen, J.S. Benton, D. Neary, D.M.A. Mann and J.S. Snowden, Catecholaminergic neurones assessed ante-mortem in Alzheimer's Disease, *Brain Res.*, 414: 365-375 (1987).
32. D.M.A. Mann, P.O. Yates and J Hawkes, The noradrenergic system in Alzheimer and multi-infarct dementias, *J. Neurol. Neurosurg. Psychiatry* 45:113-119 (1982).
33. B.E. Tomlinson, D. Irving and G. Blessed, Cell loss in locus ceruleus in senile dementia of the Alzheimer type, *J. Neurol. Sci.*, 49: 418-421 (1981).
34. T. Yamamoto and A. Hirano, Nucleus raphe dorsalis in Alzheimer's Disease: neurofibrillary tangles and loss of large neurones, *Ann. Neurol.*, 17: 573-577 (1985).
35. A.M. Palmer, P.T. Francis, J.S. Benton, N.R. Sims, D.M.A. Mann, D. Neary, J.S. Snowden and D.M. Bowen, Presynaptic serotonergic dysfunction in patients with Alzheimer's Disease, *Brain Res.*, 48: 8-15 (1987).
36. D.M.A. Mann, P.O. Yates and B. Marcyniuk, Dopaminergic neurotransmitter systems in Alzheimer's Disease and Down's Syndrome in middle age, *J. Neurol. Neurosurg. Psychiatry* 50: 341-344 (1987).
37. A.W. Proctor, A.M. Palmer, P.T. Francis, S.L. Lowe, D. Neary, D.M.A. Mann and D.M. Bowen, Evidence of glutamatergic denervation and possible abnormal metabolism in Alzheimer's Disease, *J. Neurochem.*, 50: 790-802 (1988).
38. B.T. Hyman, G.W. Van Hoesen and A. Damasio, Alzheimer's Disease: Glutamate depletion in the hippocampal perforant pathway zone, *Ann. Neurol.*, 22: 37-40 (1987).
39. B.M. Hubbard and J.M. Anderson, Age-related variation in the neuron content of the cerebral cortex in senile dementia of the Alzheimer type, *Neuropath. Appl. Neurobiol.*, 11: 309-382 (1985).

40. M.J. Ball, Neuronal loss, neurofibrillary tangles and granulovacuolar degeneration in the hippocampus with aging and dementia, *Acta Neuropath.*, 37: 111-118 (1977).
41. D.M.A. Mann, Neuropathological and neurochemical aspects of Alzheimer's Disease, in: *Psychopharmacology of the Aging Nervous System*, L.L. Iversen, S.D. Iversen and S.H. Snyder, Eds.; Plenum Press, New York, 1988, pp. 1-67.
42. C.B. Saper, B.H. Wainer and D.C. German, Axonal and transneuronal transport in the transmission of neurobiological disease: potential role in system degenerations, including Alzheimer's Disease, *Neuroscience* 23: 389-398 (1987).
43. J. Hardy, R. Adolfsson, L. Alafuzoff, G. Bucht, J. Marcusson, P. Nyberg, E. Perdahl, P. Wester and B. Winblad, Transmitter defecits in Alzheimer's Disease, *Neurochem. Int.*, 7: 545-563 (1985).
44. L. Volicer, P.J. Langlais, W.R. Matson, K.A. Mark and P.H. Gamache, Serotonergic system in dementia of the Alzheimer type. Abnormal forms of 5-hydroxytryptophan and serotonin in cerebrospinal fluid, *Arch. Neurol.*, 42: 1158-1161 (1985).
45. A. Slivka and G. Cohen, Hydroxyl radical attack on dopamine, *J. Biol. Chem.*, 260: 15466-15472 (1985).
46. M.J. Del Rio, C.V. Pardo, J. Pinxteren, W. DePotter, G. Ebinger and G. Vauquelin, Binding of serotonin and dopamine to serotonin binding proteins' in bovine frontal cortex: evidence for iron-induced oxidative mechanisms, *Eur. J. Pharmacol.*, 247: 11-23 (1993).
47. S. Udenfriend, C.T. Clark, J. Axelrod and B.B. Brodie, Ascorbic acid in aromatic hydroxylation, *J. Biol. Chem.*, 208: 731-738 (1954).
48. S. Singh, J-F. Jen and G. Dryhurst, Autoxidation of the indolic neurotoxin 5,6-dihydroxytryptamine, *J. Org, Chem.*, 55: 1484-1489 (1990).
49. S. Singh and G. Dryhurst, Further insights into the oxidation chemistry and biochemistry of the serotonergic neurotoxin 5,6-dihydroxytryptamine, *J. Med. Chem.*, 33: 3035-3044 (1990).
50. K-S. Wong and G. Dryhurst, Tryptamine-4,5-dione: properties and reactions with glutathione, *Bioorg. Chem.*, 18: 253-264 (1990).
51. H.G. Baumgarten, H.P. Klemm, L. Lachenmeyer, A Björklund, W. Lovenberg and H.G. Schlossberger, Mode and mechanism of action of neurotoxic indoleamines: a review and progress report, *Ann. New York Acad. Sci.*, 305: 3-24 (1976).
52. S. Singh, M.Z. Wrona and G. Dryhurst, Synthesis and reactivity of the putative neurotoxin tryptamine-4,5-dione, *Bioorg. Chem.*, 35: 82-93 (1992).
53. Z. Fa, R.N. Goyal, C.L. Blank and G. Dryhurst, Oxidation chemistry of the central mammalian alkaloid 1-methyl-6-hydroxy-1,2,3,4-tetrahdyro-β-carboline, *J. Med. Chem.*, 35: 82-93 (1992).
54. N. Bowery, $GABA_B$ receptors and their significance in mammalian pharmacology, *Trends Pharmacol. Sci.*, 10: 401-407 (1989).
55. R. Spector and J. Eells, Deoxynucleoside and vitamin transport into the central nervous system, *Fed. Proc.*, 43: 196-200 (1984).
56. A. Slivka, C. Mytilineou and G. Cohen, Histochemical evaluation of glutathione in brain, *Brain Res.*, 409: 275-284 (1987).
57. D.W. Choi, Ionic dependence of glutamate neurotoxicity, *J. Neurosci.*, 7: 369-379 (1987).

DOPAMINE COVALENTLY MODIFIES DNA IN A TYROSINASE-ENHANCED MANNER

Alan H. Stokes[1], Buddy G. Brown[2], Chin K. Lee[2], David J. Doolittle[2], and Kent E. Vrana[1]

[1]Department of Physiology and Pharmacology, Bowman Gray School of Medicine, Wake Forest University, Winston-Salem, NC 27157-1083
[2]Environmental and Molecular Toxicology Division, R.J. Reynolds Co., Winston-Salem, NC 27102

INTRODUCTION

Parkinson's Disease (PD) is characterized by the chronic and progressive loss of dopaminergic neurons of the nigrostriatal pathway. An important theory suggests that the dopamine neurotransmitter can itself play a role in the degeneration.[1-7] One hypothesis is that dopamine (DA), which is synthesized from the essential amino acid tyrosine, can be converted to a reactive quinone metabolite (Figure 1). This oxidation of the catecholamine can either occur spontaneously or be accelerated by the enzyme tyrosinase. The attractive aspect of this hypothesis is that tyrosinase is a known component of the DA neurons which are at risk in PD. The dopamine quinone product is destructive to macromolecules for two reasons. First, it is a highly reactive molecule which is capable of covalently modifying nucleophiles such as sulfhydryls and amino groups. In fact, dopamine quinone is known to covalently modify both DNA and proteins.[3,6-8] Second, through a series of oxidation reactions, dopamine quinone can contribute to the production of hydrogen peroxide, superoxide radical, and the hydroxyl radical. These very reactive chemical entities can in turn generate extensive macromolecular damage. In the present experiments, we investigated the role of tyrosinase in DA-mediated DNA damage. The central hypothesis is that tyrosinase catalyzes the formation of dopamine quinone and therefore enhances the covalent modification of DNA by DA.

METHODS

Salmon Sperm DNA (100 µg) was incubated with DA in the presence or absence of mushroom tyrosinase (5 µg/ml; 4400 units/mg; Sigma Chemical Co.). The reaction was conducted in phosphate-buffered saline (PBS) in a total volume of 200 µl (in the dark at room

Neurodegenerative Diseases
Edited by Gary Fiskum, Plenum Press, New York, 1996

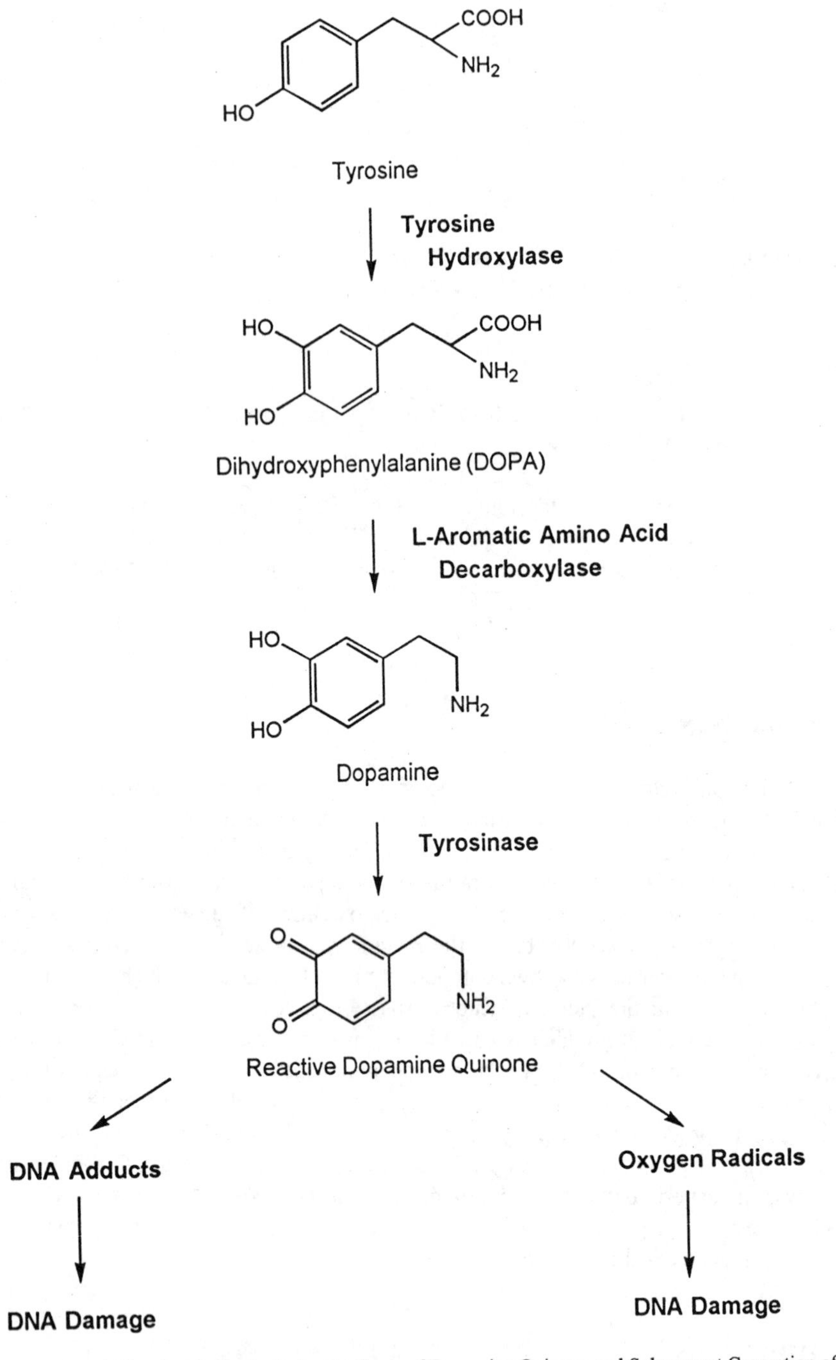

Figure 1. Proposed Chemical Pathway for the Synthesis of Dopamine Quinone and Subsequent Generation of DNA Adducts and Reactive Oxygen Species. Within dopaminergic neurons, DA (DA) is synthesized from tyrosine by the actions of tyrosine hydroxylase and aromatic amino acid decarboxylase. The pigmented neurons of the nigrostriatal pathway also contain the enzyme tyrosinase which normally converts tyrosine to DOPA and dopaquinone (in a two-step reaction). The latter of these two products then serves as a substrate for neuromelanin synthesis. However, tyrosinase can also synthesize dopamine quinone from DA. This compound is very reactive and can covalently modify DNA or contribute to the generation of reactive oxygen radicals (peroxide, superoxide and hydroxyl radicals).

temperature). DA treatments consisted of varying amounts of unlabeled DA combined with ^{3}H-DA (2 μCi). Following the incubation, samples were extracted with an equal volume of phenol/chloroform/isoamyl alcohol (25:24:1; v/v). Following centrifugation (14,000 x g; rm. temp; for 5 minutes), the DNA was precipitated from 150 μl of the supernatant by adding 5 ml ice-cold TCA (5%) and incubating on ice for fifteen minutes. The solution was then filtered through Whatman glass microfibre filters (GF/B) to collect the DNA precipitate. The filters were washed five times with five ml of ice-cold 5% TCA followed by a single five ml ethanol (70%) wash. The filters were then allowed to air dry at room temperature and were analyzed by liquid scintillation spectrometry (Model 1214 Rackbeta; LKB-Wallac Instruments Inc., Gaithersberg, MD, U.S.A.).

In several experiments, the ability of antioxidants to block tyrosinase-enhanced incorporation of ^{3}H-DA into DNA was tested. In these studies, DNA was incubated with DA and tyrosinase as described above with the inclusion of one of the following antioxidants: 0.5 mM ascorbate, 1 mM dithiothreitol (DTT), or 100 μM reduced glutathione. The DNA was then extracted, precipitated, and analyzed as described above.

RESULTS

Tyrosinase Enhances the Covalent Modification of DNA by DA

Incubation of salmon sperm DNA in the presence of increasing concentrations of ^{3}H-DA produced a small but significant dose-dependent covalent incorporation of the catecholamine into TCA-insoluble nucleic acid (Figure 2). This dose-dependent

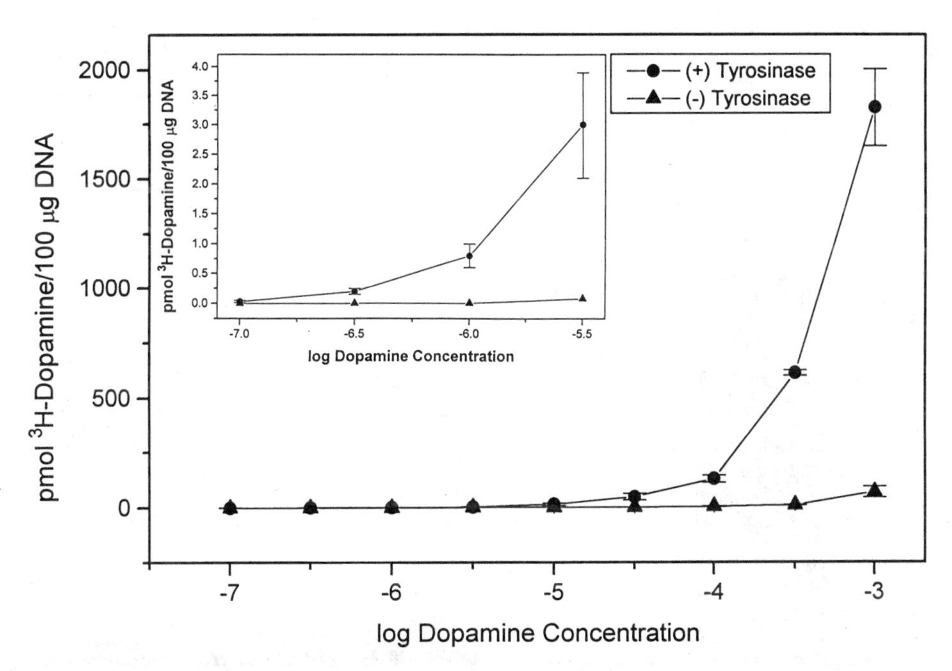

Figure 2. Tyrosinase Enhances the Dose-Dependent Incorporation of ^{3}H-DA into DNA. Salmon sperm DNA (100 μg) was incubated with various concentrations of ^{3}H-DA in the presence and absence of tyrosinase (5 μg/ml) as described in Methods. Following the *in vitro* incubation (2 hrs), the tyrosinase was extracted and the amount of covalently incorporated radioactive DA was assessed by TCA precipitation and liquid scintillation spectrometry. Each data point represents the mean ± SEM from four independent determinations. The inset represents an expanded scale for the lower concentrations. The DA incorporation was significantly increased ($P<0.05$) by tyrosinase at all concentrations of DA.

incorporation was markedly enhanced by the addition of 5 μg/ml tyrosinase. Incorporation of DA into DNA was increased as much as 160-fold by tyrosinase depending on the dose of catecholamine (*e.g.*, 0.8 pmol/μg DNA with tyrosinase *vs.* 0.005 pmol/μg DNA in the absence of tyrosinase; at 1 μM DA). Significantly enhanced modification was observed at concentrations of DA as low as 100 nM (0.03 pmol/μg DNA in the presence of tyrosinase *vs.* 0.0003 pmol/μg DNA in the absence of tyrosinase).

Antioxidants Abrogate the Tyrosinase-Enhanced DA Modification of DNA

If the covalent modification of DNA was mediated through the reactive oxidation product of DA, dopamine quinone, the presence of reducing agents should reduce or abolish the modification. Inclusion of the reducing agents ascorbate, DTT and reduced glutathione in the tyrosinase/DA incubations prevented DA-DNA adduct formation (Figure 3).

DISCUSSION

The present experiments established that DA is capable of covalently modifying DNA *in vitro* in a dose-dependent manner. This is consistent with previous reports by Moldéus and co-workers[8] and Lévay and Bodell[6]. However, the formation of DA-DNA adducts is greatly enhanced (up to 160-fold) by the presence of the oxidative enzyme, tyrosinase (Figure 2). Tyrosinase is a normal constituent of pigmented striatal dopaminergic neurons[9] and acts to convert tyrosine first to DOPA and then to dopaquinone.[1,2] Under normal circumstances, the quinone then goes on to form the pigmented melanins. While the role of tyrosinase and the neuromelanin pigments in the physiology of nigrostriatal neurons remains unclear, it is known

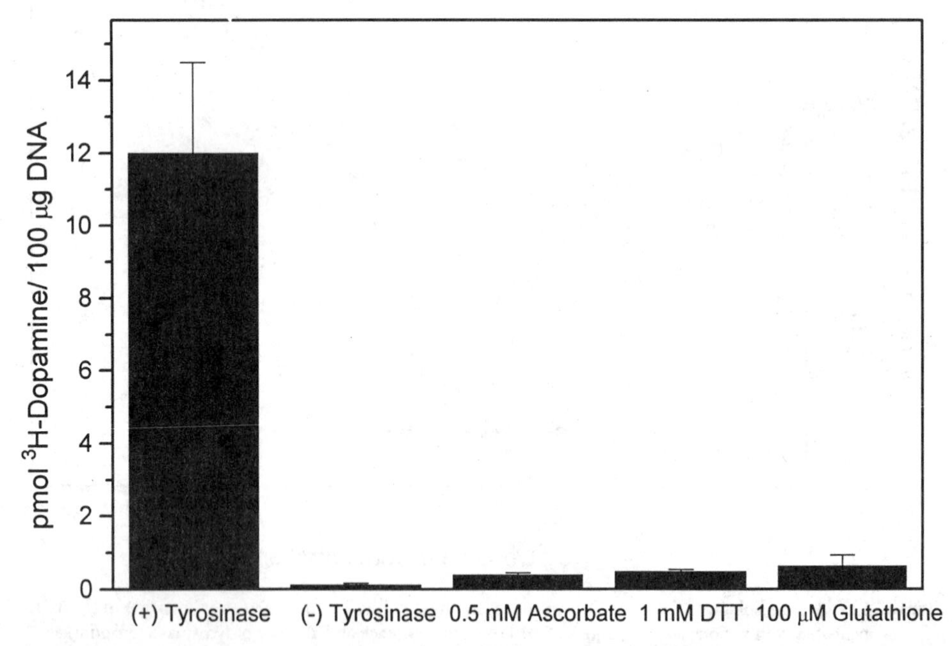

Figure 3. Antioxidants Prevent the Tyrosinase-Enhanced DA Modification of DNA. DNA was incubated with 10 μM ^{3}H-DA in the presence and absence of 5 μg/ml tyrosinase and the formation of DA-DNA adducts monitored as described in Methods. In addition, DA/tyrosinase incubations were conducted with the inclusion of 0.5 mM ascorbate, 1 mM DTT, or 100μM reduced glutathione. Each bar represents the mean ± SEM of three independent determinations. Each of the antioxidant treatments is significantly ($P<0.05$) reduced compared with the DA/tyrosinase treatment.

that the enzyme has a broad substrate specificity and can catalyze the conversion of DA to its reactive metabolite, dopamine quinone (Figure 1).[1] Therefore, conditions which would increase the levels of DA within a tissue or cell containing tyrosinase could be expected to generate dopamine quinone.

Quinones are reactive species with the propensity to modify macromolecular nucleophiles. Our presumption is that the tyrosinase-enhanced DA modification of DNA is being mediated by the dopamine quinone intermediate. While this is expected based on the enzymatic mechanism of tyrosinase, it is also supported by the findings that reducing agents and free radical scavengers such as ascorbate, DTT, and glutathione all acted to abolish the formation of DA-DNA adducts (Figure 3). This correlates with a recent observation by Hastings and Zigmond.[7] They report that the incubation of rat brain striatal slices with DA results in the formation of catechol-protein conjugates and that the covalent complex occurs at sulfhydryl moieties of cysteine. In their studies, this effect of DA was also decreased by ascorbate and glutathione.

The importance of DA-DNA adducts in the neurodegeneration in PD remains to be determined. However, modification of chromosomal DNA which exceeded the repair capacity of the cell would result in gene mutations and probably interfere with normal gene expression. There is a great deal of evidence that the direct injection of excessive levels of DA into the striatum produces neurodegeneration of the nigrostriatal neurons (see ref. 7). The fact that these neurons express tyrosinase may be a coincidence. On the other hand, the presence of tyrosinase has also been implicated in the sensitivity of pigmented melanoma cells to DA[10]. In our hands, preliminary evidence suggests that tyrosinase enhances the cytotoxicity of DA towards chinese hamster ovary cells while incubation of rat brain striatal slices also produces DA-DNA adducts (data not shown). Therefore, it seems likely that the formation of DA-DNA adducts may play a role in the DA-mediated neurodegeneration and the the presence of tyrosinase may exacerbate the pathophysiological process.

REFERENCES

1. D.G. Graham, Oxidative pathways for catecholamines in the genesis of neuromelanin and cytotoxic quinones, *Mol. Pharmacol.* 14:633-643 (1978).
2. D.G. Graham, S.M. Tiffany, W.R. Bell, Jr., and W.F. Gutknecht, Autoxidation versus covalent binding of quinones as the mechanism of toxicity of dopamine, 6-hydroxydopamine, and related compounds toward C1300 neuroblastoma cells *in vitro*, *Mol. Pharmacol.* 14:644-653 (1978).
3. R.P. Mason, Free radical metabolites of foreign compounds and their toxocological significance, *in:* "Reviews in Biochemical Toxicology," E. Hodson, J.R. Bend, and R.M. Philpot, ed., pp. 151-200, (1979).
4. C.J. Schmidt, J.K. Ritter, P.K. Sonsalla, G.R.Hanson, and J.W. Gibb, Role of dopamine in the neurotoxic effects of methamphetamine, *J. Pharmacol. Exp. Ther.* 233:539-544 (1985).
5. P.P. Michel, and F. Hefti, Toxicity of 6-hydroxydopamine and dopamine for dopaminergic neurons in culture, *J. Neurosci. Res.* 26:428-435 (1990).
6. G. Lévay, and W.J. Bodell, Detection of dopamine-DNA adducts: Potential role in Parkinson's Disease, *Carcinogenesis* 14:1241-1245 (1993).
7. T.G. Hastings, and M.J. Zigmond, Identification of catechol-protein conjugates in neostriatal slices incubated with [^{3}H]dopamine: Impact of ascorbic acid and glutathione, *J. Neurochem.* 63:1126-1132 (1994).

8. P. Moldéus, M. Nordenskjöld, G. Bolcsfoldi, A. Eiche, U. Haglund, and B. Lambert, Genetic toxicity of dopamine, *Mut. Res.* 124:9-24 (1983).
9. M. Miranda, D. Botti, A. Bonfigli, T. Ventura, and A. Arcadi, Tyrosinase-like activity in normal human substantia nigra, *Gen. Pharmacol.* 15:541-544 (1984).
10. M.M. Wick, and G. Fitzgerald, Inhibition of reverse transcriptase by tyrosinase generated quinones related to levodopa and dopamine, *Chem.-Biol. Interact.* 38:99-107 (1981).

BRAIN [52FE]-TRANSFERRIN UPTAKE IN PATIENTS WITH ALZHEIMER'S DISEASE AND HEALTHY SUBJECTS: A POSITRON EMISSION TOMOGRAPHY (PET) STUDY

Angelo Antonini, *Albert Wettstein, *Regula Schmid, Klaus L. Leenders

Paul Scherrer Institute, Villigen; *Stadtärztlicher Dienst, Zurich, Switzerland

INTRODUCTION

Alteration of cerebral iron metabolism is involved in several neurodegenerative disorders like Parkinson's disease and Alzheimer's disease. However, it is still difficult to understand whether changes in cerebral iron levels are primarily related to the pathological process or secondary to the disease state. Transferrin (Tf) is the major iron carrier protein in plasma. The transport of iron into brain involves the binding of the iron-Tf complexes to high affinity Tf receptors (TfR) which are located in the brain capillary membranes of the blood brain barrier (BBB). The complex Tf-TfR is then internalized into cells where the iron is released. Other metals like aluminium or manganese also bind to transferrin and may be transported into brain in a similar manner as iron. Neuropathological studies using [125I]-transferrin have shown increased binding in cerebral microvessels in AD patients compared to age matched healthy controls [1]. We have shown previously that blood to brain iron uptake can be measured using the tracer [52Fe]-citrate and positron emission tomography (PET) [2-3]. [52Fe] is a positron emitter which rapidly and almost completely binds to plasma Tf after intravenous injection.

In this study, we have investigated brain [52Fe]-citrate uptake in a group of patients with Alzheimer's disease (AD) and age-matched healthy subjects.

METHODS

We studied 8 patients with probable AD (age mean ± SD: 66±7; Mini Mental Scores 18 to 27) diagnosed in accordance NINDS/ADRDA criteria [4] and 10 age matched healthy controls (age mean ± SD: 65±5). AD patients were selected after neuropsychological evaluation at the Zurich memory clinic. PET scans were performed at the Paul Scherrer Institute on a CTI 933/04-16 scanner which records 7 planes simultaneously (transaxial resolution after reconstruction 8 mm full width at half maximum [FWHM]). The head was placed in an individually moulded holder to reduce movements during scanning. The scanner was aligned parallel to the orbitomeatal line (OM) using a laser beam. The gantry field of view was chosen to cover the region from OM plus 1 cm to OM plus 6.6 cm containing the temporal lobe, the lower part of the frontal cortex, the basal ganglia and the upper half of the cerebellum. A 10 minute transmission scan was performed using an external 68Ge ring source.

52Fe-citrate was prepared as previously described [5]. The purified nuclide forms with citrate buffer a $52Fe^{3+}$-citrate complex with a specific activity of 14800 MBq/mg Fe. The physical half-life of 52Fe is 8.2 hours.

52Fe-citrate (12-15 MBq) was infused intravenously in a volume of 10 ml of physiological saline, over a period of 3 minutes using a constant volume infusion pump. Scans consisted of 18 frames for a total 110 minutes. Nineteen arterial blood samples were taken during the PET-measurement to determine 52Fe plasma activity. Plasma activity curves were also logarithmic transformed to calculate 52Fe plasma half-life

Neurodegenerative Diseases
Edited by Gary Fiskum, Plenum Press, New York, 1996

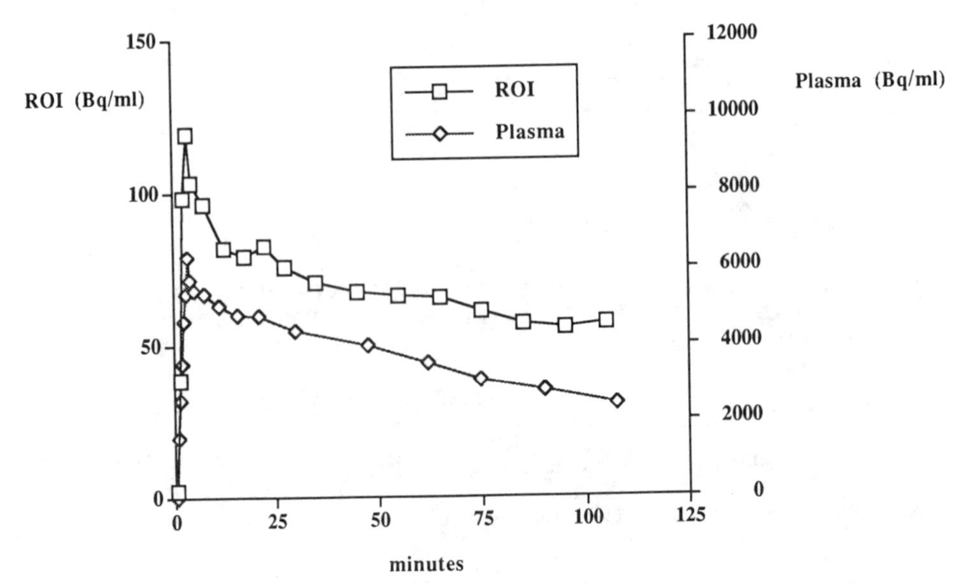

Fig 1. Time activity curves in the brain and plasma of a healthy control subject

The following regions of interest (ROI's) were defined on a visual display unit in a standard template arrangement: temporal lateral cortex, frontal cortex, striatum and cerebellum. Values of the right and left hemisphere were averaged. To facilitate identification of anatomical structures, ROI's were placed on integral images created by adding the frames between 60 and 110 minutes and then superimposed on the dynamic series for calculation.
52Fe data were analyzed using a multiple-time-graphical analysis (MTGA) approach [6] using 52Fe plasma radioactivity as input function. In the MTGA method, the gradient of the linear regression of the data reflects the unidirectional uptake rate constant (described as the net influx constant Ki (1/min)). The intercept of the regression line on the y axis is a measure of the apparent distribution volume (DV) of the tracer.

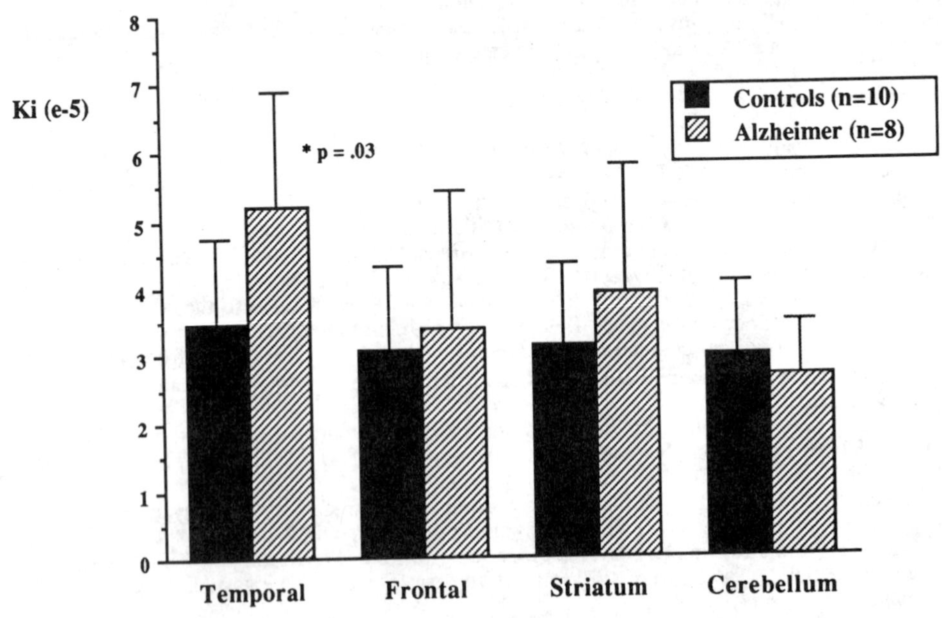

Fig 2. Mean ± SD tracer uptake constant (Ki) in AD and healthy controls

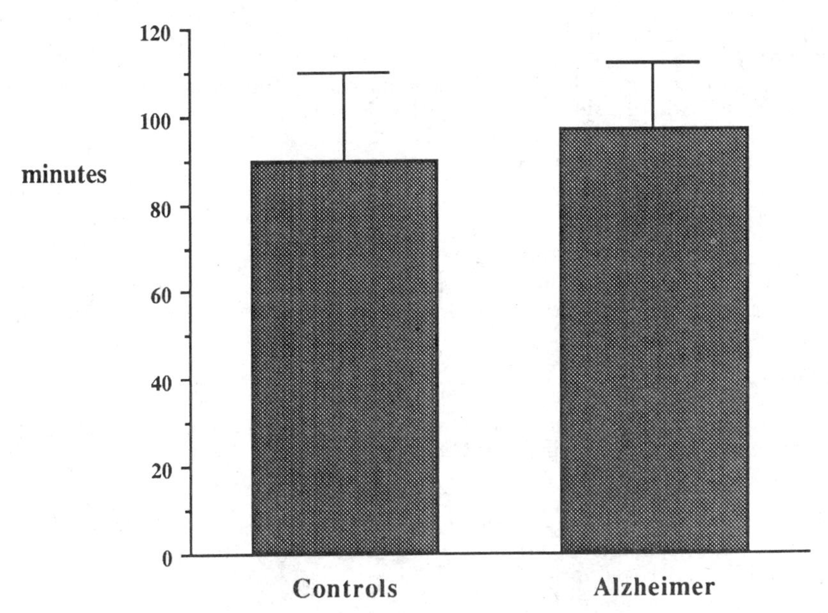

Fig 3. Mean ± SD of tracer plasma half-life

RESULTS

Fig 1 shows the time activity curve of [52Fe]-citrate in plasma and in the ROI-sample of one healthy control subject.

Mean Ki values (± SD) for the examined brain regions are displayed in fig 2. AD patients showed significantly increased values in the temporal region (p=.03; ANOVA). The frontal cortex, the striatum and the cerebellum revealed values in the control range.

No differences were found in the tracer volume of distribution (DV) between AD patients and controls (fig 3). Tracer plasma half-life in AD was also in the range of healthy subjects.

CONCLUSIONS

We reported a significant increase of [52Fe]-citrate uptake in the temporal cortex of patients with probable AD. The other examined brain regions showed uptake values in the control range.

Previous animal and patient studies have shown that [52Fe]-citrate binds rapidly after injection to plasma Tf [2-3]. Therefore, brain radioactivity accumulation mainly depends on Tf kinetic and reflects binding of the [52Fe]-Tf complexes to Tf receptors and their transport across the BBB level. Because brain iron turnover is very slow, PET measurements of [52Fe]-citrate uptake do not provide information about iron further distribution and storage within the brain. The increase of Ki values we found in the temporal region of AD patients may express increased iron turnover across the BBB secondary to changes of the Tf receptor density in the cerebral capillaries. There is evidence from neuropathological studies that [125I]Tf binding to cerebral microvessels is increased in AD patients. In particular Bmax was increased with no change of the affinity constant (Kd) suggesting increased transport of Tf bound iron (or other metals like aluminium) across the blood brain barrier endothelium in AD [1]. Conversely in the same AD patients Tf receptor density was lower particularly in the hippocampus and other cortical regions most likely secondary to degeneration of Tf containing neurons.

Metals and in particular iron can induce oxidative stress through the formation of oxygen free radicals. Free radicals are involved in the lipid peroxidation of the cell membranes resulting in cell death. It has been suggested that this mechanism of cell damage underlies neurodegenerative diseases like Parkinson's and AD [7]. Neuropathological studies have shown increased iron deposition in the cortical brain regions of AD patients [8-9]. Also increases of aluminium concentrations have been reported in the brain of AD patients [10-11]. Aluminium as well as other metals like manganese binds to plasma transferrin and is transported into brain through the Tf receptor system at the blood brain barrier level. Particularly neurofibrillary tangles show a selective accumulation of aluminium and iron [12]. Also specific accumulation of ferritin, the protein re-

sponsible for iron and other metal storage, was observed in the senile plaques and neurofibrillary tangles of AD patients.
We did not find differences in the DV of [52Fe]-citrate between AD and controls in the examined brain regions, indicating no significant changes of the plasma volume component in the patient group. Also no difference was present in [52Fe]-citrate plasma half-life between in AD and healthy controls indicating that peripheral iron metabolism is not affected in AD.
If iron turnover across the BBB is increased, AD patients may benefit from ion chelators therapy to reduce metal bioavailability for brain uptake. In a clinical study, two year treatment with the ion chelator desferrioxamine significantly slowed the rate of dementia progression in a group of 25 AD patients compared to a placebo group [13]. This finding confirms further that iron or other metals like manganese or aluminium which are transported across the BBB through the Tf receptor system play an active role in the pathological process leading to neuronal cell death in AD.

References

1. Kalaria RN, Sromek SM, Grahovac I, Harik SI Transferrin receptors of rat and human brain and cerebral microvessels and their status in Alzheimer's disease. Brain Research 1992; 585; 87-93

2. Leenders KL, Antonini A, Schwarzbach R, Smith-Jones P, Reist H, Ben-Sachar D, Youdim M, Henn V. Blood to brain iron in one monkey using [52Fe]-citrate and positron emission tomography (PET). J Neuronal Transm 1994; 43: 123-132

3. Leenders KL, Antonini A, Schwarzbach R, Smith-Jones P, Pellika R, Günther I, Psylla M, Reist H. Blood to brain iron transport in man using [52Fe]-citrate and positron emission tomography (PET). In: Uemura K (ed) Quantification of brain function. Tracer kinetics and image analysis in brain PET. Elsevier, Amsterdam 1993 pp145-150

4. McKhann G, Drachman D, Folstein M, Katzman R, Price D, Stadlan EM. Clinical diagnosis of Alzheimer's disease: report of the NINCDS-ADRDA Work Group under the auspices of the Department of Health and Human Service Task Force on Alzheimer's Disease. Neurology 1984; 34:939-944

5. Smith-Jones P, Schwarzbach R, Weinreich R The production of 52Fe by means of a medium energy proton accelerator. Radiochem Acta 1990; 50: 33-39.

6. Patlak CS, Blasberg RG. Graphical evaluation of blood-to-brain transfer constants from multiple-time uptake data. Generalizations. J Cereb Blood Flow Metabol 1985; 5:584-590

7. Gerlach M, Ben-Shacar D, Riederer P, Youdim MBH. Altered brain metabolism of iron as a cause of neurodegenerative diseases. J Neurochem 1994 63: 793-807

8. Dedman DJ, Treffry A, Candy JM, et al. Iron and aluminium in relation to brain ferritin in normal individuals and Alzheimer's disease and chronic renal-dialysis patients. Biochem J. 1992; 287; 509-514

9. Connor JR, Snyder BS, Beard JL, Fine RE, Mufson EJ Regional distribution of iron and iron regulatory proteins in the brain in aging and Alzheimer's disease. J Neurosc Research 1992; 31:327-335

10. Crapper DR, Krishnan SS, Dalton AJ. Brain aluminium distribution in Alzheimer's disease and experimental neurofibrillary degeneration. Science 1973; 180:511-513

11. Crapper DR, Krishnan SS, Quittkat. Aluminium, neurofibrillary degeneration and Alzheimer's disease. Brain 1976; 99:67-80

12. Good PF, Perl DP, Bierer LM, Schmeidler J. Selective accumulation of aluminium and iron in the neurofibrillary tangles of Alzheimer's disease: a laser microprobe (LAMMA) study. Ann Neurol 1992; 31:286-292

13. McLachlan DRC, Dalton AJ, Kruck TPA, Bell MY, Smith WL, Kalow W, Andrews DF. Intramuscular desferrioxamine in patients with Alzheimer's disease. Lancet 1991; 337:1304-1308.

βAPP METABOLITES, RADICALS, CALCIUM, AND NEURODEGENERATION: NOVEL NEUROPROTECTIVE STRATEGIES

Mark P. Mattson[1,2], Steven W. Barger[1], Katsutoshi Furukawa[1], Robert J. Mark[1,2], Virginia L. Smith-Swintosky[1], L. Creed Pettigrew[3], and Annadora J. Bruce[1].

[1]Sanders-Brown Research Center on Aging, and
[2]Departments of Anatomy & Neurobiology and
[3]Neurology
University of Kentucky, Lexington, KY 40536

INTRODUCTION

Many biochemical and molecular alterations in the brains of victims of Alzheimer's disease (AD) and other age-associated neurodegenerative disorders have been described. Unfortunately, it remains unclear which of the alterations contribute to the neuronal damage, which represent compensatory and cytoprotective responses to ongoing cell injury, and which are mere remnants of damaged cells in general. The present article describes studies that have been performed in our laboratories to help define roles for two major metabolites of the β-amyloid precursor protein (βAPP) in the pathogenesis of AD. In addition, we discuss findings concerning age-related alterations in brain metabolism (e.g., reduced glucose availability), and cellular signaling systems regulating neuronal plasticity and survival (e.g., neurotrophic factors), that are likely to impact on the biological activities of βAPP metabolites. Cellular systems regulating metabolism of calcium and reactive oxygen species (ROS) appear to be critical targets of both neurodegenerative and neuroprotective pathways. We therefore highlight the variety of both natural and synthetic compounds that can stabilize calcium homeostasis and ROS metabolism, and which may thus prove effective in reducing neuronal injury in a variety of neurodegenerative disorders. We emphasize βAPP in this chapter, not because it is the only determinant of AD, but rather because increasing data suggest it plays a pivotal role in many cases. This article is not intended to be a comprehensive review of the literature, and we refer the reader to review articles that provide more in-depth analyses of our current understanding of the molecular and cellular pathophysiology of AD[1-4]. It should be noted at the outset that many of the mechanisms of neuronal injury and neuroprotection described below have direct relevance to an array of neurodegenerative conditions ranging

Neurodegenerative Diseases
Edited by Gary Fiskum, Plenum Press, New York, 1996

from stroke to epilepsy to Parkinson's disease. Indeed, in many cases mechanisms initially elucidated in paradigms of excitotoxic, ischemic and oxidative brain injuries have subsequently been shown to apply to AD and other chronic neurodegenerative disorders.

βAPP is a glycoprotein that is expressed in many different types of cells, including neurons and glial cells. βAPP is thought to be a transmembrane protein with a single membrane-spanning region, a short cytoplasmic carboxyterminal domain, and a large extracellular region containing several biologically active domains. Several forms of βAPP have been identified (695-770 amino acids), some of which contain and some of which lack a Kunitz protease inhibitor domain near the aminoterminus. The 40-42 amino acid amyloid β-peptide (Aβ), the major component of plaques in the brains of AD victims, resides within the βAPP sequence, located partially within the plasma membrane and partially in the extracellular space. Several sites have been identified within the βAPP sequence at which enzymatic cleavages occur, including a site in the middle of the Aβ sequence (α-secretase site) and a site at the N-terminus of Aβ (β-secretase site). Alternative processing at these two sites appears to be a major determinant of whether intact Aβ (β-secretase cleavage) or secreted forms of βAPP (sAPP) (α-secretase) are released from cells.

MECHANISM OF AMYLOID β-PEPTIDE NEUROTOXICITY

Peptide Aggregation, Radical Production, and Loss of Ion Homeostasis

Aβ is released from cells in a soluble form but has a propensity to form insoluble aggregates with characteristic β-sheet structure[1,2]. The aggregation process is influenced by a variety of parameters including pH, oxidation state, metal ions (e.g., zinc), and presence of proteins that bind the peptide[5-8]. Numerous cell culture studies have shown that Aβ can be neurotoxic and that toxic activity is strongly correlated with the process of peptide aggregation[9-11]. Both full-length Aβ (1-42) and fragments thereof can aggregate and be neurotoxic, including the short 11 amino acid Aβ (25-35). Recent amino acid deletion and substitution experiments suggest that methionine 35 plays an important role in aggregation and neurotoxicity of Aβ[10].

The mechanisms underlying Aβ toxicity are still unresolved, but certain cellular pathways have been shown to be involved. We[12-15] and others[16,17] have shown that Aβ can induce accumulation of reactive oxygen ROS in cultured neurons. These ROS include hydrogen peroxide, lipoperoxides and hydroxyl radical. The major (initial) site of cellular oxidation induced by Aβ appears to be the plasma membrane[9,15,18]. Prior to identification of an oxidative process involved in Aβ neurotoxicity, we had shown that Aβ disrupts cellular calcium homeostasis resulting in elevated rest levels of intracellular free calcium $[Ca^{2+}]_i$ and markedly enhanced $[Ca^{2+}]_i$ responses to glutamate in cultured human cortical and rat hippocampal neurons[9,19]. More recently, we clarified the relationships between accumulation of ROS and loss of $[Ca^{2+}]_i$ homeostasis (Fig. 1). We found that plasma membrane ion-motive ATPase activities (Na^+/K^+-ATPase and Ca^{2+}-ATPase) are severely impaired in cultured rat hippocampal neurons and human hippocampal synaptosomes exposed to Aβ25-35 or Aβ1-40[18]. In the hippocampal cell cultures, impairment of Na^+/K^+-ATPase activity occurs within 20-30 min of exposure to Aβ and precedes elevation of $[Ca^{2+}]_i$. Experiments with ouabain, a specific inhibitor of the Na^+/K^+-ATPase, demonstrated that impairment of the sodium pump is sufficient to induce a rise in $[Ca^{2+}]_i$ and that the rise in $[Ca^{2+}]_i$ is dependent on an influx of Na^+. We also found that prevention of Na^+ influx is sufficient to attenuate the increase in $[Ca^{2+}]_i$ induced by Aβ. In contrast to the marked impairment of Na^+/K^+-ATPase and Ca^{2+}-ATPase activities by Aβ, other Mg^{2+}-dependent ATPase activities and Na^+/Ca^{2+} exchange are essentially unaffected by Aβ. ROS apparently play a key role in Aβ-induced impairment of ion-motive ATPases and elevation of $[Ca^{2+}]_i$ because both alterations are largely prevented when neuronal cultures are pretreated with the antioxidants vitamin E, propyl gallate and the spin-trapping compound PBN[18].

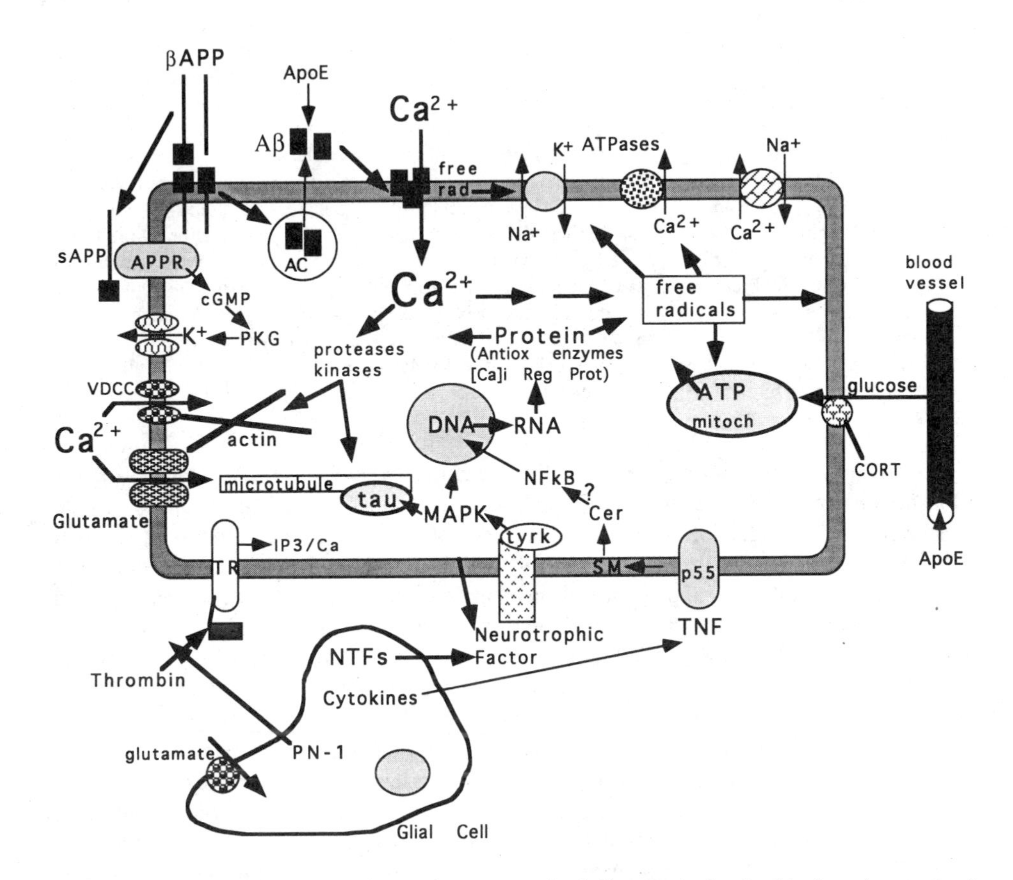

Figure 1. Mechanisms of neuronal injury and neuroprotection believed to be involved in the pathogenesis of neurodegenerative disorders. At the heart of the events leading to neuronal degeneration and death are the accumulation of free radicals and calcium within neurons. Cells have elaborate mechanisms for ensuring that levels of calcium and free radicals are maintained relatively low. The Ca^{2+} concentration outside the cell (approximately 1 mM) is normally 10,000-fold higher than inside the cell (approximately 100 nM). Calcium is normally removed from the cytoplasm by the plasma membrane Ca^{2+}-ATPase and the Na^+/Ca^{2+} exchanger (**upper right**), as well as by uptake into intracellular compartments and sequestration by Ca^{2+} binding proteins (not shown). Calcium enters cells (down the concentration gradient) through proteinaceous channels in the plasma membrane. Such Ca^{2+} channels include voltage-dependent Ca^{2+} channels (VDCC) and ligand-gated channels such as those activated by the excitatory transmitter glutamate (**left**). The sodium pump (Na^+/K^+-ATPase) plays an important role in Ca^{2+} homeostasis by maintaining the plasma membrane at resting potential. While transient elevations of $[Ca^{2+}]_i$ are involved in normal information signaling processes, sustained elevations of $[Ca^{2+}]_i$ can damage and kill cells by activating proteases and inducing free radical production (middle). A variety of potentially injurious stimuli can promote elevation of $[Ca^{2+}]_i$, including glutamate, glucose deprivation, amyloid β-peptide (Aβ), and thrombin. Glutamate induces Ca^{2+} influx through receptor channels and VDCC. The β-amyloid precursor protein (βAPP) is the source of Aβ which accumulates as plaques in the brains of AD victims (**upper left**). Aβ is liberated from βAPP by cleavage at the N-terminus of Aβ leaving behind a C-terminal βAPP fragment which is further processed, apparently in an acidic

compartment (AC) to release intact Aβ. Aβ induces accumulation of neuronal ROS, which are believed to promote Ca^{2+} influx by impairing function of ion-motive ATPases (Na^+/K^+-ATPase and Ca^{2+}-ATPase); neurons exposed to Aβ are hypersensitive to excitotoxicity induced by glutamate. Alternative processing of βAPP by α-secretase enzyme results in cleavage within the Aβ sequence, liberating secreted forms of βAPP (sAPP) from the cell. The sAPPs may play roles in plasticity and neuroprotection by activating a receptor linked to cyclic GMP (cGMP) production. Activation of cGMP-dependent protein kinase (PKG) results in activation of K^+ channels which hyperpolarizes the plasma membrane and reduces Ca^{2+} influx induced by excitatory stimuli such as glutamate. Neurotrophic factors, produced by glia and neurons activate receptor tyrosine kinases (**lower middle**)leading to phosphorylation cascades involving mitogen-activated protein kinases (MAPK) which activate transcription factors. Gene products regulated by neurotrophic factors include antioxidant enzymes and proteins involved in $[Ca^{2+}]_i$ homeostasis. Tumor necrosis factor (TNF) activates receptors linked to sphingomyelin (SM) hydrolysis, liberation of ceramide (Cer) and activation of the transcription factor NFκB. In order to maintain ATP levels, neurons rely on a constant supply of glucose which is transported into the cells via a specific high-affinity glucose transporter (**right**). In aging and AD, glucose availability to neurons may be compromised by vascular alterations and impaired glucose transport. Glucocorticoids (CORT) produced in response to stress can increase neuronal vulnerability by suppressing glucose transport. Apolipoprotein E may contribute to neuronal degeneration by influencing the development of atherosclerosis, or by affecting key events in neurodegenerative pathways such as the aggregation of Aβ. Protease - protease inhibitor interactions can influence the accumulation of Ca^{2+} and ROS in neurons. For example, thrombin induces Ca^{2+} influx and peroxide accumulation, and protease nexin-1 (PN-1) blocks the actions of thrombin. Finally, cytoskeletal elements may influence Ca^{2+} homeostasis and ROS metabolism. For example, actin depolymerization induced by Ca^{2+} influx results in suppression of further Ca^{2+} influx and protection against excitotoxicity and Aβ neurotoxicity.

A major unresolved issue concerns the mechanism whereby Aβ induces ROS in cells. One novel hypothesis is that the peptide itself can form free radical species[14]. When placed in physiological solutions, in the presence of a spin-trapping compound, synthetic Aβ generates an electron paramagnetic resonance (EPR) signal demonstrating the presence of a trapped radical. The appearance of the peptide radical signal was dependent upon the presence of molecular oxygen, but independent of the presence of metals. The kinetics of the appearance of the EPR peptide radical signal corresponded closely with the kinetics of peptide aggregation and neurotoxicity of the particular Aβs studied[14]. Moreover, Aβ (in a test tube) caused an oxygen-dependent inactivation of the enzymes glutamine synthetase and creatine kinase suggesting that Aβ radical peptides can directly damage enzymes. An attractive feature of the Aβ-radical hypothesis is that it has the potential to explain many of the properties of the peptide and its interaction with cell membrane constituents. For example, radicals are known to promote protein cross-linking and peptide-derived radicals could therefore be involved in the process of amyloid fibril formation itself. The presence of peptide radicals also provides a mechanism whereby the peptide could induce ROS in cell membranes. Available data argue against the presence of specific cell surface receptors for Aβ, and in the absence of a specific signal transduction mechanism for Aβ, the radical peptide hypothesis is consistent with much of the extant data..

The data just described suggest a scenario in which Aβ induces oxidation of membrane constituents resulting in impairment of ion-motive ATPase activities. Disruption of ion homeostasis results in elevation of $[Ca^{2+}]_i$, and both ROS and Ca^{2+} contribute to cell injury and death (Fig. 1). A prediction of this model is that agents that suppress ROS accumulation and stabilize $[Ca^{2+}]_i$ will protect neurons against Aβ toxicity. Indeed, a common feature of the various agents shown to protect neurons against Aβ toxicity is that they either (a) have intrinsic antioxidant activity, (b) enhance Ca^{2+} homeostasis, or (c) induce the expression of cellular proteins involved in regulating $[Ca^{2+}]_i$ or free radical metabolism (Table 1). Agents in the first category include: vitamin E, nordihydro-guaiaretic acid, propyl gallate, and spin-trapping compounds[12,13,14,16]. Agents in the second category include Ca^{2+} chelators, Ca^{2+} channel blockers, and actin depolymerizing

Table 1: Examples of Agents that Protect Hippocampal Neurons Against Insults Relevant to the Pathogenesis of Neurodegenerative Disorders*

Agent	Excitotoxic	Metabolic	Oxidative	Amyloid
Antioxidants				
NDGA**	+	+	+++	++
α-tocopherol	+	+	++	++
Propyl gallate	+	nd	++	++
PBN	nd	nd	++	+
$[Ca^{2+}]_i$-Stabilizers				
Calcium Channel Blockers	+	+	+/-	-
Chelators (e.g., EGTA)	++	+	+/-	+
EAA Receptor Antagonists	+++	++	+	+
Trophic Factor Mimics				
Staurosporine	+	++	+	+
K-252a	+	++	+	+
K-252b	+	++	+	+
Anticonvulsants				
Carbamazepine	+	nd	-	+
Valproate	+	nd	-	+
Phenytoin	++	nd	-	+
$K^{\pm}$ Channel Agonists				
Diazoxide	++	+	++	+
Pinacidil	++	+	++	+
Cytoskeletal Agents				
Cytochalasin D	++	nd	-	++
Taxol	+	nd	-	
Glucocortaid Suppressors				
Metyrapone	++	++	nd	nd
Neurotrophic Factors				
bFGF	+++	++	++	++
BDNF	++	++	++	+
NT-3	+	++	+	-
NT-4/5	+	++	+	nd
NGF	+/-	++	+	-
IGF-1, IGF-2	++	++	++	nd
Cytokines				
TNFα and TNFβ	++	++	+	+
TGF-β	++	++	nd	+
S100-β	+	++	nd	nd
Other				
Secreted forms of APP	++	++	+	+
Protease Nexin-1	+	++	nd	+

*Information in this table is based upon published and unpublished data from the author's laboratories and is not intended to be a comprehensive list. Metabolic insults include glucose deprivation and mitochondrial poisons. Oxidative insults include hydrogen peroxide, iron sulfate, and sodium nitroprusside. Amyloidogenic peptides include amyloid β-peptide, human amylin, and β2-microglobulin. See text for information on mechanisms of action.
**Abbreviations: nd, not determined; BDNF, brain-derived neurotrophic factor; bFGF, basic fibroblast growth factor; IGF, insulin-like growth factor; NGF, nerve growth factor; NDGA, nordihydroguaiaretic acid; NT-3, neurotrophin-3; NT-4/5, neurotrophin 4/5; PBN, N-tert-butyl-α-phenylnitrone; TGF-β, transforming growth factor-β; TNF, tumor necrosis factor.

agents [9, 20, 21, 22] (but see ref. 23 for negative data) Agents in the third category include neurotrophic factors and alkaloids that mimic neurotrophic factor signaling (see below). It should be recognized that the neuroprotective approaches just described are applicable not only to AD, but to an array of neurodegenerative conditions in which ROS and Ca^{2+} play a role including: stroke, traumatic brain injury, Parkinson's disease, amyotrophic lateral sclerosis, Huntington's disease, and severe epileptic seizures[3,24,25,26].

Different Amyloidogenic Peptides Share a Common Cytotoxic Mechanism

Alzheimer's disease can be considered, in part, an amyloid disease of the brain. Many other organ systems can be damaged by amyloidogenic peptides other than Aβ. For example, in diabetes mellitus the peptide amylin is deposited in the pancreas where it damages the insulin-producing cells. A variety of vascular amyloidoses have been described which can also involve Aβ (e.g., hereditary cerebral vascular amyloidosis, Dutch type) or other peptides, such as β2-microglobulin. While the neurotoxicity of Aβ has been the most intensively studied, recent data demonstrate that other amyloidogenic peptides can also be neurotoxic. For example, May et al.[27] reported that human amylin is neurotoxic in rat hippocampal cell cultures, whereas its non-amyloidogenic rat counterpart is not neurotoxic. Amyloidogenic scrapie prion protein[28] and β2-microglobulin[27] were also neurotoxic. As described above, accumulating data from several laboratories indicate that the mechanism of Aβ neurotoxicity involves induction of ROS, impairment of ion-motive ATPases and elevation of $[Ca^{2+}]_i$. It is therefore of considerable importance to establish whether amyloidogenic peptides involved in disorders other than AD kill cells by mechanism similar to that of Aβ.

By measuring peroxide accumulation and $[Ca^{2+}]_i$ in cultured hippocampal neurons exposed to different amyloidogenic peptides (human amylin, β2-microglobulin and Aβ) we discovered that they each induce ROS and elevated $[Ca^{2+}]_i$[29]. Control, nonamyloidogenic peptides (including rat amylin) are not neurotoxic and do not induce cellular peroxides or elevate $[Ca^{2+}]_i$. Treatment of cultures with antioxidants (vitamin E, propyl gallate or PBN) or exposure of cultures to conditions that prevent Ca^{2+} influx attenuates the neurotoxicity of each of the amyloidogenic peptides[29]. As described above, Aβ can greatly increase the vulnerability of neurons to excitotoxicity; this property of Aβ is shared with other amyloidogenic peptides: human amylin and β2-microglobulin increase neuronal vulnerability to excitotoxicity and enhance $[Ca^{2+}]_i$ responses to glutamate[29]. A better understanding of the physical and protein chemistries involved in the cytotoxic activities of amyloidogenic peptides could lead to the development of compounds that prevent the peptides from attaining a pathogenic state.

AGE-ASSOCIATED CHANGES IN BRAIN THAT MAY IMPACT ON NEURONAL VULNERABILITY TO Aβ

Because increasing age is the major risk factor for AD, it is essential to identify the age-related factors that may predispose neurons to injury and death. Here we present examples of well-documented age-related changes for which there is experimental evidence to support a role in increasing neuronal vulnerability to Aβ and excitotoxic insults. Reduced glucose availability to brain tissue has been consistently documented in brain imaging studies of aged individuals; this deficit is even more pronounced in AD patients[30,31]. Experimental evidence also supports a role for diminished glucose availability in the neuronal degeneration that occurs in AD. For example, glucose deprivation in rat hippocampal cell cultures resulted in neuronal degeneration that exhibited cytoskeletal alterations similar to those seen in the neurofibrillary tangles of AD[32]. Neuronal populations in adult rats that are vulnerable to hypoglycemic injury (e.g., CA1 hippocampal neurons) correspond to those vulnerable in AD. Importantly, reduced energy to neurons increases their susceptibility to injury and death induced by Aβ and glutamate[19,33,34]. Alterations in the brain-pituitary-adrenal cortex neuroendocrine system that mediates stress responses have been reported to occur with aging and in AD, resulting in increased levels of circulating glucocorticoids[35]. Glucocorticoids have been shown to increase the vulnerability of neurons to a variety of insults, including excitotoxicity and ischemia[36]. Sapolsky and coworkers provided evidence that the mechanism of "endangerment" by glucocorticoids involves a reduction in glucose transport ability of neurons[37]. We have found that glucocorticoids enhance excitotoxin-induced cytoskeletal alterations in the adult rat hippocampus, including tau accumulation in cell bodies and proteolysis of MAP-2 and spectrin[38,39]. Suppression of endogenous glucocorticoid production with the 11β-hydroxylase inhibitor metyrapone resulted in significant reduction in brain damage in rat models of focal and transient global ischemia, as well as seizure models[40]. Examples of metyrapone's neuroprotective action against ischemia in vivo are presented in Figure 2.

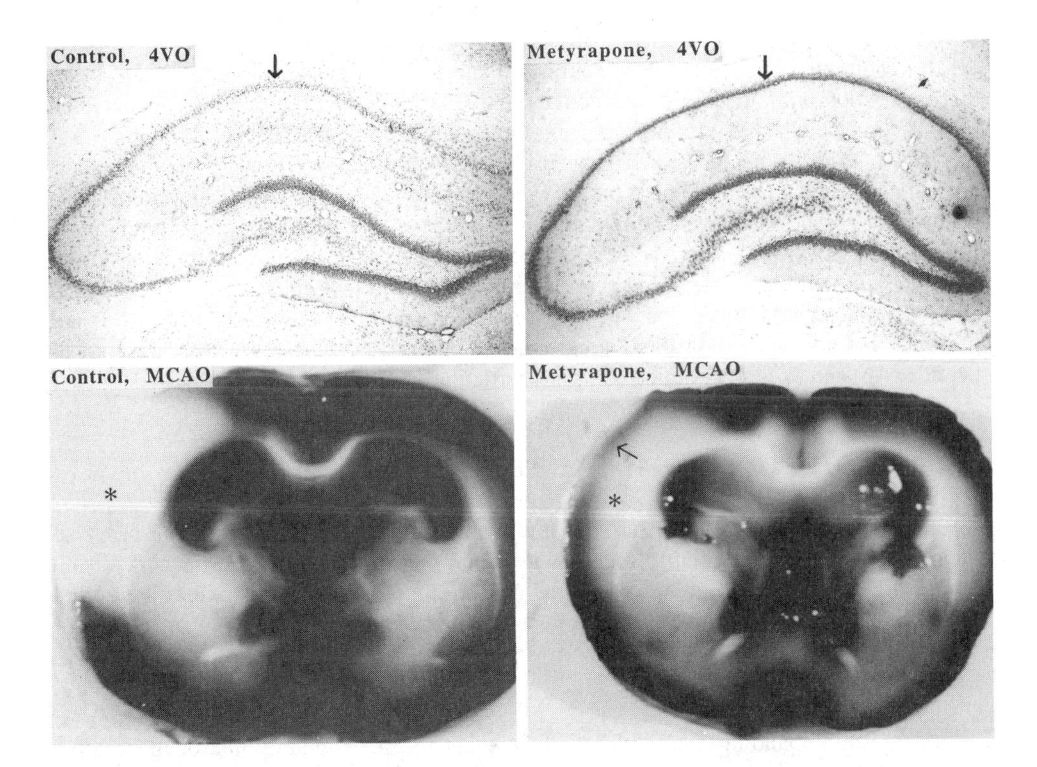

Figure 2. Metyrapone, an 11β-hydroxylase inhibitor or glucocorticoid production, protects adult rats against brain injury induced by transient global forebrain ischemia or focal cerebral ischemia. **(Upper panels)** Cresyl violet-stained coronal brain sections of hippocampus from rats killed 72 hr following 20 min of transient global forebrain ischemia induced by the 4-vessel occlusion (4VO) method. One rat was administered saline (control) and the other metyrapone (200 mg/kg b.w.) subcutaneously 30 min prior ischemia. Note marked loss of neurons in region CA1 of the control rat (arrow) and less neuron loss in the metyrapone-treated rat. (**Lower panels**) Triphenyltetrazolium chloride-stained coronal slices of

brains from rats killed 24 hr following middle cerebral artery occlusion. One rat was administered saline (control) and the other metyrapone (200 mg/kg b.w.) subcutaneously 30 min prior ischemia. Note greater infarct size (*) in brain of control rat and preservation of a band of cells in the cortex of the metyrapone-treated rat (arrow).

Evidence for age-associated increases in levels of protein oxidation and lipid peroxidation has accumulated in recent years. For example, increased levels of protein oxidation in the brain occur with increased age, and to a greater extent in AD[41]. Both amyloid plaques and tau in neurofibrillary tangles have been reported to be modified by glycation (cross-linking with sugars) to form advanced glycation end products which may promote oxidation[42]. Alterations in mitochondrial function have been reported in AD and other neurodegenerative disorders[43]. The kinds of alterations reported would be expected to result in an accumulation of oxyradicals in cells.

Additional age-related changes that may impact on the brain include alterations in immune function. Changes in inflammatory mediators and cytokines in AD suggest a state of chronic inflammation in the brain[44]. Among the alterations are the presence of microglia and T lymphocytes in regions of pathology and the presence of proteins of the classical complement pathway associated with amyloid deposits and dystrophic neurites. Epidemiological data suggest a reduced incidence of AD in persons taking non-steroidal anti-inflammatory drugs on a long-term basis and results of a clinical trial of indomethacin in AD suggested a beneficial effect[45,46].

It should be recognized that variability in age-associated changes in the brain, other than Aβ deposition, are likely to account for the variability in the relationship between plaques and neurodegeneration. For example, an individual may have abundant Aβ deposits, but have little neuronal degeneration because the supply of glucose to the cells is not compromised. On the other hand, an individual with accelerated decline in glucose availability or increased levels of oxidation may have massive neuronal loss with relatively little Aβ deposits. All other factors being equal, Aβ plays a major role in determining whether neurons live or die. However, all other things are not equal among individuals, and thus a direct relationship between Aβ deposition and manifestations of disease is not present. Indeed, it is upon this background of age-related changes where amyloid β-peptide (Aβ) and excitatory amino acids are particularly effective in wreaking havoc.

NEUROPROTECTIVE SIGNAL TRANSDUCTION

During the last 7 years it has become clear that brain cells produce an array of intercellular signals designed to preserve neuronal viability and promote recovery from injury. In many cases these injury-induced neurotrophic factors and cytokines have been shown to have neuroprotective activities in cell culture and in vivo[47,48].

Classic Neurotrophic Factors

A rapidly expanding area of research relevant to understanding neurodegenerative disorders and developing preventative and therapeutic approaches is that of neuroprotective signal transduction. This field blossomed when it was discovered that neurotrophic factors can protect neurons against excitotoxicity, metabolic impairment and oxidative insults (see refs. 47 and 48 for review). Each of these types of insults is relevant to both acute neurodegenerative conditions (stroke and traumatic brain injury) and chronic neurodegenerative disorders such as AD, Parkinsons' disease and amyotrophic lateral sclerosis. Neurons from the different CNS regions affected in these disorders can be protected against insults that involve ROS and loss of Ca^{2+} homeostasis by exposing them to one or more neurotrophic factors prior to the insult. Basic FGF, which is widely expressed in the brain, has proven to be effective in protecting many different populations of cultured neurons against an array of insults, including excitatory amino acids[49,50], glucose deprivation[51], oxidative insults[52], and mitochondrial poisons[53]. Moreover, bFGF protected hippocampal and cortical neurons against ischemic and excitotoxic insults in vivo[54].

The list of neuroprotective trophic factors is impressive and rapidly growing. For example, several different neurotrophic factors have proven to have potent neuroprotective activities towards hippocampal neurons (Table 1); they include basic fibroblast growth factor (bFGF)[49,50], nerve growth factor (NGF)[51,55], insulin-like growth factors (IGFs)[56], brain-derived neurotrophic factor (BDNF)[57], neurotrophin-3 (NT-3)[57] and neurotrophin-4/5 (NT-4/5)[58]. In each case the mechanism of action of the neurotrophic factor involves enhancement of cellular Ca^{2+} homeostasis and suppression of accumulation of ROS. For example, $[Ca^{2+}]_i$ imaging studies have shown that $[Ca^{2+}]_i$ responses to glutamate are attenuated in hippocampal neurons pretreated with bFGF[49] or BDNF[57]. Basic FGF, BDNF and NGF (but not CNTF) suppressed glutamate-induced accumulation of peroxides in cultured hippocampal neurons[59]. The specific mechanisms of action of the neurotrophic factors are beginning to be revealed and appear to include regulation of the expression of proteins involved in Ca^{2+} homeostasis and free radical metabolism. For instance, bFGF suppressed expression of an NMDA receptor protein in cultured hippocampal neurons[50], bFGF and NT-3 induced the expression of the Ca^{2+}-binding protein calbindin D28k in cultured hippocampal neurons[60], bFGF and NGF induced an increase in superoxide dismutase activity in rat hippocampal cell cultures[59], BDNF induced glutathione peroxidase in hippocampal cultures[59] and glutathione reductase activities in hippocampal cultures[59] and human neuroblastoma cells[61]. In some of these studies, the effects of the neurotrophic factor on the expression of the $[Ca^{2+}]_i$-regulating protein or antioxidant enzyme was directly linked to reduced $[Ca^{2+}]_i$ responses to glutamate or reduced accumulation of ROS, respectively[50,59].

Of particular relevance to AD are data showing that neurotrophic factors can protect neurons against Aβ toxicity. Basic FGF was the first neurotrophic factor shown to protect neurons against Aβ toxicity[9]. The elevation of rest $[Ca^{2+}]_i$ and enhancement of $[Ca^{2+}]_i$ responses to glutamate were markedly attenuated in hippocampal neurons pretreated with bFGF. Basic FGF activates receptors with intrinsic tyrosine kinase activity, and subsequent phosphorylation cascades result in activation of MAP kinases and transcription factors. As just described, bFGF affects the expression of several proteins which could affect susceptibility to Aβ toxicity (i.e., NMDA receptors, calbindin, and antioxidant enzymes). BDNF is also effective in protecting cultured hippocampal neurons against Aβ toxicity (M.P.M., unpublished data). In addition, both sAPPs and tumor necrosis factors (TNFs) suppressed Aβ-induced accumulation of ROS, elevation of $[Ca^{2+}]_i$ and neurotoxicity in hippocampal cell cultures[12,62]. The mechanism of action of sAPPs and TNFs will be considered in more detail below.

Because of the long-standing problem of lack of access of systemically administered neurotrophic factors to the brain (because of their hydrophilicity and high molecular weights), there has been considerable interest in identifying low molecular weight lipophilic compounds that can activate neurotrophic factor signal transduction pathways. We recently identified such a class of compounds. Bacteria-derived alkaloids including staurosporine, K252a and K252b were originally shown to inhibit several different kinases. Subsequently, we found that concentrations of the these compounds several orders of magnitude below those required to inhibit kinases were effective in protecting cultured hippocampal, cortical and septal neurons against a variety of insults including glucose deprivation, Aβ toxicity and radical-generating conditions[63,64]. Pretreatment of neurons with these compounds stabilized $[Ca^{2+}]_i$ and suppressed accumulation of ROS in a manner reminiscent of that previously observed with neurotrophic factors. We therefore tested the hypothesis these compounds activated neurotrophic factor signaling cascades. Exposure of cultured hippocampal neurons to 10-1000 pM staurosporine or K252 compounds resulted in a rapid increase in tyrosine phosphorylation of multiple proteins, including those with molecular weights consistent with those of neurotrophic factor receptors and MAP kinases[63]. Consistent with a mechanism of action similar to neurotrophic factors, these compounds also induced choline acetyl-transferase in spinal cord cultures[65].

Secreted Forms of β-Amyloid Precursor Protein

Recent findings suggest that βAPP normally serves beneficial roles in neuroplasticity and recovery from injury[66]. Genetic manipulation studies in species as divergent as mice[67] and *Drosophila* [68] have provided evidence that βAPP plays roles in learning and memory. Indeed, the genetic link of mutations in βAPP to rare lineages of familial AD[1] do not exclude the possibility that AD-associated deficits in memory arise from a loss of function of βAPP or its derivatives. Intuitively, such a loss of function would be more consistent with the apparent etiological diversity of AD than would the gain of function proposed by theories of Aβ toxicity. Although βAPPs are quite abundant in nervous system tissue, the concentration and localization of sAPPs are less certain. A decrease of sAPP in AD has been reported[68]. While others have had difficulty reproducing this finding in generalized AD case pools, studies focusing on specific familial lineages have been more compelling[69,70].

Speculation on the mode by which abnormalities or deficiencies in sAPPs could contribute to AD requires an understanding of their normal activities. In ironic contrast to the effects of Aβ, sAPPs attenuate $[Ca^{2+}]_i$ responses to glutamate and offer protection from excitotoxicity[71]. In fact, sAPPs also can offer protection of cultured neurons from Aβ itself[12]. The antagonism of glutamate responses by sAPPs also appears responsible for at least a portion of the neuritogenic action of sAPPs[72]. These cell culture findings have been extended in vivo with demonstrations that sAPPs protect neurons against ischemia[73] and promote synaptogenesis[74], a process which can involve glutamatergic transmission[75].

The connections of sAPPs to synapse formation or stabilization strengthens the evidence for a role in neuromodulation. More acute responses are also consistent with neuromodulation and imply that aberrant sAPP activity in AD could directly compromise memory processes independently of cell death. For instance, sAPPs rapidly activate a charybdotoxin-sensitive potassium current[76], resulting in the rapid depression of rest $[Ca^{2+}]_i$. Combined with the axonal[72,77,78] and synaptic[79] localization of βAPP and the stimulation of sAPP release by electrical activity[80], these physiological data suggest an influence of sAPPs on neurotransmission.

The mechanisms by which sAPPs exert their neuronal effects are being elucidated (Fig. 1). Ligand binding studies suggest specific binding sites in neurons[81], consistent with the striking potency of sAPPs in some of their activities. Secreted APPs were recently demonstrated to elevate cyclic GMP (cGMP) levels; the activation of cGMP-dependent kinase appears to be necessary for the effects of sAPPs on $[Ca^{2+}]_i$, neuronal survival[82], and potassium channel activation[76]. Furthermore, sAPPs may have a prolonged influence on neuronal phenotype through control of gene expression, as they activate a transcription factor similar to NFκB[83,84]. Using pharmaco-molecular disinhibition of NFκB, we provided evidence that κB-dependent transcription induces a resistance to several neuronal insults (below).

Cytokines

Although neurotrophic factors and cytokines can utilize similar signal transduction pathways, the role of cytokines in the injury response is unclear. A novel cytokine family includes tumor necrosis factors (TNFα and TNFβ), lymphotoxin-β, CD40 ligands, and CD27 ligand. TNFα and TNFβ have been shown to have a broad spectrum of activities, including promotion of vasodilation and hypotension, stimulation of osteoblast proliferation, and trophic/growth-regulating activities in lymphocytes and some tumor cell types. Specific high-affinity TNF receptors bind both TNFα and TNFβ and are widely distributed throughout most cells and tissues, including the brain. The p55 receptor has been linked specifically to the activation of a neutral sphingomyelinase (Fig. 1). This enzyme liberates ceramide, a putative second messenger which can stimulate phosphatase and kinase activities. In some systems, ceramide stimulates one of the most ubiquitous responses to TNFs: the activation of the NFκB transcription factor.

TNF is elevated in AD, leading to the proposal that it is involved in a vicious cycle of pathogenic inflammatory responses in neurodegenerative conditions. However, we found that TNFα and TNFβ both protect cultured embryonic rat hippocampal, septal, and cortical neurons against glucose deprivation-induced injury and excitatory amino acid toxicity[85]. The elevations of $[Ca^{2+}]_i$ induced by glucose deprivation, glutamate, NMDA, or AMPA are attenuated in neurons pretreated with TNFs. Given the relationship between excitotoxicity and Aβ toxicity (above), we also explored the effect of TNFs on the latter. TNFs were found to abbrogate Aβ-induced neuronal death, oxyradical generation, and calcium elevation[62]. Interestingly, TNFs activate NFκB in primary neurons[84], implying that they regulate gene expression similarly to sAPPs (above). It has been suggested that NFκB senses oxidative stress and effects a feedback elevation of antioxidant enzymes[86]. Therefore, TNFs and sAPPs may exert neuroprotective effects by elevating such enzymes preemptively (Fig. 1). To determine the contribution of κB-dependent transcription to the neuroprotective activity of TNFs, we activated NFκB by a pharmaco-molecular disinhibiton. Specifically, we applied to cultured neurons antisense oligonucleotides directed against MAD3, an endogenous inhibitory subunit of latent NFκB. This direct activation of κB-dependent transcription mimics the protection by TNFs against Aβ-evoked events and other forms of calcium-mediated neuronal damage[62,82,83]. For example, MAD3 antisense treatment ameliorates the oxyradical generation stimulated by Aβ (Fig. 3). We are currently attempting to identify candidate genes which would be relevant targets of such κB-dependent transcription. For instance, both neurons and glia respond to TNFs with an increase in the expression of the calcium-binding protein calbindin D_{28k}, which may partially explain the ability of TNFs to stabilize $[Ca^{2+}]_i$.

Transforming growth factor β (TGF-β) is often classified as a cytokine for its pleiotrophic (and often antimitotic) actions. TGF-β mRNA and protein were recently found to increase in animal brains after experimental lesions[87]. Elevations of TGF-β mRNA after lesions are prominent in microglia but are also observed in neurons and astrocytes. Moreover, TGF-β mRNA levels are autoinduced in the brain by TGF-β itself. These responses suggest models to explain the presence of TGF-β in Aβ-containing extracellular plaques of AD and Down's syndrome[88]. It has been suggested that TGF-β has organizing roles in responses to neurodegeneration and brain injury that are similar to its roles in non-neural tissues[89]. TGF-β can orchestrate a modification of extracellular matrix, exhibit brain cell chemotaxis, and regulate expression of neurotrophins. With respect to effects on neurons themselves, TGF-β has been found to protect against excitotoxicity[90], Aβ toxicicity[91], and glucose deprivation (S.W.B. and M.P.M., unpublished observations).

Interestingly, while the signal transduction systems for each of the different classes of factors described above differs, most appear to lead to common cellular defense

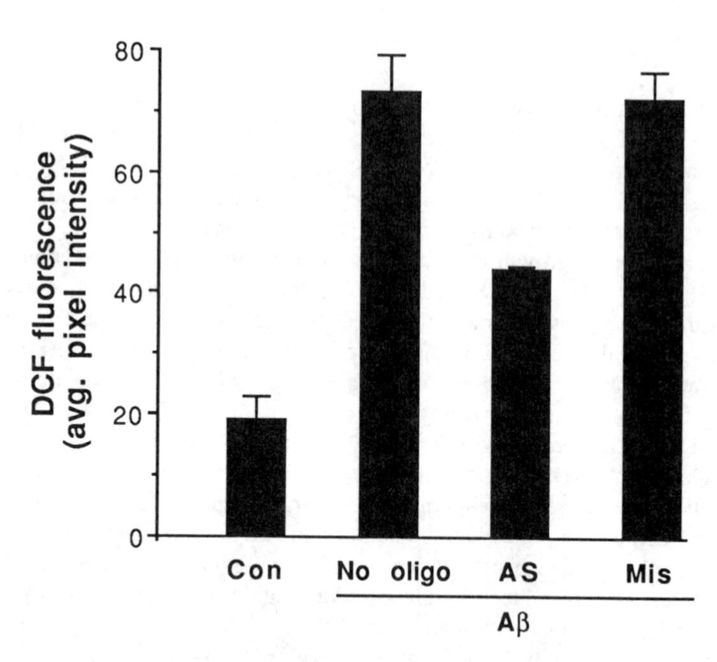

Figure 3. Amelioration of Aβ-induced oxyradicals by activation of I-κB antisense. Mouse hippocampal neurons (8 days in culture) were left untreated (Con, No oligo) or treated for 24 hrs with 30 μM of an oligonucleotide directed against the MAD3 clone of I-κB (AS) or a control oligonucleotide with two bases mismatched (Mis). Reduction of I-κB levels by the antisense oligonucleotide activates NF-κB (83). The cultures were then exposed to 20 μM of the 25-35 amino acid fragment of Aβ for 5 hrs (with the exception of "Con," which was left untreated). Relative levels of oxyradicals present in the cells were then assessed by staining with the oxidation-sensitive dye dichlorofluorescin. Data reflect the average pixel intensity in the neurons' somata obtained from confocal laser scanning microscopy images, which is proportionally related to the level of oxidation. Values are the mean ± SEM of three plates. Note that cultures treated with I-κB antisense oligonucleotide and Aβ have substantially lower levels of oxyradical signal than those treated with Aβ alone or control oligonucleotide and Aβ.

mechanisms involving regulation of the expression of proteins involved in Ca^{2+} homeostasis and free radical metabolism. Furthermore, many of the factors which have been proposed to play a role in propagating neurodegeneration experimentally exhibit excitoprotective activity. Therefore, these factors initially may be involved in compensatory responses to the incipient damage. However, additional actions or over-inductions may evolve into detrimental effects. This is especially true for some of the cytokines proposed to contribute to an inflammatory response in AD and other conditions.

THROMBIN AND NEURODEGENERATION

There is increasing interest in the role of thrombin in the CNS, particularly with regard to neurodegeneration. Over the past ten years the biological functions of thrombin in neural cells have begun to unfold. Although thrombin is primarily known as a blood coagulation factor, it appears to possess rather diverse bioregulatory activity. For example, thrombin can induce platelet activation and aggregation; stimulate proliferation and/or morphological changes in glial, sympathetic cervical ganglion, endothelial and neuroblastoma cells; cause a significant increase in intracellular calcium $[Ca^{2+}]_i$ in endothelial, smooth muscle, lymphoblastoma and neuroblastoma cells; induce the release of arachidonic acid from spinal cord cultures; and stimulate secretion of NGF from cultured astrocytes (see refs. 91,92,93 for review). Moreover, thrombin has been implicated in several injury/disease states of the nervous system. For example, thrombin is increased in severe head trauma and AD patient brains[95,96], stimulates cleavage of βAPP and secretion of either Aβ or sAPPs[97,99], and causes increased vascular endothelial permeability[93]. Both neurons and glial cells express thrombin receptor mRNA, with particular concentration in certain neuronal subpopulations such as midbrain dopaminergic neurons[99]. The presence of thrombin receptors in the brain is intriguing as thrombin is not found normally in the brain parenchyma. Nevertheless, prothrombin mRNA is expressed in the CNS where it appears to co-localize with thrombin receptor mRNA[99,100]. This suggests that cells that are a source of thrombin in the brain may also respond to thrombin in an autocrine manner.

What role does thrombin play, if any, in the pathophysiology of neurodegenerative disorders? We found thrombin to be neurotoxic to 7-10 day old cultured hippocampal neurons in a concentration-dependent manner (25 nM to 1 μM)[101]. However, similar concentrations of thrombin were considerably less toxic to embryonic hippocampal neurons at earlier stages of development in culture (1 to 3 days in culture), suggesting that thrombin differentially affects immature and mature neurons. Thrombin's toxicity is dependent on Ca^{2+} influx, as thrombin is not neurotoxic in medium lacking Ca^{2+}. Imaging studies using the calcium indicator dye, fura-2, demonstrated that thrombin induces a sustained increase of $[Ca^{2+}]_i$[101]. Moreover, pretreatment of hippocampal cultures with a submaximally toxic concentration of thrombin (25 nM) significantly exacerbates both glutamate- and glucose deprivation- induced cell damage. These data may have important implications for neuropathologies such as ischemia, severe head trauma and AD. In cerebral ischemia and head trauma, the blood brain barrier is often compromised and blood penetrates into the brain parenchyma, allowing thrombin to come in contact with neurons. In addition, it is possible that prothrombin is transported extravascularly and activated by neural cells. Deficient oxygen and glucose supply to neurons—resulting in energy failure and subsequent release of glutamate—are classic components of ischemic injury. Our data demonstrate that thrombin may worsen these events. Of particular relevance to AD is the finding that thrombin can exacerbate the neurotoxicity of Aβ[102]. In the latter study, we found that elevation of $[Ca^{2+}]_i$ induced by Aβ was enhanced by thrombin.

Protease nexin-1 (PN-1), a potent cell surface thrombin inhibitor, can block thrombin neurotoxicity in hippocampal cell cultures[101]. Moreover, hippocampal cultures pretreated with PN-1 are protected from both glutamate- and glucose deprivation-induced damage, apparently through PN-1's ability to stabilize $[Ca^{2+}]_i$[101]. PN-1 is synthesized and secreted by both neuronal and glial cells and is known to promote both cell survival and neurite outgrowth[92,103]. Interestingly, PN-1 levels have been shown to be significantly increased in several injury/disease states such as AD, cerebral ischemia, and axotomy[92]. PN-1 has similar protease inhibitor activity as some forms of βAPP and is reportedly present in the CSF in the range of approximately 2-6 nM[104]. Such basal levels of PN-1 may be sufficient to reduce neuronal injury in vivo, especially in conjunction with injury-induced increases in PN-1, although this remains to be established. Pretreatment of rat hippocampal cell cultures with PN-1 results in significant reduction in neuronal damage induced by Aβ[102]. PN-1 also attenuates Aβ-induced elevation of $[Ca^{2+}]_i$. Taken together, these findings suggest that a basal level of thrombin activity may increase neuronal vulnerability to several insults, including metabolic impairment, excitotoxicity, and Aβ toxicity. Based on our cell culture data[101,102], we believe that PN-1 secreted locally by glial cells or neurons can bind thrombin, prevent activation of thrombin receptors in neurons, and thereby reduce the vulnerability of the neurons to injury.

ACTIVE ROLES OF THE CYTOSKELETON IN NEURODEGENERATION AND NEUROPROTECTION

It has been recognized for some time that cytoskeletal elements are disrupted during the neurodegenerative process, whether that process be rapid (e.g., stroke) or slow (e.g., AD). Indeed, studies of cytoskeletal aberrancies in AD have been extensive. However, in contrast to the considerable progress made in identifying cytoskeletal alterations, very little is known about how changes in the cytoskeleton influence neuronal survival. Indeed, while it has been presumed that cytoskeletal alterations in stressed/injured neurons contribute to the cell death process, evidence supporting such a possibility is lacking. An important mediator of cytoskeletal alterations in neurons subjected to excitotoxic, ischemic and oxidative insults is calcium. Elevation of $[Ca^{2+}]_i$ promotes depolymerization of microfilaments and microtubules. We have begun to test the hypothesis that disruption of such cytoskeletal elements plays roles in determining whether or not a neuron lives or dies. The initial approach was to examine the effects of agents that selectively disrupt or stabilize actin microfilaments or microtubules on neuronal calcium homeostasis and cell survival in neurons exposed to excitotoxic insults and Aβ. In rat hippocampal cultures, the microfilament-disrupting agent cytochalasin D protects neurons against glutamate toxicity in a concentration-dependent manner and attenuates $[Ca^{2+}]_i$ responses to glutamate[105]. Cytochalasin D does not protect neurons against calcium ionophore toxicity or iron toxicity suggesting that its actions are due to specific effects on calcium influx. Moreover, cytochalasin D markedly attenuates kainate-induced damage to hippocampus of adult rats, suggesting an excitoprotective role for actin depolymerization in vivo. The actin filament stabilizing agent jasplakinolide increases the vulnerability of neurons to the $[Ca^{2+}]_i$-elevating and neurotoxic actions of glutamate. Taken together with patch-clamp data[106], our findings suggest that depolymerization of actin in response to Ca^{2+} influx may serve as a feedback mechanism to attenuate potentially toxic levels of Ca^{2+} influx. Because Ca^{2+} influx plays a role in Aβ neurotoxicity (see above), we determined whether cytochalasin D would protect neurons against Aβ toxicity and found that, indeed, this agent significantly increases neuronal survival of cultured hippocampal neurons exposed to Aβ25-35 or Aβ1-40[107]. Data from control experiments indicated that the neuroprotective mechanism of cytochalasins involves its actin depolymerizing activity rather than a less specific activity (e.g., antioxidant)[106,107].

In a parallel study we examined the effects of the microtubule-stabilizing agent taxol and the microtubule-disrupting agent colchicine on the vulnerability of cultured hippocampal neurons to excitotoxicity[108]. We found that pretreatment of cultures with taxol significantly reduces the vulnerability of neurons to glutamate toxicity. Calcium imaging studies revealed that $[Ca^{2+}]_i$ responses to glutamate are attenuated in neurons pretreated with taxol; experiments with agonists for specific glutamate receptor subtypes suggested that taxol is particularly effective in reducing Ca^{2+} influx through AMPA receptor channels. In contrast to taxol, colchicine destabilizes $[Ca^{2+}]_i$ homeostasis and exacerbates glutamate toxicity. These data suggest that the microtubule system plays a role in regulation of cellular Ca^{2+} homeostasis. These kinds of data suggest that agents that affect the neuronal cytoskeleton may prove valuable in reducing neuronal degeneration in the myriad of disorders in which loss of cellular Ca^{2+} homeostasis is part of the final common pathway of cell death (Table 1).

FAR-REACHING NEUROPROTECTIVE STRATEGIES

Compounds listed in Table 1 have proven effective in reducing neuronal injury in cell culture and/or in vivo paradigms relevant to the pathogenesis of AD. In most cases the neuronal injury paradigms are also relevant to a range of neurodegenerative conditions including stroke, traumatic brain injury, Parkinson's disease and severe epileptic seizures. The reason such compounds are likely to have broad utility is that they target events in the neurodegenerative process that are downstream from the primary cause(s). In general, the targets are involved in increasing or reducing levels of Ca^{2+} and/or free radicals in cells. The goal with this strategy is to prevent or attenuate elevation of $[Ca^{2+}]_i$ or free radicals. Although simple in concept, this goal is must be achieved while at the same time preserving the normal functions of neural circuits (i.e., synaptic transmission). The latter problem is a major one because both calcium and free radicals are involved in normal signaling mechanisms (e.g., calcium influx is a key stimulus for neurotransmitter release and nitric oxide is a normal stimulus for cyclic GMP production). While temporary disruption of brain function may be acceptable in acute neurodegenerative conditions such as stroke, it is certainly a major concern in chronic neurodegenerative conditions that require long-term therapy. It is therefore critical to continue to develop approaches aimed at prevention of the initiating events in neurodegenerative cascades. In AD, βAPP processing and Aβ aggregation are two processes being actively pursued with the hope of identifying compounds that reduce production of Aβ, or enhance clearance or prevent aggregation of soluble Aβ.

REFERENCES

1. M. Mullan, and F. Crawford, Genetic and molecular advances in Alzheimer's disease, *Trends Neurosci.* 16:398-403 (1993).
2. D. J. Selkoe, Physiological production of the β-amyloid protein and the mechanism of Alzheimer's disease, *Trends Neurosci.* 16: 403-409 (1993).
3. K. Iqbal, D.R.C. McLachlan, B. Winblad, and H. M. Wisniewski (eds). "Alzheimer's Disease: Basic Mechanisms, Diagnosis and Therapeutic Strategies," Wiley, New York (1991).
4. M. P. Mattson, Calcium and neuronal injury in Alzheimer's disease, *Ann. N.Y. Acad. Sci.* 747:50-76 (1995).
5. D. Burdick, B. Soreghan, M. Kwon, J. Kosmoski, M. Knauer, A. Henschen, J. Yates, C. Cotman and C. Glabe, Assembly and aggregation properties of synthetic Alzheimer's A4/β amyloid peptide analogs, *J. Biol. Chem.* 267: 546-554 (1992).
6. A. I. Bush, W. H. Pettingell, G. Multhaup, M. d Paradis, J. P. Vonsattel, J. F. Gusella, K. Beyreuther, C. L. Masters and R. E. Tanzi, Rapid induction of Alzheimer A β amyloid formation by zinc, *Science* 265: 1464-1467 (1994).
7. T. Dyrks, E. Dyrks, T. Hartmann, C. Masters and K. Beyreuther, Amyloidogenicity of β A4 and β A4-bearing amyloid protein precursor fragments by metal-catalyzed oxidation, *J. Biol. Chem.* 267: 18210-18217 (1992).
8. T. Wisniewski, J. Ghiso, and B. Frangione, Alzheimer's disease and soluble Aβ. *Neurobiol. Aging* 15:143-152 (1994).
9. M. P. Mattson, K. Tomaselli and R. E. Rydel, Calcium-destabilizing and neuro-degenerative effects of aggregated β-amyloid peptide are attenuated by basic FGF, *Brain Res.* 621: 35-49 (1993).
10. C. J. Pike, A. J. Walencewicz-Wasserman, J. Kosmoski, D. H. Glabe and C. W. Cotman, Structure-activity analyses of β-amyloid peptides: contributions of the

beta 25-35 region to aggregation and neurotoxicity, *J. Neurochem.* 64: 253-265 (1995).
11. J. Busciglio, A. Lorenzo, and B. A. Yankner, Methodological variables in the assessment of beta amyloid neurotoxicity, *Neurobiol. Aging* 13:609-612 (1992).
12. Y. Goodman, and M. P. Mattson, Secreted forms of β-amyloid precursor protein protect hippocampal neurons against amyloid β–peptide-induced oxidative injury, *Exp. Neurol.* 128: 1-12 (1994).
13. Y. Goodman, M. R. Steiner, S. M. Steiner, and M. P. Mattson, Nordihydroguaiaretic acid protects hippocampal neurons against amyloid β-peptide toxicity, and attenuates free radical and calcium accumulation, *Brain Res.* 654: 171-176 (1994).
14. K. Hensley, J. M. Carney, M. P. Mattson, M. Aksenova, M. Harris, J. F. Wu, R. Floyd, and D. A. Butterfield, A model for β-amyloid aggregation and neurotoxicity based on free radical generation by the peptide: relevance to Alzheimer's disease. *Proc. Natl. Acad. Sci. U.S.A.* 91:3270-3274 (1994).
15. D. A. Butterfield, K. Hensley, M. Harris, M. P. Mattson, and J. Carney, β–amyloid peptide free radical fragments initiate synaptosomal lipoperoxidation in a sequence-specific fashion: implications to Alzheimer's disease, *Biochem. Biophys. Res. Commun.* 200: 710-715 (1994).
16. C. Behl, J. B. Davis, R. Lesley, and D. Schubert, Hydrogen peroxide mediates amyloid β protein toxicity, *Cell* 77:817-827 (1994).
17. M. E. Harris, K. Hensley, D. A. Butterfield, R. A. Leedle, and J. M. Carney, Direct evidence of oxidative injury produced by the Alzheimer's β–amyloid peptide (1-40) in cultured hippocampal neurons, *Exp. Neurol.* 131:193-202 (1995).
18. R. J. Mark, K. Hensley, D. A. Butterfield, and M. P. Mattson, Mechanism of amyloid β-peptide neurotoxicity involves impairment of Na^+/K^+-ATPase activity. Submitted (1995).
19. M. P. Mattson, B. Cheng, D. Davis, K. Bryant, I. Lieberburg, and R. E. Rydel, β–amyloid peptides destabilize calcium homeostasis and render human cortical neurons vulnerable to excitotoxicity. *J. Neurosci.* 12:376-389 (1992).
20. K. Furukawa, V. L. Smith-Swintosky, and M. P. Mattson, Evidence that actin depolymerization protects hippocampal neurons against excitotoxicity by stabilizing $[Ca^{2+}]_i$, *Exp. Neurol.* In press. (1995).
21. K. Furukawa, and M. P. Mattson, Cytochalasins protect hippocampal neurons against amyloid β-peptide toxicity: evidence that actin depolymerization suppresses Ca^{2+} influx, *J. Neurochem.* In press. (1995).
22. J. H. Weiss, C. J. Pike, and C. W. Cotman, Ca^{2+} channel blockers attenuate β-amyloid peptide toxicity to cortical neurons in culture, *J. Neurochem.* 62:372-375 (1994).
23. J. S. Whitson, M. P. Mims, W. J. Strittmatter, T. Yamaki, J. D. Morrisett and S. H. Appel, Attenuation of the neurotoxic effect of Aβ amyloid peptide by apolipoprotein E, *Biochem. Biophys. Res. Commun.* 199: 163-170 (1994).
24. S. Fahn and G. Cohen, The oxidant stress hypothesis in Parkinson's disease: evidence supporting it, *Ann. Neurol.* 32: 804-812 (1992).
25. J. D. Rothstein, L. A. Bristol, B. Hosler, R. H. Brown Jr. and R. W. Kuncl, Chronic inhibition of superoxide dismutase produces apoptotic death of spinal neurons, *Proc. Natl. Acad. Sci. U. S. A.* 91:4155-4159 (1994).
26. M. K. Sutherland, M. J. Somerville, L. K. Yoong, C. Bergeron, M. R. Haussler and D. R. McLachlan, Reduction of vitamin D hormone receptor mRNA levels in Alzheimer as compared to Huntington hippocampus: correlation with calbindin-28k mRNA levels, *Mol. Brain Res.* 13: 239-250 (1992).

27. P. C. May, L. N. Boggs, and K. S. Fuson, Neurotoxicity of human amylin in rat primary hippocampal cultures: similarity to Alzheimer's disease amyloid-β neurotoxicity, *J. Neurochem.* 61: 2330-2333 (1993).
28. G. Forloni, Angeretti, N., Chiesa, R., Monzani, E., Salmona, O., Bugiani, O. and Tagliavini, F., Neurotoxicity of a prion protein fragment, *Nature* 362: 543-546 (1993).
29. M. P. Mattson, and Y. Goodman, Different amyloidogenic peptides share a common mechanism of neurotoxicity involving reactive oxygen species and calcium, *Brain Res.* 676:219-224 (1995).
30. S. Hoyer, K. Oesterreich, and O. Wagner, Glucose metabolism as the site of the primary abnormality in early-onset dementia of Alzheimer type? *J. Neurol.* 235: 143-148 (1988).
31. R. N. Kalaria, and S. I. Harik, Reduced glucose transporter at the blood-brain barrier and in cerebral cortex in Alzheimer's disease, *J. Neurochem.* 53: 1083-1088 (1989).
32. B. Cheng, and M. P. Mattson, Glucose deprivation elicits neurofibrillary tangle-like antigenic changes in hippocampal neurons: Prevention by NGF and bFGF, *Exp. Neurol.* 117:114-123 (1992).
33. J.-Y. Koh, L. L. Yang, and C. W. Cotman, β-amyloid protein increases the vulnerability of cultured cortical neurons to excitotoxic damage, *Brain Res.* 533:315-320 (1990).
34. A. Copani, J.-Y. Koh, and C. W. Cotman, β–amyloid increases neuronal susceptibility to injury by glucose deprivation, *NeuroReport* 2: 763-765 (1991).
35. B. S. Greenwald, A. A. Mathe, R. C. Mohs, M. I. Levy, C. A. Johns, and K. L. Davis, Cortisol and Alzheimer's disease II: dexamethasone suppression, dementia severity, and affective symptoms, *Am. J. Psychiatry* 143:442-446 (1986).
36. R. M. Sapolsky, A mechanism for glucocorticoid toxicity in the hippocampus: increased vulnerability to metabolic insults, *J. Neurosci.* 5:1228-1232 (1985).
37. C. Virgin, T. Ha, D. Packan, G. Tombaugh, S. Yang, H. Horner, and R. Sapolsky, Glucocorticoids inhibit glucose transport and glutamate uptake in hippocampal astrocytes: implications for glucocorticoid toxicity, *J. Neurochem.* 57:1422-1428 (1991).
38. E. Elliott, M. P. Mattson, P. Vanderklish, G. Lynch, I. Chang, and R. M. Sapolsky Corticosterone exacerbates kainate-induced alterations in hippocampal tau immunoreactivity and spectrin proteolysis in vivo, *J. Neurochem.* 61: 57-67(1993) .
39. B. Stein-Behrens, M. P. Mattson, I. Chang, M. Yeh and R. M. Sapolsky, Stress excacerbates neuron loss and cytoskeletal pathology in the hippocampus, *J. Neurosci.* 14: 5373-5380 (1994).
40. V. L. Smith-Swintosky, L. C. Pettigrew, R. M. Sapolsky, C. Phares, S. D. Craddock, S. M. Brooke, and M. P. Mattson, Metyrapone, and inhibitor of glucocorticoid production, reduces brain injury induced by focal and global ischemia and seizures, *J. Cerebral Blood Flow Metab.* Submitted (1995).
41. C. D. Smith, J. M. Carney, P. E. Starke-Reed, C. N. Oliver, E. R. Stadtman, R. A. Floyd, and W. R. Markesbery, Excess brain protein oxidation and enzyme dysfunction in normal aging and in Alzheimer's disease. *Proc. Natl. Acad. Sci. U.S.A.* 88: 10540-10543 (1991).
42. M. A. Smith, P. L. Richey, S. Taneda, R. K. Kutty, L. M. Sayre, V. M. Monnier, and G. Perry, Advanced Maillard reaction end products, free radicals, and protein oxidation in Alzheimer's disease, *Ann. N, Y. Acad. Sci.* 738: 447-54 (1994).

43. M. F. Beal, Does impairment of energy metabolism result in excitotoxic neuronal death in neurodegenerative illnesses? *Ann. Neurol.* 31:119-130 (1992).
44. P. L. McGeer, J. Rogers, and E. G. McGeer, Neuroimmune mechanisms in Alzheimer disease pathogenesis, *Alzheimer Dis .Assoc. Disord.* 8: 149-158 (1994).
45. J. Rogers, L. C. Kirby, S. R. Hempelman, D. L. Berry, P. L. McGeer, A. W. Kaszniak, J. Zalinski, M. Cofield, L. Mansukhani, and P. Wilson, Clinical trial of indomethacin in Alzheimer's disease, *Neurology* 43:1609-1611 (1993).
46. P. L. McGeer, J. Rogers, and E. G. McGeer, Neuroimmune mechanisms in Alzheimer disease pathogenesis, *Alzheimer Dis. Assoc. Disord.* 8:149-158 (1994).
47. M. P. Mattson, B. Cheng, and V. L. Smith-Swintosky, Growth factor-mediated protection from excitotoxicity and disturbances in calcium and free radical metabolism, *Seminars Neurosci.* 5:295-307 (1993).
48. M. P. Mattson, and S. W. Scheff, Endogenous neuroprotection factors and traumatic brain injury: mechanisms of action and implications for therapies, J. *Neurotrauma* 11:3-33 (1994).
49. M. P. Mattson, M. Murrain, P. B. Guthrie, and S. B. Kater, Fibroblast growth factor and glutamate: Opposing actions in the generation and degeneration of hippocampal neuroarchitecture, *J. Neurosci.* 9:3728-3740 (1989).
50. M. P. Mattson, K. Kumar, B. Cheng, H. Wang and E. K. Michaelis, Basic FGF regulates the expression of a functional 71 κDa NMDA receptor protein that mediates calcium influx and neurotoxicity in cultured hippocampal neurons, *J. Neurosci.* 13:4575-4588 (1993).
51. B. Cheng, and M. P. Mattson, NGF and bFGF protect rat and human central neurons against hypoglycemic damage by stabilizing calcium homeostasis, *Neuron* 7:1031-1041 (1991).
52. Y. Zhang, T. Tatsuno, J. Carney, and M. P. Mattson, Basic FGF, NGF, and IGFs protect hippocampal neurons against iron-induced degeneration, *J. Cerebral Blood Flow Metab.* 13:378-388 (1993).
53. M. P. Mattson, Y. Zhang and S. Bose, Growth factors prevent mitochondrial dysfunction, loss of calcium homeostasis and cell injury, but not ATP depletion in hippocampal neurons deprived of glucose, *Exp. Neurol.* 121: 1-13 (1993).
54. K. Nozaki, S. P. Finklestein, and M. F. Beal, Basic fibroblast growth factor protects against hypoxia/ischemia and NMDA neurotoxicity in neonatal rats, *J. Cereb. Blood Flow Metab.*, 13:221-228 (1993).
55. T. Shigeno, T. Mima, K. Takakura, D. I. Graham, G. Kato, Y. Hashimoto, and S. Furukawa, Amelioration of delayed neuronal death in the hippocampus by nerve growth factor, *J. Neurosci.* 11:2914-2919 (1991).
56. B. Cheng, and M. P. Mattson, IGF-I and IGF-II protect cultured hippocampal and septal neurons against calcium-mediated hypoglycemic damage, *J. Neurosci.* 12:1558-1566 (1992).
57. B. Cheng, and M. P. Mattson, NT-3 and BDNF protect hippocampal, septal, and cortical neurons against metabolic compromise, *Brain Res.* 640:56-67 (1994).
58. B. Cheng, Y. Goodman, J. G. Begley and M. P. Mattson, Neurotrophin 4/5 protects hippocampal and cortical neurons against energy deprivation- and excitatory amino acid-induced injury, *Brain Res.* 650: 331-335 (1994).
59. M. P. Mattson, M. A. Lovell, K. Furukawa and W. R. Markesbery, Neurotrophic factors attenuate glutamate-induced accumulation of peroxides, elevation of $[Ca^{2+}]_i$ and neurotoxicity, and increase antioxidant enzyme activities in hippocampal neurons, *J. Neurochem.*, In press (1995).
60. D. Collazo, H. Takahashi and R. D. McKay, Cellular targets and trophic functions of neurotrophin-3 in the developing rat hippocampus, *Neuron* 9:643-656 (1992).

61. M. B. Spina, S. P. Squinto, J. Miller, R. M. Lindsay and C. Hyman, Brain-derived neurotrophic factor protects dopamine neurons against 6-hydroxydopamine and N-methyl-4-phenylpyridinium ion toxicity: involvement of the glutathione system, *J. Neurochem.* 59: 99-106 (1992).
62. S. W. Barger, D. Horster, K. Furukawa, Y. Goodman, J. Krieglstein, and M. P. Mattson, TNFs protect hippocampal neurons against amyloid β-peptide toxicity: involvement of NFκB and attenuation of peroxide and calcium accumulation, *Proc. Natl. Acad. Sci. U.S.A.* In press (1995).
63. B. Cheng, S. W. Barger, and M. P. Mattson, Staurosporine, K-252a and K-252b stabilize calcium homeostasis and promote survival of CNS neurons in the absence of glucose, *J. Neurochem.* 62:1319-1329 (1994).
64. Y. Goodman, and M. P. Mattson, Staurosporine and K-252 compounds protect hippocampal neurons against amyloid β-peptide toxicity and oxidative injury, *Brain Res.* 650: 170-174 (1994).
65. M. A. Glicksman, J. E. Prantner, S. L. Meyer, M. E. Forbes, M. Dasgupta, M. E. Lewis and N. Neff, K-252a and staurosporine promote choline acetyltransferase activity in rat spinal cord cultures, *J. Neurochem.* 61: 210-221 (1993).
66. L. Mucke, E. Masliah, W. B. Johnson, M. D. Ruppe, M. Alford, E. M. Rockenstein, S. Forss-Petter, M. Pietropaolo, M. Mallory, and C. R. Abraham, Synaptotrophic effects of human amyloid β protein precursors in the cortex of transgenic mice, *Brain Res.* 666:151-167 (1994).
66. M. P. Mattson, S. W. Barger, B. Cheng, I. Lieberburg, V. L. Smith-Swintosky, and R. E. Rydel, β-amyloid precursor protein metabolites and loss of neuronal calcium homeostasis in Alzheimer's disease, *Trends Neurosci.* 16:409-415 (1993).
67. L. Luo, T. Tully, and K. White, Human amyloid precursor protein ameliorates behavioral deficit of flies deleted for Appl gene, *Neuron* 9:595-605 (1992).
68. W. E. Van Nostrand, S. L. Wagner, W. R. Shankle, J. S. Farrow, M. Dick, J. M. Rozemuller, M. A. Kuiper, E. C. Wolters, J. Zimmerman, and C. W. Cotman, Decreased levels of soluble amyloid β–protein precursor in cerebrospinal fluid of live Alzheimer disease patients, *Proc. Natl. Acad. Sci. U.S.A.* 89:2551-255 (1992).
69. M. Farlow, B. Ghetti, M. D. Benson, J. S. Farrow, W. E. van Nostrand and S. L. Wagner, Low cerebrospinal-fluid concentrations of soluble amyloid β–protein precursor in hereditary Alzheimer's disease. *Lancet* 340:453-454 (1992).
70. N. Nukina, K. Hashimoto, I. Kanazawa and H. Mizusawa, Soluble amyloid precursor protein in familial Alzheimer's brain (APP717 Val-Ile mutation), *Soc. Neurosci. Abstr.* 20:607 (1994).
71. M. P. Mattson, B. Cheng, A. Culwell, F. Esch, I. Lieberburg, and R. E. Rydel, Evidence for excitoprotective and intraneuronal calcium-regulating roles for secreted forms of β-amyloid precursor protein, *Neuron* 10:243-254 (1993).
72. M. P. Mattson, Secreted forms of β-amyloid precursor protein modulate dendrite outgrowth and calcium responses to glutamate in cultured embryonic hippocampal neurons, *J. Neurobiol.* 25:439-450 (1994).
73. V. L. Smith-Swintosky, L. C. Pettigrew, S. D. Craddock, A. R. Culwell, R. E. Rydel, and M. P. Mattson, Secreted forms of β-amyloid precursor protein protect against ischemic brain injury, *J. Neurochem.,* 63: 781-784 (1994).
74. J. M. Roch, E. Masliah, A. C. Roch-Levecq, M. P. Sundsmo, D. A. Otero, I. Veinbergs and T. Saitoh, Increase of synaptic density and memory retention by a peptide representing the trophic domain of the amyloid β/A4 protein precursor. *Proc. Natl. Acad. Sc.i U.S.A.* 91: 7450-7454 (1994).

75. M. P. Mattson, R. E. Lee, M. E. Adams, P. B. Guthrie and S. B. Kater, Interactions between entorhinal axons and target hippocampal neurons: a role for glutamate in the development of hippocampal circuitry. *Neuron* 1:865-876 (1988).
76. K. Furukawa, S. W. Barger, E. M. Blalock, and M. P. Mattson, Secreted βAPPs modulate neuronal excitability by activating K^+ channels,through a cGMP pathway, *Soc. Neurosci. Abstr.* 21, In press(1995).
77. E. H. Koo, S. S. Sisodia, D. R. Archer, L. J. Martin, A. Weidemann, K. Beyreuther, P. Fischer, C. L. Masters and D. L. Price, Precursor of amyloid protein in Alzheimer disease undergoes fast anterograde axonal transport, *Proc. Natl. Acad. Sci. USA* 87:1561-1565 (1990).
78. K. L. Moya, L. I. Benowitz, G. E. Schneider and B. Allinquant, The amyloid precursor protein is developmentally regulated and correlated with synaptogenesis, *Dev. Biol.* 161:597-603 (1994).
79. M. Shimokawa, K. Yanagisawa, H. Nishiye and T. Miyatake, Identification of amyloid precursor protein in synaptic plasma membrane, *Biochem. Biophys. Res. Commun.* 196:240-244 (1993).
80. M. R. Nitsch, A. S. Farber, H. J. Growdon and J. R. Wurtman, Release of amyloid β-protein precursor derivatives by electrical depolarization of rat hippocampal slices, *Proc. Natl. Acad. Sci.* 90:5191-5193 (1993).
81. H. Ninomiya, J.-M. Roch, M. P. Sundsmo, D. A. Otero, and T. Saitoh, Amino acid sequence RERMS represents the active domain of amyloid β/A4 protein precursor that promotes fibroblast growth, *J. Cell Biol.* 121:879-886 (1993).
82. S. W. Barger, R. R. Fiscus, P. Ruth, F. Hofmann, and M. P. Mattson, Role of cyclic GMP in the regulation of neuronal calcium and survival by secreted forms of β amyloid precursor, *J. Neurochem.* 64: 2087-2096 (1994).
83. S. W. Barger and M. P. Mattson, Involvement of NF-κB in neuroprotective genetic programs,*Soc. Neurosci. Abstr.* 20:687 (1994).
84. S. W. Barger and M. P. Mattson, Participation of gene expression in the protection against amyloid β-peptide toxicity by the β-amyloid precursor protein, *Ann. N.Y. Acad. Sci.* in press (1995).
85. B. Cheng, S. Christakos, and M. P. Mattson, Tumor necrosis factors protect neurons against excitotoxic/metabolic insults and promote maintenance of calcium homeostasis, *Neuron* 12: 139-153 (1994).
86. R. Schreck, K. Albermann and P. A. Baeuerle, Nuclear factor κB: an oxidative stress-responsive transcription factor of eukaryotic cells (a review),*Free Radic. Res. Commun.* 17:221-237 (1992).
87. A. Logan and M. Berry, Transforming growth factor-β1 and basic fibroblast growth factor in the injured CNS, *Trends Pharmacol. Sci.* 14:337-342 (1993).
88. E. A. van der Wal, F. Gomez-Pinilla, and C. W. Cotman, Transforming growth factor-β1 is in plaques in Alzheimer and Down pathologies,*Neuroreport* 4:69-72 (1993).
89. C. E. Finch, N. J. Laping, T. E. Morgan, N. R. Nichols and G. M. Pasinetti, TGF-b1 is an organizer of responses to neurodegeneration,*J. Cell. Biochem.* 53:314-322 (1993).
90. J. H. Prehn, B. Peruche, K. Unsicker, and J. Kriegelstein, Isoform-specific effects of transforming growth factors-β on degeneration of primary neuronal cultures induced by cytotoxic hypoxia or glutamate, *J. Neurochem.* 60:1665-1672 (1993).
91. M. F. Galindo, J. H. M. Prehn, V. P. Bindokas, and R. J. Miller, Potential role of TGF-β1 in Alzheimer's disease: regulation of APP expression and inhibition of βAP neurotoxicity, *Soc. Neurosci. Abstr.* 20:1248 (1994).
92. D. D. Cunningham, L. Pulliam, and P. J. Vaughan, Protease nexin-1 and thrombin: injury related processes in the brain,*Thromb. Haemost.* 70:168-171 (1993).

93. J. W. Fenton II, F. A. Ofosu, D. V. Brezniak, and H. I. Hassouna, Understanding thrombin and hemostasis, *Hematol. Oncol. Clin. North Am.* 7:1107-1119 (1993).
94. S. R. Coughlin, Molecular mechanisms of thrombin signaling, *Semin. Hematol.* 31: 270-277 (1994).
95. M. Suzuki, O. Motohashi, A. Nishino, V. Shiina, K. Mizoi, T.Oshimoto, M.Kameyama, and T. Onuma, Diphasic increase in thrombin-antithrombin III complex in blood from the internal jugular vein following severe head injury. *Thromb Haemost* 71:155 (1994).
96. H. Akiyama, K. Skeda, H. Kondo, and P.L. McGreer, Thrombin accumulation in brains of patients with Alzheimer's disease, *Neurosci Lett* 146:152 (1992).
97. K. Igarashi, H. Murai , and J-I. Asaka, Proteolytic processing of amyloid β protein precursor (APP) by thrombin. *Biochem Biophys Res Commun* 185:1000 (1992).
98. J. Davis-Salinas, S.M. Saporito-Irwin, F.M. Donovan, D.D. Cunningham, and W.E. Van Nostrand, Thrombin receptor activation induces secretion and nonamyloidogenic processing of amyloid beta-protein precursor, *J Biol Chem* 269:22623 (1994).
99. J. R. Weinstein, S. J. Gold, D. D. Cunningham and C. M. Gall, Cellular localization of thrombin receptor mRNA in rat brain: expression by mesencephalic dopaminergic neurons and codistribution with prothrombin mRNA, *J. Neurosci.* 15: 2906-2919 (1995).
100. M. Dihanich, M. Kaser, E. Reinhard, D.D. Cunningham, and D. Monard, Prothrombin mRNA is expressed by cells of the nervous system. *Neuron* 6:575 (1991).
101. V.L. Smith-Swintosky, S. Zimmer, J.W.Fenton II, and M.P. Mattson, Protease nexin-1 and thrombin modulate neuronal Ca^{2+} homeostasis and sensitivity to glucose deprivation-induced injury. *J. Neurosci.* in press (1995).
102. V. L. Smith-Swintosky, S. Zimmer, J. W. Fenton 2nd, and M. P. Mattson, Opposing actions of thrombin and protease nexin-1 on amyloid β-peptide toxicity, and on accumulation of peroxides and calcium in hippocampal neurons, *J. Neurochem.* Submitted.
103. E. Reinhard, H. S. Suidan, A. Pavlik, and D. Monard, Glia-derived nexin/protease nexin-1 is expressed by a subset of neurons in the rat brain, *J. Neurosci. Res.* 37:256-270 (1994).
104. B.W. Festoff, J.S. Rao, and M. Chen, Protease nexin I, thrombin- and urokinase-inhibiting serpin, concentrated in normal human cerebrospinal fluid, *Neurology* 42:1361 (1992).
105. K. Furukawa, V. L. Smith-Swintosky, and M. P. Mattson, Evidence that actin depolymerization protects hippocampal neurons against excitotoxicity by stabilizing $[Ca^{2+}]_i$, *Exp. Neurol.* In press (1995).
106. C. Rosenmund, and G. L. Westbrook, Calcium-induced actin depolymerization reduces NMDA channel activity, *Neuron* 10:805-814 (1993).
107. K. Furukawa, and M. P. Mattson, Cytochalasins protect hippocampal neurons against amyloid β-peptide toxicity: evidence that actin depolymerization suppresses Ca^{2+} influx. *J. Neurochem.* In press (1995).
108. K. Furukawa, and M. P. Mattson, Taxol stabilizes $[Ca^{2+}]_i$ and protects hippocampal neurons against excitotoxicity, *Brain Res.* In press (1995).

CHANNEL FORMATION BY A NEUROTOXIC BETA AMYLOID PEPTIDE, Aβ25-35

Meng-chin Lin,[1] Tajib Mirzabekov,[2] and Bruce Kagan[1,2,3]

[1]Neuroscience Interdepartmental Program
and [2]Department of Psychiatry and Biobehavioral Sciences
UCLA Neuropsychiatric Institute and Brain Research Institute
and [3]West Los Angeles VA Medical Center

INTRODUCTION

Alzheimer's disease (AD) is a chronic neurodegenerative disorder which is accompanied by memory loss and defects in cognitive function. The features in AD brains include extracellular senile plaques, intracellular neurofibrillary tangles, and amyloid proteins deposited in walls of cerebral and meningeal blood vessels. The major component of these pathognomonic lesions of AD has been identified as a 39-43 amino acid long peptide called beta amyloid peptide (Aβ). Aβ is derived from the amyloid β precursor protein (βAPP).[1] The thesis that Aβ deposition plays a central role in the pathogenesis of AD has garnered increasing attention. It has been demonstrated that Aβ either is directly neurotoxic to neurons in culture,[2,3] or potentiates neuronal vulnerability to excitatory neurotoxins.[4,5] Neurotoxic activity has been reported to be located in the fragment of Aβ, Aβ25-35.[2] We hypothesize that Aβ and its fragments, such as Aβ25-35, can form ion channels in neuronal membrane, which may allow calcium entry either directly or indirectly, thus leading to neuronal death and neuropathology of AD. In this study, we will characterize the ionic channels formed by Aβ25-35. Preliminary reports of this work have appeared elsewhere.[6,7,8]

MATERIALS AND METHODS

We modified the Montal and Mueller method to make solvent-free planar lipid bilayer membranes. The membranes were formed by the union of two lipid monolayers over an aperture (150-200μm) in the Teflon film that separated the two aqueous phases. Lipid monolayers were made from either 10mg/ml asolectin or a combination of phosphatidylcholine, phosphatidylethanolamine, and phosphatidylserine (2:2:1 w/w/w) dissolved in hexane. A 3% solution of hexadecane in pentane was applied to the Teflon

Neurodegenerative Diseases
Edited by Gary Fiskum, Plenum Press, New York, 1996

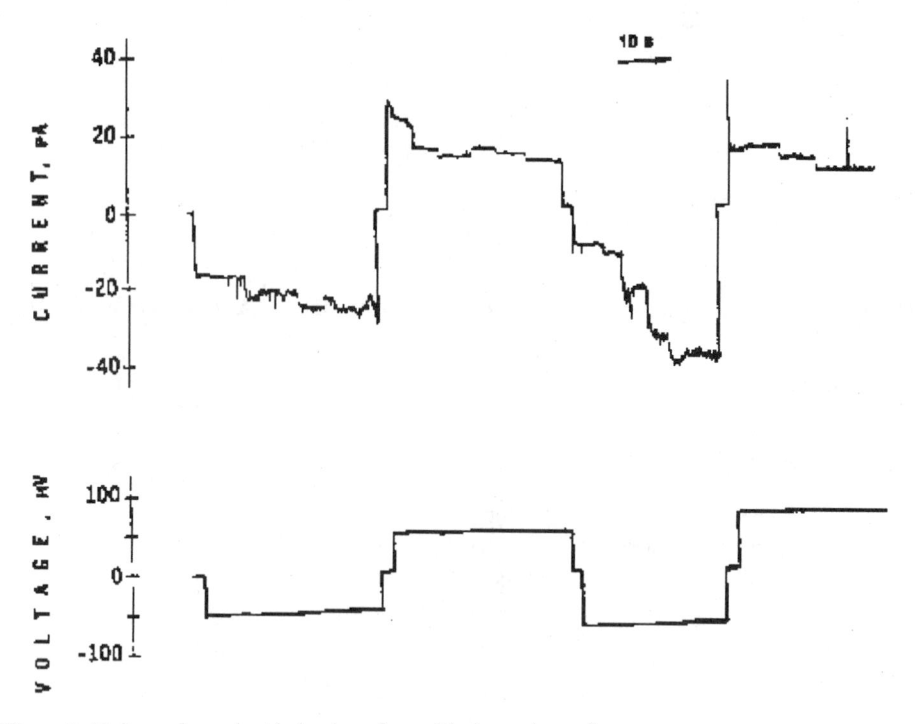

Figure 1. Voltage-dependent behavior of a multi-channel membrane.
Aβ25-35 was added to the cis compartment to the planar lipid bilayer formed from asolectin. Aqueous solutions contained 100mM NaCl, 5mM HEPES-NaOH, 2mM $MgCl_2$, and pH = 7.5.

film to facilitate bilayer formation in symmetrical salt solutions in voltage clamp conditions.[9]

Amyloid peptides (Athena Neurosciences Inc., Bachem Biosciences, Inc. Bachem California) used for the planar membrane experiments are prepared as stock solutions. The peptides were dissolved either in double-deionized H_2O, or in dimethylsulfoxide (DMSO) in concentration of 1-4mg/ml. Stock solutions were frozen immediately after preparation and stored at -20°C as small aliquots.

Liposomes were prepared from purified soybean phospholipids or 1,2-diphytanoyl-sn-glycero-3-phosphocholine (DGPC) in 100mM NaCl, 10mM NaPi, 2mM $MgCl_2$, pH 7.5, at a concentration of 10mg/ml. Lipids are dried on the walls of a glass vial under N_2 stream. The lipid-salt mixture is incubated at 4°C for 30 minutes for hydration, and then is sonicated by pulse-sonication for 10 minutes. For proteoliposome preparation, 200-400μg of lyophilized Aβ or related peptides are added to 2ml liposomes and the resulting mixture are sonicated for another 3 minutes.

A signal generator and an oscilloscope were employed to monitor membrane formation. We used a voltage clamp amplifier (Axopatch-1C, Axon Instruments, Sunnyvale, CA) with suitable head stage (CV-3B) to record single channel currents. For data acquisition, a pulse code modulation/video cassette recording digital system allow recordings which are sent through a low pass filter.[10] A storage oscilloscope and a chart recorder are useful for monitoring real-time membrane capacitance and single channel recordings.

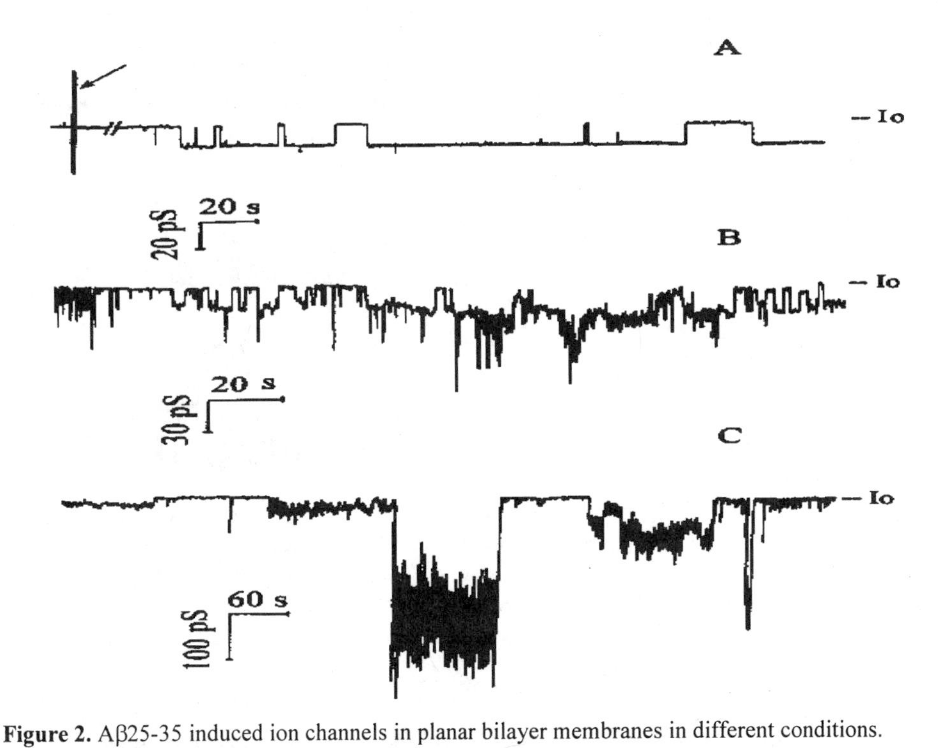

Figure 2. Aβ25-35 induced ion channels in planar bilayer membranes in different conditions.
A. The peptide was added at the time shown by arrow. Solution contained 100mM NaCl, 10mM Tris-HCl, 2mM $MgCl_2$, and pH = 7.5. The time break is 7 minutes. V_m = -40mV.
B. Aβ25-35 channels were observed at low pH. Bilayer was bathed in asymmetric solutions: 100mM NaCl, 10mM DMG-NaOH, 2mM $MgCl_2$, and pH = 4.5 in the cis-side, 100mM NaCl, 10mM Tris-HCl, 2mM $MgCl_2$, and pH = 7.5 in the trans-side. V_m = -50mV.
C. Membrane current fluctuations were induced by Aβ25-35 in high ionic strength solution which contained 1M NaCl, 10mM Tris-HCl, 2mM $MgCl_2$, and pH = 7.5. V_m = -80mV.

RESULTS AND DISCUSSION

Figure 1 exhibits ionic channels caused by Aβ25-35 in planar lipid membranes. The addition of Aβ25-35 to a voltage-clamped planar lipid bilayer induced a stepwise increase of current indicating the formation of ion-permeable channels. We suggest that these channels are relevant to the neurotoxic actions of Aβ25-35 because channels were formed at concentrations comparable to those which produced neurotoxicity *in vitro*,[11] which increased conductances in neurons[12,13] and which is expected in the central nervous system.[14]

The channels have multilevel conductances, which vary from tens of pS to hundreds of pS. In terms of conductances, Aβ25-35 forms channels that may allow a prominent ionic leakage through neuronal membranes. This leakage can disrupt calcium homeostasis. Channel activity was obtained in salt solutions of NaCl (0.1-1M), KCl (0.1-1M), and $CaCl_2$ (0.03-0.2M) and over a pH range of 4.5 to 9.9. The channel forming properties of Aβ25-35 did not change qualitatively in variations of salt concentration (from 100mM to 1M NaCl) and in variations of pH values (from 4.5 to 9.9) (Figure 2). In a ten-fold gradient of NaCl with dilute side in the trans-side where voltages are applied, a reversal potential of +9~12mV was obtained. It is inferred that the channels are weakly cation selective. Calcium, potassium, sodium, and chloride are permeable through these channels at a ratio of $Ca^{2+}:K^{+}:Na^{+}:Cl^{-}$ = 5.4:1.6:1.4:1.

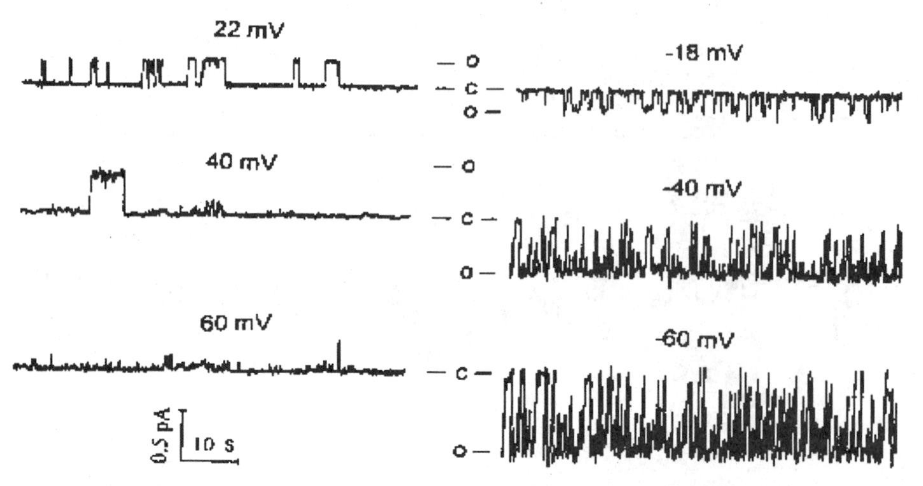

Figure 3. Recordings of membrane formed from Aβ25-35-containing proteoliposomes. Each trace shows the opening and closing of a single Aβ25-35 channels at different membrane potentials. O denotes the open state level and C denotes the closed state level. Note that the opening probability of channels was greater as the membrane potential becomes more negative.

An addition of 10-20μg/ml Aβ25-35 to the cis-side of the membrane lead to formation of voltage-dependent channels, which opened at negative voltages and closed at positive voltages (Figure 1). The voltage-dependence of channel opening is consistent with the inside negative membrane potential of neurons.

There is another method for incorporation of Aβ25-35 into planar lipid membranes. Aβ25-35-containing proteoliposomes were prepared, and planar lipid membranes were then formed directly from these proteoliposomes spread as monolayers. It was found that channel properties such as ion selectivity, single channel conductances in proteoliposome experiments were very similar to those obtained in experiments with addition of Aβ25-35 to the membrane bathing solution. However, we cannot control the orientation of Aβ25-35 insertion into the lipid in proteoliposome preparation. As a result, voltage dependence of Aβ25-35 multi-channel containing proteoliposome membrane activity is not predictable. We propose that non-potential dependent behavior of multi-channel containing membranes corresponds to the condition in which approximately the equal number of channels were incorporated in opposite directions. In one experiment in which the membrane was made from proteoliposome, single channel recordings indicated that Aβ25-35 channels have greater opening probability at negative potentials than at positive potentials (Figure 3).

Macroscopic membrane currents induced by Aβ25-35 increased slowly at negative potentials, and decreased quickly at positive potentials. Aβ25-35 channels were blocked by micromolar concentrations of Cd^{2+}, Cu^{2+}, or Zn^{2+}, but not by Ca^{2+} or Mg^{2+} (Figure 4). The blocking effect could be reversed by EGTA.

No channel formation was observed when such peptides as reverse peptide, Aβ35-25, scrambled peptide, Aβ25-35scr, and smaller fragments Aβ28-35, Aβ25-31, Aβ25-28, Aβ34-42 were added to the planar lipid bilayer membranes. Indeed, we were unable to obtain consistently reproducible channel activity with full length versions of Aβ (1-40, 1-42) at concentrations up to 150μg/ml. Different states of aggregation for Aβ, or different

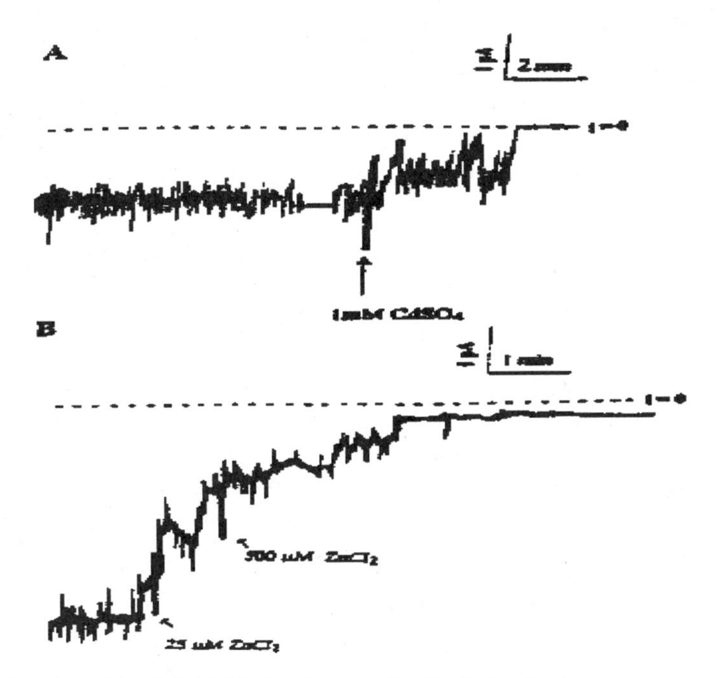

Figure 4. Inhibition of the Aβ25-35 induced current by divalent cations.
A. Membrane was formed from proteoliposomes. Salt solution contained 100mM NaCl, 2mM HEPES-NaOH, 1mM $MgCl_2$, and pH = 7.2. 1mM $CdSO_4$ was added to the side with proteoliposomes at the time shown by arrow.
B. Membrane made from PC:PE:PS = 2:2:1 was held at -40mV. Salt solution contained 100mM NaCl, 5mM Tris-HCl, 1mM EDTA, and pH = 7.5. Aβ25-35 was added to a final concentration of 35μg/ml. $ZnCl_2$ was added to the cis-side and decreased the Aβ25-35 induced current by more than 95%.

methodology might underlie the discrepancy of our preliminary results from Arispe et al's.[15,16] More intensive investigation on the effect of Aβ on planar lipid bilayer system is required to ascertain its channel forming activity.

Therefore, we have observed that Aβ25-35 can form channels in lipid bilayer membranes. The results suggest that Aβ25-35 can form ion channels in neuronal membrane, which may allow calcium entry either directly or indirectly, thus leading to neuronal death.

REFERENCES

1. L. Berg, and J.C. Morris, Aging and dementia, *in*: "Neurobiology of disease," A.L. Pearlman, and R.C. Collins, ed., Oxford University Press, New York, Oxford (1990).
2. B.A. Yankner, L.K. Duffy, and D.A. Kirschner, Neurotrophic and neurotoxic effects of amyloid β protein: reversal by tachykinin neuropeptides. *Science*, 25:279-282 (1990).
3. N.W. Kowall, M.F. Beal, J. Busciglio, L.K. Duffy, and B.A. Yankner, An *in vivo* model for the neurodegenerative effects of β amyloid and protection by substance P. *Proc. Natl. Acad. Sci. USA*, 88:7247-7251 (1991).
4. J. Koh, L.L. Yang, and C.W. Cotman, β amyloid protein increases the vulnerability of cultured cortical neurons to excitotoxic damage. *Brain Res.*, 533:315-320 (1990).

5. M.P. Mattson, B. Cheng, D. Davis, K. Bryant, I. Lieberburg, and R.E. Rydel, β-amyloid peptides destabilize calcium homeostasis and render human cortical neurons vulnerable to excitoxicity. *J. Neurosci.*, 12:376-389 (1992).
6. P. Marshall, T. Mirzabekov, W.L. Yuan, M. Carman, I. Lieberburg, and B.L. Kagan, Channels in lipid bilayers induced by a fragment of the β-amyloid peptide. *Biophys. J.*, 64(pt2):A94 (1993).
7. T. Mirzabekov, M. Lin and B.L. Kagan, Ion permeability properties of lipid bilayers containing amyloid beta-peptide and its fragments. *Biophys. J.*, 66(pt2):A430 (1994).
8. T. Mirzabekov, M. Lin, W. Yuan, P.J. Marshall, M. Carman, K. Tomaselli, I. Lieberburg, and B.L. Kagan, Channel formation in planar lipid bilayers by a neurotoxic fragment of the beta-amyloid peptide. *Biochem. Biophys. Res. Commun.*, 202:1142-1148 (1994).
9. B.L. Kagan, and Y. Sokolov, The use of lipid bilayer membranes to detect pore formation by toxins. *Methods in Enzymol.*, 235:699-713 (1994).
10. O. Alvarez, How to set up a bilayer system, *in*: "Ion channels reconstitution," C. Miller, ed., Plenum Press, New York (1986).
11. C.J. Pike, A.J. Walencewicz, C.G. Glabe, and C.W. Cotman, *In vitro* aging of β-amyloid protein causes peptide aggregation and neurotoxicity. *Brain Res.*, 563:311-314 (1991).
12. M.A. Simmons and C.R. Schneider Amyloid β peptides act directly on single neurons. *Neurosci. Lett.*, 150:133-136 (1993).
13. B. Carette, P. Poulain, and A. Delacourte, Electrophysiological effects of 25-35 amyloid-β-protein on guinea-pig lateral septal neurons. *Neurosci. Lett.*, 151:111-114 (1993).
14. P. Seubert, C. Vigo-Pelfrey, F. Esch, M. Lee, H. Dovey, D. Davis, S. Sinha, M. Schlossmacher, J. Whaley, C. Swindlehurst, R. McCormack, R. Wolfert, D. Selkoe, I. Lieberburg, and D. Schenk, Isolation and quantification of soluble Alzheimer's beta-peptide from biological fluids. *Nature*, 359:325-327 (1992).
15. N. Arispe, E. Rojas, and H.B. Pollard, Alzheimer disease amyloid β protein forms calcium channels in bilayer membranes: blockade by tromethamine and aluminum. *Proc. Natl. Acad. Sci. USA*, 90:567-571 (1993).
16. N. Arispe, H.B. Pollard, and E. Rojas, Giant multilevel cation channels formed by Alzheimer disease amyloid β-protein[AβP-(1-40)] in bilayer membranes. *Proc. Natl. Acad. Sci. USA*, 90:10573-10577 (1993).

CYSTEINE STRING PROTEINS: PRESYNAPTIC FUNCTION AND DYSFUNCTION

Cameron B. Gundersen, Joy A. Umbach, and Alessandro Mastrogiacomo

Department of Molecular & Medical Pharmacology
UCLA School of Medicine
Los Angeles, CA 90095

INTRODUCTION

The observation that is the basis of this chapter is the finding by Zinsmaier and colleagues[1] that neuronal degeneration is present in organisms with mutations of the gene encoding cysteine string proteins (csps). Since there is still a very limited number of instances in which neurodegenerative disorders have been traced to alterations of single genes, it is plausible that studies of *csp* mutants will provide useful insights into the etiology of neurodegenerative diseases. To develop this thesis, we will review what is known about csps and then consider scenarios by which mutation of the csp gene could have neurodegenerative sequelae.

DISCOVERY AND CHARACTERIZATION OF CSPS

Csps were identified using a synapse-specific monoclonal antibody in *Drosophila*[2]. This antibody bound selectively to all neuropil regions, as well as motor nerve terminals in *Drosophila*[2]. Western analysis revealed two immunoreactive proteins of 32kDa and 34kDa which were shown by cDNA cloning and sequencing to derive from alternative splicing of mRNA transcribed from a single gene on the third chromosome[2]. The most striking feature of the deduced amino acid sequence of the encoded proteins was a string of 11 consecutive cysteine residues; indeed it was this cysteine string that gave these proteins their name[2]. It was concluded that csps were likely to be presynaptic proteins, but there was no evidence regarding their function[2].

The first clue to the possible function of csps came somewhat serendipitously. We were using a suppression cloning strategy to search for cDNAs encoding subunits or modulators of presynaptic calcium channels in the electric fish, *Torpedo*[3]. The basis of this assay was the inhibition by antisense cRNA of the expression of presynaptic (ω-conotoxin-sensitive) calcium channels in *Xenopus* oocytes[3]. We isolated and sequenced a cDNA of about 7.4kDa, and it encoded a 195 residue protein with high homology to *Drosophila* csps. In fact, it represented a *Torpedo* csp (Tcsp). By virtue of the fact that antisense Tcsp cRNA inhibited the expression of presynaptic calcium channels in frog oocytes, it seemed plausible that Tcsp was a subunit or modulator of these channels. Indeed, such a conclusion was supported by the fact that *Drosophila* csps were almost exclusively associated with presynaptic elements. However, in spite of the progress outlined below, the precise link between csps and presynaptic calcium channels remains elusive.

Advances in understanding the contribution of csps to presynaptic function have come from several sources. By raising antibodies against the *Torpedo* protein, we were able to establish that csps were almost exclusively membrane-associated antigens in *Torpedo*[4]. Moreover, relative to the protein produced by *in vitro* translation of the Tcsp cRNA, we found[4] that the native csp in electric organ migrated as a higher mass protein (by the equivalent of about 7kDa). Interestingly, this increase in mass was accompanied by a dramatic change in the capacity of Tcsp to partition into Triton X-114 (a measure of its hydrophobicity)[4]. *In vitro* translated Tcsp did not partition into Triton X-114, whereas more than 95% of the native protein did extract into Triton X-114 (ref 4). These results indicated to us that Tcsp was post-translationally modified in a manner that simultaneously altered its electrophoretic mobility and enhanced its hydrophobicity. Based on these observations we subsequently established that Tcsp is extensively fatty acylated(ref 5).

Treatment of native Tcsp with neutral hydroxylamine or with 0.1M KOH in methanol resulted in a complete shift in the mobility of Tcsp from its native (34kDa) mass to 27kDa (ref 5). Since hydroxylamine and methanolic KOH have been used widely to remove fatty acyl groups from cysteine residues of proteins[5], it seemed likely that we had displaced one or more fatty acyl groups from Tcsp, thus causing a shift in its electrophoretic mobility. Concomitantly, we showed that the 27kDa form of Tcsp, just like its *in vitro* translated counterpart, did not extract into Triton X-114 (C.G., unpublished). Thus, deacylating reagents caused a loss of mass and a loss of hydrophobicity of native Tcsp. By analyzing the fatty acyl hydroxamates of hydroxylamine-treated Tcsp, we confirmed that it was predominantly palmitic acid that was esterified to Tcsp. However, two important questions remained: how many cysteine residues are fatty acylated, and which residues, if any, escape fatty acylation? Preliminary data[5] indicate that as many as 11 or 12 of the 13 cysteine residues of Tcsp are fatty acylated. We are seeking independent confirmation of this result, as well as the answer to question two.

At virtually the same time that evidence emerged that Tcsp is fatty acylated, we found that csps are synaptic vesicle proteins[6]. This result came from analysis of highly enriched membrane fractions from *Torpedo* electric organ. While Tcsp immunoreactivity is prominent in fractions enriched (to better than 95%) in synaptic vesicles, it is undetectable in fractions that include presynaptic plasma membrane[6]. This was an unexpected result, because, owing to the csp-calcium channel link implied by our suppression cloning efforts, we had expected to find csps associated with presynaptic calcium channels. How does one resolve this problem? To date, it has not been resolved, but we are looking for evidence of a direct or indirect (mediated by some other protein) regulatory interaction between vesicle-associated csps and calcium channels. A discussion of possible interactions is in press (ref 7).

The nearly simultaneous observations that csps are synaptic vesicle proteins and that they harbor a large number of fatty acylated cysteine residues had an additional impact on our thinking. The domains of csp that flank the cysteine string are very hydrophilic. By adding several fatty acyl residues to the cysteine string, csps become highly amphipathic. Indeed, in unpublished experiments, we have found that three or more fatty acyl groups are sufficient to allow Tcsp to partition with the Triton X-114 bead in Triton X-114 partitioning experiments. Thus, if we assume from this result that three fatty acyl groups are adequate to tether Tcsp to the external face of a synaptic vesicle (where immunoprecipitation data indicate csps are localized[6]), then what is the purpose of all of the additional 8 or 9 fatty acyl groups esterified to Tcsp? While we do not know the answer to this question, we have proposed that csps act as interfacial catalysts of exocytosis at the boundary between synaptic vesicles and the presynaptic plasma membrane[5]. A detailed description of this proposal is presented in reference 8. In addition, compelling evidence for a role of csps in neurotransmitter release has emerged from studies of *Drosophila csp* mutants. These results are considered next.

Zinsmaier and colleagues[1] obtained hypomorphic and null csp alleles, as well as two EMS (ethyl methane sulfonate) alleles. The percent viability of all of these alleles is quite low. The highest percent viability (about 16%) is seen in the hypomorphs and a low of 1.6% viability is seen for one of the EMS alleles[1]. All adult mutants die prematurely[1]. However, the presence of survivors, especially in the null allele, indicates that csps are not indispensable for survival, at least not at room temperature. This qualification reflects an interesting temperature-sensitive phenotype of the *csp* mutants: at 29°C, *csp* mutations are 100% lethal and mutant adults that are transferred from 18°C to 29°C show temperature-sensitive paralysis and death.

We investigated the cellular basis of this behavioral paralysis of *csp* mutants. Our results[9] showed that larvae of the null and hypomorphic allele have diminished release of neurotransmitter relative to controls at room temperature. When the temperature is raised to 30°C, evoked transmitter release fails completely in *csp* mutants[9]. (Neuromuscular transmission is largely unaffected in controls between 20°C and 30°C). This effect on the mutants is due purely to a failure of depolarization (invading nerve endings as a consequence of the stimulation of the nerve) to elicit the quantal discharge of transmitter[9]. However, spontaneous release events persist in *csp* mutants[9]. So, what can one conclude from this? One could achieve this kind of an uncoupling of the normal depolarization-secretion process either by inhibiting the function of presynaptic calcium channels, or by inhibiting a step further downstream in the exocytotic cascade. Since csps have been implicated both in regulating presynaptic calcium channels, and in catalyzing exocytosis, the available results do not allow us to distinguish between a primary failure of presynaptic calcium channels or an effect on the release apparatus in the heat-treated mutants. However, some additional corollaries do emerge from this work.

First, the continued presence of spontaneous quantal events in *csp* mutants (even at 30°C) indicates that csps do not play a vital role in regulating this release process. It will be of some interest to determine whether another csp-like protein participates in such constitutive secretory events.

Second, we sought evidence to exclude the possibility that the failure of transmitter release in the *csp* mutants reflects the transient loss of a releasable population of vesicles. While this is a relatively unlikely explanation based on the rapid recovery of evoked transmitter release when the *csp* mutants are cooled back to 20°C (ref 9), experiments with α-latrotoxin also exclude this explanation. Preliminary experiments (J.U., unpublished and see ref 9) indicate that α-latrotoxin venom gland extract from black widow spiders induces a comparable discharge of quanta from control and mutant terminals. The persistence of releasable quanta in the mutants still argues for a principal role for csps in the release arm of the life cycle of synaptic vesicles. In other words, the results obtained with vonom gland extract indicate that one can circumvent the lesion in the release process and still drive the release of transmitter-containing vesicles. However, these data do not exclude the possibility that csps also play a role in recycling of synaptic vesicles. Additional experiments are needed to address this possibility.

We now come to the observations that are central to the issue of neurodegeneration in *Drosophila csp* mutants. Electron microscopic examination of nerve terminals of photoreceptor axons revealed a paucity of synaptic vesicles relative to controls[1]. Instead of vesicles, there was abundant electron-dense debris in the nerve terminals[1]. It could not be conclusively established that there were no vesicles, because it is possible that vesicles in adult *csp* mutants are more fragile than in control or larval preparations. It is likely, at least in adult *csp* mutants, that this change in nerve terminal ultrastructure contributes to the inhibition of synaptic transmission that is seen by recording of the electroretinogram[1]. However, based both on the electrophysiological data discussed above, as well as electron microscopic examination of larval neuromuscular junctions (M. Kreman, J. Umbach, unpublished), there are still vesicles present at this developmental stage. From these results, it appears that the degenerative changes seen in adult nerve terminals may in part be age or activity related. In any circumstance, because *Drosophila* are well suited to a wide range of analytical investigations, it may be very providential to pursue further the cause(s) and consequence(s) of nerve terminal degeneration in these organisms.

MUTATION OF THE CSP GENE AND NERVE TERMINAL DEGENERATION IN DROSOPHILA: CAUSES AND CONSEQUENCES

While the available data suggest that *Drosophila csp* mutants may be useful models to investigate degenerative changes in nerve terminals, a number of additional issues needs to be resolved. For instance, we need to know if nerve terminals at sites other than the laminar neuropil (where photoreceptor axons make synapses) show similar degenerative changes. It is also important to know whether other parts of the neuron are affected. The developmental time course of these changes is also of interest. Once these issues are resolved, it should be more clear how one might proceed toward establishing the cellular and molecular events that culminate in the observed neurodegenerative changes.

In spite of the fact that evidence is lacking with respect to several of the issues raised in the preceding paragraph, it is evident even at this stage that there are two distinct hypotheses that can be formulated to explain why *csp* mutants display neurodegeneration. First, these changes (loss of vesicles and accumulation of debris in photoreceptor nerve terminals) may be a direct consequence of the absence of csps. In other words, csps may be necessary for preserving synaptic vesicle structure and nerve terminal function. Alternatively, these changes may be indirect. They may come about as an indirect effect of the reduction in synaptic efficacy already documented in *csp* mutants[1,9]. Thus, either the reduced synaptic activity itself, the disruption of a trophic interaction, or an over-compensating change of function in the mutants may have deleterious consequences. It should be of considerable interest to distinguish among these possibilities, and to ascertain whether similar arguments prevail in vertebrates.

Acknowledgments: We thank Judy Amos for preparing this manuscript. Supported in part by NIH grants to CG (NS31517) and JU (NS31934).

REFERENCES

1. K. Zinsmaier, K.K. Eberle, E. Buchner, K. Walter, and S. Benzer, Paralysis and early death in cysteine string protein mutants of Drosophila, *Science*, 263:977-980 (1994).
2. K. Zinsmaier, A. Hofbauer, G. Heimbeck, G.O. Pflugfelder, S. Buchner, and E. Buchner, A cysteine-string protein is expressed in retina and brain of *Drosophila*, *J. Neurogenet.* 7:15-29 (1990).
3. C.B. Gundersen and J.A. Umbach, Suppression cloning of the cDNA for a candidate subunit of a presynaptic calcium channel, *Neuron*, 9:527-537 (1992).
4. A. Mastrogiacomo, C.J. Evans, and C.B. Gundersen, Antipeptide antibodies against a *Torpedo* cysteine string protein, *J. Neurochem.* 62:873-880 (1994).
5. C.B. Gundersen, A. Mastrogiacomo, K. Faull, and J.A. Umbach, Extensive lipidation of a *Torpedo* cysteine string protein, *J. Biol. Chem.* 269:19197-19199, (1994).
6. A. Mastrogiacomo, S.M. Parsons, G.A. Zampighi, D.J. Jenden, J.A. Umbach, and C.B. Gundersen, Cysteine string proteins: a potential link between synaptic vesicles and presynaptic calcium channels, *Science*, 263:981-982 (1994).
7. J.A. Umbach, A. Mastrogiacomo, and C.B. Gundersen, Cysteine string proteins and presynaptic function, *J. Physiol.* (Paris) in press.
8. C.B. Gundersen, A. Mastrogiacomo and J.A. Umbach, Cysteine-string proteins as templates for membrane fusion: models of synaptic vesicle exocytosis, *J. Theor. Biol.* 172:269-277 (1995).
9. J.A. Umbach, K.E. Zinsmaier, K.K. Eberle, E. Buchner, S. Benzer, and C.B. Gundersen, Presynaptic dysfunction in *Drosophila* csp mutants, *Neuron* 13:899-907 (1994).

AIDS DEMENTIA AS A FORM OF EXCITOTOXICITY: POTENTIAL THERAPY WITH NMDA OPEN-CHANNEL BLOCKERS AND REDOX CONGENERS OF NITRIC OXIDE

Stuart A. Lipton

From the Laboratory of Cellular & Molecular Neuroscience, Department of Neurology, Children's Hospital, and the Departments of Neurology, Beth Israel Hospital, Brigham & Women's Hospital, Massachusetts General Hospital; Program in Neuroscience, Harvard Medical School, Boston, Massachusetts 02115

ABSTRACT

The neurological manifestations of AIDS include dementia, encountered even in the absence of opportunistic superinfection or malignancy. The AIDS Dementia Complex appears to be associated with several neuropathological abnormalities, including astrogliosis and neuronal injury or loss. How can HIV-1 result in neuronal damage if neurons themselves are only rarely, if ever, infected by the virus? *In vitro* experiments from several different laboratories have lent support to the existence of HIV- and immune-related toxins. In one recently defined pathway to neuronal injury, HIV-infected macrophages/microglia as well as macrophages activated by HIV-1 envelope protein gp120 appear to secrete excitants/neurotoxins. These substances may include arachidonic acid, platelet-activating factor, free radicals (NO· and $O_2^{\cdot -}$), glutamate, quinolinate, cysteine, cytokines (TNF-α, IL1-β, IL-6), and as yet unidentified factors emanating from stimulated macrophages and possibly reactive astrocytes. A final common pathway for neuronal susceptibility appears to be operative, similar to that observed in stroke, trauma, epilepsy, and several neurodegenerative diseases, including Huntington's disease, Parkinson's disease, and amyotrophic lateral sclerosis. This mechanism involves excessive activation of *N*-methyl-D-aspartate (NMDA)

Neurodegenerative Diseases
Edited by Gary Fiskum, Plenum Press, New York, 1996

receptor-operated channels, with resultant excessive influx of Ca^{2+} leading to neuronal damage, and thus offers hope for future pharmacological intervention.

This chapter reviews two clinically-tolerated NMDA antagonists, memantine and nitroglycerin: *(i)* Memantine is an open-channel blocker of the NMDA-associated ion channel and a close congener of the anti-viral and anti-parkinsonian drug amantadine. Memantine blocks the effects of escalating levels of excitotoxins to a greater degree than lower (physiological) levels of these excitatory amino acids, thus sparing to some extent normal neuronal function. *(ii)* Nitroglycerin acts at a redox modulatory site of the NMDA receptor/channel complex to downregulate its activity. The neuroprotective action of nitroglycerin at this site is mediated by a chemical species related to nitric oxide, but in a higher oxidation state, resulting in transfer of an NO group to a critical cysteine on the NMDA receptor. Because of the clinical safety of these drugs, they have the potential for trials in humans. As the structural basis for redox modulation is further elucidated, it may become possible to design even better redox reactive reagents of clinical value.

To this end, redox modulatory sites of NMDA receptors have begun to be characterized at a molecular level using site-directed mutagenesis of recombinant subunits (NMDAR1, NMDAR2A-D). Two types of redox modulation can be distinguished. The first type gives rise to a persistent change in the functional activity of the receptor, and we have identified two cysteine residues on the NMDAR1 subunit (#744 and #798) that are responsible for this action. A second site, presumably also a cysteine(s) because $\leq$1 mM N-ethylmaleimide can block its effect in native neurons, underlies the other, more transient redox action. It appears to be at this, as yet unidentified, site on the NMDA receptor that the NO group acts, at least in recombinant receptors.

INTRODUCTION

Approximately a third of adults and half of children with the acquired immunodeficiency syndrome (AIDS) eventually suffer from neurological manifestations, including dysfunction of cognition, movement, and sensation, that are a direct consequence of HIV-1 infection of brain.[1,2] These neurological problems occur even in the absence of superinfection with opportunistic pathogens or secondary malignancies.[3] Clinical manifestations include difficulty with mental concentration and slowness of hand movements and gait. This malady was initially termed the AIDS dementia complex by Price and colleagues,[3] but more recently has been placed under the rubric HIV-1-associated cognitive/motor complex. Pathologically, HIV-1 infection in the CNS (or HIV encephalitis) is characterized by widespread reactive astrocytosis, myelin pallor, and infiltration by monocytoid cells, including blood-derived macrophages, resident microglia and multinucleated giant cells.[4] In addition, subsets of neurons display a striking degree of injury, including dendritic pruning and simplification of synaptic contacts, as well as frank cell loss, which may herald the onset of cognitive and motor deficits in affected individuals,[5-9] but for another view also see Ref. 10. Such neuronal injury may result in reversible dysfunction rather than inevitable demise. Unlike any other encephalitis studied in recent times, the progressive clinical sequelae occur without direct infection of neurons by HIV-1 or significant autoimmunity induced by virus. Surprisingly, mononuclear phagocytes (brain macrophages, microglia and multinucleated giant cells) in the CNS represent the cell type that is predominantly infected.[11] Although infection can occur in astrocytes, it is highly restricted, and perhaps limited to the pediatric population.[12,13] An emerging body of evidence strongly supports the idea that activated HIV-1 infected brain macrophages secrete neurotoxins that are largely responsible for the pathological alterations of brain tissue seen following viral infection. These toxins may include HIV-1 proteins (gp120, *tat*, *nef* and possibly others) and cell-encoded substances (glutamate-like neurotoxic molecules, free radicals, cytokines, and eicosanoids).[14-25] The mechanism underlying this indirect form of neuronal injury is related to excessive influx of Ca^{2+} into neurons in response to the noxious factors released from immune activated HIV-infected or gp120-stimulated brain macrophages/microglia.[15,26,27]

BRAIN MACROPHAGE-MEDIATED NEURONAL INJURY: TOXIC SUBSTANCES RELEASED AFTER HIV INFECTION OR gp120 STIMULATION

Production of Cytokines, Quinolinate, Cysteine, PAF, and Free Radicals

A paradox exists between the small numbers of productively HIV-infected brain macrophages/microglia and the severe clinical cognitive and motor deficits that patients with the acquired immunodeficiency syndrome (AIDS) experience. This suggests that some sort of cellular amplification and/or activation is necessary for the generation of viral and cellular toxins that lead to tissue injury and sustained viral infection. Indeed, there is ample evidence for diffuse CNS immune-related activation in HIV-1 associated neurological impairments.[28-30] Moreover, the secretion of neurotoxins by HIV-1 infected macrophages is likely to be regulated by a complex series of intracellular interactions between several different brain cell types including mononuclear phagocytes, astrocytes and neurons.[27] HIV-infected brain mononuclear phagocytes, especially after immune activation, secrete substances that contribute to neurotoxicity (Fig. 1).[19,31-33] These include, but are not limited to, eicosanoids, e.g., arachidonic acid, its metabolites and platelet-activating factor (PAF), proinflammatory cytokines (tumor necrosis factor-alpha [TNF-α] and interleukin-1beta [IL-1β]), free radicals such as nitric oxide (NO·) and superoxide anion ($O_2{\cdot}^-$), and the glutamate-

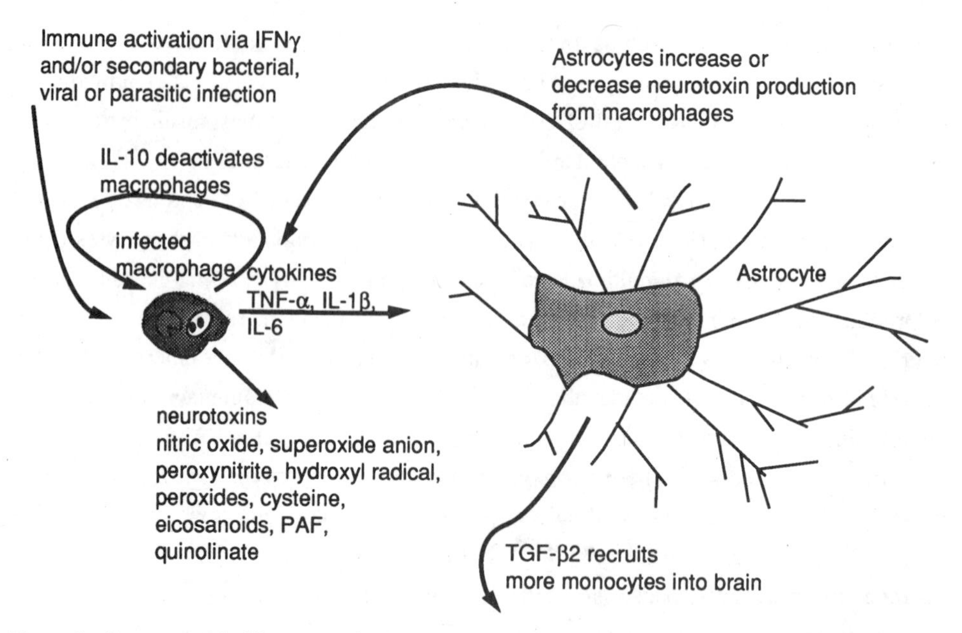

Figure 1. Current model of immune activation of HIV-infected brain macrophages/microglia to enhance the release of potentially neurotoxic substances. Astrocytes can modulate up or down the production of these substances by macrophages depending on their state of immune activation. See text for explanation.

like agonist, cysteine.[19,23-25] Similarly, macrophages activated by HIV-1 envelope protein gp120 release arachidonic acid and its metabolites, TNF-α, IL-1β, and cysteine.[17,25] Some eicosanoids and free radicals can lead to increased release or decreased re-uptake of glutamate, which can contribute to neuronal damage.[22] PAF also elicits neuronal death in in vitro systems by a mechanism probably involving increased neuronal Ca^{2+} and the release of glutamate.[23] TNF-α and IL-1β stimulate astrocytosis.[34] Chronic immune stimulation of the brain, with widespread CNS (microglial and astroglial) activation, can result from interferon-gamma (IFN-γ) production.[28] This immune activation continues the process of neuronal injury initiated by HIV infection and its protein product, gp120. IFN-γ induces production of macrophage PAF[35] and quinolinate, a tryptophan metabolit found in high concentrations in the cerebrospinal fluid of HIV-infected patients with dementia; quinolinate can also act as an glutamate-like agonist to injure neurons.[36] Cytokines participate in this cellular network in several additional ways. TNF-α may also increase voltage-dependent calcium currents in neurons.[37] In conjunction with IL-1β, IFN-γ can induce nitric oxide synthase (NOS) expression with consequent NO· (nitric oxide) production in cultured astrocytes.[38] Importantly, most of these factors (cytokines, quinolinate, PAF and products of arachidonic acid metabolism) have been shown to be elevated in brain and/or cerebrospinal fluid of AIDS patients with clinical neurological deficits including dementia.[18,23,28,30]

A final common pathway for neuronal susceptibility appears to be operative, similar to that observed in stroke, trauma, epilepsy, neuropathic pain and several neurodegenerative diseases, possibly including Huntington's disease, Parkinson's disease, and amyotrophic lateral sclerosis. This mechanism involves overactivation of voltage-dependent Ca^{2+} channels and *N*-methyl-D-aspartate (NMDA) receptor-operated channels, which are also permeable to Ca^{2+}, with resultant generation of free radicals.[22] Ultimately, the macrophage-synthesized toxins lead to overstimulation of this receptor and increased levels of neuronal Ca^{2+}, with consequent release of glutamate. In turn glutamate overexcites neighboring neurons leading to further increases in intracellular Ca^{2+}, neuronal injury, and thus more glutamate release (Fig. 2). For many neurons, this pathway to toxicity can be blocked by antagonists of the NMDA receptor.[31,39,40] For some neurons, this form of damage can also be ameliorated to some degree by calcium channel antagonists or non-NMDA receptor antagonists, perhaps depending on the repertoire of ion channel types in a specific population of neurons.[41] Thus, the elucidation of HIV-1-induced neurotoxins and their mechanism of action(s) offers hope for future pharmacological intervention.[27,42]

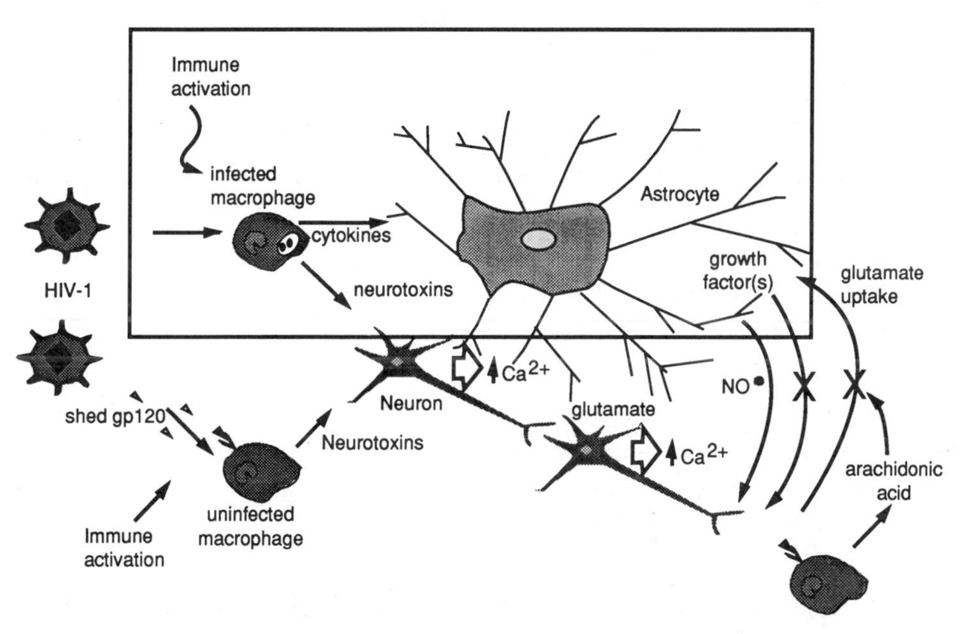

Figure 2. Current model of neuronal injury by HIV-infected or gp120-stimulated brain macrophages/microglia. To simplify the diagram, the contribution of the glial elements (HIV-infected macrophages and astrocytes) from Fig. 1 are represented in the box. Substances from immune-activated macrophages and possibly from reactive astrocytes contribute to neuronal injury and astrocytosis. Neuronal injury is primarily mediated by overactivation of NMDA receptors with a resultant increase in intracellular Ca^{2+} levels. This in turn leads to overactivation of a variety of potentially harmful enzyme systems and release of the neurotransmitter, glutamate. Glutamate then overstimulates NMDA receptors on neighboring neurons, resulting in further injury. See text for additional details of the model.

HIV-1 COAT PROTEIN gp120 AND NEURONAL INJURY

gp120 Inhibits Glutamate Re-uptake into Astrocytes via Arachidonic Acid

The coat protein gp120 is thought to exert its predominant deleterious effect on neurons indirectly, via the induction of macrophage toxins, some of which have been elaborated above.[43,44] However, direct effects on neurons have not been entirely ruled out. In addition, gp120 may have direct or indirect effects on astrocytes, e.g., to decrease growth factor production[45] or to inhibit glutamate re-uptake (Fig. 2).[46] Recently we found that some of these effects may be mediated by arachidonate production of gp120-stimulated macrophages. Arachidonic acid had previously been shown to inhibit re-uptake of excitatory amino acids, such as glutamate, by astrocytes and nerve ending preparations (synaptosomes).[47,48] Since gp120-stimulated macrophages, similar to HIV-infected macrophages, had been shown to produce arachidonate and its metabolites,[17,19] it seemed plausible that this arachidonate might affect the re-uptake of excitatory amino acids. In fact, picomolar concentrations of gp120 inhibit excitatory amino acid uptake into cultured astrocytes.[16] Purified neonatal rat "astrocyte cultures" also contain approximately 5% monocytoid cells. After depletion of the monocytic cells from these cultures, gp120 no longer inhibits the uptake of excitatory amino acids into astrocytes. This finding implies that this effect of gp120 is predominantly mediated via the monocytic cells in the culture.[16]

One mechanism for the generation of free arachidonic acid is via the activity of the enzyme phospholipase A_2. Inhibitors of phospholipase A_2, such as aristolochic acid and quinacrine, prevent the decreased re-uptake of excitatory amino acid into astrocytes engendered by gp120. Taken together, these results suggest that a gp120-induced increase in arachidonate production by macrophages is responsible for the inhibition of glutamate uptake into astrocytes.[16] Moreover, this mechanism may account at least in part for the observation that gp120 enhances glutamate neurotoxicity in culture.[40,49] Arachidonate has also been recently reported to enhance NMDA-evoked currents in neurons.[50] Therefore arachidonate could contribute to neurotoxicity not only by inhibiting the re-uptake of glutamate but also by increasing the effectiveness of glutamate in evoking NMDA receptor-mediated responses.

gp120-Stimulated Human Macrophages Release the NMDA Agonist Cysteine

Recently, we have shown that gp120 activation or frank HIV infection enhances cysteine secretion from human macrophages [25]. Cysteine is a known NMDA agonist[51,52] and could therefore represent at least one of the neurotoxic substances released from gp120-stimulated or HIV-infected macrophages. The release of cysteine by gp120-activated macrophages is greatly attenuated by TNF-α blocking antibodies. Thus, it appears that macrophage secretion of cysteine is mediated by gp120-stimulated production of TNF-α

(Yeh and Lipton, manuscript in preparation). These results may serve to link, at least in part, the previously recognized pattern of TNF-α elevation in the brains of patients with AIDS dementia.[28,29]

Undoubtedly, other noxious substances are released by HIV-infected or gp120-activated macrophages, and the identity of these potential neurotoxins are currently under intensive investigation.[44] However, the common denominator is that this form of neuronal injury appears to be mediated by excessive increases in intracellular Ca^{2+} levels and can be largely inhibited by NMDA antagonists.[15,31,39,40,44,49,53] For these reasons, studies in AIDS patients with dementia of a clinically-tolerated NMDA antagonist as an adjunctive agent to anti-retroviral therapy has been considered prudent by the AIDS Clinical Trials Group (ACTG) of the NIH. In addition, under some conditions, antagonists of voltage-dependent calcium channels (another mode of entry of Ca^{2+} into neurons) can alleviate neuronal damage due to excessive activation of NMDA receptors, occurring, for example, with exposure to gp120.[15,54] Along these lines, a phase 1-2 clinical trial for AIDS dementia with the calcium channel antagonist nimodipine was recently completed by the ACTG, and the data indicate that the drug is safe in this population of patients and that a larger trial needs to be performed to assess the possibility of efficacy.

POTENTIAL CLINICAL UTILITY OF NMDA ANTAGONISTS FOR AIDS DEMENTIA: OPEN-CHANNEL BLOCKERS AND REDOX CONGENERS OF NITRIC OXIDE

As detailed elsewhere,[22,55] many NMDA antagonists are not clinically tolerated, while some others appear to be tolerated by humans at concentrations that are effective neuroprotectants. Several NMDA antagonists have been found to prevent neuronal injury associated with HIV-infected macrophages, gp120, PAF, cysteine, or quinolinate.[27,31,39,40,44,49,53,56-59] Among these, two of the most promising, because of their long experience in patients with other diseases, are memantine and nitroglycerin (Fig. 3). Memantine blocks the NMDA receptor-associated ion channel only when it is open. Unlike other NMDA open-channel blockers, such as dizocilpine (MK-801), memantine does not remain in the channel for an excessively long time, and this kinetic parameter correlates with its safe use in humans for over a dozen years in Europe as a treatment for Parkinson's disease.[60] Increasing concentrations of glutamate or other NMDA agonists cause NMDA channels to remain open on average for a greater fraction of time. Under such conditions, an open-channel blocking drug such as memantine has a better chance to enter the channel and block it. Because of this mechanism of action, the untoward effects of greater (pathologic) concentrations of glutamate are prevented to a greater extent than the effects of lower (physiologic) concentrations.[22] Moreover, in model systems memantine can ameliorate gp120-associated neuronal injury.[56,57,61]

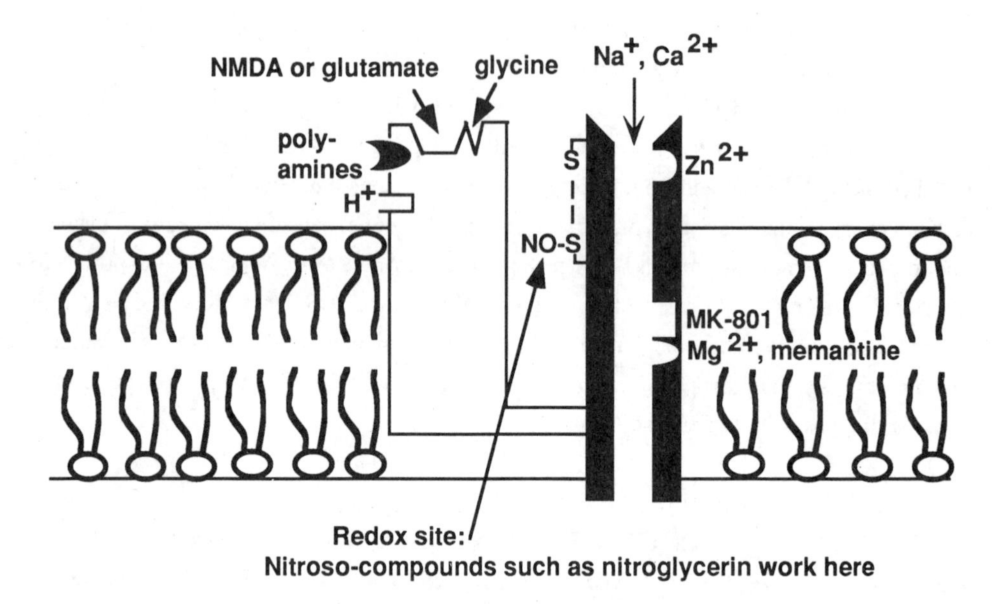

Figure 3. Sites of potential antagonist action on the NMDA receptor-channel complex. Competitive antagonists such as CGS-19755 can compete with NMDA or glutamate for binding to the agonist site [63]. Several antagonists to the glycine co-agonist site have been described that are chlorinated and sulfated derivatives of kynurenic acid.[64-67] It is not yet definitively known if any will prove to be tolerated clinically at concentrations effective in combating neurotoxicity. H^{+} effects are transmitted through a noncompetitive site; decreasing pH acts to downregulate channel activity.[68-70] Other modulatory sites for polyamines and Zn^{2+} can also be used to affect receptor-channel function.[71-76] Sites that inhibit channel activity by binding Mg^{2+} or drugs such as MK-801, phencyclidine, and memantine are within the electric field of the channel and are only exposed when the channel is previously opened by agonist (termed uncompetitive antagonism).[60,77-82] A redox modulatory site, comprised or several thiol groups from our site-directed mutagenesis work on recombinant subunits,[83] is affected by chemical reducing and oxidizing agents. Oxidation may possibly favor disulfide bond formation (S-S) over free thiol (-SH) groups and thus downregulate channel activity.[84,85] At least one of the NMDA receptor's redox modulatory sites can also be downregulated by NO group transfer (*S*-nitrosylation with NO^{+} equivalents to form RS-NO), which may facilitate disulfide bond formation (represented by the dashed line).[52] This reaction, which leads to less NMDA-evoked Ca^{2+} influx and thus neuroprotection, involves NMDA receptor thiol because it is blocked by specific sulfhydryl alkylating agents such as *N*-ethylmaleimide (NEM).[62]

As discussed above, nitric oxide (NO·, where the dot represents one unpaired electron in the outer molecular orbital) can contribute to neuronal damage, and one of these pathways to neurotoxicity involves the reaction of NO· with $O_2^{\cdot -}$ to form peroxynitrite ($ONOO^-$). In contrast, NO· can be converted to a chemical state that has just the opposite effect, i.e., that protects neurons from injury due to NMDA receptor-mediated overstimulation. The change in chemical state is dependent on the removal or addition of an electron to NO·. This change in the chemical redox state can be influenced by the presence or absence of electron donors such as ascorbate and cysteine. With one less electron, NO· becomes nitrosonium ion (NO^+), which can react with critical thiol group(s) comprising a redox modulatory site on the NMDA receptor-channel complex to decrease channel activity. This reaction can afford neuronal protection from overstimulation of NMDA receptors which would otherwise result in excessive Ca^{2+} influx.[52] One such drug that can react with NMDA receptors in a manner resembling nitrosonium is the common vasodilator nitroglycerin.[22,52,55,62] Chronic use of nitroglycerin induces tolerance to its effects on the cardiovascular system, but the drug still appears to work in the brain to attenuate NMDA receptor-mediated neurotoxicity.[55] Nonetheless, the exact dosing regimen has yet to be worked out for the neuroprotective effects of nitroglycerin in the brain; therefore, caution has to be exercised before attempting to implement this form of therapy.

In the coming months, as these and other clinically-tolerated NMDA antagonists are tested in clinical studies in an attempt to ameliorate AIDS dementia, we hope to be able to offer our patients better adjunctive therapy to their anti-retroviral medicines to treat the cognitive and other neurologic manifestations of AIDS.

Acknowledgments

I am indebted to the members of my laboratory who made this work possible. Research detailed in this review was supported in part by grants from the National Institutes of Health, the American Foundation for AIDS Research, and the Pediatric AIDS Foundation.

REFERENCES

1. H. Bacellar, A. Muñoz, E.N. Miller, B.A. Cohen, D. Besley, O.A. Selnes, J.T. Becker, and J.C. McArthur, Temporal trends in the incidence of HIV-1-related neurologic diseases: multicenter AIDS cohort study, 1985-1992, *Neurology* 44:1892-1900 (1994).
2. S.A. Lipton, Neurobiology: HIV displays its coat of arms, *Nature* 367:113-114 (1994).

3. R.W. Price, B. Brew, J. Sidtis, M. Rosenblum, A.C. Scheck, and P. Clearly, The brain and AIDS: central nervous system HIV-1 infection and AIDS dementia complex, *Science* 239:586-92 (1988).
4. H. Budka, Neuropathology of human immunodeficiency virus infection, *Brain Pathology* 1:163-175 (1991).
5. S. Ketzler, S. Weis, H. Haug, and H. Budka, Loss of neurons in frontal cortex in AIDS brains, *Acta Neuropathol (Berlin)* 80:90-92 (1990).
6. I.P. Everall, P.J. Luthbert, and P.L. Lantos, Neuronal loss in the frontal cortex in HIV infection, *Lancet* 337:1119-21 (1991).
7. C.A. Wiley, E. Masliah, M. Morey, C. Lemere, R.M. DeTeresa, M.R. Grafe, L.A. Hansen, and R.D. Terry, Neocortical damage during HIV infection, *Ann Neurol* 29:651-7 (1991).
8. E. Masliah, C.L. Achim, N. Ge, R. DeTeresa, R.D. Terry, and C.A. Wiley, Spectrum of human immunodeficiency virus-associated neocortical damage, *Ann Neurol* 32:321-9 (1992).
9. W.N. Tenhula, S.Z. Xu, M.C. Madigan, K. Heller, W.R. Freeman, and A.A. Sadun, Morphometric comparisons of optic nerve axon loss in acquired immunodeficiency syndrome, *Am J Ophthalmol* 113:14-20 (1992).
10. D. Seilhean, C. Duyckaerts, R. Vazuex, F. Bolgert, P. Brunet, C. Katlama, M. Gentilini, and J.-J. Hauw, HIV-1-associated cognitive/motor complex: Absence of neuronal loss in the cerebral neocortex, *Neurology* 43:1492-1499 (1993).
11. S. Koenig, H.E. Gendelman, J.M. Orenstein, M.C. Dal Canto, G.H. Pozeshkpour, M. Yungbluth, F. Janotta, A. Kasmit, M.A. Martin, and A.S. Fauci, Detection of AIDS virus in macrophages in brain tissue from AIDS patients with encephalopathy, *Science* 233:1089-1093 (1986).
12. T. Saito, L.R. Sarer, L.G. Epstein, J. Michales, M. Mintz, M. Louder, K. Goldring, T.A. Cvetkovich, and B.M. Blumberg, Overexpression of *nef* as a marker for restricted HIV-1 infection of astrocytes in postmortem pediatric central nervous tissues, *Neurology* 44:474-481 (1994).
13. C. Tornatore, R. Chandra, J.R. Berger, and E.O. Major, HIV-1 infection of subcortical astrocytes in the pediatric central nervous system, *Neurology* 44:481-487 (1994).
14. D.E. Brenneman, G.L. Westbrook, S.P. Fitzgerald, D.L. Ennist, K.L. Elkins, M. Ruff, and C.B. Pert, Neuronal cell killing by the envelope protein of HIV and its prevention by vasoactive intestinal peptide, *Nature* 335:639-42 (1988).
15. E.B. Dreyer, P.K. Kaiser, J.T. Offermann, and S.A. Lipton, HIV-1 coat protein neurotoxicity prevented by calcium channel antagonists, Science 248:364-7 (1990).
16. E.B. Dreyer andS.A. Lipton, Toxic neuronal effects of the HIV coat protein gp120 may be mediated through macrophage arachidonic acid, *Soc Neurosci Abstr* 20:1049 (1994).

17. L.M. Wahl, M.L. Corcoran, S.W. Pyle, L.O. Arthur, A. Harel-Bellan, and W.L. Farrar, Human immunodeficiency virus glycoprotein (gp120) induction of monocyte arachidonic acid metabolites and interleukin 1, *Proc Natl Acad Sci USA* 86:621-5 (1989).
18. M.P. Heyes, B.J. Brew, A. Martin, R.W. Price, A.M. Salazqr, J.J. Sidtis, J.A. Yergey, M.M. Mouradian, A. Sadler, J. Keilp, D. Rubinow, and S.P. Markey, Quinolinic acid in cerebrospinal fluid and serum in HIV-1 infection: relationship to clinical and neurological status, *Ann Neurol* 29:202-9 (1991).
19. P. Genis, M. Jett, E.W. Bernton, T. Boyle, H.A. Gelbard, K. Dzenko, R.W. Keane, L. Resnick, T. Mizrachi, D.J. Volsky, L.G. Epstein, and H.E. Gendelman, Cytokines and arachidonic acid metabolites produced during human immunodeficiency virus (HIV)-infected macrophage-astroglia interactions: implications for the neuropathogenesis of HIV disease, *J Exp Med* 176:1703-18 (1992).
20. M. Hayman, G. Arbuthnott, G. Harkiss, H. Brace, P. Filippi, V. Philippon, D. Thompson, R. Vigne, and A. Wright, Neurotoxicity of peptide analogies of the transactivating protein tat from Maedi-Visna virus and human immunodeficiency virus, *Neuroscience* 53:1-6 (1993).
21. S.M. Toggas, E. Masliah, E.M. Rockenstein, G.F. Rall, C.R. Abraham, and L. Mucke, Central nervous system damage produced by expression of the HIV-1 coat protein gp120 in transgenic mice, *Nature* 367:188-193 (1994).
22. S.A. Lipton andP.A. Rosenberg, Excitatory amino acids as a final common pathway for neurologic disorders, *N Engl J Med* 330:613-622 (1994).
23. H.A. Gelbard, H.S.L.M. Nottet, S. Swindells, M. Jett, K.A. Dzenko, P. Genis, R. White, L. Wang, Y.-B. Choi, D. Zhang, S.A. Lipton, W.W. Tourtellotte, L.G. Epstein, and H.E. Gendelman, Platelet-activating factor: a candidate human immunodeficiency virus type 1-induced neurotoxin, *J Virol* 68:4628-4635 (1994).
24. M.I. Bukrinsky, H.S.L.M. Nottet, H. Schmidtmayerova, L. Dubrovsky, C.R. Flanagan, M.E. Mullins, S.A. Lipton, and H.E. Gendelman, Regulation of nitric oxide synthase activity in human immunodeficiency virus type 1 (HIV-1)-infected monocytes: Implications for HIV-associated neurological disease, *J Exp Med* 181:735-745 (1995).
25. M.W. Yeh, H.L.M. Nottet, H.E. Gendelman, and S.A. Lipton, HIV-1 coat protein gp120 stimulates human macrophages to release L-cysteine, an NMDA agonist, *Soc Neurosci Abstr* 20:451 (1994).
26. T.-M. Lo, C.J. Fallert, T.M. Piser, and S.A. Thayer, HIV-1 envelope protein evokes intracellular calcium oscillations in rat hippocampal neurons, *Brain Res* 594:189-196 (1992).

27. S.A. Lipton, Models of neuronal injury in AIDS: another role for the NMDA receptor?, *Trends Neurosci* 15:75-79 (1992).
28. W.R. Tyor, J.D. Glass, J.W. Griffin, S. Becker, J.C. McArthur, L. Bezman, and D.E. Griffin, Cytokine expression in the brain during the acquired immunodeficiency syndrome, *Ann Neurol* 31:349-60 (1992).
29. S.L. Wesselingh, C. Power, J.D. Glass, W.R. Tyor, J.C. McArthur, J.M. Farber, J.W. Griffin, and D.E. Griffin, Intracerebral cytokine messenger RNA expression in acquired immunodeficiency syndrome dementia, *Ann Neurol* 33:576-582 (1993).
30. D.E. Griffin, S.L. Wesselingh, and J.C. McArthur, Elevated central nervous system protaglandins in human immunodeficiency virus-associated dementia, *Ann Neurol* 35:592-597 (1994).
31. D. Giulian, K. Vaca, and C.A. Noonan, Secretion of neurotoxins by mononuclear phagocytes infected with HIV- 1, *Science* 250:1593-6 (1990).
32. L. Pulliam, B.G. Herndler, N.M. Tang, and M.S. McGrath, Human immunodeficiency virus-infected macrophages produce soluble factors that cause histological and neurochemical alterations in cultured human brains, *J Clin Invest* 87:503-12 (1991).
33. H.S.L.M. Nottet, M. Jett, Q.-H. Zhai, C.R. Flanagan, A. Rizzino, P. Genis, T. Baldwin, J. Schwartz, and H.E. Gendelman, A regulatory role for astrocytes in HIV-encephalitis: an overexpression of eicosanoids, platelet-activating factor and tumor necrosis factor-α by activated HIV-1-infected monocytes is attenuated by primary human astrocytes, *J Immunol*, in press.
34. K.N. Selmaj, M. Farooq, T. Norton, C.S. Raine, and C.F. Brosman, Proliferation of astrocytes in vitro in response to cytokines, *J Immunol* 144:129-35 (1990).
35. F.H. Valone andL.B. Epstein, Biphasic platelet-activating factor synthesis by human monocytes stimulated with IL-β, tumor necrosis factor, or IFN-γ, *J Immunol* 141:3945-50 (1988).
36. M.P. Heyes, K. Saito, and S.P. Markey, Human macrophages convert L-tryptophan to the neurotoxin quinolinic acid, *Biochem J* 283:633-5 (1992).
37. B. Soliven andJ. Albert, Tumor necrosis factor modulates Ca^{2+} currents in cultured sympathetic neurons, *J Neurosci* 12:2665-71 (1992).
38. M.L. Simmons andS. Murphy, Cytokines regulate L-arginine-dependent cyclic GMP production in rat glial cells, *Eur J Neurosci* 5:825-831 (1993).
39. S.A. Lipton, P.K. Kaiser, N.J. Sucher, E.B. Dreyer, and J.T. Offermann, AIDS virus coat protein sensitizes neurons to NMDA receptor-mediated toxicity, *Soc Neurosci Abstr* 16:289 (1990).
40. S.A. Lipton, N.J. Sucher, P.K. Kaiser, and E.B. Dreyer, Synergistic effects of HIV coat protein and NMDA receptor-mediated neurotoxicity, Neuron 7:111-8 (1991).

41. S.A. Lipton, Calcium channel antagonists in the prevention of neurotoxicity, *Adv Pharmacol* 22:271-91 (1991).
42. S.A. Lipton andH.E. Gendelman, The dementia associated with the acquired immunodeficiency syndrome, *N Engl J Med* 332:934-40 (1995).
43. S.A. Lipton, Requirement for macrophages in neuronal injury induced by HIV envelope protein gp120, *NeuroReport* 3:913-5 (1992).
44. D. Giulian, E. Wendt, K. Vaca, and C.A. Noonan, The envelope glycoprotein of human immunodeficiency virus type 1 stimulates release of neurotoxins from monocytes, *Proc Natl Acad Sci USA* 90:2769-2773 (1993).
45. D.E. Brenneman, T. Nicol, D. Warren, and L.M. Bowers, Vasoactive intestinal peptide: a neurotrophic releasing agent and an astroglial mitogen, *J Neurosci Res* 25:386-94 (1990).
46. D.J. Benos, B.H. Hahn, J.K. Bubien, S.K. Ghosh, N.A. Mashburn, M.A. Chaikin, G.M. Shaw, and E.N. Benveniste, Envelope glycoprotein gp120 of human immunodeficiency virus type 1 alters ion transport in astrocytes: Implications for AIDS dementia complex, *Proc Natl Acad Sci USA* 91:494-498 (1994).
47. B. Barbour, M. Szatkowski, N. Ingledew, and D. Attwell, Arachidonic acid induces a prolonged inhibition of glutamate uptake into glial cells, *Nature* 342:918-20 (1989).
48. A. Volterra, D. Trotti, P. Cassutti, C. Tromba, A. Salvaggio, R.C. Melcangi, and G. Racagni, High sensitivity of glutamate uptake to extracellular free arachidonic acid levels in rat cortical synaptosomes and astrocytes, *J Neurochem* 59:600-6 (1992).
49. V.L. Dawson, T.M. Dawson, G.R. Uhl, and S.H. Snyder, Human immunodeficiency virus-1 coat protein neurotoxicity mediated by nitric oxide in primary cortical cultures, *Proc Natl Acad Sci USA* 90:3256-3259 (1993).
50. B. Miller, M. Sarantis, S.F. Traynelis, and D. Attwell, Potentiation of NMDA receptor currents by arachidonic acid, *Nature* 355:722-5 (1992).
51. J.W. Olney, C. Zorumski, M.T. Price, and J. Labryuere, L-cysteine, a bicarbonate-sensitive endogenous excitotoxin, *Science* 248:596-9 (1990).
52. S.A. Lipton, Y.-B. Choi, Z.-H. Pan, S.Z. Lei, H.-S.V. Chen, N.J. Sucher, J. Loscalzo, D.J. Singel, and J.S. Stamler, A redox-based mechanism for the neuroprotective and neurodestructive effects of nitric oxide and related nitroso-compounds, *Nature* 364:626-632 (1993).
53. T. Savio andG. Levi, Neurotoxicity of HIV coat protein gp120, NMDA receptors, and protein kinase C: a study with rat cerebellar granule cell cultures, *J Neurosci Res* 34:265-272 (1993).
54. S.Z. Lei, D. Zhang, A.E. Abele, and S.A. Lipton, Blockade of NMDA receptor-mediated mobilization of intracellular Ca^{2+} prevents neurotoxicity, *Brain Res* 598:196-202 (1992).

55. S.A. Lipton, Prospects for clinically tolerated NMDA antagonists: open-channel blockers and alternative redox states of nitric oxide, *Trends Neurosci* 16:527-532 (1993).

56. S.A. Lipton, Memantine prevents HIV coat protein-induced neuronal injury *in vitro*, *Neurology* 42:1403-5 (1992).

57. W.E.G. Müller, H.C. Schröder, H. Ushijima, J. Dapper, and J. Bormann, gp120 of HIV-1 induces apoptosis in rat cortical cell cultures: prevention by memantine, *Eur J Pharmacol - Molec Pharm Sect* 226:209-214 (1992).

58. S.A. Lipton, 7-Chlorokynurenate ameliorates neuronal injury mediated by HIV envelope protein gp120 in retinal cultures, *Eur J Neurosci* 4:1411-1415 (1992).

59. S.A. Lipton, Human immunodeficiency virus-infected macrophages, gp120, and *N*-methyl-D-aspartate receptor-mediated neurotoxicity, *Ann Neurol* 33:227-228 (1993).

60. H.-S.V. Chen, J.W. Pellegrini, S.K. Aggarwal, S.Z. Lei, S. Warach, F.E. Jensen, and S.A. Lipton, Open-channel block of NMDA responses by memantine: therapeutic advantage against NMDA receptor-mediated neurotoxicity, *J Neurosci* 12:4427-36 (1992).

61. S.A. Lipton andF.E. Jensen, Memantine, a clinically-tolerated NMDA open-channel blocker, prevents HIV coat protein-induced neuronal injury in vitro and in vivo, *Soc Neurosci Abstr* 18:757 (1992).

62. S.Z. Lei, Z.-H. Pan, S.K. Aggarwal, H.-S.V. Chen, J. Hartman, N.J. Sucher, and S.A. Lipton, Effect of nitric oxide production on the redox modulatory site of the NMDA receptor-channel complex, *Neuron* 8:1087-99 (1992).

63. D.E. Murphy, A.J. Hutchison, S.D. Hurt, M. Williams, and M.A. Sills, Characterization of the binding of [^{3}H]CGS-19755: a novel *N*-methyl-D-aspartate antagonist with nanomolar affinity in rat brain, *Br J Pharmacol* 95:932-938 (1988).

64. J.W. Johnson andP. Ascher, Glycine potentiates the NMDA response in cultured mouse brain neurons, *Nature* 325:529-31 (1987).

65. N.W. Kleckner andR. Dingledine, Requirement for glycine in activation of NMDA-receptors expressed in Xenopus oocytes, *Science* 241:835-7 (1988).

66. J.A. Kemp, A.C. Foster, P.D. Leeson, T. Priestley, R. Tridgett, L.L. Iversen, and G.N. Woodruff, 7-Chlorokynurenic acid is a selective antagonist at the glycine modulatory site of the N-methyl-D-aspartate receptor complex, *Proc Natl Acad Sci USA* 85:6547-50 (1988).

67. F. Moroni, A. Marina, L. Facci, E. Fadda, S.D. Skaper, A. Galli, G. Lombardi, F. Mori, M. Ciuffi, B. Natalini, and R. Pellicciari, Thiokynurenates prevent excitotoxic neuronal death in vitro and in vivo by acting as glycine antagonists and as inhibitors of lipid peroxidation, *Eur J Pharmacol* 218:145-151 (1992).

68. C. Tang, M. Dichter, and M. Morad, Modulation of the *N*-methyl-D-asparate channel by extracellular H^{+}, *Proc Natl Acad Sci USA* 87:6445-6449 (1990).

69. S.F. Traynelis andS.G. Cull-Candy, Proton inhibition of *N*-methyl-D-aspartate receptors in cerebellar neurons, *Nature* 345:347-350 (1990).
70. L.J. Vyklicky, V. Vlachova, and J. Krusek, The effect of external pH changes on responses to excitatory amino acids in mouse hippocampal neurones, *J Physiol (Lond)* 430:497-517 (1990).
71. R.W. Ransom andN.L. Stec, Cooperative modulation of [^{3}H]MK-801 binding to the N-methyl-D-aspartate receptor-ion channel complex by L-glutamate, glycine, and polyamines, *J Neurochem* 51:830-6 (1988).
72. J.F. McGurk, M.V. Bennett, and R.S. Zukin, Polyamines potentiate responses of N-methyl-D-aspartate receptors expressed in xenopus oocytes, *Proc Natl Acad Sci USA* 87:9971-4 (1990).
73. K. Williams, V.L. Dawson, C. Romano, M.A. Dichter, and P.B. Molinoff, Characterization of polyamines having agonist, antagonist, and inverse agonist effects at the polyamine recognition site of the NMDA receptor, *Neuron* 5:199-208 (1990).
74. K. Williams, C. Romano, M.A. Dichter, and P.B. Molinoff, Modulation of the NMDA receptor by polyamines, *Life Sci* 48:469-98 (1991).
75. G.L. Westbrook andM.L. Mayer, Micromolar concentrations of Zn^{2+} antagonize NMDA and GABA responses of hippocampal neurons, *Nature* 328:640-3 (1987).
76. S. Peters, J. Koh, and D.W. Choi, Zinc selectively blocks the action of N-methyl-D-aspartate on cortical neurons, *Science* 236:589-593 (1987).
77. B. Ault, R.H. Evans, A.A. Francis, D.J. Oakes, and J.C. Watkins, Selective depression of excitatory amino acid induced depolarization by magnesium ions in isolated spinal cord preparations, *J Neurophysiol* 32:424-442 (1980).
78. L. Nowak, P. Bregestovski, P. Ascher, A. Herbert, and A. Prochiantz, Magnesium gates glutamate-activated channels in mouse central neurones, *Nature* 307:462-5 (1984).
79. M.L. Mayer, G.L. Westbrook, and P.B. Guthrie, Voltage-dependent block by Mg^{2+} of NMDA responses in spinal cord neurones, *Nature* 309:261-3 (1984).
80. J.F. MacDonald, Z. Miljkovic, and P. Pennefather, Use-dependent block of excitatory amino acid currents in cultured neurons by ketamine, *J Neurophysiol* 58:251-66 (1987).
81. J.E. Huettner andB.P. Bean, Block of N-methyl-D-aspartate-activated current by the anticonvulsant MK-801: selective binding to open channels, *Proc Natl Acad Sci USA* 85:1307-1311 (1988).
82. J. Bormann, Memantine is a potent blocker of N-methyl-D-aspartate (NMDA) receptor channels, *Eur J Pharmacol* 166:591-2 (1989).
83. J.M. Sullivan, S.F. Traynelis, H.-S.V. Chen, W. Escobar, S.F. Heinemann, and S.A. Lipton, Identification of two cysteine residues that are required for redox modulation of the NMDA subtype of glutamate receptor, *Neuron* 13:929-936 (1994).

84. E. Aizenman, S.A. Lipton, and R.H. Loring, Selective modulation of NMDA responses by reduction and oxidation, *Neuron* 2:1257-63 (1989).

85. J.W. Lazarewicz, J.T. Wroblewski, M.E. Palmer, and E. Costa, Reduction of disulfide bonds activates NMDA sensitive glutmate receptors in primary cultures of cerebellar granule cells, *Neurosci Res Commun* 4:91-97 (1989).

CALCIUM, AMPA/KAINATE RECEPTORS, AND SELECTIVE NEURODEGENERATION

John H. Weiss, Hong Z. Yin, Sean G. Carriedo and You M. Lu

Department of Neurology
University of California, Irvine
Irvine, CA 92717-4290

INTRODUCTION

Neurotoxic effects of the excitatory amino acid (EAA) glutamate play an important role in the pathogenesis of acute diseases of the nervous system such as stroke;[1] additionally, several lines of evidence lend support to a possibility that slower EAA toxicity may contribute to neurodegeneration associated with certain neurodegenerative disease.[2-5] The purpose of this chapter is twofold. First, I will review several lines of evidence supporting the idea that activation of AMPA/kainate-type glutamate receptors might play a particular role in the selective pattern of neurodegeneration seen in Alzheimer's disease (AD) and amyotrophic lateral sclerosis (ALS). Secondly, I will discuss certain recent results bearing on the role of Ca^{2+} ions in selective AMPA/kainate receptor-mediated degeneration.

AMPA/KAINATE RECEPTORS IN ALZHEIMER'S DISEASE AND ALS

Environmental Toxic Syndromes in Neurodegenerative Diseases

Evidence for an excitotoxic component to ALS derives most directly from observation of abnormalities in the metabolism or uptake of glutamate in the disease.[4-7] While there is little comparably direct evidence for an excitotoxic component in AD, studies of three environmental illnesses, lathyrism, domoic acid toxicity and, the ALS-Parkinsonism-dementia complex of Guam (ALS-PD), have provided direct evidence that excitotoxic mechanisms can produce syndromes in humans or primates with certain clinical and pathologic features of both ALS and AD.

The domoic acid toxicity syndrome occurred in individuals after eating mussels contaminated with a marine microorganism that produced large amounts of the potent EAA, domoic acid.[8] Long lasting symptoms observed in survivors included memory impairment (probably on the basis of hippocampal damage) and some muscular weakness and atrophy, likely reflecting motor neuronal damage.[9] Lathyrism is a spastic leg weakness due to upper motor neuronal degeneration, induced by prolonged consumption of the chickling pea; the toxic agent is another potent EAA, β-N-oxalylamino-L-alanine (BOAA).[10] ALS-PD comprises pathologic and clinical features of ALS, AD and Parkinson's disease. While the cause is still controversial, considerable attention has been focused on toxic components of the cycad plant, the seed of which has long served as a food source in indigenous populations of Guam and neighboring islands. The finding that primates fed large amounts of the cycad seed EAA toxin β-N-methylamino-L-alanine (BMAA) develop a syndrome with certain similarities to the human disease suggests that BMAA may be a contributory factor.[11] Whether or not BMAA does play a role in the human disease, its effects in primates merits its inclusion on this list of disease-causing EAA neurotoxins.

While domoic acid and BOAA are well established as potent AMPA/kainate receptor agonists,[12-13] early studies of BMAA toxicity suggested that it acted primarily at NMDA receptors, thus failing to suggest a common receptor mechanism for the motor/cognitive effects of these three toxins.[14] Studies of BMAA toxicity in cortical neuronal cultures provided a potential resolution of this dilemma. Consistent with prior studies, the widespread neuronal damage induced by intense BMAA exposure could be attenuated by a selective NMDA antagonist. However at low concentrations perhaps more relevant to disease, BMAA unlike NMDA preferentially destroyed a subpopulation of cortical neurons containing the enzyme NADPH-diaphorase, a pattern of injury consistent with predominant AMPA/kainate receptor mediation.[15]

While the link between AMPA/kainate receptors and neurodegenerative disease provided by study of these toxic syndromes is intriguing, it is very indirect. For instance, toxins taken systemically have variable access to the CNS, they could have effects other than activation of glutamate receptors, or they could induce excitotoxicity indirectly by triggering release of glutamate from excitatory pathways.

Intrinsic Vulnerability of Neuronal Subpopulations

A distinctive feature of both AD and ALS is a characteristic pattern of selective neuronal degeneration. Among the populations of neurons that have been reported to preferentially degenerate in AD are cortical somatostatin (SS)[16-18] and parvalbumin (PV)[19-20] containing neurons, basal forebrain cholinergic (BFC)[21] neurons, and cortical and hippocampal pyramidal neurons, while in ALS, motor neurons selectively degenerate. We hoped that use of the simplified dissociated culture system would circumvent some of the interpretive difficulties with animal studies and permit direct assessment of the intrinsic EAA vulnerability of these neuronal populations (all cultures were used after 14-17 days *in vitro*).

The first study compared the survival of SS- and PV-immunoreactive neurons in murine cortical cultures with that of the majority of neurons in the cultures, after prolonged (24 h) exposures to AMPA, kainate or NMDA. We found the small (<2%) subpopulations of neurons labeled by SS or PV immunostaining to be largely destroyed by low AMPA (2 μM) and kainate (3-5 μM) exposures that caused little damage to the overall cortical neuronal population. Conversely, these populations were somewhat more resistant to NMDA than most neurons.[22]

We subsequently used three assays, immunocytochemical staining for choline acetyltransferase (ChAT), assay of the cultures for ChAT activity, and histochemical staining for acetylcholine esterase (AChE), to study the BFC population in basal forebrain cultures. Staining revealed BFC neurons to comprise a small subpopulation (< 3%) of neurons in the cultures. Consistent with studies in animals intracerebrally injected with agonists,[23] all three assays found the BFC neurons to be considerably more vulnerable to

AMPA/kainate receptor-mediated injury, and somewhat less vulnerable to NMDA toxicity than most neurons in the cultures.[24]

Culture studies of pyramidal neurons and motor neurons have in the past been hindered by a paucity of reliable, robust markers for these neurons, which are most readily identified *in vivo* by anatomic and morphologic criteria. In an attempt to identify pyramidal neurons in culture, we employed immunostaining with SMI-32, a monoclonal antibody against non-phosphorylated neurofilament epitopes that selectively labels subsets of pyramidal neurons *in vivo*, many of which may be especially prone to degeneration in AD.[25-28]

In dissociated cultures of murine cortex as well as hippocampus, SMI-32 labeled small subpopulations (<4%) of distinctly stained neurons (SMI-32(+) neurons), which were of larger than average size, and were usually multipolar with prominent neurites.[29] Double staining revealed substantial numbers of SMI-32(+) neurons (about 82% in cortex; 54% in hippocampus) to be immunoreactive for gamma-amino butyric acid (GABA),[29] suggesting the possibility that under present culture conditions either some GABAergic interneurons express unusually high levels of neurofilaments, or, as has been reported,[30] pyramidal neurons express GABA immmunoreactivity. In both cortex and hippocampus, SMI-32(+) neurons were preferentially injured by kainate exposures that caused little damage to the overall neuronal population, while NMDA exposures triggered equal injury to SMI-32(+) neurons and other neurons.[29]

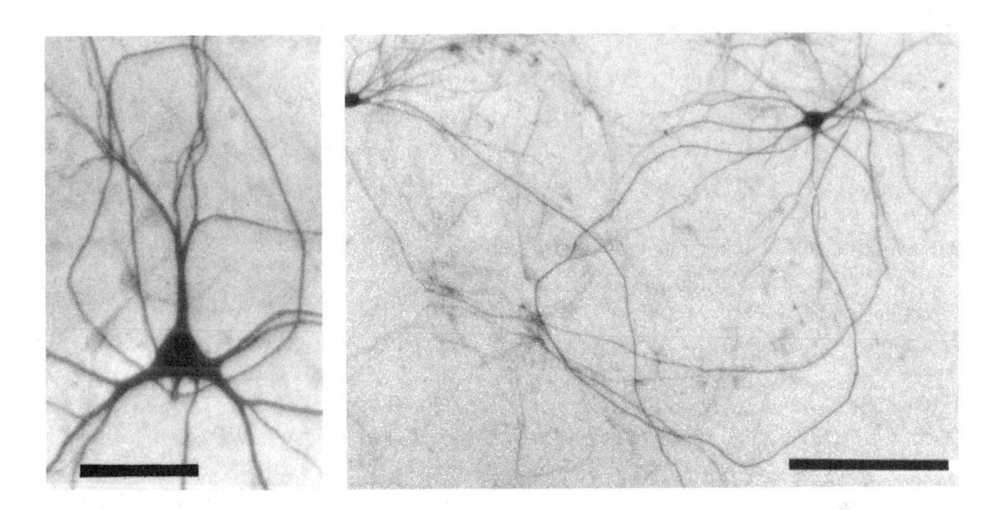

Figure 1. Large SMI-32 immunoreactive neurons in spinal cord culture. Note large neuronal size, extensive dendritic arborization, and single long axon-like neurite projecting from each neuron. Bar = 100 microns (left photo); 200 microns (right photo).

As SMI-32 has also been reported to label motor neurons in spinal cord slice,[31] we next examined SMI-32 immunostaining in both spinal cord slice and dissociated culture. We were impressed by the apparent selectivity of the stain for ventral horn neurons in slice; little or no staining was evident in other neurons. In culture, a small minority of neurons (<2%) were immunoreactive to SMI-32. A subpopulation (about 30%) of these SMI-32(+) neurons ("large SMI-32(+) neurons") have morphologic features that are characteristic of motor neurons; they are strongly stained, have cell bodies over 20 μm in diameter, a large and complex dendritic tree, and a single apparent axon which often projects over several millimeters (Fig. 1).[32] Studies with other motor neuronal markers, including antibodies against acetylcholine, calcitonin gene related peptide, and peripherin, lend support to the presumed motor neuronal identity of these neurons. Consistent with studies *in vivo* (after intrathecal injection[33]), and in spinal cord slice (treated with glutamate uptake blockers[34]) these large SMI-32(+) neurons were selectively vulnerable to AMPA/kainate-, but not NMDA- receptor mediated injury.

CALCIUM AND SELECTIVE AMPA/KAINATE VULNERABILITY

Studies in culture have revealed that brief periods of NMDA receptor activation are capable of triggering lethal injury to most neurons. This rapidly triggered NMDA receptor mediated injury is Ca^{2+}-dependent and probably reflects the high Ca^{2+} permeability of the NMDA receptor-gated channel.[35] In contrast, longer (several hour) periods of activation of AMPA/kainate channels, which are generally Ca^{2+}-impermeable, are necessary for comparable widespread neuronal injury to occur.[36] However, small subpopulations of neurons in the central nervous system have been recently found to possess AMPA/kainate receptor gated channels with direct Ca^{2+} permeability,[37] a factor that, by analogy to NMDA channels, might be expected to confer an enhanced vulnerability to AMPA/kainate receptor-mediated injury.

Ca^{2+} Permeable AMPA/kainate Channels in Selective Vulnerability

Neurons possessing Ca^{2+} permeable AMPA/kainate receptor-gated channels can be identified by a histochemical stain based on kainate stimulated uptake of Co^{2+} ions[38] (Co^{2+}(+) neurons; the specificity of the stain is indicated by the failure of NMDA or high-K^+ to substitute for kainate in triggering Co^{2+} uptake). Furthermore, supporting a possible role of such Ca^{2+} permeable channels in selective injury, Co^{2+}(+) neurons in both cerebellum and cortex are unusually vulnerable to Ca^{2+} dependent kainate toxicity.[39-40]

To examine the role of Ca^{2+} in selective vulnerability of identified neuronal subpopulations, cultures were exposed to kainate for brief ($\leq$ 45 min.) periods in the presence of defined Ca^{2+} concentrations and injury evaluated the next day. In the presence of physiologic (1.8 mM) Ca^{2+}, these exposures induced substantial and preferential injury to BFC neurons,[41] to cortical and hippocampal SMI-32(+) neurons[29], and to spinal SMI-32(+) neurons.[32] Injury to all these populations was Ca^{2+} dependent; removal of Ca^{2+} from the media during the brief exposures decreased subsequent degeneration, while raising Ca^{2+} enhanced injury. Furthermore, implicating possession of Co^{2+} permeable AMPA/kainate channels as central to this selective vulnerability, double staining techniques revealed that a majority ($\geq$ 60%) of each of these neuronal populations was Co^{2+}(+).[29,32,41] In preliminary studies, it appears likely that a substantial majority of cortical SS and PV neurons are Co^{2+}(+) as well.

Thus, studied *in vitro*, cortical SS and PV neurons, BFC neurons, cortical and hippocampal SMI-32(+) neurons, and large spinal cord SMI-32(+) neurons are all unusually vulnerable to AMPA/kainate receptor-mediated injury. Furthermore, the vulnerability of these neurons likely reflects at least in part possession of AMPA/kainate receptors gating channels with direct Ca^{2+} permeability.

Calcium Influx and Intracellular Homeostasis in Selective Vulnerability

While the simplest explanation for the unusual vulnerability of Co^{2+}(+) neurons to AMPA/kainate receptor-mediated injury would be that the Ca^{2+} permeable channels permitted a high rate of agonist triggered Ca^{2+} influx, rapidly overwhelming the cells homeostatic capabilities, other factors could be important. For instance, the injuriousness of Ca^{2+} ions could vary depending on the precise locations (or type of channels) through which they enter the neuron;[42] additionally, factors intrinsic to the neuron unrelated to either route or rate of Ca^{2+} entry, such as metabolic rate or Ca^{2+} buffering capacity could be relevant.

We first set out to compare kainate-triggered Ca^{2+} influx rates in Co^{2+}(+) neurons with those in other neurons (where influx is likely indirect, due to depolarization and activation of voltage sensitive Ca^{2+} channels (VSCC)). As Co^{2+}(+) neurons only comprise a small subpopulation of cortical neurons in culture (about 13%), direct measurement of this influx rate is not possible. However, Co^{2+}(+) neurons can be selectively destroyed by one hour kainate exposures.[40,43] Thus, comparing $^{45}Ca^{2+}$ influx rates in untreated cultures with those in cultures in which the Co^{2+}(+) neurons had been destroyed (by a kainate exposure the prior day) we estimated kainate triggered $^{45}Ca^{2+}$ influx into Co^{2+}(+) neurons to represent over 50% of total kainate triggered Ca^{2+} influx. Normalization for the small size of the Co^{2+}(+) population and comparison to the much greater NMDA triggered Ca^{2+} influx rates[44] suggests that kainate triggered Ca^{2+} influx rates into Co^{2+}(+) neurons are much greater than those into other neurons, and are comparable with NMDA triggered influx rates.[44]

Subsequent experiments used fura-2 Ca^{2+} imaging to compare kainate induced changes in intracellular free Ca^{2+} concentration ($[Ca^{2+}]i$) in Co^{2+}(+) neurons with those in other neurons. With low level (10 μM) exposures, consistent with their greater kainate vulnerability, Co^{2+}(+) neurons showed greater mean $[Ca^{2+}]i$ elevations than other neurons. These greater elevations could be explained if the weak kainate exposures caused poor depolarization and VSCC activation, thus allowing little influx in the absence of directly Ca^{2+} permeable AMPA/kainate channels. In contrast, with more intense (100 μM) exposures, mean $[Ca^{2+}]i$ elevations displayed by Co^{2+}(+) neurons were only slightly greater than those seen in other neurons.[43]

Thus, kainate appears to trigger an exceptionally rapid rate of Ca^{2+} entry into Co^{2+}(+) neurons that parallels their unusual vulnerability to Ca^{2+}-dependent AMPA/kainate receptor-mediated injury. Surprisingly, this rapid influx rate was not reflected in correspondingly greater $[Ca^{2+}]i$ elevations. Such lack of correspondence could be explained if rapid buffering and sequestration limited $[Ca^{2+}]i$ rises. However, as kainate damage to Co^{2+}(+) neurons does depend critically upon Ca^{2+} influx, this sequestered and "invisible" Ca^{2+} may play a critical role in the development of the injury. For instance, with intense NMDA exposures Ca^{2+} sequestration by mitochondria may trigger excess production of damaging reactive oxygen species.[45-48]

CONCLUSIONS

Antibodies which identify neurons that preferentially degenerate in AD or ALS were used in murine cultures, with the aim of examining the intrinsic excitotoxic vulnerability of specific neuronal subsets. Each of the small subpopulations of neurons identified by these markers demonstrated an unusual high sensitivity to AMPA/kainate receptor-mediated injury, likely reflecting at least in part possession of AMPA/kainate receptors gating channels with direct Ca^{2+} permeability. Further studies suggest that the unusual vulnerability of the entire population of neurons possessing Ca^{2+} permeable AMPA/kainate channels (identified by kainate triggered Co^{2+} uptake staining) reflects an unusually high rate of agonist triggered Ca^{2+} influx; additional factors may well contribute to the selective vulnerability of specific subpopulations.

The potential applicability of these findings to human disease depends upon the extent to which behavior of the neurons studied *in vitro* parallels the behavior of their counterparts in the mature human central nervous system. The high AMPA/kainate vulnerability of at least some of the populations does appear to carry over to animal models: studies suggest that motor neurons,[33,34] SS neurons,[49] and BFC neurons[23] all are selectively vulnerable to AMPA/kainate receptor-mediated injury. Also, hippocampal pyramidal neurons in slice are Co^{2+}(+).[38] Thus, present results lead us to consider the possibility that possession of Ca^{2+} permeable AMPA/kainate receptor-gated channels may be one factor that helps to predispose certain neurons to degenerate in Alzheimer's disease and ALS.

In summary, the *in vitro* system we have employed may permit certain mechanistic studies of neuronal vulnerability that are impossible in whole animals. However, as cultured rodent neurons may behave very differently from neurons in mature human brain, hypotheses developed in the simple *in vitro* system will need to be further evaluated by studies in more complex systems.

REFERENCES

1. S.M. Rothman and J.W. Olney, Glutamate and the pathophysiology of hypoxic-ischemic brain damage, *Ann. Neurol.* 19:105 (1986).
2. M.F. Beal, Does impairment of energy metabolism result in excitotoxic neuronal death in neurodegenerative illnesses?, *Ann. Neurol.* 31:119 (1992).
3. J.T. Greenamyre and A.B. Young, Excitatory amino acids and Alzheimer's disease, *Neurobiol. Aging* 10:593 (1989).
4. A. Plaitakis and J.T. Caroscio, Abnormal glutamate metabolism in amyotrophic lateral sclerosis, *Ann. Neurol.* 22:575 (1987).
5. J.D. Rothstein, L.J. Martin, and R.W. Kuncl, Decreased glutamate transport by the brain and spinal cord in amyotrophic lateral sclerosis, *New Engl. J. Med.* 326:1464 (1992).
6. J. Hugon, F. Tabaraud, M. Rigaud, J.M. Vallat, and M. Dumas, Glutamate dehydrogenase and aspartate aminotransferase in leukocytes of patients with motor neuron disease, *Neurology* 39:956 (1989).
7. J.D. Rothstein, G. Tsai, R.W. Kuncl, L. Clawson, D.R. Cornblath, D.B. Drachman, A. Pestronk, B.L. Stauch, and J.T. Coyle, Abnormal excitatory amino acid metabolism in amyotrophic lateral sclerosis, *Ann. Neurol.* 28:18 (1990).
8. T.M. Perl, L. Bedard, T. Kosatsky, J.C. Hockin, E.C. Todd, and R.S. Remis, An outbreak of toxic encephalopathy caused by eating mussels contaminated with domoic acid, *New Engl. J. Med.* 322:1775 (1990).
9. J.S. Teitelbaum, R.J. Zatorre, S. Carpenter, D. Gendron, A.C. Evans, A. Gjedde, and N.R. Cashman, Neurologic sequelae of domoic acid intoxication due to the ingestion of contaminated mussels, *New Engl. J. Med.* 322:1781 (1990).
10. P.S. Spencer, A. Ludolph, M.P. Dwivedi, D.N. Roy, J. Hugon, and H.H. Schaumberg, Lathyrism: evidence for the role of the neuroexcitatory amino acid BOAA, *Lancet* 2:1066 (1986).
11. P.S. Spencer, P.B. Nunn, J. Hugon, J. Ludolph, A.C. Ross, S.M. Ross, D.N. Roy, and R.C. Robertson, Guam amyotrophic lateral sclerosis- Parkinsonism-dementia linked to a plant excitant neurotoxin, *Science* 237:517 (1987).
12. R.J. Bridges, D.R. Stevens, J.S. Kahle, P.B. Nunn, M. Kadri, and C.W. Cotman, Structure-function studies on N-oxalyl-diamino-dicarboxylic acids and excitatory amino acid receptors: evidence that beta-L-ODAP is a selective non-NMDA agonist, *J. Neurosci.* 9:2073 (1989).
13. G. Debonnel, L. Beauchesne, and C. de Montigny, Domoic acid, the alleged "mussel toxin," might produce its neurotoxic effect through kainate receptor activation: an electrophysiological study in the dorsal hippocampus, *Can. J. Physiol. Pharmacol.* 67:29 (1989).
14. S.M. Ross, M. Seelig, and P.S. Spencer, Specific antagonism of excitotoxic action of "uncommon" amino acids assayed in organotypic mouse cortical cultures, *Brain Res.* 425:120 (1987).
15. J.H. Weiss, J. Koh, and D.W. Choi, Neurotoxicity of beta-N-methylamino-L-alanine (BMAA) and beta-N-oxalylamino-L-alanine (BOAA) on cultured cortical neurons, *Brain Res.* 497:64 (1989).

16. V. Chan-Palay, Somatostatin immunoreactive neurons in the human hippocampus and cortex shown by immunogold/silver intensification on vibratome sections: coexistence with neuropeptide Y neurons, and effects in Alzheimer-type dementia, *J. Comp. Neurol.* 260:201 (1987).
17. P. Davies, R. Katzman, and R.D. Terry, Reduced somatostatin-like immunoreactivity in cerebral cortex from cases of Alzheimer disease and Alzheimer senile dementia, *Nature.* 288:279 (1980).
18. M.N. Rossor, P.C. Emson, C.Q. Mountjoy, M. Roth, and L.L. Iversen, Reduced amounts of immunoreactive somatostatin in the temporal cortex in senile dementia of Alzheimer type, *Neurosci. Lett.* 20:373 (1980).
19. H. Arai, P.C. Emson, C.Q. Mountjoy, L.H. Carassco, and C.W. Heizmann, Loss of parvalbumin-immunoreactive neurones from cortex in Alzheimer-type dementia, *Brain Res.* 418:164 (1987).
20. J. Satoh, T. Tabira, M. Sano, H. Nakayama, and J. Tateishi, Parvalbumin-immunoreactive neurons in the human central nervous system are decreased in Alzheimer's disease, *Acta Neuropathol.* 81:388 (1991).
21. P.J. Whitehouse, D.L. Price, A.W. Clark, J.T. Coyle, and M.R. DeLong, Alzheimer disease: evidence for selective loss of cholinergic neurons in the nucleus basalis, *Ann. Neurol.* 10:122 (1981).
22. J.H. Weiss, J. Koh, K.G. Baimbridge, and D.W. Choi, Cortical neurons containing somatostatin or parvalbumin-like immunoreactivity are atypically vulnerable to excitotoxic injury in vitro, *Neurology* 40:1288 (1990).
23. K.J. Page, B.J. Everitt, T.W. Robbins, H.M. Marston, and L.S. Wilkinson, Dissociable effects on spatial maze and passive avoidance acquisition and retention following AMPA- and ibotenic acid-induced excitotoxic lesions of the basal forebrain in rats: differential dependence on cholinergic neuronal loss, *Neuroscience* 43:457 (1991).
24. J.H. Weiss, H. Yin, and D.W. Choi, Basal forebrain cholinergtic neurons are selectively vulnerable to AMPA/kainate receptor-mediated neurotoxicity, *Neuroscience* 60:659 (1994).
25. M.J. Campbell and J.H. Morrison, Monoclonal antibody to neurofilament protein (SMI-32) labels a subpopulation of pyramidal neurons in the human and monkey neocortex, *J. Comp. Neurol.* 282:191 (1989).
26. J.H. Morrison, D.A. Lewis, M.J. Campbell, G.W. Huntley, D.L. Benson, and C. Bouras, A monoclonal antibody to non-phosphorylated neurofilament protein marks the vulnerable cortical neurons in Alzheimer's disease, *Brain Res.* 416:331 (1987).
27. P.R. Hof, K. Cox, and J.H. Morrison, Quantitative analysis of a vulnerable subset of pyramidal neurons in Alzheimer's disease: I. Superior frontal and inferior temporal cortex, *J. Comp. Neurol.* 301:44 (1990).
28. P.R. Hof and J.H. Morrison, Quantitative analysis of a vulnerable subset of pyramidal neurons in Alzheimer's disease: II. Primary and secondary visual cortex, *J. Comp. Neurol.* 301:55 (1990).
29. S.J. Burke, H.Z. Yin, and J.H. Weiss, Ca^{2+} and in vitro kainate damage to cortical and hippocampal SMI-32(+) neurons, *NeuroReport* 6:629 (1995).
30. M.P. Mattson and S.B. Kater, Development and selective neurodegeneration in cell cultures from different hippocampal regions, *Brain Res.* 490:110 (1989).
31. T. Gotow and J. Tanaka, Phosphorylation of neurofilament H subunit as related to arrangement of neurofilaments, *J. Neurosci. Res.* 37:691 (1994).
32. S.G. Carriedo, H.Z. Yin, R. Lamberta, and J.H. Weiss, In vitro kainate injury to large, SMI-32 spinal neurons is Ca^{2+} dependent, *NeuroReport* 6:945 (1995).
33. J. Hugon, J.M. Vallat, P.S. Spencer, M.J. Leboutet, and D. Barthe, Kainic acid induces early and delayed degenerative neuronal changes in rat spinal cord, *Neurosci. Lett.* 104:258 (1989).
34. J.D. Rothstein, L. Jin, M. Dykes-Hoberg, and R.W. Kuncl, Chronic inhibition of glutamate uptake produces a model of slow neurotoxicity, *Proc. Natl. Acad. Sci. USA* 90:6591 (1993).
35. A.B. MacDermott, M.L. Mayer, G.L. Westbrook, S.J. Smith, and J.L. Barker, NMDA-receptor activation increases cytoplasmic calcium concentration in cultured spinal cord neurones, *Nature* 321:519 (1986).
36. D.W. Choi, Excitotoxic cell death, *J. Neurobiol.* 23:1261 (1992).
37. M. Iino, S. Ozawa, and K. Tsuzuki, Permeation of calcium through excitatory amino acid receptor channels in cultured rat hippocampal neurones, *J. Physiol.* 424:151 (1990).
38. R.M. Pruss, R.L. Akeson, M.M. Racke, and J.L. Wilburn, Agonist-activated cobalt uptake identifies divalent cation-permeable kainate receptors on neurons and glia, *Neuron* 7:509 (1991).
39. J.R. Brorson, P.A. Manzolillo, and R.J. Miller, Ca^{2+} entry via AMPA/KA receptors and excitotoxicity in cultured cerebellar purkinje cells, *J. Neurosci.* 14:187 (1994).
40. D.M. Turetsky, L.M.T. Canzoniero, S.L. Sensi, J.H. Weiss, M.P. Goldberg, and D.W. Choi, Cortical neurons exhibiting kainate-activated Co^{2+} uptake are selectively vulnerable to AMPA/kainate receptor-mediated toxicity, *Neurobiol. Dis.* 1:101 (1994).

41. H. Yin, A.D. Lindsay, and J.H. Weiss, Kainate injury to cultured basal forebrain cholinergic neurons in Ca^{2+} dependent, *NeuroReport* 5:1477 (1994).
42. M. Tymianski, M.P. Charlton, P.L. Carlen, and C.H. Tator, Source specificity of early calcium neurotoxicity in cultured embryonic spinal neurons, *J. Neurosci.* 13:2085 (1993).
43. Y.M. Lu, H.Z. Yin, and J.H. Weiss, Ca^{2+} permeable AMPA/kainate channels permit rapid injurious Ca^{2+} entry, *NeuroReport* 6 (1995), in press.
44. D.M. Hartley, M.C. Kurth, L. Bjerkness, J.H. Weiss, and D.W. Choi, Glutamate receptor-induced $45Ca^{2+}$ accumulation in cortical cell culture correlates with subsequent neuronal degeneration, *J. Neurosci.* 13:1993 (1993).
45. L.L. Dugan, S.L. Sensi, L.M.T. Cazoniero, M.P. Goldberg, S.D. Handran, S.M. Rothman, and D.W. Choi, Imaging of mitochondrial oxygen radical production in cortical neurons exposed to NMDA, *Soc. Neurosci. Abstr.* 20:1532 (1994).
46. M. Lafon-Cazal, S. Pietri, M. Culcasi, and J. Bockaert, NMDA-dependent superoxide production and neurotoxicity, *Nature* 364:535 (1993).
47. I.J. Reynolds, T.G. Hastings, and K.R. Hoyt, Studies on the role of mitochondria in NMDA receptor-mediated excitotoxicity, *Soc. Neurosci. Abstr.* 20:1530 (1994).
48. J.L. Werth and S.A. Thayer, Mitochondria buffer physiological calcium loads in cultured rat dorsal root ganglion neurons, *J. Neurosci.* 14:348 (1994).
49. M.F. Beal, K.J. Swartz, S.F. Finn, M.F. Mazurek, and N.W. Kowall, Neurochemical characterization of excitotoxin lesions in the cerebral cortex, *J. Neurosci.* 11:147 (1991).

DILTIAZEM AND MK-801 BUT NOT APV ACT SYNERGISTICALLY TO PROTECT RAT HIPPOCAMPAL SLICES AGAINST HYPOXIC DAMAGE

Avital Schurr*, Ralphiel S. Payne and Benjamin M. Rigor

Department of Anesthesiology
University of Louisville, School of Medicine
Louisville, KY 40292

INTRODUCTION

The noncompetitive N-methyl-D-aspartate (NMDA) receptor antagonist MK-801 {(+)-5-methyl-10,11-dihydro-5H-dibenzo[a,d]cyclohepten-5,10-imine maleate, dizocilpine} potently protects hippocampal slices against hypoxic neuronal damage, while the competitive NMDA receptor antagonist 2-amino-5-phosphonovalerate (APV) has a much weaker neuroprotective potency.[1,2] In addition, MK-801 protects hippocampal slices against the combined effects of hypoxia + NMDA, and hypoxia + kainate (KA) or α-amino-3-hydroxy-5-methyl-4-isoxazolpropionate (AMPA). Furthermore, MK-801 was found to be a better antagonist of KA (or AMPA)-enhanced hypoxic neuronal damage than the specific KA/AMPA antagonist GYKI 52466 [1-(4-aminophenyl)-4-methyl-7,8-methylenedioxy-5H-2,3-benzodiazepine HCl]. The combination of APV and 7-chloro-kynurenate could not abolish the protective effect of MK-801 against hypoxia, a effect one should expect if MK-801 exerts its protective action via an open NMDA receptor.[3] Thus, at least in the continuously-perfused hippocampal slice preparation, where hypoxia-induced glutamate accumulation cannot reach excitotoxic levels, the antihypoxic effect of MK-801 does not appear to be mediated through the NMDA receptor. Several other recent studies also raise questions in regard to MK-801 mode of action. One study[4] concluded that ketamine, an anesthetic known to compete for MK-801 binding sites,[5] is an inhibitor of L-type calcium channels, while another study[6] points to both similarities and differences between these two compounds depending on the method of administration of ketamine. Yet, another study[7] demonstrated the ability of diltiazem (DILT), an L-type calcium channel blocker,[8-10] to inhibit [^{3}H]MK-801 binding to hippocampal synaptic membranes at IC_{50} value which is very similar to the IC_{50} of unlabelled MK-801. These studies, along with our own results that showed DILT to provide a significant protection against hypoxia-induced neuronal damage,[1] led us to suggest that MK-801 may possess L-type calcium channel blocking properties.

The present study tested the combined effects of MK-801 and DILT in protecting the rat hippocampal slice preparation against neuronal damage produced by a prolonged (20 min) hypoxic episode.

METHODS

Preparation and Maintenance of Slices

Adult (200-350 g) male Sprague-Dawley rats were used. For each experiment, one rat was decapitated, its brain rapidly removed and rinsed with cold (6-8°C) artificial cerebrospinal fluid (ACSF, see composition below), and dissected. Isolated hippocampi were sliced transversely at 400 *u*m with a McIlwain tissue chopper, and the resulting slices (10-15 slices from each hippocampus) were placed in a dual linear-flow incubation /recording chamber.[11]. Each of the two compartments of the dual chamber was supplied with a humidified gas mixture (95% O_2/5% CO_2) through separate flow meters (1.5 liter/min) and ACSF via a dual peristaltic pump (60 ml/h). Hypoxia was produced by replacing O_2 in the gas mixture with N_2. The ACSF contained the following (in mM): NaCl, 124; KCl, 5; NaH_2PO_4, 3; $CaCl_2$, 3.0; $MgSO_4$, 2.0; $NaHCO_3$, 23; D-glucose, 10. The pH of the ACSF was 7.3-7.4. The temperature of the incubation chamber was held at 34°C $\pm$ 0.3°C.

Assessment of Neuronal Function

Extracellular recordings of evoked population spikes (neuronal function) in the stratum pyramidale of the CA1 region were made from one slice in each compartment of the dual chamber using borosilicate micropipettes filled with ACSF (impedence 2-5 Mohms). A two-channel preamplifier (X100) and two field-effect transistor headstages were used. The output was digitized and stored on a floppy disk for later analysis. Bipolar stimulating electrodes were placed in the Schaffer collaterals (orthodromic stimulation), and stimulus pulses of 0.1 ms in duration and of an amplitude 8-9 V (x2 threshold) were applied once/min. At the end of each experiment (following a washout, reoxygenation period) the rest of the slices (9-14) in each compartment were tested for the presence of neuronal function by stimulating the Schaffer collaterals and by recording from the CA1 cell body layer. A waveform analysis program was used to determine the amplitude of the evoked population spike. Any slice exhibiting a population spike of 3 mV or larger was considered neuronally functional. Slices in which a population spike could not be evoked or in which the spike amplitude was smaller than 3 mV at the end of the experiment were considered to be neuronally damaged. Monitoring of neuronal function in one slice in each compartment was begun 90-120 min after decapitation and was continued minute-by-minute throughout the experiment. Each experiment consisted of a 10-min baseline recording period, a 45-min drug treatment period (the last 20 min of which was combined with hypoxia), and a 30-min washout, reoxygenation (recovery) period. Frequently, the experimental manipulations were performed on slices in one compartment of the dual chamber while slices in the other compartment served as "controls".[12] Statistical analysis was done using the X^2-test for significant differences. Where indicated, the ACSF was supplemented with one or more of the following compounds: MK-801 (Wyeth-Ayerst Research Laboratories, Princeton, NJ), DL-APV, diltiazem hydrochloride (Sigma Chemical Co., St. Louis, MO).

RESULTS

Only 34 out of 242 slices (14%) recovered their neuronal function after exposure to 20-min hypoxia (see control value, Figs. 1 and 2). Both MK-801 and DILT had a significant, dose-dependent neuroprotective effect on hippocampal slices when administered during the hypoxic period (Fig. 1). MK-801 was approximately 50 times more potent than DILT in protecting slices against hypoxic damage. Nevertheless, MK-801, a noncompetitive antagonist of the NMDA receptor,[5,13-16] and DILT, an L-type calcium channel blocker,[8,9,17] are two agents with supposedly different modes of action and thus, should not interact with each other when added together; at most, they should have an additive neuro-protective effect. However, if the combination of the two should have an effect larger than the expected additive value i.e., synergy, it could indicate an interaction between these two agents.[18]

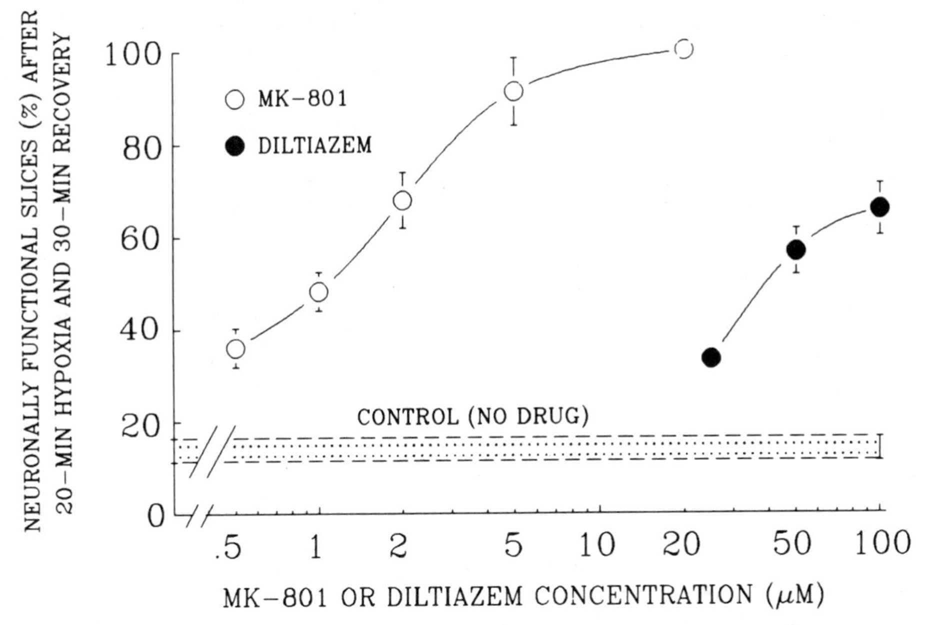

Fig. 1. The dose-dependent neuroprotective effects of MK-801 and diltiazem (DILT) against neuronal damage in hippocampal slices induced by 20-min hypoxia. The effects are expressed as the rate (%) of recovery of neuronal function at the end of a 30-min recovery period. Only 13.8±2.6% of control (no drug) slices recovered their neuronal function following the 30-min reoxygenation period. Each data point is the mean of at least three replications. The bars represent standard deviation of the mean. The drugs were perfused through the ACSF 30 min before and 20 min during the hypoxic period and were washed out during the reoxygenation/recovery period.

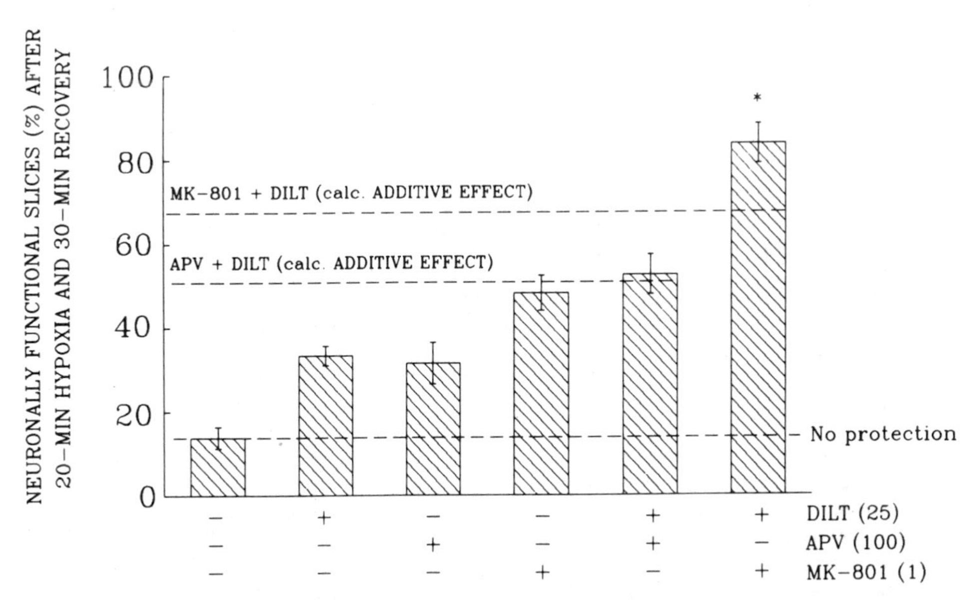

Fig. 2. The rate (%) of recovery of neuronal function in hippocampal slices exposed to 20-min hypoxia in the absence (no protection, control) or presence of diltiazem (DILT), DL-2-amino-5-phosphonovalerate (APV), MK-801 or the combinations [APV + DILT] and [MK-801 + DILT]. Numbers in parentheses are doses in uM. Each data point was replicated at least 4 times. Bars are standard deviation of the mean. Other technical details are as in Fig. 1. *Significantly different from the calculated additive effect of [MK-801 + DILT] (0.025<P<0.05).

Figure 2 shows the percentage of hippocampal slices that recover their neuronal function after 20-min hypoxia and 30-min reoxygenation in the presence of DILT, APV or MK-801. Of the three, APV (100 μM) had the weakest neuroprotective effect (17/54 slices recovered after hypoxia; less than 18% improvement over the control value) while MK-801 (1 μM) was the most potent neuroprotectant (30/60 slices recovered after hypoxia, a 36% improvement over the control value). Figure 2 also shows the effect of two drug combinations, [APV + DILT] and [MK-801 + DILT]. The first (100 μM + 25 μM, respectively) produced an additive effect (44/84 slices recovered after hypoxia, a 39% improvement over the control value as compared to a calculated additive value of 37%). The second combination (1 μM + 25 μM, respectively) resulted in a significant synergistic effect (48/57 slices recovered after hypoxia, a 70% improvement over the control value as compared with a calculated additive value of 54%). A significant ($P < 0.05$) synergistic effect also prevailed when the combination [0.5 μM MK-801 + 25 μM DILT] was used (42/69 slices recovered after hypoxia, a 48% improvement over the control value as compared with a calculated additive value of 36%).

DISCUSSION

These results strongly support our notion that the noncompetitive NMDA receptor antagonist MK-801 expresses its neuroprotective effect *via* additional mechanism(s) including the blockade of L-type calcium channels. The one study that attempted to measure the effect of MK-801 on calcium channels could not demonstrate any direct effect.[19] However, such measurements were not made under hypoxic conditions where L-type calcium channels may undergo changes that could be driven by lower ATP levels. Alternatively, it is possible that a separate class of calcium channels exists, mainly inactive under normal physiological conditions, but fully active under stressful conditions such as long-term hypoxia or ischemia. Such channels could be blocked by the L-type calcium channel blocker DILT, and even more efficiently by MK-801. Since both drugs exert their protective effect via the same protein component, an interaction between them that lead to synergistic effect is possible. Additional support of the notion that MK-801 interacts with L-type calcium channels was reported recently[20] where the dihydropyridine L-type calcium channel blocker nitrendipine potently displaced [^{3}H]MK-801 binding to mouse brain sections.

Interestingly, Uematsu et al.[10] reported on the combined protective effect of MK-801 and nimodipine against ischemic brain damage in the cat. Unfortunately, the study compared the combined effect against the effect of MK-801 alone but not against the effect of nimodipine alone and, thus, synergism between the two could not be evaluated. Nevertheless, combining nimodipine with a relatively low dose of MK-801 significantly improved the histologic damage and reduced Ca^{2+} entry as compared with MK-801 alone. This finding was interpreted by the investigators to be the outcome of a dual blockade of both ligand-, and voltage-activated calcium channels.

An *in vivo* study is underway in our laboratory to determine the neuroprotective potential of the combination MK-801 + diltiazem in global cerebral ischemia.

REFERENCES

1. A. Schurr, R.S. Payne and B.M. Rigor, Protection by MK-801 against hypoxia-, excitotoxin-, and depolarization-induced neuronal damage in vitro, *Neurochem. Int.* (in press).
2. P.G. Aitken, M. Balestrino and G.G. Somjen, NMDA antagonists: lack of protective effect against hypoxic damage in CA1 region of hippocampal slices, *Neurosci. Lett.* 89:187-192 (1988).
3. J.E. Huettner and B.P. Bean, Block of N-methyl-D-aspartate-activated current by the anticonvulsant MK-801: selective binding to open channels, *Proc. Natl. Acad. Sci. USA* 85:1303-1311 (1988).
4. P.H. Ratz, P.E. Callahan and F.A. Lattanzio, Ketamine relaxes rabbit femoral arteries by reducing $[Ca^{2+}]_i$ and phospholipase C activity, *Eur. J. Pharmacol.* 236:433-441 (1993).
5. E.H.F. Wong, J.A. Kemp, T. Priestley, A.R. Knight, G.N. Woodruff and L.L. Iversen, The anticonvulsant MK-801 is a potent N-methyl-D-aspartate antagonist, *Proc. Natl. Acad. Sci. USA* 83:7104-7108 (1986).

6. M.D. Kelland, R.P. Soltis, R.C. Boldry and J.R. Walters, Behavioral and electrophysiological comparison of ketamine with dizocilpine in the rat, *Physiol. Behav.* 54:547-554 (1993).
7. M. Maruyama and K. Takeda, Electrophysiologically potent non-competitive glutamate antagonists at crayfish neuromuscular junctions are also potent inhibitors of [^{3}H]MK-801 binding to synaptic membranes from rat central nervous system, *Comp. Biochem. Physiol.* 107C:105-110 (1994).
8. S. Adachi-Akahane and T. Nagao, Binding site for diltiazem is on the extracellular side of the L-type Ca^{2+} channel, *Circulation* 88: I-230 (1993).
9. K.J. Brooks and R.A. Kauppinen, Calcium-mediated damage following hypoxia in cerebral cortex *ex vivo* studied by NMR spectroscopy. Evidence for direct involvement of voltage-gated Ca^{2+}-channels, *Neurochem. Int.* 23:441-450 (1993).
10. D. Uematsu, N. Araki, J.H. Greenberg, J. Sladky and M. Reivich, Combined therapy with MK-801 and nimodipine for protection of ischemic brain damage, *Neurology* 41:88-94 (1991).
11. A. Schurr, K.H. Reid, M.T. Tseng, H.L. Edmonds, Jr. and B.M. Rigor, A dual chamber for comparative studies using the brain slice preparation, *Comp. Biochem. Physiol.* 82A:701-705 (1985).
12. A. Schurr, P. Lipton, C.A. West and B.M. Rigor, The role of energy metabolism and divalent cations in the neurotoxicity of excitatory amino acids in vitro, *in*: Pharmacology of Cerebral Ischemia 1990, J. Krieglstein and H. Oberpichler, Eds., Wissenschaftliche Verlagsgesellschaft, Stuttgart (1990) pp. 217-226.
13. A.C. Foster and E.H.F. Wong, The novel anticonvulsant MK-801 binds to the activated state of the N-methyl-D-aspartate receptor in rat brain, *Br. J. Pharmacol.* 91:403-409 (1987).
14. A.C. Foster, R. Gill, J.A. Kemp and G.N. Woodruff, Systemic administration of MK-801 prevents N-methyl-D-aspartate-induced neuronal degeneration in rat brain, *Neurosci. Lett.* 76:307-311 (1987).
15. A.C. Foster, R. Gill and G.N. Woodruff, Neuroprotective effects of MK-801 in vivo: selectivity and evidence for delayed degeneration mediated by NMDA receptor activation, *J. Neurosci.* 8:4745-4754 (1988).
16. G.N. Woodruff, A.C. Foster, R. Gill, J.A. Kemp, E.H.F. Wong and L.L. Iversen, The interaction between MK-801 and receptors for N-methyl-D-aspartate: functional consequences, *Neuropharmacology* 26:903-909 (1987).
17. T. Watanabe, H. Kalasz, H. Yabana, A. Kuniyasu, J. Mershon, K. Itagaki, P.L. Vaghy, K. Naito, H. Nakayama and A. Schwartz, Azidobutyryl clentiazem, a new photoactivatable diltiazem analog, labels benzothiazepine binding sites in the c_1 subunit of the skeletal muscle calcium channel, *FEBS Lett.* 334:261-264 (1993).
18. M. C. Berenbaum, What is synergy? *Pharmacol. Rev.* 41: 93-141 (1989).
19. R. Netzer, V. Graf, G. Pfimlin and G. Trube, Effects of some NMDA receptor channel antagonists on voltage-dependent Ca^{2+} channels in cultured cortical neurons, *in:* Pharmacology of Cerebral Ischemia 1990, J. Krieglstein and H. Oberpichler, Eds., Wissenschaftliche Verlagsgesellschaft, Stuttgart (1990) pp 129-137.
20. F.M. Filloux, R.C. Fitts, G.A. Skeen and H.S. White, The dihydropyrydine nitrendipine inhibits [^{3}H]MK-801 binding to mouse brain sections, *Eur. J. Pharmacol.* 269:325-330 (1994).

INTRACELLULAR CALCIUM STORES AND ISCHEMIC NEURONAL DEATH

David C. Perry, [1] Huafeng Wei[1], Wenlin Wei,[1] and Dale E. Bredesen[2]

[1]Department of Pharmacology
The George Washington University Medical Center
2300 I St NW
Washington, DC 20037
[2]Program on Aging
The La Jolla Cancer Research Foundation
10901 N. Torrey Pines Blvd.
La Jolla, CA 92037

INTRODUCTION

Disruption of cellular regulation of calcium has long been implicated in the mechanism of neruonal death following cerebral ischemia and reperfusion[9,17]. While much research has been directed at the potential role of calcium influx in ischemia, less attention has been paid to other mechanisms of calcium regulation that may be altered. We are investigating the possibility that altered storage and release of Ca^{2+} from the endoplasmic reticulum (ER) contributes to ischemic neuronal death.

Free cytosolic concentrations of Ca^{2+} are normally maintained at 50-100 nM, several orders of magnitude below extracellular level, by three mechanisms: exchange with extracellular Ca^{2+}(via Ca^{2+}-ATPases and Na^+ /Ca^{2+} exhangers), binding to a variety of Ca^{2+} binding proteins, and sequestration in subcellular organelles[3]. The ER is the primary organelle for routine Ca^{2+} sequestration and release. It relies on a high affinity Ca^{2+}ATPase to pump Ca^{2+} from the cytosol (this enzyme is substantially different from the plasma membrane form). Evidence indicates that the ER does more than serve as a passive storage site for Ca^{2+}; in fact, this ion is crucial for ER structure[11] and for a number of ER-specific cellular functions, including protein synthesis[3,16]. ER Ca^{2+} can be released into the cytosol by two different types of receptor/channels: the ryanodine receptor (RYR) and the IP_3 receptor (IPR). Both receptors can be stimulated by increases in Ca^{2+} in a positive feedback system known as calcium-induced calcium release (CICR)[4], as well as the intracellular second messengers inositol trisphosphate (IP_3) and cyclic ADP-ribose (cADPR).

Although little data exists on the role of intracellular Ca^{2+} release in ischemic death, several studies have found that it contributes significantly to neuronal death from excitotoxicity, possibly by a mechanism whereby initial Ca^{2+} influx triggers further Ca^{2+} release from ER stores[8,13,15]. Dantrolene, a drug that blocks Ca^{2+} release from ER ryanodine receptor-channels, inhibits excitotoxic cell death in these models[8,13], and inhibits the increase in cytosolic Ca^{2+} caused by chemical hypoxia induced by NaCN in cultured hippocampal neurons[7]. Few *in vivo* studies have addressed the issue of intracellular Ca^{2+}

Neurodegenerative Diseases
Edited by Gary Fiskum, Plenum Press, New York, 1996

release. A recent report demonstrated that dantrolene administered intracerebroventricularly (but not i.p.) inhibited hippocampal neuronal death after global cerebral ischemia in gerbils[21]. Another study demonstrated that preadministration of the intracellular Ca^{2+} chelator, BAPTA-AM, was neuroprotective in a rat model of focal cerebral ischemia[19].

Although ischemic neuronal death has long been presumed to be necrotic in nature, recent evidence indicates that apoptosis may also be involved[5]. Increases in cytosolic Ca^{2+} have been shown to induce apoptosis in a variety of models[12,14]. The protooncogene *bcl-2* can protect against apoptosis induced by a large variety of insults[10,12,22]. For instance, drugs such as thapsigargin that stimulate ER Ca^{2+} release cause apoptosis in mouse thymocytes which is prevented by overexpression of *bcl-2*[12].

STUDIES OF THE EFFECT OF INTRACELLULAR CALCIUM RELEASE

In Vitro Studies

To study the effects of intracellular Ca^{2+} release on neuronal viability we chose as a model the GT1-7 line, an immortalized mouse hypothalamic neuronal line[14]. Ca^{2+} release was stimulated by treatment with either thapsigargin (TG) or caffeine. TG is a highly specific inhibitor of ER Ca^{2+}-ATPase[18], whereas caffeine stimulates release of ER

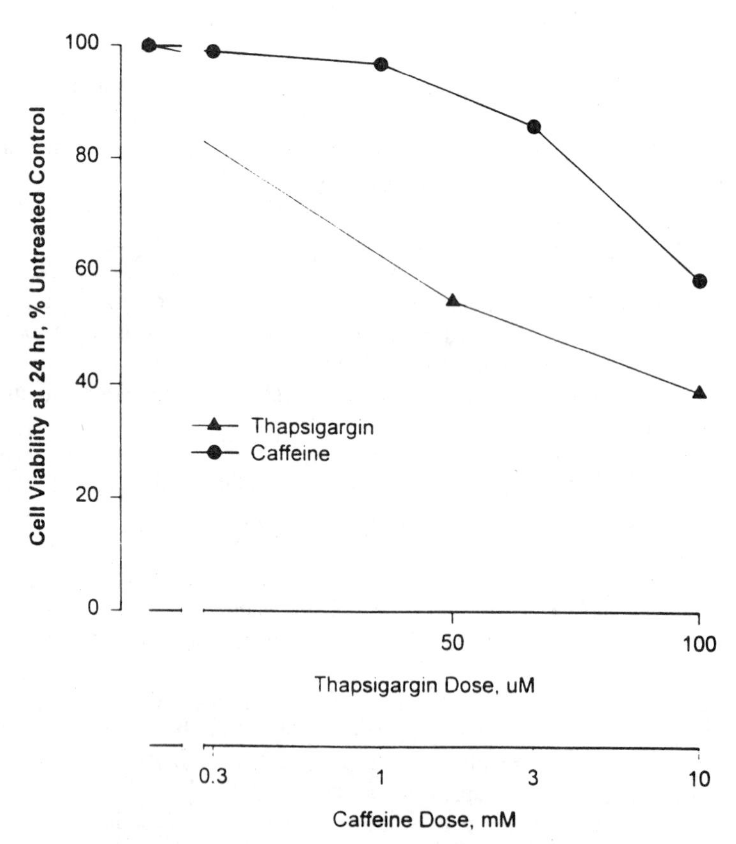

Figure 1 Effect of 24 hr exposure to thapsigargin or caffeine on viability of GT1-7 cells.

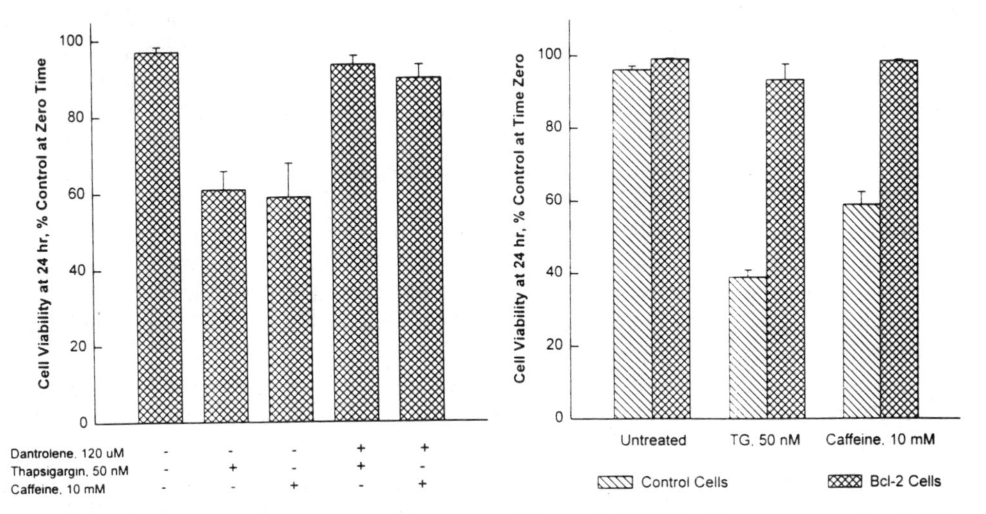

Figure 2 Dantrolene added 30 min prior protects against TG & caffeine toxicity.

Figure 3 GT1-7 cells overexpressing *bcl-2* are protected from TG or caffeine toxicity.

Ca^{2+} from the ryanodine receptor channel . Cell viability was assessed by the trypan blue exclusion method. The effects of 24 hr exposure to TG or caffeine on GT1-7 viability are shown in Figure 1. Both 50 nM and 100 nM doses of TG were highly cytotoxic. Caffeine had little effect at lower doses (0.3 and 1 mM) but showed a dose-related cytotoxicity at 3 and 10 mM.

For further evidence that the cytotoxic effect of these two drugs was due to release of ER Ca^{2+}, we added 120 μM dantrolene to cells 30 min prior to addition of either TG or caffeine. Dantrolene inhibits Ca^{2+} release from ER via ryanodine receptors. As can be seen in Figure 2, dantrolene almost completely blocked the cytotoxicity of each drug at 24 hr.

Cell death can be due to necrosis or apoptosis[20]. Many insults known to produce apoptosis can be can be blocked by the expression of the protooncogene *bcl-2*. We used GT1-7 cells that were transfected with a plasmid containing the *bcl-2* gene which have been shown to be resistant to death from a variety of different insults[10,22]. Figure 3 demonstrates that overexpression of this gene offers almost complete protection against 24 hr exposure to 50 nM TG or 10 mM caffeine. Protection by *bcl-2* is not necessarily limited to apoptosis[22]. Therefore, we examined the cells for histopathological markers of apoptosis[20] by staining with the nuclear dye acridine orange. Apoptotic nuclei show a characteristic pattern of nuclear fragmentation which can be detected by fluorescence microscopy. We quantified the number of cells demonstrating this form of damage after 24 hr of different treatments; the results are in Figure 4. Greater than 40% of cells showed evidence of apoptotic damage with 24 hr of treatment with 50 nM TG or 10 mM caffeine, whereas these same treatments in cells expressing *bcl-2* produced no more damage than in untreated cells.

In Vivo Studies

To demonstrate that ER Ca^{2+} release contributes to ischemic neuronal damage, we chose to study the neuroprotective effects of dantrolene using the gerbil model of global cerebral ischemia. Male Mongolian gerbils (60-70g) were anesthetized with 2-3%

halothane in 70%N_2O:30%CO_2, while rectal temperature was maintained at 37-38°C by a heated pad and lamps. The common carotid arteries were exposed and clamped for 5 min with microclips. Following resumption of normal blood flow, dantrolene was injected into the jugular vein at a dose of 50 mg/kg. Sham controls were operated on without arterial clamping. Ischemic controls received arterial clamping but vehicle injections. Following removal from anesthesia, animals were allowed to recover while being kept warm for at least four hr. Seven days following surgery, animals were anesthetized with 100 mg/kg pentobarbital i.p., then perfused through the heart with formaldehyde. Brains were removed and frozen, and coronal sections through the dorsal hippocampus were cut, mounted and stained with cresyl violet acetate. Hippocampal CA1 pyramidal cells were counted in at least two separate sections by two separate blinded observers, and the number of viable cells / mm was calculated

Figure 5 demonstrates that 5 min global ischemia produces a profound loss of CA1 pyramidal neurons (reduced to 39% of sham controls). However, i.v. dantrolene at 50 mg/kg immediately post-ischemia showed substantial protection: the number of viable neurons was 84% of sham control, which was highly significantly greater than those in the untreated ischemic animals ($p<0.01$), although still less than sham controls ($p<0.05$).

DISCUSSION

Although it has long been postulated that disruption of cellular Ca^{2+} regulation plays an important role in ischemic neuronal death, the precise nature of the disruption has not been clear. The present studies have focussed on release of ER stores as a potential source. Using an *in vitro* cell culture model, we have shown that drugs known to cause release of ER Ca^{2+} stores (TG and caffeine) are cytotoxic, and that dantrolene, which blocks ER Ca^{2+} release at the ryanodine receptor channel, will inhibit the effects of these drugs. Although caffeine has a variety of neurochemical effects, the dose required to produce the effects seen here are consistent with its acting via the ryanodine receptor. This conclusion is greatly strengthened by the fact that its cytotoxicity is prevented by dantrolene. TG, as noted, is highly selective at inhibiting ER Ca^{2+}ATPase. The source of the Ca^{2+} efflux from the ER after TG inhibition is not known for certain, but again the ability of dantrolene to inhibit TG toxicity argues that the ryanodine receptor channel plays a major role. The potential contribution of the IP_3 receptor channel remains unclear; unfortunately, few drugs are selective for this site, and those that exist have various problems (i.e. inability to penetrate cell membranes) that limit their utility in these types of experiments.

The mechanism of cell death arising from ER Ca^{2+} release is also uncertain. Numerous cellular activities can be stimulated by increases in cytosolic [Ca^{2+}], including activation of proteases, lipases, kinases and nucleases, many of which may contribute to cell death. Another important feature of ER Ca^{2+} release is the fact that not only do cytosolic concentrations of Ca^{2+} increase, but ER concentrations of Ca^{2+} decrease. This decrease may have detrimental effects independent of cytosolic changes[11,16]. Release of ER Ca^{2+} may be coupled to influx of extracellular calcium[15]. TG toxicity has been found to be dependent upon extracellular Ca^{2+} in many but not all cases studied[18]. We have preliminary results suggesting that adding EGTA to the medium to chelate extracellular Ca^{2+} does not inhibit the toxicity of TG; however, EGTA itself may be causing toxicity, obscuring our interpretation. Studies are currently ongoing to resolve this question.

Controversy still exists over whether ischemia induces neuronal death by necrosis or apoptosis. Manipulation of Ca^{2+} in cells has been shown to produce both types of

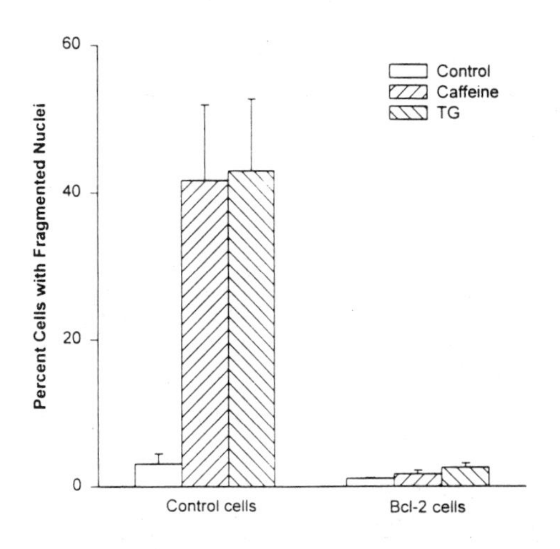

Figure 4 *Bcl-2* protects against nuclear fragmentation in GT1-7 cells caused by 24 hr of 50 nM TG or 10 mM caffeine.

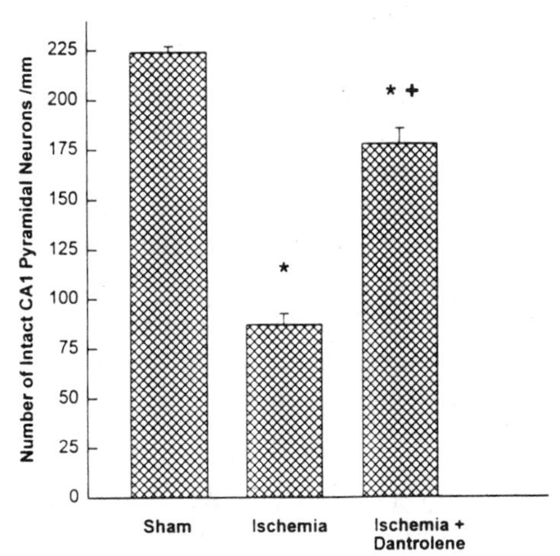

Figure 5 Dantrolene (50 mg/kg i.v.) protects against ischemic damage to CA1 hippocampal neurons.

death. If both processes are activated, differences in the speed of the processes may mean that the faster one appears to be the only process stimulated. Typically the cytosolic Ca^{2+} concentrations following ER Ca^{2+} release are in the range of 1-250 nM; while high, this is considerably lower than the 1-2 μM concentrations achieved by stimulation of influx via NMDA receptors or with ionophores such as A23187[4,22]. It is possible that qualitatively different results are achieved at these two different concentration ranges: the high range may be associated with rapid, necrotic death (which is known to activated by Ca^{2+} in some systems[6]), while perhaps a sustained but moderate increase may be associated with a slower, apopotic death, possibly due in part to the additional effects of ER Ca^{2+} depletion.

Our finding that dantrolene is neuroprotective in the gerbil model is important for both theoretical and practical reasons. Because the primary cellular effect of dantrolene is to block release of Ca^{2+} from the ER and sarcoplasmic reticulum, this finding lends support to the hypothesis that release of ER Ca^{2+} contributes to ischemic neuronal death. Also, because this drug has been used in humans for over 30 years, it represents a promising clinical approach to treatment of patients recovering from cardiac arrest and stroke.

REFERENCES

1. M.J. Berridge, Inositol trisphosphate and calcium signaling, *Nature* **361**: 315-25 (1993).
2. E. Bonfoco, D. Krainc, P. Nicotera, & S.A. Lipton, NMDA or nitric oxide ($NO^{\bullet}$)/superoxide ($O_2^{\bullet}$) induces apoptosis in rat cortical culture, *Soc. Neurosci. Abst.* **20**: 650 (1994).
3. C.O. Brostrom & M.A. Brostrom, Calcium-dependent regulation of protein synthesis in intact mammalian cells, *Annu. Rev. Physiol.* **52**: 577-90 (1990).
4. E. Carafoli, Intracellular calcium homeostasis, *Annu. Rev. Biochem.* **56**: 395-433 (1987).
5. C. Charriatu-Marlangue, H. Pollard, A. Heron, I. Margaill, M. Plotkine, B. Zivkovic & Y. Ben-Ari, Apoptotic laddered DNA in global and focal cerebral ischemia in rats, *Soc. Neurosci. Abst.* **20**: 250 (1994).

6. D.W. Choi, J. Koh, M.B. Wie, B.J. Gwag, D. Lobner, L.M.T. Canzoniero, S.L. Sensi, Y.J. Oh & K.L. O'Malley, BCL-2 transfection protects MN9D cells against β-amyloid toxicity and glucose deprivation but not against several other insults, *Soc. Neurosci. Abst.* **20**: 604 (1994).
7. J.M. Dubinsky & S.M. Rothman, Intracellular calcium concentrations during "chemical hypoxia" and excitotoxic neuronal injury, *J. Neurosci.* **11**: 2545-51 (1991).
8. A. Frandsen & A. Schousboe, Excitatory amino acid-mediated cytotoxicity and calcium homeostasis in cultured neurons, *J. Neurochem.* **60**: 1202-11 (1993).
9. M.P. Goldberg & D.W. Choi, Combined oxygen and glucose deprivation in cortical cell culture: calcium-dependent and calcium-independent mechanisms of neuronal injury, *J. Neurosci.* **13**: 3510-24 (1993).
10. D.J. Kane, T.A. Sarafian, R. Anton, H. Hahn, E. Butler Gralla, J. Selverstone Valentine, T. Ord & D.E. Bredesen, Bcl-2 inhibition of neural death: decreased generation of reactive oxygen species, *Science* **262**: 1274-7 (1993).
11. G.L.I.. Koch, C. Booth & F.B.P. Wooding, Dissociation and re-assembly of the endoplasmic reticulum in live cells, *J. Cell Sci.* **91**: 511-22 (1988).
12. M. Lam, G. Dubyak, L. Chen, G. Nunez, R.L. Miesfeld & C.W. Distelhorst, Evidence that BCL-2 represses apoptosis by regulating endoplasmic reticulum-associated Ca^{2+} fluxes, *Proc. Natl. Acad. Sci. USA* **91**: 6569-73 (1994).
13. S.L. Lei, D. Zhang, A.E. Abele & S.A. Lipton, Blockade of NMDA receptor-mediated mobilization of intracellular Ca^{2+} prevents neurotoxicity, *Brain Res.* **598**: 196-202 (1992).
14. P.L. Mellon, J.J. Windle, P.C. Goldsmith, C.A. Padula, J.L. Roberts & R.I. Weiner, Immortalization of hypothalamic GnRH neurons by genetically targeted tumorigenesis, *Neuron*, **5**: 1-10 (1990).
15. J.W. Putney & G. St.J. Bird, The signal for capacitative calcium entry, *Cell* **75**: 199-201 (1993).
16. J.F. Sambrook, The involvement of calcium in transport of secretory proteins from the endoplasmic reticulum, *Cell* **61**: 197-9 (1990).
17. B.K. Siesjo & F. Bengtsson, Calcium fluxes, calcium antagonists an calcium-related pathology in brain ischemia, hypoglycemia and spreading depression: a unifying hypothesis, *J. Cereb. Blood Flow Metab.* **9**: 127-40 (1989).
18. O. Thastrup, P.J. Cullen, B.K. Drobak, M.R. Hanley & A.P. Dawson, Thapsigargin, a tumor promoter, discharges intracellular Ca^{2+} stores by specific inhibition of the endoplasmic reticulum Ca^{2+}-ATPase, *Proc. Natl. Acad. Sci USA* **87**: 2466-70 (1990).
19. M. Tymianski, M.C. Wallace, I. Spigelman, M. Uno, P.L. Carlen, C.H. Tator & M.P. Charlton, Cell-permeant Ca^{2+} chelators reduce early excitotoxic and ischemic neuronal injury in vitro and in vivo, *Neuron* **11**: 221-235 (1993).
20. A.H. Wyllie, J.F.R. Ker & A.R. Currie, Cell death: the significance of apoptosis, *Int. Rev. Cytology* **68**: 251-306 (1980).
21. L. Zhang, Y. Andou, S. Masuda, K. Mitani & K. Kataoka, Dantrolene protects against ischemic, delayed neuronal death in gerbil brain, *Neurosci. Lett.* **158**: 105-8 (1993).
22. L.-T. Zhong, T. Sarafian, D.J. Kane, A.C. Charles, S.P. Mah, R.H. Edwards & D.E. Bredesen, *bcl-2* inhibits death of central neural cells induced by multiple agents. *Proc. Natl. Acad. Sci. USA* **90**, 4533-7 (1993).

CALPAIN, A CATABOLIC MEDIATOR IN SPINAL CORD TRAUMA

Naren L. Banik, Denise Lobo-Matzelle, Gloria Gantt-Wilford, and
Edward L. Hogan

Department of Neurology
Medical University of South Carolina
Charleston, SC 29425

INTRODUCTION

Spinal cord trauma causes tissue damage and necrosis with neurological dysfunction proportional to the severity of injury. The traumatic morphology in the cord lesion correlates with the extent of the neurochemical alterations. The changes include axonal granular degeneration, myelin vesiculation and phagocytosis, losses of axonal and myelin protein and lipids and an accumulation of calcium. Subsequently, there is an increased activity of lysosomal and neutral proteinase in the lesion with the greatest change an elevated activity of neutral proteinase. Although no single factor has been implicated in the pathogenesis of secondary injury, calcium is clearly one of the most important players in spinal cord injury tissue degeneration. Calcium plays many roles in cell function; an important one being enzyme activation of both proteinase and lipase. Lipase activation releases arachidonic acid and initiates a cascade of reactions that are also observed following vascular injury. Calcium also activates the neutral proteinase, calpain, which degrades endogenous substrate cytoskeletal and myelin protein and is associated with neuronal death, axonal degeneration and myelinolysis. This tightly linked series of events suggest a pivotal role for calpain in the tissue destruction of spinal cord trauma. This short chapter deals with the importance of calpain as a catabolic mediator in spinal cord injury. This may serve as a model for events occurring in other brain degenerative disorders.

Changes in Morphology

Light microscopic studies have revealed edema, inflammation, hemorrhage and necrosis in the lesion: these changes appear first in the gray matter and extend to white matter with time and severity of injury.[1,2] Electron microscopic studies of the acute severe cord lesion show changes in axon and myelin at 15 to 30 minutes after trauma. Myelinated axons are swollen with reduction of neurofilaments at 15 minutes and the axon undergoes degeneration; later - by 30 minutes - microtubules and neurofilaments granulate and myelin lamellae loosen.[3] There is splitting of myelin lamellae and vacuolation of myelin as well

Neurodegenerative Diseases
Edited by Gary Fiskum, Plenum Press, New York, 1996

as myelin phagocytosis by inflammatory cells, many of which contain ingested myelin debris.[4-7] Importantly, an accumulation of calcium (hydroxyapatite crystals) and calcification of axonal mitochondria was recognized in the lesion.[1,6,8-10]

Alterations in Protein and Proteolytic Enzymes

There is a progressive loss of myelin and axonal protein in the lesion following injury. The myelin marker enzyme CNPase (adenosine 2',3'-cyclic nucleotide 3'-phosphohydrolase) is significantly decreased (55%) at 72 hours.

The neurofilament proteins (200kD, 150kD and 68kD) show the greatest losses in the lesion. This is observed as early as 15 minutes after injury and correlates with granular axonal degeneration seen ultrastructurally.[3,11] In vitro, these proteins of spinal cord are preferentially degraded in the presence of calcium. They are excellent substrates of purified calpain.[12] The losses of myelin proteins in the lesion progress with time following trauma with myelin basic protein most susceptible. It is extensively degraded (80-90%) at 24 hours after injury compared to proteolipid protein, the loss of which is less dramatic and is decreased by only 30% at one day. A complete loss of myelin and axonal protein accompanies structural dissolution of the axon-myelin unit.

The breakdown of protein in the lesion implicates an increased activity of proteolytic enzymes. In the lesion, there is a progressive increase of lysosomal (cathepsins B and D) and neutral proteinase activity.[13] The greatest increase is in neutral proteinase activity with lesser rises of cathepsin B and D. The degradation in the lesion of neurofilament proteins, the increased neutral proteinase activity and the elevated level of calcium suggests to us that the breakdown of protein is caused by the Ca^{2+} activated neutral proteinase, calpain. These studies led us to propose that calpain is a key determinant in the tissue degeneration of spinal cord trauma.

Calpain in Trauma

In examining calpain's pivotal role in traumatic spinal cord tissue damage and degeneration several features of the enzyme have been defined. These include its immunocytochemical localization, the relation of tissue damage and calpain activity and its tissue level.

Immunocytochemical studies of calpain have been made in the lesion and in the adjacent cord at different times after trauma in both compression and weight drop injury. Calpain immunoreactivity is elevated in the lesion and the increased immunoreactivity is proportional to the extent of trauma and duration after injury.[2] The immunoreactivity also increases adjacent to the lesion. Astrocytes and microglial cells in dorsal and ventral funiculi manifest increased calpain immunostaining. The proliferation in the lesion of astrocytes and microglia has been identified by GFAP (glial fibrillary acidic protein) and GSA (Griffonica Simplifolia Agglutinin) lectin binding staining, respectively. The first appearance of microglia and astrocytes in the lesion is observed after one and three days, respectively. The neurons in ventral horn atrophy and show increased calpain staining. Neuronal losses are evident in the lesion and may stem from intraneuronal calpain increase causing destruction of cytoskeletal protein and neuronal death. The elevated glial calpain suggests that there is glial participation as well in the tissue degeneration. The greater calpain immunoreactivity strongly implies that an increased calpain expression (both mRNA and protein) underlies the release of activity in the lesion. This of course will be clarified by studies at the transcriptional and translational levels that we are currently pursuing.

We have examined calpain activity and content in the lesion at intervals following trauma. The degeneration of endogenous cytoskeletal protein in the lesion has been determined by SDS-PAGE taking the extent of loss of protein (neurofilaments) compared to control as a measure of calpain activity. Protein loss was quantitated from the immunoblot by chemiluminescence. Progressive losses of neurofilament proteins (68kD and 200kD) and MAP_2 (microtubule associated protein) or MAG (myelin associated glycoprotein) are evident in the lesion and correlate with time after injury. There is almost complete loss of 68kD protein at 24 to 72 hours after trauma indicating an increased calpain activity in the lesion (Table 1). Similar changes are also found with MAP_2 and MAG proteins. The loss of cyto-skeletal proteins and increased calpain activity in the cord lesion is similar to that found in brain after head trauma.[14] The latter workers also concluded that the loss of protein was the result of increased calpain activity.

Table 1. Decreased Levels of 68kD Neurofilament Protein in Lesion after Spinal Cord Injury

Time after injury (hours)	% Loss compared to control (sham)
0.5	43 (2)
1	55 (2)
24	65 ± 9.3 (3)
72	92 ± 7.9 (3)

Proteins were separated by SDS-PAGE and the extent of protein loss was quantitated from the immunoblot by chemiluminescence compared to control. Numbers in parentheses indicate number of experiments. Results are expressed as average ± SD where applicable.

Although calpain expression (mRNA) and content have not yet been directly demonstrated in the lesion, the increased immunocytochemical staining implicates increased expression of calpain. In this context, it is noteworthy that calpain expression both at the mRNA and protein levels is increased in lymphoid cells following activation by cytokines and calcium ionophore A23187.[15] In addition to increased calpain immuno-reactivity in the endogenous cells (e.g., glial) in the lesion, inflammatory cells, both neutrophils and monocytes, also showed strong calpain staining. Hence, calpain expression may be regulated by cytokines or other factors in both glial and inflammatory cells.

In order to determine the content of calpain in spinal cord injury, we have examined lesion calpain by western blotting and chemiluminescence. The intensity of the calpain in the lesion progressively increases with time following injury. The increased content of the calpain as quantitated by chemiluminescence also reflects elevated enzyme activity in the lesion.

Since calpain activation is believed to occur at the cell membrane, we wonder whether calpain synthesized in the lesion is translocated from cytosol to the membrane in the course of activation. Lesion and control cords have been fractionated into cytosolic and membrane fractions and the extent of 68kD neurofilament protein loss compared between subcellular fractions. Our preliminary data indicates that there is increased translocation and activation of calpain at the membrane in the cord lesion.

The control cytosol degraded more 68kD neurofilament proteins than that from the lesion. More 68kD neurofilament was degraded by the lesion membrane than control. An elevated calpain activation in the lesion membrane would, of course, degrade membrane protein and foster membranolysis: an hypothesis in keeping with myelin lamellae splitting etc., found in the lesion histology.

The finding of increased calpain activity in the lesion implicates it as potentially pivotal in the tissue degeneration of spinal cord injury (Figure 1).

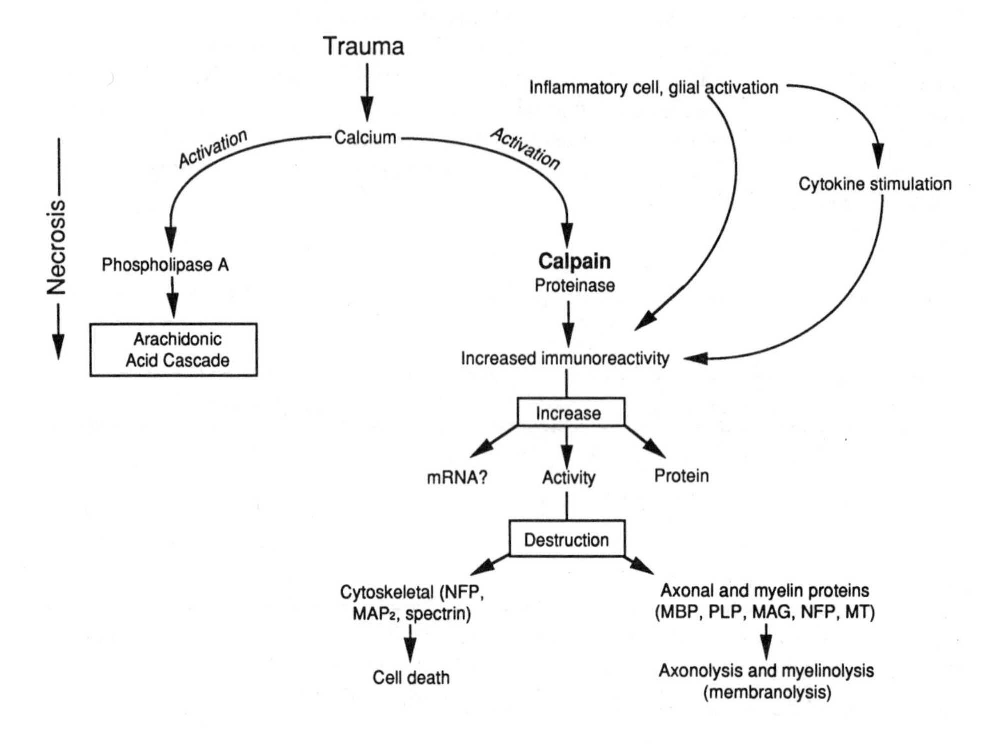

The calpain activity in the adjacent tissue also suggests a "spill over" effect of calpain from the lesion into this area. Such would be entirely consistent with diffusion of cytokines secreted in cells resident in or infiltrating the lesion. In the adjacent tissue neurons may be partially damaged. Whether they will recover their functional capabilities is not known but certainly is plausible, and if the tissue injury is less in the penumbra than the lesion, cell may be preserved by treatment with proteinase inhibitors (neuroprotection). The inhibitors now being examined as therapeutic agents in spinal cord injury may ameliorate tissue damage by preventing protein degradation, thereby preserving cells and their structures.

ACKNOWLEDGEMENT

This work was partly supported by grants NINDS/NS-31622 and NINDS/NS-31767 from the National Institutes of Health and SCRF-1238 from the Paralyzed Veterans of America. We also thank Ms. Lois Godfrey for her skillful secretarial assistance.

REFERENCES

1. J.D. Balentine, E.L. Hogan, N.L. Banik, and P.L. Perot, Jr., Calcium and the pathogenesis of spinal cord injury, *in:* "Trauma of the Central Nervous System," R.G. Cacey, H.R. Winn, R.W. Rimel, and J.A. Jane, eds., Raven Press, New York (1985).

2. Z. Li, E.L. Hogan, and N.L. Banik, Role of calpain in spinal cord injury: Increased mcalpain immunoreactivity in spinal cord after compression in the rat. *Neurochem. Int.* (In press).

3. N.L. Banik, Degradation of cytoskeletal proteins in experimental spinal cord injury. Neurochem. Res. 7:1465(1982).

4. J.D. Balentine, Pathology of experimental spinal cord trauma. Part 2: Ultrastructure of axon and myelin. *Lab. Invest.* 39:254(1978).

5. J. Bresnahan, An electron microscopic analysis of axonal alterations following blunt contusion of the spinal cord of the Rhesus monkey. *J. Neurol. Sci.* 37:59(1978).

6. N.L. Banik, J.M. Powers, and E.L. Hogan, Effects of spinal cord trauma on myelin. *J. Neuropathol. Res.* 39:232(1980).

7. G.J. Dohrman, F.C. Wagner, Jr., P.C. Bucy, The microvasculature in transitory traumatic paraplegia. An electron microscopic study in the monkey. *J. Neurosurg.* 35:263(1971).

8. R.D. Happel, K. Smith, N.L. Banik, J.M. Powers, E.L. Hogan, and J.D. Balentine, Calcium accumulation in experimental spinal cord trauma. *Brain Res.* 211:476(1981).

9. W. Young, V. Yen, and A. Blight, Extracellular ionic activity in experimental spinal cord contusion. *Brain Res.* 253:105(1982).

10. B.T. Stokes, P. Fox, and G. Hollinden, Extracellular calcium activity in the injured spinal cord. *Exp. Neurol.* 80:561(1983).

11. J.M. Braughler, and E.D. Hall, Effects of multidose methylprednisolone sodium succinate administration on injured cat spinal cord neurofilament degradation and energy metabolism. *J. Neurosurg.* 61:290(1984).

12. N.L. Banik, A.K. Chakrabarti, and E.L. Hogan, Role of calcium-activated neutral proteinase in myelin: Its role and function, *in:* "Myelin, Biology and Chemistry," R. Martenson, ed., Telford Press, Caldwell (1992).

13. N.L. Banik, E.L. Hogan, J.M. Powers, L.J. Whetstine, Proteolytic enzymes in spinal cord injury. *J. Neurol. Sci.* 73:245(1986).

14. R. Postmantur, R.L. Hayes, C.E. Dixon, W.C. Taft, Neurofilament 68 and neurofilament 200 protein levels decrease after traumatic brain injury. *J. Neurotrauma* 11:533(1994).

15. R.V. Deshpande, J.M. Goust, A.K. Chakrabarti, E. Barbosa, E.L. Hogan, N.L. Banik, Calpain expression in lymphoid cells. Increased mRNA and protein levels after cell activation. *J. Biol. Chem.* 270(6):2497(1995).

DETECTION OF SINGLE AND DOUBLE STRAND DNA BREAKS DURING EXCITOTOXIC OR APOPTOTIC NEURONAL CELL DEATH

Michel Didier[1,3], Sherry Bursztajn[1,2,3] and Stephen A. Berman[1,2,3]

[1]Department of Psychiatry and [2]Program in Neuroscience,
Harvard Medical School, Boston and
[3]Laboratories for Molecular Neuroscience, McLean Hospital,
Belmont MA 02178.

INTRODUCTION

Excitotoxins have been implicated in a number of neurodegenerative diseases.[1,2] These excitatory amino acids can cause neuronal death by several mechanisms, but their effect on the genome is not well defined. Damage of the genomic as well as mitochondrial DNA may appear during chronic excitotoxicity leading to a gradual and progressive accumulation of long term macromolecular damage.[3,4] Until now, two major categories of genomic DNA degradation have been distinguished in cells. Apoptotic DNA damage results from a specific molecular and biochemical program mediating a cellular death without the manifestations of cytoplasmic alterations such as swelling and lysosomal activation found in the more commonly described necrotic death.[5-7] In most cases, apoptotic cell death involves a specific endonucleic cleavage of double stranded DNA and the cell nuclei adopt a characteristic morphology with fragmentation and perinuclear condensation of chromatin.[5,6,8-10] Aside from apoptosis, less specific DNA strand breaks have many possible causes such as radiation,[11,12] free radicals and other chemical agents.[13-16] Defects in one or more DNA repair enzyme can also lead to the accumulation of DNA damage.[17-19]

Indirect observations have strongly suggested that during glutamate excitotoxicity, early DNA damage may initiate pathological intracellular mechanisms responsible for a later neuronal cell death.[20-21] In particular, a precocious and sustained activation of the poly(ADP-ribose) polymerase by nicked DNA could deplete intracellular energy and lead to ca!cium homeostasis deregulation. Recent reports have primary focused on the potential induction of apoptotic DNA degradation by activation of glutamate receptors *in vitro* or during ischemic injury.[22-28] However, these studies, involving acute overstimulation of glutamate receptors, did not investigate early DNA alterations during the excitotoxic cell

Neurodegenerative Diseases
Edited by Gary Fiskum, Plenum Press, New York, 1996

death. The potential importance of both apoptotic and non-apoptotic DNA damage in the nervous system relates to the possibility that such damage might occur during an early and reversible period of the degenerative processes. For this purpose, we decided to develop a new *in vitro* approach to quantify and visualize the effects of moderate, but not complete, glutamate toxicity on the genome of neuronal cells. Moreover, we compared the characteristics of DNA damage detected in this glutamate-mediated neuronal death to the degradation of genomic DNA observed during apoptosis.

THE NEURONAL CULTURE MODEL

Cerebellar granule cells in primary cultures represent the model of choice for studying excitatory amino acid-dependent neurotoxicity. First, they express several functional receptors for glutamate during their *in vitro* development.[29-31] Second, they differentiate and establish glutamatergic synaptic contacts.[30,32] Third, in particular culture conditions, they represent at least 95% of the cells surviving *in vitro*.[32] During their maturation, granule neurons display progressively higher glutamate sensitivities.[30,31] After 10 days in culture, the major part of glutamate excitotoxicity is mediated by the activation of NMDA receptors. At this time, short NMDA or glutamate exposure, is largely responsible for a dramatic intracellular calcium overload which leads to an extensive neuronal death in the next 18-24 hours.[33,34] To produce a model of moderate and chronic excitotoxicity, we treated granule cells with L-α-Amino Adipate (LAA), a competitive blocker of the cerebellar glutamate transporter.[35,36] Because D-adipic acid is a potential glutamate receptor ligand, we verified by fura2 calcium video imaging that elevated concentration (up to 500 μM) of the L-isomer did not acutely affect either neuronal or glial calcium homeostasis. Moreover, as this compound has been used as gliotoxin in various models, we have also verified that the survival of the few glial cells present in two week-old cultures, was not affected by LAA treatment. The depolarization-induced vesicular release of endogenous glutamate in cerebellar cultures has been shown to reach a plateau after 11 days *in vitro* suggesting a complete maturation of excitatory synaptic contacts at this developmental period.[32] At this culture stage, the basal glutamate concentration in the culture medium does not exceed 5 μM.[34] This endogenous glutamate maintains spontaneous depolarizations of granule neurons in culture.[30,31] After 24 hrs exposure to LAA, there was a 2-3 fold increase in endogenous glutamate levels (Didier et al., submited) which produced a gradual excitotoxicity. Minor morphological transformations in most of the LAA-treated granular cells becomed evident (Figure 1). These alterations consist mainly of cytoplasmic swelling and can be prevented by the addition of a NMDA receptor antagonist such as MK801 (Figure 1). Further experiments, using specific dyes which discriminate between living and dead cells, showed that the neuronal survival was unaffected at least 24 hrs after LAA exposure, and that the removal of extracellular glutamate also reversed LAA excitotoxicity of neuronal cells. Taken together, these results demonstrate that LAA-induced moderate neurotoxicity results from NMDA receptor activation by endogenous glutamate. In the same cultures, a 10 min incubation with 50 μM glutamate killed most of granular cells in few hours (Figure 1). Finally, besides their high sensitivity to excitotoxic cell death, cerebellar granule cells in primary cultures can undergo apoptosis.[37] Indeed, the reduction of the chronic neuronal depolarization in chemically-defined culture medium during the first week *in vitro* results in a dramatic neuronal death in few hours. Granular cells are rescued by depolarizing agents, growth

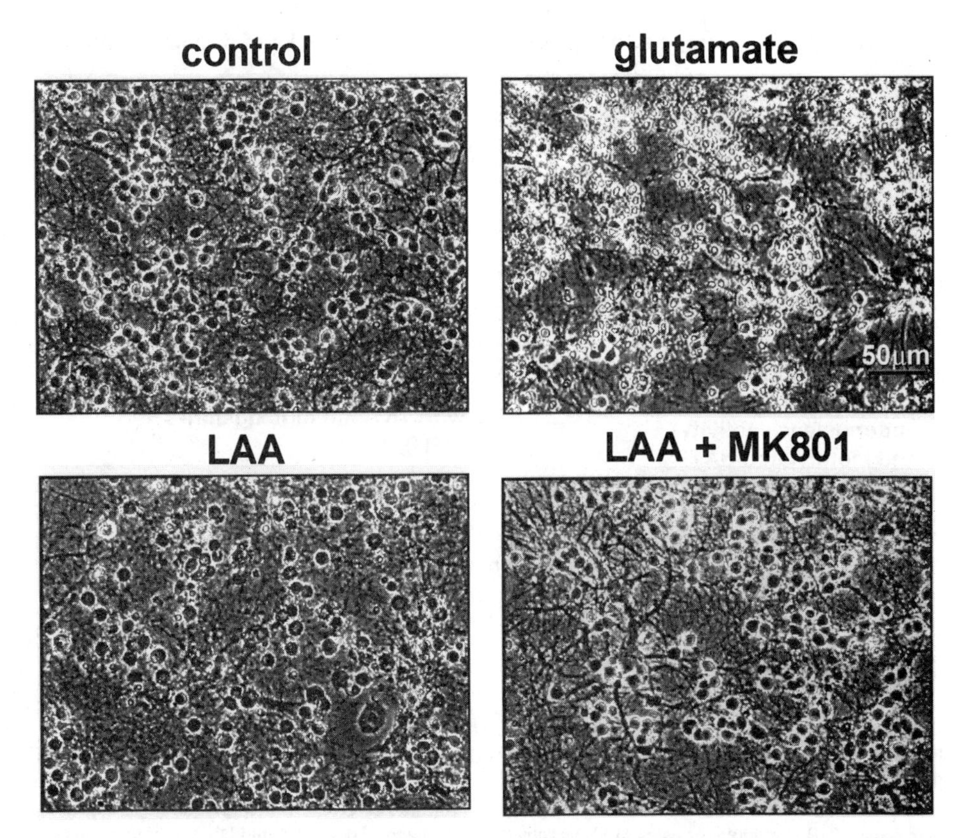

Figure 1. Phase contrast microscopic observations of cerebellar granule cells exposed to LAA or glutamate. A 24 hr 500 μM LAA treatment induced morphological changes of granular cells without affecting their survival. The LAA-mediated cellular swelling was prevented by a NMDA receptor inhibitor (MK801). In comparison, a 10 min incubation with 50 mM glutamate reduced dramatically the number of living neurons.

factors or by the inhibition of RNA or protein synthesis. Moreover, the morphology of these dying neurons shares a number of similarities with well defined apoptotic cells, and a characteristic genomic DNA laddering is observed in electrophoretic gel analysis. Two distinct neuronal cell death types may thus be induced and studied in the same neuronal population developing *in vitro*. Interestingly, both apoptotic and necrotic deaths of cerebellar granule neurons seem also to occur naturally *in vivo* during distinct periods of their development.[38]

THE DNA LABELING PROCEDURES

Two methods were used to reveal predominantly single strand or double strand DNA breaks. Figure 2 shows schematic representation of these procedures. In order to detect the nicks in neuronal cell DNA, we applied the nick-translation reaction using DNA polymerase I. This enzyme is regularly used to label DNA when combined with DNase I, an enzyme that catalyses production of single strand nicks. Employing the polymerase in the absence of exogenous DNase gives a sensitive measure of the number of single strand nicks present in the genomic DNA.[12,13,39] In such experimental conditions, radiolabeled

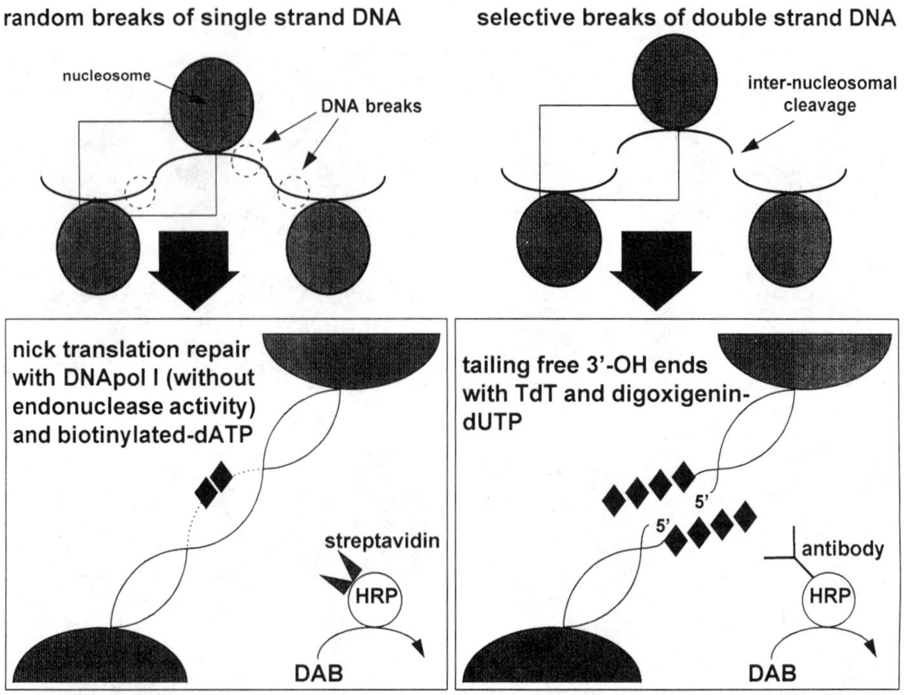

Figure 2. Schema of the experimental procedures for the labeling of single and double DNA strand breaks. Random nicks of genomic DNA were detected by a nick translation reaction with DNA polymerase I and biotinylated deoxynucleotide. The nucleotide incorporation was detected by streptavidin-horse radish peroxydase (HRP) complexes using DAB as substrate. Cleavage of double strand DNA, which occurs between nucleosomes during apoptosis, was visualized by a modified TUNEL method. Free 3'-OH ends were labeled with terminal deoxynucleotidyl transferase and digoxigenin-dUTP. The DNA tailing was revealed with an anti-digoxigenin antibody linked to horse radish peroxidase.

or biotinylated-labeled nucleotides may be used for the nick-translation DNA repair. The incorporated radioactivity can be determined by liquid scintillation counting and biotinylated-dATP is detected *in situ* by the binding of streptavidin-horse radish peroxidase complex (HRP) with 3,3' diaminobenzidine tetrahydrochloride (DAB) as substrate. Double stranded cleavage of DNA can be visualized by a TUNEL method.[40] This experimental procedure has now been extensively applied and commercialized protocols are available. Here, we used the Apoptag kit (ONCOR, MD) to perform modified TUNEL reactions. The TUNEL method takes advantage of the property of an unsual DNA polymerase, namely terminal deoxynucleotidyl transferase (TdT). In the presence of a divalent cation, the TdT preferentially catalyses the addition of deoxynucleotides to protruding 3'-OH termini of DNA molecules. Thus, cleaved double strand DNA may be tailed with up to thousands of deoxynucleotides. In the Apoptag procedure, apoptotic DNA is labeled with digoxigenin-dUTP and *in situ* detected with an antibody linked to HRP followed by a DAB enzymatic degradation.

In a first step to validate the nick translation repair method, we assessed DNA nicking on isolated cerebellar granule cell nuclei. We adapted a nick translation protocol which was developed for detection of carcinogen-induced DNA breaks in fibroblasts.[13] In

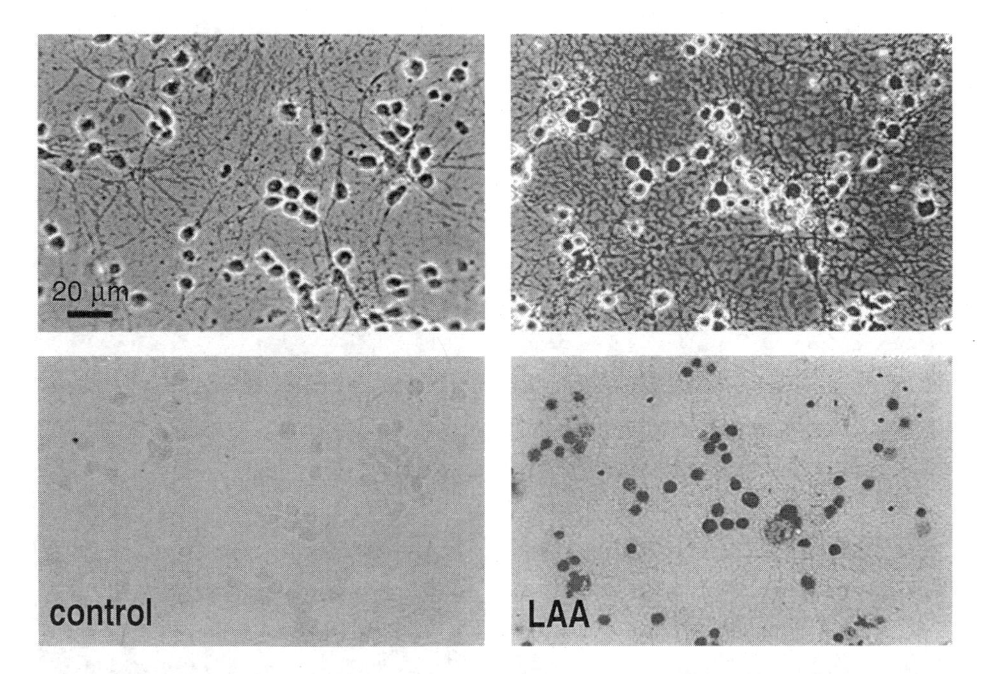

Figure 3. Repair of nick translation-sensitive sites from cerebellar cell nuclei after LAA treatment. Nuclei from normal or treated (24 hr with 500 µM LAA) cerebellar cultures were isolated and then incubated in nick translation mixture (right panel). The radioactive nucleotide incorporation was measured by liquid scintillation counting. An *in situ* enzymatic detection of incorporated biotinylated nucleotide into neuronal nuclei was observed when cells were exposed to LAA (left panel). For both culture conditions (control or LAA treatment), pictures represent the same field under phase contrast or bright field microscopic observations.

brief, cultured granule cells were gently scrapped in ice-cold phosphate buffer saline and lysed with Nonidet P40. After centrifugation on a sucrose cushion, nuclei were counted under phase contrast microscopy.[41] Nick translation experiments were carried out on a given number of isolated nuclei with ^{32}P-dCTP and DNA polymerase I for 20 min at 20°C. The reaction was terminated by the addition of cold 5% trichloroacetic acid (TCA). Then, samples were filtered through glass filters and washed extensively with TCA solution. The radioactivity retained on filters was measured by liquid scintillation counting. Figure 3 shows a typical experiment where a significant increase of incorporation of radioactive nucleotides occurs in nuclei isolated from cells treated with LAA. Nuclei displaying DNA damage can be also specifically labeled by *in situ* nick translation. For this, cerebellar cells grown on coverslips were fixed in a paraformaldehyde-methanol/acetic acid (20:1 v/v) solution and subjected to the nick translation procedure using biotinylated-dATP and a high concentration of DNA polymerase I. As described above, the incorporated biotin was detected with streptavidin-HRP complexes and a final enzymatic transformation of DAB. The nick translation signal is usually very low in non-treated cultures (Figure 3). When cells are exposed for 24 hr to LAA, a strong nuclear labeling in most neuronal cells is observed although no dramatic change in their morphology is appearent (Figure 3). Further experiments clearly showed that only nuclei from neurons but not glial cells were labeled and that some variations in the labeling intensity were present among granule cells.

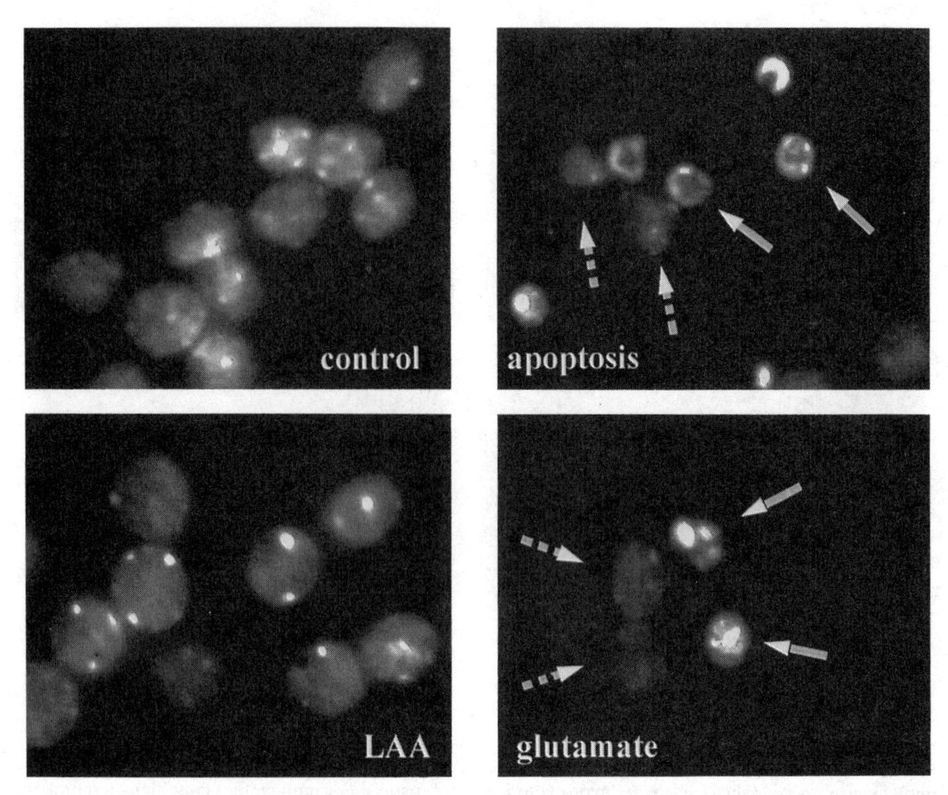

Figure 4. Nuclear structure of cerebellar granule cells during LAA or glutamate excitotoxicity or apoptosis. Cerebellar cultures were fixed with paraformaldehyde/glutaraldehyde solutions and incubated with bisBenzimide. When observed with a microscope equiped with fluorescence optics, nuclei from normal granule cells appear bright and round with a regular shape. No particular modifications of the nuclear morphology were visualized after 24 hr of treatment with 500 μM LAA. In contrast, apoptotic nuclei show severe nuclear transformations such as shrinkage of nuclei with decreased bisBenzimide staining (dashed arrows) and marginated condensed chromatin (arrows). Cellular exposure to 50 μM glutamate for 24 hr also induces similar nuclear morphological alterations except that a spherical condensation of chromatin is ultimitaly observed (arrows).

DETECTION OF SINGLE AND DOUBLE STRAND DNA BREAKS DURING LAA OR GLUTAMATE-MEDIATED NECROSIS AND APOPTOSIS

By combining morphological studies, nick-translation and Apoptag labelings, we compared the characteristics of DNA degradation in chronic or acute excitotoxicity and apoptosis. As already described above, 24 hr LAA treatment produced minor cytoplasmic morphological transformations. Moreover, when observed at higher magnifications with bisBenzimide staining, nuclei of granular neurons display normal morphological characteristics. No chromatin condensation, modifications of the nuclear volume nor shape is evident (Figure 4). In these LAA treated or non-treated cerebellar granule cells, nuclei are regular, round and represent approximately 80% of the cellular volume. In apoptotic neuronal cells, two major nuclear transformations are observed: 1) a nuclear shrinkage characterized by smaller nuclei displaying irregular shapes and a light bisBenzimide staining (Figure 4, dashed arrows) and 2) pyknotic nuclei with marginated chromatin (Figure 4, white arrows). Incubation with high glutamate concentration (50 μM) induces a similar initial nuclear degradation, but a spherical condensation of chromatin is ultimately observed in most dead neuronal cells.

In situ nick translation and Apoptag labelings were performed on cerebellar granule cells exposed to these various treatments (Figure 5). Normal granular neurons are negative to both of these DNA damage staining methods (figure 5, control). When cells display moderate excitotoxicity under LAA treatment, a clear labeling is detected by *in situ* nick translation as previously shown (Figure 3). However, a very low nuclear signal is visualized with the Apoptag procedure. In apoptotic nuclei, intense staining is observed with the two DNA labeling procedures. A uniform (dashed arrows) or a perinuclear (normal arrows) signal is present in these nuclei. These particular distributions correlate with the nuclear localization of condensed chromatin (Figure 4). As shown above, nuclei of neurons display distinct morphologies when undergoing acute glutamate excitotoxicity (Figure 4). When nick translation repair labeling is applied, both types of damaged nuclei have a positive signal. However, pyknotic nuclei display a more intense labeling (Figure 5, normal arrow). In these cells, the TUNEL method reveals a different labeling pattern (Figure 5). A moderate signal occurs in nuclei showing early morphological transformations (dashed arrows). In contrast, the extensively condensed chromatin in pyknotic nuclei is intensely labeled with the Apoptag procedure (normal arrows).

CONCLUSIONS

Numerous recent studies have investigated the potential DNA fragmentation in glutamate-dependent neurotoxicity.[22-28] One major aim for such research is to determine whether excitotoxic cell death may share similar properties with apoptosis. On the other hand, does glutamate neurotoxicity involve an active suicide process common to apoptosis and different from that described for necrosis?. Apoptotic cellular death is characterized by an early double-stranded cleavage of genomic DNA followed by a typical morphological criteria (chromatin condensation and formation of pyknotic nuclei).[6,8-10] It is likely that different endonucleases sequentially digest genomic DNA during this process. In fact, DNA fragmentation might occur in two stages: the first involves double-stranded DNA cleavages into long fragments. This is sufficient to cause the chromatin condensation. A second internucleosomal DNA digestion seems to appear later producing the characteristic laddering of DNA fragments.[10] According to these observations, the TUNEL procedure revealed the presence of double-strand DNA breaks not only in pyknotic nuclei of apoptotic cerebellar granule cells but also in neurons displaying the early minor transformations in their nuclear structures. The time course of molecular and cellular pathological events should be thus considered when investigating the glutamate effects on neuronal DNA. Indeed, double-stranded cleavage of genomic DNA could be detected in excitotoxic granular cells by the TUNEL *in situ* labeling. However, such a staining was only clearly present in pyknotic nuclei which exhibited an advanced and final stage of the neuronal death. This suggested that double strand DNA degradation was not an initial event of excitotoxicity but occured later with the appearance of extended chromatin condensation. Therefore, using another approach for the visualization of damaged DNA, we clearly showed that single strand DNA breaks could be detected in dying neuronal cells during excitotoxicity. Because neurotoxicity mediated by exogenous addition of glutamate is massive and mostly irreversible, we also investigated the potential production of DNA breaks in a slow and moderate excitotoxicity. A chronic treatment of cerebellar granule cell cultures with LAA, a blocker of the membranal glutamate transport, raised extracellular endogenous glutamate sufficiently to activate the NMDA receptor and to induce minor cytoplasmic morphological alterations. Granular neurons were still alive after LAA

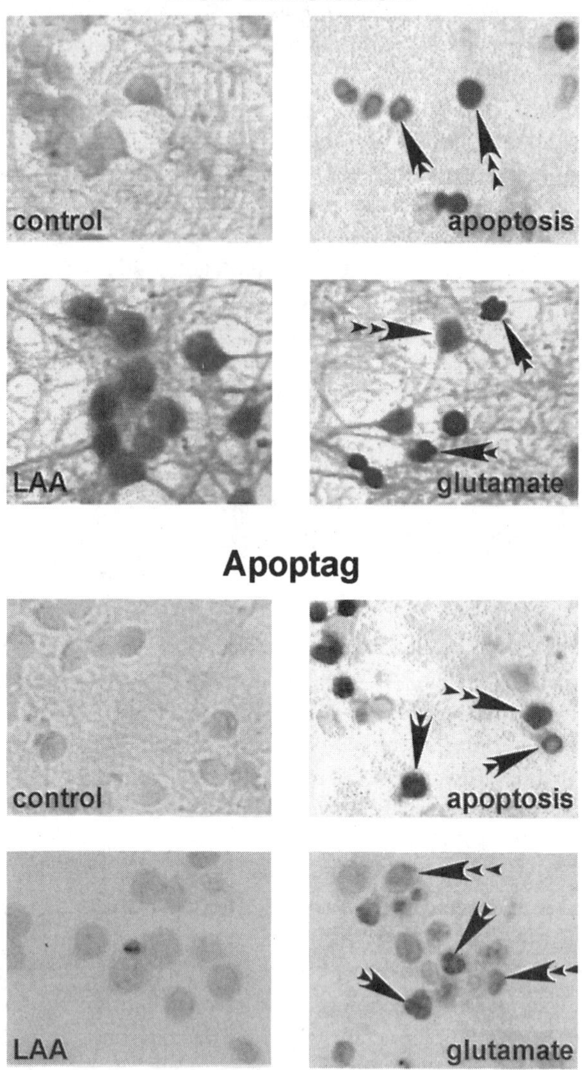

Figure 5. *In situ* nick-translation and Apoptag labelings of genomic DNA damage occuring during LAA or glutamate excitotoxicity and apoptosis. LAA-treated neuronal cells showed a positive signal for DNA damage only with nick-translation repair labeling. Apoptotic nuclei were intensely stained with both of the labeling procedures. Nuclei of cerebellar granule cells exposed to high glutamate concentration displayed strong signals after the nick translation reaction. An intense labeling of condensed chromatin was also seen in these nuclei by the Apoptag procedure.

treatment and the morphology of their nuclei was not affected. However, single-stranded DNA breaks were detected as evidenced by nick translation repair assays. In contrast, the Apoptag procedure did not detect DNA alterations produced by exposure to the glutamate transport inhibitor. Taken together, these results suggest that single strand DNA damage could constitute an early step in a slow but still reversible glutamate excitotoxicity.

Neurotoxicity with similar characteristics could also occur in certain diseases *in vivo*, such as ALS where a perturbation of glutamate transporter has been hypothesized to play a role in motoneuron death.[42] In other neurodegenerative illnesses, a moderate and slow neuronal cell death may also involve chronic glutamate receptor activation.[1,2,7,43] Currently, most *in vivo* studies which detected double stranded breaks of genomic DNA did so in cases where a massive neuronal death was induced in particular brain structures.[23-28] It is likely that such a DNA degradation represent a final step in the neurodegeneration processes. In our study, we demonstrated that the *in situ* nick translation repair could detect early alterations of single strand DNA during a moderate glutamate neurotoxicity. Such a procedure could also help to visualize similar DNA damage occuring during chronic excitotoxicity *in vivo*.

REFERENCES

1. M. Flint Beal, "Mechanisms of excitotoxicity in neurologic diseases" FASEB J. 6: 3338 (1992).
2. J.T. Coyle and P. Puttfarcken, "Oxidative stress, glutamate, and neurodegenerative disorders" *Science* 262: 689 (1993).
3. M.E. Boerrigter, J.Y. Wei and J. Vijg, "DNA repair and Alzheimer's disease" *J. Gerontol.* 47: B177 (1992).
4. V.M. Mann, J.M. Cooper and A.H. Schapira, "Quantification of a mitochondrial DNA deletion in Parkinson's disease" *FEBS. Lett.* 299: 218 (1992).
5. W. Bursh, F. Oberhammer and R. Schulte-Hermann, "Cell death by apoptosis and its protective role against disease" *TIPS* 13: 245 (1992).
6. M.K.L. Collins and A.L. Rivas, "The control of apoptosis in mammalian cells" TIBS 18: 307 (1993).
7. B. Meldrum and J. Garthwaite, "Excitatory amino acid neurotoxicity and neurodegenerative disease" *TIPS* 11: 379 (1990).
8. Y.A. Lazebnick, S. Cole, C.A. Cooke, W. G. Nelson and W.C. Earnshaw, "Nuclear events of apoptosis *in vitro* in cell-free mitotic extracts: a model system for analysis of the active phase of apoptosis" *J. Cell. Biol.* 123: 7 (1993).
9. D. Y. Sun, S. Jiang, L-M. Zheng, D.M. Ojcius and J. D-E. Young, "Separate metabolic pathways leading to DNA fragmentation and apoptotic chromatin condensation" *J. Exp. Med.* 179: 559 (1994).
10. P.R. Walker, V. M. Weaver, B. Lach, J. LeBlanc and M. Sikorska, "Endonuclease activities associated with high molecular weight and internucleosomal DNA fragmentation in apoptosis" *Exp. Cell Res.* 213: 100 (1994).
11. N.J. Sargentini and K.C. Smith, "Involvement of RecB-mediated repair of double strand breaks in the gamma-radiation production of long deletions in Escherichia Coli" *Mutation Res.* 265: 83 (1992).
12. Y. Maehara, H. Anai, T. Kusumoto, Y. Sakaguchi and K. Sakaguchi, "Nick translation detection *in situ* of cellular DNA strand break induced by radiation" *Am. J. Pathol.* 134: 7 (1989).
13. K. Nose and H. Okamoto, "Detection of carcinogen-induced DNA breaks by nick translation in permeable cells" *Biochem. Biophys. Res. Com.* 111: 383 (1983).
14. T.M. Masuck, A.R. Taylor and J. Lough, "Arabinosylcytosine-induced accumulation of DNA nicks in myotube nuclei detected by *in situ* nick translation" *J. Cell. Physio.* 144: 12 (1990).
15. T. Nguyen, D. Brunson, C.L. Crespi, B.W. Penman, J.S. Wishnok and S.R. Tannenbaum, "DNA damage and mutation in human cells exposed to nitric oxide *in vitro*" Proc. *Natl. Acad. Sci. USA* 89: 3030 (1992).
16. A.C. Povey, V.L. Wilson, J.L. Zweier, P. Kuppusamy, I.K. O'Neill and C.C. Harris, "Detection by 32P-postlabeling of DNA adducts induced by free radicals and unsaturated aldehydes formed during the aerobic decomposition of fecapentaene-12" *Carcinogenesis* 13: 395 (1992).
17. S. Seki and T. Oda, "An exonuclease possibly involved in the initiation of repair of bleomycin-damaged DNA in mouse ascites sarcoma cells" *Carcinogenesis* 9: 2239 (1988).
18. W.B. Mattes, "Lesion selectivity in blockage of lambda exonuclease by DNA damage" *Nucleic Acid Res.* 18: 3723 (1990).
19. B.M. Hannigan, S.A. Richardson and P.G. McKenna, "DNA damage in mammalian cell lines with different antioxidant levels and DNA repair capacities" *Exs* 62: 247 (1992).
20. C. Cosi, H. Suzuki, D. Milani, L. Facci, M. Menegazzi, G. Vantini, Y. Kanai and S.D. Skaper, "Poly(ADP-ribose) polymerase: early involvement in glutamate-induced neurotoxicity in cultured cerebellar granule cells" *J. Neurosci. Res.* 39: 38 (1994).
21. J. Zhang, V.L. Dawson, T.M. Dawson and S.H. Snyder, "Nitric oxide activation of poly(ADP-ribose) synthetase in neurotoxicity" *Science* 263: 687 (1994).

22. F. Dessi, C. Charriaut-Marlangue, M. Khrestchatisky and Y. Ben-Ari, "Glutamate-induced neuronal death is not a programmed cell death in cerebellar culture" *J. Neurochem.* 60: 1953 (1993).
23. A. Heron, H. Pollard, F. Dessi, J. Moreau, F. Lasbennes, Y. Ben-Ari and C. Charriaut-Marlangue, "Regional variability in DNA fragmentation after global ischemia evidenced by combined histological and gel electrophoresis observations in the rat brain" *J. Neurochem.* 61: 1973 (1993).
24. J.P. MacManus, A.M. Buschan, I.E. Hill, I. Rasquinha and E. Preston, "Global ischemia can cause DNA fragmentation indicative of apoptosis in rat brain" *Neurosci. Lett.* 164: 89 (1993).
25. T. Tominaga, S. Kure, K. Narisawa and T. Yoshimoto, "Endonuclease activation following focal ischemic injury in the rat brain" *Neurosci. Lett.* 608: 21 (1993).
26. S-I. Kihara, T. Shiraishi, S. Nakagawa, K. Toda and K. Tabuchi, "Visualization of DNA double strand breaks in the gerbil hippocampal CA1 following transient ischemia" *Neurosci. Lett.* 175: 133 (1994).
27. H. Pollard, C. Charriaut-Marlangue, S. Cantagrel, A. Represa, O. Robain, J. Moreau and Y. Ben-Ari, "Kainate-induced apoptotic cell death in hippocampal neurons" *Neurosci.* 63: 7 (1994).
28. Y. Sei, D.K.J.E. Von Lubitz, A.S. Basile, M.M. Borner, R.C-S. Lin, P. Skolnick and L.H. Fossom, "Internucleosomal DNA fragmentation in gerbil hippocampus following forebrain ischemia" *Neurosci. Lett.* 171: 179 (1994).
29. M.J. Courtney, J.L. Lambert and D.G. Nicholls, "The interactions between plasma membrane depolarization and glutamate receptor activation in the regulation of cytoplasmic free calcium in cultured cerebellar granule cells" *J. Neurosci.* 10: 3873 (1990).
30. M. Didier, M. Heaulme, N. Gonalons, P. Soubrie and J-P. Pin, "35 mM K+-stimulated 45Ca2+-uptake in cerebellar granule cell cultures mainly result from NMDA receptor activation" *Eur. J. Pharmacol.* 244: 57 (1993).
31. M. Didier, J-M. Mienville, P. Soubrie, J. Bockaert, S. Berman, S. Bursztajn and J-P. Pin, "Plasticity of NMDA receptor expression during mouse cerebellar granule cell development" *Eur. J. Neurosci.* 6: 1536 (1994).
32. B.J. Van-Vliet, M. Sebben, A. Dumuis, J. Gabrion, J. Bockaert, J-P. Pin, "Endogenous amino acid release from cultured cerebellar neuronal cells: effect of tetanus toxin on glutamate release" *J. Neurochem.* 52: 1229 (1989).
33. H. Manev, M. Favaron, A. Guidotti and E. Costa, "Delayed increase of Ca2+ influx elicited by glutamate: role in neuronal death" *Mol. Pharmacol.* 36: 106 (1989).
34. M. Didier, M. Heaulme, P. Soubrie, J. Bockaert and J-P. Pin, "Rapid, sensitive and simple method for the quantification of both neurotoxic and neurotrophic effects of NMDA on cultured cerebellar granule cells" *J. Neurosci. Res.* 27: 25 (1990).
35. M.B. Robinson, M. Hunter-Ensor and J. Sinor, "Pharmacologically distinct sodium-dependent L-[3H]glutamate transport processes in rat brain" *Brain Res.* 544: 196 (1991).
36. Y. Kanai, C.P. Smith and M.A. Hediger, "The elusive transporters with a high affinity for glutamate" *TINS* 16: 365 (1993).
37. S.R. D'Mello, C. Galli, T. Ciotti and P. Calissano, "Induction of apoptosis in cerebellar granule neurons by low potassium: Inhibitionof death by insulin-like growth factor I and cAMP" *Proc. Natl. Acad. Sci. USA* 90: 10989 (1993).
38. K.A. Wood, B. Dipasquale and R.J. Youle, "In situ labeling of granule cells for apoptosis-associated DNA fragmentation reveals different mechanisms of cell loss in developing cerebellum" *Neuron* 11: 621 (1993).
39. S. Iseki, "DNA strand breaks in rat tissues as detected by in situ nick translation" *Exp. Cell Res.* 167: 311 (1986).
40. Y. Gavrieli, Y. Sherman and S.A. Ben-Sasson, "Identification of programed cell death in situ via specific labeling of nuclear DNA fragmentation" *J. Cell. Biol.* 119: 493 (1992).
41. M. Didier, P. Roux, M. Piechaczyk, P. Mangeat, B. Verrier, G. Devilliers, J. Bockaert and J-P. Pin, "Long-term expression of c-fos protein during the *in vitro* maturation of cerebellar granule cells induced by potassium or NMDA" *Mol. Brain Res.* 12: 249 (1992).
42. S.H. Appel, "Excitotoxic neuronal cell death in amyotrophic lateral sclerosis" *TINS* 16: 3 (1993).
43. G.J. Lees, "Contributory mechanisms in the causation of neurodegenerative disorders" Neurosci. 54: 287 (1993).

SIGNALLING FOR SURVIVAL: POTENTIAL APPLICATIONS OF SIGNAL-TRANSDUCTION THERAPIES FOR SUPPRESSION OF APOPTOSIS IN THE NERVOUS SYSTEM

Aviva M Tolkovsky
Department of Biochemistry
University of Cambridge
Tennis Court Road
Cambridge CB2 1QW, UK.

INTRODUCTION

One of the prime aims of therapies for neurodegenerative disorders is to find ways of preventing neuronal cell death. At the basis of this quest lie two major questions: (i) what is the molecular basis of death, with particular reference to death commitment point, and (ii) how can this process be prevented or slowed down? Some clues to these mechanisms can be obtained by studying neurons from the developing peripheral nervous system, where a process of limited death occurs when the amount and location of innervation are being matched to the requirements of the targets[1]. It was within this context that the discovery of the first nerve cell survival factor, NGF, was made by Levi-Montalcini and colleagues[2]. Since that time, several other neuronal survival factors have been discovered and it is now well established that neuronal survival factors include not only the NGF-family of neurotrophins (NGF, BDNF, NT3, and NT4/5) but also a range of cytokines and other polypeptide growth factors[3]. Moreover, some adult neurons may produce their own survival factors by autocrine or paracrine mechanisms. Despite these exciting advances, and the breakthroughs that have come with the identification of the *trk* family of receptors which mediate the actions of each of the neurotrophins[4], it is still not known how survival factors signal for survival. Here, I summarise some of our studies on the mechanisms used by NGF-dependent superior cervical ganglion (SCG) neurons to prepare for death commitment point and outline possible mechanisms that may be utilised for preventing their death due to survival-factor-deprivation. Three main concepts are discussed:

1. That rescue by NGF and other survival factors may be mediated by post-translational mechanisms.
2. That parallel survival signalling pathways exist, some of which may not require the activation of cell-surface receptors.
3. That it may take longer for NGF to rescue neurons than the time its takes for NGF to initiate signalling cascades.

SUPPRESSION OF THE DEATH PROGRAMME

About 1/3 of the rat SCG neurons die between 3-6 days postnatally[5] by a stochastic process in the sense that no specific neurons are pre-selected to die. This is unlike the nematode paradigm, where a genetic programme specifies that the same sets of particular neurons (and other cells at other times) will always die[6]. Why do SCG neurons die? The model outlined by Barde[1], that death is due to a limiting supply of survival factors which are competed for at the target by the innervating neurons, has intrinsic logic in that a delicate balance is thereby established where each neuron that dies leaves more NGF to ensure that the remaining neurons are viable. SCG neurons removed from neonate rats are thus poised to die, allowing the study of neurons at their peak of vulnerability in relation to NGF-withdrawal. Each day thereafter, however, the surviving neurons develop a decreased dependence on NGF although the dependence on NGF for survival is never totally abrogated[7]. Whether this independence is induced as part of the intrinsic mechanism of action of NGF or other survival factors is not yet clear. Understanding this phenomenon should help reveal whether there is more than one strategy which can be used by a single cell type to prevent cell death.

Inhibition of death is a post-translational phenomenon

SCG neurons die by apoptosis, including membrane blebbing, cytoplasmic granulation and vacuolisation, and what is sometimes considered the hallmark of apoptosis, DNA condensation and its fragmentation into 180-200 base pair 'ladders', all of which occur before the plasma membrane becomes permeable[8,9]. All neurons showing signs of apoptosis are already committed to die. The key question is therefore what happens immediately prior to the onset of apoptosis, when the neurons can still be rescued? Two phases have been identified between the time of NGF withdrawal and the time of death commitment, (i) a phase when death is prevented either by the neurotrophins or by inhibitors of RNA and protein synthesis[10] and (ii) a phase when such inhibitors fail to rescue the neurons, but one which allows NGF and other survival factors to continue to do so for a further 5-10 hours[8].

Because rescue by NGF during this second phase can occur even in the absence of protein synthesis, rescue within this phase is most likely to be mediated by activating post-translational mechanisms. Moreover, if the end of the period of rescue by inhibitors indicates the expression of 'death protein(s)'[10], then, since NGF, cAMP and high K^+ can still rescue these neurons, these 'death proteins' do not absolutely commit the neurons to die because their actions, or those of downstream effectors, can be directly suppressed by post-translational actions of survival factors.

Younger neurons have different requirements compared to older neurons

Is death prevention by inhibitors of RNA and protein synthesis a hallmark of neuronal cell death? The answer is still unclear. The phenomenon of prevention of cell death by protein synthesis inhibitors was first stipulated to be a part of the apoptotic process by Wyllie et al[11] who were examining cells that were induced to die by steroids (whose receptors are transcription factors) and was first described for SCG neurons by Martin et al.[10]. Indeed, some genes whose expression increases during the period of NGF-deprivation and which may be causally linked to the process of preparation for death due to the production of oxygen free radicals are beginning to be identified[12]. However, oxygen radical damage may be less of a problem *in vivo*, where the environment is much more anaerobic.

Moreover, while this gene-expression-dependent phase of rescue is easily demonstrated in established 7-day SCG neuronal cultures where there is a 12-16 hr lag time between the removal of NGF and the death of the first neurons, it is hardly apparent in newly isolated neurons, which display a shorter lag time of about 3-6 hr before the onset of death[9]. Yet, these 'young' neurons share with the older neurons the competence to die by apoptosis, and they are efficiently rescued by survival factors. That the younger neurons may pre-express a competent, but latent, apoptotic machinery, is supported by the finding that some inhibitors (such as teniposide or mitoxanthrone, which stabilise topoisomerase II with DNA in a cleavable complex, or nucleoside analogues commonly used for cancer treatment) promote the activation of apoptosis in these neurons in the presence of NGF[13]. These 'younger' neurons are thus similar to non-neuronal cells where inhibition of *de novo* gene expression not only does not prevent death, but may actively induce death[14]. The mechanisms by which drugs (many of which are used to treat cancer) activate the execution of apoptosis in the presence of NGF is of great interest because understanding their action may help to pin point crucial mediators of death, and explain why a neurotrophin may lose its potential to act as a survival factor.

Another difference is that young neonatal neurons also cannot be rescued by depolarization with high K^{+}[15], the age-gap phenomenon for high K^{+} rescue being previously observed in chick sensory neurons[16]. Since the 'young' and older neurons appear to contain the same types of voltage-dependent calcium channels[17], it may be that either the sensory mechanisms for transducing the calcium signals are unavailable in the younger neurons, or that the calcium-induced survival signals are over-ridden by inappropriately localised elevated calcium due to impaired calcium homeostasis. That high K^{+} is not non-specifically toxic to the neurons is shown by their dying by apoptosis with normal kinetics (S N Edwards, unpublished data).

A labile-activator of death as an alternative model

One action of inhibitors of RNA and protein synthesis of potential interest in SCG neurons is their induction over several hours of a 3-fold increase in peak influx of calcium through voltage-dependent calcium channels[18]. This must be due to a change in the conductance of the pre-existing calcium channels, since no new proteins are being synthesised during the period of exposure to the inhibitors. It appears therefore that calcium channel permeability is kept normally under tonic inhibition by a labile inhibitor with a half-life of about 15 hr. Aside from the possibility that the inhibitors of protein and RNA synthesis alter calcium homeostasis, which by itself may alter the sensitivity of the neurons to NGF-deprivation (as discussed above), these observations highlight the putative role of labile proteins, and perhaps that of proteolytic processing, in maintaining key physiological functions in the neurons. Thus, it is possible that NGF-maintained neurons which already express a competent cell death programme do not die because rapidly-turning over proteins that orchestrate the actions of more stable inhibitor(s) of cell death are inhibited post-translationally by NGF. According to this scenario, and using a protease as a model for the labile protein, withdrawal of NGF would allow the labile proteases to be activated and remove these inhibitors, but inhibition of protein synthesis after NGF withdrawal would cause the disappearance of labile proteases before they had the opportunity to degrade the inhibitors of cell death. In younger neurons, then, inhibition of protein synthesis would not rescue because the proteases might be more stable relative to the death inhibitors. Since the discovery that the death gene *ced* 3 codes for a protease with homology to ICE[18] (the interleukin 1 beta converting enzyme) many novel ICE-like proteases (enzymes that cleave substrates after the amino acid aspartate) have been implicated in the apoptotic process induced by a wide variety of insults.

Older Neurons may develop cooperative mechanisms to prevent accidental cell death

It has already been mentioned that older neurons die more slowly upon NGF-deprivation compared to younger neurons. One mechanism by which the older neurons may develop more resistance to NGF deprivation is to build more control over the rate limiting steps that intervene between NGF withdrawal and the process of death commitment (here defined as the time at which neurons can no longer be rescued by NGF). Clearly, the larger the number of conditions that need to be fulfilled to activate the death-effector processes, the more resistant the cell becomes to cell death by accident. An analysis of the shape of the rescue curves[9] reveals that the controlling steps of death commitment are highly cooperative. One way to implement such cooperativity is to require that a group of separate reactions be active at the same moment in order to affect cell death. These proteins would thus constitute a 'death set'. It is also possible, however, that a rate-limiting death-committing protein is activated by an allosteric modulator, or that the cooperativity is due to an autocatalytic mechanism (such as that which might be predicted for ICE, which is activated by its self-cleavage at an aspartate residue).

In summary, the differences between younger and older neurons demonstrate that a single type of neuron may alter the rate limiting step of death depending on its stage of development, or which neurotrophin is used for its survival[19]. Clearly, the most urgent task is to define the molecular identity of the processes underlying death commitment point. Whether the proteins that are required for the gene expression-dependent type of death are those that are already expressed in neurons that do not require *de novo* gene expression remains to be resolved, as does the identification of the genes involved. When these become known, it may become possible to predict which cells are likely to die by one mechanism or the other, the differences being crucial for evolving strategies for protecting against accidental cell death.

MECHANISMS THAT SUPPRESS IMPLEMENTATION OF THE CELL DEATH PROGRAMME

The finding that SCG neurons *in vitro* can be rescued by several survival factors which begin their signalling from different starting points potentially provides a means of identifying several components of the survival signalling pathways. It is of course possible that survival signals converge on a central 'hub'. Alternatively, if death commitment is driven by a 'death set', survival factors may operate non-convergent parallel pathways since inactivation of any one member of the set would be sufficient to disable the entire set from activating apoptosis.

p21Ras Sits At A Node Of Several Survival Signal Transducing Pathways.

In SCG neurons, NGF activates the homodimerisation and tyrosine phosphorylation of TrkA (of which there are two forms of 140 and 115 kDa in SCG neurons[20]) which in turn mediates the phosphorylation of two prominent proteins of 94 and 74 kDa. Activation of TrkA causes the activation of p21Ras in SCG and other primary neurons, demonstrated by enhanced exchange of GTP for GDP. In SCG neurons, the activation of p21Ras appears to be both necessary for the survival response to NGF and sufficient for survival *per se* because (i) oncogenic or cellular p21Ras promote survival (and neurite regeneration) (C.D. Nobes, J. Reppas, A. Markus and A.M. Tolkovsky, submitted) and (ii) blocking antibodies to p21Ras in the presence of NGF cause the neurons to die[20]. Similar results were found previously by Borasio et al.[21,22] for chick sensory neurons but, surprisingly, not for chick sympathetic neurons[22]. Elucidating the mechanisms which account for the differences between these two cell types in terms of signalling in response to NGF will be of immense interest. Interestingly, in the PC12 cell line which has been used extensively for studies of NGF-action, the forced expression of a dominant negative Ras analogue, N17Ras, was found to actively promote survival after serum and NGF withdrawal[23]. Taking these results at face value, it might then be predicted that activation of Ras would cause the cells to die. Since the latter experiment has been done many times, surely, such death would not have gone unnoticed. Thus, the question of why N17Ras, and also p21Ras, promote survival in the same cell is not clear. One hypothesis[23] first put forward by Rubin and colleagues[24] is that PC12 cells die by attempting to enter the cell cycle at an inappropriate stage, and that N17Ras rescues them from this fate. Prevention of death in PC12 cells and SCG neurons may thus be fundamentally different.

The cell-permeant non-hydrolysable analogue of cyclic AMP (CPTcAMP) is also an excellent long-term survival factor for SCG neurons in vitro[25], especially when the neurons are plated on laminin[20]. Despite substantial neurite regeneration, the cell bodies remain small compared to the hypertrophy found with NGF. This might be reflecting the lack of induction of MAP kinase by cAMP[26] which in turn is an upstream regulator of the PHAS-1-eIF4e initiation of protein synthesis cascade[27] (see further below). The survival effect of CPTcAMP is not dependent on neurite regeneration, however, since neurons in suspension maintained by CPTcAMP for two days are all rescued when plated on laminin in the presence of NGF or CPTcAMP (K Virdee, unpublished data). The survival activity of CPTcAMP, but not that of NGF, is directly mediated by PKA, shown by the ability of the inactive Rp stereoisomer of thio-cyclic AMP (Rp-cAMPS) to block exclusively the survival induced by CPTcAMP, but not those induced by NGF or the cytokines CNTF or LIF[20]. Additional evidence for PKA as an independent starting point in the signalling cascade is our finding that forced expression of the catalytic unit of PKA cloned into a disabled genomic Herpes Simplex vector can promote survival[28]. One interesting aspect of these experiments was that as a result of overabundant PKA activation, the cells responded by making more of the PKA inhibitory regulatory subunit R(I). This counter-response emphasises the point that therapy by overexpression of any endogenous gene might be short-lived. For this reason, overexpression of any protein should be followed by an inspection of whether functional compensation is occuring.

CNTF (a survival factor first discovered for its potent survival actions on ciliary ganglion parasympathetic neurons[29] and LIF (whose beta receptor subunit is also part of the CNTF signalling-receptor complex[30]) activate the IL-6 receptor signalling subunit gp130. CNTF and LIF are dual-function cytokines, promoting short term (<3 days) survival of purified SCG neurons but killing SCG neurons after longer exposures, even in the presence of NGF[31,32]. Like CPTcAMP, survival and regeneration in neonatal neurons in response to CNTF or LIF are also enhanced when the neurons are plated on laminin, although it is clear that the receptors for CNTF and LIF are capable of transmitting signals even when the neurons are plated on collagen (for example, over 85% of the neurons plated on collagen express c-Fos protein in response to LIF, C Nobes, unpublished data). Why the cytokines induce cell death in the older neurons in the presence of NGF is obviously of prime interest because it may pinpoint some aspect of the NGF-survival signal transduction machinery downstream from TrkA that is only found in older neurons. Thus here we have the opposite effect of high K^+ in young neurons, namely that only the older neurons are vulnerable to the toxic effects of the cytokines. Surprisingly, the survival-promoting actions of all these survival factors is blocked by neutralising antibodies to p21Ras[20], suggesting that p21Ras may indeed form a point of convergence (or 'hub') for several survival signal transduction pathways in SCG neurons.

Multiple non convergent kinase cascades may be utilised for survival of SCG neurons.

In many cell systems p21Ras activates a kinase cascade that leads to an activation of ERKs, which are members of the MAP kinase family. We therefore examined whether MAP kinases are activated in SCG neurons in response to each of the survival factors[26]. However, we found no correlation between ERK activity and survival. Thus, although NGF activated ERK1 (p44) and ERK2 (p42) robustly and persistently for as long as it remained present in the medium (up to 2 days), we found that CPTcAMP had no effect on the ERKs in the absence of NGF, nor was it inhibitory in the presence of NGF. Furthermore, although the cytokines briefly activated the ERKs, this effect subsided within 30 min but survival was extended for 3 days. Finally, when NGF was added briefly to mimic the brief activation of the ERKs induced by the cytokines, no survival extension was obtained. Taken together, these results suggest that there exists a bifurcation in the signalling pathway downstream from p21Ras. The branch leading to MAP kinase activation may have important cellular functions, for example it may be involved in mediating the long-term, hypertrophic effects of NGF, but its activation is clearly not necessary for survival. Therefore, another signalling pathway which branches out from p21Ras appears to mediate the survival signals, perhaps in conjunction with co-activation of other kinases. Thus, we propose that SCG neurons, and by extension other types of neurons, may contain multiple signalling pathways that can be utilised to promote survival. The challenge now is to define the identity of these kinases and, most importantly, to define their intracellular targets.

HOW LONG DOES IT TAKE NGF TO RESCUE NEURONS FROM DEATH?

One feature of experiments in which addition of NGF is used to rescue neurons, is the finding that some neurons that look healthy cannot be rescued by NGF. By performing experiments we call 'punctuated rescue' where the period of NGF withdrawal is punctuated by brief additions of NGF, we have found that even though the NGF signalling pathway down to the ERKs is activated, this is not sufficient to promote survival. The intriguing possibility is that although signalling by NGF is very rapid, the operation of rescue takes longer, perhaps because the post-translational actions of NGF must be accompanied by a disappearance of one of its protein targets. The search for these targets is underway.

CONCLUSIONS AND FUTURE PROSPECTS

(i) We clearly need to discover more about the molecular processes that lead up to and control death commitment point. In this context, signs of apoptosis may be of least interest for studying the problem of neuronal cell death because once apoptosis has begun, neurons cannot be rescued.

(ii) It is clear that apoptosis in some neurons can be suppressed by activation of well characterised signalling pathways, and we suggest that day-to-day death-suppression most likely involves post-translational mechanisms. However, the act of death-suppression may not occur as quickly as the rapid signalling events initiated by NGF may suggest.

(iii) The therapeutic implications are that multiple signalling pathways may be available for survival in any one cell and therefore inactivation of any one pathway will not necessarily cause these neurons to become irreversibly committed to die. Thus, even if a supply of NGF became limiting, it might still be possible to extend survival of these neurons by utilising alternative signalling mechanisms. How these findings extend to other mammalian systems is now of prime interest.

Acknowledgements

I am indebted to my colleagues Anne Buckmaster, Earl Clarke, Sue Edwards, Ruth Murrell, Kate Nobes, Chris Tomkins and Kanwar Virdee, whose experiments have been described here, and to Bob Amess, for their seminal contributions to the lab, and to Viv Wilkins and Molly Sheldon for unfailing technical support. The work on Ras could not have been done without the help of Gian Borasio, Ludwig-Maximilians University, Munich, and Annette Markus and Rolf Heumann, Ruhr University, Bochum, Germany. I thank The Wellcome Trust, the MRC and Action Research for their support.

REFERENCES

1. Y.-A. Barde, Trophic factors and neuronal survival. *Neuron* 2:1525 (1989).
2. R. Levi-Montalcini, The nerve growth factor: its mode of action on sensory and sympathetic nerve cells. *Harvey Lect.* 60: 217(1965).
3. P. Ernfors, K.F. Lee, and R. Jaenisch, Target-derived and putative local actions of neurotrophins in the peripheral nervous-system. *Prog. Brain Res.* 103:43 (1994).
4. D.R. Kaplan and R.M. Stephens, Neurotrophin signal-transduction by the trk receptor, *J. Neurobiol.* 25: 1404 (1994).
5. I.A. Hendry and J. Campbell, Morphometric analysis of rat superior cervical ganglion after axotomy and NGF treatment, *J. Neurocytol.* 5: (1976).
6. R.E. Ellis, J.Y. Yuan, and H.R. Horvitz, Mechanisms and functions of cell-death, *Ann. Rev. Cell Biol.* 7: 663 (1991).

7. K.G. Ruit, P.A. Osborne, R.E., Schmidt, E.M. Johnson, And W.D. Snider, Nerve Growth-Factor Regulates sympathetic-ganglion cell morphology and survival in the adult-mouse, *J. Neurosci.* 10:2412 (1990).
8. S.N. Edwards, A.E. Buckmaster, and A.M. Tolkovsky, The death programme in sympathetic neurons can be suppressed at the post-translational level by NGF, cyclic AMP and depolarization. *J. Neurochem.* 57: 2140 (1991).
9. S.N. Edwards and A.M. Tolkovsky, Characterisation of apoptosis in cultured rat sympathetic neurons after nerve growth factor (NGF) withdrawal, *J. Cell Biol.* 124:537 (1994).
10. D.P. Martin, R.E. Schmidt, P.S. DiStephano, O.H. Lowry, J.G. Carter, and E.M. Johnson Jr, Inhibitors of protein synthesis and RNA synthesis prevent neuronal cell death caused by nerve growth factor deprivation. *J. Cell Biol.* 106:829 (1988).
11. A.H. Wyllie, J.F.R. Kerr, and A.R. Currie, Cell Death: the significance of apoptosis. *Int. Rev. Cytol.* 68:251 (1980).
12. I Silos-Santiago, L.S.J. Greenlund, E.M. Johnson Jr, and W.D. Snider, Molecular genetics of neuronal survival, *Curr Opinion Neurobiol* 5:42 (1995).
13. C.E. Tomkins, S.N. Edwards, and A.M. Tolkovsky, Apoptosis is induced in postmitotic neurons by arabinosides and topoisomerase II inhibitors in the presence of NGF. *J. Cell Sci.* 107:1499 (1994).
14. D.L. Vaux and I.L. Weissman, Neither macromolecular synthesis not Myc is required for cell death via the mechanism that can be controlled by Bcl2, *Mol. Cell Biol.* 13:7000 (1993).
15. J.L. Franklin and E.M. Johnson, Suppression of programmed neuronal death by sustained elevation of cytoplasmic calcium. *Trends Neurosci.* 15:501 (1992).
16. A. Acheson, Y.-A. Barde, and H. Thoenen, High K-mediated survival of spinal sensory neurons depends on developmental age. *Exp. Cell Res.* 170:56 (1987).
17. R.D. Murrell and A.M. Tolkovsky, Role of Ca^{2+} channels and intracellular Ca^{2+} in rat sympathetic neuron survival and function promoted by high K^+ or cyclic AMP in the presence or absence of NGF. *Eur. J. Neuroscience* 5:1261 (1993).
18. J.Y. Yuan, S. Shaham , S. Ledoux, H.M. Ellis, and H.R. Horvitz, The c-elegans cell-death gene ced-3 encodes a protein similar to mammalian interleukin-1-beta-converting enzyme. *Cell* 75:641 (1993).
19. T.E. Allsopp, S. Wyatt, H.F. Paterson, and A.M. Davies, The proto-oncogene bcl2 can selectively rescue neurotrophic factor-dependent neurons from apoptosis. *Cell* 73:295 (1993).
20. C.D. Nobes and A.M. Tolkovsky, Neutralising anti-$p21^{ras}$ Fabs suppress rat sympathetic neuron survival induced by NGF, LIF, CNTF and cyclic AMP. Eur. J. Neurosci. 7:340 (1995).
21. G.D. Borasio, J. John, A. Wittinghofer, Y.-A. Barde, M. Sendtner, and R. Heumann Ras p21 protein promotes survival and fibre outgrowth of cultured embryonic neurons. *Neuron* 2:1087 (1989).
22. G.D. Borasio, A. Markus, A. Wittinghofer, Y.-A., and Heumann, Involvement of ras p21 in neurotrophin-induced response of sensory but not sympathetic neurons. *J. Cell Biol.* 121:665 (1993).
23. G. Ferrari and L.A. Greene, Proliferative inhibition by dominant-negative ras rescues naive and neuronally differentiated pc12 cells from apoptotic death. *EMBO J.* 13:5922 (1994).
24. L.L. Rubin, K.L. Philpott, and S.F. Brooks, Apoptosis - the cell-cycle and cell-death, *Current Biol.* Vol.3 No.6 pp.391 (1993).
25. R.E. Rydel and L.A. Greene, cAMP analogs promote survival and neurite outgrowth in cultures of rat sympathetic and sensory neurons independently of nerve growth factor. *Proc. Natl. Acad. Sci. USA.* 85:1257 (1988).
26. K. Virdee and A.M. Tolkovsky, Activation of p42 and p44 MAP kinases is not essential for suppressing apoptosis in NGF-dependent sympathetic neurons. *Eur J Neurosci,* in press (1995).
27. T.A. Lin, X.M. Kong, T.A.J. Haystead, A. Pause, G. Belsham, N. Sonenberg, and J.C. Lawrence, Phas-1 as a link between mitogen-activated protein-kinase and translation initiation, *Science* 266: 653 (1994).
28. A.E. Buckmaster and A.M. Tolkovsky, Expression of the cyclic AMP-dependent protein kinase (PKA) catalytic subunit from a herpes simplex vector extends the survival of sympathetic neurons in the absence of NGF, *Eur J Neuroscience* 6:1316 (1994).
29. G. Barbin, M. Manthorpe, and S. Varon, Purification of the eye ciliary neurotrophic factor, *J. Neurochem.* 43:1468 (1984).
30. N.Y. Ip, S.H. Nye, T.G. Boulton, S. Davis, T. Taga, L. Yanping, S.J. Birren, K. Yasukawa, T. Kishimoto, D.J. Anderson, N. Stahl, and G.D. Yancopoulos, CNTF and LIF act on neuronal cells via shared signalling pathways that involve the IL-6 signal transducing receptor component gp130, *Cell* 69:1121 (1992).
31. J.A. Kessler, W.H. Ludlam, M.M. Freidin, D.H. Hall, M.D. Michaelson, and D.K. Batter, Cytokine-induced programmed death of cultured sympathetic neurons. *Neuron* 11:1123 (1993).
32. P. Burnham, J.-C. Louis, E. Magal, and S. Varon, Effects of ciliary neurotrophic factor on the survival and response to nerve growth factor of cultured rat sympathetic neurons. *Dev. Biol.* 161:96 (1994).

CILIARY NEUROTROPHIC FACTOR (CNTF): POSSIBLE IMPLICATIONS IN THE PATHOGENESIS OF AMYOTROPHIC LATERAL SCLEROSIS

Richard W. Orrell[1,2], Russell J.M. Lane[2], and Jackie S. de Belleroche[1]

[1]Department of Biochemistry
[2]Academic Unit of Neuroscience
Charing Cross and Westminster Medical School
London W6 8RF, England

INTRODUCTION

Amyotrophic lateral sclerosis (ALS) is a neurodegenerative disorder, primarily affecting both upper and lower motor neurons. The median age of onset is around 68 years, and median survival approximately 2 years, although with significant variability[1]. The pathogenesis is unknown, but the recognition of mutations in the gene for copper/zinc superoxide dismutase (SOD-1) [2] in a proportion of patients (around 1-2% of all cases), almost all of whom have a family history of the condition, has provided fruitful avenues of research[3]. Nevertheless, the mechanism by which these mutations lead to the expression of the disease is not yet established[4,5].

Mutations of SOD-1 are present in only around 20% of the familial form of ALS, which represents only 5-10% of all ALS [6]. We are interested in determining the other risk factors in the remainder of these families, and also any additional features which may explain the variable presentation and penetrance of those with the SOD-1 mutation[7].

GROWTH FACTORS

Growth factors have been implicated in a number of neurological disorders. They have been defined[8] as "signalling proteins required for normal cellular growth and division, and hence important in the regulation of developmental processes". The group includes nerve growth factor (NGF) and structural homologues brain-derived neurotrophic factor (BDNF), neurotrophin 3 (NT-3), and neurotrophin 4 (NT-4), and the structurally unrelated ciliary neurotrophic factor (CNTF), insulin growth factor (IGF-1) and glial-cell-line-derived neurotrophic factor (GDNF) [9]. Therapeutic trials have recently been undertaken with these agents in diseases including ALS [8,10].

CILIARY NEUROTROPHIC FACTOR (CNTF)

In vitro studies of CNTF have demonstrated it to have potent survival effects on neurons in culture, including chick ciliary, dorsal root sensory, and sympathetic ganglia [11,12]. Local application of CNTF to transected neonatal rat facial nerve axons prevents the degeneration of the neuron[13], and it was suggested that CNTF may have a general effect on neuronal survival. Using a stable transfection of mouse D3 stem cells containing mouse CNTF genomic DNA, implanted intraperitoneally, a continuous high systemic level of CNTF was produced in a mouse model of progressive motor neuronopathy[14]. A prolongation of survival was observed in the treated mice. (The mice were intolerant of repetitive daily injection of CNTF, which also has a short intravenous half life). CNTF may induce local sprouting of motor neurons when injected in mice, and in higher doses more generalised sprouting in adult mice[15]. (These high doses also caused significant weight loss and cachexia). Further evidence that CNTF may have a maintenance function in adult motor neurons was suggested in a study of abolition of CNTF gene expression in mice[16]. There appeared to be no effect on embryonic or neonatal motor neurons, but progressive atrophy and degeneration of spinal motor neurons occurred in the adult mouse.

CNTF GENE

The CNTF protein consists of 200 amino acids, encoded by a gene located on chromosome 11 [17,18]. The gene has two exons and one intron. A recent report[19,20] described a frequent mutation, which is a single base substitution in the intron. This creates a new splice receptor site, and consequent frameshift, leading to a new stop codon. In a Japanese population, 2.3% were homozygotes for the mutation and did not express the CNTF protein. The absence of CNTF protein did not appear to be significantly associated with neurological disease[19].

The point mutation abolishes a restriction site, which allows detection in restricion enzyme digests. Using polymerase chain reaction amplification of this region of the CNTF gene and restriction enzyme digestion we studied this null mutation as a contributing factor in patients with a family history of ALS, where the genetic contribution might be expected to be strongest[21]. Investigating one affected member from each of 49 European families with ALS, 65% were normal homozygotes, and 35% were heterozygotes for the mutation. No mutant homozygotes were detected.

CONCLUSION

The variable determining factors in both sporadic and familial ALS remain to be defined. The role of neurotrophins has largely been investigated in animal models, and relevance to the human disease of ALS remains undetermined. Reduced levels of CNTF in the anterior horn of the spinal cords of humans with ALS may be secondary to the disease process[22]. The absence of CNTF protein expression associated with the homozygote mutation does not appear to be of major significance in the development of ALS. The use of CNTF as a therapeutic agent is not necessarily dependent on any direct implication in the pathogenesis of the disease. However, therapeutic trials of CNTF have so far appeared to be disappointing, including problems with drug delivery and clinical side effects as suggested by animal studies[23]. It may be that CNTF will have an adjunctive role with other neurotrophins such as BDNF [24], or compounds acting on complimentary mechanisms such as riluzole which is thought to act as an inhibitor of glutamate release[25].

REFERENCES

1. S. Yoshida, D.W. Mulder, L.T. Kurland, C. Chu, and H. Okazaki, Follow-up study on amyotrophic lateral sclerosis in Rochester, Minn., 1925 through 1984, *Neuroepidemiology* 5:61-70 (1986).
2. D.R. Rosen, T. Siddique, D Patterson, D.A. Figlewicz, P. Sapp, A. Hentati, D. Donaldson, J. Goto, J.P. O'Regan, H.-X. Deng, Z. Rahmani, A. Kirzus, D. McKenna-Yasek, A. Cayabyab, S.M. Gaston, R. Berger, R.E. Tanzi, J.J. Haperin, B. Herzfeldt, R. Van den Bergh, W.-Y. Hung, T. Bird, G. Deng, D.W. Mulder, C. Smyth, N.G. Laing, E. Soriano, M.A. Pericack-Vance, J. Haines, G.A. Rouleau, J.S. Gusella, H.R. Horvitz, and R.H. Brown, Mutations in Cu/Zn superoxide dismutase are associated with familial amyotrophic lateral sclerosis, *Nature* 362:59-62 (1993).
3. R.W. Orrell and J. de Belleroche, Superoxide dismutase and ALS, *Lancet* 344:1651-1652 (1994).
4. R.H. Brown. Amyotrophic lateral sclerosis: recent insights from genetics and transgenic mice. Cell 80:687-692 (1995).
5. R. Orrell, J. de Belleroche, S. Marklund, F. Bowe, and R. Hallewell, A novel SOD mutant and ALS, *Nature* 374:504-505 (1995)
6. L.P. Rowland, Amyotrophic lateral sclerosis: human challenge for neuroscience, *Proc Natl Acad Sci USA* 92:1251-1253 (1995).
7. J. de Belleroche, R. Orrell, and A. King, Familial amyotrophic lateral sclerosis/motor neurone disease (FALS): a review of current developments, *J Med Genet* (in press).
8. J. Drago, T.J. Kilpatrick, S.A. Kolbar, and P.S. Talman, Growth factors: potential therapeutic applications in neurology, *J Neurol Neurosurg Psychiatry* 57:1445-1450 (1994).
9. Q. Yan, C, Matheson, and O.T. Lopez, In vivo neurotrophic effects of GDNF on neonatal and adult facial motor neurons, *Nature* 373:341-344 (1995).
10. R.W. Orrell, R.J.M. Lane, and R.J. Guiloff, Recent developments in the drug treatment of motor neurone disease, *Br Med J* 309:140-141 (1994)
12. Y. Arakawa, M. Sendtner, and H. Thoenen, Survival effect of ciliary neurotrophic factor (CNTF) on chick embryonic motoneurons in culture: comparison with other neurotrophic factors and cytokines, *J Neuroscience* 10:3507-3515 (1990).
12. G. Barbin, M. Manthorpe, and S. Varon. Purification of the chick eye ciliary neurotrophic factor, *J Neurochem* 43:1468-1478 (1984).
13. M. Sendtner, G.W. Kreutzberg, and H. Thoenen. Ciliary neurotrophic factor prevents the degeneration of motor neurons after axotomy. *Nature* 345:440-441 (1990).
14. M.Sendtner, H. Schmalbruch, K.A. Stockli, P. Carroll, G.W. Kreutzberg, and H. Thoenen, Ciliary neurotrophic factor prevents degeneration of motor neurons in mouse mutant progressive motor neuronopathy, *Nature* 358:502-504.
15. Y.W. Kwon, and M.E. Gurney, Systemic injections of ciliary neurotrophic factor induce sprouting by adult motor neurons, *NeuroReport* 5:789-792 (1994).
16. Y. Masu, E. Wolf, B. Holtmann, M. Sendtner, G. Brem, and H. Thoenen, Disruption of the CNTF gene results in motor neuron degeneration, *Nature* 365:27-32 (1993).
17. A. Lam, F. Fuller, J. Miller , J. Kloss, M. Manthorpe, S. Varon, and B. Cordell, Sequence and structural organization of the human gene encoding ciliary neurotrophic factor, *Gene* 102:271-276 (1991)..
18. A. Negro, E. Tolosano, S.D. Skaper, I. Martinin, L. Callegaro, L. Silengo , F. Fiorini, and F. Altruda, Cloning and expression of human ciliary neurotrophic factor, *Eur J Biochem* 201:289-294 (1991).
19. R. Takahashi, H.Yokoji, H.Misawa, M. Hayashi, J. Hu, and T. Deguchi, A null mutation in the human CNTF gene is not causally related to neurological diseases, *Nature Genetics* 7:79-84 (1994).
20. R. Takahashi, H. Yokoji, H. Misawa, M. Hayashi, J. Hu, and T. Deguchi, A null mutation in the human CNTF gene is not causally related to neurological diseases, *Nature Genetics* 7:215 (1994).
21. R.W. Orrell, A.W. King, R.J.M. Lane, and J.S. de Belleroche, Investigation of a null mutation of the CNTF gene in familial amyotrophic lateral sclerosis, *J Neurol Sci* (in press).
22. P. Anand, A. Parrett, J. Martin, S. Zeman, P. Foley, M. Swash, P.N. Leigh, J.M. Cedarbaum, R.M. Lindsay, R.E. Williams-Chestnut, and D.V. Sinicropi, Regional changes of ciliary neurotrophic factor and nerve growth factor levels in post mortem spinal cord and cerebral cortex from patients with motor neuron disease, *Nature Medicine* 1:168-172 (1995).
23. F. Dittrich, H. Thoenen, and M. Sendtner, Ciliary neurotrophic factor: pharmacokinetics and acute-phase response in rat, *Ann Neurol* 35:151-163 (1994).
24. H. Mitsumoto, K. Ikeda, B. Klinkosz, J.M. Cedarbaum, V. Wong, and R.M. Lindsay, Arrest of motor neuron disease in wobbler mice cotreated with CNTF and BDNF, *Science* 265:1107-1110 (1994).
25. G. Bensimon, L. Lacomblez, V. Meininger, and the ALS/Riluzole Study Group, A controlled trial of riluzole in amyotrophic lateral sclerosis, *N Engl J Med* 330:613-22 (1994).

NEURONAL DIFFERENTIATION OF PC12 CELLS IN THE ABSENCE OF EXTRACELLULAR MATRIX ADHESION INDUCES APOPTOSIS

Herbert W. Harris., M.D., Ph.D

Laboratory of Neurosciences
National Institute on Aging
National Institutes of Health
Bethesda, Maryland, 20892

INTRODUCTION

In recent years, evidence has accumulated implicating programmed cell death as a major mechanism underlying neurodegenerative processes. Programmed cell death has been described in a wide range of cell types in response to diverse factors including steroids, cytokines, ionizing radiation, virus infection, and many other agents[1]. It is an active process that is often dependent on RNA and protein synthesis and probably involves a number of complex pathways that ultimately result in endonuclease activation, DNA fragmentation, and cell death. Programmed cell death may be distinguished form "accidental" cell death that results from hypoxia, trauma, or toxic agents, although all of these processes may involve common pathways[2]. In the brain, programmed cell death has been well established as a part of normal development[3], following excitotoxic death of targets of innervation[4], and as a consequence of amyloid toxicity[5,6]. These findings suggest that programmed cell death may be a central common pathway in numerous neurodenegerative processes including Alzheimer's disease. This concept is supported by observations that neurons are strictly dependent on growth factors to prevent programmed cell death and promote survival[7] and by evidence that growth factor response mechanisms are attenuated in aging[8].

Although there are many known causes of programmed cell death, one mechanism that has been extensively studied in nonneuronal cells involves the extracellular matrix. An extensive literature exists on the induction of programmed cell death in epithelial and endothelial cells by blockade of contact with extracellular matrix[9]. It has become evident that integrin-mediated adhesion and signalling is an important determinant of cell survival. However, little is currently known about the role of adhesion in the survival of neuronal cells. The present studies were undertaken to investigate whether neuronal cells undergo programmed cell death when deprived of extracellular matrix contact.

RESULTS AND DISCUSSION

Undifferentiated PC12 cells can be successfully cultured on plastic surfaces without prior coating with matrix proteins such as collagen or laminin. Under these conditions, the cells adhere weakly to the plastic, but show high levels of viability when cultured in appropriate media. Addition of Nerve Growth Factor to PC12 cells cultured on plastic results in neurite extension during the first 2-4 days comparable to that seen in collagen-adherent PC12 cells. However, following this initial period of differentiation, PC12 cells grown on plastic quickly detach and can be found floating in the media. Viability counting with trypan blue reveals that a substantial number of these floating cells (40-60%) are dead. In contrast, PC12 cells differentiated on collagen-coated surfaces retain nearly 100% viability. Agarose electrophoresis of DNA extracted from these cells shows a nucleosomal "ladder" pattern characteristic of programmed cell death.

Dependence on extracellular matrix adhesion is a well documented phenomenon most frequently observed in cells of endothelial or epithelial origin[9]. Human Umbilical Vein Endothelial cells undergo programmed cell death when matrix adhesion is blocked but these cells survive when cultured on immobilized anti-β_1 integrin antibodies[10]. This process was delayed by the inhibition of protein synthesis with cycloheximide[11]. Apoptosis of adhesion-blocked epithelial cells was prevented by transformation with *ras* or *src* oncogenes or treatment with phorbol ester[12]. Mammary epithelial cells undergo programmed cell death when cultured in the absence of matrix which is associated with enhanced expression of the apoptosis-associated protease, interleukin-1β converting enzyme[13].

The mechanism of programmed cell death induced by blockade of extracellular matrix adhesion is poorly understood. It is known that integrin-mediated adhesion elicits a signalling cascade that involves activation of the focal adhesion kinase, a membrane associated tyrosine kinase[14]. Focal adhesion kinase interacts with several membrane-associated signalling molecules that contain *src* homology domains including Grb2, PI-3 kinase, and phospholipase C[15]. Integrin-mediated signalling has also been shown to activate the MAP kinase cascade[16]. This indicates that cellular signalling associated with matrix adhesion extensively overlaps signalling pathways associated with growth factors. Therefore, loss of attachment to matrix proteins may have effects that are similar to growth factor withdrawal. In addition, there is a great deal of evidence indicating that extracellular matrix proteins directly interact with growth factors such as fibroblast growth factor modifying their biological activity[17]. In the absence of extracellular matrix proteins, growth factors may lose their trophic effects and may even acquire toxic properties.

The extensive "cross-talk" and overlap between growth factor signalling and integrin-mediated adhesion signalling may help to explain the interaction between nerve growth factor and apoptosis observed in the present study. We noted that undifferentiated PC12 cells are relatively insensitive to matrix deprivation. Upon stimulation with nerve growth factor, the PC12 cells require matrix adhesion for survival. Therefore, growth and differentiation of PC12 cells requires both integrin-mediated and nerve growth factor-associated signals. In the absence of the integrin component, the nerve growth factor signal results in apoptosis. The exact mechanism of this interaction of growth and adhesion signalling as it influences cell survival is not well understood. The mechanisms are currently under active investigation.

REFERENCES

1. L. Fesus, Biochemical Events in Naturally Occurring Forms of Cell Death. *FEBS.* 328:1 (1993).
2. R.A. Schwartzman and J.A.Cidlowski, Apoptosis: the biochemistry and molecular biology of programmed cell death. *Endocr Rev.* 14:133 (1993).

3. K.A. Wood, B. Dipasquale, and R.J.Youle, In situ labeling of granule cells for apoptosis-associated DNA fragmentation reveals different mechanisms of cell loss in developing cerebellum. *Neuron.* 11:621 (1993).
4. A. Macaya, F. Munell, R.M. Gubits, and R.E. Burke, Apoptosis in substantia nigra following developmental striatal excitotoxic injury. *Proc Natl Acad Sci U S A.* 91: 8117 (1994).
5. D.T. Loo, A. Copani, C.J. Pike, E.R. Whittemore A.J. Walencewicz and C.W. Cotman, Apoptosis is induced by beta-amyloid in cultured central nervous system neurons. *Proc Natl Acad Sci U S A.* 90: 7951 (1993).
6. G. Forloni, Beta-Amyloid neurotoxicity. *Funct Neurol.* 8: 211 (1993).
7. M.C. Raff, B.A. Barres, J.F. Burne, H.S. Coles, Y. Ishizaki, and M.D. Jacobson, Programmed cell death and the control of cell survival: lessons from the nervous system, *Science.* 262: 695 (1993).
8. L. Olson, NGF and the treatment of Alzheimer's disease, *Exp Neurol.* 124: 5 (1993)
9. E. Ruoslahti,and J.C. Reed, Anchorage dependence, integrins, and apoptosis, *Cell.* 77: 477 (1994).
10. J.J. Meredith, B. Fazeli, and M.A. Schwartz, M. A., The extracellular matrix as a cell survival factor, *Mol Biol Cell.* 4: 953 (1993).
11. F. Re, A. Zanetti, M. Sironi, N. Polentarutti, L. Lanfrancone, E. Dejana, and F. Colotta, Inhibition of anchorage-dependent cell spreading triggers apoptosis in cultured human endothelial cells, J Cell Biol. 127: 537 (1994).
12. S.M. Frisch, and H. Francis, Disruption of epithelial cell-matrix interactions induces apoptosis. *J Cell Biol.* 124: 619 (1994).
13. N. Boudreau, C.J. Sympson, Z. Werb, and M.J. Bissell, Suppression of ICE and apoptosis in mammary epithelial cells by extracellular matrix. *Science.* 267: 891 (1995).
14. J.T. Parsons, M.D. Schaller, J. Hildebrand, T.H. Leu, A. Richardson, and C. Otey, Focal adhesion kinase: structure and signalling. *J Cell Sci Suppl,* 18:109 (1994).
15. E.A. Clark, and J.S. Brugge, Integrins and signal transduction pathways: the road taken, *Science.* 268: 233 (1995).
16. Q. Chen, M.S. Kinch, T.S. Lin, K. Burridge, and .L. Juliano, Integrin-mediated cell adhesion activates mitogen-activated kinases. *J. Biol. Chem.* 269: 26602 (1994).
17. E. Ruoslahti, Y Yamaguchi, A. Hildebrand, and W.A. Border, Extracellular matrix/growth factor interactions, *Cold Spring Harbor Symp. Quant.Biol.* 57:309 (1992).

EXCITABLE MEMBRANE-DERIVED LIPID MEDIATORS: GLUTAMATE RELEASE AND REGULATION OF GENE EXPRESSION

Nicolas G. Bazan, Miriam Kolko, and Geoffrey Allan
Neuroscience Center of Excellence and Department of Opthamology
Louisiana State University Medical Center, School of Medicine
2020 Gravier Street, Suite B
New Orleans, LA 70112, USA

ABSTRACT

Ischemia and seizures lead to the accumulation of free arachidonic (AA) and docosahexaenoic acids (DHA); diacylglycerol, due to activation of phospholipase A_2; and phospholipase C. The platelet-activating factor (PAF) precursor, 1-alkyl-2-acyl-sn-glycero-phosphocholine, is enriched in polyunsaturated fatty acids such as the 2-acyl group. Under resting conditions PAF is undetectable in brain, but it accumulates after injury. The PAF antagonist BN 52021 elicits neuroprotection during brain ischemia-reperfusion and inhibits AA accumulation and inositol lipid degradation. This work led to the discovery of presynaptic PAF binding sites upon which the neuroprotective BN 52021 acts. PAF plays at least two major roles in the nervous system: as a modulator of synapse function and as a mediator of inflammation/injury. Depending upon the degree of neuronal activity, there appear to be roles, on the one hand,for PAF as a retrograde messenger in long-term potentiation, and, on the other hand, for PAF acting on neural responses to injury. PAF activates immediate-early gene expression through an intracellular PAF binding site. An early increase in inducible prostaglandin G/H synthase (TIS-10, COX-2) expression takes place after brain injury. The intracellular PAF antagonist BN 50730 inhibits this effect, as does dexamethasone. These PAF-mediated synaptic and gene expression events may represent intracellular pathways of signal transduction in repair, neuronal plasticity, and cell death. Moreover, these events may

Neurodegenerative Diseases
Edited by Gary Fiskum, Plenum Press, New York, 1996

be involved in epileptogenesis and the remodeling of synaptic circuitry. Thus, excitable membrane-derived lipid mediators as new targets may be explored for pharmacological intervention in stroke and other cerebrovascular diseases, epileptic damage, neurotrauma, and neurodegenerative diseases.

INTRODUCTION

The function of excitable membranes includes the storage of bioactive lipids and their precursors within the structure of phospholipids. Physiological signals such as those elicited by receptor occupancy trigger, via the activation of phospholipases, the release of phospholipid moieties. Figure 1 illustrates the sites of action of the different classes of phospholipases on a typical phospholipid molecule. These enzymes are under control of receptors, calcium ions, kinase/phosphatase systems, and of other signaling mechanisms which still need to be better characterized. The best defined regulatory events of these enzymes are those controlled by seven transmembrane domain superfamily of receptors that, through G proteins, activate phospholipase C (PLC). This enzyme predominantly hydrolyzes phosphatidylinositol 4', 5' bisphosphate (PIP_2), resulting in the production of two second messengers, diacylglycerol (DAG) and inositol 1,4,5-trisphosphate (IP_3). The synaptic inositol signaling system is activated by neurotransmitters, growth factors and drugs. Cerebral ischemia[1] and seizures[2] result in a rapid activation of PLC and accumulation of PIP_2-derived products. IP_3 promotes intracellular Ca^{2+} mobilization by binding to a specific receptor in the endoplasmic reticulum, which in turn enhances calcium-dependent cellular processes. Overactivation of PLC and accumulation of IP_3 may be an important contributory factor of calcium-induced neurotoxicity and cell death in neurodegenerative diseases.

The other product of inositol lipid degradation, DAG, is an activator of protein kinase C (PKC). PKC plays a modulatory role by catalyzing the phosphorylation of specific proteins that contribute to cell function. The modulation of cell function by PKC is complex because a) there are several isoforms of this enzyme, some brain specific[3]; b) also PLA_2 products such as free arachidonic and docosahexaenoic acids modulate some protein kinases C and c) there is a high degree of synergy and feedback modulation between various intracellular pathways of cell signal transduction, including also other protein kinases. Synaptic plasticity involves changes in PKC activity[4,5], hence phospholipases may play a role in regulating the formation of neuronal pathways.

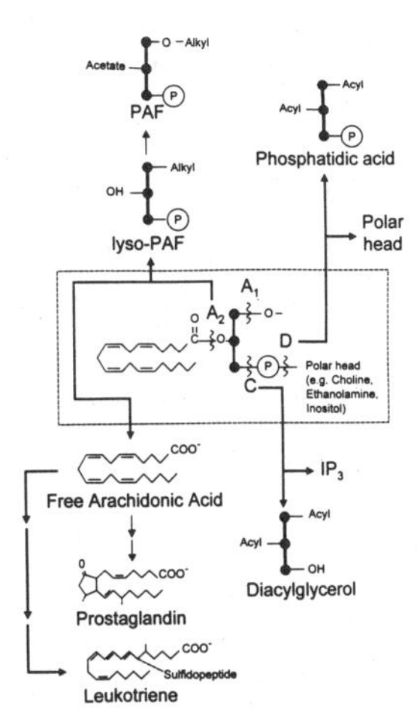

Figure 1. Sites of action of the different classes of phospholipases involved in the generation of lipid second messengers and their precursors.

Phospholipases A_1 and A_2 (PLA_1 and PLA_2) cleave fatty acid residues from the first and second carbons respectively of the glycerol-3-phosphate backbone. The PLA_2 class has been much more extensively characterized. It consists of two subtypes, type II secretory, calcium-independent PLA_2 (found, for instance in snake venom and pancreatic secretions) and type I intracellular PLA_2, which are primarily important in hydrolysis of membrane phospholipids. Of particular importance in neuronal signal transduction is a cytosolic PLA_2 which, upon stimulus is translocated to the plasma membrane and hydrolyzes a small, but metabolically highly active pool of phosphatidylcholines in excitable membranes which have a saturated, ether-linked fatty alkyl chain in the first position (usually 16:0 or 18:0) and a polyunsaturated fatty acyl chain (often arachidonoyl) in the second position. PLA_2 activity on this molecule yields arachidonic acid and lyso-PAF. Arachidonic acid is a substrate for cyclooxygenation and subsequent conversion into biologically active prostaglandins, leukotrienes and HETES. Lyso-PAF can be acetylated by an acetyl transferase to generate PAF by the remodeling pathway.

Phospholipases C and D (PLC and PLD) both cleave phospholipids at the third position. PLC-mediated hydrolysis of phosphatidyl inositol yields two important signaling molecules, diacylglycerol (DAG) which activates protein kinase C, and inositol triphosphate (IP_3) which, when bound to it's receptor, mobilizes intracellular calcium from calcium-storage protein complexes in the endoplasmic reticulum. PLD cleaves the other side of the phosphate group from PLC to yield phosphatidic acid and either choline, ethanolamine or inositol.

There is growing information on the heterogeneity of PLA_2. These enzymes cleave the fatty acyl chain at the C2 position of a phospholipid. In excitable membranes, the C2 acyl chain of phospholipid is often polyunsaturated and, when cleaved, can provide a substrate for oxidative metabolism to biologically active lipid mediators. In addition, there is, in excitable membranes, a small, but metabolically active pool of 1-alkyl-2-acyl-glycero-3-phosphocholine. The lyso-phospholipid PLA_2 product of this phospholipid is a precursor for the potent lipid mediator platelet-activating factor (PAF). The focus of this chapter is on the potential significance of PLA_2 in neurodegenerative disorders, since various of the products of its activity on membrane phospholipids accumulate in brain during ischemia and seizures and these products are precursors of potent bioactive lipids affecting neurotransmission, gene expression, and neuronal plasticity.

Bioactive Lipids as Injury Mediators

There is growing body of knowledge about the roles of membrane-derived lipid mediators in physiological processes. However, it is as mediators of inflammatory and immune responses that the products of PLA_2 -catalyzed hydrolysis of membrane phospholipids have been most extensively characterized. There is, in excitable membranes, a small but metabolically highly active pool of 1-alkyl-2-acyl-glycero-3-phosphocholine (alkylacyl GPC) in which the acyl residue is either arachidonoyl (20:4) or docosahexanoyl (22:6). A cytosolic PLA_2 ($cPLA_2$) which responds to increases in intracellular calcium concentration by membrane translocation, and which is selective for the hydrolysis of arachidonoyl residues has been found in brain[6]. $cPLA_2$ -catalyzed hydrolysis of alkylacyl GPC provides substrate for bifurcating pathways of lipid mediator synthesis, the remodeling pathway of PAF synthesis, and the arachidonic acid cascade.

PAF is the most potent bioactive lipid known and it is the only phospholipid so far known to elicit its actions through a specific receptor . As its name suggests, it was first described as a potent activator of platelet aggregation, but is now known to have much wider roles as an active mediator of inflammatory and immune responses [7,8]. PAF accumulates in the brain during neurotrauma, including electrically and chemically-induced seizures in rat brain[9] and cerebral ischemia in rabbit[10] and fetal rat[11] brain. There are specific PAF binding sites in rat cerebral cortex[12,13] and hippocampus (V.L.Marcheselli and N.G.Bazan, unpublished observations).The synaptosomal and microsomal sites are kinetically and pharmacologically distinct. The microsomal binding activity appears to be composed of two sites, one with a Kd = 22.5pM, B_{max} = 82.5 fmol/mg protein (the highest affinity PAF binding site yet reported), and a lower affinity site of Kd = 25.0 nM, B_{max} = 60.0 pmol/mg protein. Synaptosomal PAF binding activity is composed of a single site with a relatively lower affinity for PAF compared with the primary microsomal site (Kd = 1.2nM, B_{max} = 0.96 pmol/mg protein). The synaptosomal binding site has a strong affinity for the PAF antagonists BN 50726, BN 50727 and BN 52021, and a somewhat lower affinity for CV 3988 and CV 6209. While the microsomal activity has some affinity for the ginkgolide BN 52021, this is much lower than the synaptosomal site. Of all the PAF antagonists tested, BN 52021 is the only one not to display competitive inhibition of PAF binding (i.e. direct competition for the same site on the receptor), but appears to act via a noncompetitive or uncompetitive mechanism. Of the antagonists which compete for microsomal PAF binding activity, the most important is the hetrazepine BN 50730, which has a strong affinity for the microsomal site, but which does not displace PAF from the synaptosomal site to any significant degree. The selectivity of BN 52021 and BN 50730 for the two types of PAF binding sites have been exploited to assign biological activities of PAF in the central nervous system to each of these putative receptors. The synaptosomal binding site appears to correspond with the cell surface PAF receptor that has been cloned[14-16]; however, the identity of the intracellular PAF receptor, as well as its possible relationships with the cell surface receptor, has not been defined.

PAF is responsible for some of the neurotoxic consequences of cerebral ischemia. Systemic pretreatment with PAF antagonists confers neuroprotection in animal models of cerebral ischemia-reperfusion Pretreatment by intraperitoneal injection with the PAF antagonist BN 52021 improves post-ischemic blood flow concomitant with decreased free fatty acid and diacylglycerol accumulation in gerbil brain following 10 minutes of bilateral carotid artery ligation[17]. The antagonist also exerts similar effects on free fatty acid accumulation, but not PLC activity, in mouse brain following 1 minute of postdecapitation ischemia[18]. In a global ischemia-reperfusion model using isolated dog brains, the PAF antagonists BN 50726 or BN 50739, in combination with the free radical scavenger dimethyl sulphoxide, significantly improves postischemic recovery of mitochondrial function and energy metabolism[19].

During reperfusion after brain ischemia, and in the post-seizure phase of multiple seizures, there is a surge of oxidative metabolism. Some of the free AA released by PLA_2 activation during the ischemic phase is now converted into biologically active eicosanoids[20,21]. These second messengers elicit various actions at the microvasculature and on parenchymal neural cells. Mediators released from infiltrating polymorphonuclear leukocytes and microglia[22,23] may play a major role in limiting neuronal survival. A side effect of this sudden increase in peroxidative metabolism is the generation of large amounts of free radicals, initially superoxide[24], which can be converted to the extremely reactive hydroxyl radical[25]. Oxygen radicals are normally removed by efficient scavenging systems. However, during post-ischemic reperfusion, the levels of superoxide dismutase and catalase in brain may be inadequate to cope with the elevated free radical production[26]. While oxygen radicals may contribute to brain damage by reacting with proteins, DNA, lipids and other cellular molecules[27,28], there is also evidence that they impair cerebral arteriole function[29] and open the blood-brain barrier[30,31].

Physiological significance of lipid mediators in glutamate neurotransmission, neural plasticity and memory.

PAF is an important physiological modulator of glutamate neurotransmission and neuronal plasticity. The following three lines of evidence support this conclusion:

1) The PAF analog methylcarbamyl PAF (MC-PAF), but not biologically inactive lyso-PAF, when applied to synaptically connected hippocampal neurons at a concentration of 1μM, specifically augments endogenous excitatory synaptic responses, but not inhibitory GABA-mediated responses[32]. It does not, however, augment the effects of exogenously-added glutamate. The PAF receptor antagonist BN 52021 (at a concentration of 1μM) blocks the MC-PAF-enhanced neurotransmission. Since this antagonist is specific for the synaptosomal PAF receptor (12), this implies that PAF under these conditions acts as a retrograde messenger on a presynaptic site. Incidentally, the neuroprotective effect of BN 52021 in ischemic-reperfusion in gerbil brain (17) may be due partly to an inhibition of the PAF-induced release of excessive amounts of glutamate which results in excitotoxic damage.

2) PAF is a potential retrograde messenger in LTP. When applied to the CA1 subfield of the hippocampus it seems, as in the hippocampal cell studies described above, to be acting presynaptically via the cell-surface PAF receptor. Bath application of BN 52021 (2μM) to hippocampal slices 25 minutes before or 10 minutes after stimulation blocks the sustained enhancement of excitatory postsynaptic potentials that follows tetanic stimulation of the Schaffer collateral pathway[33]. This is not due to interference with NMDA receptor activity because the NMDA component of synaptic currents remains altered under all conditions. Thus PAF must be acting downstream of the NMDA component of the LTP cascade. In the pretreated slices although an initial potentiation is seen, this declines to basal levels within 10 minutes, indicating that the effect of the antagonist is not on transmission of the initial stimulus. Indeed, there is a window of 5-10 minutes after commencement of stimulation in which BN52021 can still inhibit establishment of LTP. Very high concentrations of exogenous MC-PAF (40μM) can overcome this inhibition, indicating that the antagonist is acting directly on the synaptic PAF receptor. Supporting this idea is the

inability of the highly selective intracellular PAF receptor antagonist BN 50730[13] to affect the establishment of LTP.

3) PAF enhances memory in rats performing an inhibitory avoidance task[34] and in a water maze task[35]. In these instances specificity for the synaptic PAF receptor was shown, as was neuroanatomical and temporal selectivity. For instance, in the inhibitory avoidance task PAF only had an effect when infused at specific times after training into defined areas of the limbic system, and in the water maze task PAF was only effective when infused into the striatum.

Arachidonic acid is another potential retograde messenger in hippocampal LTP[36] but slices require long-term exposure to high concentrations of the fatty acid. Furthermore inhibition of this form of LTP by the NMDA receptor antagonist D-2-amino-5-phosphopentanoic acid (APV)[37,38] indicates that arachidonic acid acts via a different mechanism than PAF, which acts independently of the NMDA receptor. There is a report of PAF eliciting LTP which is sensitive to APV[39]. However, this was after prolonged exposure, and since arachidonic acid can potentiate NMDA responses[40] and PAF promotes the release of arachidonic acid[41], then this form of LTP could arise solely from the effects of arachidonic acid.

PAF Activates Gene Transcription

Many signals trigger the rapid and transient expression of immediate-early genes (primary genomic response genes, early response genes). In the brain physiological and pathological events (such as long-term potentiation, ischemia, seizures and NMDA receptor activation) initiate transcription of genes encoding transcription factors and thus have the potential to initiate cascades of gene expression. As a result of gene activation, long-term cellular responses such as neuronal plasticity occur.

PAF is a potential mediator of extracellular and intracellular signal/gene transcription coupling since it is known to stimulate immediate-early gene expression in neuronal and other cells in culture, and the use of specific PAF antagonists blocks gene expression in animal models of brain trauma. PAF rapidly and transiently augments levels of the c-*fos* and c-*jun* transcription factor mRNAs in a neuronal cell line[42]. Phorbol esters and PAF synergistically stimulate c-*fos* expression suggesting that the transcriptional effects of PAF are not mediated by protein kinase C. The effect of PAF is at the transcriptional level, as opposed to increasing the stability of the mRNA 5′ deletion mutagenesis studies of the c-*fos* promoter show that the calcium-response element is necessary for the PAF-induced response. There are other examples of PAF-induced gene expression in neuronal and non-neuronal cells including the heparin-binding epidermal growth factor in monocytes[43], *c-fos* and egr-2 in lymphoblastoid cell lines[44], the transcription factor NF-kappa B and immunoglobulins[45] in human B cell lines, *c-fos* and zif/268 in rat astroglia[46], and *c-fos* and TIS 1 in A-431 epidermoid carcinoma cells[47]. There also reports that PAF is able to up-regulate gene expression of its own cell-surface receptor[48,49]. The intracellular PAF binding site antagonist BN 50730 blocks *c-fos* and zif/268 transcription factor expression induced in rat hippocampus and cerebral cortex by a single electroconvulsive shock [13], or zif/268 and inducible cyclooxygenase (PGHS-2, COX-2, TIS 10) in cryogenically-induced brain edema (V.L.Marcheselli and N.G. Bazan, unpublished observations).

PAF Activates Expression of the Inducible Prostaglandin Synthase Gene

The rate-limiting step of prostaglandin sysnthesis is conversion of arachidonic acid to prostaglandin H_2, catalyzed by prostaglandin G/H synthase. There are now known to be two forms of the enzyme, a "constitutive" enzyme (PGHS-1), and an "inducible" form (PGHS-2). The two enzymes have 61% amino acid sequence homology and share highly conserved domains essential for enzyme function, including glycosylation sites, heme binding sites, the aspirin acetylation sequence, and a region with homology to epidermal growth factor . They are, however, encoded by separate genes. PGHS-2 was first described as an immediate early gene induced by mitogens, cytokines, including by a number of mitogens and cytokines including serum, bFGF, TNFα, IL-1β, and pp60$^{v\text{-}src}$, and by phorbol esters, and down-regulated by glucocorticoids[50-52]. The rationale behind having two different forms of the enzyme is not entirely clear because, in cells which express both forms, up-regulation of PGHS-2 is not matched by an increase in prostaglandin synthesis. Nevertheless, recent findings suggest different subcellular localizations for the two enzymes: PGHS-1 is localized in the endoplasmic reticulum, while PGHS-2 is enriched in the nuclear envelope. One line of thought is that PGHS-2 is responsible for the synthesis of PGs which act as intracellular signals, while PGHS-1 activity leads to the generation of PGs which act as extracellular mediators[53]. PGHS-2 has been found in rat brain, where it is expressed throughout the forebrain in discrete populations and is enriched in the cerebral cortex and hippocampus [54]. Its expression is rapidly and transiently induced by seizure or NMDA-dependent synaptic activity, while basal expression in the developing and adult appears to be regulated by synaptic activity. For instance, there is detectable expression in the developing animal from post-natal day 5 onwards, a period during which the rat brain is undergoing massive synaptic plasticity and developmental changes. The inducibilty of PGHS-2 expression in response to synaptic stimuli implies a role in neuronal plasticity.

Using constructs of the promoter of the murine inducible prostaglandin synthase gene transfected to neural and non-neural cell it was found that PAF stimulates expression of the luciferase reporter gene[55]. Retinoic acid modulates gene expression through nuclear receptors that are members of a superfamily of ligand-dependent transcription factors. Since PAF may also affect gene expression through an intracellular site, the effect of retinoic acid was studied. Retinoic acid induces the expression of constructs of the TIS-10/PGS-2 promoter with a luciferase reporter transfected into neuroblastoma cells. Using the calcium phosphate co-precipitation transfection procedure in the presence of retinoic acid, there is a PAF-dependent (from 1 to 50 nM) activation of luciferase reporter constructs driven by regulatory regions of the TIS-10/PGHS-2 gene. The effect of PAF in indicating that a preexisting latent transcription factor(s) is engaged in the effect. Deletion studies of the TIS-10/PGHS-2 promoter/luciferase reporter constructs showed the deletion of sequences between -371 and -300 reduced the PAF inductive effect from 31-fold to 4.1-fold. BN 50730 inhibited the PAF inductive effect when incubated for 1 h with cultured cells before transfection. This indicates that a PAF receptor is involved in the expression of the inducible prostaglandin synthase and that a major PAF-responsive element lies within this region. Further deletions showed no response to added PAF.

Membrane-Derived Lipid Second Messengers in Neurodegeneration

The membrane-derived lipids accumulated during brain injury represent an overactivation of processes that regulate normal synaptic function[56]. Some of the membrane-derived lipid second messengers generated by injury may participate in the coupling of injury either with plasticity responses and repair/regenerative responses, or with cell death. PAF links injury response with transcriptional activation of immediate-early genes. Several key mediators of nervous system function may also play role in pathological conditions, depending upon the degree of impairment of neuronal activity. This duality is a feature of glutamate, an excitatory amino acid, that, by far, is the most abundant neurotransmitter of the mammalian brain. Glutamate plays a critical role in such neurophysiological processes as developmental plasticity and memory formation. However, in stroke, seizures, neurotrauma and some neurodegenerative diseases, there is an abnormally high synaptic accumulation of glutamate, making it a critical effector of excitotoxic damage. Other examples of neuronal mediators with this duality are interleukin-1 and, perhaps, amyloid peptide, coexist with neural cells under physiological conditions but may, when overproduced, be engaged in abnormal actions in diseases.

PAF also elicits physiological effects and, when overproduced, may become an endogenous neurotoxin. Although PAF is most often referred to as a mediator of the inflammatory and immune responses as well as of cell injury, low PAF concentrations elicits sprouting in PC12 cells, whereas neuropathological changes occur when these cells are exposed to high PAF concentrations[57]. The presynaptic PAF binding site linked to glutamate release is a target through which PAF participates in excitotoxicity. However, the same presynaptic site involves PAF as a potential retrograde messenger in long-term potentiation. The very high affinity binding site, localized intracellularly and linked to immediate-early gene expression, may be a bridge between the initial response to injury and gene expression[58]. It is not yet known if two or more of these actions are interlinked or if there are different effects among the diverse cellular responses stimulated by PAF. The significance of the effect of PAF on gene expression may be related either to long-term adaptive responses or to nuclear events or apoptosis. PAF may participate in neurodegeneration by its ability to increase intracellular Ca^{2+} [59,57] likely by activating a Ca^{2+} channel, by activating phospholipase C with resultant intracellular Ca^{2+} mobilization by inositol trisphosphate and by other actions.

PAF stimulation of the inducible prostaglandin synthase gene establishes a link between cell injury-generated PAF and the gene encoding the enzyme that catalyzes the cyclooxygenation of arachidonic acid. Arachidonic acid is readily available since cell injury promotes its accumulation[60,61]. The inducible prostaglandin synthase utilizes only a small proportion of the free arachidonic acid available under these conditions. PGE_2 synthesis during overstimulation (e.g., seizure) or ischemia-reperfusion or neurotrauma, may require enhanced expression of inducible prostaglandin synthase. PGE_2 action may comprise a feedback loop of the inflammatory response. The signal transduction pathways affected by membrane-derived lipid mediators may permit the identification of movel sites to target drugs that slow down neurodegenerative diseases.

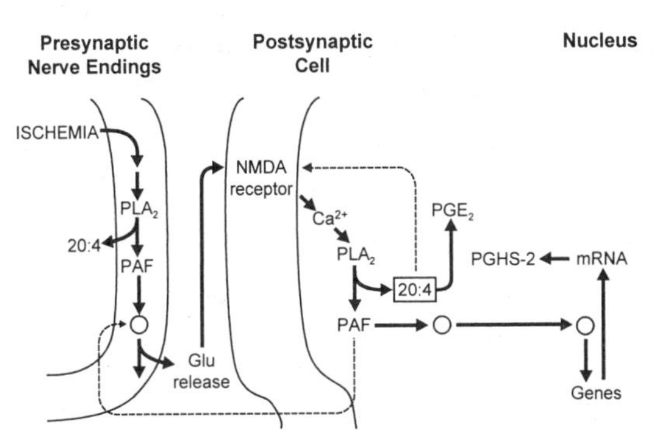

Figure 2. Schematic representation of the proposed role of PAF in the regulation of excitatory neurotransmitter release and how, during cerebral trauma, it may play a role in the excessive release of neurotransmitter. Under normal physiological conditions, glutamate is released from a presynaptic terminal and stimulates NMDA receptors at the postsynaptic terminal. The resultant elevation of intracellular calcium activates cytoplasmic PLA_2 and initiates accumulation of PAF and 20:4. Upon sufficient stimulation there is enough PAF to diffuse back across the synapse and provide a sufficient concentration at the presynaptic terminal to stimulate the BN 52021-sensitive, presynaptic PAF receptor and, via a mechanism as yet undefined, stimulate glutamate release. At the same time, 20:4 accumulates in the postsynaptic terminal and have an positive effect on glutamate neurotransmission involving the NMDA receptor. Under pathophysiological conditions such as during cerebral ischemia, glutamate neurotransmission could be elevated by a superinduction of these pathways where an initially elevated signal at the presynaptic terminal could initiate a positive feedback loop.

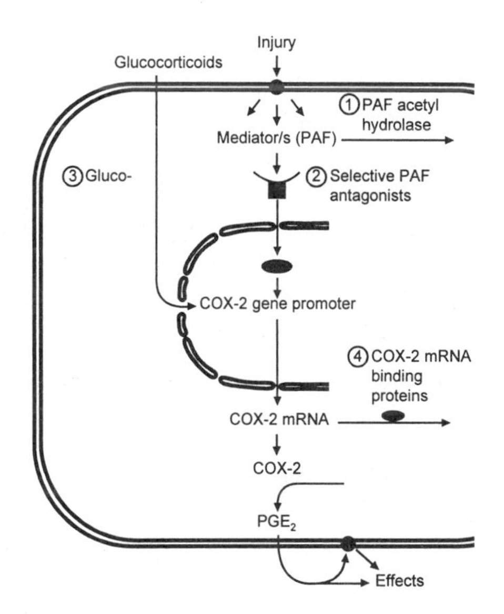

Figure 3. Sites of regulation of PAF-induced COX-2 (PGHS-2) expression and prostaglandin synthesis.. PLA_2 activation generates lyso-PAF, which is then transaectylated to form PAF. PAF has a short biological half-life. It is quickly deacetylated by PAF acetylhydrolase ①and only under stimulated conditions does sufficient PAF accumulate to activate the intracellular PAF receptor ②, the target for the PAF antagonist BN 50730. Receptor

activation triggers a series of events which allows one or more transcription factors to bind to the "PAF-responsive" domain on the COX-2 gene promoter. Glucorticoids, bound to their nuclear receptor may inhibit COX-2 gene induction by sterically hindering binding of the PAF-induced transcription factor ③. Immediate-early gene mRNAs are inherently unstable, containing 3' sequences which direct their rapid degradation. Although COX-2 induction is longer-lived than, for instance, induction of transcription factor immediate-early genes such as *c-fos*, nevertheless, post-transcriptional regulation of COX-2 expression via proteins which bind to the mRNA ④ needs to be explored.

Acknowledgement:

This research was sponsored by the National Institutes of Health, NINDS, NS 2300. This does not necessarily reflect the position or the policy of the NIH, and no official endorsement should be inferred.

References

1. M.I. Aveldaño and N.G. Bazan, Rapid production of diacylglycerols enriched in arachindonate and stearate during early brain ischemia, *J. Neurochem.* 25: 919-920 (1975b).
2. N.G. Bazan and T.S. Reddy, Arachidonic acid, stearic acid and diacylglycerol accumulation correlates with the loss of phosphatidylinositol 4,5-bisphosphate in cerebrum 2 seconds after electroconvulsive shock. Complete reversion of changes 5 minutes after stimulation, *J. Neurosci. Res.* 18:449-455 (1987).
3. Y. Nishizuka, M.S. Shearman, T. Oda, N. Berry, T. Shinomura, Y. Asaoka, K Ogita, H. Koide, U.Kikkawa, A. Kishimoto, Protein kinase C family and nervous function. *Prog Brain Res* 89:125-41 (1991).
4. P. Pasinelli, G.M. Ramakers, I. J. Urban, J.J. Hens, A.B. Oestreicher, P.N. de Graan, W. H. Gispen, Long-term potentiation and synaptic protein phosphorylation, *Behav Brain Res* 66:53-9 (1995).
5. T. Suzuki, Protein kinases involved in the expression of long-term potentiation, *Int J Biochem* 26:735-44 (1995).
6. J.D. Clark, L.L. Lin, R.W. Kriz, C.S. Ramesha, L.A. Sultzman, A.Y. Lin, N. Milona, J.L. Knopf, A novel arachidonic acid-selective cytosolic PLA2 contains a Ca^{2+}-dependent translocation domain with homology to PKC and GAP, *Cell* 65: 1043-51(1991).

7. P. Braquet, L.Touqui, T.Y. Shen, and B.B. Vargaftig, Perspectives in platelet-activating factor research, *Pharmacol. Rev.* 39: 97-145 (1987).
8. S.M. Prescott, G.A. Zimmerman, T.M. McIntyre, Platelet-activating factor, *J Biol Chem.* 265:17381-4 (1990).
9. R. Kumar, S. Harvey, N. Kester, D. Hanahan, M. Olson, Production and effects of platelet-activating factor in the rat brain, *Biochim Biophys Acta.* 963:375-83 (1988).
10. P.J. Lindsberg, K. Freirichs, J.M. Hallenbeck, and G.Z. Feuerstein, Evidence for platelet-activating factor as a novel mediator in experimental stroke in rabbits. *Stroke* 21:1452-1457 (1990).
11. B. Kunievsky, E.Yavin, Production and metabolism of platelet-activating factor in the normal and ischemic fetal rat brain, *J. Neurochem.* 63:2144-51 (1991).
12. V.L. Marcheselli, M. Rossowska, M.T. Domingo, P. Braquet , N.G. Bazan, Distinct platelet-activating binding sites in synaptic endings and in intracellular membranes of rat cerebral cortex, *J Biol Chem.* 265:9140-5 (1990).
13. V.L. Marcheselli, N.G. Bazan, Platelet-activating factor is a messenger in the electroconvulsive shock-induced transcriptional activation of c-*fos* and *zif*-268 in hippocampus, *J Neurosci Res.* 37:54-61 (1994).
14. Z. Honda, M. Nakamura, I. Miki, M. Minami, T. Wantabe, Y. Seyama, H. Okado, H. Toh, K. Ito, T. Miyamoto, and T. Shimizu, Cloning by functional expression of platelet-activating factor receptor from guinea-pig lung, *Nature* 394:342-346 (1991).
15. D. Kunz, N.P. Gerard, C. Gerard, The human leukocyte PAF receptor. cDNA cloning, cell surface expression and construction of a novel epitopr-bearing analog, *J. Biol. Chem.* 267:9101-6 (1992).
16. T. Sugimoto, H. Tsuchimochi, C.G. McGregor, H. Mutoh, T. Shimizu, Y. Kurachi, Molecular cloning and characterization of the platelet-activating factor receptor from human heart, *Biochem. Biophys. Res. Commun.* 189:617-24 (1992).
17. T. Panetta, V.L. Marcheselli, P. Braquet, B. Spinnewyn, N.G. Bazan, Effects of diacylglycerols, polyphosphoinositides and blood flow in the gerbil brain: Inhibition of ischemia-reperfusion induced cerebral injury, *Biochem. Biophys. Res. Comm.* 149:580-7 (1987).
18. D.L. Birkle, P. Kurian, P. Braquet, N.G. Bazan, Platelet-activating factor antagonist BN 52021 decreases accumulation of free polyunsaturated fatty acid in mouse brain during ischemia and electroconvulsive shock, *J. Neurochem.* 51:1900-5 (1990).

19. D.D. Gilboe, D. Kintner, J.H. Fitzpatrick, S.E. Emoto, A. Esanu, P.G. Braquet, and N.G. Bazan, Recovery of postischemic brain metabolism and function following treatment with a free radical scavenger and platelet activating factor, *J. Neurochem.* 56: 311-319 (1991).

20. R.J. Gaudet, J. Alam, and L. Levine, Accumulation of cyclooxygenation products of arachidonic acid metabolism in gerbil brain during reperfusion after bilateral common carotid atrtery occlusion, *J. Neurochem.* 35:653-658 (1990).

21. M.A. Moskowitz, K.J. Kiwak, K. Hekiman, and L. Levine, Synthesis of compounds with the properties of leukotrienes C_4 and D_4 in gerbil brains after ischemia and reperfusion, *Science* 224:886 (1984).

22. A.-L. Siren, R.M. McCarron, Y. Liu et al., Perivascular macrophage signaling of endothelium via cytokines: Mechanism by which stroke risk factors operate to increase stroke likelihood, *in*: "Pharmacology of Cerebral Ischemia," J. Krieglstein and H. Oberpichler-Schwenk H, eds., Wissenschaftliche Verlagsgesellschaft, Stuggart (1992).

23. D. Giulian, K. Vaca, Inflammatory glia mediate delayed neuronal damage after ischemia in the central nervous system, *Stroke* 24 (Suppl):84-90 (1993).

24. W.M. Armstead, R. Mirro, D.W. Busija, and C.W. Leffler, Postischemic generation of superoxide anion by newborn pig brain, *Am J. Physiol.* 255:401-403 (1988).

25. E.L. Cerchiari, T.M. Hoel, P. Safar, and R.J. Schlabassi, Protective effects of combined superoxide dismutase and deferoxamine on recovery of cerebral blood flow after cardiac arrest in dogs, *Stroke* 18:869-878 (1987).

26. B.C. White, J.F. Hildenbrandt, A.T. Evans, I. Aronson, R.J. Indrieri, T. Hoehner, L. Fox, R. Huang, and D. Johns, Prolonged cardiac arrest in resuscitation in dogs: brain mitochondrial function with different artificial perfusion methods, *Ann. Emerg, Med.* 14: 383-388 (1985).

27. B. Halliwell and J.M. Gutteridge, Oxygen toxicity, oxygen radicals, transition metals, and disease, *Biochem J.* 219:1-14 (1984).

28. S.Yoshida, R. Busto, M. Santiso, and M.D. Ginsberg, Brain lipid peroxidation induced by postischemic reoxygenation in vitro: effect of Vitamin E, *J. Cereb. Blood Flow Metab.* 4:466-469 (1980).

29. E.P. Wei, C.W. Christman, H.A. Kontos, and J.T. Povlishock, Effects of oxygen radicals on cerebral arterioles, *Am. J. Physiol.* 248:H157-H162 (1985.)

30. C.W. Nelson, E.P. Wei, J.T. Povlishok, H.A. Kontos, and M.A. Moskowitz, Oxygen radicals in cerebral ischemia, *Am. J. Physiol.* H1356- H1362 (1992).

31. A. Shukla, R. Shukla, M. Dikshit, and R.C. Srimal R C, Alterations in free radical scavenging mechanisms following blood-brain barrier disruption, *Free Radic.Biol. Med.* 15:97-100 (1993).

32. G.D. Clark, L.T. Happel, C.F. Zorumski, N.G. Bazan, Enhancement of hippocampal excitatory synaptic transmission by platelet- activation factor, *Neuron.* 9:1211-1216 (1992).

33. K. Kato, G.D. Clark, N.G. Bazan, C.F. Zorumski, Platelet-activating factor as a potential retrograde messenger in CA1 hippocampal long-term potentiation, *Nature* 367:175-179 (1994).

34. I. Izquierdo, C. Fin, P.K. Schmitz, R.C. Da Silva, O. Jerusalinsky, J.A. Quillfeldt, M.B. Ferreira, J.H. Medina, and N.G. Bazan, Memory enhancement by intrahippocampal, intraamygdala, or intraentorhinal infusion of platelet-activating factor measured in an inhibitory avoidance task, *Proc. Natl. Acad. Sci. U S A* 92:5047-51 (1995).

35. L.Teather, M.G. Packard, and N.G. Bazan, Effects of intra-caudate nucleus injections of platelet-activating factor and BN52021 on memory (abstract), *Soc. Neurosci.* 21: 1230 (1995).

36. J.H. Williams, M.L. Errington, M.A. Lynch, and T.V.P. Bliss, Arachidonic acid induces long-term activity-dependent enhancement of synaptic transmission in the hippocampus, *Nature* 341: 739-42 (1989).

37. T.J. O'Dell, R.D. Hawkins, E.R. Kandel, O. Arancio, Tests of the roles of two diffusible substances in long term potentiation: evidence for nitric oxide as a possible early retrograde messenger, *Proc. Natl Acad. Sci. U S A* 88: 11285-9 (1991).

38. K. Kato, K. Uruno, K. Saito, H. Kato, Both arachidonic acid and 1-oleoyl-2-acetyl glycerol in low magnesium solution induce long-term potentiation in hippocampal CA1 neurons in vitro, *Brain Res* 563(1-2): 94-100 (1991).

39. A. Wieraszko, G. Li, E. Kornecki, M.V. Hogan, and Y.H. Ehrlich, Long-term potentiation in the hippocampus induced by platelet-activating factor, *Neuron* 10: 553-557 (1993).

40. B.Miller, M. Sarantis, S.F. Traynelis, D. Attwell, Potentiation of NMDA receptor currents by arachidonic acid, *Nature* 355:722-5 (1992).

41. B. Kunievsky, E.Yavin, Platelet-activating factor stimulates arachidonic acid release

and enhances thromboxane B2 production in intact fetal rat brain ex vivo, *J Pharmacol Exp Ther* 263:562-8 (1992).

42. S.P. Squinto, A.L. Block, P. Braquet, and N.G. Bazan, Platelet-activating factor stimulates a Fos/Jun/AP-1 transcriptional signaling system in human neuroblastoma cells, *J Neurosci Res* 24:558-66 (1981).
43. Z. Pan, V.V. Kravchenko, and R.D. Ye, Platelet activating factor stimulates transcription of the heparin-binding epidermal growth factor-like growth factor in monocytes. Correlation with an increased kappa B binding activity, *J. Biol. Chem.* 270: 7787-7790 (1995).
44. B. Mazer, J. Domenico, H. Sawami, and E.W. Gelfand, Platelet-activating factor induces an increase in intracellular calcium and expression of regularity genes in human B lymphoblastoma cells, *J Immunol.* 146:1914-20 (1991).
45. C.S. Smith, and W.T. Shearer, Activation of NF-kappa B and immunoglobulin expression in response to platelet-activating factor in a human B cell line, *Cell. Immunol.* 155:292-303 (1994).
46. P. Dell' Albani, D.F. Condorelli, G. Mudo, C. Amico, M.Bindoni, and N. Belluardo, Platelet-activating factor and its methoxy analogue ET-18-OCH3 simullate immediate-early gene expression in rat astroglial cultures, *Neurochem. Int.* 22: 567-74 (1993).
47. Y. Tripathy, J. Kandala, R. Guntaka, R. Lim, and S. Shukla, Platelet-activating factor induces expression of early response genes c-*fos* and Tis-1 in human epidermoid carcinoma A-431 cells, *Life Sci.* 49:1761-7 (1991).
48. H. Mutoh, S. Ishii, T. Izumi, S. Kato, and T. Shimizu, Platelet activating factor (PAF) positively auto-regulates the expression of human PAF receptor transcript 1 (leukocyte type) through NF-kappa B, *Biochem. Biophys. Res. Commun.* 205:1137-1142 (1994).
49. H. Shirasaki, I. M. Adcock, O.J. Kwon, M. Nishikawa, J.C. Mak, and P.J. Barnes, Agonist-induced up-regulation of platelet-activating factor receptor messenger RNA, *Eur. J. Pharmacol.* 268:263-266 (1994).
50. D.A. Kujubu, B.S. Fletcher, B.C. Varnum, R.W. Lim, and H.R. Herschman, TIS10, a phorbol ester tumor promoter-inducible mRNA from Swiss 3T3 cells, encodes a novel prostaglandin synthase/cyclooxygenase homologue, *J. Biol. Chem.* 266:12866-12872 (1991).

51. M.K. O'Banion, H.B. Sadowski, V.Winn, and D.A.Young, A serum- and glucocorticoid-regulated 4-kilobase mRNA encodes a cycooxygenase-related protein, *J. Biol. Chem.* 266:23261-23267 (1991).

52. H.R. Herschman, Regulation of prostaglandin synthase-1 and prostaglandin synthase-2, *Cancer Metastasis Rev.* 13:241-56 (1994).

53. E.J. Goetzl, W.L. An S. Smith, Specificity of expression of eicosanoid mediators in normal physiology and human disease, *FASEB J.* 9:1051-58 (1995).

54. K. Yamagata, K.I. Andeasson, W.E. Kaufmann, C.A. Barnes, P.F. Worley, Expression of a mitogen-inducible cyclooxygenase in brain neurons: Regulation by synaptic activity and glucocorticoids, *Neuron* 11:371-86 (1993).

55. N.G. Bazan, B.S. Fletcher, H.R. Herschman, P.K. Mukherjee, Platelet-activating factor and retinoic acid synergistically activate the inducible prostaglandin synthase gene, *Proc. Natl Acad Sci U S A* 91:5252-6 (1993).

56. N.G. Bazan, C.F. Zorumski, G.D. Clark, The activation of phospholipase A_2 and release of arachidonic acid and other lipid mediators at the synapse: The role of platelet-activating factor, *J Lipid Med* 6:421-7 (1993).

57. E. Kornecki and Y.H. Ehrlich, Neuroregulatory and neuropathological actions of the ether phospholipid platelet-activating factor, *Science* 240:1792-4 (1988).

58. N.G. Bazan, S.P. Squinto, P. Braquet, T. Panetta, V.L. Marcheselli, Platelet-activation factor and polyunsaturated fatty acids in cerebral ischemia or convulsions: Intracellular PAF-binding sites and activation of a Fos/Jun/Ap-1 transcriptional signaling system, *Lipids* 26:1236-42 (1991).

59. H. Bito, M. Nakamura, A. Honda, et al, Platelet-activating factor (PAF) receptor in rat brain: PAF mobilizes intracellular Ca^{2+} in hippocampal neurons, *Neuron.* 9:1-10 (1992).

60. N.G. Bazan, Effects of ischemia and electroconvulsive shock on free fatty acid pool in the brain, *Biochim Biophys Acta* 218:1-10 (1970).

61. N.G. Bazan, Arachidonic acid (AA) in the modulation of excitable membrane function and at the onset of brain damage, *Ann N Y Acad Sci* 559:1-16 (1989).

BCL-2 PROTECTION OF MITOCHONDRIAL FUNCTION FOLLOWING CHEMICAL HYPOXIA/AGLYCEMIA

Anne N. Murphy,[1] Dale E. Bredesen,[2] and Gary Fiskum[1]

[1]Department of Biochemistry and Molecular Biology
The George Washington University Medical Center
Washington, D.C. 20037

[2]The Program on Aging
La Jolla Cancer Research Foundation
La Jolla, CA 92037

Neuronal death following cardiac arrest or stroke is a primary cause of delayed morbidity and mortality. In cardiac arrest, neurologic compromise is primarily the result of delayed neuronal death which develops over 24 to 72 h following resuscitation. In the case of stroke, the cells at the core of the lesion die acutely. However, the cells at the penumbra are at risk in subsequent days, and it is their fate that can determine survival or the degree of debilitation of the victim. The mechanisms underlying delayed neuronal death following cerebral ischemia and reperfusion are not fully understood. Amelioration of *in vivo* damage through the administration of excitatory amino acid antagonists of the NMDA and especially non-NMDA type have met with success. As well, inhibitors of free-radical induced damage such as antioxidants or heavy metal chelators have been found to inhibit delayed neuronal death. However, no treatment has been found to completely prevent the deleterious effects of ischemia/reperfusion, due either to a complex interplay of multiple degradative mechanisms[1], or to a lack of appreciation of the sequence and relative importance of events in the death pathway.

APOPTOSIS IN DELAYED NEURONAL DEATH FOLLOWING ISCHEMIC INJURY

Recent data has suggested that apoptosis is involved in the delayed neuronal death associated with ischemia/reperfusion,[2-8] especially in those areas of the brain most susceptible to delayed neuronal injury (i.e. the pyramidal neurons of the CA1 region of the hippocampus[2,5-8], however, see Deshpande *et al.*[9]). MacManus *et al.*[7] have provided evidence that transient forebrain ischemia in rats leads to apoptotic-like death; whereas ischemia with

Neurodegenerative Diseases
Edited by Gary Fiskum, Plenum Press, New York, 1996

no reperfusion (by decapitation) results in necrotic death. The evidence therefore suggests that neuronal loss may result from processes leading to either necrotic or apoptotic death, depending on the severity of the insult and the neuronal type. If this is so, then effective therapies for ameliorating ischemia/reperfusion-induced injury must address both modes of cell death. Alternatively, it is possible that the cellular sequence of events leading to either necrosis or apoptosis share a common pathway for which a therapy may be designed. This concept is supported by data indicating that either necrosis or apoptosis can result from a variety of cellular insults, and the mode of cell death is dependent on the dose of the injuring agent.[10] A common pathway for necrosis and apoptosis may involve increases in cell [Ca^{2+}],[10] or oxidative damage,[11] both of which are mechanisms of cell injury in ischemia/reperfusion and glutamate excitotoxicity for which there is significant data.[1] Therefore, an appropriately designed inhibitor of the step(s) triggering the cell death pathway induced by oxidative stress and Ca^{2+} overload may inhibit both apoptotic and necrotic death in response to ischemia/reperfusion injury.

A number of recent studies provide evidence that a prototype for such an inhibitor could be the product of the anti-death gene *bcl-2*.[12] *In vitro*, Bcl-2 is known to inhibit cell death in a variety of cell types in response to diverse insults, including a number which induce oxidative stress,[12-16] or an increase in Ca^{2+} induced by an ionophore.[15,17,18] However, direct evidence for the potential of Bcl-2 to inhibit ischemia/reperfusion neuronal loss has recently been provided. Specifically, Dubois-Dauphin *et al.*[19] have discovered in a transgenic mouse model that *bcl-2* overexpression can decrease the infarct volume in a model of stroke by approximately 50%. As well, Linnik *et al.*[20] have shown that infection of neurons with a viral vector containing *bcl-2* protects them from death in permanent focal ischemia. Others have shown that Bcl-2 protein levels increase in the brain following ischemia[21,22] as though part of the stress response includes up-regulation of self-protective agents. This increase in Bcl-2 levels is associated with tolerance to subsequent ischemia,[21] and occurs in neurons that are destined to survive.[22] In addition, Krajewski *et al.*[23] have provided evidence that selectively vulnerable regions of the rat brain following cardiac arrest express high levels of the death promoting Bcl-2 antagonist, Bax, and low levels of Bcl-2.

THEORIES ON THE MECHANISM OF ACTION OF BCL-2

The specific mechanism of action of Bcl-2 remains a mystery. Two separate theories have been proposed for which there is experimental evidence. Recent studies on different cell types responding to different forms of insult suggest that Bcl-2 can protect cells through antioxidant activity.[14,16] Specifically, Kane *et al.*[16] have shown that the neural cell line, GT1-7, is protected by *bcl-2* expression from death induced by GSH depletion. The *bcl-2* expressing cells demonstrate lower levels of reactive oxygen species (ROS) and lipid peroxides, suggesting that Bcl-2 either decreases the generation or increases the detoxification of free-radicals. Hockenbery *et al.*[14] have provided evidence that lymphocytes expressing Bcl-2 are protected from H_2O_2, menadione (an agent that generates $O^{\cdot-}_2$), and dexamethasone treatment. Dexamethasone treated cells expressing *bcl-2* demonstrated lower levels of lipid peroxides; but not lower levels of ROS, suggesting that Bcl-2 operates by detoxifying free-radical species. This theory as the sole mechanism of action of Bcl-2 has recently been questioned by two studies proposing that reactive oxygen species generation is not required for apoptotic death under conditions of very low oxygen tension, and these conditions do not preclude the protective effect of Bcl-2.[24,25] It should be kept in mind that these two studies were performed under conditions of prolonged, continuous hypoxia in the presence of glucose (conditions which are substantially different from those during ischemia/reperfusion injury in which cellular deenergization and oxidative stress are known to play a role). Another proposed mechanism of action for which there is experimental evidence has been provided by Lam *et al.*[18], suggesting that Bcl-2 inhibits Ca^{2+} efflux from the endoplasmic reticulum in response to thapsigargin, an inhibitor of the endoplasmic reticular Ca^{2+}-ATPase. It is not known whether these two mechanisms are related, and specifically how they might operate under conditions of ischemia/reperfusion injury.

Bcl-2 is a 26 kDa protein[26] known to be membrane-bound, but its primary localization is somewhat controversial. Studies have found it localized to the inner mitochondrial membrane,[13] but also to the endoplasmic reticulum, nuclear envelope, and outer mitochondrial membrane.[22,27] The mitochondrial localization of Bcl-2 is particularly intriguing in light of evidence by Newmeyer *et al.*[28] using a reconstitution assay for apoptosis that the mitochondrial fraction is required for induction of changes in nuclei consistent with apoptosis. These data suggest a pro-active role of altered mitochondria in the death pathway.

The intracellular location of Bcl-2 protein would be consistent with an antioxidant mechanism, as it is localized to membranes which contain electron-transporting systems which have a capacity for $O_2^{\cdot-}$ generation. A primary site of $O_2^{\cdot-}$ generation which Bcl-2 may act to inhibit or detoxify is the mitochondrial electron transport chain.[29] It is estimated that approximately 1-2% of the electrons flowing through the respiratory chain leak from the pathway to produce $O_2^{\cdot-}$ which is readily converted to hydrogen peroxide by mitochondrial superoxide dismutase. However, this proportion may increase under certain conditions of mitochondrial inhibition.[29] Other potential sites of cellular free-radical production include the cytochrome P-450 system of the endoplasmic reticulum, xanthine oxidase in the cytoplasmic fraction, and an NAD(P)H oxidase of the plasma membrane of a few (non-neuronal) cell types.[29]

In addition, the localization of Bcl-2 would be consistent with the potential to affect Ca^{2+} distribution as mitochondrial Ca^{2+} transporters are inner membrane proteins. Mitochondrial Ca^{2+} transport properties allow for rapid respiration-dependent uptake of Ca^{2+} and relatively slow release mediated primarily in electrically excitable cells by a Na^+/Ca^{2+} exchanger.[30,31] Ca^{2+} sequestration occurs via a uniporter that is driven directly by the electrochemical potential. Under physiological conditions, modest accumulation likely serves to stimulate Ca^{2+}-sensitive dehydrogenases of the TCA cycle, thereby enhancing provision of reducing equivalents to the electron transport chain for ATP synthesis.[30,31] Under pathological conditions when Ca^{2+}_i is supraphysiological, mitochondria have the capacity to accumulate significant quantities of Ca^{2+} (far in excess of the endoplasmic reticulum), functioning as a large capacity sink. If this accumulation is excessive, however, the consequences can include respiratory inhibition,[32,33] opening of a non-specific megachannel (undergoing the membrane permeability transition,[30,31]) or osmotic lysis of the mitochondria. Therefore, the capacity for Ca^{2+} uptake under pathological conditions is relevant to continued mitochondrial function and ATP production. A number of newly published works suggest that the process of Ca^{2+} redistribution following significant increases in Ca^{2+}_i, and the resulting potentiation of mitochondrial free radical production are key to understanding the triggers for delayed cell death. Specifically, White and Reynolds,[34] as well as others, have shown that mitochondria buffer the increase in Ca^{2+}_i following NMDA stimulation of neurons; that is, mitochondria sequester Ca^{2+} under these conditions. As important, Reynolds and Hastings[35] and Dugan *et al.*[36] have provided evidence that this Ca^{2+} sequestration results in the potentiation of mitochondrial reactive oxygen species generation.

Cell damage following cerebral ischemia is closely associated with the loss of ionic homeostasis as a result of ATP depletion and activation of ligand- and voltage-gated Ca^{2+}

channels. Mitochondria are a prime subcellular target for hypoxic injury and the degree to which mitochondrial damage occurs can limit the rate and extent of cell recovery during reoxygenation because of the important roles this organelle plays in ATP formation, Ca^{2+} homeostasis, and free-radical detoxification. The ability of respiring mitochondria to buffer increases in cytosolic Ca^{2+} (i.e. sequester Ca^{2+}) such that increases in extramitochondrial concentrations of Ca^{2+} are dampened[34] may relate to mechanisms involved in delayed mitochondrial alterations and increased free radical production. Therefore, delayed neuronal death may result from a time dependent collapse of mitochondrial function following excessive Ca^{2+} accumulation.

We have detected significant differences in mitochondrial respiratory properties between *bcl-2* and non-*bcl-2*-expressing cells early in the development of delayed cell death in an *in vitro* model designed to mimic the events in ischemia/reperfusion injury.[37] In this model, we have used chemical hypoxia/aglycemia [3 mM potassium cyanide (KCN) in the presence of serum and absence of glucose] followed by reenergization (removal of KCN and return to growth medium) to model the rapid deenergization during ischemia and reenergization that occurs during reperfusion *in vivo.*[38] Using control and *bcl-2* transfectants of a murine hypothalamic GT1-7 cell line established by Kane *et al.*,[16] a model was established in which there was no cell death evident immediately following treatment (as there is no evidence of acute neuronal death either immediately or within several hours following resuscitation from cardiac arrest), but significant delayed cell death that develops over the ensuing days. The conditions were achieved by a 30 minute treatment of the cells with chemical hypoxia/aglycemia followed by 24-72 hr of incubation in growth media. Control transfectants were approx. 30-40% less viable than *bcl-2* expressors at 24 hr and this difference persisted up to the longest time point, 72 hr. We found that this regimen of chemical hypoxia/aglycemia resulted in respiratory impairment that was evident in the control transfectants immediately following the treatment. This evidence was provided by experiments measuring in intact cells the rates of the endogenous phosphorylating respiration (State 3), resting respiration induced by the addition of an inhibition of the ATP synthase (State 4), and uncoupler-stimulated respiration (an indication of the maximal rate of electron flow through the electron transport chain). This impairment was evident when expressing either the acceptor control ratios (the State 3 rate/the State 4 rate) or the respiratory control ratios (the uncoupler-stimulated rate /the State 4 rate). Both the acceptor control and respiratory control ratios were significantly inhibited (by approx. 25% and 41%, respectively) in the control cells immediately following treatment, whereas these ratios were unaffected in the *bcl-2* expressors. It was found, however, that the rate of decline of cellular ATP concentration during chemical hypoxia/aglycemia as well as the rate of recovery of ATP following replacement of the treatment media with normal growth media was found to be quite similar in GT1-7*puro* and *bcl-2* transfectants. Therefore, the mechanism by which Bcl-2 protects these cells under these conditions is not via preservation, or accelerated return of cellular ATP. These data suggest either that the mitochondrial impairment evidenced in the control cells by lower respiratory ratios is not severe enough to affect mitochondrial ATP production, or that glycolytic ATP production can compensate for the mitochondrial deficit. Using an HPLC technique for separation of oxidized lipids, we also found that chemical hypoxia/aglycemia and reenergization induced lipid oxidation in viable control transfectants to a level that was 150% of untreated controls at 24 hr following the 30 min. treatment. Bcl-2 completely inhibited this increase in lipid oxidation, thereby providing additional evidence for a correlated role of oxidative stress in the death pathway which Bcl-2 can inhibit.

In addition, we have found profound differences in mitochondrial Ca^{2+} transport properties that likely have functional significance with regard to cell injury under conditions

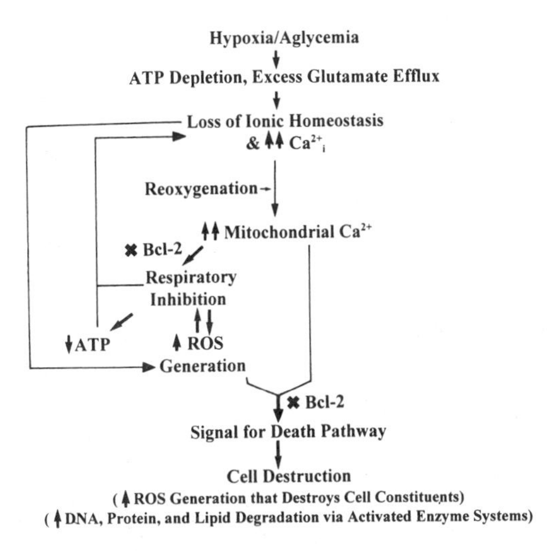

Fig. 1. Proposed scheme of events leading to delayed death in response to ischemia/reperfusion injury and potential sites of inhibition by Bcl-2. As a result of ATP depletion and excitotoxic mechanisms, the cytoplasmic Ca^{2+} concentration rises to pathological levels. Due to intrinsic Ca^{2+} transport properties, respiring mitochondria sequester a large Ca^{2+} load, which can result in potentiation of mitochondrial free radical production. The added stresses of increased matrix Ca^{2+} and free radical generation lead to triggering the signal (as yet unidentified) that leads to cell destruction, which is mediated by downstream effector systems that ultimately degrade DNA, proteins, and lipids.

in which the loss of ionic homeostasis is a triggering factor. Using digitonin-permeabilized GT1-7 transfectants, we have found that the mitochondrial Ca^{2+} uptake capacity (the maximal amount of Ca^{2+} that respiring mitochondria can sequester) is significantly greater in the cells expressing *bcl-2*[39]. As well, this enhancement correlates with greater resistance to Ca^{2+}-induced respiratory inhibition.[39] Therefore, if Bcl-2 potentiates mitochondrial Ca^{2+} uptake capacity thereby avoiding respiratory inhibition and possibly mitochondrial free radical production, then this activity could be the mechanism by which Bcl-2 protects cells from delayed cell death following ischemia/reperfusion injury.

These data have led to the development of a theoretical scheme for the events of ischemia/reperfusion injury and potential sites of Bcl-2 intervention (Fig. 1). Hypoxia/aglycemia induces a pathological increase in Ca^{2+}_i that results in significant mitochondrial Ca^{2+} sequestration upon cellular reenergization. This accumulation can result in respiratory inhibition which potentiates mitochondrial ROS generation. The combined effects of increased mitochondrial Ca^{2+} and enhanced ROS generation can trigger the death pathway (although either effect alone, if intense enough, may do the same), and this triggering step can be inhibited by Bcl-2. The identity of this triggering step remains elusive, but could involve the opening of the membrane permeability transition pore,[30,31] potentially releasing an activator of a cascade of enzyme systems that ultimately degrade the cell. Evidence that activity of homologues of the cysteine protease Il-1β converting enzyme lie close to this trigger in the apoptotic pathway is compelling.[40] Yet, significant experimental work explaining the biochemical linkage of these events is still required to complete our understanding of the critical steps in delayed neuronal death. In addition to global cerebral ischemia and stroke, other neurodegenerative disorders may share a similar mechanism of injury. By gaining a better understanding of Bcl-2 function in this regard, novel potential therapeutic approaches to these pathologies may ultimately be developed.

REFERENCES

1. Siesjö, B.K., Pathophysiology and treatment of focal cerebral ischemia. *J. Neurosurg.* 77:337-354 (1992).
2. Héron, A., Pollard, H., Dessi, F., Moreau, J., Lasbennes, Ben-Ari, Y., and Charriaut-Marlangue, C., Regional variability in DNA fragmentation after global ischemia evidenced by combined histological and gel electrophoresis observations in the rat brain. *J. Neurochem.* 61:1973-1976 (1993).
3. Linnik, M.D., Zobrist, R.H., and Hatfield, M.D., Evidence supporting a role for programmed cell death in focal cerebral ischemia in rats. *Stroke* 24:2002-2009 (1993).
4. Okamoto, M., Matsumoto, M., Ohtsuki, T., Taguchi, A., Mikoshiba, K., Yanagihara, T., and Kamada, T., Internucleosomal DNA cleavage involved in ischemia-induced neuronal death. *Biochem. Biophys. Res. Commun.* 196:1356-1362 (1993).
5. Sei Y., Von Lubitz D.K.J.E., Basile A.S., Borner M.M., Lin R.C.-S., Skolnick P., and Fossom L.H., Internucleosomal DNA fragmentation in gerbil hippocampus following forebrain ischemia. *Neurosci. Lett.* 171, 179-182 (1994).
6. Nitatori, T., Sato, N., Waguri, S., Karasawa, Y., Araki, H., Shibanai, K., Kominami, E., and Uchiyama, Y., Delayed neuronal death in the CA1 pyramidal cell layer of the gerbil hippocampus following transient ischemia is apoptosis. *J. Neurosci.* 15:1001-1011 (1995).
7. MacManus, J.P., Hill, I.E., Preston, E., Rasquinha, I., Walker, T., and Buchan, A.M., Differences in DNA fragmentation following transient cerebral or decapitation ischemia in rats. *J. Cereb. Blood Flow Metab.* 15:728-737 (1995).
8. Kihara S., Shiraishi T., Nakagawa S., Toda K., and Tabuchi K., Visualization of DNA double strand breaks in the gerbil hippocampal CA1 following transient ischemia. *Neurosci. Lett.* 175, 133-136 (1994).
9. Deshpande, J., Bergstedt, K., Linden, T., Kalimo, H., and Wieloch, T., Ultrastructural changes in the hippocampal CA1 region following transient cerebral ischemia: evidence against programmed cell death. *Exp. Brain Res.* 88:91-105 (1992).
10. Lennon, S.V., Martin, S.J., and Cotter, T.G., Dose-dependent induction of apoptosisin human tumour cell lines by widely diverging stimuli. *Cell Prolif.* 24:203-214 (1991).
11. Buttke, T.M., and Sandstrom, P.A., Oxidative stress as a mediator of apoptosis. *Immun. Today* 15:7-10 (1994).
12. Reed, J.C., Bcl-2 and the regulation of programmed cell death. *J. Cell Biol.* 124:1-6 (1994).
13. Hockenbery, D.M., Nuñez, G., Milliman, C., Schreiber, R.D., and Korsmeyer, S.J., Bcl-2 is an inner mitochondrial membrane protein that blocks programmed cell death. *Nature* 348:334-336 (1990).
14. Hockenbery, D.M., Oltvai, Z.N., Yin, X.-M., Milliman, C.L., and Korsmeyer, S.J. Bcl-2 functions in an antioxidant pathway to prevent apoptosis. *Cell* 75:241-251 (1993).
15. Zhong, L.-T., Sarafian, T., Kane, D.J., Charles, A.C., Mah, S.P., Edwards, R.H., and Bredesen, D.E., bcl-2 inhibits death of central neural cells induced by multiple agents. *Proc. Natl. Acad. Sci. USA*, 90:4533-4537 (1993).
16. Kane, D.J., Sarafian, T.A., Anton, R., Hahn, H., Gralla, E.B., Valentine, J.S., Örd, T., and Bredesen, D.E., Bcl-2 inhibition of neural death: Decreased generation of reactive oxygen species, *Science* 262:1274-1277 (1993).
17. Mah, S.P., Zhong, L.T., Liu, Y., Roghani, A., Edwards, R.H., and Bredesen, D.E. The protooncogene bcl-2 inhibits apoptosis in PC12 cells. *J. Neurochem.* 60:1183-1186 (1993).
18. Lam M., Dubyak G., Chen L., Nunez G., Miesfeld R.L., and Distelhorst C.W., Evidence

that BCL-2 represses apoptosis by regulating endoplasmic reticulum-associated Ca^{2+} fluxes. *Proc. Natl. Acad. Sci. USA.* 91, 6569-6573 (1994).

19. Dubois-Dauphin M., Frankowski H., Tsujimoto Y., Huarte J., and Martinou J.-C., Neonatal motoneurons overexpressing the *bcl-2* protooncogene in transgenic mice are protected from axotomy-induced cell death. *Proc. Natl. Acad. Sci. USA* 91, 3309-3313 (1994).
20. Linnik M.D., Zahos P., Geschwind M.D., Federoff H.J., Expression of bcl-2 from a defective herpes simplex virus-1 vector limits neuronal death in focal cerebral ischemia. *Stroke* 26, 1670-1675 (1995).
21. Shimazaki, K., Ishida, A., and Kawai, N., Increase in bcl-2 oncoprotein and the tolerance to ischemia-induced neruonal death in the gerbil hippocampus. *Neurosci. Res.* 20:95-99 (1994).
22. Chen, J., Graham, S.H., Chan, P.H., Lan, J., and Simon, R.P., bcl-2 is expressed in neurons that survive focal ischemia in the rat. *Neuroreport* 6:394-398. (1995).
23. Krajewski, S., Mai, J.K., Krajewska, M., Sikorska, M., Mossakowski, M.J., and Reed, J.C., Upregulation of Bax protein levels in neurons following cerebral ischemia. *J. Neurosci.* 15:6364-6367 (1995).
24. Jacobson M.D., and Raff M.C., Programmed cell death and Bcl-2 protection in very low oxygen. *Nature* 374, 814-816 (1995).
25. Shimizu S., Eguchi Y., Kosaka H., Kamiike W., Matsuda H., and Tsujimoto Y., Prevention of hypoxia-induced cell death by Bcl-2 and Bcl-x_L. *Nature* 374, 811-813 (1995).
26. Chen-Levy, Z., Nourse, J., and Cleary, M.L., The Bcl-2 candidate proto-oncogene product is a 24-kilodalton integral-membrane protein highly expressed in lymphoid cell lines and lymphomas carrying the t(14;18). *Mol. Cell. Biol.* 9:701-710 (1989).
27. Krajewski, S., Tanaka, S., Takayama, S., Schibler, M.J., Fenton, W., and Reed, J.C., Investigation of the subcellular distribution of the *bcl-2* oncoprotein: residence in the nuclear envelope, endoplasmic reticulum, and outer mitochondrial membranes. *Cancer Res.* 53:4701-4714 (1993).
28. Newmeyer, D.D., Farschon, D.M., and Reed, J.C., Cell-free apoptosis in *Xenopus* egg extracts: inhibition by Bcl-2 and requirement for an organelle fraction enriched in mitochondria. *Cell* 79, 353-364 (1994).
29. Cross, A.R., and Jones, O.T.G., Enzymic mechanisms of superoxide production. *Biochim. Biophys. Acta* 1057:281-298 (1991).
30. Gunter, T.E., Gunter, K.K., Sheu, S.-S., and Gavin, C.E., Mitochondrial calcium transport: physiological and pathological relevance. *Am. J. Physiol.* 267:C313-C339 (1994).
31. Gunter, T.E., and Pfeiffer, D.R., Mechanisms by which mitochondria transport calcium. *Am. J. Physiol.* 258:C755-C786 (1990).
32. Sciammanna, M.A., Zinkel, J., Fabi, A.Y., Lee, C.P., Ischemic injury to rat forebrain mitochondria and cellular calcium homeostasis, *Biochem Biophys. Acta.* 1134:223-232, 1992.
33. Sun D., and Gilboe D.D., Ischemia-induced changes in cerebral mitochondrial free fatty acids, phospholipids, and respiration in the rat. *J. Neurochem.* 62, 1921-1928 (1994).
34. White, R.J., and Reynolds, I.J., Mitochondria and Na^+/Ca^{2+} exchange buffer glutamate-induced calcium loads in cultured cortical neurons. *J. Neurosci.* 15:1318-1328 (1995).
35. Reynolds, I.J., and Hastings, T.G., Glutamate induces the production of reactive oxygen species in cultured forebrain neurons following NMDA receptor activation. *J. Neurosci.* 15:3318-3327 (1995).
36. Dugan, L.L., Sensi, S.L., Canzoniero, L.M.T., Handran, S.D., Rothman, S.M., T.-S. Lin, Goldberg, M.P., and Choi, D.W., Mitochondrial production of reactive oxygen species

in cortical neurons following exposure to *N*-methyl-D-aspartate. *J. Neurosci.* 15:6377-6388 (1995).

37. Myers, K.M., Fiskum, G., Liu, Y., Simmens, S.J., Bredesen, D.E., and Murphy, A.N., Bcl-2 protects neural cells from cyanide/aglycemia induced lipid oxidation, mitochondrial injury, and loss of viability, *J. Neurochem.* (1995, in press).
38. Nishijima, M.K., Koehler, R.C., Hurn, P.D., Eleff, S.M., Norris, S., Jacobus, W.E., and Traystman, R.J., Postischemic recovery rate of cerebral ATP, phosphocreatine, pH, and evoked potentials. *Am. J. Physiol.* 257, H1860-H1870 (1989).
39. Murphy, A.N., Bredesen, D.E., and Fiskum, G., Bcl-2 protects neural cell mitochondria from Ca^{2+} overload and Ca^{2+} -induced respiratory inhibition. *Soc. for Neurosci. Abst.* 21(3):1728 (1995).
40. Wang L., Miura M., Bergeron L., Zhu H., and Yuan J., *Ich-1*, an *Ice/ced-3*-related gene, encodes both positive and negative regulators of programmed cell death. *Cell* 78, 739-750 (1994).

BIOCHEMICAL CHARACTERISTICS OF OXYGEN-INDUCED AND LOW K^+ MEDIUM-INDUCED APOPTOTIC NEURONAL DEATH

Yasushi Enokido[1], Takekazu Kubo[1], Noboru Sato[2] Yasuo Uchiyama[2] and Hiroshi Hatanaka[1]

[1]Division of Protein Biosynthesis Institute for Protein Research, Osaka University, 3–2 Yamadaoka, Suita, Osaka 565, [2]1st Department of Anatomy, Osaka University school of Medicine, 2–2 Yamadaoka, Suita, Osaka 565, Japan.

INTRODUCTION

Neuronal cell death is thought to be an important phenomenon not only in the developmental but also in the post–matured stages of PNS and CNS. In the developmental stages, massive neuronal death is observed at specific periods, e.g. the immediately after the arrival of axons to the postsynaptic target fields. This phenomenon is precicely programmed in the developing nervous system and called programmed (or apoptotic) neuronal death. Recent studies also suggest that similer apoptotic neuronal death occurs in the matured nervous system, e.g. neurodegenerative diseases, such as Alzheimer's and Parkinson's diseases, and aging. Thus, it is expected that the study of apoptosis in the nervous system give us useful clues to know how the neuronal net work are constructed in the development and why the neurodegenerateve diseases and brain aging occur.

To study the molecular mechanisms of apoptosis in CNS neurons, we investigated the biochemical characteristics of neuronal death induced by oxygen toxicity and low K^+ medium. Furthermore, we examined the survival–promoting effect of BCL–2 protein using *bcl–2*–transfected PC12 cells.

MATERIALS AND METHODS

Cell Culture

Primary cultures of dissociated hippocampal and cerebellar granule neurons were prepared from the brains of embryonic day 20～21 and postnatal day 9 rats, respectively, as previously described[1,2]. Cells were plated at a cell density of $4{\sim}5\times10^5$ cells/cm^2 on a polyethyleneimine–coated culture plates. Hippocampal neurons and PC12 cells[3] were cultured in the chambers with 20% or 50% O_2 and a constant 5% CO_2 atmosphere in a N_2–O_2–CO_2 gas incubator. Cerebellar granule neurons were

Neurodegenerative Diseases
Edited by Gary Fiskum, Plenum Press, New York, 1996

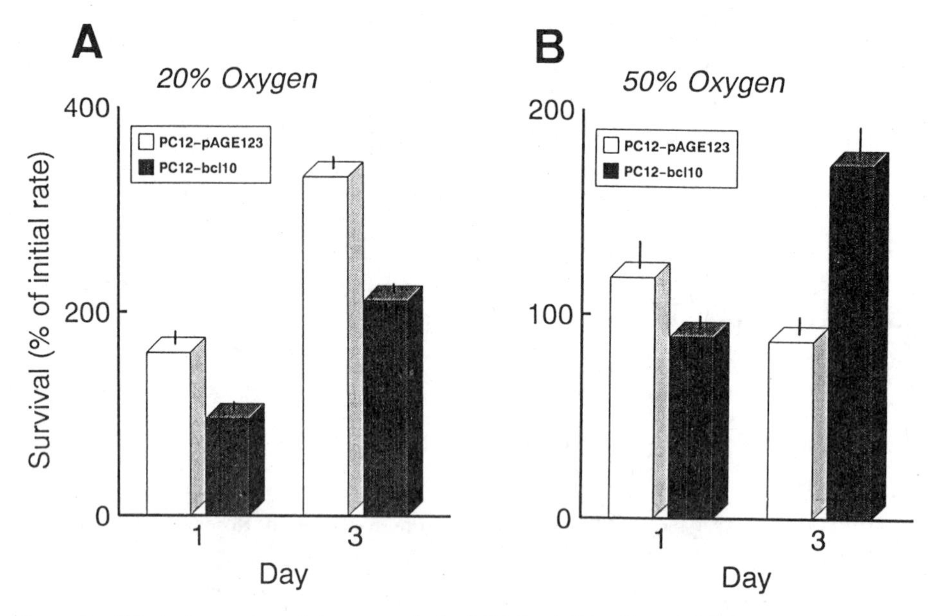

Figure 1. Effect of oxygen toxicity on survival of PC12-pAGE cells and PC12-bcl10 cells. Control cells (PC12-pAGE123) and *bcl-2* expressing cells (PC12-bcl10) were cultured in a 20% (A) or 50% (B) oxygen atmosphere for the days indicated. Cell survival was determined by MTT assay. Initial value of A_{570} were 0.280 ± 0.024 (PC12-pAGE123) and 0.255 ± 0.027 (PC12-bcl10), respectively. Values = means $\pm$ S.D. (n=3~4).

cultured in high K^+ (26 mM) MEM containing 5% horse serum for 7 days, and then the medium were changed to serum-free high K^+ or low K^+ (5.4 mM) MEM. The high K^+ MEM was prepared by increasing the $KHCO_3$ concentration from normal value of 5.4 to 26 mM with the omission of a corresponding concentration $NaHCO_3$.

Assays of neuronal survival

Neuronal survival was determined by the MTT assay. Briefly, the tetrazolium salt MTT (3-(4,5-dimethyl-2-thiazolyl)-2,5-diphenyl-2H tetrazolium bromide) was added to cultures (1 mg/ml final concentration). After a 2 hr incubation at 37° C, the assay was stopped by adding lysing buffer (20% SDS in 50% N,N-dimethyl formamide, pH 4.7). The absorbance was measured photometrically at 570 nm after a further overnight incubation at 37℃. The value obtained from cells maintained in high K^+ medium was the daily control. The percent survival was defined as [Abs.(experimental-blank)/Abs.(control-blank)] $\times$ 100 and the blank was the value taken from wells without cells.

When indicated, neuronal survival was quantified by counting the number of MAP2-positive neurons.

RESULTS AND DISCUSSION

When embryonic rat hippocampal neurons were cultured in a 50% O_2 atmosphere, neurons were gradually died after 20 h in culture. This death pattern was found to be mediated by apoptosis, as follows: (1) RNA and protein synthesis inhibitors prevented cell death. (2) DNA fragmentation was detected during the course of cell death. (3) Depolarization with high K^+ medium (26-50 mM) prevented cell death.

(4) Overexpression of *bcl-2* gene in PC12 cells prevents oxygen-induced cell death (Fig.1). Since cell death was also prevented by some antioxidants, such as α-tocopherol and N-acetylcysteine, it was expected that oxidative stress deeply related to the cell death. We also examined the effects of several kinase inhibitors to prevent the neuronal death. As a result, H-7, H-8 and H-89 effectively prevented the cell death but genistein and herbimycin A, both tyrosine kinase inhibitors, did not. Furthermore, 3-aminobenzamide which inhibits ADP-ribosylation, delayed the cell death. This result suggest that DNA damage involved in the death cascade. In contrast, the low K^+ medium-induced death of cerebellar granule neurons was specifically prevented by H-7.

These results suggest that above two types of neuronal death posses different intracellular death cascades and induce ultimate apoptotic CNS neuronal death.

REFERENCES

1. Y. Enokido and H. Hatanaka, Apoptotic cell death occurs in hippocampal neurons cultured in a high oxygen atmosphere., *Neuroscience*, **57**:965(1993).
2. T. Kubo, T. Nonomura, Y. Enokido and H. Hatanaka, Brain-derived neurotrophic factor (BDNF) can protect apoptotis of rat cerebellar granule neurons in culture., *Dev. Brain Res*, **85**:249(1995).
3. N. Sato, K. Hotta, S. Waguri, T. Nitatori, K. Tohyama, Y. Tsujimoto and Y. Uchiyama, Neuronal differentiation of PC12 cell as a result of prevention of cell death by *bcl-2*., *J. Neurobiol.*, **25**:1227(1994).
4. M.B. Hansen, S.E. Nielsen and K. Berg, Re-examination and further development of aprecice and rapid dye method for measureing cell growth/cell kill., *J. Immunol. Methods*, **119**:203(1989).

MECHANISMS OF SELECTIVE NEURONAL VULNERABILITY TO 1-METHYL-4-PHENYLPYRIDINIUM (MPP+) TOXICITY

Patricia A. Trimmer[1], Jeremy B.Tuttle[2], Jason P. Sheehan[3], and James P. Bennett, Jr.[4]

[1]Departments of Neurology and Neuroscience
[2]Departments of Urology and Neuroscience
[3]Department of Neurosurgery
[4]Department of Neurology
University of Virginia
Charlottesville, VA 22908

A common feature of neurodegenerative diseases such as Parkinson's disease is the selective, inappropriate death of specific populations of central neurons. In Parkinson's disease, dopaminergic neurons are lost in the zona compacta of the substantia nigra.[1] This selective neuronal death gives rise to a progressive movement disorder whose cause is unknown and for which there is no known cure.

Insight into how and why neurons in the substantia nigra die increased dramatically after the discovery of the compound 1-methyl-4-phenyl-1,2,3,6-tetrahydropyridine (MPTP).[2] In human and primates, exposure to MPTP induces a syndrome with symptoms that mimic idiopathic Parkinson's disease.[3] MPTP itself is not toxic (for a review see Johannessen[4]). After it enters the brain, MPTP is taken up largely by astrocytes where it is converted to 1-methyl-4-phenylpyridinium ion (MPP+), the active metabolite, by monoamine oxidase B.[5] MPP+ is specifically accumulated in neurons by the dopamine transporter.[6,7] However, if the concentration is sufficiently high, MPP+ can enter cells without the aide of the transporter.[4] MPP+ accumulates in vesicles in monoaminergic terminals but the primary site of action is the mitochondria. In mitochondria, MPP+ inhibits complex I of the mitochondrial respiratory chain and severely decreases the output of ATP.[7,8] Other cellular consequences of MPTP toxicity include altered calcium homeostasis[9] and excessive free radical formation[10] and ultimately, cell death.

In order to better understand what specific factors cause the death of nigral neurons in Parkinson's disease or as a consequence of MPP+ exposure, we must first determine the mechanism of cell death. Cell death can occur as consequence of necrosis or apoptosis.[11] Necrotic cell death is associated with ischemia, anoxia or drug toxicity and is characterized by cell swelling, membrane disruption and random DNA fragmentation.[11,12] In contrast, apoptosis is an active process that occurs in response to a variety of physiological events such as tissue remodeling, growth factor removal or in response to various pathological condition.[11,13,14] Hallmarks of apoptotic cell death include chromatin collapse and the

Neurodegenerative Diseases
Edited by Gary Fiskum, Plenum Press, New York, 1996

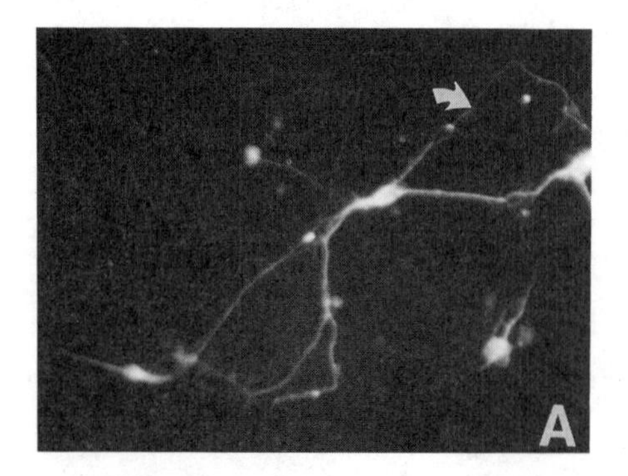

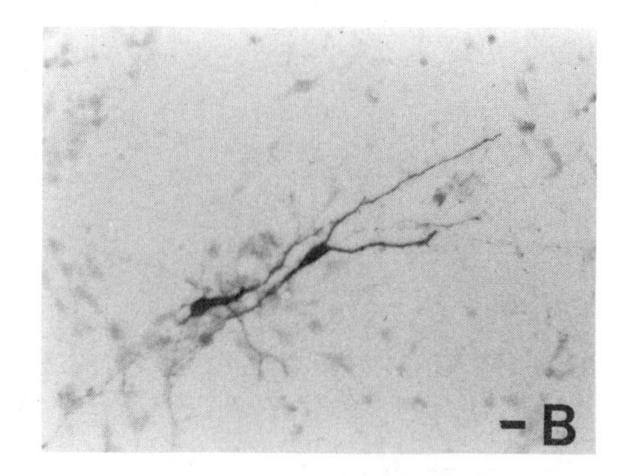

Figure 1- Immunocytochemically-stained neurons after 5 days in culture. 1A shows a neuron stained with TuJ1 antibodies followed by an FITC-conjugated secondary antibody. TuJ1 staining is present in the cell body, dendrites and axon (arrow) of this cell. 1B shows a culture stained with tyrosine hydroxylase antibodies followed by visualization with peroxidase. Two positively-stained neurons are present in the field. Bar = 10μM.

condensation of the nucleus and the cleavage of DNA into double-stranded fragments by a Ca++ activated endogenous endonuclease.[12,15] Cells undergoing apoptosis can be specifically labeled by tailing biotinylated dUTP onto the 3'-OH ends of DNA using terminal deoxynucleotidyl transferase.[16,17]

The focus of this paper is the development of tissue culture models that can be used to study the factors and mechanisms involved in selective neuronal cell death, especially in relation to neurodegenerative diseases. Defining the molecular mechanisms of cell death, both in vitro and in vivo, is an important avenue for the development of relevent therapeutic treatments for Parkinson's and other neurodegenerative diseases.

METHODS

Primary cultures of fetal rat mesencephalon were prepared from E14-16 embryos. Briefly, the mesencephalic tissue was collected in sterile saline, minced and dissociated with trypsin (0.25% for 15 min at 37°C). Using methods previously described[18], the neurons were seeded at low density on polylysine coated coverslips. These coverslips were then inverted over a bed layer of neonatal rat astrocytes growing in defined medium. Small paraffin feet separated the neurons on the coverslips from the astrocytes in the bed layer which serve as a source of trophic factors.

Human neuroblastoma cells (SH-SY5Y) were maintained using standard techniques in culture medium consisting of Dulbecco's modified Eagle's medium with high glucose, 10% fetal bovine serum and antibiotics. Serum-free medium was added to the neuroblastoma cells one day prior to MPP+ treatment.

Immunocytochemical staining of fixed cultures was carried out as previously described.[19]

Cultures of primary neurons and neuroblastoma cells were exposed to a 20μM concentration of MPP+ in defined or serum-free medium. Details of the exposure periods are described in the results and figure legends.

To identify neurons that undergo apoptosis as a consequence of MPP+ exposure, cultures were fixed with 4% paraformaldehyde in phosphate buffered saline and terminal deoxynucleotidyl transferase was used to tail digoxigenin-labeled dUTP onto the 3' ends of

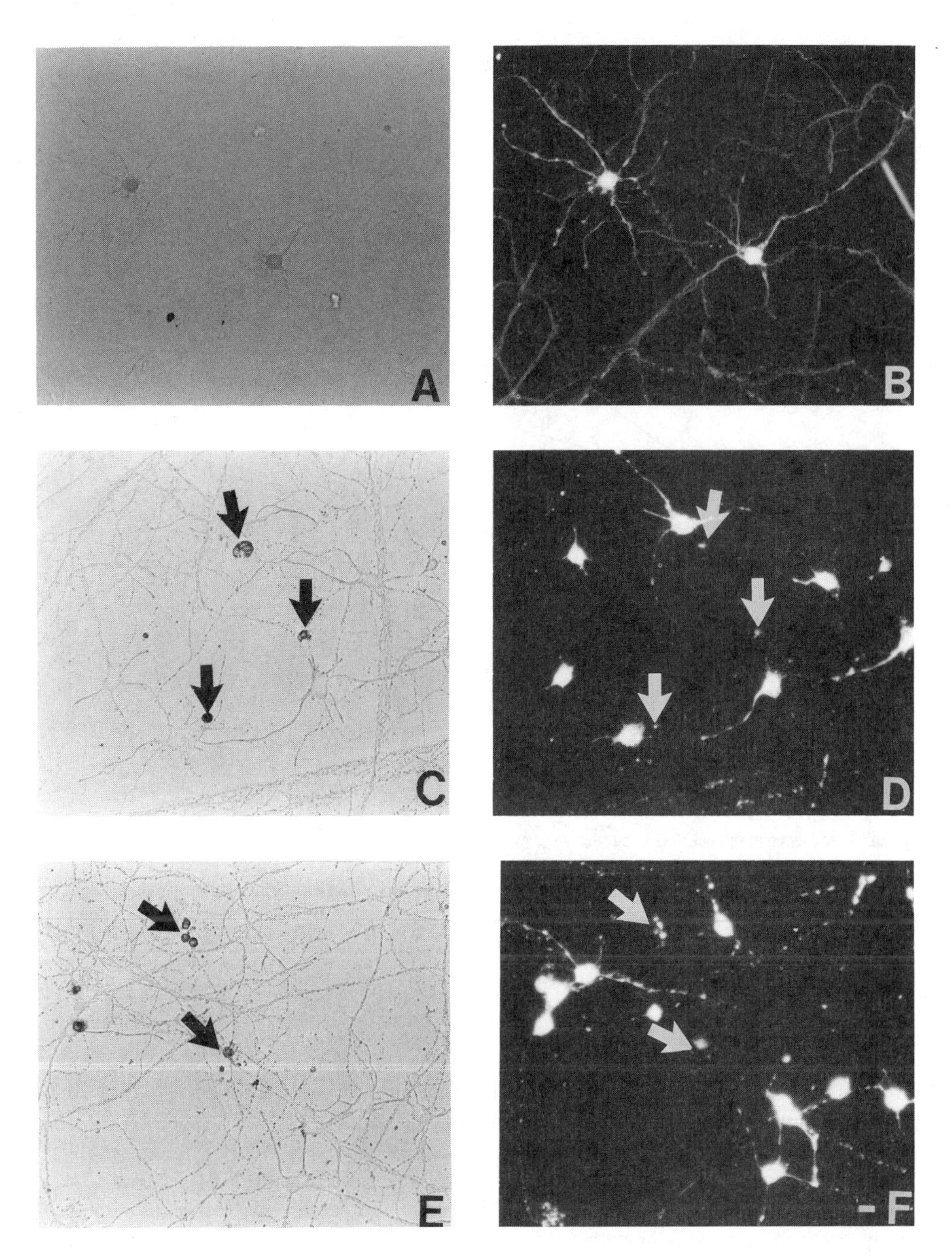

Figure 2- Mesencephalic neurons in culture after double staining with TuJ1 antibodies and the ApopTag kit to visualize apoptotic nuclei. 2A and 2B show two neurons after a 3 hour exposure to MPP+. These cells have normal, non-apoptotic nuclei (2A) and the pattern of TuJ1 staining (2B) in the cell bodies and processes is normal. 2C and 2D show several neurons after a 6 hour exposure to MPP+. Of the ten neurons in this field, three have apopototic nuclei (arrows). Arrows indicate the same three neurons after TuJ1 staining (2D). These cells have no remaining TuJ1 staining. The other neurons in the field (2D) are abnormal in appearance and have shrunked, beaded dendrites. 2E and 2F show several neurons after a 48 hour exposure to MPP+. A number of the neurons (2E) have apoptotic nuclei (arrows). Arrows indicate several of these same neurons after TuJ1 staining (2F). The apoptotic cells have no residual TuJ1 staining . The other neurons in the field (2F) are abnormal in appearance and have shrunken, beaded dendrites. Bar = 10μM.

any DNA fragments present using the ApopTag kit (Oncor). The labeled DNA is then visualized with either a peroxidase or FITC-conjugated anti-digoxigenin antibody according to kit directions. To generate a positive control, intact DNA in neurons was nicked with the enzyme, deoxyribonuclease I.[12]

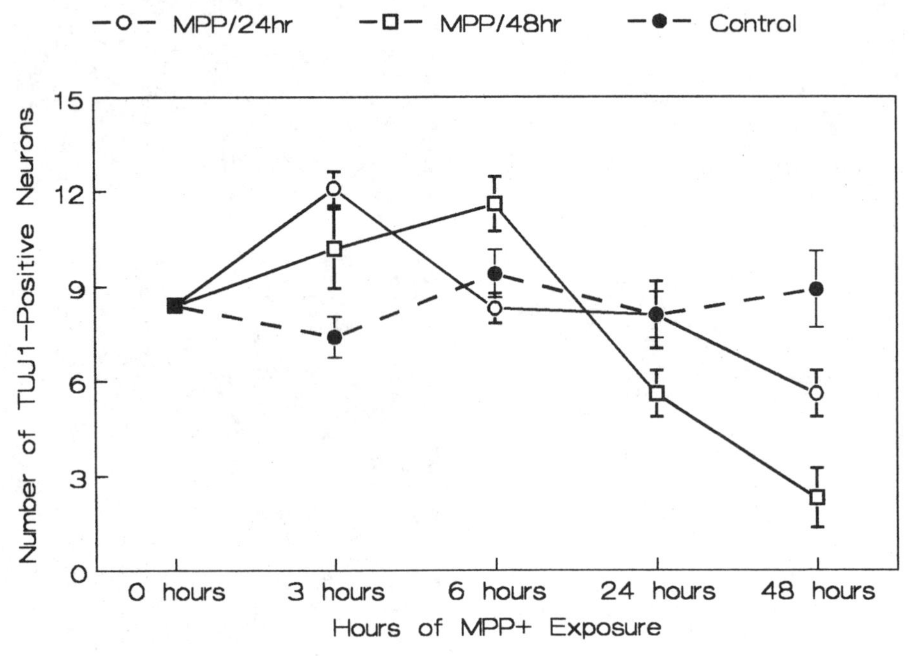

Graph 1- Mesencephalic neurons in culture were exposed to 20 μM MPP+ for 0, 3, 6, 24 or 48 hours. The toxin was then removed and the cells were allowed to survive for an additional 24 or 48 hours before fixation. TuJ1-stained neurons in ten fields per culture and time point were counted. The error bars represent standard error.

RESULTS

Mesencephalic Neuronal Cultures

The overwhelming majority of cells in the primary mesencephalic cultures were identified as neurons based on specific staining with antibodies to TuJ1, a neuron-specific isoform of beta-tubulin and the astrocyte specific protein, glial fibrillary acidic protein.[19,20] Figure 1A shows a neuron in culture that has been stained with TuJ1 followed by a fluorescein-conjugated secondary antibody (Vector). Immunoreactivity is present in the cell body, dendrites and axons. Dopaminergic neurons were identified by staining with antibodies to tyrosine hydroxylase (Eugene Tech International, Inc). The tyrosine hydroxylase staining in the neurons shown in Figure1B has been visualized using the ABC technique (Vector) and peroxidase. The proportion of tyrosine hydroxylase-positive neurons in our cultures was small, (<5%).

This plate illustrates the appearance of stained neurons from primary cultures of fetal rat mesencephalon that have been exposed to a 20 uM concentration of MPP+ for 3, 6 or 48 hours prior to fixation. The neurons were double-stained with TuJ1 antibodies and with the ApopTag kit to label fragmented DNA in apoptotic neurons. After 3 hrs exposure to MPP+

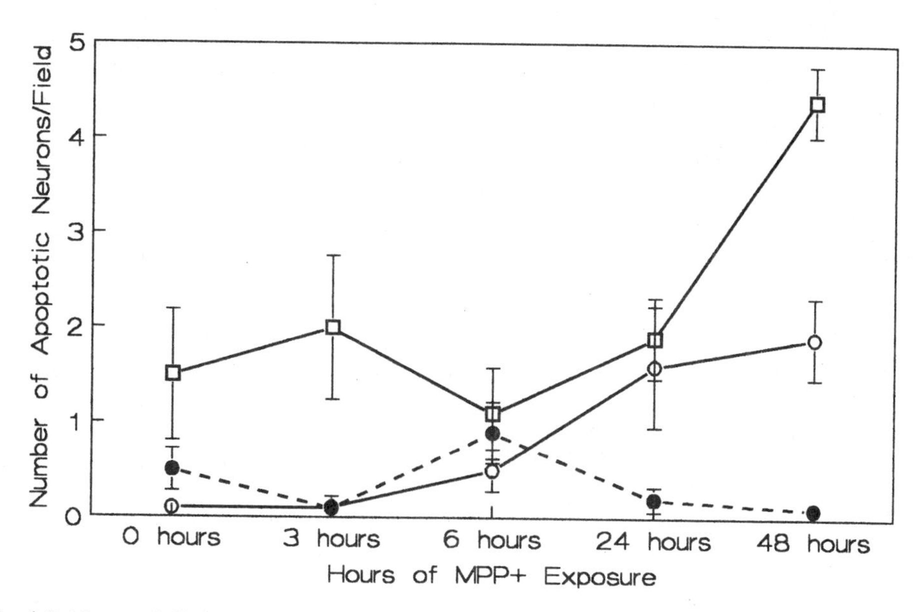

Graph 2- Mesencephalic neurons in culture were exposed to 20 μM MPP+ for 0, 3, 6, 24 or 48 hours. The toxin was then removed and the cells were allowed to survive for an additional 24 or 48 hours before fixation. Apoptotic neurons in ten fields per culture and time point were counted. The error bars represent standard error.

(Figure 2A,B), there were few apoptotic neurons and TuJ1 staining was normal and distributed into the dendrites and axons. In contrast, after the 6 hours exposure to MPP+ (Figures 2C,D), brown-stained, apoptotic nuclei were clearly identifiable. These apoptotic neurons no longer exhibited TuJ1 staining in the cell bodies or processes. Other non-apoptotic neurons retained TuJ1 staining but had neurites that were beaded and fragmented. Similarly, after a 48 hour exposure to MPP+, brown-stained apoptotic neurons were detectable. These cells exhibited no residual TuJ1 staining. Non-apoptotic neurons in these cultures had TuJ1 staining that was restricted to beaded, retracted neurites (Figure 2E,F).

Graph 1 depicts the cell counts of TuJ1-stained neurons in primary mesencephalic cultures exposed to a 20μM concentration of MPP+. Exposure to MPP+ for 6 and 24 hr resulted in a substantial decline in the number of TuJ1-stained neurons at 48 hrs post-exposure. Graph 2 shows the number of apoptotic neurons per field in the same cultures. Increased numbers of apoptotic neurons were detected after 6-24 hrs exposure to MPP+ in both the 24 and 48 hr survival groups.

We have also conducted similar experiments with cultures of SH-SY5Y neuroblastoma cells. These cells possess tyrosine hydroxylase and store dopamine.[21] Figure 3A illustrates a field of undifferentiated neuroblastoma cells that were exposed to 20 μM MPP+. Two apoptotic nuclei are present.

The SY5Y cells can be differentiated with phorbol ester or retinoic acid,[22] causing the cells to assume a more neuronal morphology. Preliminary results suggest that differentiated SH-SY5Y neuroblastoma cells are also sensitive to MPP+ toxicity and respond by undergoing apoptosis (Figure 3B). In fact, differentiated SH-SY5Y neuroblastoma cells may be somewhat more sensitive to MPP+ toxicity than undifferentiated neuroblastoma cells.

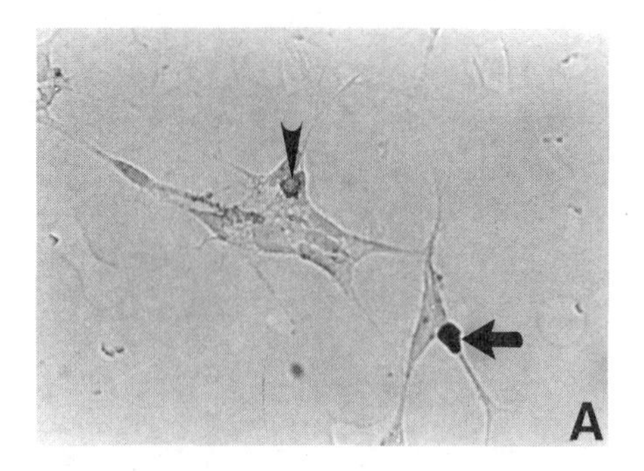

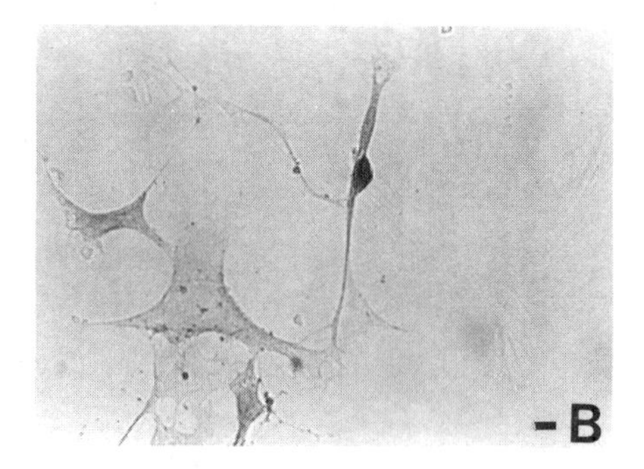

Figure 3- SH-SY5Y neuroblastoma cells after exposure to 20 μM MPP+ and visualization of apoptotic nuclei. 3A shows several undifferentiated neuroblastoma cells in culture. There are two apoptotic nuclei in this field. One is condensed and darkly stained (arrow). The other apoptotic nucleus is in an early stage of the process and has dark apopototic bodies present around the perimeter of the nucleus (arrowhead). 3B shos neuroblastoma cells that have been differentiated with retinoic acid. The neuron-like cell in the center of this field has undergone apoptosis. The nucleus is heavily stained. Bar = 10μM.

DISCUSSION

Our results show that MPP+ caused dose-dependent cell death in primary cultures of mesencephalic neurons. These findings support results of previous studies with cerebellar neurons[23,24] as well as dopaminergic neurons.[25,26] In agreement with these studies, our results showed that MPP+ exposure causes neurite degeneration and disruption of the neuritic network.[24,25,26] Our study also confirms the report by Mochizuki et al[27] who showed that MPP+ induces apoptotic cell death in ventral mesencephalon-striatal co-cultures. We have carried this study one step further to determine the time course of MPP+ induced apoptosis. Loss of TuJ1-positive neurons was detectable after a 6 hour exposure to a 20μM concentration of MPP+. A parallel increase in the number of apoptotic nuclei had a similar time course. Furthermore, our study showed that apoptotic neurons cease to express fundamental proteins such as beta-tubulin based on staining with the neuron-specific TuJ1 antibodies.

To our knowledge this is the first demonstration of apoptosis in SH-SY5Y neuroblastoma cells that is the result of MPP+ toxicity. This is not unexpected because these cells possess tyrosine hydroxylase and concentrate dopamine.[21] Further study is needed to determine whether differentiated SH-SY5Y cells are more sensitive to MPP+ and the basis for this increased vulnerability.

The observation that MPP+ induces apoptotic cell death suggests that apoptosis may be the mechanism of cell death in neurodegenerative diseases like Parkinson's. To date, we have been unable to demonstrate apoptotic cell death the the substantia nigra of C57BL/6 mice treated with MPTP (unpublished observations). This may be due to the fact that apoptotic cells are rapidly removed in vivo and MPTP toxicity occurs over a longer period of time in vivo. It will be important to establish the mechanisms of cell death in neurodegenerative diseases. We plan to use these culture models to examine the molecular events that trigger apoptosis in neurons and to define the temporal sequence of cellular changes associated with apopotosis. This work will help elucidate the biological basis of selective vulnerability that appears to play a role in the pathogenesis of neurodegenerative diseases such as Parkinson's disease.

ACKNOWLEDGMENTS

The authors would like to thank Graciela Gamez-Torre for her help in preparation and maintenance of the cultures. This work was supported by NIH grant P01 NS 30024.

REFERENCES

1. L. S. Forno, Pathology of Parkinson's disease, *in:* "Movement Disorders," C. D. Marsden and S. Fahn, eds., Butterworths, London (1982). p25.
2. P. A. Ballard, J. W. Tetrud, and J.W. Langston, Permanent human parkinsonism due to 1-methyl-4-phenyl-1,2,3,6-tetrahydropyridine (MPTP): seven cases, *Neurol.* 35:949 (1985).
3. R. S. Burns, Subclinical damage to the nigrostriatal dopamine system by MPTP as a model of preclinical Parkinson's disease: A review, *Acta Neurol. Scand. Suppl.* 136:29 (1991).
4. J. N. Johannessen, A model of chronic neurotoxicity: Long-term retention of the neurotoxin 1-methyl-4-phenyl pyridinium (MPP+) within catecholaminergic neurons, *Neurotoxicology* 12:285 (1991).
5. B. R. Ransom, D. M. Kunis, I. Irwin and J. W. Langston, Astrocytes convert the parkinsonism inducing neurotoxin, MPTP, to its active metabolite, MPP+, *Neurosci. Lett.* 75:323 (1987).
6. R. R. Ramsey, J. Dadgar, A. Trevor and T.P. Singer, Energy-driven uptake of N-methyl-4-phenylpyridine by brain mitochondria mediates the neurotoxicity of MPTP, *Life Sci.* 39:581 (1986).
7. K. F. Tipton and T. P. Singer, Advances in our understanding of the mechansims of the neurotoxicity of MPTP and related compounds, *J. Neurochem.* 61:1191 (1993).
8. Y. Mizuuno, K. Suzuki, N. Sone and T. Saitoh, Inhibition of ATP synthesis by 1-methyl-4-phenylpyridinium ion (MPP+) in isolated mitochondria from mouse brains, *Neurosci. Lett.* 81:204 (1987).
9. G. E. N. Kass, J. M. Wright, P. Nicoreta and S. Orrenius, The mechanism of MPTP toxicity: Role of intracellular calcium, *Arch. Biochem. Biophys.* 260:789 (1988).
10. E. Hasegawa, K. Takeshige, T. Oishe, Y. Murai, and S. Nimakami, 1-methyl-4-phenyl pyridinium (MPP+) induces NADH dependent superoxide formation and enhances NADH- dependent lipid peroxidation in bovine heart submitochondrial particles, *Biochem. Biophys. Res. Commun.* 170:1049 (1990).
11. A. H. Wyllie, Cell death: a new classification separating apoptosis from necrosis, *in:* "Cell Death in Biology and Pathology," I. D. Bowen and R. A. Lockshen, eds., Chapman and Hall, London (1981), p.9.
12. M. J. Arends, R. G. Morris and A. H. Wyllie, Apoptosis. The role of endonuclease, *Am. J. Pathol.* 136:593 (1990).
13. M. K. L. Collins and A. L. Rivas, The control of apoptosis in mammalian cells, *TIBS* 18:307 (1993).
14. C. Portera-Cailliau, J. C. Hedreen, D. L. Price and V.E. Koliatsos, Evidence for apoptotic cell death in Huntington disease and excitotoxic animal models, *J. Neurosci.* 15:3775 (1995).
15. J. Searle, J. F. R. Kerr and C. J. Bishop, Necrosis and apoptosis: distinct modes of cell death with fundamentally different significance, Pathol. Annu.17(pt2):229 (1982).
16. Y. Gavrieli, Y. Sherman and S. A. Ben-Sasson, Identification of programed cell death in situ via specific labeling of nuclear DNA fragmentation, *J. Cell Biol.* 119:493 (1992).
17. R. Gold, M. Schmied, G. Gregerich, H. Breitschopf, H. P. Hartung, K. V. Yoyka and H. Lassmann, Differentiation between cellular apoptosis and necrosis by the combined use of in situ tailing and nick translation techniques, *Lab. Invest.* 71:219 (1994).
18. W. P. Bartlett, and G. A. Banker, An electron microscopic study of the development of axons and dendrites by hippocampal neurons in culture I. J. Neurosci. 4:1944 (1984).

19. P. A. Trimmer, L. L. Phillips and O. Steward, Combination of in situ hybridization and immunocytochemistry to detect messenger RNAs in identified CNS neurons and glia in tissue culture, *J. Histochem. Cytochem.* 39:891 (1991).
20. M. Lee, J. B. Tuttle, L. Rebhun, D. W. Cleveland and A. Frankfurter, The expression and post-translational modification of a neuron-specific beta-tubulin isotype during chick embryogenesis, *Cell Motil. & Cytoskel.* 17:118 (1990).
21. J. M. Willets, D. G. Lambert and H. R. Griffiths, Suitability of B65 and SH-SY5Y neuroblastoma cells as models for 'in vitro' neurotoxicity testing, *Biochem. Soc. Trans.* 22:452S (1993).
22. S. Pahlman, A. I. Ruusala, L. Abrahamsson, M. E. K. Mattson and T. Esscher, Retinoic acid-induced differentiation of cultured human neuroblastoma cells: a comparison with phorbol ester-induced differentiation, *Cell Differ.* 14:135 (1984).
23. B. Dipasquale, A. M. Marini, and R. J. Youle, Apoptosis and DNA degradation induced by 1-methyl-4-phenylpyridinium in neurons, *Biochem. Biophys. Res. Commun.* 181:1442 (1991).
24. A.M. Marini, J. P. Schwartz and I. J. Kopin, The neurotoxicity of 1-methyl-4-phenylpyridinium in cultured cerebellar granule cells, *J. Neurosci.* 9:3665 (1989).
25. K. Nishi, H. Mochizuki, Y. Furukawa, Y. Mizuno and M. Yoshida, Neurotoxic effects of 1-methyl-4-phenylpyridinium (MPP+) and tetrahydroisoquinoline derivatives on dopaminergic neurons in ventral mesencephalic-striatal co-culture, *Neurodegeneration* 3:33 (1994).
26. Mytilineou, G. Cohen and R. E. Heikkila, 1-methyl-4-phenylpyridine (MPP+) is toxic to mesencephalic dopamine neurons in culture, Neurosci. Lett 57:19 (1985).
27. H. Mochizuki, N. Nakamura, K. Nishi and Y. Mizuno, Apoptosis is induced by 1-methyl-4-phenylpyridinium ion (MPP+) in ventral mesencephalic-striatal co-culture in rat, *Neurosci Lett.* 170:191 9194).

POTENTIAL ROLE OF HYPERACTIVATION OF SIGNAL TRANSDUCTION PATHWAYS IN ALZHEIMER'S DISEASE: PROTEIN KINASE C REGULATES PHF-LIKE PHOSPHORYLATION OF TAU WITHIN NEURONAL CELLS

Thomas B. Shea[1,2], John J. Boyce[1,2], and Corrine M. Cressman[1]

[1]Center for Cellular Neurobiology and Neurodegeneration Research
[2]Department of Biological Sciences
University of Massachusetts at Lowell
Lowell, MA 01854

INTRODUCTION

Paired helical filaments (PHF) that accumulate in affected neurons in Alzheimer's disease are comprised of hyperphosphorylated tau that exhibits electrophoretic and antigenic properties distinct from that of normal adult CNS tau (for reviews, see refs. 1-3). Accordingly, one approach towards understanding the onset and progression of Alzheimer's disease is to determine the kinase(s) responsible for phosphorylating tau.

A number of kinases, including Ca^{2+}-calmodulin kinase, cyclin-dependent kinases, mitogen-activated protein (MAP) kinases, glycogen synthase kinases, and proline-directed kinases[4-14], have been shown to phosphorylate tau in a manner that confers the altered antigenicity and electrophoretic migration characteristically observed in AD brains. By contrast, which kinase(s) are responsible for these phenomena within intact neurons remains unclear. Moreover, kinases *in situ* do not necessarily operate in isolation; many kinases may influence the activity of other kinases to varying degrees by means of kinase cascades. Accordingly, this line of reasoning underscores the possibility that the activity of signal transduction kinases may influence tau hyperphosphorylation, although they may not directly phosphorylate the tau molecule.

Protein kinase C (PKC) plays a pivotal role in neuronal differentiation and homeostasis. PKC inhibition promotes neurite outgrowth, while activation restricts it[15-23]. Two modes of PKC activation have been described. PKC is activated by the transient generation of 1,2-diacylglycerol, coupled with the release of Ca^{2+} from intracellular stores, following receptor-mediated hydrolysis of inositol phospholipids (for review, see ref. 24). PKC can also be irreversibly activated by limited proteolysis: under cell-free conditions, limited calpain-mediated proteolysis cleaves the catalytic and regulatory subunits to generate a free catalytic subunit, termed PKM[25-30]. The physiological significance of this latter potential mode of PKC activation is the subject of controversy (e.g., see ref. 24). However, since PKM, unlike the parent enzyme, is neither co-factor dependent nor bound to the plasma membrane[27-29, 31], it therefore has potential access to additional substrates versus those accessible to PKC. Recent studies from this laboratory indicate that PKM, once formed, is itself rapidly degraded by calpain-mediated proteolysis at a faster rate than the intact enzyme[32,33], suggesting that if indeed any PKM is generated *in situ*, it is normally rapidly eliminated. However, it has been

Neurodegenerative Diseases
Edited by Gary Fiskum, Plenum Press, New York, 1996

considered that excessive PKM generation, e.g., by hyperactivation of calpain following calcium influx[34], could promote hyperphosphorylation of a distinct array of non-membrane associated proteins, including cytoskeleton-associated proteins (e.g., see ref. 35).

In cell free analyses, PKC has been reported to phosphorylate tau, and, in doing so, to alter its association with microtubules, but has not been demonstrated to induce PHF-like alterations in electrophoretic migration or antigenicity[36,37]. However, the pivotal nature of PKC in signal transduction prompted the consideration that an inappropriate increase in PKC activity could induce "downstream" effects on the phosphorylation state of tau, perhaps by activation of one or more kinase cascades. We approached this possibility by utilizing a variety of techniques to manipulate PKC activity within intact neuronal cells, including cell-permeant inhibitors, calcium influx, antisense oligonucleoties and microinjection of functional enzyme. In addition, we investigated whether or not calpain-mediated limited PKC proteolysis represented an "upstream" influence in this process.

MATERIALS AND METHODS

Culture Conditions and Treatment with Calpain and Kinase Inhibitors and Activators

SH-SY-5Y cells (originally obtained from the stocks of Dr. June L. Biedler, Memorial Sloan-Kettering Cancer Center, Rye, New York) were cultured in Dulbecco's modified Eagle's medium (DMEM) containing 10% fetal calf serum (FCS) at 37°C in a humidified atmosphere containing 10% CO_2. Twenty-four hours later, the medium was replaced with the same medium containing no further additions or one or more of the following: 1μM N-acetyl-leucyl-leucyl-norleucinal ("CI", also referred to in other studies as "AllNal"; Boehringer-Mannheim, Inc., Indiannapolis, IN); 1μM calpeptin (generous gift of Mitsubishi-Kasei Corporation, Japan; Tsujinaka et al., 1988); 1μM 1-(5-isoquinolinesulfonyl)-2-methylpiperazine dihydrochloride (H7), 1μm staurosporine (generous gift of Dr. Yazuru Matsuda; Kyowa Hakko Kogyo Co, Tokyo, Japan); 100μM N-(2-guanidinoethyl)-5-isoquinolinesulfonamide (HA-1004; Seikagaku Kogyo (Tokyo, Japan); 2-0-Tetradecanoylphorbol-13-acetate (TPA; 1μM); N-(2-guanidinoethyl)-5-isoquinolinesulfonamide (HA-1004); N-(6-aminohexyl)-5-chloro-1-napthalenesulfonamide (W-7; Research Biochemicals, Inc., Natick, MA) for 4 -24hrs. Cell culture reagents, HA-1004 and TPA were purchased from Sigma Chemical Co. (St. Louis, MO). Stock solutions of inhibitors were prepared in 100% dimethyl sulfoxide (DMSO), then diluted in culture medium so that the final concentration of DMSO did not exceed 0.1%; this concentration did not induce morphological differentiation by 6hr, but did induce the outgrowth of short neurites by 24hrs of continuous treatment[21]. Influx of calcium into intact cells was achieved by addition of 3μM calcium ionophore A23187 (Sigma) directly into cultures in DMEM, which contains 1.8mM $CaCl_2$[38]. Previous analyses under cell-free conditions have demonstrated specific inhibition of PKA by HA-1004, PKC by H-7 and staurosporine, and calpain by C1, calpeptin and the I2 antibody and activation of PKC by TPA; moreover, none of these kinase inhibitors were effective against calpain, and calpain inhibitors were not effective against kinases[39]. PKC translocation and functional activation has also been demonstrated following TPA treatment of intact SH-SY-5Y neuroblastoma cells[32,39].

Intracellular Delivery of a Synthetic Calpain Inhibitory Peptide and Anti-Calpain Antibodies

Influx of calcium from the culture medium was also achieved by transient permeabilization of cells in calcium-containing medium as described previously[39]; this second method afforded the simultaneous intracellular delivery of a non-permeant calpain inhibitor and anti-calpain antibodies. Cultures plated at least 24hr previously in 8-well Lab-Tek Chamber slides (PGC Scientific, Inc., Gaithersburg, MD) were rinsed with serum-free DMEM. The cultures received 100μl of pre-warmed (37°C) 1.2M glycerol in PBS and were placed directly on ice for 10min, after which 4μl lysophosphatidyl choline (LPC; egg-white derivative; Sigma) was aliquoted into cultures at a final concentration of 40μg/ml, and the incubation continued on ice for an additional 8min. Alternate control cultures were "mock-permeabilized" by witholding LPC. Cultures then received 100μl of pre-warmed serum-free DMEM containing either no further additives or 200μg of a 14kDa recombinant peptide (Takara Biochemicals, Berkeley, CA) corresponding to the inhibitory domain of human calpastatin; a calpain neutralizing IgG ("anti-calpain I2") raised against a 20-amino acid peptide corresponding to the catalytic domain of chicken calpain[40]; or a non-neutralizing anti-calpain IgG ("anti-calpain C-7") raised in this laboratory against a synthetic peptide corresponding to the N-terminus of human calpain[39]. Cell-free analyses confirm inhibition of calpain activity by this synthetic calpastatin peptidee[39] and anti-calpain I2[41]. After a 10min incubation at 37°C, an additional 100μl of DMEM containing serum was added, and the cultures were then incubated for an additional 4hr after which they were fixed for 15 min with 4% paraformaldehyde in 0.1M phosphate buffer (pH 7.4). Internalization of calpastatin by permeabilized cells was confirmed by immunocytochemical reaction of loaded and unloaded fixed cultures with a rabbit polyclonal antibody raised against this synthetic peptide. Internalization of anti-calpain IgG was confirmed by immunocytochemical reaction of loaded and unloaded fixed cultures with peroxidase-conjugated secondary antibody (not shown; see refs. 39, 42).

Preparation and Use of "Sense-" and "Antisense"-Oriented Oligonucleotides

DNA oligonucleotides spanning the translation initiation codon were synthesized according to the published PKC sequence[43] as follows (presented in 5'-3' orientation): α "antisense" (AAG ACG TCA GCC ATG GTC CCT CGC CGC CTC CT), α "sense" (AGG A GG CGG CGA GGG ACC ATG GCT GAC GTC TT), ε "antisense" (AAG GCC ATT GAA CAC TAC CAT GGT CGG GGC GGG GGT CA), ε "sense" (TGA CCC CCG CCC CGA CCA TGG TAG TGT TCA ATG GCC TT), β "antisense" (CCG GGT CAG CCA TCT TGC GCG CGG GGA GCC GGA), and β "sense" (TCC GGC TCC CCG CGC GCA AGA TGG CTG ACC CGG). Oligonucleotides were dispensed at a final concentration of 12.5μM in serum-containing culture medium at 24hr intervals along with replacement of medium for 5 days total (a length of time determined to be required to achieve reproducible down-regulation in preliminary studies), after which cultures were immediately fixed and subjected to immunocytochemical analyses (below).

Production of PKM

PKCα (1μg; UBI, Lake Placid, NY) was diluted in 50 mM Tris HCl (pH 7.4) containing 2 mM PMSF and incubated with purified human erythrocyte μ calpain (generous gift of Drs. P. S. Mohan and R. A. Nixon, McLean Hospital) at 30°C with 7 mM $CaCl_2$, 0.4 μg/μl phosphatidylserine (PS; Serdary Research Laboratory, Canada) as described previously (Shea et al., 1994b). The reaction was terminated by the addition of 10 mM of EDTA to chelate Ca^{2+}. PKM (migrating at approx. 46kDa) was separated from μ calpain (76-80kDa) and residual PKC (80kDa) by centrifugation in a centricon-50 concentrator (Amicon, Beverly, MA) with a 50,000mw cut-off. Fractions were stored at -70°C pending immunoblot analysis or preparation for microinjection.

Cell-free Phosphorylation of Tau by PKC and PKM

Purified human brain tau (generous gift of Dr. Marc Mercken, McLean Hospital) (190 ng) was incubated with PKC or PKM (25 ng each) at 37°C for 20 hr. in 20 mM Tris (pH 7.5) containing 100 μM ATP, 250 nCi g-AT[P^{32}] (DuPont, Boston, MA), 5 mM $CaCl_2$, 100 mM $MgCl_2$, 0.4 μg/μl PS, 15 μM TPA (phorbol 12-myristate 13-acetate). Reactions terminated by addition of Laemmli sample buffer on ice.

Immunocytochemical and Immunoblot Analysis

For immunocytochemical analyses, fixed cells were rinsed with Tris-buffered saline (TBS; pH 7.4), blocked with 20% normal goat serum in TBS at room temperature for 20min, then incubated overnight with a 1:100 dilution of monoclonal antibodies directed against tau from AD brain ("ALZ-50"; generous gift of Dr. Peter Davies) in TBS containing 1% normal goat serum. For immunoblot analysis of calpain and PKC, particulate (e.g., membrane) and soluble fractions were prepared as described[39]. For immunoblot analysis of tau, heat-stable proteins were derived from Triton-soluble fractions. Briefly, cells were scraped from the plate in the presence of 1% Triton in 50mM Tris HCl (pH 6.8) containing 2mM PMSF and 5mM EDTA at 4°C, homogenized in a glass-teflon homogenizer (50 strokes), then Triton-insoluble material was sedimented by centrifugation at 13,000 x g for 15min. The resulting supernatant was then heated at 60°C for 15min, then the above centrifugation repeated to yield a heat-labile pellet and the heat-stable (Triton-soluble) fraction. In some experiments, this fraction was incubated as described[32] at 30°C for 0-15min with 85 μg/ml m calpain with 7mM $CaCl_2$ (to neutralize EDTA and generate 2mM free $CaCl_2$) of purified from human brain[44]. Various cellular fractions were subjected to SDS-gel electrophoresis, transferred to nitrocellulose and reacted with ALZ-50 and with a monoclonal antibodies directed against a phosphate-independent non-phosphorylated epitope of tau ("5E2"[45]; generous gift of Dr. Ken Kosik, Brigham and Women's Hospital, Brookline, MA), a non-phosphorylated epitope of tau ("Tau-1", which reacts with ser-202 when this site is nonphosphorylated; generous gift of Dr. Lester Binder), a monoclonal antibody directed against μM calpain[46] or a monoclonal antibody directed against the catalytic domain of PKCα (UBI, Lake Placid, NY)[32,39]. Cultures and nitrocellulose replicas were then rinsed and reacted with the appropriate secondary antibody conjugated to peroxidase or alkaline phosphatase and visualization as described[39]. Replicas containing radiolabeled preparations were subsequently placed against X-ray film to generate multiple autoradiographs.

Microinjection

Under the above incubation conditions, μ calpain rapidly cleaves the PKC regulatory and catalytic subunits, but does not further degrade either subunit[28]. Accordingly, following centrifugation in a Centricon-50 filter (with a 50,000mw cut-off; above), free regulatory (36kDa) and catalytic (46kDa) subunits passed through this filter, while intact PKC (80kDa) and calpain (76-80kDa) were retained. In order that identical amounts of each kinase could be microinjected, the fraction which passed through the filter (confirmed to be devoid of PKC or calpain; e.g., see Fig. 1) was normalized according to total protein with PKC diluted in the identical buffer but not incubated with calpain. Samples were prepared for microinjection by mixing these kinases with Dextran tetramethylrhodamine ("Fluoro Ruby"; Molecular Probes, Eugene, OR) as tracer to identify injected cells, followed by dilution with Dulbecco's phosphate buffer saline (Sigma) to a final concentration of 3.2 ng/ml of kinase and 3.0 mg/ml of tracer, and clarified immediately prior to injection by centrifugation at 100,000 x g for 20 min. in a Beckman Airfuge. The resulting supernatant was immediately microinjected into mouse NB2a/d1 neuroblastoma cells cultured with 10% serum as described previously[47] using a Narishige hydraulic micromanipulator model MO-203 (Narishige USA, INC., NY) driven by an Eppendorf 5342 microinjector (Eppendorf, Germany). Within each culture, 20-40 cells were injected with either tracer alone or tracer mixed with kinase, and the injected area marked on the culture plate.

Immunofluoresence

Cultures were rinsed with cold TBS (pH 7.4) and fixed with 4% paraformaldehyde in Dulbecco's phosphate buffer saline at room temperature for 15 minutes. The fixed cultures were rinsed in TBS, blocked with 3% BSA in TBS and incubated with 1:10 dilutions of ALZ-50, Tau-1, or PHF-1, or a 1:100 dilution of AT8 (which reacts with ser-202 when this site is phosphorylated and therefore provides complementary information to Tau-1) in TBS containing 1% BSA overnight at room temperature. Cultures were then rinsed three times with TBS, incubated for 2 hrs at room temperature with goat anti-mouse secondary antibody conjugated to fluorescein and viewed under phase contrast and UV optics.

Microdensitometric analyses

Microdensitometry was carried out using the public-domain NIH "IMAGE" analysis software (version 1.57; obtained via internet by anonymous FTP to zippy.nimh.nih.gov). Briefly, images of 10 representative microscopic fields obtained at 60X on an Olympus inverted microscope were imported via a Sony CCD camera into Power PC Macintosh 7100AV (equipped with video import capabilities), digitized and subjected to microdensitometry. Identical magnification, illumination, capture and record settings were maintained for all cells in all cultures in these analyses. Images were subjected to the program's "threshold" function, and the scale adjusted to insure that all images fell within a linear range; the identical threshold setting was utilized for all images from all cultures. Images were subjected to the "analyze particle" function, with the minimal particle size adjusted to 50 pixels, which reports the optical density, area, and background for each "particle" (in this case, each cell); background values were subtracted from the densitometric values obtained for each cell, and this net densitometric value was divided by the respective area reported for each cell. These corrected densitometric values were exported into Excel

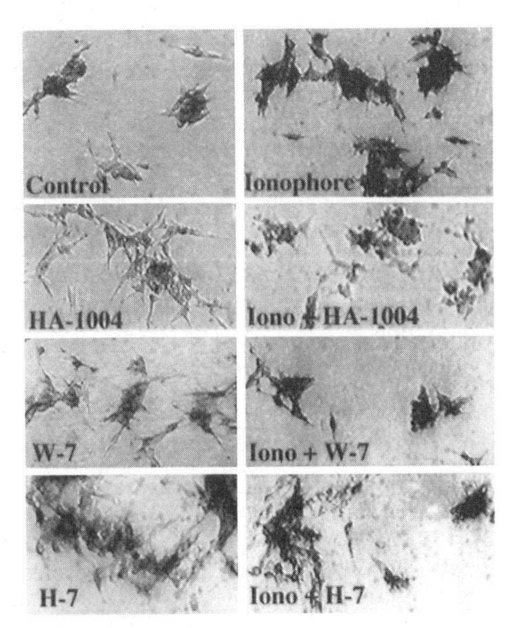

Fig. 1: Immunocytochemical analyses of PHF-1 levels in cells treated with various kinase inhibitors in the presence and absence of calcium ionophore A23187. Representative digitized images of microscopic fields obtained under bright-field illumination following treatment as indicated, followed by immunostaining with PHF-1.

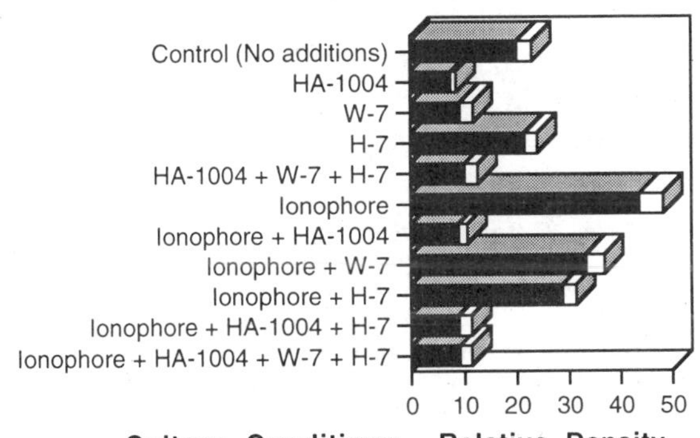

Fig. 2: Densitometric analysis of PHF-1 immunoreactivity. Images of cells were digitized and quantitated by microdensitometry as described in Materials and Methods. Values presented represent the mean ± standard error of the mean of net densitometric values of all cells (derived by subtracting associated background levels and dividing by the area of the cell) from 10 representative microscopic field from each treatment.

spreadsheet software for statistical analyses and graphics. Figures contain representative digitized images, annotated with PowerPoint software and printed with a Fargo dye-sublimation printer, that were utilized in densitometric quantitations.

RESULTS

Multiple Classes of Kinases Phosphorylate Tau within Intact Neuronal Cells: Calcium Recruits an Additional Class of Kinases to Phosphorylate Tau

The "basal" level of PHF-1 immunoreactivity displayed by undifferentiated SH-SY-5Y neuroblastoma cells was reduced by approximately 50% following a 6hr treatment with 1μM HA-1004 W-7 (Fig. 1, 2). By contrast, treatment with 1μM H-7 did not affect PHF-1 immunoreactivity (Fig. 1, 2).

Exposure of cells to 3μM calcium ionophore A23187 in DMEM culture medium (which contains 1.8mM calcium) resulted in an approximate doubling of PHF-1 immunoreactivity. Concurrent treatment with HA-1004 completely prevented this ionophore-mediated increase in PHF-1 immunoreactivity (Fig. 1, 2). W-7 reduced the ionophore-mediated increase in PHF-1 immunoreactivity to a similar relative extent (approximately 45%) as it did for cells not exposed to ionphore. However, unlike the situation observed for cells not treated with ionophore, H-7 inhibited (by approximately 22%) the ionophore-mediated increase in PHF-1 immunoreactivity (Fig. 1, 2). This ionophore-mediated increase in PHF-1 immunoreactivity was not due simply to an increase in total tau, since this treatment did not induce an increase in immunoreactivity for a monoclonal antibody (5E2) directed against an epitope of tau not affected by phosphorylation (32.9 ± 1.4 in non-treated and 29.8 ± 1.6 in ionophore-treated cells).

Increasing the concentration of either W-7 or H-7 to 2μM did not further inhibit the ionophore-mediated increase in PHF-1 immunoreactivity. Simultaneous treatment of cells with 1μM concentrations of HA-1004 and H-7, or all 3 kinase inhibitors yielded identical results as did treatment with 1μM HA-1004 alone in the presence and absence of ionophore (Fig. 1, 2). By contrast, HA-1004 did not inhibit overall protein phosphorylation, since immunoreactivity with an antibody (H3) specific for phosphorylated isoforms of the high molecular weight neurofilament subunit was not reduced following treatment of cells with this kinase inhibitor; densitometric analyses of H3 immunoreactivity was 208.8 ± 8.9 in untreated cells and 201.9 ± 7.7 in HA-1004-treated cells.

Since H-7 is reported to be active against PKC, and induces effects in these cells similar to those obtained following inhibition of PKC activity, we examined whether or not PKC activation, induced by TPA treatment, increased PHF-1 immunoreactivity (Fig. 3). Densitometric analyses indicated that TPA treatment significantly (p>0.0005) increased PHF-1 immunoreactivity (40.1 ± 2.7 following TPA treatment as compared to 24.9 ± 1.9 in non-treated cells).

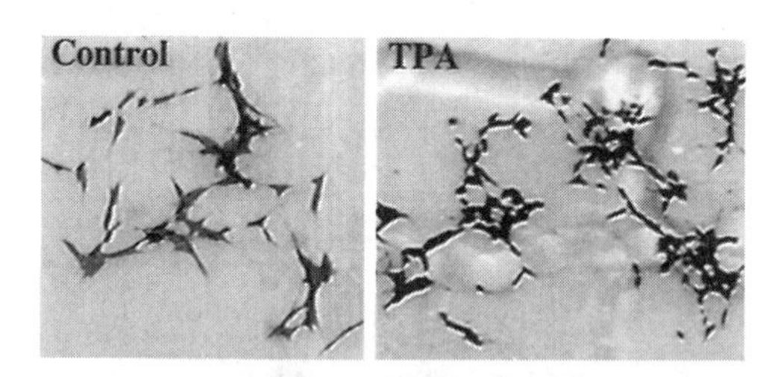

Fig. 3: Immunocytochemical analyses of PHF-1 levels in cells treated with the protein kinase C activator, TPA Representative digitized images of microscopic fields obtained under bright-field illumination following treatment as indicated, followed by immunostaining with PHF-1.

PKC-Mediated Tau Phosphorylation is Isoform-Specific: PKCε Mediates Tau Phosphorylation to Generate the PHF-1 and ALZ-50 epitopes

We next examined whether or not certain PKC isoforms were involved in mediation of tau phosphorylation to yield "AD-like" immunoreactivity. This was accomplished by down-regulation of individual isoforms by continuous treatment with antisense oligonucleotides directed at isofoirm-specific sequences. Our examinations in this area were initially restricted to the α and ε isoforms of PKC, since these isoforms have been demonstrated to regulate phenotypic differentiation (Leli et al., 1992a,b). Additional cultures received antisense oligonucleotides directed towards the β isoform as an additional control.

Specific down-regulation of "target" isoforms, achieved following 5 days of continuous treatment with 12.5μM oligonucleotide, was confirmed by immunocytochemistry with isoform-specific antibodies (Fig. 4, 5). Treated cultures were next examined for the effect of PKC isoform down-regulation on steady-state levels of PHF-1 and ALZ-50 immunoreactivity. Both the PHF-1 and ALZ-50 epitopes were specifically reduced in cultures that had been treated with antisense oligonucleotides directed against PKCε; neither PKCα nor PKCβ antisense oligonucleotides, nor any "sense" oriented olionucleotides, altered PHF-1 or ALZ-50

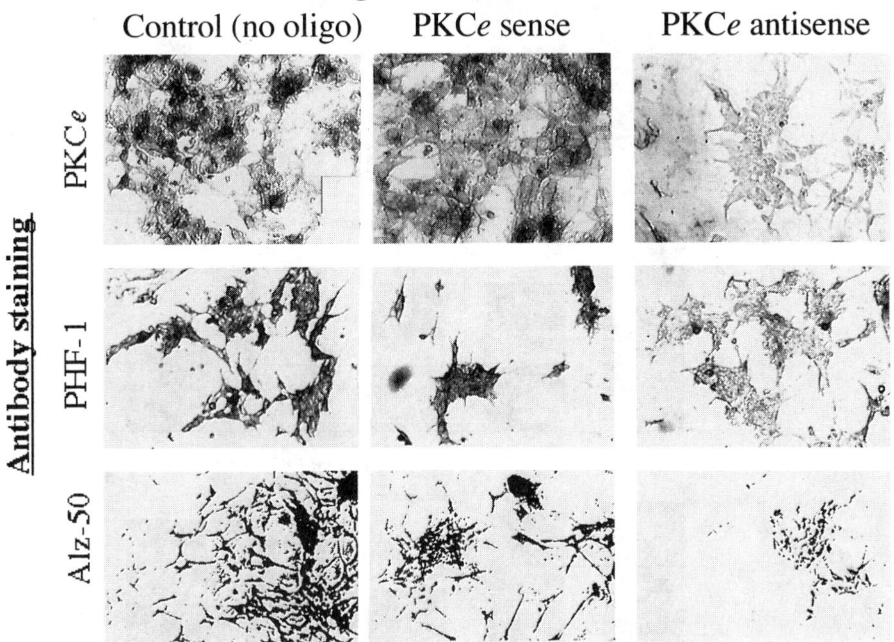

Fig. 4. Downregulation of target PKC isoforms and corresponding downregulation of PHF-1 and Alz-50 immunoreactivity. Cells were treated with sense and antisense oligonucleotides corresponding to PKCε as indicated, then fixed and processed for immunocytochemistry. Representative digitized images of cells are presented.

immunoreactivity (Fig. 4, 6). Treatment with a combination of PKCε and PKCβ antisense oligonucleotides had the same effect as PKCε antisense oligonucleotides alone (Fig. 6).

Hyperphosphorylation of Tau and Filopodial Retraction Following Microinjection of PKM, the Free Catalytic Subunit of PKC

The above findings suggested that up-regulation and down-regulation of PKC could influence tau phosphorylation. We further examined these phenomena by raising the intracellular concentration of constitutively active PKC. This was accomplished by microinjection of free PKC catalytic subunits.

Generation of PKM by limited proteolysis: As described previously (see refs. in Introduction), μ calpain carries out limited proteolysis of 80kDa PKC to generate free the regulatory and 46kDa catalytic subunits (Fig. 1) and 36kDa regulatory subunits (not shown, see ref. 28); no further proteolysis of either the catalytic or regulatory subunit was observed under these conditions (see also ref. 28). Free catalytic and regulatory subunits were separated from intact PKC and calpain (76-80kDa) by centrifugation in a Centericon filter with a 50,000mw cut-off (Fig. 7).

PKC and PKM generate the PHF-1 and ALZ-50 epitopes in vitro: Purified human brain tau (Fig. 8) was incubated for 20hr with PKC in the presence of PS, DAG, Ca^{2+} and TPA or with PKM in the absence of cofactors. The samples were subjected to SDS gel-electrophoresis, transferred to nitrocellulose, and the resulting nitrocellulose replicas were probed with polyclonal antiserum (JM) that reacts with tau regardless of tau phosphorylation

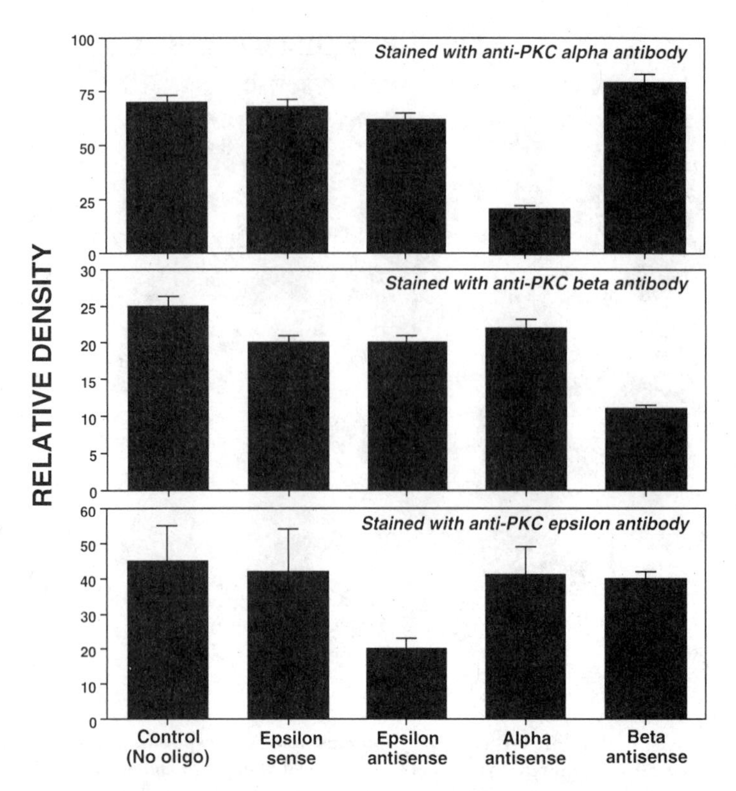

Fig. 5. Antisense oligonucleotide-mediated downregulation of PKC isoforms. Cells treated with sense and antisense oligonucleotides corresponding to various PKC isoforms as indicated, then fixed and processed for immunocytochemistry. Micordensitometric analyses confirmed specific downregulation of "target" isoforms only.

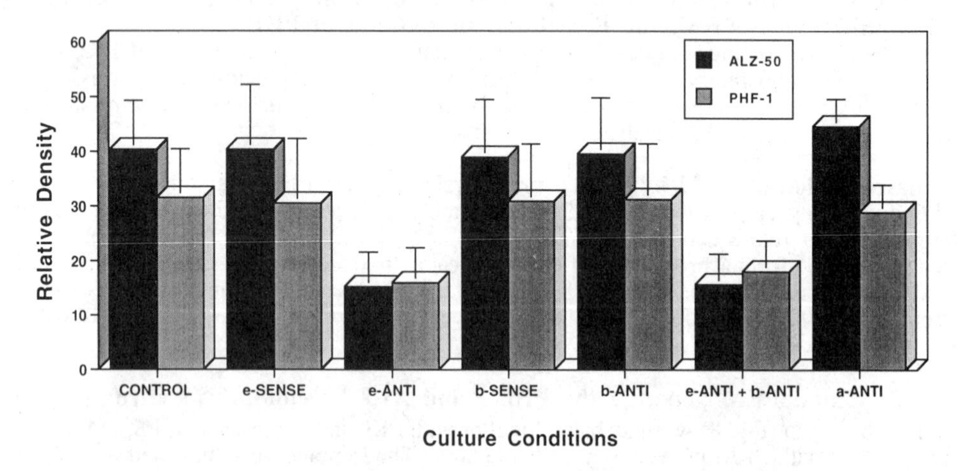

Fig. 6. Downregulation of PKCe specifically reduces ALZ-50 and PHF-1 levels. Micodensitometric analyses revealed that ALZ-50 and PHF-1 levels were reduced only in those cultures treated with PKCe antisense oligonucleotides

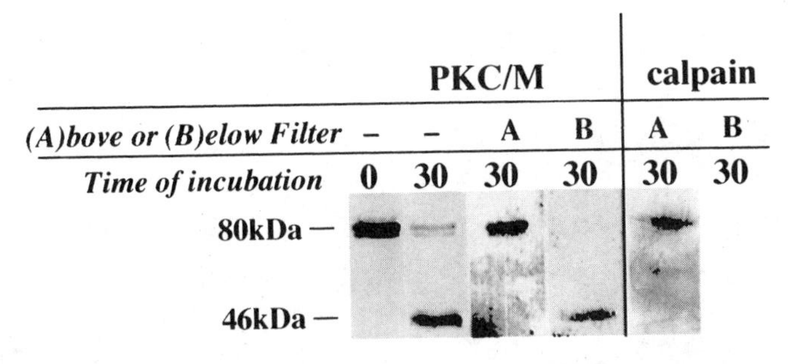

<u>Fig. 7</u>: Generation of PKM by calpain-mediated limited PKC proteolysis PKCα was incubated for 30 min. with purified μ calpain in the presence of calcium, then subjected to immunoblot analysis. Note the elimination of most immunoreactivity at 80kDa, and the generation an immunoreactive species (46kDa) after 30 min., indicating that PKM was generated by limited proteolysis of PKC by μ calpain. Following 30 min. incubation, samples were centrifuged in a Centericon-50 filter (50,000mw cut-off), and fractions recovered either (A)bove or (B)elow the filter were also subjected to immunoblot analysis with anti-PKC and anti-calpain antibodies. Note the complete recovery of PKM within the filtrate, and the complete retention of intact PKC and calpain by the filter.

state, PHF-1, ALZ-50, Tau-1, or AT8. The immunoreactive profile of tau before and after incubation with PKC or PKM (Fig. 8), exhibited minor differences, including the *de novo* appearance of a minor immunoreactive polypeptide at 68kDa. Prior to incubation with PKC or PKM, purified human brain tau exhibited no cross-reactivity with PHF-1 or AT8 and minor reactivity with ALZ-50; by contrast, however, incubation with PKC and PKM for 20hr induced PHF-1, and increased ALZ-50, immunoreactivity with multiple isoforms including the novel 68kDa polypeptide. Tau-1 immunoreactivity was not diminished, and the AT-8 epitope was not generated, following incubation with either PKC or PKM. These results confirm the ability of PKC and PKM to phosphorylate the tau molecule in a manner that generates the PHF-1 and ALZ-50 epitopes under cell-free conditions. The failure to generate the AT-8 epitope, and to diminish the Tau-1 epitope, (phosphate-dependent epitopes which are increased and decreased, respectively, by phosphorylation; see ref. 48 and refs. therein) suggests that tau phosphorylation by PKC and PKM is site-specific.

	CBB	JM			ALZ-50			PHF-1			AT8			Tau-1		
PKC	–	–	+	–	–	+	–	–	+	–	–	+	–	–	+	–
PKM	–	–	–	+	–	–	+	–	–	+	–	–	+	–	–	+

68kDa→

<u>Fig. 8</u>: Phosphorylation of purified human brain tau by purified PKC or PKM under cell-free conditions Prior to incubation with PKC or PKM, major tau isoforms migrated as a series of isoforms from approximately 42-56kDa, with several lower molecular weight breakdown products, as visualized by Coomassie staining (CBB) and immunoblot analysis with a polyclonal anti-tau antibody JM as indicated. Note the *de novo* appearance of a 68kDa polypeptide that is recognized by anti-tau polyclonal antibody JM, the appearance of PHF-1 and ALZ-50 immunoreactivity, by tau following incubation with purified PKC or PKM for 20 hr. Note also that after incubation for 20hr with PKC or PKM, Tau-1 immunoreactivity is not diminished, and the AT8 epitope is not detected. These analyses confirm that PKC and PKM are capable of phosphorylating tau in a manner to generate epitopes common to paired helical filaments.

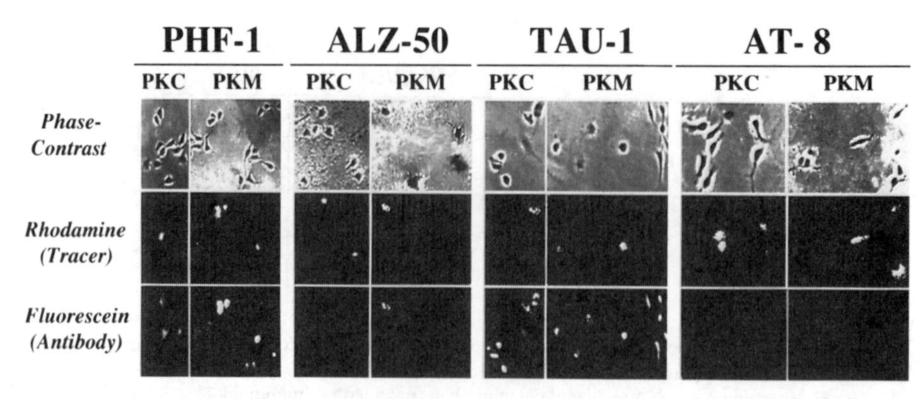

Fig. 9: Generation of PHF-1 and ALZ-50 immunoreactivity *in situ* following microinjection of PKM. Images obtained by CCD camera from uninjected and microinjected cultures as indicated. The identical microscopic field is presented under phase-contrast, rhodamine illumination (for identification of injected cells) and fluorescein illumination. Arrowhead in phase-contrast images of PKC- and PKM-injected cultures denote a representative microinjected cell as ascertained by visualization of co-injected rhodamine-conjugated tracer under rhodamine illumination. Subsequent immunofluorescent analyses using PHF-1, ALZ-50, AT8, and Tau-1 antibodies and fluorescein-conjugated secondary antibody demonstrates a specific increase in PHF-1 and ALZ-50 immunoreactivity, and the absence of filopodia-like neurites, in PKM-injected, but not PKC-injected cells.

An Increase in PHF-1, and a Decrease in Filopodia-Like Neurites, is Observed Following Microinjection of PKM but not PKC: Cells were microinjected with rhodamine-conjugated tracer and either PKC or PKM, the injected regions were circled on the bottom of the culture plate, and the cultures were processed for immunofluorescence. The circled areas of plates were examined under rhodamine illumination, and injected cells were located by the presence of rhodamine-conjugated tracer (e.g., Fig. 9). Uninjected cells did not exhibit appreciable antibody immunoreactivity or any rhodamine immunoreactivity as ascertained by examination of areas outside of the circle, or in plates in which no injection was performed (e.g., Fig. 9). Considerable heterogeneity was observed during densitometric analyses of "basal" antibody immunoreactivity levels observed in uninjected cells, and cells injected with PKC fell within this range (Fig. 9, 10). By contrast, cells injected with PKM demonstrated an average 3- and 2-fold increase in PHF-1 and ALZ-50 immunofluoresence, respectively (Fig. 9, 10), while no change was observed in AT8 or Tau-1 immunofluoresence levels.

Undifferentiated NB2a/d1 cells lack axonal or dendritic neurites, but continuously elaborate and retract relatively short, filopodia-like neurites (e.g., Shea et al., 1991). When uninjected and microinjected cells were scored for the presence or absence of one or more neurites of any length, we observed that similar percentages of uninjected and PKC-injected cells possessed filopodia-like neurites [65±5% (n=126) and 59±5% (n=17), respectively]. By contrast, we observed a drastic reduction (28±9%, n=40) in the percentage of PKM-injected cells with neurites.

Role of Limited Calpain-Mediated PKC Hydrolysis, and Generation of free Catalytic Subunits ("PKM"), in Tau Hyperphosphorylation

Calpain Inhibition Prevents Ionophore-Mediated Tau Hyperphosphorylation: Treatment of cells with calcium ionophore A23187 in the presence of 1.8mM extracellular calcium induced a dramatic increase in ALZ-50 immunoreactivity (Fig. 11). This increase was prevented by reduction of extracellular calcium by inclusion of 1.8mM EDTA in the culture medium, and by the addition of the calpain inhibitor C1. The specificity of these cell-permeant inhibitors for PKC and calpain, respectively has been demonstrated[39]. As shown above for PHF-1 immunoreactivity, PKC inhibition (by H-7) also prevents the ionopnore-mediated increase in ALZ-50, while PKC activation (by TPA treatment) also increased ALZ-50 immunoreactivity in the absence of

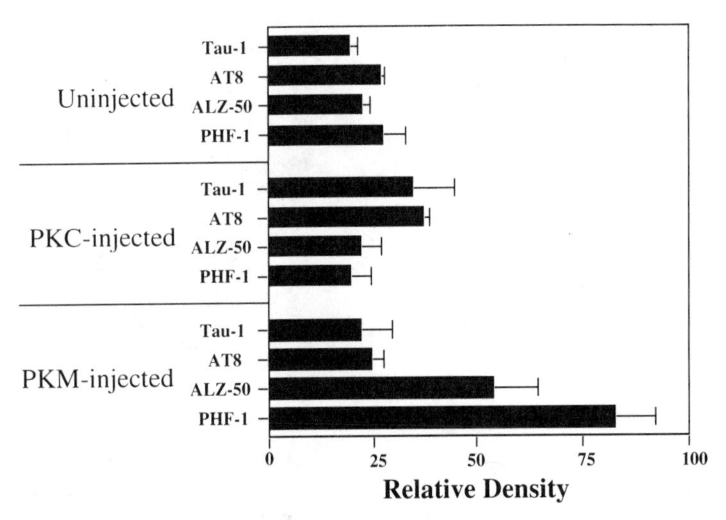

Fig. 10: Densitometric analysis of antibody immunoreactivity in uninjected and microinjected cells Microscopic images of cells at 200X were digitized and quantitated by microdensitometry using NIH "Image" analysis software (version 1.52) on a Power PC Macintosh equipped with audio-visual import capabilities. Microinjected and uninjected cells were outlined on-screen with the Image program's freehand selection tool and the relative density within each cell recorded. For each microscopic field, 5 areas devoid of cells were designated as "background", and these values along with those obtained for a total of 17 PKC-injected, 40 PKM-injected and 220 uninjected cells from 40 microscopic fields, were exported into Excel spreadsheet software. The average background density was subtracted from all recorded cells from each respective field, and statistical analyses for the individual antibodies were carried out on the resultant "corrected" cell density values using Student's *t* test. Note the 3- and 2-fold increase in relative density of PKM-injected cells stained with PHF-1 and ALZ-50, respectively, compared to uninjected and PKC-injected cells.

ionophore (Fig. 11). Transient permeabilization of cells in the presence of calcium increased ALZ-50 immunoreactivity, while intracellular delivery of a synthetic peptide corresponding to the inhibitory domain of calpastatin during this transient permeabilization prevented this increase; this inhibition by intracellular delivery of calpastatin was overcome by TPA treatment (Fig. 12).

Calpain Generates PKM *In Situ*: Immunoblot analyses confirmed calpain activation following calcium influx, and limited PKC proteolysis to generate free catalytic subunits, i.e., PKM (see also refs. 32, 39); calpain inhibition during calcium influx prevented PKM generation (Fig. 13).

Calpain Regulates Steady-state levels of Tau: Immunoblot analysis confirmed the immunocytochemical demonstration of increased ALZ-50 immunoreactivity following calcium influx. Heat-stable Triton-soluble fractions derived from ionophore-treated cells demonstrated increased ALZ-50 as compared to cells not treated with ionophore; this ALZ-50-immunoreactive tau migrated on SDS-gels at 68kDa as opposed to the typical migration of 5E2- or Tau-1-immunoreactive tau at 42-58kDa (Fig. 13). The ionophore-mediated increase in ALZ-50 was prevented by treatment with calpain inhibitor C1, indicating the involvement of calpain activity in this phenomenon (Fig. 13). Probing of additional replicas with the phosphate-independent monoclonal antibody 5E2 demonstrated rapid depletion of total tau following ionophore-mediated calcium influx, and prevention of this depletion by C1 (Fig. 13). Inhibition of calpain in intact cells by intracellular delivery of a neutralizing anti-calpain antibody resulted in a dramatic increase in total tau levels as ascertained by immunoblot analysis, but, by contrast, resulted in only a modest increase in ALZ-50 immunoreactive tau (Fig. 14).

ALZ-50 immunoreactive tau isoforms are selectively resistant to proteolysis: That total tau immunoreactivity declines following calcium influx initially

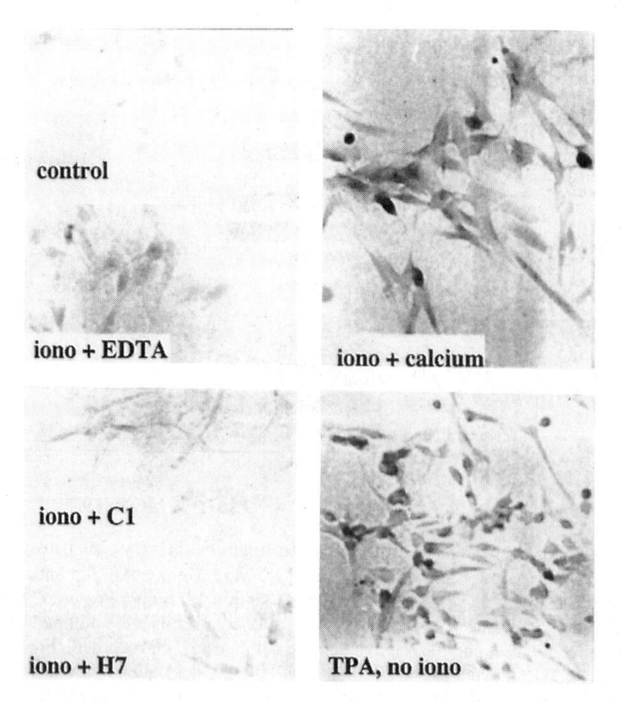

Fig. 11: Calcium influx increases ALZ-50 immunoreactivity Bright field micrographs of SH-SY-5Y cells immunostained with ALZ-50 under the following conditions: Untreated control cells (panel A); calcium ionophore A23187 with 1.8mM EDTA to reduce extracellular calcium (panel B); calcium ionophore A23187 (panel C); calcium ionophore A23187 and the calpain inhibitor C-I (panel D); calcium ionophore A23187 and the protein kinase C inhibitor H-7 (panel E); the protein kinase C activator TPA (panel F). Note that ALZ-50 immunoreactivity is increased following ionophore-mediated calcium influx; however, this increase is prevented by C-I and H-7. TPA treatment alone also increased ALZ-50 immunoreactivity.

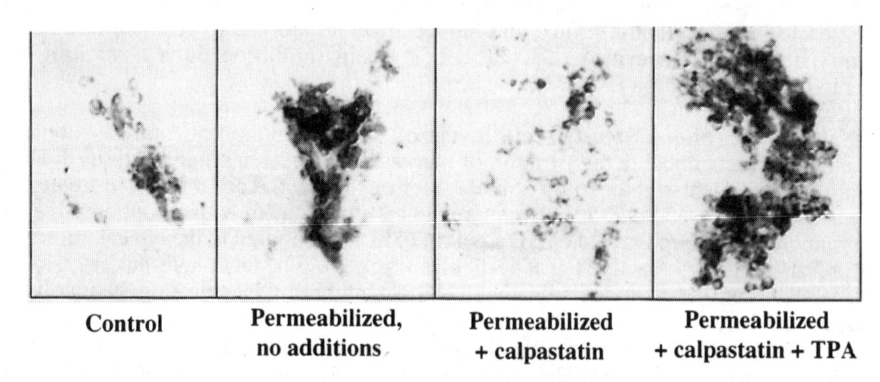

Fig. 12: Calpastatin prevents the ALZ-50 increase induced by calcium influx, while TPA treatment overcomes this inhibition Panel A presents mock-permeabilized cells as described in Materials and Methods. Transient permeabilization of cells in the presence of calcium increases ALZ-50 (panel B). Intracellular delivery of a synthetic calpastatin peptide during this transient permeabilization prevents this increase into SH-SY-5Y cells prevents the increase in ALZ-50 resulting from transient permeabilization of cells ialcium influx induces ALZ-50, loading of calpastatin prevents this increase. Note also that TPA treatment results in an increase in ALZ-50 immunoreactivity despite the intracellular delivery of calpastatin.

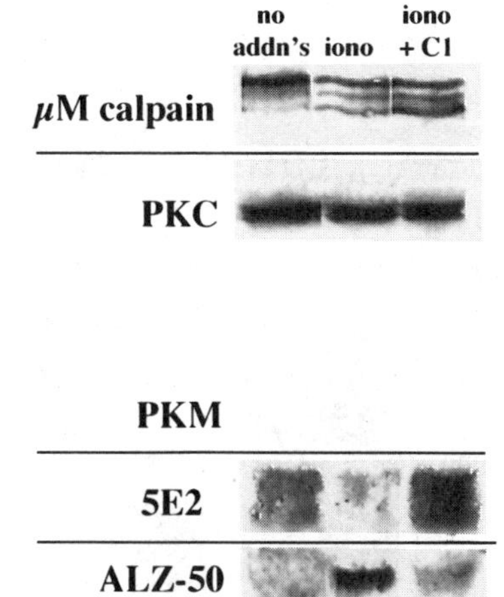

Fig. 13: Calcium influx induces calpain and PKC activation and alters tau immunoreactivity Cells were treated with ionophore A23187 in calcium-containing medium with or without calpain inhibitor C1 as indicated. Replicate immunoblots were probed with monoclonal antibodies against µM calpain, the catalytic domain of PKCα, 5E2 and ALZ-50; particulate fractions were assayed for calpain and PKC, while soluble fractions were assayed for 5E2 and ALZ-50. Calcium influx induced activation of µM calpain as evidenced by autolysis (see text). Calcium influx induced limited PKC proteolysis as evidenced by the appearance of a 42kDa immunoreactive peptide (corresponding to free catalytic subunits, or PKM; see text). Total tau was markedly decreased following calcium influx as evidenced by 5E2 immunoreactivity, while C1 inhibited this decrease. In contrast to total tau, ALZ-50 immunoreactivity was increased following calcium influx, and this increase was prevented by C1.

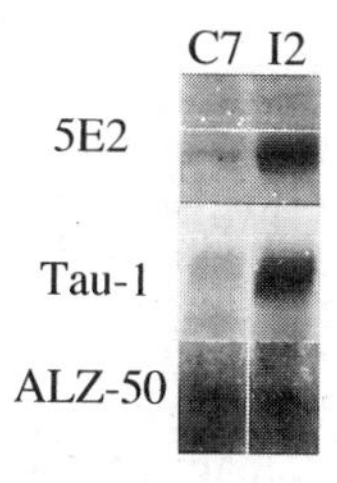

Fig. 14: Calpain inhibition increases steady-state levels of tau within intact cells Cells were transiently permeabilized in the presence of either non-neutralizing anti-capain antibody C-7 or neutralizing anti-calpain antibody I2 as described in Materials and Methods. Cells were incubated for 4hr, then harvested and heat-stable Triton-soluble fractions were subjected to immunoblot analysis. Only the relevent portions of replicas are presented. Note the dramatic increase in the 42-58kDa 5E2 and Tau-1 immunoreactive forms of tau, and the marginal increase in the 68kDa ALZ-50 immunoreactive form.

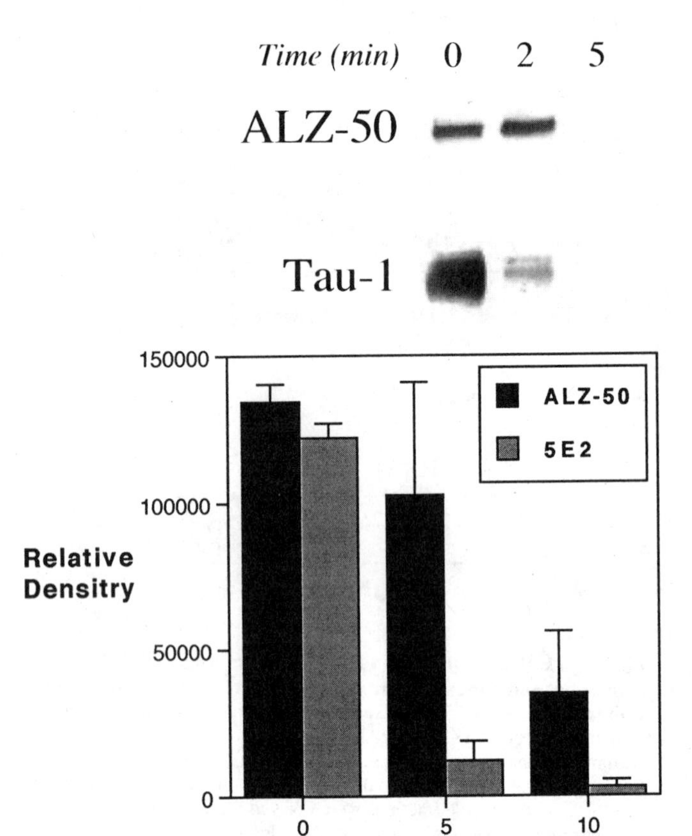

Fig. 15: ALZ-50-immunoreactive tau isoforms demonstrate increased resistance to calpain-dependent proteolysis. Heat-stable Triton-soluble fractions were incubated for the indicated times in the presence of 85 μg/ml purified m calpain and 2mM excess $CaCl_2$ as described in Materials and Methods, harvested and subjected to immunoblot analysis. Note the slower loss of 68kDa ALZ-50 immunoreactivity versus the multiple Tau-1 immunoreactive bands.

seems inconsistent with the observation of increased ALZ-50 immunoreactivity following calcium-mediated calpain activation (above). To address this issue, we examined whether or not ALZ-50-immunoreactive tau isoforms were selectively resistant to calpain-mediated proteolysis. Incubation of heat-stable Triton-soluble cellular fractions with purified mM calpain in the presence of calcium followed by immunoblot analysis with ALZ-50 and Tau-1 demonstrated that ALZ-50-immunoreactive tau isoforms were indeed more resistant to degradation than were total tau isoforms (Fig. 15). One interpretation of these collective findings is that PKM, generated by calpain activation, modifies (perhaps via a kinase cascade) a percentage of total tau and renders it resistant to calpain-mediated proteolysis; although overall tau levels decline as a consequence of calpain activation, ALZ-50-immunoreactive tau is selectively increased (e.g., Fig. 16).

DISCUSSION

We have utilized multiple approaches to demonstrate a role for signal transduction pathways, specifically those regulated by PKC, in the generation of the A68 tau isoforms that accumulate in AD. These findings collectively underscore the possibility that minor amplification of homeostatic mechanisms may, over time, lead to the development of neuropathology characteristic of AD.

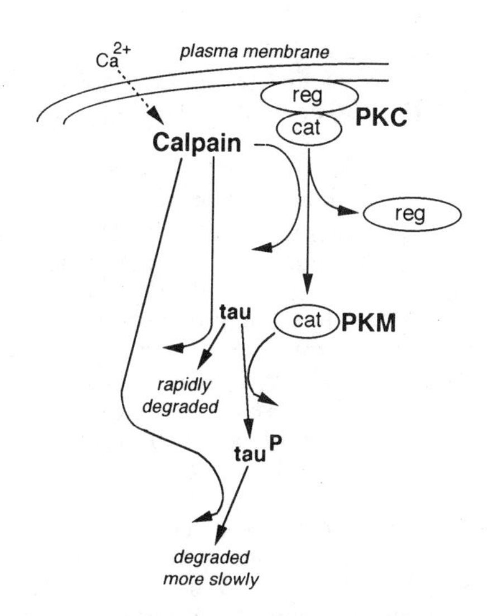

Fig. 16: Hypothetical scheme for the consequences of intracellular calpain activation on PKC and tau This working hypothesis is based on the collective findings of the present and previous studies (see text). Excessuve calcium influx, achieved in this case by ionophore A23187, activates calpain. Calpain then initiates degradation of cytoskeletal proteins, including tau. However, a simultaneous, additional consequence of calpain activation is PKM generation by limited PKC proteolysis. PKM, either directly or through one or more kinase cascades, then initiates phosphorylation of a percentage of tau molecules, altering their antigenicity and conferring selective resistance to calpain-mediated degradation.

Multiple Kinases Contribute to Tau Phosphorylation Under "Resting Conditons" in Intact Neuronal Cells, and Calcium Influx Recruits an Additional Class of Kinases in this Process

Our use of multiple kinase inhibitors demonstrate that it is indeed possible to modulate PHF-1 immunoreactivity within intact neuronal cells via cell-permeant kinase inhibitors. Moreover, the substantial increase in PHF-1 immunoreactivity following ionophore-mediated calcium influx further indicates at least one class of PHF-1-generating kinases to be calcium-dependent. Recently, it has been shown that the PHF-1 epitope can be routinely detected in normal brain[48]; and furthermore may undergo transient increases during development[49,50]; accordingly, the observation of PHF-1 immunoreactivity in undifferentiated SH-SY-5Y cells is not surprising. However, the disparate results observed in the presence of H-7 prior to and following calcium influx indicate the recruitment of an H-7-sensitive class of kinases in tau phosphorylation, and indicate that the *de novo* recruitment of additional kinases, beyond those that generate this epitope under normal conditions, may contribute to PHF accumulation observed in AD. While the H-7-sensitive kinase(s) may indeed also phosphorylate tau under "steady-state" conditions (e.g., in the absence of experimental pathological conditions such as ionophore-induced calcium influx), the relative levels of phosphorylation leading to the generation of the PHF-1 epitope are apparently minimal relative to those induced in the absence of ionophore by the other classes of kinases observed in this study (e.g., those inhibited by HA-1004 and W-7). That HA-1004 and W-7 each inhibit PHF-1 immunoreactivity to a similar respective extent both in the absence and presence of ionophore suggests that they inhibit constitutively-active kinases, and that the extent of activity of each of these classes of kinases is not affected by ionophore-mediated calcium influx. The different results observed for H-7 prior to and following calcium influx clearly distinguish the H-7-sensitive kinases from those inhibited by HA-1004 and W-7. In addition, the class of kinases inhibited by HA-1004 apparently also differed from that inhibited by W-7, since increasing the concentration of W-7 to 2μM induced no further inhibition beyond that observed with 1μM, yet the addition of 1μM each of W-7 and HA-1004 inhibited PHF-1 immunoreactivity to the same extent as 1μM HA-1004 alone. The possibility remains that sequential phosphorylation events confer PHF-1 immunoreactivity to tau, and HA-1004 inhibits a class of kinases that carry out initial events in any such sequence.

Although HA-1004, W-7 and H-7 are effective inhibitors of protein kinase A, calcium/calmodulin kinase and protein kinase C in cell-free analyses[51] kinase inhibitors, regardless of their specificity in cell-free analyses, may inhibit additional kinases other than their purported "target" kinase *in situ*. Accordingly, the findings of the present study cannot be interpreted to indicate the direct involvement of purported "target" kinases of these inhibitors. It is furthermore not appropriate to compare the relative potency of these various inhibitors to each other; rather, in these analyses, the only comparisons made are those regarding the relative efiicacy of each inhibitor to itself in the presence and absence of ionophore. Nevertheless, these findings do clearly demonstrate the recruitment of an additional kinase, or class of kinases, in tau phosphorylation following calcium influx.

Despite the above caveat, PKC represents one potential candidate responsible for these phenomena, since it is both activated following calcium influx and TPA treatment and inhibited under cell-free conditions by H-7 (for reviews, see refs. 24,25,51). Protein kinase C plays a pivotal role in neuronal differentiation and homeostasis (for review, see 51). This pivotal role of PKC underscores the possibility that even slight alterations in its activation may result in downstream protein hyperphosphorylation in a variety of neurological disorders.

PKCε Mediates Tau Hyperphosphorylation in Intact Neuronal Cells

The potential involvement of PKC in tau hyperphosphorylation was supported by decreases in PHF-1 and ALZ-50 immunoreactivity following antisense oligonucleotide-mediated down-regulation of PKCε; moreover, these findings also demonstrate the involvement of PKC in this phenomena to be isoform-specific, since similar down-regulation of PKCα or PKCβ reduced PHF-1 or ALZ-50 immunoreactivity.

PKC Phosphorylates Tau in a Site-Specific Manner under Cell-Free Conditions and within Intact Neuronal Cells

PKC has been reported to phosphorylate tau, but not to induce a molecular weight shift or to generate PHF-1 or ALZ-50 epitopes[36,37]; our use of an extended incubation time is likely to be one factor contributing to our demonstration of altered tau electrophoretic migration and antigenicity following incubation with PKC, since these earlier studies used incubations of 2hr or less. Indeed, our analyses under cell-free conditions demonstrate that, although radiolabeled phosphate was incorporated by tau following as little as 30min incubation of tau with PKC, a minimum of 15 hr was required before PHF-1 immunoreactivity, ALZ-50 immunoreactivity, and the 68kDa isoform were detectable, and 20hr of incubation markedly increase these alterations. In addition, the present study was confined to PKCα and to free catalytic subunits derived from this particular PKC isoform. PKC, by contrast, consists of a family of related kinases, many of which have been demonstrated in neurons[52]. Our exclusive use of this isoform, as opposed to previous investigations, may also have contributed to our demonstration of altered electrophoretic migration and the generation PHF-1 or ALZ-50. Further such studies with other PKC isoforms, and their respective free catalytic subunits, are warranted. While PKC is capable of phosphorylating tau under cell-free conditions, it apparently does not do so following microinjection, while, by contrast, free PKC catalytic subunits (PKM) readily do so following microinjection. Phosphorylation of tau *in situ* by PKM but not PKC is consistent with the previous observation of phosphorylation of a distinct set of cytoskeletal proteins following PKM generation in intact neutrophils[29], and suggests that hyperactivation of PKC may be required to mediate accumulation of "AD-like" tau isoforms.

PKM Microinjection Induces Neurite Retraction

Retraction of filopodia following PKM injection is consistent with our previous demonstration that PKC activation induces retraction of existing neurites[16,17,22]; since PKM is constitutively active, it is not unexpected that PKM injection would effect similar morphological changes; whether or not neurite retraction is a consequence of microtubule disassembly, brought about by phosphorylation-induced dissociation of tau from microtubules, has not been examined, although this possibility is consistent with previous studies of PKC-mediated tau phosphorylation[36] and our demonstrated requirement of tau for neuritogenesis[53]. Moreover, it is entirely possible that increased PKM-mediated phosphorylation has induced filopodial retraction by means completely independent of tau and/or microtubules.

Influence of Calpain-Mediated PKM Generation on Tau Hyperphosphorylation

In support of the results following microinjection of PKM, we observed that calpain activation in situ also influenced tau phosphorylation, apparently by PKM generation. Along with our previous studies[17,22,39], these findings suggest that this mode of PKC activation exerts a crucial regulatory influence on neuronal homeostasis: calpain inhibition apparently serves to promote and maintain neuronal differentiation by suppressing PKC activation[39] while the extent of activation induced by calcium influx, by promoting PKC activation, induces "AD-like" alterations in tau as a consequence of PKC activation. That TPA could induce an increase in ALZ-50 in the absence of calcium influx, and in the presence of a calpain inhibitor, suggests that the influence of PKC is downstream from that of calpain.

While calcium influx can reversibly activate PKC by non-proteolytic means (for review, see ref. 24), calpain inhibition by specific inhibitor peptides or antibodies was equally as effective in preventing any calcium-meditated increase in ALZ-50 as was PKC inhibition. These data suggest that, although PKC-mediated phosphorylation apparently plays a role in tau phosphorylation to generate a degree of ALZ-50 and PHF-1 immunoreactivity in the absence of calcium influx, calpain-mediated PKM generation may be an obligate step to mediate the extent of the increase in ALZ-40 induced by calcium-influx in the present study. This possibility is supported by our finding that microinjection of PKM, but not PKC, generates ALZ-50 and PHF-1 immunoreactivity. The observation that such epitopes are present in SH-SY-5Y cells under normal conditions, but are significantly increased following calcium influx, is consistent with recent findings that several "AD-like" epitopes of tau (including PHF-1 and ALZ-50) are developmentally regulated[1,54], and are detectable within normal adult brain but are accentuated in AD brain[48].

The present findings support previous studies indicating that PKC activation as a consequence of calcium influx may play a role in tau hyperphosphorylation[55]. They also provide novel evidence, in accord with previous hypotheses[35,56], for the involvement of calpain in this process, and suggest that calpain regulates tau steady-state levels and the relative distribution of phosphorylated isoforms, both by direct proteolysis of the tau molecule and by activation of PKC; this line of reasoning is consistent with previous studies that indicate that phosphorylation renders tau resistant to calpain-mediated proteolysis[57-59]. The physiological consequences of excessive PKM generation are further underscored by the observation that antisense oligonucleotide-mediated PKCα downregulation did not reduce steady-state levels of tau hyperphosphorylation, while microinjection of PKM derived from PKCα induced a dramatic increase in hyperphosphorylated tau.

PKM Formation and Elimination is Under Regulation by Calpains

Our previous results suggest that PKM, once formed, is rapidly degraded by mM calcium-requiring calpain (m calpain) at a rate faster than intact PKC, indicating that, if it is indeed generated within intact cells (discussed more fully below), it is likely to exist only transiently under normal conditions [32-34,39]. Our further studies indicate that lipids inhibit m calpain mediated PKC proteolysis[60]. By contrast, μ calpain cleaves PKC to yield free catalytic and regulatory subunits, but does not further degrade PKM, and this limited proteolysis is lipid-insensitive. These findings leave open the possibility that physiological functions of the calpains may include generation of PKM at the cell membrane by μ calpain-mediated limited PKC proteolysis, and m calpain activity may be required for elimination of PKM that enters the cytosolic compartment. A substantial increase in μ calpain activity, perhaps as a consequence of inappropriate calcium influx, not compensated for by a similar increase in m calpain activity, could result in PKM persistence for sufficient length of time to effect phosphorylation of inappropriate substrates, including tau. In this regard, the differing calcium requirements of the calpains, as demonstrated at least in cell-free studies[44], are consistent with the possibility that influx of certain concentrations calcium could stimulate μ but not m calpain activation. The results of the present study do not provide any information regarding the rate or extent of PKM formation *in situ*, the physiological importance of which remains controversial (e.g., see ref.

24). They do, however, provide a model for the potential consequences of increased PKM generation and/or extended PKM half-life, and demonstrate that PKC and PKM can exhibit markedly different behaviors within intact neuroblastoma cells, including phosphorylation of tau to generate the PHF-1 and ALZ-50 epitopes.

PKC-Mediated Tau Phosphorylation May be the Result of Kinase Cascades

These findings should not be necessarily be interpreted to indicate that PKC directly phosphorylates tau, although Hoshi et al.[36] have demonstrated that PKC can do so, and we have demonstrated herein that both PKC and PKM can do so, under cell-free conditions, and that PKM can do so following microinjection into these cells; rather, calpain-mediated activation of PKC may initiate tau hyperphosphorylation via kinase cascades, potentialy including those kinases reported to induce normal tau to exhibit PHF-like characteristics under cell-free conditions, including Ca^{2+}-calmodulin kinase, cyclin-dependent kinases, MAP kinases, glycogen synthase kinases, and proline-directed kinases[4-14]. This possibility is further underscored by our observation that, following microinjection, neither PKC nor PKM were able to phosphorylate tau within cells or under cell-free conditions at ser 202, which generates the AT-8 epitope, and masks the Tau-1 epitope (e.g., ref. 48 and refs. therein), suggesting that an additional kinase activity, or activities, must be recruited to generate all AD-like changes in tau. This finding underscores the likelihood that the development of AD neuropathology is multifactoral, and suggests the potential involvement of PKM generation in early stages of this process. This line of reasoning is consistent with a recent study demonstrating that initial phosphorylation of tau by PKC hastened the generation of AD-like phosphorylation by glycogen synthase kinase[61].

SUMMARY AND CONCLUSIONS

The findings of the present study, and in particular, the relative resistance of ALZ-50-immunoreactive tau isoforms to subsequent degradation, leaves open the possibility that calcium influx may contribute to pathological levels of conversion of tau into A68 by hyperactivation of the calpain/PKC system. Downstream consequences of continued hyperphosphorylation, even at low levels, could eventually contribute to neuronal degeneration. In these analyses, we have only examined the development of certain immunoreactivities, and have not examined whether or not PHFs were generated; it remains possible that the phenomena we have described herein are relatively or completely reversible, and that additional events are required to achieve altered tau associations giving rise to PHF. Moreover, we have not addressed the possibility that altered tau immunoreactivity and/or PHF formation may be a by-product of, rather than a contributing factor in, AD-like neurodegeneration; nevertheless, the accumulation of altered tau immunoreactivity as observed herein remains an index of altered phosphorylation/dephosphorylation cycles, and a potentially useful indicator for the initial development of certain therapeutic approaches.

ACKNOWLEDGEMENTS

The writing of this chapter, and the original research presented herein, were supported by grants from the National Institute for Aging to T.B.S.

REFERENCES

1. Goedert, M., Jakes, R., Crowther, R. A., Six, J., Lubke, U., Vandermeeren, M., Cras, P., Trojanowski, J. Q., and Lee, V. M.-Y. (1993) The abnormal phosphorylation of tau proteins at Ser-202 in Alzheimer disease recapitulates phosphorylation during development. Proc. Natl. Acad. Sci. USA 90:5066-5070.
2. Trojanowski, J. Q., Schmidt, M. L., Shin, R.-W., Bramblett, G. T., Rao, D., and Lee, V. M.-Y. (1993) Altered tau and neurofilament proteins in neurodegenative diseases: diagnostic implications for Alzheimer's disease and Lewy body dementias. Brain Pathol. 3:45-54.
3. Trojanowski, J. Q., Schmidt, M. L., Shin, R.-W., Bramblett, G. T., Goedert, M., and Lee, V. M.-Y. (1993) PHF tau (A68): from pathological marker to potential mediator of neuronal dysfunction and degeneration in Alzheimer's disease. Clin. Neurosci. 1:184-191.
4. Baudier, J. and Cole, D. R. (1987) Phosphorylation of tau proteins to a state like that in Alzheimer's brain is catalyzed by a calcium/calmodulin-dependent kinase and modulated by phospholipids. J. Biol. Chem. 262:17577-17583.
5. Baumann, K., Mandelkow, E.-M., Biernat, J., Piwnica-Worms,H., and Mandelkow, E. (1993) Abnormal Alzheimer-like phosphorylation of tau-protein by cyclin-dependent kinases cdk2 and cdk5. FEBS Lett. 336:417-424.

6. Drewes, G., Lichtenberg-Kragg, B., Doring, F., Mandelkow, E.-M., Bienart, J., Doree, M., and Mandelkow, E. (1992) Mitogen activated protein (MAP) kinase transform tau protein into an Alzheimer-like state. EMBO J. 11:2131-2138.
7. Goedert, M., Cohen, E. S., Jakes, R., and Cohen, P. (1992) p42 MAP kinase phosphorylation sites in microtubule-associated protein tau are dephosphorylated by protein phosphatase 2A1. FEBS Lett. 312:95-99.
8. Hanger, D. P., Hughes, K., Woodgett, J. R., Brion, J. P., and Anderton, B. H. (1992) Glycogen synthase kinase-3 induces Alzheimer's disease-like phosphorylation of tau: generation of paired helical filament epitopes and neuronal localization of the kinase. Neurosci. Lett. 147:58-62.
9. Kobayahi S, Ishiguro K, Omori A, Takamatsu M, Arioka M, Imahora K, Uchida T (1993) A cdc-related kinase PSSALRE/cdk5 is homologous with the 30kDa subunit of tau protein kinase II, a proline-directed kinase associated with microtubules. FEBS Lett 335: 171-175.
10. Ledesma, M. D., Correas, L., Avila, J., and Diaz-Nido, J. (1992) Implication of brain cdc2 and MAP kinases in the phosphorylation of tau protein in Alzheimer's disease, FEBS Lett. 308: 218-224.
11. Mandelkow, E.-M., Drewes, G., Biernat, J., Gustke, N., Van Lint, J., Vandenheede, J. R., and Mandelkow, E. (1992) Glycogen synthase kinase 3 and the Alzheimer's disease-like state of microtubule-associated protein tau. FEBS Lett. 314:315-321.
12. Mulot, S. F. C., Hughes, K., Woodgett, J. R., Anderton, B. H., and Hanger, D. P. (1994) PHF-tau a from Alzheimer's brain comprises four species on SDS-PAGE which can be mimicked by in vitro phosphorylation of human brain tau by glycogen synthase kinase-3b. FEBS Lett. 349: 359-364.
13. Paudel, H. K., Lew, J., Zenobia, A., and Wang, J. H. (1993) Brain proline-directed kinase phosphorylates tau on sites that are abnormally phosphorylated in tau associated with Alzheimer's paired helical filaments. J. Biol. Chem. 268: 23512-23518.
14. Vulliet, R., Halloran, S. M., Braun, R. K., Smith, A. J., and Lee, G. (1992) Proline-directed phosphorylation of human tau protein. J. Biol. Chem. 267: 22570-22574.
15. Heikkila JE, Akerlind G and Akerman KEO (1991) Protein kinase C activation and down-regulation in relation to phorbol ester-induced differentiation of SH-SY-5Y human neuroblastoma. J Cell Physiol 140: 593-600.
16. Leli U, Cataldo AM, Shea TB, Nixon RA and Hauser G (1992a) Distinct mechanisms of differentiation of SH-SY-5Y neuroblastoma cells by protein kinase C activators and inhibitors. J Neurochem 58: 1191-1198.
17. Leli U, Parker PJ and Shea TB (1992b) Intracellular delivery of protein kinase C-a or -e isoform-specific antibodies promotes acquisition of a morphologically differentiated phenotype in neuroblastoma cells. FEBS Lett 297: 91-94.
18. Minana M-D, Felipo V and Grisolia S (1989) Inhibition of protein kinase C induces differentiation of neuroblastoma cells. FEBS Lett 255: 184-186.
19. Ono K, Katayama N, Yamagata Y, Tokunaga A and Tsuda M (1991) Morphology of neurites from N18TG2 cells induced by H-7 and by cAMP. Brain Res Bull 26: 605-612.
20. Ono K, Tokunaga A and Tsuda M (1993) Neurite outgrowth from N18TG2 neuroblastoma induced by H-7, a protein kinase inhibitor, in the presence of colchicine. Brain Res Bull 31: 209-215.
21. Shea TB and Beermann ML (1991a) Stauroporine-induced morphological differentiation of human neuroblastoma cells. Cell Biol Internat Rep 15: 161-168.
22. Shea TB, Beermann ML, Leli U and Nixon RA (1992a) Opposing influences of protein kinase activities on neurite outgrowth in human neuroblastoma cells: Initiation by kinase A and restriction by kinase C. J Neurosci Res 33: 398-407.
23. Tsuda M, Ono K, Katayama N, Yamagata Y, Kikuchi K and Tsuchiya T (1989) Neurite outgrowth from mouse neuroblastoma and cerebellar cells induced by the protein kinase inhibitor H-7. Neurosci Lett 105: 241-245.
24. Murray, A. W., Fournier, A., and Hardy, S. J. (1987) Proteolytic activation of protein kinase C: a physiological reaction? Trends Neurol. Sci. 12:53-54.
25. Hashimoto K, Mikawa K, Kuroda T, Ase K and Kishimoto A (1990) Calpains and regulation of protein kinase C. In: Intracellular calcium-dependent proteolysis (Mellgren RL, Murachi T, eds.) CRC Press Boca Raton, FA pp 181-190.
26. Inoue M, Kishimoto A, Takai Y and Nishizuka Y (1977) Studies on a cyclic nucleotide-independent protein kinase and its proenzyme in mammalian tissues. II. Proenzyme and its activation by calcium-dependent protease from rat brain. J Biol Chem 252: 7610-7616.
27. Kishimoto A, Kajikawa N, Shiota M and Nishizuka Y (1983) Proteolytic activation of calcium-activated, phospholipid-dependent protein kinase by calcium-dependent neutral protease. J Biol Chem. 258: 1156-1160.
28. Kishimoto A, Mikawa K, Hashimoto K, Yasuda I, Tanaka S-I, Tominaga M, Kurode T and Nishizuka Y (1989) Limited proteolysis of protein kinase C subspecies by calcium-dependent neurtral protease (Calpain) J Biol Chem 264: 4088-4092.
29. Pontremoli, S., Michetti, M., Melloni, E., Sparatore, B., Salamino, F. and Horecker B. L. (1990) Identification of the proteolytically activated form of protein kinase C in stimulated human neurtrophils. Proc. Natl. Acad. Sci. USA 87: 3705-3707.

30. Young, S., Parker, P. J., Ulrich, A. and Stable, S. (1987) Down-regulation of protein kinase C is due to an increased rate of degradation. Biochem. J. 244: 775.
31. Al, Z. and Cohen, C. M. (1993) Phorbol 12-myristate 13-acetate-stimulated phosphorylation of erythrocyte membrane skeletal proteins is blocked by calpain inhibitors: possible role of protein kinase M. Biochem J. 296: 675-683.
32. Shea, T. B., Beermann, M. L., Griffin, W. R., Anthony, P. K., and Leli, U. (1994a) Calpain degrades the free PKC catalytic subunit (PKM) faster than intact PKC. Trans. Am. Soc. Neurochem. 25: 262.
33. Shea, T. B., Beermann, M. L., Griffin, W. R., and Leli, U. (1994b) Degradation of protein kinase Cα and its free catalytic subunit, protein kinase M, in intact human neuroblastoma cells and under cell-free conditions: Evidence that PKM is degraded by calpain-mediated proteolysis at a faster rate than PKC. FEBS Lett. 350:223.
34. Shea, T. B., Cressman, C. M., Beermann, M. L., and Nixon, R. A. (1994c) Hyperphosphorylation of tau following calcium influx into human neuroblastoma: Role of calpain and PKC. Mol. Biol. Cell: 5: 168a.
35. Nixon, R. A. (1989) Calcium-activated neutral proteinases as regulators of cellular function. Implications for Alzheimer's disease Pathogenesis. Ann. N. Y. Acad. Sci. 568: 198-208.
36. Hoshi, M., Nishida, E., Miyata, Y., Sakai, H., Miyoshi,T., Ogawara, H. and Akiyama, T. (1987) Protein kinase C phosphorylates tau and induces its functional alterations. FEBS Lett. 217: 237-241.
37. Steiner, B., Mandelkow, E.-M., Biernat, J., Gustke, N., Meyer, H. E., Schmidt, B., Mieskes, G., Söling, H. D., Dreschsel, D., Kirschner, M. W., Goedert, M., and Mandelkow, E. (1990) Phosphorylation of microtubule-associated protein tau: identification of the site for Ca^{2+}-calmodulin dependent kinase and relationship with tau phosphorylation in Alzheimer tangles. EMBO J. 9:3539-3544.
38. Shea TB. (1990) Neuritogenesis in mouse NB2a/d1 neuroblastoma cells: Triggering by calcium influx and involvement of actin and tubulin dynamics. Cell Biol Internat Reports 14:967-79.
39. Shea TB, Cressman CM, Spencer MJ, Beermann ML and Nixon RA (1995) Enhancement of neurite outgrowth following calpain inhibition is mediated by protein kinase C. J Neurochem 65: in press
40. Spencer MJ and Tidball JG (1992) Calpain concentration is elevated although net calcium-dependent proteolysis is suppressed in dystrophin-deficient muscle. Exp Cell Res 203: 107-114.
41. Uemori T, Shimojo T, Asada K, Asano T, Kimizuka F, Kato I, Maki M, Hatanaka M, Murachi T, Hanzawa H (1990) Characterization of a functional domain of human calpastatin. Biochem Biophys Res Comm 166: 1485-1493.
42. Shea TB and Beermann ML (1991b) A method for phospholipid-mediated delivery of specific antibodies into adherent cultured cells. Biotechniques 10: 288-291.
43. Parker P, Coussens L, Totty N, Rhee L, Young S, Chen E, Stabel S, Waterfield M, Ullrich A (1986): The Complete Primary Structure of Protein Kinase C-the Major Phorbol Ester Receptor. Science 233:853-859.
44. Vitto A. and Nixon R.A. (1986). Calcium-activated neurtral proteases of human brain: subunit structure and enzymatic properties of multiple molecular forms. J. Neurochem. 47, 1039-1051.
45. Kosik KS, Orecchio LD, Binder L, Trojanowski JQ, Lee VM-Y, and Lee G (1988) Epitopes that span the tau molecule are shared with paired helical filaments. *Neuron* **1**: 817-825.
46. Samis J.A., Zboril G. and Elce J.S. (1987). Calpain I remains intact and intracellular during platelet activation. Biochem. J. 246, 481-488.
47. Shea TB, Beermann ML and Nixon RA (1991) Multiple proteases regulate neurite outgrowth in NB2a/d1 neuroblastoma. J Neurochem 56:842-851.
48. Matsuo, E. S., Shin, R.-W., Billingsley, M. L., Van deVoorde, A., O'Connor, M., Trojanowski, J. Q. and Lee, V. M.-Y. (1994) Biopsy-derived adult human brain tau is phosphorylated at many of the same sites as Alzheimer's disease paired helical filament tau. Neuron 13: 989-1002.
49. Pope, W. B., Enam, S. A., Bawa, N., Miller, B. E., Ghanbari, H. A., and Klein, W. L. (1993) Phosphorylated tau epitope of Alzheimer's disease is coupled to axon development in the avian central nervous system. Exp. Neurol. 120: 106-113.
50. Pope, W. B., Lambert, M. P., Leypold, B., Seupaul, R., Sletten, L., Krafft, G., and Klein, W. L. (1994) Microtubule-associated protein tau is hyperphosphorylated during mitosis in the human neuroblastoma cell line SH-SY-5Y. Exp. Neurol. 126: 185-194.
51. Nishizuka, Y. (1989) Studies and prospectives of the protein kinase C family for cellular regulation. Cancer 63: 1892-1903.
52. Leli, U., Shea, T. B., Cataldo, A., Hauser, G., Grynspan, F., Beermann, M. L., Liepkalns, V. A., Nixon, R. A, and Parker, P. A. (1993) Differential expression and subcellular localization of protein kinase C α,β,δ,ε and γ isoforms in SH-SY-5Y neuroblastoma cells: Modifications during differentiation. J. Neurochem. 60:289-298.
53. Shea TB, Beermann ML, Nixon RA. (1992b) Microtubule-associated protein tau is required for axonal neurite elaboration by neuroblastoma cells. J Neurosci Res. 32:363-374.
54. Lee, J. H., Goedert, M., Hill, W. D., Lee. V. M.-Y., and Trojanowski, J. Q. (1993) Tau proteins are abnormally expressed in olfactory epithelium of Alzheimer's disease and developmentally regulated in fetal spinal cord. Exp. Neurol. 121: 93-105.
55. Mattson MP (1991) Evidence for the involvement of protein kinase C in neurodegenerative changes in cultured human cortical neurons. Exp Neurol 112:95-103.

56. Nixon RA, Saito K-I, Grynspan F, Griffin WR, Katayama S, Honda T, Mohan PS, Shea TB, Beermann ML. (1994) The calcium-activated neurotral proteinase (calpain) system in aging and Alzheimer's disease. Ann NY Acad Sci 747:77-91.
57. Johnson GVW, Jope RS and Binder LI (1989) Proteolysis of tau by calpain. Biochem Biophys Res Commun 163: 1505-1511.
58. Johnson GVW and Foley VG (1993) Calpain-mediated proteolysis of microtubule-associated protein 2 (MAP2) is inhibited by phosphorylation by cAMP-dependent protein kinase, but not by Ca^{2+}/calmodulin-dependent protein kinase II. J Neurosci Res 34: 642-647.
59. Litersky JM, Scott CW and Johnson GVW (1993) Phosphorylation, calpain proteolysis and tubulin binding of recombinant human tau isoforms. Brain Res 604: 32-40.
60. Lang, D., Hauser, G. H., Cressman, C. M., Mohan, P. S., Nixon, R. A., and Shea, T. B. (1995) Lipids inhibit mM calpain-mediated proteolysis of PKC in vitro. J. Neurochem., 64, 25a
61. Singh TJ, Zaidi T, Grundke-Iqbal I and Iqbal K(1995) Modulation of GSK-3-catalyzed phosphorylation of microtubule-associated protein tau by non-proline-dependent protein kinases. FEBS Lett 358:4-8.

INDEX